Processes Controlling the Composition of Clastic Sediments

Edited by

Mark J. Johnsson*
U.S. Geological Survey
345 Middlefield Road, MS 999
Menlo Park, California 94025

and

Abhijit Basu
Department of Geological Sciences
Indiana University
Bloomington, Indiana 47405

SPECIAL PAPER

284

*Present address: Department of Geology, Bryn Mawr College, Bryn Mawr, Pennsylvania 19010-2899.

1993

Published by The Geological Society of America, Inc.
3300 Penrose Place, P.O. Box 9140, Boulder, Colorado 80301

Printed in U.S.A.

GSA Books Science Editor Richard A. Hoppin

Library of Congress Cataloging-in-Publication Data

Processes controlling the composition of clastic sediments / edited by
Mark J. Johnsson and Abhijit Basu.
p. cm. — (Special paper ; 284)
Based on papers presented at the 1991 annual meeting of the
Geological Society of America.
Includes bibliographical references and index. ISBN 0-8137-2284-5
1. Sedimentation and deposition—Congresses. 2. Weathering-
-Congresses. 3. Sediments (Geology)xx—Composition—Congresses.
I. Johnsson, Mark J. II. Basu, Abhijit. III. Series: Special
papers (Geological Society of America) ; 284.
QE571.P76 1993
551.3—dc20 93-28881
CIP

Cover photo: The Río Mendoza carries detritus from the rising high Andes of central Argentina and Chile to the Andean foreland basin near Mendoza, Argentina. The river is reworking alluvial deposits stored in alluvial fans as it struggles to maintain its grade during continued uplift of the mountains; note the truncated alluvial fan on the left side of the photograph. More detritus is temporarily stored in the meander belt of the river, awaiting remobilization by channel migration and cutbank growth. **Background photo:** Compositional sorting of sand by hydrodynamic processes; segregation of denser minerals (principally ilmenite and magnetite) from lighter minerals (principally quartz) in Año Nuevo Creek at Año Nuevo beach, central California. Photographs by Mark J. Johnsson.

10 9 8 7 6 5 4 3 2 1

Contents

Preface

Henry Clifton Sorby studied mountains with a microscope, for which he was ridiculed by his critics and tormentors. Little did they know that some 150 years later sedimentary petrologists would be studying not only the skins, deep interiors, and roots of mountains, but also lower crustal and mantle material with various microanalytical tools. Our understanding of tectonic processes, of interaction of the lithosphere with the atmosphere and the biosphere, and of the global interrelationship of Earth processes have increased dramatically in the last half century. Students of sedimentology have benefited from the seminal teachings of Krynine in relating sediments and their mineralogical compositions with tectonics and climate. The pivotal discovery of Pettijohn that intrastratal solutions preferentially destroy different mineralogic entities in sandstones forever changed the tenor of sandstone petrology. As now seems historically inevitable given the plate tectonic revolution, Dickinson and his coworkers in the early 1970s related the framework petrography of sandstones to the plate tectonic setting of the rocks from which they were derived. Sandstone petrology now plays an integral part in reconstructing the history of the Earth's crust and in understanding the Earth as an integrated system. Sediment provenance studies, at present, include the interpretation of the processes that generate the parent rocks, of the processes that emplace the parent rocks to be subjected to erosion, and of the processes of weathering, erosion, transportation, deposition, burial, diagenesis, and subsequent uplift.

This integrated concept of sediment provenance is relatively new. The concept has been shaped, in part, through two international meetings. Gian G. Zuffa organized a NATO Advanced Studies conference in 1984 on the "Provenance of Arenites." This was the first meeting specifically focused on the factors controlling sandstone composition. In 1989, Andrew C. Morton organized a special meeting of the Geological Society of London to consider further the "Developments in Sedimentary Provenance Studies." It became obvious at this meeting that provenance interpretation now extended to the precursors and the protoliths of the immediate parent rocks of sands and sandstones, and possibly to reconstructing the paleogeography of basins and source areas that are now irrevocably detached from each other.

In this spirit, we convened a theme session at the 1991 annual meeting of the Geological Society of America explicitly to appraise our understanding of the interrelated processes controlling the composition of clastic materials. This volume grew out of that session. Our goal has been to assemble a group of papers that, taken together, sketch an outline of the integrated processes shaping the composition of clastic materials.

The first two papers (Johnsson; McLennan et al.) deal with general topics and set the stage for our exploration of the processes controlling the composition of clastic sediments. Respectively, they give an overview of the processes shaping sediment composition and a discussion of recent advances in geochemical studies in sedimentology. The subsequent six papers discuss aspects of source rock control on the composition of clastic material. The papers by Marsaglia and by Ridgway and DeCelles present baseline information concerning sediments derived from two tectonic environments that have largely escaped examination under earlier tectonic-provenance investigations: isolated basaltic islands and strike-slip basins. In the papers by Ridgway and DeCelles and by Devaney and Ingersoll, sandstone composition is used to chart tectonic evolution in regions where clastic materials provide the best clues to the former disposition of source rocks.

Girty et al. and Linn and DePaolo offer new ways in which some of the geochemical methods discussed by McLennan et al. can be put to use in the evaluation of provenance. The paper by Heins illustrates the potential effects of an often overlooked factor—source rock texture—in controlling the composition of sands derived from similar source rocks. The three following papers discuss ways in which chemical weathering affects clastic sediment composition. Johnsson et al. show that even in geographically restricted regions, varying precipitation can produce distinct clay mineral assemblages in soils. The paper by Morton and Johnsson illustrates the limited extent to which heavy mineral suites from an active tectonic region can be modified by chemical weathering during alluvial storage—in marked contrast to the greater modification of overall sandstone composition. Dutta and Wheat formulate a hypothesis integrating the effects of temporal changes in both tectonic activity and climate to explain compositional variations in foreland basin sandstones from the Sydney Basin. Mechanical controls on sediment composition are explored in the following three papers. Dutta et al. present theoretical arguments for the depletion of mechanically less durable minerals in eolian sand through their disintegration by ballistic impact. An experimental study of the mechanical degradation of granitic detritus by Osborne et al. extends the classic work of Kuenen by quantifying the results through Fourier grain-shape analysis. The last paper in this section, by Picard and McBride, brings such studies to the field by illustrating the effects of abrasion on modern beach sands. The three following papers evaluate processes operating in integrated depositional systems. Cavazza et al. illustrate how sand composition is affected by sedimentary recycling in a steep, temperate, fluvial setting. Kairo and Suttner examine the effects of depositional processes in deltaic, fluvial, and littoral facies, and Soreghan and Cohen show how basin asymmetry can serve as an important "master variable" in controlling processes affecting sediment composition in the east African rift valleys. The final section of the volume deals with one of the thorniest subjects in sedimentology—the estimation of rates and fluxes from compositional data. Molinaroli et al. point out the profound importance of Saharan dust to Mediterranean depositional settings. Palomares and Arribas present a "sand generation index" by which the propensity of different source rocks to produce sand-size detritus can be quantified. Finally, Molinaroli and Basu offer a first attempt at quantifying the relative rate of denudation of various source materials based on preserved sediments.

This selection of papers reflects, in part, our bias toward submissions directly dealing with processes that modify sediment composition, calculate fluxes, quantify source rock contribution, and go beyond the identification of parent rocks to yield understanding of the processes responsible for sediment composition. Only the readership will tell how far we have been successful in achieving those goals.

Mark J. Johnsson
Abhijit Basu
October 1992

Acknowledgments

Many people have given freely toward the publication of this volume. Gary H. Girty encouraged us to convene a theme session dealing with controls on the composition of clastic sediments at the 1991 annual meeting of GSA and assisted us throughout the planning phases. Due to enthusiastic response, the session swelled to morning and afternoon oral sessions, and a poster session continued into the early evening. We wholeheartedly thank Gary Girty, the organizers of the meeting, and all those who submitted abstracts for making the theme session a success. The foundations for this volume were laid at that meeting, and authors worked within a strict time frame to produce quality manuscripts in short order. We owe special thanks to the more than 50 reviewers who took time to criticize constructively and to improve the quality of the final product immensely. Some reviewers were called on several times, either for multiple reviews or for re-reviews of the same manuscript—we thank them for their dedication to this volume. Lorie Canada of Indiana University assisted with the editorial process. Finally, we thank the U.S. Geological Survey—especially David G. Howell—and Indiana University for allowing us the time and resources to devote to this volume.

List of Reviewers

Richard April, *Colgate University*
Asish R. Basu, *University of Rochester*
Sam Boggs, Jr., *University of Oregon*
John A. Breyer, *Texas Christian University*
Mark J. Cintala, *National Aeronautics and Space Administration*
Harvey A. Cohen, *Princeton University*
Rónadh Cox, *Stanford University*
Robert L. Cullers, *Kansas State University*
Dennis A. Darby, *Old Dominion University*
Peter G. DeCelles, *University of Rochester*
John Decker, *ARCO Alaska*
Rebecca Dorsey, *Northern Arizona University*
Steven Dreise, *University of Tennessee at Knoxville*
Prodip K. Dutta, *Indiana State University*
Robert Ehrlich, *University of South Carolina*
Peter Flemmings, *Massachusetts Institute of Technology*
Robert L. Folk, *University of Texas at Austin*
Gary H. Girty, *San Diego State University*
Santosh Ghosh, *Petroscan, Inc.*
Stephen A. Graham, *Stanford University*
Jeffrey D. Grigsby, *University of Texas at Austin*
Barry R. Hill, *U.S. Geological Survey*
Karen Houck, *University of Colorado at Boulder*
David G. Howell, *U.S. Geological Survey*
Ralph E. Hunter, *U.S. Geological Survey*
Raymond V. Ingersoll, *University of California at Los Angeles*
W. Calvin James, *Ohio State University at Columbus*
Gary Kocurek, *University of Texas at Austin*
Stephen K. Kennedy, *R.J. Lee Group*
Eric Kvale, *Indiana Geological Survey*
Timothy F. Lawton, *New Mexico State University at Las Cruces*
Anne M. Linn, *University of California at Berkeley*
David Loope, *University of Nebraska*
Gretchen Luepke, *U.S. Geological Survey*
Neil Lundberg, *Florida State University*
Gregory H. Mack, *New Mexico State University at Las Cruces*
Barry J. Maynard, *University of Cincinnati*
James M. Mazzullo, *Texas A&M University*
Hugh McLean, *U.S. Geological Survey*
Scott M. McLennan, *State University of New York at Stony Brook*
Douglas W. Ming, *National Aeronautics and Space Administration*
Emanuela Molinaroli, *University of Venice*
Richard W. Murray, *University of Rhode Island*
Richard W. Ojakangas, *University of Minnesota at Duluth*
Zell E. Peterman, *U.S. Geological Survey*
Albert Schultz, *CONOCO, Inc.*
Garrison Sposito, *University of California at Berkeley*
Christopher A. Suczek, *Western Washington University*
Lee J. Suttner, *Indiana University*
Peter van de Kamp, *Cornelius Corporation*
Stephen Young, *Exxon International*

List of Contributors

José Arribas, *Universidad Complutense de Madrid, Spain*
Robin W. Barber, *San Diego State University*
Abhijit Basu, *Indiana University*
Edwin J. Bomer III, *University of Southern California*
Claudio Camporesi, *University of Bologna, Italy*
William Cavazza, *University of Bologna, Italy*
Andrew S. Cohen, *University of Arizona*
Peter G. DeCelles, *University of Rochester*
Donald J. DePaolo, *University of California at Berkeley*
Kathleen A. Devaney, *University of California at Los Angeles*
Paulo R. dos Santos, *University of São Paulo, Brazil*
Prodip K. Dutta, *Indiana State University*
Stephen D. Ellen, *U.S. Geological Survey*
Carlo Ferretti, *University of Bologna, Italy*
Gary H. Girty, *San Diego State University*
Stefano Guerzoni, *National Research Council, Italy*
Gilbert N. Hanson, *State University of New York at Stony Brook*
William A. Heins, *University of California at Los Angeles*
Sidney Hemming, *State University of New York at Stony Brook*
Raymond V. Ingersoll, *University of California at Los Angeles*
Mark J. Johnsson, *U.S. Geological Survey*
Suzanne Kairo, *Indiana University*
Charles Knaack, *Washington State University*
Anne M. Linn, *University of California at Berkeley*
Yi Lu, *University of Southern California*
Kathleen M. Marsaglia, *University of Texas at El Paso*
Earle F. McBride, *University of Texas at Austin*
Diane K. McDaniel, *State University of New York at Stony Brook*
Mary Anne McKittrick, *U.S. Geological Survey*
Scott M. McLennan, *State University of New York at Stony Brook*
Emanuela Molinaroli, *University of Venice, Italy*
Andrew C. Morton, *British Geological Survey*
Robert H. Osborne, *University of Southern California*
Marta Palomares, *Universidad Complutense de Madrid*
M. Dane Picard, *University of Utah*
Giancarlo Rampazzo, *University of Venice, Italy*
Kenneth D. Ridgway, *University of Rochester*
Michael J. Soreghan, *University of Arizona*
Lee J. Suttner, *Indiana University*
Yeon-Chang Wang, *University of Southern California*
Russel W. Wheat, *Indiana State University*
Ziye Zhou, *Indiana State University*
Gian Gaspare Zuffa, *University of Bologna, Italy*

Geological Society of America
Special Paper 284
1993

The system controlling the composition of clastic sediments

Mark J. Johnsson*
U.S. Geological Survey, 345 Middlefield Road, MS 999, Menlo Park, California 94025

ABSTRACT

The composition of clastic sediments and rocks is controlled by a complex suite of parameters operating during pedogenesis, erosion, transport, deposition, and burial. The principal first-order parameters include source rock composition, modification by chemical weathering, mechanical disaggregation and abrasion, authigenic inputs, hydrodynamic sorting, and diagenesis. Each of these first-order parameters is influenced to varying degrees by such factors as the tectonic settings of the source region, transportational system and depositional environment, climate, vegetation, relief, slope, and the nature and energy of transportational and depositional systems. These factors are not independent; rather a complicated web of interrelationships and feedback mechanisms causes many factors to be modulated by others. Accordingly, processes controlling the composition of clastic sediments are best viewed as constituting a *system*, and in evaluating compositional information the dynamics of the system must be considered as whole.

INTRODUCTION

Sedimentary rocks are our principal sources of information concerning past conditions on the Earth's surface. Clastic rocks may preserve detritus from orogenic settings now obscured by tectonic overprinting, dismemberment or erosion. At times, the composition of such clastic materials provides the only available clues to the composition of long-eroded source rocks, and thus is an invaluable tool in paleogeologic reconstructions. Indeed, clastic sediments provide us with our only windows into Earth's earliest past—the oldest geologic materials on Earth are detrital zircons from Archean sandstones (Compston et al., 1985; Compston and Pidgeon, 1986; Liu et al., 1992; Mueller et al., 1992).

Accordingly, there is a long history of the use of compositional data from clastic sediments to evaluate the nature of the terranes from which they were derived. This history extends back to the time of Hutton, who concluded that "if this part of the earth which we now inhabit had been produced, in the course of time, from the materials of a former earth, we should, in the examination of our land, find data from which to reason, with regard to the nature of that world . . ." (Hutton, 1785, p. 23). In the ensuing two centuries, hundreds of studies have been undertaken in which clastic rocks, particularly sandstones, have been utilized as clues to their source regions. Compositional data have been used to chart orogenic progression, unroofing, and most recently, plate tectonic evolution.

The composition of clastic materials does not, however, correlate on a one-to-one basis with source rock composition, implying that factors other than source rock composition also are important in determining the ultimate composition of clastic sediments. During the formation of clastic rocks, material passes through several evolutionary stages, including pedogenesis, erosion, transport, deposition, and burial. Processes acting on sediment composition during this evolution include chemical weathering, physical breakdown, abrasion, hydrodynamic sorting, modification by authigenic inputs and, in the case of lithified sedimentary rocks, burial diagenesis. Erosion and sedimentation are essentially partitioning processes whereby the components of the source rock are differentially preserved. If the materials eroded from a group of source rocks were in physical and chemical equilibrium, then their proportions in the resulting sediment would be dictated merely by their abundance in the source rocks and by hydrodynamic sorting effects operating during transport and deposition. However, mineral phases differ in chemical and physical stability and accordingly are modified by physical and chemical weathering in different ways. The effects of physical and chemical weathering processes are in turn controlled by such parameters as climate and relief of source terrain, distance of

*Present address: Department of Geology, Bryn Mawr College, Bryn Mawr, Pennsylvania 19010-2899.

Johnsson, M. J., 1993, The system controlling the composition of clastic sediments, *in* Johnsson, M. J., and Basu, A., eds., Processes Controlling the Composition of Clastic Sediments: Boulder, Colorado, Geological Society of America Special Paper 284.

transport, time spent in transport, and energy of the transportational and depositional environments. Finally, the sediment is likely to experience further compositional modification during diagenesis. All of these factors, rarely easily separated, are important in establishing the composition of clastic sediments.

Modifications that occur during transport and deposition tend to obscure information regarding parent materials, but may, at the same time, provide a wealth of information concerning the environment in which sediment was formed. Modifications subsequent to deposition may provide clues to the diagenetic environment of a sedimentary basin. For these reasons, the processes modifying sediment composition both prior to and following deposition are currently the focus of much interest, and a picture is beginning to emerge relating sediment composition to environmental conditions in addition to source rocks. An understanding of these processes and how they interact is desirable for the development of more refined provenance interpretation schemes, to provide a basis for evaluating past environmental conditions, and to evaluate the roles of erosion and sedimentation within the tectonic and hydrologic cycles. The goal of this paper is to provide an overview of the system regulating the composition of clastic materials and to suggest potential feedback mechanisms operating within that system.

CONCEPTUAL BASELINE

What is composition?

Sediment composition can be defined in many ways, depending on the nature of the sediment and on the questions addressed by the study. Perhaps most basic is a definition based on the relative abundance of the elements or their oxides in the bulk sample. This approach is especially valuable when making comparisons between sediment composition and the composition of solid or dissolved weathering products, but it is of limited utility in studying sedimentary rocks because whole-rock analyses include both detrital and diagenetic components. Mineralogical composition provides an alternate measure. In shales, in which composition is routinely evaluated through x-ray diffraction, mineralogical composition is the standard measure of composition. In sandstones, however, fine-grained minerals within polymineralic grains present identification difficulties (Boggs, 1968). Furthermore, schemes defining sandstone composition on the basis of either elemental or mineralogical compositions ignore the textural information available in thin section. Accordingly, sandstone composition is most commonly defined on the basis of grain type, a procedure that includes a great deal of textural information as well.

The petrographic microscope and the x-ray diffractometer thus remain the principal tools for the evaluation of the mineralogical composition of sandstone and shale composition. Bulk chemical data may be collected by a wide variety of techniques, including traditional atomic and plasma absorption and emission spectroscopic techniques, x-ray fluorescence, and neutron activation analysis. Chemical data of individual phases are commonly acquired through use of the electron microprobe, backscatter scanning electron microscopy, energy-dispersive spectral analysis and, indirectly, through cathodoluminescence. Finally, "composition" may be taken broadly to include detailed information about sediment components, such as varieties of quartz grains, U-Pb dating of individual zircon grains, and stable isotope signatures of the bulk sediment.

Effects of methods

While between-sample variations can be adequately described by any definition of composition, comparisons among data sets require scrupulous attention to the methods employed in collecting compositional data (Zuffa, 1985). This is particularly important when applying previously derived provenance models to new data sets (Ingersoll, 1990).

Irrespective of other factors affecting overall sediment composition, composition is strongly correlated with grain size. Obviously, shales and sandstones derived under identical conditions from identical source rocks are markedly different in composition simply because clay minerals make up the bulk of finer grain sizes and detritus consisting principally of primary minerals is found in coarser fractions. More subtle is the effect of grain size within the sand fraction; numerous studies have demonstrated a strong dependence of sand composition on grain size (e.g., Füchtbauer, 1964; Boggs, 1968; Odom et al., 1976; Basu, 1985a; Cather and Folk, 1991; Savage and Potter, 1991).

Since grain-size effects can obscure meaningful trends in samples, a means of normalizing for grain size must be achieved if samples are to be compared. In unconsolidated sands, normalization may be achieved by restricting analysis to a particular grain size that has been isolated by sieving. In sandstones, such a procedure is impractical, so an effort is generally made to collect samples of nearly identical grain size (most commonly, the medium size fraction, 0.25-0.50 mm). Another means to help normalize for compositional dependence on grain size is to adopt point-counting conventions such that sand-size (>63 μm) subgrains within larger polymineralic grains are assigned to the category of the subgrain rather than to the category of the larger fragment (Gazzi, 1966; Dickinson, 1970; Gazzi and Zuffa, 1970). In fact, no point-counting technique can correct for actual variation of monomineralic grains with grain size (Ingersoll et al., 1984; Decker and Helmold, 1985). The Gazzi-Dickinson method, however, by assuming that polycrystalline components will disaggregate into constituent grains, provides improved clustering of data obtained from sandstones of diverse grain sizes and places an increased emphasis on provenance (Ingersoll et al., 1984). More traditional methods, in which the entire detrital grain is used to arrive at grain classification (Suttner and Basu, 1981, 1985) emphasize the textural and compositional information found in lithic fragments, and place emphasis on paleoenvironmental conditions during erosion and transport. Clearly, the point-counting method adopted depends on the goals of the study (Ingersoll et al., 1984, 1985a,b; Decker and Helmold, 1985; Suttner and Basu, 1985;

Zuffa, 1985), but Zuffa (1985) demonstrated that significantly disparate results are obtained by different point-counting procedures. Furthermore, sampling scale can affect the interpretation of compositional data. Ingersoll (1990) pointed out that provenance interpretation schemes such as those of Dickinson and Suczek (1979) or Dickinson (1985), derived as they are from mean values representing many data sets, are not applicable to data sets representing regions on the scale of drainage basins or individual mountains.

SYSTEM APPROACH TO CLASTIC COMPOSITION

Parameters influencing the composition of clastic sediments are intimately interrelated; to fully understand what information is contained within composition data, all such parameters must be considered together. Thus, a system approach is advocated, whereby an emphasis is placed on the interrelationships among parameters influencing composition and on their feedback mechanisms.

The study of parameters affecting sediment composition and how they interact tends to be highly uniformitarianistic. Given the difficulties inherent in establishing past values for many of these variables, sedimentary petrologists have turned to modern sediments and the principles of comparative sedimentology to cast light on these questions. It must be borne in mind, however, that modern systems may be atypical of past Earth surface conditions in a number of ways. Of particular significance may be the effects of Pleistocene glaciation and of anthropogenic perturbation of erosion and soil production rates.

SOURCE ROCK CONTROLS ON SEDIMENT COMPOSITION

Sediment composition is obviously strongly controlled by the composition of the rocks from which the sediment is derived. Under many conditions, sediment composition correlates well with the composition of source rocks. Because source rock composition is in turn controlled to a great degree by tectonic setting, an attempt has been made to classify sands and sandstones on the basis of the tectonic environment of the source terrain. The modern movement in this direction was foreshadowed by Krynine (1941a–c; 1942) when he linked stratigraphic variations in sandstones with progressing geosynclinal evolution. Krynine envisioned the upper portion of the Earth's crust to be divisible into an upper layer of dominantly sedimentary rocks, a middle layer composed of metamorphic rocks, and a lower layer of plutonic rocks. He reasoned that these layers would give rise to quartzose sands, lithic arenites, and arkoses, respectively, and that such a succession of sands would reflect progressive unroofing of the source terrane during orogenesis.

Correlation with tectonic setting

With the advent of the plate tectonic model, it was noted that the global distribution of rock types could be explained in terms of tectonic setting, suggesting that sandstone composition could be used to infer ancient tectonic setting as well. In 1970, Dickinson formulated a classification scheme making possible quantitative evaluation of sandstone composition. Dickinson and his coworkers, as well as other researchers, next used this classification scheme to relate sandstone composition to the tectonic environments of source terranes (e.g., Dickinson and Rich, 1972; Crook, 1974; Schwab, 1975; Dickinson and Suczek, 1979; Dickinson, 1982; Dickinson et al., 1983a). Sand composition shows a particularly strong correlation with provenance among sands of modern arc-trench systems (e.g., Harrold and Moore, 1975; Ingersoll and Suzcek,1979; Moore, 1979; Dickinson and Valloni, 1980; Valloni and Maynard, 1981; Enkeboll, 1982; Maynard et al., 1982; Valloni and Mezzadri, 1984; Yerino and Maynard, 1984; Valloni, 1985; Gergen and Ingersoll, 1986; Packer and Ingersoll, 1986; Cawood, 1991; Pirrie, 1991), probably because of the steep slopes, short transport times and distances, and rapid sedimentation rates usually associated with such regions. Narrow ranges of sand composition were successfully correlated with equally specific tectonic settings, and in some cases spatial changes in sand composition could be related to derivation from different sources (Enkeboll, 1982; Johnsson, 1990).

In 1979, Dickinson and Suczek formalized the correlation of sandstone composition with tectonic setting by examining petrographic data from 88 sandstone suites reported in the literature. By plotting the means of each suite on various ternary diagrams, distinct clusterings emerged. They found that these clusters correlated with the tectonic environment of the inferred source terranes. Three main categories were recognized: continental block, magmatic arc, and recycled orogen. The limits of these categories on the most frequently used ternary diagrams (Q_tFL and Q_mFL_t) were refined by Dickinson et al. (1983a) and Dickinson (1985). This work has since seen broad application in the interpretation of the tectonic evolution of sedimentary basins on a regional scale (e.g., Graham et al., 1976; Dickinson et al., 1979, 1986; Schwab, 1981; DeCelles, 1986; Lawton, 1986; Girty, 1987; Dorsey, 1988; Dumoulin, 1988; Jett and Heller, 1988; Ingersoll et al., 1990; Ingersoll and Dickinson, 1990; Graham et al., 1991; Ridgeway and DeCelles, this volume; Devaney and Ingersoll, this volume).

Although the groupings on the ternary diagrams as first defined by Dickinson and Suczek (1979) and later refined by Dickinson (1985) do correlate remarkably well with tectonic environment, important exceptions and misclassifications can occur. Molanaroli et al. (1991) have shown that the 1983 refinement (Dickinson et al., 1983a) correctly classifies only 74% of the samples from which the provenance model was defined. These misclassifications, as well as much of the scatter within groupings, probably relate to the use of diverse methods (Wolf, 1971; Mack, 1984; Zuffa, 1985; Ingersoll, 1990), sediment recycling (Blatt, 1967), and transport of sediment across tectonic boundaries (Mack, 1984; Velbel, 1985; Saccani, 1987), as well as to factors modifying sediment during transport, deposition, and diagenesis (Suttner, 1974; Johnsson, 1992).

Attention has turned more recently to the use of bulk chemical data, rather than the petrographic identification of framework grains, to classify sandstones by tectonic setting (e.g., Bhatia, 1983, 1985; van de Kamp and Leake, 1985; Bhatia and Crook, 1986; Roser and Korsch, 1986, 1988; Argast and Donnelly, 1987). Chemical analyses can be performed to greater precision that petrographic modal analyses, and are certainly easier to perform, but the textural information that is frequently critical to the interpretation of provenance is lost. Furthermore, authigenic material such as cement is included in bulk-chemical analyses, such that different diagenetic histories may make sediment from similar tectonic settings appear compositionally distinct. Nevertheless, the bulk-chemical approach may prove useful in instances where the influence of weathering or diagenesis has destroyed the original phases; if the sediment behaved as a closed system, elemental ratios may be relatively unaffected even when phases have reequilibrated to the diagenetic environment (Argast and Donnelly, 1987).

Petrofacies

The concept of sedimentologic "petrofacies"—facies defined in terms of sediment composition rather than lithology—was introduced in the early 1970s by workers studying the Great Valley Group in California (Gilbert and Dickinson, 1970; Swe and Dickinson, 1970; Mansfield, 1971; Dickinson and Rich, 1972). In northern California, petrofacies are much more laterally persistent than are lithofacies based on grain size and bedding style (Dickinson and Rich, 1972). Consideration of clastic sedimentary packages based on their compositional similarity rather than on gross lithologic character has proven useful both in the evaluation of source regions and their evolution over time, and in regional correlation—especially where biostratigraphic control is poor (e.g., Stanley, 1976; Ingersoll, 1978, 1983, 1990; Dickinson et al., 1982, 1983b; Thornburg and Kulm, 1987; Ingersoll et al., 1990; Ingersoll and Dickinson, 1990; Short and Ingersoll, 1990; Ingersoll and Cavazza, 1991).

Nevertheless, petrofacies do not provide an unambiguous window on tectonic setting and source rock composition any more than any other aspect of compositional information. Petrofacies may reflect not only similarities in source rock, but in any of the factors influencing sediment composition (Suttner, 1974; Suttner and Dutta, 1986; Johnsson et al., 1991). Furthermore, petrologically defined facies do not necessarily imply common provenance. Johnsson et al. (1991), for example, defined a quartz-arenite petrofacies among modern sands of the Orinoco River drainage basin that was the result of diverse source rock and chemical weathering controls on sand composition: compositional similarity in these sands did not reflect similar provenance, tectonic setting, or transportation history.

Assessing provenance

The extent to which the overall composition of clastic sediments may be modified by environmental factors has led to the use of specific sediment components, rather than overall composition, as indicators of provenance. Most obvious is the use of an unusual component that fortuitously is found in a clastic sequence and may constitute a unique tracer, such as the lawsonite and "blue sodic amphibole" encountered in the Alpine molasse of France (Mange-Rajetzky and Oberhänsli, 1982). Even the principal sandstone constituents, however, may provide provenance clues that are little affected by weathering and diagenetic processes. These include plagioclase composition (Pittman, 1970; Trevena and Nash, 1979, 1981; Maynard, 1984), feldspar twinning (Helmold, 1985; Devaney and Ingersoll, this volume), inclusions and vesicles in quartz (Folk, 1974), nature of quartz polycrystallinity and undulose extinction (Blatt and Christie, 1963; Folk, 1974; Basu et al., 1975; Arribas et al., 1985; Basu, 1985b; Girty et al., 1988; Johnsson et al., 1991), opaque mineral texture and composition (Reizebos, 1979; Molinaroli and Basu, 1987, this volume; Basu and Molinaroli, 1989, 1991; Grigsby, 1990, 1992), and composition of the heavy mineral suite (van Andel, 1959; Reizebos, 1979; Morton, 1985; Statteggar, 1987; Hurst and Morton, 1988; Morton et al., 1989; Morton, 1991; Grigsby, 1992).

More recently, attention has turned to dating of individual detrital grains as an aid in provenance discrimination. Such techniques offer a fundamentally new type of information than has previously been applied to provenance studies: that of time. Fission-track dating of apatite and zircon (Baldwin et al., 1986; Hurford and Carter, 1991), ^{40}Ar-^{39}Ar dating of amphiboles, micas, and feldspars (Kelly and Bluck, 1989), and U-Pb dating of zircons (Compston et al., 1985; Compston and Pidgeon, 1986; Drewey et al., 1987; Liu et al.,1992; Mueller et al., 1992; Rainbird et al., 1992), and even quartz (Hemming et al., 1990, 1991; McLennan et al., this volume) have each provided important constraints on provenance. As more precise dating techniques become available, better detrital age distributions will allow not only for less ambiguous provenance determinations, but also for the development of mixing models important to understanding the evolution of the crust.

Further attention is being directed at the distribution of trace elements, particularly the rare earth elements, as constraints on provenance. Rare earth element distributions appear to be particularly sensitive to tectonic setting (e.g., Balashov et al., 1964; Bhatia, 1985; Cullers et al., 1987; McLennan, 1989; McLennan et al., 1990, this volume; Floyd et al., 1991; McCann, 1991; McLennan and Taylor, 1991; Milodowski and Zalasiewicz, 1991; Girty et al., this volume). An advantage of using rare earth element distributions to evaluate provenance is that they are usually concentrated in finer grain sizes; thus shale fragments and matrix material in sandstones can be evaluated with the same set of criteria as shales. Furthermore, because most of the rare earth elements are relatively immobile during chemical weathering and diagenesis (Nesbitt, 1979; Nesbitt et al., 1980; Humphris, 1984; Maksimovic and Pantó, 1991), provenance information may not be lost even after considerable alteration of framework grains.

^{147}Sm decays to ^{143}Nd with a half-life of 1.063×10^{11} yr, providing another rare earth index of provenance. Because the lighter Nd is segregated from Sm during crustal differentiation, the Nd-Sm "model age" is generally taken to indicate the time of fractionation of crustal material from the mantle (McCulloch and Wasserburg, 1978; Taylor and McLennan, 1985; DePaolo, 1988). In practice, Nd-Sm systematics are generally expressed in terms of ϵ_{Nd}, the chondrite-normalized ratio of radiogenic ^{143}Nd to non-radiogenic ^{144}Nd, a quantity that increases with increased time since derivation from the mantle. Because of the refractory nature of Nd, ϵ_{Nd} and Nd-Sm model ages appear to be nearly immune to chemical weathering or diagenesis (Awwiller and Mack, 1989, 1991; McLennan et al., 1989). Furthermore, since model ages do not reflect crystallization times, but rather the time of extraction from the mantle reservoir, closed system metamorphism and even melting will not affect model ages (DePaolo, 1988). The Nd-Sm age of a sediment thus records the mean age of the crustal components from which it is derived, and has been used to help constrain tectonic provenance in both sandstones and shales (Michard et al., 1985; Andre et al., 1986; Frost and Winston, 1987; Frost and Coombs, 1989; Todd, 1989; Basu et al., 1990; McLennan et al., 1990; Evans et al., 1991; Floyd et al., 1991; Graham et al., 1991; Linn et al., 1991; Ross et al., 1991). See McLennan and others, this volume, for a thorough review of rare earth element geochemistry and its application to provenance studies.

CHEMICAL WEATHERING CONTROLS ON SEDIMENT COMPOSITION

Most minerals found in rocks were formed under different temperature, pressure, and chemical conditions than those at the Earth's surface. Goldich (1938) showed that the common silicate minerals form a weathering stability series analogous to the Bowen's reaction series: minerals that crystallize at higher temperatures (e.g., olivine, amphiboles, pyroxenes, and calcium plagioclase) are markedly less stable under chemical weathering conditions than minerals that crystallize at lower temperatures (e.g., sodium plagioclase, potassium feldspar, micas, and quartz). In general, nonsilicate minerals such as carbonates, sulfides, sulfates, and halides are more susceptible to chemical weathering than are the silicate minerals. The effect that chemical weathering has on sediment composition is to deplete sediments of the more unstable minerals, causing a relative increase in the proportion of the more stable minerals. Rock and mineral textures have strong effects on the level of alteration produced by chemical weathering. Fractures in bedrock, which can serve as conduits for fluids, become important loci for weathering. On a microscopic scale, mineral cleavages play a similar role. Polymineralic grains are particularly susceptible to chemical weathering because the boundaries between the mineral subgrains serve as efficient conduits for water.

The degree of alteration attained depends both on the intensity and the duration of the weathering. Intensity of weathering appears to be controlled primarily by climate (precipitation and temperature) and vegetation (particularly the nature and activity of organic acids). The duration of weathering is controlled by many factors, including relief, slope, sediment storage prior to ultimate deposition, and sedimentation rate. Complicated feedback mechanisms occur among many of these parameters.

Mechanisms of chemical weathering

A large body of experimental data has been amassed regarding the mechanisms of weathering at an atomic scale, but a thorough review of these data is beyond the scope of this review. The mechanisms of protonation, complexation, and dissolution have been explored by laboratory dissolution studies (e.g., Berner, 1978; Eggleton and Buseck, 1980; White, 1983; Holdren and Speyer, 1985; Schott and Berner, 1985; Wollast and Chou, 1985), and through atomic scale imaging with scanning tunneling and atomic force microscopy (e.g., Hochella, 1988, 1990; Johnsson et al., 1992). Studies of the surface textures of minerals from soils, including quartz (Crook, 1968; Douglas and Platt, 1977; Darmody, 1985; Brantley et al., 1986a,b), feldspars (Berner and Holdren, 1977; Nixon, 1979; Velbel, 1986), and mafic minerals (Berner et al., 1980; Berner and Schott, 1982; Velbel, 1984a, b, 1989) indicate that mineral dissolution is controlled largely by crystallographic imperfections, especially dislocations. Despite the progress made in understanding the mechanisms of weathering, applying laboratory data to nature has remained difficult. Mineral breakdown rates do not necessarily control rates of natural weathering where slope processes and tectonic controls appear to be more important (Stallard, 1988). Possibly for this reason, laboratory dissolution rates vastly exceed natural weathering rates (Paçes, 1973; Velbel, 1992).

Intensity versus duration of chemical weathering during pedogenesis

Weathering intensity. Even before leaving the source area, chemical weathering processes act on the composition of detrital material during pedogenesis. Clastic sediments generally are not produced directly from the bedrock, but instead represent reworked soils.

Chemical weathering reactions take place in the presence of water, and the intensity of chemical weathering is controlled to a large extent by the quantity and composition of water in the weathering environment. Climate plays a leading role in determining the quantity of water entering the weathering environment through precipitation, and the correlation of climate with chemical weathering intensity has been long-established (e.g., Woolnough, 1930; Krynine, 1936; Basu, 1976; Potter, 1978a; James et al., 1981; Mack, 1981; Suttner and Basu, 1981). Tropical environments, for example, may be characterized by chemical weathering so intense that the nature of the source rock cannot be determined from sediment composition; quartz arenites may result from the weathering of diverse rocks such as granites, schists, and gneisses (Potter, 1978a; Franzinelli and Potter, 1983; Johnsson et al., 1988, 1991). In more temperate areas, chemical weathering may be less intense; on the North Carolina Piedmont, Mann and Cavaroc (1973) were able to distinguish sands derived from igneous, metamorphic, and sedimentary sources. In arid environments, a generally good correlation exists between source rock composition and sediment composition (Girty et al., 1988). Climatic differences in weathering intensity across the United States have been graphically demonstrated by Meirding (1981), who discovered systematic differences in the weathering of marble tombstones.

The effects of weathering intensity on sediment composition are perhaps best evaluated by comparing sediment produced from similar parent material, under identical transport/depositional regimes, but under different climatic regimes. This approach was adopted by Basu (1976) and Suttner and Basu (1981). These authors concluded that sands derived from granitic rocks in a semi-arid environment were compositionally distinct from sands derived from similar rocks in a humid/temperate environment, although factors other than climate may also have varied between their study areas (e.g., Heins, this volume). Within regions underlain by a similar suite of source rocks, Breyer and Ehlmann (1981) showed that local variations in precipitation can have a profound impact on sediment composition, and Johnsson and others (this volume) show how such variations in sediment composition reflect different weathering processes in the soils from which they were derived.

Such studies have provided a baseline for the use of compositional data to infer paleoclimatology in clastic rocks. For example, Baker (1978) and Baker and Penteado-Orellana (1978) inferred a changing climate to be the cause of temporally changing sandstone compositions in Colorado River deposits in Texas. Suttner and Dutta (1986) quantitatively used framework mineralogy to evaluate changing paleoclimatic conditions during the deposition of sandstones produced from crystalline basement.

In addition to overall framework mineralogy, particular sedimentary components have been proposed as paleoclimatic indicators. Feldspar has long been regarded as particularly sensitive to climatic variability, and Krynine (1950) suggested that the presence of unaltered feldspars could be used as a criterion for the recognition of arid climates. Further, the coexistence of fresh and weathered feldspar may indicate humid weathering conditions: fresh feldspar could be eroded from canyon walls and buried prior to significant alteration, while more altered feldspars could be the product of weathering in soil profiles on the interfluves. Damuth and Fairbridge (1970) attributed Pleistocene arkoses on the South American continental shelf to a less intense chemical weathering environment corresponding to an arid climate during the last glacial maximum. Following the arguments of Krynine (1950), however, these arkoses could equally well have been derived from lower, less mature soil horizons in response to deeper downcutting of the Amazon and its tributaries during the sea-level lowstand at the last glacial maximum. Todd (1968) proposed that the amount of weathering of orthoclase relative to plagioclase may be indicative of climate. Citing the relative stability of potassium feldspars over plagioclase, he proposed that, although plagioclase will usually be more altered than orthoclase in temperate climates, heavy leaching in a low-relief setting characterized by heavy nonseasonal rains and subtropical temperatures could reverse this relationship. These predictions were not, however, borne out by Johnsson et al., (1991) from just such an environment in southern Venezuela. The use of feldspar as a paleoclimatic indicator was also questioned by James et al. (1981), who showed that there is a broad overlap in the level of alteration of detrital feldpar from wet and dry climates, and that the amount of alteration of feldspar probably is an inadequate paleoclimatic indicator, perhaps because the duration of weathering may have as large a role as climate. These authors did find, however, that the types of feldspar alteration *products* may show some climatic dependence.

The effects of vegetation on weathering intensity are complex. Probably most important is the release of CO_2 by the decay of organic matter in soils to form H_2CO_3, a powerful reactant during weathering. In addition, plants have evolved organic acids that are released through root mycorrhizae specifically to dissolve soil particles, aiding root penetration in soils. Such acids may increase weathering rates in the upper levels of soils (Knoll and James, 1987), but probably more important are the much more abundant humic and fulvic acids. These organic acids, produced through the degradation of more complex organic materials, have been shown to greatly increase mineral dissolution rates (Eckhardt, 1985; Krumbein and Dyer, 1985; Stumm et al., 1985; Bennett and Siegel, 1987; Hansley, 1987; Mast and Drever, 1987; Bennett et al., 1988; Hinman, 1990). Jackson and Keller (1970) showed that lichens could accelerate the weathering of bare basalt surfaces on Hawaii through chelation of metals. Vegetation may also increase weathering rates by increasing soil capacity to hold water and thereby enhancing soil/water interaction. Perhaps as important as these direct effects, however, are the complex effects vegetation may have on slope stability. On steep slopes, vegetation may reduce erosion over short time intervals (Moore, 1984). Vegetation may, however, increase weathering rates by anchoring soils on otherwise bare surfaces. Soil moisture and organic acids will then accelerate bedrock decomposition, leading to instability, landslides, and a new cycle of soil accumulation (Stallard, 1985).

Because the nature and extent of vegetation on the Earth's surface has changed dramatically over geologic time, the effect of vegetation on weathering may have been much different in the past. Pre-Silurian chemical weathering (e.g., before the advent of widespread vascular land plants) was almost certainly slower than at present (Basu, 1981a,b; Graustein and Velbel, 1981; Knoll and James, 1987). Basu (1981b) suggested that uptake of potassium and other nutrient elements by plants might have lowered their activity in soil waters, promoting chemical weathering of phases containing such elements. Nutrient elements may be cycled back into the soil by throughfall or decay, however, and a nearly closed system may cycle nutrients through the biosphere without involving soil minerals at all (Graustein and Velbel,

1981). Stallard (1985, 1988) pointed out that the extent of chemical weathering in the tributaries of the Amazon River, as measured by the concentrations of dissolved major ions in surface waters, show no correlation with vegetation. Nevertheless, the ultimate source of potassium and other nutrient elements must be soils, and potassium, which is removed from the soil-biosphere system through runoff, must be replenished by mineral weathering (Basu, 1981a).

Weathering duration. Equally as important as the intensity of chemical weathering in determining sediment composition is the amount of time over which weathering occurs. The greatest effects of chemical weathering occur in environments that are characterized not only by intense chemical weathering, but also by long duration of weathering.

The effects of weathering duration were vividly demonstrated by Barton (1916) in his classic study of archeologically dated granitic materials in Egypt. Conversely, the level of alteration of granitic materials, as expressed by the thickness of weathering rinds on cobbles, the degrees of pitting of cobbles and the level of grusification, is now a standard method for dating glacial deposits (e.g., Birkeland, 1964, 1973; Porter, 1975; Burke and Birkeland, 1979; Birkeland et al., 1980; Chinn, 1981; Colman and Pierce, 1981; Colman, 1982).

During pedogenesis, the duration of weathering is controlled by a complex set of factors, of which physiography is particularly important (Carson and Kirkby, 1972; Stallard and Edmond, 1983; Stallard, 1985, 1988; Grantham and Velbel, 1988; Johnsson and Stallard, 1989; Johnsson et al., 1991). Erosion can be viewed in terms of transport-limited and weathering-limited denudation regimes. Material loosened by weathering moves downslope by transport processes. If the transport processes removing weathered material from an area are potentially more rapid than the weathering processes generating the material, then erosion is said to be weathering limited. If, in contrast, the maximum weathering rate exceeds the ability of transport processes to remove material, then erosion is said to be transport limited.

Only recently have the implications of weathering-limited versus transport-limited denudation regimes become apparent to those studying chemical weathering and its products. Stallard (Stallard and Edmond, 1983, 1987; Stallard, 1985, 1988; Stallard et al., 1991) tracked differences in chemical weathering reactions by examining the dissolved load of tropical rivers draining areas of contrasting erosional regimes. These concepts have since been extended to the solid products of weathering (Johnsson et al., 1988, 1991; Johnsson and Stallard, 1989; Stallard et al., 1991). Weathering products have a much longer time to react with the soil and ground waters under transport-limited conditions. Exposed to the weathering environment for long periods of time, soils are extensively leached and weathering products are correspondingly poor in soluble cations. The extent to which leaching proceeds also depends on the composition of the weathering fluids, particularly with respect to Eh conditions (Retallack, 1990), but leaching will in any case be greater than in weathering-limited erosion regimes with similar fluid conditions. Incompletely chemically weathered residual detritus is greatly enriched in resistant phases, especially quartz. Relatively immobile elements, such as Fe and Al, may be retained in soils in the form of pisolitic horizons or lateritic crusts (Stallard, 1985; Johnsson and Stallard, 1989). Weathering in low-relief regions of tropical climate may lead to such extreme chemical weathering that nearly monomineralic quartz arenites are produced, regardless of the composition of the bedrock (Potter, 1978a,b; 1984; Franzinelli and Potter, 1983, 1985; Johnsson et al., 1988, 1991). Under weathering-limited conditions, where potential sediment transport rates exceed the rate at which weathering can generate loose material, detritus spends much less time in the soil profile. Under these conditions, weathering products are incompletely leached, cation-rich phases, and immobile elements do not accumulate in the soil. Residual detritus consists of a varied suite of unstable phases, and the proportions of these phases reflect the composition of the bedrock (Johnsson, 1990; Johnsson et al., 1991).

Chemical weathering during transport and deposition

In most large depositional systems, the amount of time sediment spends in transport is probably insignificant compared to time spent in temporary storage in alluvial sequences such as floodplains, point bars, and terraces. Fluvial sediment is stored in such deposits through lateral and vertical accretion, and previously deposited sediment is reincorporated into river bedload during meander growth, channel migration, and avulsion. The result is a net replacement of younger, freshly eroded sediments by older, previously stored material. When sediments are stored in alluvial sequences in an intense weathering environment, profound alteration of sediment composition may result (Rolfe and Hadley, 1964; Krook, 1969; Macaire, 1986; Savage et al., 1988; Johnsson and Meade, 1990; Morton and Johnsson, this volume). This process gives rise to first-cycle quartz sands in the Venezuelan Andean foreland basin (Johnsson et al., 1988, 1991). Mills and Wagner (1985) used differential weathering of stored alluvium to chart changes in the history of the New River, an approach advocated by Retallack (1986) for the study of ancient rivers as well.

Vegetation plays a major role in stabilizing fluvial systems (Evans, 1991), and is important in determining the extent to which material is stored in alluvial sequences and the rate at which it is reincorporated into river bedload. As the nature and extent of vegetation evolved over geological time, its effects on fluvial dynamics and on the extent of alluvial storage likely evolved as well (Schumm, 1968).

Sediment is ultimately deposited in a depositional setting where it is buried and isolated from the weathering environment. The rapidity with which sediment is isolated from weathering may be important in determining its composition, especially where exposure to the weathering environment has been brief during pedogenesis and transport. Wascher et al. (1947) and Leighton and Willman (1950) noticed a positive correlation between calcium carbonate content and sedimentation rate in Mississippi Valley loess deposits, and attributed this correlation to dissolution of calcium carbonate when sedimentation rates were sufficiently slow to allow weathering to proceed during deposition. Conversely, Krynine (1935, 1936, 1941b,c) suggested that rapid sedimentation rates overwhelm the effects of climate in determining the amount of weathering undergone by a sediment, even in the tropics. This conclusion is supported by Ruxton (1970) and Basu (1985a), who found modern fluvial sands to be exceptionally quartz-poor in tropical, high-relief settings in Papau. Such settings have been shown by Milliman and Meade (1983) to be sources for a disproportionate amount of sediment reaching the world ocean, suggesting that tropical environments characterized by only minor chemical weathering may be particularly important in the geologic record.

Role of slope

Slope is an important variable controlling the duration of weathering during pedogenesis, transport, and deposition. During each of these processes, steep slopes work to reduce the time sediment is exposed to weathering (Stallard, 1985, 1988; Retallack, 1990). Steep slopes correlate with rapid downslope transport rates, favoring weathering-limited erosion conditions. Steep canyons afford little opportunity for alluvial storage, decreasing transport times and limiting sediments exposure to chemical weathering during transport. Finally, steep slopes generally tend to lead to high sedimentation rates in adjacent sedimentary basins, limiting the exposure of sediments to weathering in the depositional environment. Sediments will be least affected by chemical weathering in tectonically active areas where rapid uplift not only maintains steep slopes but also continually brings a fresh supply of unweathered bedrock to the surface.

Chemical weathering, the carbon cycle, and sediment composition

The role of mineral weathering in the global cycling of carbon is of interest because CO_2 is a primary determinant of the intensity of chemical weathering and, at the same time, is an important greenhouse gas. CO_2 allows short-wavelength solar radiation to enter the atmosphere but traps most of this energy when it is reradiated as heat at longer wavelengths. The level of atmospheric CO_2 is thus an important control on secular variation in global climate. Berner, Lasaga, and Garrels have modeled the modern carbon cycle and estimated partitioning among various carbon reservoirs through geologic time (Berner et al., 1983; Lasaga et al., 1985; Berner and Lasaga, 1989); for alternate views, see Volk (1987) and Tans et al. (1990).

Sedimentary rocks represent the largest reservoir of carbon, containing more than 1,500 times the carbon found in the atmosphere, ocean, and biosphere combined (Berner et al., 1983; Lasaga et al., 1985; Berner and Lasaga, 1989). This sedimentary carbon reservoir is divided roughly equally between CO_3^{-2} in limestones and dolostones, and organic carbon in sedimentary rocks (Berner et al., 1983; Lasaga et al., 1985; Berner and Lasaga, 1989).

According to the Berner-Lasaga-Garrels model, CO_2 is absorbed from the atmosphere by plants through photosynthesis and stored in the soil in the form of organic matter; there it reacts with water to form H_2CO_3. Chemical weathering reactions involving H_2CO_3 release HCO_3^- and soluble metallic cations, including Ca. These weathering products are transported by ground and river waters to the ocean, where marine zoo- and phytoplankton utilize the Ca and HCO_3^- to precipitate tests consisting of $CaCO_3$, which ultimately sink and accumulate to form new carbonate rocks. The biological precipitation of $CaCO_3$ releases CO_2 into the ocean and, because the ocean and atmosphere maintain a balance through gas exchange, this CO_2 ultimately is released back into the atmosphere, completing the cycle. The stoichiometry of the weathering reactions of carbonate minerals dictates that carbonate weathering results in no net loss of atmospheric CO_2. However, only half the CO_2 consumed from the atmosphere during silicate weathering is returned to the atmosphere subsequent to marine precipitation of $CaCO_3$. A balance is maintained through the release of CO_2 from the sedimentary reservoir during metamorphism of carbonate and silicate minerals; CO_2 released by these reactions enters the atmosphere principally through volcanic eruptions. The rate of volcanic degassing of CO_2 is controlled by plate tectonics; variations in global seafloor spreading rates control long-term temporal variations in volcanism associated with the creation and destruction of oceanic crust at mid-ocean ridges and subduction zones, respectively.

Chemical weathering largely determines the rate of CO_2 consumption from the atmosphere. Thus, variations in the intensity of chemical weathering and tectonism govern the level of dissolved CO_2 in the atmosphere and ocean. Because the reservoir of carbon in the atmosphere is small relative to the fluxes among reservoirs, even small variations in these fluxes can have a profound effect on the concentration of atmospheric CO_2. Owing to the importance of CO_2 as a greenhouse gas, such variations in atmospheric CO_2 may be a dominant control on temperatures at the Earth's surface over geologic time. Thus the system controlling the composition of clastic sediments is intimately linked with the carbon cycle: chemical weathering, itself strongly influenced by climate, plays an important role in determining global climate. The composition of clastic sediments provides a detailed record of chemical weathering through geologic time. Clastic sediments may thus provide a means of directly observing long-term changes in the intensity of chemical weathering, providing a better understanding of the interactions among the atmosphere, hydrosphere, and biosphere over geologic time.

TRANSPORT CONTROLS ON SEDIMENT COMPOSITION

In addition to further chemical weathering, a number of other processes may act on sediment composition during transport and deposition. These processes—which include abrasion, mechanical breakdown, hydrodynamic sorting, and mixing effects—are commonly especially difficult to separate in natural systems.

Abrasion and mechanical breakdown

Kuenen's classic tumbler and flume experiments (Kuenen, 1955, 1956, 1959, 1960, 1962, 1964) suggest that, while abrasion during fluvial transport may be important in rounding gravel and cobble-size material, it does not result in a significant reduction in angularity or a volumetric loss of softer constituents in sand. These experimental results have recently been more rigorously quantified by Osborne et al. (this volume). Mechanical abrasion does appear to be important, however, in rounding eolian and beach sands (Kuenen, 1960, 1964), and perhaps in quantitatively removing less mechanically stable components, especially lithic fragments (Dutta and Zhou, this volume) from eolian deposits. In many fluvial systems, eolian transport of sediment is an important and commonly overlooked process (Johnsson et al., 1991) implying that mechanical breakdown may be of more importance in fluvial systems than is commonly believed.

In natural systems, results are somewhat ambiguous. This is largely due to the difficulty of separating the effects of chemical weathering during protracted alluvial storage, mechanical breakdown, and hydrodynamic sorting. Some workers have found downstream enrichment of durable minerals relative to less resistant materials to be insignificant (Russell, 1937; Pollack, 1961; Breyer and Bart, 1978), while others have noted increases in quartz/feldspar ratios (Mackie, 1896; Martens, 1931; Plumley, 1948; Hayes, 1962), reduction of unstable lithic fragments (Cameron and Blatt, 1971; Shukis and Ethridge, 1975), increase in zircon, tourmaline, and rutile relative to other heavy minerals (Hubert, 1962), and progressive downstream destruction of less mechanically stable plagioclase grains (Pittman, 1969). In a study of a short, high-gradient stream in northwestern Italy, McBride and Picard (1987) compared the outcrop areas of different rock types with corresponding percentages of different grain types in stream sediments, and found that more durable rock types were overrepresented in the sediments. Estimates of source rock abundance based on the composition of sands in beach environments similarly may be in error (Picard and McBride, this volume). Garzanti (1986) called on mechanical breakage of feldspar along cleavages or twin planes to explain the elimination of feldspar from coarser grain sizes and its accumulation in finer grain sizes over time in deltaic sandstones of northern Italy. The effect chemical weathering may have played in this process was not, however, immediately addressed.

The fact that most of the less durable constituents common in fluvial sands are also chemically the least stable makes the effects of mechanical erosion and abrasion difficult to separate from the effects of chemical weathering. Slatt and Eyles (1981) addressed this problem by examining first- and second-cycle glacial sands in Arctic environments, where the influence of chemical weathering could be presumed to be slight. These authors noted a dramatic decrease in lithic fragment abundance and an increase in feldspar abundance between the first- and second-cycle sands, which they explained by the mechanical breakdown of lithic fragments. Mechanically weak and strongly foliated lithic fragments were most severely affected. Cather and Folk (1991) examined various size fractions of volcaniclastic sands in an arid environment, and demonstrated that sedimentary components were redistributed between different size fractions, rather than quantitatively removed, through mechanical breakdown.

Finally, emphasis should be placed on the interrelated nature of the processes of physical and chemical weathering. Just as abrasion opens fresh mineral surfaces to chemical weathering, steadily increasing the ratio of surface area to mineral volume, particles which are more severely weathered are more susceptible to abrasion (Bradley, 1970). Such a feedback mechanism makes separation of the two processes difficult.

Sorting and mixing effects

When different grain sizes differ markedly in composition, hydrodynamic sorting may be of paramount importance in determining local sediment composition (Garzanti, 1986). Hydrodynamic sorting is responsible for drastic differences in composition between the suspended load and bedload of most rivers (cf. Koehnken, 1990; Johnsson et al., 1991); the former typically is dominated by clay minerals and monomineralic silt grains, whereas the latter usually consists of more varied primary minerals and lithic fragments. Within the bedload, compositional sorting resulting from density differences is most pronounced where rapid changes in current velocity occur, such as on point bars, transverse bars, and behind obstructions. Local concentrations of dense minerals may drastically affect local sediment composition, but the overall composition of the material within the system is not quantitatively changed (Garzanti, 1986).

Overall sediment composition does, of course, vary as a result of addition of new sources, such as incoming tributaries in a fluvial system. Secondary sources may carry sediment of different provenance and history than that of the principal source, thus complicating interpretation of compositional information. Columbia River sands, for example, show a downstream *reduction* in the quartz/feldspar ratio, coupled with an *increase* in grain size; this clearly suggests that the effects of downstream dilution or sorting overwhelm any abrasion or chemical weathering effects operating in that river (Kelley and Whetten, 1969; Whetten et al., 1969).

Subsequent to deposition, sediment may be subjected to significant modification in the depositional environment. Mechanical and chemical weathering continue to be of great importance prior to final burial. Davies and Ethridge (1975) suggested that composition reflects characteristics of the depositional environment; they found that recent and ancient fluvial, deltaic, and beach sands could be distinguished on the basis of composition alone. Kairo and others (this volume) show a correlation between framework composition and depositional setting within the Fountain and Minturn Formations (Pennsylvanian of Colorado). These authors attribute compositional differences among depositional environments to mechanical disaggregation and hydrodynamic sorting. Such studies suggest that framework composition, perhaps coupled with studies of grain-size distribution, may aid in interpretation of depositional environments in ancient sedimentary rocks of more ambiguous setting.

DIAGENETIC CONTROLS ON SEDIMENT COMPOSITION

Following burial, clastic sediments are likely to be subjected to an entirely different set of modification processes. Burial diagenesis commonly results in the destruction of feldspar and lithic grains, dissolution of unstable phases, the growth of authigenic phases, and cementation (Scholle and Schluger, 1979; McDonald and Surdam, 1984). Like chemical weathering, the principal changes to sediment composition brought about by burial diagenesis result from reequilibration of original phases to their current chemical and physical environment. Much greater variability exists in diagenetic environments than in weathering environments, however, because of correspondingly greater temperature and pressure ranges. Accordingly, progressive diagenetic alteration results in a more complex set of compositional modifications as compared to chemical weathering.

In coarse clastic rocks, diagenesis may result in a net removal of unstable constituents through dissolution in an open system, or in a rearrangement of constituents through a dissolution-precipitation mechanism in a closed system. Perhaps the best studied aspect of sandstone diagenesis is the creation of "secondary porosity" through the dissolution of unstable framework constituents (principally feldspar) in an open system (e.g., Bjørlykke, 1984; Boles, 1984; Franks and Forester, 1984; Markert and Al-Shaieb, 1984; Moncure et al., 1984; Surdam et al., 1984; Nedkvitne and Bjørlykke, 1992), which is of obvious importance in the evaluation of petroleum reservoir rocks. In addition to outright dissolution, feldspars are vulnerable to calcitization, albitization, and zeolitization through dissolution-precipitation processes (e.g., Land and Milliken, 1981; Boles, 1982, 1984; Siebert et al., 1984; Walker, 1984; Walker, 1984; Helmold, 1985; McBride, 1985; Morad et al., 1990). Diagenesis of sandstones also commonly results in chert dissolution (McBride, 1985; Shanmugam and Higgins, 1988), dissolution and dolomitization of carbonate rock fragments (McBride, 1985), chloritization and zeolitization of volcanic rock fragments (Helmold and van de Kamp, 1984; Helmold, 1985; McBride, 1985), and alteration of heavy mineral assemblages (van Andel, 1959; Nickel, 1973; Morton, 1984, 1986; Morton et al., 1989; Cavazza and Gandolfi, 1992).

Diagenesis of shales presents particular problems because clay minerals are generally more reactive than constituents of coarser clastic sediments, and because the modification of porosity and permeability is more important in controlling the extent to which diagenesis proceeds. In general, shales are nearly impermeable, which tends to force diagenesis toward closed system reactions and which minimizes the importance of external chemical controls on diagenesis. Accordingly, diagenetic reactions in shales are potentially valuable indicators of the physical controls on diagenesis. The transformation of smectite to illite, through various "mixed layer" intermediaries, for example, is relatively well calibrated to diagenetic temperature (e.g., Hower et al., 1976; Hower, 1981; Nadeau and Reynolds, 1981; Pollastro, 1985; Bethke and Altaner, 1986; Pollastro and Barker, 1986; Inoue et al., 1990; Freed and Peacor, 1992).

Cementation of clastic rocks is a direct result of reprecipitation of constituents that are dissolved by diagenetic pore waters. Cementation "stratigraphy" can thus be used to evaluate changing diagenetic conditions. Because much cementation may take place within the reach of meteoric waters (Longstaffe, 1984), authigenic mineralogy and cementation history may provide clues to the climatic conditions prevailing during early diagenesis (Dutta and Suttner, 1986). Diagenesis is also linked with tectonic setting, through the control tectonics may exert on the thermal structure of, and fluid flow within, the upper crust (Siever, 1979).

RECYCLING

About 66% of the Earth's land surface is covered by sedimentary rock (Blatt and Jones, 1975) and, because sedimentary rocks supply a disproportionate volume of detritus, perhaps as much as 80% of the material within the global sedimentary system has been derived from preexisting sedimentary rocks (Blatt and Jones, 1975; Garrels, 1986). Clastic sedimentary rocks that have already been processed through the system described above are unique source rocks for subsequent sediments in that their composition has been "premodified" according to the tectonic, climatic, geomorphic, and biologic conditions prevailing during the rock's derivation and deposition. Clastic sediments derived from preexisting clastic rocks thus contain a complex compositional signal reflecting each previous episode of erosion, weathering, transport, deposition, and diagenesis.

The most obvious effect of polycyclicity on sediment composition is an enrichment in chemically and mechanically stable phases and textures at the expense of more labile ones. Even in settings where chemical weathering, mechanical breakdown, and diagenesis have little effect on sediment composition, the effect they do have will be amplified in the next cycle of sedimentation. Thus, polycyclic sandstones tend to be progressively enriched in quartz, zircon, and other stable phases. Nevertheless, recycling can also result in a sediment rich in phases unstable at Earth-surface conditions, as authigenic and diagenetic phases are released through erosion. Such effects may be of greatest importance in fine-grained clastic rocks, such as shales, because the chemical reactivity of clay minerals is more likely to force them to reequilibrate with diagenetic conditions than phases found principally in coarser fractions. Recycling of fine-grained clastic rocks may yield sediments that are petrographically "immature" in that they are composed principally of lithic fragments,

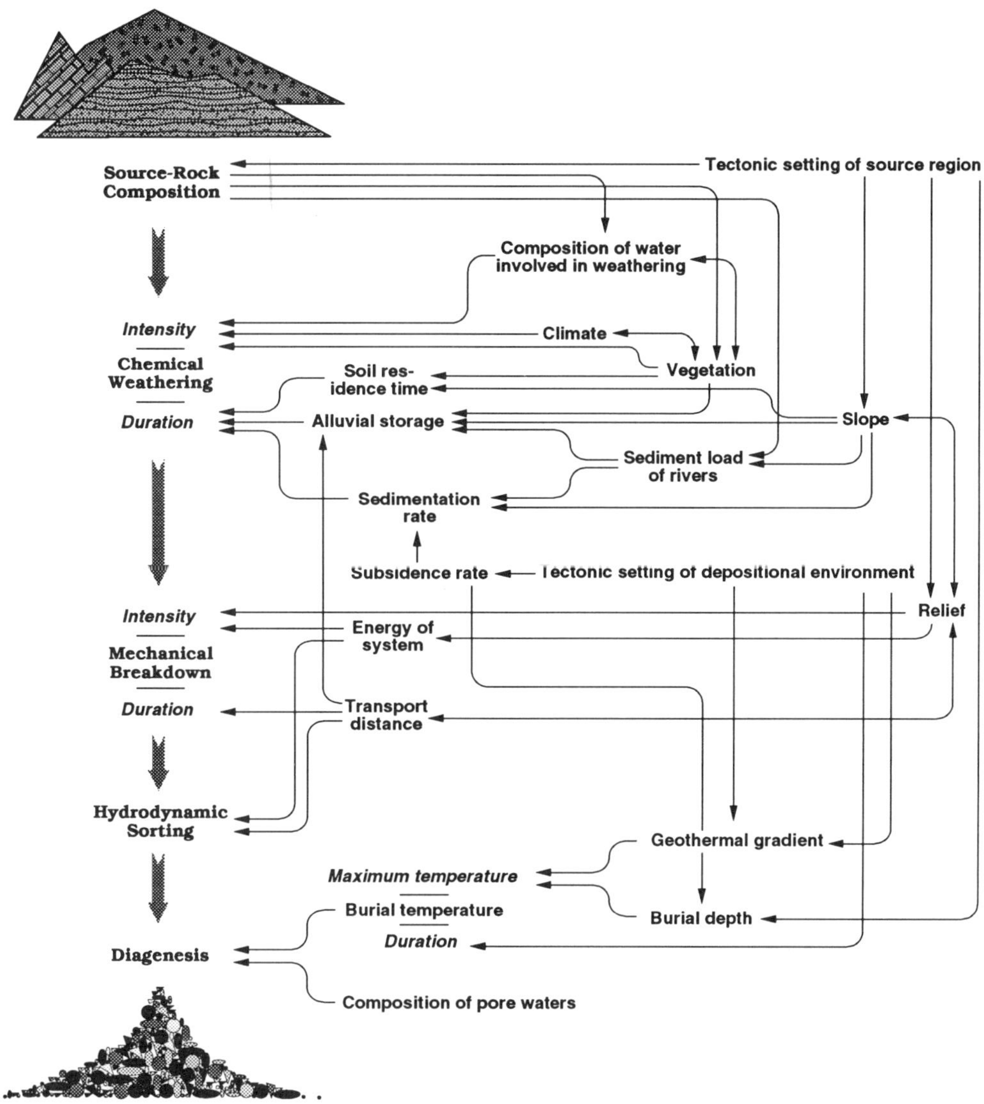

Figure 1. Schematic representation of the system controlling the composition of clastic sediments. Arrows indicate parameters that exert influence on each other.

but nonetheless are depleted in the more mobile phases or elements (Johnsson et al., 1991).

The effect of recycling on the composition of clastic sediments is thus to complicate the detrital compositional signal by overprinting environmental influences on composition from successive sedimentary cycles. In the process, sediments tend to be progressively depleted in labile elements. In addition, the tectonic imprint on sand composition may be passed on through successive sedimentary cycles, even as tectonic settings change (Graham et al., 1993).

SYNTHESIS

The composition of clastic rocks is thus controlled by a complex suite of factors. The fact that an empirical correlation nonetheless exists between sediment composition and plate tectonic setting suggests that, in addition to controlling the global distribution of rock types, tectonic setting also influences some of the other parameters controlling sediment composition. Most notable among these is slope; active tectonism is responsible for maintaining steep slopes, and in intense chemical weathering environments, where differences in source rock composition may be obliterated by chemical weathering, the role of tectonic setting in controlling sediment composition may lie as much in the maintenance of steep slopes as in the control of dominant rock types.

Processes operating during erosion, transport, deposition, and burial are intimately interlinked, creating a complex web of feedback mechanisms within the system controlling the composition of clastic sediments (Fig. 1). For example, as discussed above, the duration of chemical weathering may be extended by temporary storage of sediment on alluvial plains. The extent to which such storage occurs depends, however, on the sediment load of the river; which is in turn dependent on source rock composition, tectonism in the source area, and climate. The amount of reworking of stored alluvium depends on the stability of the fluvial system, again dependent on the sediment load of the river, but also influenced by the extent of stabilization by vegetation (itself controlled by climate) and the geometry of the alluvial system. Finally, the ultimate limit of storage, before sediments are removed from the fluvial system by deep burial, is controlled by the sedimentation rate within the alluvial system. This rate is determined not only by the sediment influx, but also by the tectonic environment of the alluvial system and the ultimate depositional environment. Altering any of these parameters—tectonics in the source area, alluvial system or depositional setting, source rock composition, climate, vegetation, or physiography—affects other parameters within the system. A similar web of interrelationships could be constructed for other processes discussed above that affect sediment composition (Fig. 1).

A firm understanding of the controls on clastic composition is a prerequisite to accurate interpretation of provenance. In addition, potentially valuable information concerning the sedimentary environment may be obtained if the interactions among these processes can be understood. Information about ancient climate, vegetation, and the geometry and nature of the fluvial system are all contained in the composition of clastic materials.

ACKNOWLEDGMENTS

My research into the system regulating the composition of clastic sediments has been supported by the U.S. Geological Survey, Princeton University, the National Science Foundation (Grant EAR-8407651 to R. F. Stallard, and Grant EAR-8616904 to R. F. Stallard and N. S. Lundberg), the Smithsonian Tropical Research Institute, the Venezuelan Ministry of the Environment and Renewable Natural Resources, the Geological Society of America, the American Association of Petroleum Geologists, and Sigma Xi. The idea that the composition of clastic sediments was controlled by an interrelated system of parameters, and the potential feedback mechanisms among these parameters, grew out of many years of valuable conversations with individuals too numerous to list. Special acknowledgment, however, must be given to A. Basu, G. D. Girty, R. V. Ingersoll, P. A. Johnsson, N. S. Lundberg, R. H. Meade, R. F. Stallard, L. J. Suttner, and M. A. Velbel, each of whom was instrumental in the development of the ideas expressed here. The manuscript benefitted from critical reviews by A. Basu, R. V. Ingersoll, and W. C. James.

REFERENCES CITED

Andre, L., Deutsch, S., and Hertogen, J., 1986, Trace-element and Nd isotopes in shales as indexes of provenance and crustal growth: The early Paleozoic from the Brabant Massif (Belgium): Chemical Geology, v. 57, p. 101–115.

Argast, S., and Donnelly, T. W., 1987, The chemical discrimination of clastic sedimentary components: Journal of Sedimentary Petrology, v. 57, p. 813–823.

Arribas, J., Marfil, R., and De La Peña, J., 1985, Provenance of Triassic feldspathic sandstones in the Iberian Range (Spain): Significance of quartz types: Journal of Sedimentary Petrology, v. 55, p. 864–868.

Awwiller, D. N., and Mack, L. E., 1989, Diagenetic resetting of Sm-Nd isotope systematics in Wilcox Group sandstones and shales, San Marcos Arch, south-central Texas: Transactions—Gulf Coast Association of Geological Societies, v. 39, p. 321–330.

Awwiller, D. N., and Mack, L. E., 1991, Diagenetic modification of Sm-Nd model ages in Tertiary sandstones and shales, Texas Gulf Coast: Geology, v. 19, p. 311–314.

Baker, V. R., 1978, Adjustment of fluvial systems to climate and source terrain in tropical and subtropical environments, *in* Miall, A. D., eds., Fluvial sedimentology: Canadian Society of Petroleum Geologists Memoir 5, p. 211–230.

Baker, V. R., and Penteado-Orellana, M. M., 1978, Fluvial sedimentation conditioned by Quaternary climatic change in central Texas: Journal of Sedimentary Petrology, v. 48, p. 433–451.

Balashov, Y. A., Ronov, A. B., Migdisov, A. A., and Turanskaya, N. V., 1964, The effect of climate and facies environment on the fractionation of the rare earth elements during sedimentation: Geochemistry International, v. 1, p. 951–969.

Baldwin, S. L., Harrison, T. M., and Burke, K., 1986, Fission track evidence for the source of accreted sandstones, Barbados: Tectonics, v. 5, p. 457–468.

Barton, D. C., 1916, The disintegration of granite in Egypt: Journal of Geology, v. 24, p. 382–393.

Basu, A., 1976, Petrology of Holocene fluvial sand derived from plutonic source rocks: Implications to paleoclimatic interpretation: Journal of Sedimentary Petrology, v. 46, p. 694–709.

Basu, A., 1981a, Reply to Comment on "Weathering before the advent of land plants: Evidence from unaltered K-feldspars in Cambrian-Ordovician arenites": Geology, v. 9, p. 505–506.

Basu, A., 1981b, Weathering before the advent of land plants: Evidence from unaltered K-feldspars in Cambrian-Ordovician arenites: Geology, v. 9, p. 132–133.

Basu, A., 1985a, Influence of climate and relief on compositions of sands released at source areas, *in* Zuffa, G. G., ed., Provenance of arenites: Dordrecht, D. Reidel, p. 1–18.

Basu, A., 1985b, Reading provenance from detrital quartz, *in* Zuffa, G. G., ed., Provenance of arenites: Dordrecht, D. Reidel, p. 231–248.

Basu, A., and Molinaroli, E., 1989, Provenance characteristics of detrital opaque Fe-Ti oxide minerals: Journal of Sedimentary Petrology, v. 59, p. 922–934.

Basu, A., and Molinaroli, E., 1991, Reliability and application of detrital opaque Fe-Ti oxide minerals in provenance determination, *in* Morton, A. C., Todd, S. P., and Haughton, P.D.W., eds., Developments in sedimentary provenance studies: Geological Society of London Special Publication 57, p. 55–65.

Basu, A., Young, S. W., Suttner, L. J., James, W. C., and Mack, G. H., 1975, Re-evaluation of the use of undulatory extinction for polycrystallinity in detrital quartz for provenance interpretation: Journal of Sedimentary Petrology, v. 45, p. 873–882.

Basu, A. R., Sharma, M., and DeCelles, P. G., 1990, Nd, Sr-isotopic provenance and trace element geochemistry of Amazonian foreland basin fluvial sands, Bolivia and Peru: Implications for ensialic Andean orogeny: Earth and Planetary Science Letters, v. 100, p. 1–17.

Bennett, P., and Siegel, D. I., 1987, Increased solubility of quartz in water due to complexing by organic compounds: Nature, v. 326, p. 684–687.

Bennett, P. C., Melcer, M. E., Siegel, D. I., and Hassett, J. P., 1988, The dissolution of quartz in dilute aqueous solutions of organic acids at 25°C: Geochimica et Cosmochimica Acta, v. 52, p. 1521–1530.

Berner, R. A., 1978, Rate control of mineral dissolution under earth surface conditions: American Journal of Science, v. 278, p. 1235–1252.

Berner, R. A., and Holdren, G. R., 1977, Mechanisms of feldspar weathering, II. Observations of soil grains: Geochimica et Cosmochimica Acta, v. 43, p. 1173–1186.

Berner, R. A., and Lasaga, A. C., 1989, Modeling the geochemical carbon cycle: Scientific American, v. 260, p. 74–81.

Berner, R. A., and Schott, J., 1982, Mechanisms of pyroxene and amphibole weathering II. Observations of soil grains: American Journal of Science, v. 282, p. 1214–1231.

Berner, R. A., Sjöberg, E. L., Velbel, M. A., and Krom, M. D., 1980, Dissolution of pyroxenes and amphiboles during weathering: Science, v. 207, p. 1205–1206.

Berner, R. A., Lasaga, A. C., and Garrels, R. M., 1983, The carbonate-silicate geochemical cycle and its effect on atmospheric carbon dioxide over the past 100 million years: American Journal of Science, v. 283, p. 641–683.

Bethke, C. M., and Altaner, S. P., 1986, Layer-by-layer mechanism of smectite illitization and application to a new rate law: Clays and Clay Minerals, v. 34, p. 136–145.

Bhatia, M. R., 1983, Plate tectonics and geochemical composition of sandstones: Journal of Geology, v. 91, p. 611–627.

Bhatia, M. R., 1985, Rare earth element geochemistry of Australian Paleozoic graywackes and mudrocks: Provenance and tectonic controls: Sedimentary Geology, v. 45, p. 97–113.

Bhatia, M. R., and Crook, K.A.W., 1986, Trace element characteristics of graywackes and tectonic setting discrimination of sedimentary basins: Contributions to Mineralogy and Petrology, v. 92, p. 181–193.

Birkeland, P. W., 1964, Pleistocene glaciation of the northern Sierra Nevada north of Lake Tahoe, California: Journal of Geology, v. 72, p. 810–825.

Birkeland, P. W., 1973, Use of relative age-dating methods in a stratigraphic study of rock glacier deposits, Mt. Sopris, Colorado: Arctic and Alpine Research, v. 5, p. 401–416.

Birkeland, P. W., Burke, R. M., and Walker, A. L., 1980, Soils and subsurface rock-weathering features of Sherwin and pre-Sherwin glacial deposits, eastern Sierra Nevada, California: Geological Society of America Bulletin, v. 91, p. 238–244.

Bjørlykke, K., 1984, Formation of secondary porosity: How important is it? *in* McDonald, D. A., and Surdam, R. C., eds., Clastic diagenesis: American Association of Petroleum Geologists Memoir 37, p. 277–286.

Blatt, H., 1967, Provenance determinations and recycling of sediments: Journal of Sedimentary Petrology, v. 37, p. 1031–1044.

Blatt, H., and Christie, J. M., 1963, Undulatory extinction in quartz of igneous and metamorphic rocks and its significance in provenance studies of sedimentary rocks: Journal of Sedimentary Petrology, v. 33, p. 559–579.

Blatt, H., and Jones, R. L., 1975, Proportions of exposed igneous, metamorphic, and sedimentary rocks: Geological Society of America Bulletin, v. 86, p. 1085–1088.

Boggs, S., Jr., 1968, Experimental study of rock fragments: Journal of Sedimentary Petrology, v. 38, p. 1326–1339.

Boles, J. R., 1982, Active albitization of plagioclase, Gulf Coast Tertiary: American Journal of Science, v. 282, p. 165–180.

Boles, J. R., 1984, Secondary porosity reactions in the Stevens Sandstone, San Joaquin Valley, California, *in* McDonald, D. A., and Surdam, R. C., eds., Clastic diagenesis: American Association of Petroleum Geologists Memoir 37, p. 217–224.

Bradley, W. C., 1970, Effect of weathering on abrasion of granitic gravel, Colorado River (Texas): Geological Society of America Bulletin, v. 81, p. 61–80.

Brantley, S. L., Crane, S. R., Crerar, D. A., Hellmann, R., and Stallard, R., 1986a, Dislocation etch pits in quartz *in* Davis, J. A., and Nayes, K. F., eds., Geochemical processes at mineral surfaces: Washington, D.C., American Chemical Society, p. 635–649.

Brantley, S. L., Crane, S. R., Crerar, D. A., Hellmann, R., and Stallard, R., 1986b, Dissolution at dislocation etch pits in quartz: Geochimica et Cosmochimica Acta, v. 50, p. 2349–2361.

Breyer, J. A., and Bart, H. A., 1978, The composition of fluvial sands in a temperate semiarid region: Journal of Sedimentary Petrology, v. 48, p. 1311–1320.

Breyer, J. A., and Ehlmann, A. J., 1981, Mineralogy of arc-derived sediment: Siliciclastic sediment on the insular shelf of Puerto Rico: Sedimentology, v. 28, p. 61–74.

Burke, R. M., and Birkeland, P. W., 1979, Reevaluation of multiparameter relative dating techniques and their application to the glacial sequence along the eastern escarpment of the Sierra Nevada, California: Quaternary Research, v. 11, p. 21–51.

Cameron, K. L., and Blatt, H., 1971, Durabilities of sand-size schist and "volcanic" rock fragments during fluvial transport, Elk Creek, Black Hills, South Dakota: Journal of Sedimentary Petrology, v. 41, p. 565–576.

Carson, M. A., and Kirkby, M. J., 1972, Hillslope form and process: Cambridge, Cambridge University Press, 475 p.

Cather, S. M., and Folk, R. L., 1991, Pre-diagenetic sedimentary fractionation of andesitic detritus in a semi-arid climate: An example from the Eocene Datil Group, New Mexico, *in* Fisher, R. V., and Smith, G. A., eds., Sedimentation in volcanic settings: Tulsa, Oklahoma, SEPM Special Publication 45, p. 211–226.

Cavazza, W., and Gandolfi, G., 1992, Diagenetic processes along a basin-wide marker bed as a function of burial depth: Journal of Sedimentary Petrology, v. 62, p. 261–272.

Cawood, P. A., 1991, Nature and record of igneous activity in the Tonga arc, SW Pacific, deduced from the phase chemistry of derived detrital grains, *in* Morton, A. C., Todd, S. P., and Haughton, P.D.W., eds., Developments in sedimentary provenance studies: Geological Society of London Special Publication 57, p. 305–321.

Chinn, T.J.H., 1981, Use of rock weathering-rind thickness for Holocene absolute age-dating in New Zealand: Arctic and Alpine Research, v. 13, p. 33–45.

Colman, S. M., 1982, Chemical weathering of basalts and andesites; Evidence from weathering rinds: U.S. Geological Survey Professional Paper 1246, 43 p.

Colman, S. M., and Pierce, K. L., 1981, Weathering rinds on andesitic and basaltic stones as Quaternary age indicators, western United States: U.S. Geological Survey Professional Paper 1210, 56 p.

Compston, W., and Pidgeon, R. T., 1986, Jack Hills, evidence of more very old detrital zircons in western Australia: Nature, v. 321, p. 766–769.

Compston, W., and 6 others, 1985, The age of (a tiny part of) the Australian continent: Nature, v. 317, p. 559–560.

Crook, K.A.W., 1968, Weathering and roundness of quarts sand grains: Sedimentology, v. 11, p. 171–182.

Crook, K.A.W., 1974, Lithogenesis and geotectonics: The significance of compositional variations in flysch arenites (graywackes), *in* Dott, R. H., and Shaver, R. H., eds., Modern and ancient geosynclinal sedimentation: Society of Economic Paleontologists and Mineralogists Special Publication 19, p. 304–310.

Cullers, R. L., Barrett, T., Carlson, R., and Robinson, R., 1987, Rare-earth distributions in size fractions of Holocene soil and stream sediment, Wet Mountains region, Colorado, USA: Chemical Geology, v. 63, p. 275–297.

Damuth, J. E., and Fairbridge, R. W., 1970, Equatorial Atlantic deep-sea arkosic sands and ice-age aridity in tropical South America: Geological Society of America Bulletin, v. 81, p. 189–206.

Darmody, R. G., 1985, Weathering assessment of quartz types: A semiquantitative approach: Soil Science Society of America Journal, v. 49, p. 1322–1324.

Davies, D. K., and Ethridge, F. G., 1975, Sandstone composition and depositional environment: American Association of Petroleum Geologists Bulletin, v. 59, p. 239–264.

DeCelles, P. G., 1986, Sedimentation in a tectonically partitioned, nonmarine foreland basin: The Lower Cretaceous Kootenai Formation, southwestern Montana: Geological Society of America Bulletin, v. 97, p. 911–931.

Decker, J. and Helmold, K. P., 1985, The effect of grain size on detrital modes: A test of the Gazzi-Dickison point-counting method—Discussion: Journal of Sedimentary Petrology, v. 55, p. 619–619.

DePaolo, D. J., 1988, Neodymium isotope geochemistry: Berlin, Springer-Verlag, 187 p.

Dickinson, W. R., 1970, Interpreting detrital modes of graywacke and arkose: Journal of Sedimentary Petrology, v. 40, p. 695–707.

Dickinson, W. R., 1982, Compositions of sandstones in circum-Pacific subduction complexes and fore-arc basins: American Association of Petroleum Geologists Bulletin, v. 66, p. 121–137.

Dickinson, W. R., 1985, Interpreting provenance from detrital modes of sandstones, *in* Zuffa, G. G., ed., Provenance of arenites: Dordrecht, D. Reidel,

p. 333–362.

Dickinson, W. R., and Rich, E. I., 1972, Petrologic intervals and petrofacies in the Great Valley Sequence, Sacramento Valley, California: Geological Society of America Bulletin, v. 83, p. 3007–3024.

Dickinson, W. R., and Suczek, C. A., 1979, Plate tectonics and sandstone compositions: American Association of Petroleum Geologists Bulletin, v. 63, p. 2164–2182.

Dickinson, W. R., and Valloni, R., 1980, Plate settings and provenance of sands in modern ocean basins: Geology, v. 8, p. 82–86.

Dickinson, W. R., Helmold, K. P., and Stein, J. A., 1979, Mesozoic lithic sandstones in central Oregon: Journal of Sedimentary Petrology, v. 49, p. 501–516.

Dickinson, W. R., Ingersoll, R. V., Cowan, D. S., Helmold, K. P., and Suczek, C. A., 1982, Provenance of Franciscan graywackes in coastal California: Geological Society of America Bulletin, v. 93, p. 95–107.

Dickinson, W. R., and 8 others, 1983a, Provenance of North American Phanerozoic sandstones in relation to tectonic setting: Geological Society of America Bulletin, v. 94, p. 222–235.

Dickinson, W. R., Harbaugh, D. W., Saller, A. H., Heller, P. L., and Snyder, W. S., 1983b, Detrital modes of Upper Paleozoic sandstones derived from Antler orogen in Nevada: Implications for nature of Antler orogeny: American Journal of Science, v. 283, p. 481–509.

Dickinson, W. R., Lawton, T. F., and Inman, K. F., 1986, Sandstone detrital modes, central Utah foreland region: Stratigraphic record of Cretaceous-Paleogene tectonic evolution: Journal of Sedimentary Petrology, v. 56, p. 276–293.

Dorsey, R. J., 1988, Provenance evolution and unroofing history of a modern arc-continent collision: Evidence from petrography of Plio-Pleistocene sandstones, eastern Taiwan: Journal of Sedimentary Petrology, v. 58, p. 208–218.

Douglas, L. A., and Platt, D. W., 1977, Surface morphology of quartz and age of soils: Soil Science Society of America Journal, v. 41, p. 641–645.

Drewey, S., Cliff, R. A., and Leeder, M. R., 1987, Provenance of Carboniferous sandstones from U-Pb dating of detrital zircons: Nature, v. 325, p. 51–53.

Dutta, P. K., 1987, The role of variables in controlling sandstone composition: Indiana State University Department of Geography and Geology Professional Paper 18, p. 30–40.

Dutta, P. K., and Suttner, L. J., 1986, Alluvial sandstone composition and paleoclimate, II. Authigenic mineralogy: Journal of Sedimentary Petrology, v. 56, p. 346–358.

Eckhardt, F.E.W., 1985, Solubilization, transport, and deposition of mineral cations by microorganisms—Efficient rock weathering agents, *in* Drever, J. I., ed., The chemistry of weathering: Dordrecht, D. Reidel, p. 161–173.

Eggleton, R. A., and Buseck, P. R., 1980, High resolution electron microscopy of feldspar weathering: Clays and Clay Minerals, v. 28, p. 173–178.

Enkeboll, R. H., 1982, Petrology and provenance of sands and gravels from the Middle America Trench and trench slope, southwestern Mexico and Guatemala: Initial Reports of the Deep Sea Drilling Project, v. 66, p. 521–530.

Evans, J. A., Stone, P., and Floyd, J. D., 1991, Isotopic characteristics of Ordovician graywacke provenance in the southern uplands of Scotland, *in* Morton, A. C., Todd, S. P., and Haughton, P.D.W., eds., Developments in sedimentary provenance studies: Geological Society of London Special Publication 57, p. 161–172.

Evans, J. E., 1991, Paleoclimatology and paleobotany of the Eocene Chumstick Formation, Cascade Range, Washington (USA): A rapidly subsiding alluvial basin: Palaeogeography, Palaeoclimatology, Palaeoclimatology, v. 88, p. 239–264.

Floyd, P. A., Shail, R., Leveridge, B. E., and Franke, W., 1991, Geochemistry and provenance of Rhenohercynian synorogenic sandstones: Implications for tectonic environment discrimination, *in* Morton, A. C., Todd, S. P., and Haughton, P.D.W., eds., Developments in sedimentary provenance studies: Geological Society of London Special Publication 58, p. 173–188.

Folk, R. L., 1974, Petrology of sedimentary rocks: Austin, TX, Hemphill, 182 p.

Franks, S. G., and Forester, R. W., 1984, Relationships among secondary porosity, pore-fluid chemistry and carbon dioxide, Texas Gulf Coast, *in* McDonald, D. A., and Surdam, R. C., eds., Clastic diagenesis: American Association of Petroleum Geologists Memoir 37, p. 63–79.

Franzinelli, E., and Potter, P. E., 1983, Petrology, chemistry, and texture of modern river sands, Amazon river system: Journal of Geology, v. 91, p. 23–39.

Franzinelli, E., and Potter, P. E., 1985, Areias Recentes do rios da bacia Amazônica: Composicões petrográfica, textural e química: Revista Brasileira de Geociências, v. 15, p. 213–220.

Freed, R. L., and Peacor, D. R., 1992, Diagenesis and the formation of authigenic illite-rich I/S crystals in Gulf Coast shales: TEM study of clay separates: Journal of Sedimentary Petrology, v. 62, p. 220–234.

Frost, C. D., and Coombs, D. S., 1989, Nd isotope character of New Zealand sediments: Implications for terrane concepts and crustal evolution: American Journal of Science, v. 289, p. 744–770.

Frost, C. D., and Winston, D., 1987, Nd isotope systematics of coarse- and fine-grained sediments: Examples from the Middle Proterozoic Belt–Percell Supergroup: Journal of Geology, v. 95, p. 309–329.

Füchtbauer, H., 1964, Sedimentpetrographische Untersuchungen in der älteren Molasse nördlich der Alpen: Eclogae Geologicae Helvetiae, v. 57, p. 157–298.

Garrels, R. M., 1986, Sediment cycling and diagenesis: U.S. Geological Survey Bulletin, v. 1578, p. 1–11.

Garzanti, E., 1986, Source rock versus sedimentary control on the mineralogy of deltaic volcanic arenites (Upper Triassic, northern Italy): Journal of Sedimentary Petrology, v. 56, p. 267–275.

Gazzi, P., 1966, Le arenarie del flysch sopracretaceo dell'Appennino modenese; correlazioni con il flysch di Monghidoro: Mineralogica et Petrographica Acta, v. 12, p. 69–97.

Gazzi, P., and Zuffa, G. G., 1970, Le arenarie paleogeniche dell'Appennino emiliano: Mineralogica et Petrographica Acta, v. 16, p. 97–137.

Gergen, L. D., and Ingersoll, R. V., 1986, Petrology and provenance of deep sea drilling project and sandstone from the north Pacific Ocean and the Bering Sea: Sedimentary Geology, v. 51, p. 29–56.

Gilbert, W. G., and Dickinson, W. R., 1970, Stratigraphic variations in sandstone petrology, Great Valley Sequence, central California coast: Geological Society of America Bulletin, v. 81, p. 949–954.

Girty, G. H., 1987, Sandstone provenance, Point Lomo Formation, San Diego, California: Evidence for uplift of the Peninsular Ranges during the Laramide orogeny: Journal of Sedimentary Petrology, v. 57, p. 839–844.

Girty, G. H., Mossman, B. J., and Pincus, S. D., 1988, Petrology of Holocene sand, Peninsular Ranges, California and Baja Norte, Mexico: Implications for provenance-discrimination models: Journal of Sedimentary Petrology, v. 58, p. 881–887.

Goldich, S. S., 1938, A study in rock-weathering: Journal of Geology, v. 46, p. 17–58.

Graham, J. R., Wrafter, J. P., Daly, J. S., and Menuge, J. F., 1991, A local source for the Ordovician Derryveeny Formation, western Ireland: Implications for the Connemara Dalradian, *in* Morton, A. C., Todd, S. P., and Haughton, P.D.W., eds., Developments in sedimentary provenance studies: Geological Society of London Special Publication 57, p. 199–213.

Graham, S. A., Hendrix, M. S., Wang, L. B., and Carroll, A. R., 1993, Collisional successor basins of western China: Impact of tectonic inheritance on sand composition: Geological Society of America Bulletin, v. 105, p. 323–344.

Grantham, J. H., and Velbel, M. A., 1988, The influence of climate and topography on rock-fragment abundance in modern fluvial sands of the southern Blue Ridge Mountains, North Carolina: Journal of Sedimentary Petrology, v. 58, p. 219–227.

Graustein, W. C., and Velbel, M. A., 1981, Comment on "Weathering before the advent of land plants: Evidence from unaltered detrital potassium feldspar": Geology, v. 9, p. 505–506.

Grigsby, J. D., 1990, Detrital magnetite as a provenance indicator: Journal of Sedimentary Petrology, v. 60, p. 940–951.

Grigsby, J. D., 1992, Chemical fingerprinting in detrital ilmenite: A viable alternative in provenance research?: Journal of Sedimentary Petrology, v. 62, p. 331–337.

Hansley, P. L., 1987, Petrologic and experimental evidence for the etching of garnets by organic acids in the Upper Jurassic Morrison Formation, northwestern New Mexico: Journal of Sedimentary Petrology, v. 57, p. 666–681.

Harrold, P. J., and Moore, C., 1975, Composition of deep-sea sands from marginal basins of the northwestern Pacific: Initial Reports of the Deep Sea Drilling Project, v. 31, p. 507–514.

Hayes, J. R., 1962, Quartz and feldspar content in South Platte, Platte, and Missouri River sands: Journal of Sedimentary Petrology, v. 32, p. 793–800.

Helmold, K. P., 1985, Provenance of feldspathic sandstones—The effect of diagenesis on provenance interpretations, *in* Zuffa, G. G., ed., Provenance of arenites: Dordrecht, D. Reidel, p. 139–164.

Helmold, K. P., and van de Kamp, P. C., 1984, Diagenetic mineralogy and controls on albitization and laumontite formation in Paleogene arkoses, Santa Ynez Mountains, California, *in* McDonald, D. A., and Surdam, R. C., eds., Clastic diagenesis: American Association of Petroleum Geologists Memoir 37, p. 239–276.

Hemming, S., McLennan, S. M., Hanson, G. N., Krogstad, E. J., and Mezger, K., 1990, Pb isotope systematics in quartz [abs.]: EOS, Transactions of the American Geophysical Union, v. 71, p. 654.

Hemming, S., McLennan, S. M., Hanson, G. N., and Sharp, W. D., 1991, U-Pb isotope systematics in detrital quartz, feldspar and zircon [abs.]: Geological Society of America Abstracts with Programs, v. 23, p. A108.

Hinman, N. W., 1990, Chemical factors influencing the rates and sequences of silica phase transitions: Effects of organic constituents: Geochimica et Cosmochimica Acta, v. 54, p. 1563–1574.

Hochella, M. F., Jr., 1988, Auger electron and x-ray photoelectron spectroscopies, *in* Hawthorne, F. C., eds., Spectroscopic methods in mineralogy and geology: Mineralogical Society of America Reviews in Mineralogy 18, p. 87–132.

Hochella, M. F., Jr., 1990, Atomic structure, microtopography, composition, and reactivity of mineral surfaces, *in* Hochella, M. F., Jr., and White, A. F., eds., Mineral-water interface geochemistry: Mineralogical Society of America Reviews in Mineralogy 23, p. 87–132.

Holdren, G. R., and Speyer, P. M., 1985, Reaction rate-surface area relationships during the early stages of weathering—I. Initial observations: Geochimica et Cosmochimica Acta, v. 49, p. 675–681.

Hower, J., 1981, Shale diagenesis, *in* Longstaffe, F. J., eds., Clays and the resource geologist: Mineralogical Association of Canada Short Course Handbook 7, p. 60–80.

Hower, J., Eslinger, E. V., Hower, M. E., and Perry, E. A., 1976, Mechanisms of burial metamorphism of argillaceous sediment: 1. Mineralogical and chemical evidence: Geological Society of America Bulletin, v. 87, p. 725–737.

Hubert, J. F., 1962, A zircon-tourmaline-rutile maturity index and the interdependence of the composition of heavy mineral assemblages with the gross composition and texture of sandstones: Journal of Sedimentary Petrology, v. 32, p. 440–450.

Humphris, S. E., 1984, The mobility of rare earth elements in the crust, *in* Henderson, P., ed., Rare earth element geochemistry: Amsterdam, Elsevier, p. 317–342.

Hurford, A. J., and Carter, A., 1991, The role of fission track dating in discrimination of provenance, *in* Morton, A. C., Todd, S. P., and Haughton, P.D.W., eds., Developments in sedimentary provenance studies: Geological Society of London Special Publication 57, p. 67–78.

Hurst, A. R., and Morton, A. C., 1988, An application of heavy mineral analysis to lithostratigraphy and reservoir modeling in the Oseberg Field, northern North Sea: Marine and Petroleum Geology, v. 5, p. 157–169.

Hutton, J., 1785, Abstract of a dissertation read in the Royal Society of Edinburgh, upon the seventh of March, and fourth of April, M,DCC,LXXXV, concerning the system of the Earth, its duration, and stability, 30 p., *reprinted in* White, G. W., ed., 1973, Contributions to the history of geology, Volume 5, James Hutton: New York, Hafner Press, 203 p.

Ingersoll, R. V., 1978, Petrofacies and petrologic evolution of the Late Cretaceous fore-arc basin, northern and central California: Journal of Geology, v. 86, p. 335–352.

Ingersoll, R. V., 1983, Petrofacies and provenance of Late Mesozoic forearc basin, northern and central California: American Association of Petroleum Geologists Bulletin, v. 67, p. 1125–1142.

Ingersoll, R. V., 1990, Actualistic sandstone petrofacies: Discriminating modern and ancient source rocks: Geology, v. 18, p. 733–736.

Ingersoll, D. V., and Cavazza, W., 1991, Reconstruction of Oligocene-Miocene volcaniclastic dispersal patterns in north-central New Mexico using sandstone petrofacies, *in* Fisher, R. V., and Smith, G. A., eds., Sedimentation in volcanic settings: Tulsa, OK, SEPM Special Publication 45, p. 227–236.

Ingersoll, R. V., and Dickinson, W. R., 1990, Great Valley Group (sequence), Sacramento Valley, California, *in* Ingersoll, V., and Nilsen, T. H., eds., Sacramento Valley symposium and guidebook: Bakersfield, CA, Pacific Section, SEPM (Society for Sedimentary Geology), p. 183–215.

Ingersoll, R. V., and Suczek, C. A., 1979, Petrology and provenance of Neogene sand from Nicobar and Bengal fans, DSDP sites 211 and 218: Journal of Sedimentary Petrology, v. 49, p. 1217–1228.

Ingersoll, R. V., Bullard, T. F., Ford, R. L., Grimm, J. P., Pickle, J. D., and Sares, S. W., 1984, The effect of grain size on detrital modes: A test of the Gazzi-Dickinson point-counting method: Journal of Sedimentary Petrology, v. 54, p. 103–116.

Ingersoll, R. V., Bullard, T. F., Ford, R. L., and Pickle, J. D., 1985a, The effect of grain size on detrital modes: A test of the Gazzi-Dickinson point-counting method—Reply to discussion of John Decker and Kenneth P. Helmold: Journal of Sedimentary Petrology, v. 55, p. 620–621.

Ingersoll, R. V., Bullard, T. F., Ford, R. L., and Pickle, J. D., 1985b, The effect of grain size on detrital modes: A test of the Gazzi-Dickinson point-counting method—Reply to discussion of Lee J. Suttner and Abhijit Basu: Journal of Sedimentary Petrology, v. 55, p. 617–618.

Ingersoll, R. V., Cavazza, W., Baldridge, W. S., and Shafiqullah, M., 1990, Cenozoic sedimentation and paleotectonics of north-central New Mexico: Implications for initiation and evolution of the Rio Grande rift: Geological Society of America Bulletin, v. 102, p. 1280–1296.

Inoue, A., Watanabe, T., Kohyama, N., and Brusewitz, A. M., 1990, Characterization of illitization of smectite in bentonite beds at Kinnekulle, Sweden: Clays and Clay Minerals, v. 38, p. 241–249.

Jackson, T. A., and Keller, W. D., 1970, A comparative study of the role of lichens and "inorganic" processes in the chemical weathering of recent Hawaiian lava flows: American Journal of Science, v. 269, p. 446–466.

James, W. C., Mack, G. H., and Suttner, L. J., 1981, Relative alteration of microcline and sodic plagioclase in semi-arid and humid climates: Journal of Sedimentary Petrology, v. 51, p. 151–164.

Jett, G. A., and Heller, P. L., 1988, Tectonic significance of polymodal compositions in mélange sandstones, Western Mélange Belt, North Cascades Range, Washington: Journal of Sedimentary Petrology, v. 58, p. 52–61.

Johnsson, M. J., 1990, Tectonic versus chemical-weathering controls on the composition of fluvial sands in tropical environments: Sedimentology, v. 37, p. 713–726.

Johnsson, M. J., 1992, Chemical weathering controls on sand composition, *in* Nierenber, W. A., ed., Encyclopedia of earth system science: Orlando, FL, Academic Press, p. 455–466.

Johnsson, M. J., and Meade, R. H., 1990, Chemical weathering of fluvial sediments during alluvial storage: The Macuapanim Island point bar, Solimões River, Brazil: Journal of Sedimentary Petrology, v. 60, p. 827–842.

Johnsson, M. J., and Stallard, R. F., 1989, Physiographic controls on the composition of sediments derived from volcanic and sedimentary terrains on Barro Colorado Island, Panama: Journal of Sedimentary Petrology, v. 59, p. 768–781.

Johnsson, M. J., Stallard, R. F., and Meade, R. H., 1988, First-cycle quartz arenites in the Orinoco River basin, Venezuela and Colombia: Journal of Geology, v. 96, p. 263–277.

Johnsson, M. J., Stallard, R. F., and Lundberg, N., 1991, Controls on the composition of fluvial sands from a tropical weathering environment: Sands of the Orinoco River drainage basin, Venezuela and Colombia: Geological Society of America Bulletin, v. 103, p. 1622–1647.

Johnsson, P. A., Blum, A. E., Hochella, M. F., Jr., Parks, G. A., and Sposito, G., 1992, Direct observation of muscovite basal-plane dissolution and secondary phase formation: An XPS, LEED, and SFM study, *in* Kharaka, Y. F., and Maest, A. S., eds., Water-rock interaction, Volume 1, Low temperature environments: Rotterdam, A. A. Balkema, p. 159–162.

Kelley, J. C., and Whetten, J. T., 1969, Quantitative statistical analyses of Columbia River sediment samples: Journal of Sedimentary Petrology, v. 39, p. 1167–1173.

Kelly, S., and Bluck, B. J., 1989, Detrital mineral ages from the Southern Uplands using ^{40}Ar-^{39}Ar laser probe: Journal of the Geological Society of London, v. 146, p. 401–403.

Knoll, M. A., and James, W. C., 1987, Effect of the advent and diversification of vascular land plants on mineral weathering through geologic time: Geology, v. 15, p. 1099–1102.

Koehnken, L., 1990, The composition of fine-grained weathering products in a large tropical river system, and the transport of metals in fine-grained sediments in a temperate estuary [Ph.D. dissertation]: Princeton, New Jersey, Princeton University, 246 p.

Krook, L., 1969, Investigations on the mineralogical composition of the Tertiary and Quaternary sands in northern Surinam: Verhandelingen van het Koninklijke Nederlandse Geologische Mijnbouwkundig Genootschap, v. 27, p. 89–100.

Krumbein, W. E., and Dyer, B. D., 1985, This planet is alive—Weathering and biology, a multi-faceted problem, *in* Drever, J. I., ed., The chemistry of weathering: Dordrecht, D. Reidel, p. 143–160.

Krynine, P. D., 1935, Arkose deposits in the humid tropics. A study of sedimentation in southern Mexico: American Journal of Science, v. 29, p. 353–363.

Krynine, P. D., 1936, Geomorphology and sedimentation in the humid tropics: American Journal of Science, v. 32, p. 297–306.

Krynine, P. D., 1941a, Differentiation of sediments during the life history of a landmass [abs.]: Geological Society of America Bulletin, v. 52, p. 1915.

Krynine, P. D., 1941b, Paleogeographic and tectonic significance of graywackes [abs.]: Geological Society of America Bulletin, v. 52, p. 1916.

Krynine, P. D., 1941c, Paleogeographic and tectonic significance of sedimentary quartzites [abs.]: Geological Society of America Bulletin, v. 52, p. 1915–1916.

Krynine, P. D., 1942, Differential sedimentation and its products during one complete geosynclinal cycle, *in* Primer Congresso Panamericano de Ingenieria de Minas y Geología: Santiago, Chile, p. 537–561.

Krynine, P. D., 1950, Petrology, stratigraphy and origin of the Triassic rocks of Connecticut: Connecticut Geological Survey Bulletin 73, 239 p.

Kuenen, P. H., 1955, Experimental abrasion of pebbles 1. Wet sandblasting: Leidse Geologische Mededlingen, v. 20, p. 142–147.

Kuenen, P. H., 1956, Experimental abrasion of pebbles 2. Rolling by current: Journal of Geology, v. 64, p. 336–368.

Kuenen, P. H., 1959, Experimental abrasion 3. Fluviatile action on sand: American Journal of Science, v. 257, p. 172–190.

Kuenen, P. H., 1960, Experimental abrasion 4: Eolian action: Journal of Geology, v. 68, p. 427–449.

Kuenen, P. H., 1962, Experimental abrasion 5. Frosting and defrosting of quartz grains: Journal of Geology, v. 70, p. 648–658.

Kuenen, P. H., 1964, Experimental abrasion: 6. Surf action: Sedimentology, v. 3, p. 29–43.

Land, L. S., and Milliken, K. L., 1981, Feldspar diagenesis in the Frio Formation, Brazoria County, Texas Gulf Coast: Geology, v. 9, p. 314–318.

Lasaga, A. C., Berner, R. A., and Garrels, R. M., 1985, An improved geochemical model of atmospheric CO_2 fluctuations over the past 100 million years, *in* Sundquist, E. T., and Broecker, W. S., eds., The carbon cycle and atmospheric CO_2: Natural variations Archean to present: American Geophysical Union Monograph 32, p. 397–411.

Lawton, T. F., 1986, Compositional trends within a clastic wedge adjacent to a fold-thrust belt, Indianola Group, central Utah, USA, *in* Allen, P. A., and Homewood, P., eds., Foreland basins: International Association of Sedimentologists Special Publication 8, p. 411–423.

Leighton, M. M., and Willman, H. B., 1950, Loess formations of the Mississippi Valley: Journal of Geology, v. 58, p. 599–622.

Linn, A. M., DePaolo, D. J., and Ingersoll, R. V., 1991, Nd-Sr isotopic provenance analysis of Upper Cretaceous Great Valley fore-arc sandstones: Geology, v. 19, p. 803–806.

Liu, D. Y., Nutman, A. P., Compston, W., Wu, J. S., and Shen, Q. H., 1992, Remnants of $\geqslant$3800 Ma crust in the Chinese part of the Sino-Korean craton: Geology, v. 20, p. 339–342.

Longstaffe, F. J., 1984, The role of meteoric water in diagenesis of shallow sandstones: Stable isotope studies of the Milk River aquifer and gas pool, southeastern Alberta, *in* McDonald, D. A., and Surdam, R. C., eds., Clastic diagenesis: American Association of Petroleum Geologists Memoir 37, p. 91–98.

Macaire, J. J., 1986, Apport de l'alteration superficielle à la stratigraphie—exemple des formations alluviales et eoliennes Plio-Quaternaires de Touraine (France): Bulletin de l'Association française pour l'étude du Quaternaire, v. 1986, p. 233–245.

Mack, G. H., 1981, Composition of modern stream sand in a humid climate derived from a low-grade metamorphic and sedimentary foreland fold-thrust belt of north Georgia: Journal of Sedimentary Petrology, v. 51, p. 1247–1258.

Mack, G. H., 1984, Exceptions to the relationship between plate tectonics and sandstone composition: Journal of Sedimentary Petrology, v. 54, p. 212–220.

Mackie, W., 1896, Sand and sandstones of the eastern Moray: Edinburgh Geological Society Transactions, v. 7, p. 148–172.

Maksimovic, Z., and Pantó, G., 1991, Contribution to the geochemistry of the rare earth elements in the karst-bauxite deposits of Yugoslavia and Greece: Geoderma, v. 51, p. 93–109.

Mange-Rajetzky, M. A., and Oberhänsli, R., 1982, Detrital lawsonite and blue sodic amphibole in the Molasse of Savoy, France, and their significance in assessing Alpine evolution: Schweizer Mineralogische und Petrographisches Mitteilungen, v. 62, p. 415–436.

Mann, W. R., and Cavaroc, V. V., 1973, Composition of sand released from three source areas under humid, low-relief weathering in the North Carolina piedmont: Journal of Sedimentary Petrology, v. 43, p. 870–881.

Mansfield, C. F., 1971, Stratigraphic variation in sandstone petrology of the Great Valley Sequence in the southern Coast Ranges west of Coalinga, California [abs.]: Geological Society of America Abstracts with Programs (Cordilleran Section), v. 3, p. 157.

Markert, J. C., and Al-Shaieb, Z., 1984, Diagenesis and evolution of secondary porosity in Upper Minnelusa Sandstones, Powder River Basin, Wyoming, *in* McDonald, D. A., and Surdam, R. C., eds., Clastic diagenesis: American Association of Petroleum Geologists Memoir 37, p. 367–389.

Martens, J.H.C., 1931, Persistence of feldspars in beach sand: American Mineralogist, v. 16, p. 526–531.

Mast, M. A., and Drever, J. I., 1987, The effect of oxalate on the dissolution rates of oligoclase and tremolite: Geochimica et Cosmochimica Acta, v. 51, p. 2559–2568.

Maynard, J. B., 1984, Composition of plagioclase feldspar in modern deep-sea sands: Relationship to tectonic setting: Sedimentology, v. 31, p. 493–501.

Maynard, J. B., Valloni, R., and Yu, H. S., 1982, Composition of modern deep-sea sands from arc-related basins, *in* Leggett, J. K., ed., Trench-forearc geology: Geological Society of London Special Publication 10, p. 551–560.

McBride, E. F., 1985, Diagenetic processes that affect provenance determination in sandstone, *in* Zuffa, G. G., eds., Provenance of arenites: Dordrecht, D. Reidel, p. 95–114.

McBride, E. F., and Picard, M. D., 1987, Downstream changes in sand composition, roundness, and gravel size in a short-headed high-gradient stream, northwestern Italy: Journal of Sedimentary Petrology,v. 57, p. 1018–1026.

McCann, T., 1991, Petrological and geochemical determination of provenance in the southern Welsh Basin, *in* Morton, A. C., Todd, S. P., and Haughton, P.D.W., eds., Developments in sedimentary provenance studies: Geological Society of London Special Publication 57, p. 215–230.

McCulloch, M. T., and Wasserburg, G. J., 1978, Sm-Nd and Rb-Sr chronology of continental crust formation: Science, v. 200, p. 1003–1011.

McDonald, D. A., and Surdam, R. C., 1984, Clastic diagenesis: American Association of Petroleum Geologists Memoir 37, 434 p.

McLennan, S. M., 1989, Rare earth elements in sedimentary rocks: Influence of provenance and sedimentary processes: Mineralogical Society of America Reviews in Mineralogy, v. 21, p. 169–200.

McLennan, S. M., and Taylor, S. R., 1991, Sedimentary rocks and crustal evolution: Tectonic setting and secular trends: Journal of Geology, v. 99, p. 1–21.

McLennan, S. M., McCulloch, M. T., Taylor, S. R., and Maynard, J. B., 1989, Effects of sedimentary sorting on neodymium isotopes in deep-sea turbidites: Nature, v. 337, p. 547–549.

McLennan, S. M., Taylor, S. R., McCulloch, M. T., and Maynard, J. B., 1990, Geochemical and Nd-Sr isotopic composition of deep-sea turbidites: Crustal evolution and plate tectonic associations: Geochimica et Cosmochimica Acta, v. 54, p. 2015–2050.

Meirding, T. C., 1981, Marble tombstone weathering rates: A transect of the United States: Physical Geography, v. 2, p. 1–18.

Michard, A., Gurriet, P., Soudant, M., and Albarede, F., 1985, Nd isotopes in French Phanerozoic shales: External vs. internal aspects of crustal evolution: Geochimica et Cosmochimica Acta, v. 49, p. 601–610.

Milliman, J. D., and Meade, R. H., 1983, World-wide delivery of river sediment to the oceans: Journal of Geology, v. 91, p. 1–21.

Mills, H. H., and Wagner, J. R., 1985, Long-term change in regime of the New River indicated by vertical variation in extent and weathering intensity of alluvium: Journal of Geology, v. 93, p. 131–142.

Milodowski, A. E., and Zalasiewicz, J. A., 1991, Redistribution of rare earth elements during diagenesis of turbidite/hemipelagite mudrock sequences of Llandovery age from central Wales, *in* Morton, A. C., Todd, S. P., and Haughton, P.D.W., eds., Developments in sedimentary provenance studies: Geological Society of London Special Publication 57, p. 101–124.

Molinaroli, E., and Basu, A., 1987, Studio di minerali opachi in sabbie fluviali oloceniche e nelle corispondenti rocce madri di zone sottoposte a climi diversi (Montagne Rocciose e Monti Appalachi in USA): Rendiconto della Societa Italiana di Mineralogia e Petrologia, v. 42, p. 271–283.

Molinaroli, E., Blom, M., and Basu, A., 1991, Methods of provenance determination tested with discriminant function analysis: Journal of Sedimentary Petrology, v. 61, p. 900–908.

Moncure, G. K., Lahann, R. W., and Siebert, R. M., 1984, Origin of secondary porosity and cement distribution in a sandstone/shale sequence from the Frio formation (Oligocene), *in* McDonald, D. A., and Surdam, R. C., eds., Clastic diagenesis: American Association of Petroleum Geologists Memoir 37, p. 151–161.

Moore, G. F., 1979, Petrography of subduction zone sandstones from Nias Island, Indonesia: Journal of Sedimentary Petrology, v. 49, p. 71–84.

Moore, P. D., 1984, Clues to past climate in river sediment: Nature, v. 308, p. 316.

Morad, S., Bergan, M., Knarud, R., and Nystuen, P., 1990, Albitization of detrital plagioclase in Triassic reservoir sandstones from the Snorre field, Norwegian North Sea: Journal of Sedimentary Petrology, v. 60, p. 411–425.

Morton, A. C., 1984, Stability of detrital heavy minerals in Tertiary sandstones from the North Sea basin: Clay Minerals, v. 19, p. 287–308.

Morton, A. C., 1985, Heavy minerals in provenance studies, *in* Zuffa, G. G., ed., Provenance of arenites: Dordrecht, D. Reidel, p. 249–278.

Morton, A. C., 1986, Dissolution of apatite in North Sea Jurassic sandstones: Implications for the generation of secondary porosity: Clay Minerals, v. 21, p. 711–733.

Morton, A. C., 1991, Geochemical studies of detrital heavy minerals and their application to provenance research, *in* Morton, A. C., Todd, S. P., and Haughton, P.D.W., eds., Developments in sedimentary provenance studies: Geological Society of London Special Publication 57, p. 31–45.

Morton, A. C., Stiberg, F. P., Hurst, A., and Quale, H., 1989, Use of heavy minerals in lithostratigraphic correlation, with examples from Brent sandstones of the northern North Sea, *in* Collinson, J. D., ed., Correlation in hydrocarbon exploration: Dordrecht, Graham and Trotman, p. 217–230.

Mueller, P. A., Wooden, J. L., and Nutman, A. P., 1992, 3.96 Ga zircons from an Archean quartzite, Beartooth Mountains, Montana: Geology, v. 20, p. 327–330.

Nadeau, P. H., and Reynolds, R. C., Jr., 1981, Burial and contact metamorphism in the Mancos Shale: Clays and Clay Minerals, v. 29, p. 249–259.

Nedkvitne, T., and Bjølykke, 1992, Secondary porosity in the Brent Group (Middle Jurassic), Huldra field, North Sea: Implication for predicting lateral continuity of sandstones: Journal of Sedimentary Petrology, v. 62, p. 23–34.

Nesbitt, H. W., 1979, Mobility and fractionation of rare earth elements during weathering of a granodiorite: Nature, v. 279, p. 206–210.

Nesbitt, H. W., Markovics, G., and Price, R. C., 1980, Chemical processes affecting alkalis and alkaline earths during continental weathering: Geochimica et Cosmochimica Acta, v. 44, p. 1659–1666.

Nickel, E., 1973, Experimental dissolution of light and heavy minerals in comparison with weathering and intrastratal solution: Contributions to Sedimentology, v. 1, p. 1–68.

Nixon, R. A., 1979, Differences in incongruent weathering of plagioclase and microcline—Cation leaching versus precipitates: Geology, v. 7, p. 221–224.

Odom, I. E., Doe, T. W., and Dott, J. H., Jr., 1976, Nature of feldspar-grain size relations in some quartz-rich sandstones: Journal of Sedimentary Petrology, v. 46, p. 862–870.

Paçes, T., 1973, Steady-state kinetics and equilibrium between ground water and granitic rocks: Geochimica et Cosmochimica Acta, v. 37, p. 2641–2663.

Packer, B. M., and Ingersoll, R. V., 1986, Provenance and petrology of deep sea drilling project sands and sandstones from the Japan and Mariana forearc and backarc regions: Sedimentary Geology, v. 51, p. 5–28.

Pirrie, D., 1991, Controls on the petrographic evolution of an active margin sedimentary sequence: the Larsen basin, Antarctica, *in* Morton, A. C., Todd, S. P., and Haughton, P.D.W., eds., Developments in sedimentary provenance studies: Geological Society of London Special Publication 57, p. 231–249.

Pittman, E. D., 1969, Destruction of plagioclase twins by stream transport: Journal of Sedimentary Petrology, v. 39, p. 1432–1437.

Pittman, E. D., 1970, Plagioclase feldspar as an indicator of provenance in sedimentary rocks: Journal of Sedimentary Petrology, v. 40, p. 591–598.

Plumley, W. J., 1948, Black Hills terrace gravels: A study in sediment transport: Journal of Geology, v. 56, p. 526–577.

Pollack, J. M., 1961, Significance of compositional and textural properties of South Canadian River channel sands, New Mexico, Texas, and Oklahoma: Journal of Sedimentary Petrology, v. 31, p. 15–37.

Pollastro, R. M., 1985, Mineralogical and morphological evidence for the formation of illite at the expense of illite/smectite: Clays and Clay Minerals, v. 33, p. 265–274.

Pollastro, R. M., and Barker, C. E., 1986, Application of clay-mineral, vitrinite reflectance, and fluid inclusion studies to the thermal and burial history of the Pinedale anticline, Green River basin, Wyoming, *in* Gautier, D. L., ed., Roles of organic matter in sediment diagenesis: Society of Economic Paleontologists and Mineralogists Special Publication 38, p. 73–83.

Porter, S. C., 1975, Weathering rinds as relative age criteria: Application to subdivisions of glacial deposits in the Cascade Range: Geology, v. 3, p. 101–104.

Potter, P. E., 1978a, Petrology and chemistry of modern big river sands: Journal of Geology, v. 86, p. 423–449.

Potter, P. E., 1978b, Significance and origin of big rivers: Journal of Geology, v. 86, p. 13–33.

Potter, P. E., 1984, South African [sic] modern beach sand and plate tectonics: Nature, v. 311, p. 645–648.

Rainbird, R. H., Heaman, L. H., and Young, G., 1992, Sampling Laurentia: Detrital zircon geochronology offers evidence for an extensive Neoproterozoic river system originating from the Grenville orogen: Geology, v. 20, p. 351–354.

Reizebos, P. A., 1979, Compositional downstream variation of opaque and translucent heavy residues in some modern Río Magdalena sands (Colombia): Sedimentary Geology, v. 24, p. 197–225.

Retallack, G. J., 1986, Fossil soils as grounds for interpreting long-term controls on ancient rivers: Journal of Sedimentary Petrology, v. 56, p. 1–18.

Retallack, G. J., 1990, Soils of the past: An introduction to paleopedology: Boston, Unwin Hymen, 520 p.

Rolfe, B. N., and Hadley, R. F., 1964, Weathering and transport of sediment in the Cheyenne River basin, eastern Wyoming, *in* Bradley, W. F., ed., Clays and clay minerals: Proceedings of the 12th National Conference on Clays and Clay Minerals: New York, Macmillan, p. 649–670.

Roser, B. P., and Korsch, R. J., 1986, Determination of tectonic setting of sandstones-mudstone suites using SiO_2 content and K_2O/Na_2O ratio: Journal of Geology, v. 94, p. 635–650.

Roser, B. P., and Korsch, R. J., 1988, Provenance signatures of sandstone-mudstone suites determined using discriminant function analysis of major element data: Chemical Geology, v. 67, p. 119–139.

Ross, G. N., Parrish, R. R., and Dudas, F. O., 1991, Provenance of the Bonner Formation (Belt Supergroup), Montana: Insights from U-Pb and Sm-Nd analyses of detrital minerals: Geology, v. 19, p. 340–343.

Russell, R. D., 1937, Mineral composition of Mississippi River sands: Geological Society of America Bulletin, v. 48, p. 1307–1348.

Ruxton, B. P., 1970, Labile quartz-poor sediments from young mountain ranges in northeast Papua: Journal of Sedimentary Petrology, v. 40, p. 1262–1270.

Saccani, E., 1987, Double provenance of sand-size sediments in the southern Aegean forearc basin: Journal of Sedimentary Petrology, v. 57, p. 736–745.

Savage, K. M., and Potter, P. E., 1991, Petrology of modern sands of the Ríos Gauviare and Inírida, southern Colombia: Tropical climate and sand composition: Journal of Geology, v. 99, p. 289–298.

Savage, K. M., De Cesero, P., and Potter, P. E., 1988, Mineralogic maturity of modern sand along a high-energy tropical coast, Baixada de Jacarepaguá, Rio de Janeiro, Brazil: Journal of South American Earth Sciences, v. 1, p. 317–328.

Scholle, P. A., and Schluger, P. R., 1979, Aspects of diagenesis: Society of Economic Paleontologists and Mineralogists Special Publication 26, 443 p.

Schott, J., and Berner, R. A., 1985, Dissolution mechanisms of pyroxenes and olivines during weathering, *in* Drever, J. I., ed., Chemistry of weathering: Dordrecht, D. Reidel, p. 35–54.

Schumm, S. A., 1968, Speculations concerning paleohydrologic controls of terrestrial sedimentation: Geological Society of America Bulletin, v. 79, p. 1573–1588.

Schwab, F. L., 1975, Framework mineralogy and chemical composition of continental margin-type sandstones: Geology, v. 3, p. 487–490.

Schwab, F. L., 1981, Evolution of the western continental margin, French-Italian Alps: Sandstone mineralogy as an index of plate tectonic setting: Journal of Geology, v. 89, p. 349–368.

Shanmugam, G., and Higgins, J. B., 1988, Porosity enhancement from chert dissolution beneath Neocomian unconformity: Ivishak Formation, North Slope, Alaska: American Association of Petroleum Geologists Bulletin, v. 72, p. 523–535.

Short, P. F., and Ingersoll, R. V., 1990, Petrofacies and provenance of the Great Valley Group, southern Klamath Mountains and northern Sacramento Valley, *in* Ingersoll, R. V., and Nilsen, T. H., eds., Sacramento Valley Symposium and Guidebook: Bakersfield, CA, Pacific Section, SEPM (Society for Sedimentary Geology), p. 39–52.

Shukis, P. S., and Ethridge, F. G., 1975, A petrographic reconnaissance of sand size sediment upper St. Francis River, southeastern Missouri: Journal of Sedimentary Petrology, v. 45, p. 115–127.

Siebert, R. M., Moncure, G. K., and Lahann, R. W., 1984, A theory of framework grain dissolution in sandstones, *in* McDonald, D. A., and Surdam, R. C., eds., Clastic diagenesis: American Association of Petroleum Geologists Memoir 37, p. 163–175.

Siever, R., 1979, Plate tectonic controls on diagenesis: Journal of Geology, v. 87, p. 127–155.

Slatt, R. M., and Eyles, N., 1981, Petrology of glacial sand: implications for the origin and mechanical durability of lithic fragments: Sedimentology, v. 28, p. 171–183.

Stallard, R. F., 1985, River chemistry, geology, geomorphology, and soils in the Amazon and Orinoco basins, *in* Drever, J. I., ed., The chemistry of weathering: Dordrecht, D. Reidel, p. 293–316.

Stallard, R. F., 1988, Weathering and erosion in the humid tropics, *in* Lerman, A., and Meybeck, M., eds., Physical and chemical weathering in geochemical cycles: Dordrecht, Kluwer, p. 225–246.

Stallard, R. F., and Edmond, J. M., 1983, Geochemistry of the Amazon 2. The influence of geology and weathering on the dissolved load: Journal of Geophysical Research, v. 88, p. 9671–9688.

Stallard, R. F., and Edmond, J. M., 1987, Geochemistry of the Amazon 3. Weathering chemistry and limits to dissolved input: Journal of Geophysical Research, v. 92, p. 8293–8302.

Stallard, R. F., Koehnken, L., and Johnsson, M. J., 1991, Weathering processes and the composition of inorganic material transported through the Orinoco River system, Venezuela and Colombia: Geoderma, v. 51, p. 133–165.

Stanley, K. O., 1976, Sandstone petrofacies in the Cenozoic High Plains sequence, eastern Wyoming and Nebraska: Geological Society of America Bulletin, v. 87, p. 297–309.

Statteggar, K., 1987, Heavy minerals and provenance of sands: Modeling of lithological end members from river sands of northern Austria and from sandstones of the Austroalpine Gosau Formation (Late Cretaceous): Journal of Sedimentary Petrology, v. 57, p. 301–310.

Stumm, W., Furrer, G., Wieland, E., and Zinder, B., 1985, The effects of complex-forming ligands on dissolution of oxides and aluminosilicates, *in* Drever, J. I., ed., The chemistry of weathering: Dordrecht, D. Reidel, p. 55–74.

Surdam, R. C., Boese, S. W., and Crossey, L. J., 1984, The chemistry of secondary porosity, *in* McDonald, D. A., and Surdam, R. C., eds., Clastic diagenesis: American Association of Petroleum Geologists Memoir 37, p. 127–161.

Suttner, L. J., 1974, Sedimentary petrographic provinces: An evaluation, *in* Ross, C. A., ed., Paleogeographic provinces and provinciality: Society of Economic Paleontologists and Mineralogists Special Publication 21, p. 75–84.

Suttner, L. J., and Basu, A., 1981, Climate and the origin of quartz arenites: Journal of Sedimentary Petrology, v. 51, p. 1235–1246.

Suttner, L. J., and Basu, A., 1985, The effect of grain size on detrital modes: A test of the Gazzi-Dickinson point-counting method—Discussion: Journal of Sedimentary Petrology, v. 55, p. 616–617.

Suttner, L. J., and Dutta, P. K., 1986, Alluvial sandstone composition and paleoclimate, I. Framework mineralogy: Journal of Sedimentary Petrology, v. 56, p. 329–345.

Swe, W., and Dickinson, W. R., 1970, Sedimentation and thrusting of Late Mesozoic rocks in the Coast Ranges near Clear Lake, California: Geological Society of America Bulletin, v. 81, p. 165–188.

Tans, P. P., Fung, I. Y., and Takahashi, T., 1990, Observational constraints on the global atmospheric CO_2 budget: Science, v. 247, p. 1431–1438.

Taylor, S. R., and McLennan, S. M., 1985, The continental crust: its composition and evolution: Oxford, Blackwell Scientific, 312 p.

Thornburg, T. M., and Kulm, L. D., 1987, Sedimentation in the Chile Trench: Petrofacies and provenance: Journal of Sedimentary Petrology, v. 57, p. 55–74.

Todd, S. P., 1989, Role of the Dingle Bay Lineament in the evolution of the Old Red Sandstone of southwest Ireland, *in* Arthurton, R. S., Gutteridge, P., and Nolan, S. C., eds., The role of tectonics in Devonian and Carboniferous sedimentation in the British Isles: Yorkshire Geological Society Occasional Publication 6, p. 35–54.

Todd, T. W., 1968, Paleoclimatology and the relative stability of feldspar minerals under atmospheric conditions: Journal of Sedimentary Petrology, v. 38, p. 832–844.

Trevena, A. S., and Nash, W. P., 1979, Chemistry and provenance of detrital plagioclase: Geology, v. 7, p. 475–478.

Trevena, A. S., and Nash, W. P., 1981, An electron microprobe study of detrital plagioclase: Journal of Sedimentary Petrology, v. 51, p. 137–150.

Valloni, R., 1985, Reading provenance from modern marine sands, *in* Zuffa, G. G., ed., Provenance of arenites: Dordrecht, D. Reidel, p. 309–332.

Valloni, R., and Maynard, J. B., 1981, Detrital modes of recent deep-sea sands and their relation to tectonic setting: A first approximation: Sedimentology, v. 28, p. 75–83.

Valloni, R., and Mezzadri, G., 1984, Compositional suites of terrigenous deep-sea sands of the present continental margins: Sedimentology, v. 31, p. 353–364.

van Andel, T. H., 1959, Reflections on the interpretation of heavy mineral analyses: Journal of Sedimentary Petrology, v. 29, p. 153–163.

van de Kamp, P. C., and Leake, B. E., 1985, Petrography and geochemistry of feldspathic and mafic sediments of the northeastern Pacific margin: Transactions of the Royal Society of Edinburgh, v. 76, p. 411–449.

Velvel, M. A., 1984a, Natural weathering mechanisms of almandine garnet: Geology, v. 12, p. 631–634.

Velbel, M. A., 1984b, Weathering processes of rock-forming minerals, *in* Fleet, M. E., ed., Environmental geochemistry: London, Ontario, Mineralogical Association of Canada, p. 67–111.

Velbel, M. A., 1985, Mineralogically mature sandstones in accretionary prisms: Journal of Sedimentary Petrology, v. 55, p. 685–690.

Velbel, M. A., 1986, Influence of surface area, surface characteristics, and solution composition on feldspar weathering rates, *in* Davis, J. A., and Kayes, K. F., eds., Geochemical processes at mineral surfaces: American Chemical Society Symposium Series 323, p. 615–634.

Velbel, M. A., 1989, Weathering of hornblende to ferruginous products by a dissolution-reprecipitation mechanism: Petrography and stoichiometry: Clays and Clay Minerals, v. 37, p. 515–524.

Velbel, M. A., 1992, Geochemical mass balances and weathering rates in forested watersheds of the southern Blue Ridge, III. Cation budgets in an amphibolite watershed, and weathering rates of plagioclase and hornblende: American Journal of Science, v. 292, p. 58–78.

Volk, T., 1987, Feedbacks between weathering and atmospheric CO_2 over the last 100 million years: American Journal of Science, v. 287, p. 763–779.

Walker, T. R., 1984, Diagenetic albitization of potassium feldspar in arkosic sandstones: Journal of Sedimentary Petrology, v. 54, p. 3–16.

Wascher, H. L., Humbert, R. P., and Cady, J. G., 1947, Loess in the southern Mississippi Valley: Identification and distribution of loess sheets: Soil Science Society Proceedings, p. 389–399.

Whetten, J. T., Kelley, J. C., and Hanson, L. G., 1969, Characteristics of Columbia River sediment and sediment transport: Journal of Sedimentary Petrology, v. 39, p. 1149–1166.

White, A. F., 1983, Surface chemistry and dissolution kinetics of glassy rocks at 25°C: Geochimica et Cosmochimica Acta, v. 47, p. 805–815.

Wolf, K. H., 1971, Textural and compositional transitional stages between various lithic grain types (with a comment on "Interpreting detrital modes of graywacke and arkose"): Journal of Sedimentary Petrology, v. 41, p. 328–332.

Wollast, R., and Chou, L., 1985, Kinetic study of the dissolution of albite with a continuous flow-through fluidized bed reactor, *in* Drever, J. I., ed., Chemistry of weathering: Dordrecht, D. Reidel, p. 75–96.

Woolnough, W. G., 1930, The influence of climate and topography in the formation and distribution of products of weathering: Geological Magazine, v. 67, p. 123–132.

Yerino, L. N., and Maynard, J. B., 1984, Petrography of modern marine sands from the Peru-Chile Trench and adjacent areas: Sedimentology, v. 31, p. 83–89.

Zuffa, G. G., 1985, Optical analyses of arenites: Influence of methodology on compositional results, *in* Zuffa, G. G., ed., Provenance of arenites: Dordrecht, D. Reidel, p. 165–190.

Manuscript Accepted by the Society December 24, 1992

Printed in U.S.A.

Geological Society of America
Special Paper 284
1993

Geochemical approaches to sedimentation, provenance, and tectonics

S. M. McLennan, S. Hemming, D. K. McDaniel, and G. N. Hanson
Department of Earth and Space Sciences, State University of New York at Stony Brook, Stony Brook, New York 11794-2100

ABSTRACT

Geochemical and isotopic approaches to constraining provenance of sedimentary rocks complement the information inferred from petrography. Geochemical approaches have several advantages, including applicability to both matrix-rich sandstones and shales and ability to constrain provenance age and geochemical history. Five provenance components, or terrane types, have been defined on the basis of whole-rock chemical and Nd-isotopic composition, including *Old Upper Continental Crust, Recycled Sedimentary Rocks, Young Undifferentiated Arc, Young Differentiated Arc,* and various *Exotic Components,* such as ophiolites. Among the most important geochemical characteristics that define these provenance types are Nd isotopic composition (reflecting average provenance age), europium anomalies (reflecting intracrustal igneous differentiation processes), large-ion lithophile element enrichments (provenance composition), alkali and alkaline earth depletions (weathering and alteration), Zr and Hf enrichments (heavy mineral enrichments), and high Cr abundances (ultramafic sources).

Pb isotopic compositions of whole rocks and framework (quartz, feldspar) and accessory (e.g., zircon, monazite) grains constrain the age and crustal history of sediment sources. The U-Pb system may be used to date the time of crystallization of small populations and, for favorable circumstances, individual sand-sized grains of detrital quartz. The Early Proterozoic Pokegama Quartzite contains detrital quartz populations that give a Pb-Pb age of 2647 ± 16 Ma, consistent with detrital zircon ages and with the age of the Archean Superior Province, from which this formation is mainly derived. The initial Pb isotopic composition may be approximated by the Pb isotopic composition of leached feldspars, due to their low U/Pb and Th/Pb ratios. Pb isotopic compositions of detrital feldspars may also provide information about sedimentary provenance. K-feldspars from the Pokegama Quartzite and Early Proterozoic Chelmsford Formation fall within the field of Archean Superior Province igneous K-feldspars. They are distinct from K-feldspars found in other potential provenance terranes, including the Penokean Orogen or, in the case of the Pokegama, the Minnesota River Valley gneisses.

INTRODUCTION

The notion that there is a record of tectonic history within the composition of sedimentary rocks and their components is now well established (e.g., Blatt, 1967; Dickinson, 1970). Much effort has been expended in discriminating the tectonic setting of the provenance of sedimentary rocks, mainly using sandstone petrography but more recently employing various geochemical and isotopic techniques. A preliminary evaluation of some of these geochemical and isotopic approaches is the primary focus of this work. To date, there are exceedingly few studies that have combined the various petrographical, geochemical, and isotopic techniques. Nor have many studies applied such approaches to the study of individual components. Although we have drawn on the available integrated studies wherever possible, many of our inferences come from studies using far more limited data bases.

McLennan, S. M., Hemming, S., McDaniel, D. K., and Hanson, G. N., 1993, Geochemical approaches to sedimentation, provenance, and tectonics, *in* Johnsson, M. J., and Basu, A., eds., Processes Controlling the Composition of Clastic Sediments: Boulder, Colorado, Geological Society of America Special Paper 284.

Accordingly, our approach and many of our conclusions are certainly tentative and will no doubt require extensive revision as more data become available.

Using relative abundances of major framework components of sandstones, sedimentologists have endeavored to discriminate the tectonic setting of the provenance (e.g., Dickinson and Suczek, 1979; Valloni and Maynard, 1981; Dickinson, 1985, 1988). There has been some interest in combining petrographical data with major and trace element geochemical data, again with an emphasis on discrimination of tectonic setting (e.g., Schwab, 1971; Bhatia, 1983, 1985; Bhatia and Crook, 1986; Roser and Korsch, 1986; Floyd and Leveridge, 1987). A number of studies have also applied isotopic data, mainly neodymium isotopes, to sedimentary rocks. Studies of Nd isotopes have centered primarily on issues of crustal evolution (e.g., McCulloch and Wasserburg, 1978; Taylor et al., 1983; Miller et al., 1986), although some have also attempted to address tectonic issues (e.g., Nelson and DePaolo, 1988; Barovich et al., 1989; Frost and Coombs, 1989; McLennan et al., 1990; Linn et al., 1991; McLennan and Hemming, 1992).

Applying radiogenic isotope techniques to sedimentary provenance studies is especially alluring because it provides a fundamentally different perspective, that of time. In the case of neodymium isotopes, the most important process that fractionates Sm from Nd, thus providing the conditions necessary for significant isotopic variations over time, is melting of the mantle to form the more evolved plutonic and volcanic rocks of the exposed continental crust. The Nd model age of sedimentary rocks is generally interpreted to reflect the mean age of mantle extraction for the various provenance components (McCulloch and Wasserburg, 1978). In practice, such data appear most useful in evaluating the relative role of new (mantle derived) crust and old recycled upper crust at the time of deposition. Other conventional isotopic approaches, including Rb-Sr and K-Ar dating, have also proven useful (e.g., Peterman et al., 1981; Heller et al., 1985).

Analyses of individual components also hold great promise. K-feldspar, and often plagioclase, have very low U/Pb and Th/Pb ratios. Consequently, Pb isotope ratios typically approach the initial composition, which in many cases appears to be characteristic of specific tectonic age terranes. Pb isotope data also allow an evaluation of the geochemical history of U, Th, and Pb in the source rocks. Improvements in conventional U/Pb isotopic measurements and development of ion microprobe techniques allow for precise dating on portions of single grains of zircon, and thus it is possible to establish age distributions for this mineral. Improvements in conventional techniques further permit analysis of the isotopic composition and, under favorable circumstances, the age of other major components, notably quartz (Hemming et al., 1990, 1991) and many other minor mineral components.

In this work we examine the application of geochemical and isotopic data to the evaluation of the provenance and tectonic history of sedimentary rocks. It is not our purpose to attempt to provide new methods for discrimination of tectonic settings; with currently available data it is not possible to offer significant further advances in this area. In fact, a major conclusion of our studies has been that geochemical and isotopic approaches best provide a complementary perspective to the origin of sedimentary rocks. Instead, our intent is to show that geochemical data can provide insight into the sedimentary processes affecting the rocks, the nature of the provenance, and the geochemical history of certain provenance components. In turn, such insight may provide constraints on the tectonic conditions and tectonic history affecting the sedimentary rocks.

ADVANTAGES OF GEOCHEMICAL APPROACHES

There are a number of intrinsic advantages that geochemical approaches to sedimentary provenance have over the more traditional petrographical approaches. Some of these are summarized below.

1. In general, geochemical approaches are equally applicable to coarse- and fine-grained sedimentary rocks. This contrasts with petrographical approaches where provenance studies for fine-grained as well as very coarse grained sediments are difficult. An important caveat is that fine-grained sediments typically represent more homogeneously mixed sources, but it is compositional variability that commonly leads to insights about provenance and attendant sedimentary processes. There is growing evidence that even intimately associated sands and muds may be derived from quite different sources with different sedimentary histories and so both require evaluation (e.g., McLennan et al., 1990).

2. In principle, it is possible to apply such techniques to mineralogically altered samples in cases where alteration does not affect bulk chemistry or by concentrating on elements and isotope systems not substantially affected by such processes. Accordingly, there is great potential for evaluating the origin of sandstones with substantial amounts of secondary matrix. In turn, geochemistry in some cases may be used to quantify the occurrence and/or extent of some secondary processes, such as weathering, sorting, and diagenesis (e.g., McDaniel and McLennan, 1991).

3. Key trace elements or isotope systems may be very sensitive in identifying minor components not readily recognized petrographically (e.g., Hiscott, 1984; Nelson and DePaolo, 1988). In some cases, recognizing even trivial amounts of certain exotic components, e.g., ophiolites, may be of considerable importance in understanding the tectonic history.

4. In cases where certain geochemical characteristics of provenance components can be inferred, there is the potential to further evaluate the origin and geochemical history of that component. For example, recognition that a volcanic component has originated from either a large-ion lithophile element (LILE)–depleted or enriched mantle source may have tectonic implications.

5. Using various isotopic approaches, there is great potential for constraining the age of the provenance and its temporal history. It is possible to evaluate both the mean age of the provenance, using techniques such as Sm-Nd whole-rock data, or estimating the crystallization or resetting (cooling) age of the

TABLE 1. PETROGRAPHICAL AND GEOCHEMICAL PERSPECTIVES OF SEDIMENTARY PROCESSES AND PROVENANCE

Process/Provenance	Petrographic Perspective	Geochemical Perspective
Sedimentary Processes		
1. Weathering	Ambiguous. Some mineralogical effects (e.g., feldspar/clay ratios)	Possible to quantify in favorable conditions. Major elements (CIA), Rb/Sr (also Sr-isotopes)
2. Diagenesis	Relative chronology and mineral reactions. Overgrowth, albitization, smectite/illite transformation	Potential but few studies. Major elements (CIA), redox elements (e.g., Fe^{3+}/Fe^{2+}), various isotopes (Rb/Sr, U/Pb)
3. Sorting	Textural maturity, heavy mineral content and variety	Major elements (Si/Al). Quantify heavy mineral fractionation such as zircon (Zr, Hf) and monazite (REE)
4. Sedimentary recycling	Quartz content, sedimentary rock fragments	Various geochemical approaches (CIA, Zr, Hf, Th/U); Nd-isotopes
Sedimentary Provenance		
1. Rock types	Rock fragments; quartz morphology	Little or no direct information
2. Terrane type	Q-F-L (continental block, magmatic arc, recycled orogen)	Combined trace element/Nd-isotopes (Old Upper Crust, Young Differentiated Arc, Young Undifferentiated Arc)
3. Terrane identification	Little or no direct information	Pb isotopic composition of feldspars and whole rocks
4. Provenance age	Relative age of components in favorable circumstances	Mean age (Nd-model ages); component ages (U/Pb ages of zircon, quartz, etc.)
5. Crust/mantle character	Little or no information	Nature of crust/mantle melting (Eu-anomalies, HREE-depletion). Nature of crust/mantle sources (Th/U, μ_i)

individual components (e.g., U-Pb dating of zircon, monazite, sphene, and quartz; Rb-Sr dating of detrital muscovite). In addition, it may be possible to identify specific tectonic terranes as provenance components using Pb isotopic data for feldspars (as an estimate of the initial Pb isotopic composition) and further to evaluate the geochemical history of U, Th, and Pb in those components.

SEDIMENTARY PROCESSES: PERSPECTIVES FROM GEOCHEMISTRY

To identify geochemical approaches relevant to characterizing provenance, it is first important to separate as much as possible the geochemical "signals" due predominately to processes from those due to provenance. In addition, sedimentary rocks associated with certain tectonic regimes in some cases appear more or less affected by such processes. In Table 1 a comparison of the types of information that may be obtained from both petrographic and geochemical/isotopic data is summarized.

Weathering and diagenesis

Using petrography to evaluate the weathering history preceeding the deposition of a sediment is ambiguous at best. Weathering involves the conversion of volcanic glass and unstable minerals, mainly feldspars and mica, to clay. Although mineralogical data and textural information regarding the alteration of feldspar can be useful, attempts at quantification have not been forthcoming. Similarly, diagenesis also commonly involves the alteration of feldspars and other unstable minerals to clays; accordingly, it may be difficult to distinguish these processes geochemically (e.g., Hower et al., 1976; Land, 1984; Helmhold and van de Kamp, 1984).

During weathering (and in many cases, diagenesis), there is

a substantial increase in the Rb/Sr ratio of most rocks. This is because Rb^+, a large alkali trace element (1.72 Å for 12-fold coordination), is more readily retained on exchange sites of clays than the smaller Sr^{2+} (1.26 Å for eight fold coordination). This major change in Rb/Sr provides opportunities for applying Sr isotopes to the evaluation of such history (e.g., Ohr et al., 1991). There is evidence that under certain conditions the rare earth elements (REE) may be fractionated substantially in the weathering profile (Nesbitt, 1979; Banfield and Eggleton, 1989). However, it appears that there is only rarely any significant gain or loss of Sm or Nd from the weathering profile during subsequent erosion and sedimentation and, accordingly, the Nd isotope system is less likely to provide insights into weathering history of sedimentary rocks. Diagenetic reactions in some cases also appear to redistribute REE among various diagenetic minerals, without significant movement of REE into or out of the system (Ohr and others, 1991; although see the alternate view of Anwiller and Mack, 1991). Accordingly, diagenesis does not appear to significantly affect bulk-rock REE patterns (also see McLennan, 1989a). It may be possible to evaluate the timing of diagenesis by examining the isotopic systematics of co-existing diagenetic phases (Ohr et al., 1991).

The bulk chemical changes that take place during weathering have also been used to quantify the weathering history of sedimentary rocks, primarily to understand past climatic conditions (Nesbitt et al., 1980; Nesbitt and Young, 1982, 1984). The Chemical Index of Alteration (CIA) has been established as a general guide to the degree of weathering (Nesbitt and Young, 1982), although chemical changes resulting from other processes, such as diagenesis and metamorphism, have not been fully evaluated. Using molecular proportions:

$$CIA = Al_2O_3/[Al_2O_3 + K_2O + Na_2O + CaO^*] * 100$$

or, abbreviating,

$$CIA = A/[A + K + N + C] * 100,$$

where CaO* represents that residing only in silicate minerals (i.e., corrected for calcium residing in carbonates and phosphates). For unconsolidated sediments, it is sometimes also necessary to make a correction on Na_2O for seawater salt (McLennan et al., 1990). On such a scale, unweathered igneous rocks have values of 50 or below, residual clays have values of near 100, and typical shales average about 70 to 75.

The petrological rationale behind this index, and for the relationship between these elements and Fe and Mg, is illustrated in Figure 1A. The bulk-rock chemistry is dominated by the conversion of feldspars (and for volcanic rocks, glass) to clay minerals during weathering. The expected pathways for increasing degrees of weathering for a mafic and felsic igneous rock are illustrated. On the A-CN-K diagram, these paths are confirmed both from weathering profiles and thermodynamic/kinetic calculations. In the case of the A-CNK-FM diagram, appropriate kinetic data are unavailable and trends are derived from weathering profiles alone. There are several complexities that must be kept in mind when employing such an approach. The first is that many mineral reactions that take place during diagenesis involve these same minerals, and for a system where any of these chemical components are added or lost there will be comparable effects. Studies are not yet available to evaluate this in any systematic manner (although see Nesbitt and Young, 1989). An additional complexity comes from mixing provenance components of different compositions and/or weathering histories, illustrated in Figure 1B. In some cases the trajectory of the pathway will help in distinguishing mixing from weathering trends (H. W. Nesbitt, personal communication). For example, a very flat trend that intersects the A-CN boundary is inconsistent with weathering being the sole control and mixing may be indicated. Mass balance requires that weathering (or most water-rock interaction) trends should lie on a line extrapolated from groundwater compositions (that lie on the CN-K join) and through the parent material composition.

Sedimentary sorting

There are several petrographic approaches to examine sorting processes in sedimentary rocks. The most straightforward is to evaluate the textural maturity of the sediment, using characteristic grain sizes and shapes, and mineralogy (e.g., Folk, 1974). The chemical composition of a sedimentary rock may also be helpful in evaluating the influence of sedimentary sorting.

With increasing textural maturity in sandstones, there is typically an increase in quartz at the expense of primary clay-sized material, resulting in an elevation of the SiO_2/Al_2O_3 ratio and a decrease in the abundances of most trace elements. Of special interest is the recognition that, on a local scale, in some mineralogically immature sediments, there can be an enrichment in the major framework mineral plagioclase due to sedimentary sorting processes. This may have serious consequences both for tectonic interpretations of sandstone petrography and chemical composition in samples where plagioclase is abundant (Nance and Taylor, 1977; McLennan, 1989a). A well-documented example of this phenomenon comes from the Devonian Baldwin Formation (Nance and Taylor, 1977). In this volcanogenic turbidite of probable forearc origin, Nance and Taylor (1977) noted a strong enrichment in Eu, a rare earth element typically enriched in feldspars, for some samples. These enrichments correlate well with framework plagioclase content (Chappell, 1968; Bhatia, 1981), and accordingly have been interpreted as resulting from plagioclase enrichments due to sedimentary sorting processes (Fig. 2). Any variations in REE abundance patterns of sedimentary rocks that can be assigned to sedimentary processes are of significance because REE patterns are used widely for provenance characterization (Taylor and McLennan, 1985; McLennan, 1989a). In a study of deep-sea turbidites from a variety of tectonic settings, McLennan et al. (1990) noted that many sands from active tectonic settings had elevated Eu/Eu* (where Eu* is

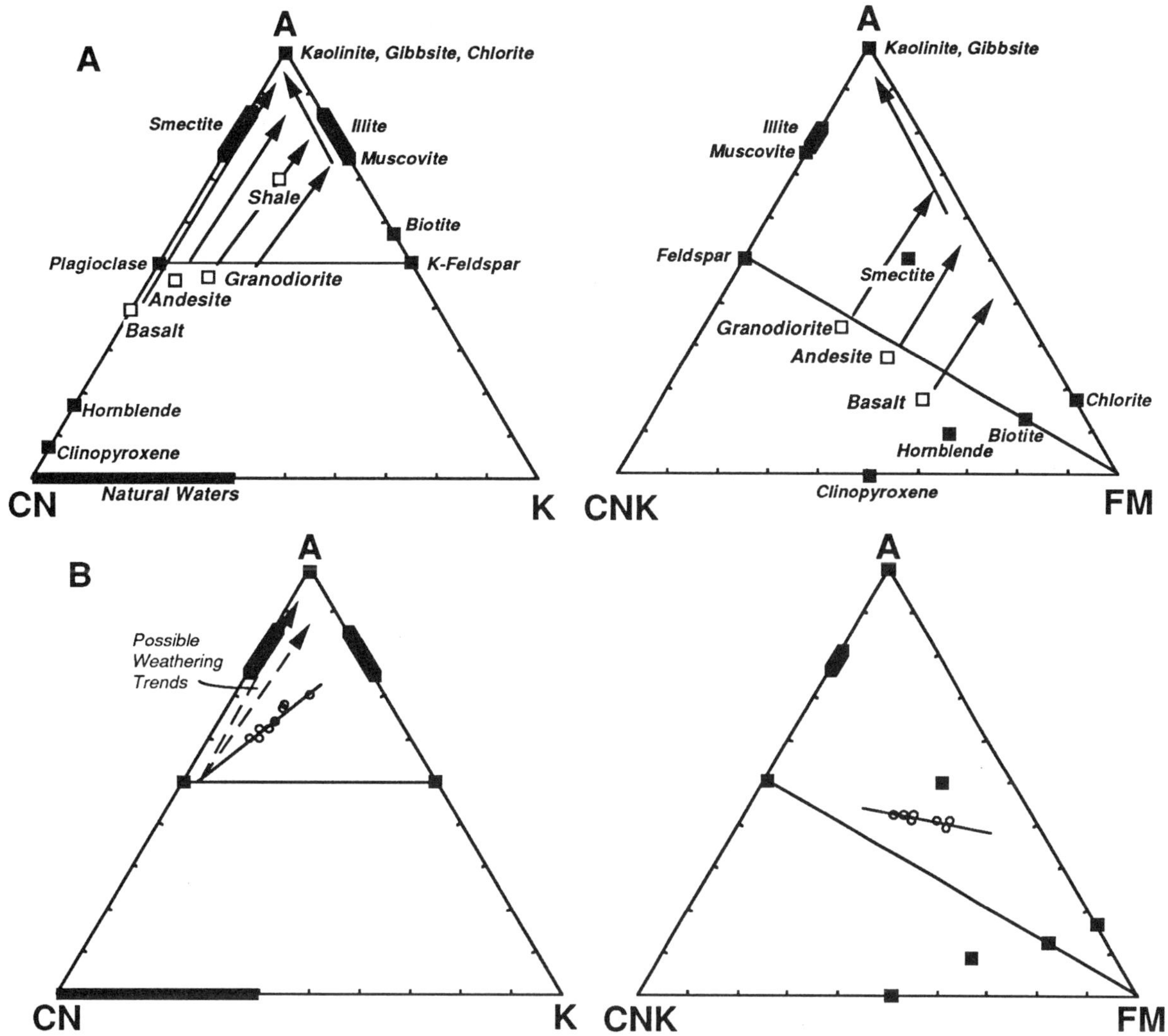

Figure 1. A, Ternary plots of A-CN-K and A-CNK-FM where, in mole fraction, A = Al_2O_3, C = CaO (silicate fraction only), N = Na_2O, K = K_2O, F = total Fe as FeO, and M = MgO (after Nesbitt and Young, 1984, 1989). Plotted are simplified compositions of major minerals, typical rock types, and natural waters. Arrows indicate the general trends of weathering exhibited by various rock types. B, Ternary plots of A-CN-K and A-CNK-FM showing sandstones and argillites from the Proterozoic Chelmsford Formation, Sudbury Basin, Ontario (McDaniel, 1992). Two argillite samples are the two most evolved samples in terms of CIA. These samples show a linear trend that is inconsistent with simple weathering being the sole control of the composition (compare with Fig. 1A). For example, on the A-CN-K diagram the predicted weathering trend is shown for a hypothetical composition derived from extrapolating the sediment trend to an unweathered composition. Possible explanations include mixing of a relatively weathered source with an unweathered source of differing primary composition or perhaps some influence from secondary sedimentary process that resulted in redistribution (gains or losses depending on composition) of Ca, Na, and/or K in the silicate fraction (note that plagioclase in these sandstones is albitized).

a theoretical value for Eu, assuming no chondrite-normalized Eu anomaly), by as much as 0.10 over coexisting muds. This was also ascribed as being possibly due to sorting processes, and suggests that such a process may be more important than commonly thought. In tectonic settings where abundant plagioclase is not found there is no evidence of systematic Eu enrichment in sands over coexisting muds (Nathan, 1976; Bhatia, 1985).

Since a number of heavy minerals are dominated by elements that are trace elements in most sedimentary rocks (e.g., Zr in zircon, REE in monazite and allanite), it is possible to evaluate the role of heavy mineral concentration during sedimentary sorting (McLennan, 1989a). An example of this can be illustrated for modern deep-sea turbidites. In Figure 3, Th/Sc is plotted against Zr/Sc. The ratio Zr/Sc is a useful index of zircon enrichment, since Zr is strongly enriched in zircon, whereas Sc is not enriched but generally preserves a signature of the provenance similar to

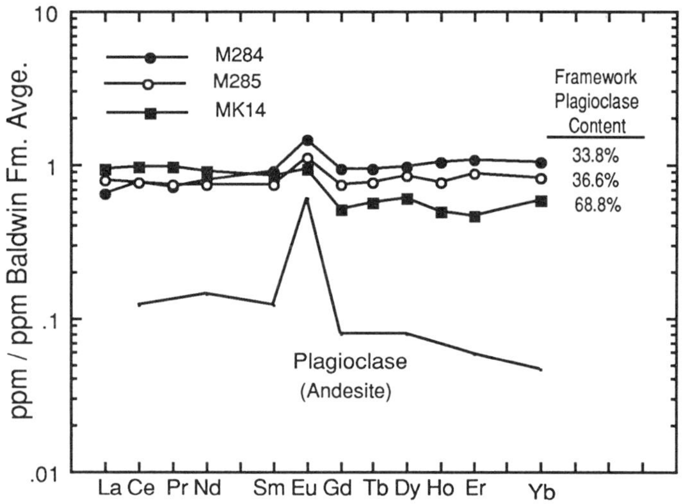

Figure 2. REE diagram for Eu-enriched turbidite sandstones from the Devonian Baldwin Formation, normalized to the Baldwin Formation average (of non-Eu-enriched samples). The Baldwin Formation is a volcanogenic turbidite thought to have been deposited in a forearc setting. The amount of framework plagioclase in these samples is shown on the right. The formation average is about 16%. Also shown is a typical REE pattern for plagioclase from an andesite. The Eu enrichment in some of the Baldwin Formation sandstones is likely due to plagioclase enrichments formed during sedimentary sorting. Data from Schnetzler and Philpotts (1970), Chappell (1968), Nance and Taylor (1977), Bhatia (1981, 1985).

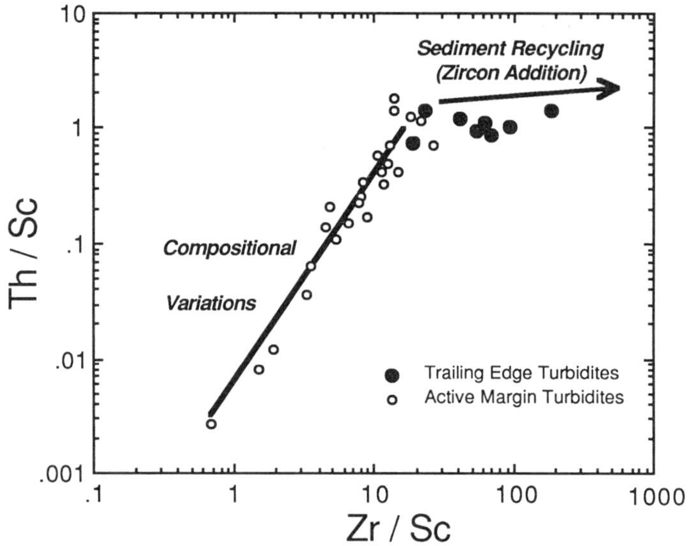

Figure 3. Plot of Th/Sc versus Zr/Sc for modern turbidites from various tectonic settings showing the enrichments of zircon (high Zr/Sc ratio) in trailing edge settings resulting from sedimentary sorting and recycling. Active margin samples, which are less affected by this sedimentary process, show a simple correlation for these ratios, and this relationship is interpreted to be due to compositional variations of the provenance. Data from McLennan et al. (1990).

other REE (McLennan, 1989a). In contrast, Th/Sc is a good overall indicator of igneous chemical differentiation processes since Th is typically an incompatible element, whereas Sc is typically compatible in igneous systems. It can be seen that both Th/Sc and Zr/Sc vary sympathetically for turbidite sands from active margin settings, following a trend consistent with igneous differentiation being the primary control (i.e., provenance). In contrast, turbidite sands from passive margins show Zr/Sc increasing substantially, with Th/Sc increasing far less, consistent with zircon enrichment. In the case of monazite, an REE-enriched heavy mineral, it is also possible to estimate enrichments from REE patterns (McLennan, 1989a; Fig. 4). Since monazite has very high REE abundances and a very steep chondrite-normalized heavy REE pattern, even small amounts (<0.01%) result in significant increases in the Gd_N/Yb_N ratio (where subscript N refers to chondrite-normalized values). Monazite enrichments may be recognizable since post-Archean sedimentary REE patterns and most upper crustal igneous rocks rarely have Gd_N/Yb_N outside of the range of 1.0 to 2.0 (McLennan, 1989a; McLennan and Taylor, 1991).

Sedimentary recycling

Quantifying sedimentary recycling processes, or the cannibalistic turnover of the sedimentary mass, is becoming increasingly important in understanding issues such as crustal evolution (e.g., Veizer and Jansen, 1979; McLennan, 1988) and the tecton-

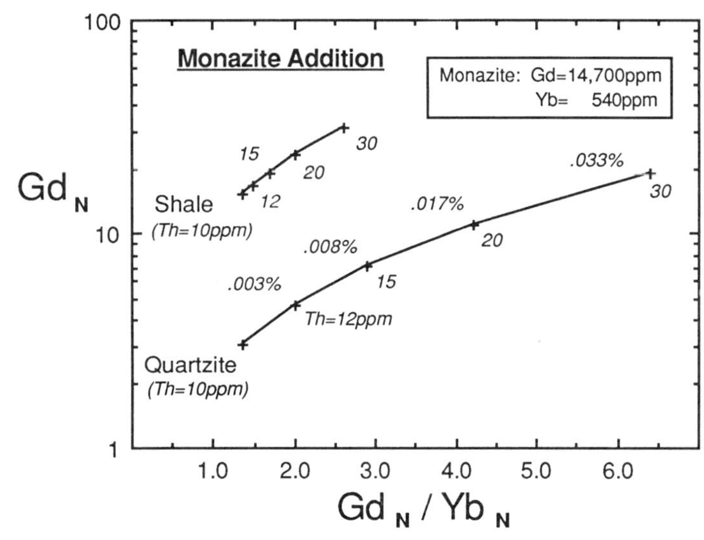

Figure 4. Plot of Gd_N versus Gd_N/Yb_N to illustrate the effect on REE patterns of concentrating monazite in sedimentary rocks. Calculations for two hypothetical sediments are shown, one a low REE sediment (e.g., quartzite) and the other a high REE sediment (e.g., shale). For both, an REE pattern parallel to typical post-Archean shales (Taylor and McLennan, 1985) is assumed. Both are assumed to have 10 ppm Th. Monazite REE data are from Lee and Bastron (1967), and a typical monazite Th abundance of 6% is assumed. Adding even small amount of monazite (as little as 0.005% to the quartzite) can result in significant increases in the Gd/Yb ratio of the sediment. The Gd_N/Yb_N ratio of sedimentary rocks is usually rather constant and in the range of 1.0 to 2.0.

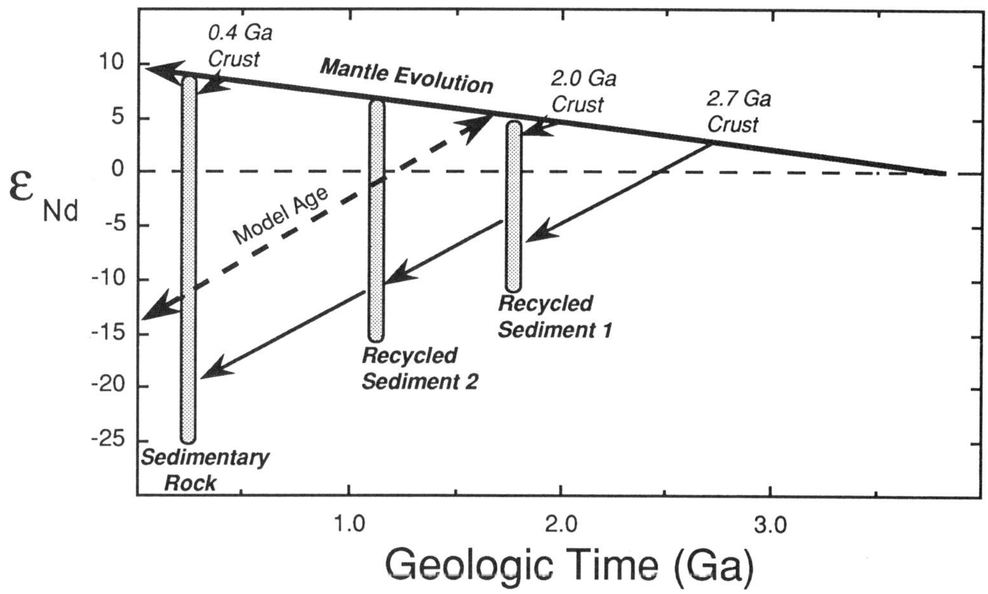

Figure 5. Plot of ϵ_{Nd} versus time to illustrate the interpretation of sedimentary Nd model ages in terms of sedimentary recycling. A sediment with a 250-Ma stratigraphic age has a present day ϵ_{Nd} of −14 and, for a typical sedimentary Sm/Nd ratio, gives a depleted mantle model age (T_{DM}) of 1.7 Ga. Shown is a highly simplified model of the sediment's origin. The 2.7-Ga upper crust is mixed with new crustal material of 2.0-Ga age and incorporated into a sedimentary rock (Recycled Sediment 1) at about 1.8 Ga. This sediment is then recycled into another sedimentary basin (Recycled Sediment 2) at about 1.2 Ga with no addition of new crust material. At the time of sedimentation, Recycled Sediment 2 and new crust, formed during the Early Paleozoic, act as the provenance for the analyzed sediment. In practice, the history is likely to be even more complicated. Also, it is not possible to distinguish, solely on the basis of Nd isotopes, between old igneous/metamorphic crust acting directly as the provenance and old crust recycled through the sedimentary record.

ics of sedimentary basins (e.g., Ingersoll, 1988; Johnsson et al., 1988; Johnsson and Meade, 1990; Steidtmann and Schmitt, 1988). Recycled sediment is likely the dominant source of preserved sedimentary rocks, but in detail the extent of any recycled component is highly variable and in many cases is related to tectonic history (Veizer and Janson, 1985). In Figure 5, the Nd-isotopic evolution of a sedimentary rock is shown in terms of a highly simplified sedimentary recycling model. The model age of the sediment (in this example, 1.7 Ga) has no bearing on any particular provenance component. Instead, it represents a complex mix of differing ages of crust that has gone through various cycles of sedimentary processes.

SEDIMENTARY PROVENANCE: PERSPECTIVES FROM GEOCHEMISTRY

Terrane types

It is recognized that sediment compositions are characteristic of certain assemblages of igneous/metamorphic/sedimentary source rocks that have specific styles of sedimentary history (e.g., Dickinson, 1985, 1988), or what is here termed terrane (or provenance) types. In turn, some inferences about the tectonic conditions can be made. In this section, we describe terrane types that can be identified from geochemical data, mainly combining major and trace elements with Nd-isotopic compositions of whole rocks. Of special importance is the observation that these terrane types do not precisely correspond to those recognized from petrographical approaches but instead provide a different perspective of the plate tectonic association. It is also important to note that, although a given terrane type may dominate the provenance of certain sedimentary rocks (and some types are more likely than others), it is just as likely that the provenance will be a mix. Unravelling the contributions of each terrane type is, in effect, a major part of the task at hand. Some of the chemical and isotopic characteristics of these terrane types are summarized in Table 2.

These concepts are largely based on combined geochemical and Nd isotope studies of the provenance of young continental margin sediments (e.g., White et al., 1985; Ben Othman et al., 1989; McLennan et al., 1990; Basu et al., 1990). Similar studies of Phanerozoic sequences are rare but the general approach appears to apply (Andre et al., 1986; Frost and Coombs, 1989; Chen et al., 1990). In Precambrian terranes, additional complexities arise, largely related to different lithologies being present in the Archean upper crust. These are discussed in greater detail below.

Two of the major factors used in characterizing terrane types

TABLE 2. SUMMARY OF GEOCHEMICAL AND Nd-ISOTOPIC CHARACTERISTICS OF TERRANE (PROVENANCE) TYPES

Terrane Type	ε_{Nd}†	Eu/Eu*	Th/Sc	Th/U	Other
Old Upper Continental Crust (OUC)	≤-10	≈0.60-0.70	≈1.0	>3.8 (shales)	Evolved major element composition (e.g., high Si/Al, CIA); high LILE abundances; uniform compositions
Recycled Sedimentary Rocks (RSR)	≤-10	≈0.60-0.70	≥1.0		Evidence of heavy mineral concentration from trace elements (e.g., Zr, Hf for zircon, REE for monazite)
Young Undifferentiated Arc (YUA)	≥+5	≈1.0	<1.0	<3.0	Unevolved major element compositions (e.g., low Si/Al, CIA); low LILE abundances; variable compositions
Young Differentiated Arc (YDA)	≥+5	≈0.50-0.90	Variable	Variable	Evolved major element composition (e.g., high Si/Al, CIA); high LILE abundances; variable compositions
Exotic components	Chemical and/or isotopic signature depends on the nature of the component; e.g., very high Mg, Cr, Ni, V, and Cr/V would be distinctive of ophiolite sources.				

†ε_{Nd} values are for modern sediments only. Distinctions between terrane types would persist for older sedimentary rocks but would be increasingly less pronounced with age. Exact values would depend on sedimentation age.

are bulk chemical composition and Nd-isotopic composition at the time of sedimentation. The influence of these factors can be summarized in Figure 6 where modern turbidites (McLennan et al., 1990) are plotted on a diagram of ϵ_{Nd} versus Th/Sc ratio, where ϵ_{Nd} represents deviations of the $^{143}Nd/^{144}Nd$ ratio, in parts per 10^4, from average chondritic meteorites (representative of the bulk earth). The Th/Sc ratio is a convenient and sensitive index of bulk composition (Taylor and McLennan, 1985). Th is a relatively incompatible element in most igneous processes whereas Sc behaves compatibly; both appear to be transported in the terrigenous component during sedimentary processes. The samples are divided into *quartzose* (Recycled Orogen, Craton, Transitional Continental in terms of QFL) and *nonquartzose.* Quartzose samples include those from trailing edge margins and continental collisions with an additional few from active settings in Japan and Java where old continental crust is present within or near the arcs. The main features to note are that nonquartzose turbidites (mainly active settings) have higher ϵ_{Nd} (reflecting young provenance components) and highly variable and commonly low Th/Sc ratios (reflecting heterogeneous and generally less differentiated provenance). In contrast, quartzose turbidites have generally lower ϵ_{Nd} (reflecting older provenance) and very uniform Th/Sc ratios of about 1.0, the value of the upper crust (Taylor and McLennan, 1985). Also shown are data for the Andean foreland (Basu et al., 1990) interpreted to be derived from Paleozoic-Mesozoic upper crust. The extension of these data toward high Th/Sc is likely related to heavy mineral enrichments since samples with highest Th/Sc have exceptionally high Zr content (up to 2,500 ppm).

An entirely comparable approach, using only Nd-isotopic data, is shown in Figure 7, an $f^{Sm/Nd}$ - ϵ_{Nd} diagram. The value $f^{Sm/Nd}$ is the fractional deviation of the $^{147}Sm/^{144}Nd$ ratio from that in chondritic meteorites ($f^{Sm/Nd} = (^{147}Sm/^{144}Nd)_{sample}/(^{147}Sm/^{144}Nd)_{chondrite} - 1$). Since REE patterns tend to become more light REE (LREE; La-Sm)–enriched (i.e., lower Sm/Nd) during igneous differentiation processes, $f^{Sm/Nd}$ monitors the general extent of differentiation, albeit in a far less sensitive manner than Th/Sc (McLennan and Hemming, 1992).

1. Old upper continental crust (OUC). This provenance component constitutes old igneous/metamorphic/sedimentary terranes that have been influenced by intracrustal geochemical differentiation. Old stable cratons and the old continental foundations of active tectonic settings are examples. Intracrustal fractionation includes processes of both intracrustal partial melting and/or crystal fractionation, but in either case involves considerable separation of plagioclase (stable essentially only at crustal depths). It is this intracrustal fractionation that imparts the negative Eu anomaly to the upper crust (Taylor and McLennan, 1985). The relatively great age of this component is reflected in the nonradiogenic Nd isotopic character (i.e., low $^{143}Nd/^{144}Nd$ ratio), with ϵ_{Nd} typically being less than about −10 for modern sediments and appropriately low for older sedimentary rocks.

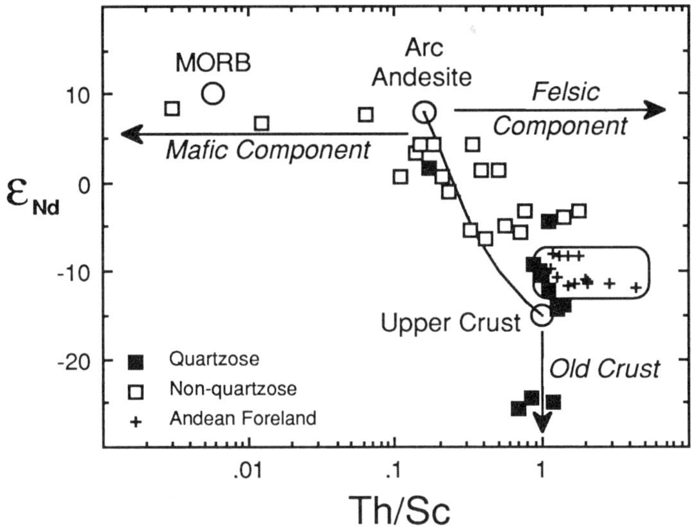

Figure 6. Plot of ϵ_{Nd} versus Th/Sc ratio for various modern sediments. The Nd isotopic composition is strongly controlled by the mean provenance age whereas Th/Sc ratio is a sensitive index of the bulk composition of the provenance. Plotted are modern turbidites (McLennan et al., 1990) divided into quartzose (mainly trailing edge) and nonquartzose (mainly active margins). Note that the nonquartzose turbidites tend to have variable but higher ϵ_{Nd} (reflecting a younger provenance) and more variable but commonly lower Th/Sc ratio. The quartzose turbidites are of more uniform composition and generally from an older provenance. Also plotted are modern sands from the Andean Foreland (Basu et al., 1990) considered to be derived from old recycled upper crust. In this case, high Th/Sc is not due to derivation from more felsic sources but is related to concentration of heavy mineral during sedimentary recycling and sorting.

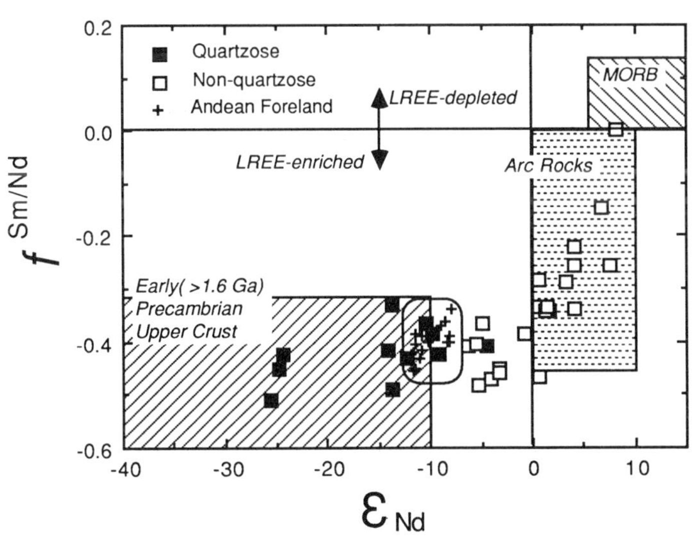

Figure 7. Plot of $f^{Sm/Nd}$ versus ϵ_{Nd} for the same samples shown in Figure 6. The degree of LREE enrichment in a sedimentary rock is also sensitive to bulk compositions; such a diagram conveys similar information to that in Figure 6. Note that this diagram is essentially a slightly recast "isochron" diagram (see Shirey and Hanson, 1986, for an explanation).

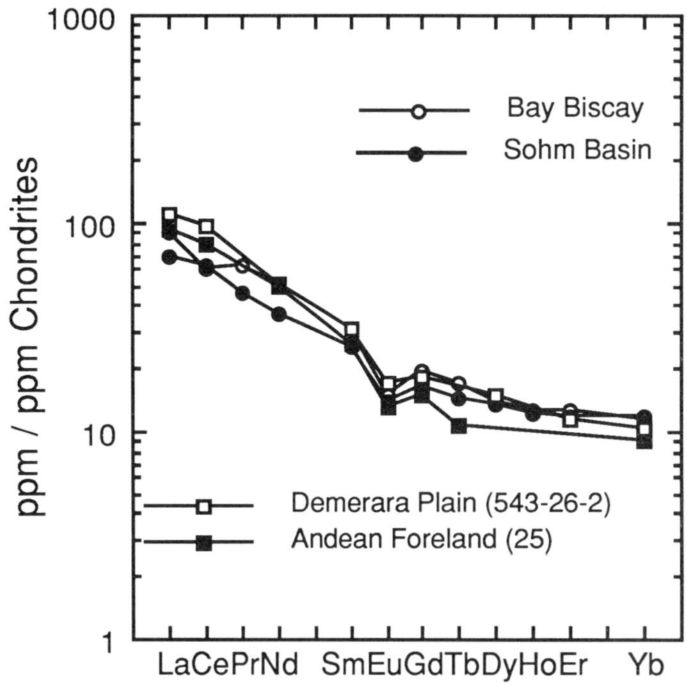

Figure 8. Chondrite-normalized REE plot of selected modern sediments that have an old upper continental crustal (OUC) provenance. The ϵ_{Nd} for each of these samples is less than –10. Note the similarity of the patterns with LREE enrichment, fairly flat HREE (i.e., Gd_N/Yb_N = 1.0 to 2.0) and the negative Eu anomaly, characteristic of the upper continental crust. Data from White et al. (1985), McLennan et al. (1990), and Basu et al. (1990).

The geochemistry of sediments derived from this component is normally characterized by relatively uniform compositions, reflecting a broad, well-mixed provenance (Potter, 1978) and the greater level of recycling expected for such environments (Veizer and Jansen, 1979, 1985). In addition to a substantial negative Eu anomaly (Eu/Eu* about 0.60 to 0.70; Taylor and McLennan, 1985), sediments typically have evolved major element compositions (e.g., high SiO_2/Al_2O_3, K_2O/Na_2O, CIA), reflecting a dominance of upper crustal granitic sources and a relatively severe weathering (and recycling) history. Trace element features include enrichments of normally incompatible over compatible elements (e.g., LREE enrichment, high Th/Sc, La/Sc) reflecting relatively felsic average provenance compositions, and high Rb/Sr (>0.5) and Th/U (>3.8 for shales) reflecting weathering and sedimentary recycling.

Examples of sedimentary rocks dominated by OUC include abundant cratonic shales (Nance and Taylor, 1976), most passive margin turbidites (McLennan et al., 1990), and foreland basin sediments derived from the older exposed continental crust (including, for example, the exposed continental bulge). Characteristic REE patterns are shown for sediments dominated by this terrane type in Figure 8. Each displays the negative Eu anomaly and LREE enrichment characteristic of the upper continental crust and each has ϵ_{Nd} <–10. Most sediments derived from tectonically active settings also contain variable amounts of such

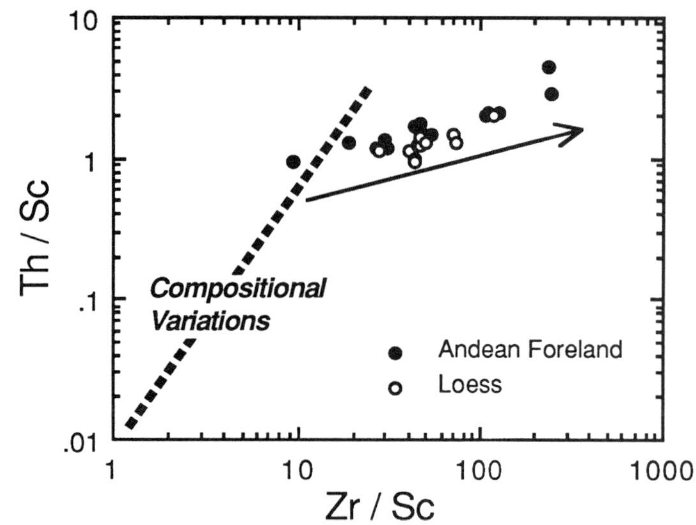

Figure 9. Plot of Th/Sc versus Zr/Sc for modern sands from the Andean foreland and for loess. Sands are enriched in heavy minerals, notably zircon, due to sedimentary sorting and recycling. Similar enrichments are seen for the aeolian loess (silt), due to the efficient concentration of zircon during wind transport. Data from Taylor et al. (1983) and Basu et al. (1990).

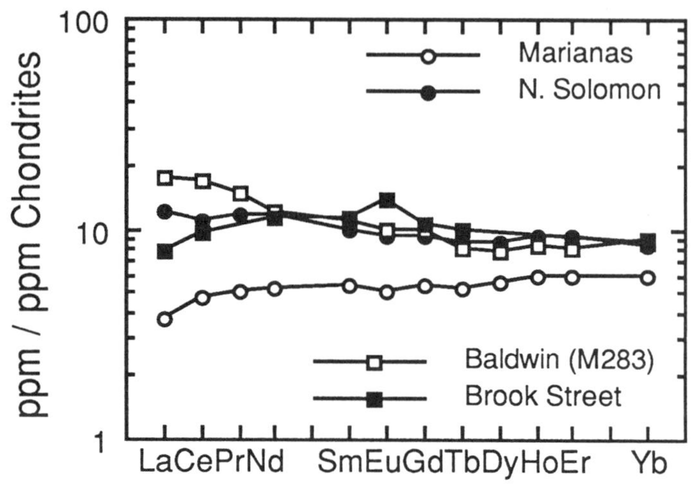

Figure 10. Chondrite-normalized REE plot of selected modern and Phanerozoic turbidites that are dominated by young undifferentiated arc (YUA) provenance. Except for the Baldwin sample, for which there are no Nd isotopic data, each sample has $\epsilon_{Nd} > +6$ at the time of deposition. Compared to samples with OUC provenance (Fig. 8), these have lower REE abundances, variable but less LREE enrichment (in some cases, LREE depletion) and lack significant negative Eu anomalies. Data from Nance and Taylor (1977), Frost and Coombs (1989), and McLennan et al. (1990).

material. In general, cannibalistic sedimentary recycling is considerably more efficient at preserving sedimentary rocks associated with stable cratonic tectonic settings than those from tectonically active settings (Veizer and Jansen, 1985). Accordingly, sedimentary rocks dominated by this provenance component are preferentially preserved in the geologic record.

*2. **Recycled sedimentary rocks (RSR).*** Although old upper continental crust (OUC) includes recycled sedimentary and metasedimentary rocks, in favorable cases it may be possible to distinguish recycled sediment as a separate component. Petrographically, identification of recycled sedimentary components is straightforward either by identifying sedimentary rock fragments or abraded quartz overgrowths and may be suggested by the quartz-rich nature of sands (although note Johnsson et al., 1988). Sedimentary recycling may also be accompanied by fractionation and enrichment of heavy minerals, notably zircon. As noted above, it may be possible to identify such enrichments using trace element relationships. It is likely inappropriate to use geochemical data alone because most well-sorted sediment and highly reworked first-cycle sediments could also possess enrichments in heavy minerals (Johnsson et al., 1988). The latter effect may be seen in data for loess, where Zr and Hf are strongly enriched over upper crustal abundances (Fig. 9). It appears that aeolian processes are particularly efficient at transporting zircon (e.g., Taylor et al., 1983; Brimhall et al., 1991), presumably due to some optimum combination of specific gravity and typical grain size (and perhaps shape).

An example where a recycled sedimentary component may be identified comes from the Andean foreland basin sands of Bolivia and Peru (Basu et al., 1990). In these sands, a major recycled metasedimentary component may be directly identified from the petrography. Such a component is also observed through the trace element relationships where zircon and other trace minerals are concentrated due both to the sedimentary provenance and further enrichments related to sedimentary transport (Fig. 9).

*3. **Young undifferentiated arc (YUA).*** This component constitutes young (dominantly mantle derived) volcanic and plutonic igneous rocks from island and continental arcs that have not undergone intracrustal differentiation to the degree that separates significant amounts of plagioclase. Accordingly, significant negative Eu anomalies are absent. Although lithological relations in arc terranes are complex, they are typically less evolved, with basaltic and andesitic rocks (and plutonic equivalents) being far more common than in stabilized upper continental crust. A young provenance is demonstrated by radiogenic neodymium isotopic compositions such that ϵ_{Nd} is typically above about +5 for modern sediments. In addition to lacking negative Eu anomalies (Fig. 10), such sediments typically possess rather variable major and trace element compositions reflecting relatively localized provenance and general lack of sedimentary recycling. Thus, major elements typically have relatively low, but variable, SiO_2/Al_2O_3, K_2O/Na_2O, and CIA, reflecting less differentiated and less weathered sources, and lower ratios of incompatible to compatible elements, such as Th/Sc (<1.0), and less LREE enrichment than old upper crustal sources (see Figs. 3, 6, 7, 10).

For Phanerozoic sedimentary rocks, at least, information regarding the nature of mantle sources may be extracted from the geochemical and isotopic data. For example, Pb isotopic data for potassium feldspar (or plagioclase) should constrain the U/Pb

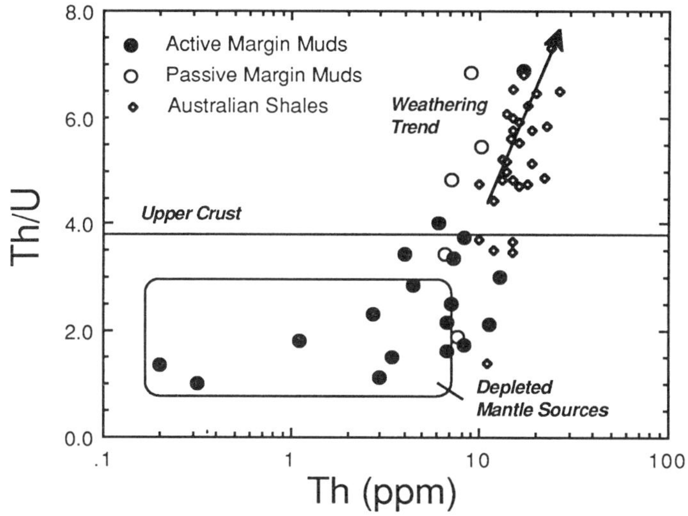

Figure 11. Plot of Th/U versus Th for modern turbidite muds from various tectonic settings and for Australian post-Archean shales mainly of cratonic provenance. During weathering, there is a tendency for an elevation of Th/U above upper crustal igneous values of 3.5 to 4.0 that are reflected in shales. Sedimentary processes that lower Th/U usually are the result of U enrichments. It is suggested that low Th/U, commonly seen in active margin sediments, especially when accompanied by low Th and U abundances, reflects geochemically depleted mantle sources of the arc provenance.

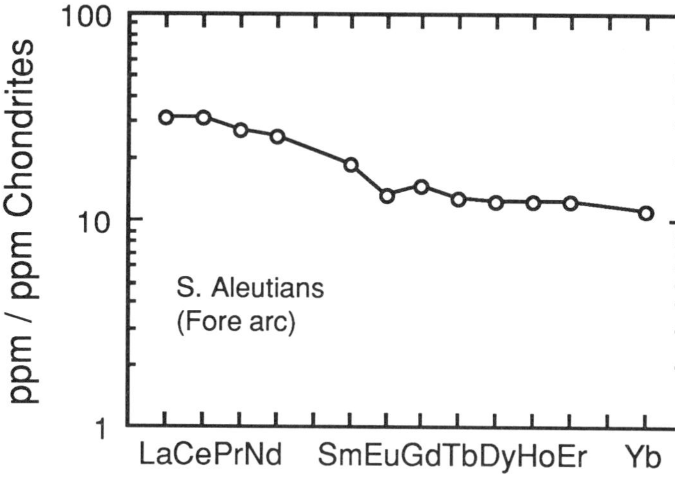

Figure 12. Chondrite-normalized REE plot of a turbidite sample from the Aleutians forearc. This sample has an ϵ_{Nd} of +7.7, suggestive of a provenance dominated by the island arc. The negative Eu anomaly suggests that igneous rocks in the young arc provenance were formed by intracrustal differentiation that fractionated plagioclase. This is an example of a young differentiated arc (YDA) provenance. Data from McLennan et al. (1990).

ratio and Pb isotopic history of the mantle sources as it does for mantle-derived igneous rocks themselves (e.g., Sun, 1980). For most upper crustal rocks, Th/U is typically about 3.5 to 4.0. In most cases, weathering and sedimentary recycling under oxidizing conditions typically results in oxidation of U^{4+} to U^{6+}, the latter being more soluble. Dissolution and loss during sedimentation follows, thus elevating the Th/U ratio, especially for shales, where heavy minerals are less likely to be a complicating factor (e.g., McLennan and Taylor, 1980; Taylor and McLennan, 1985). In other cases where low Th/U exists due to sedimentary processes, it is likely that it results from U enrichment and accordingly this will be accompanied by high U content reflecting the greater mobility of this element. Many sediments from active margin tectonic settings, with major components of young undifferentiated crust, have Th/U significantly below 3.5 accompanied by low Th and U contents (Fig. 11) and this is interpreted as dominantly reflecting a low ratio in the source rocks (McLennan, 1989b, McLennan and Taylor, 1991). Low Th/U ratios are rather common in mantle-derived volcanic rocks and reflect the geochemically depleted nature of such reservoirs (e.g., Newman et al., 1984).

This provenance component is likely to dominate in forearc basins. It is also likely to be found in most other volcanically active tectonic settings; however, data for modern turbidites indicate that most of these settings also have significant but variable amounts of old upper continental crust. Examples where YUA dominates is the Devonian-Carboniferous volcanogenic sediments of eastern Australia (Chappell, 1968; Nance and Taylor, 1977; Bhatia, 1981, 1985; Hensel et al., 1985) and the Permian Brook Street terrane of southwesternmost New Zealand (Frost and Coombs, 1989) (Fig. 10).

4. Young differentiated arc (YDA). This provenance component represents young (mantle derived) volcanic and plutonic igneous rocks from island and continental arcs that have been affected by intracrustal differentiation such that they possess significant negative Eu anomalies. It is dominated by differentiated felsic volcanics and ashes and plutons that form within the arc and are exposed as the arc becomes dissected. This component is distinguished from young undifferentiated arc (YUA) sources by the presence of significant negative Eu anomalies. It is distinguished from old upper crustal sources (OUC and RSR) by its relatively high $^{143}Nd/^{144}Nd$ ratio (ϵ_{Nd}>+5 for modern sediment). In effect, this component represents "young" upper continental crust since the intracrustal differentiation essentially removes heat-producing elements to the upper crust and stabilizes the continental crust. In terms of major and trace element compositions, this component appears to be generally similar but of much greater variability than old upper crustal provenance, due to the more localized sources. Because of the first-cycle character and rapid erosion rates of high-standing arc terranes, in most cases this component is likely to be less affected by weathering processes (i.e., low CIA, Rb/Sr) than is old upper crust (OUC).

Such a component should be important in mature arc environments, such as the Aleutians (Fig. 12), where the arc is dissected (McLennan et al., 1990) or where abundant felsic volcanism occurs. The origin of this component suggests that it is unlikely that it would commonly be dominant, compared to OUC or YUA. Instead, mixtures with OUC and YUA are to be expected. Floyd and Leveridge (1987) have suggested a dissected continental arc provenance for much of the Devonian Gramsca-

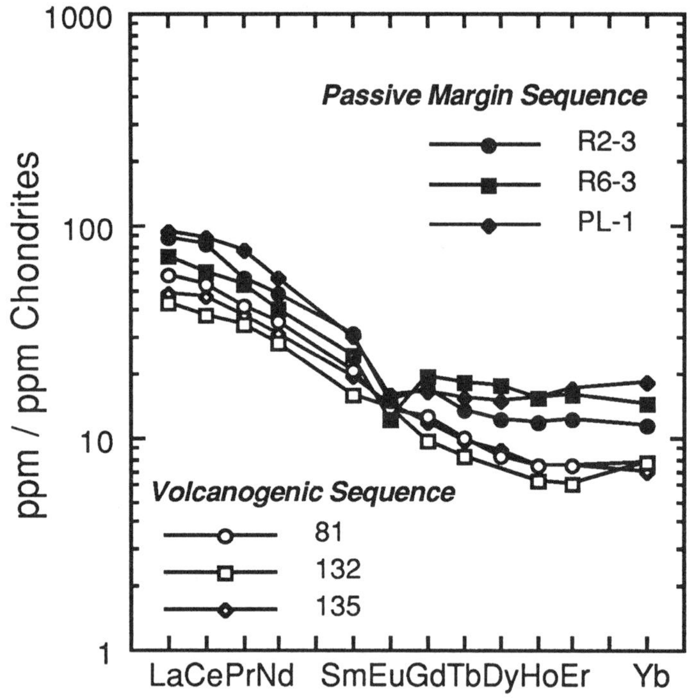

Figure 13. Chondrite-normalized REE plot showing selected Early Proterozoic sedimentary rocks from northern New Mexico and southern Colorado. Examples from two sequences are shown. One is a lower volcanogenic turbidite sequence, lacking substantial negative Eu anomalies, typical of an undifferentiated arc provenance. The second, younger, sequence is characterized by LREE-enriched patterns with negative Eu anomalies, characteristic of the upper continental crust. Unpublished data.

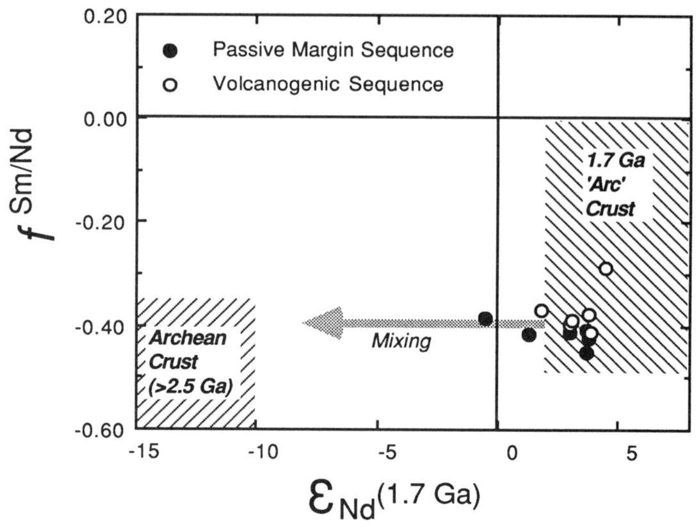

Figure 14. Plot of $f^{Sm/Nd}$ versus ϵ_{Nd} for sedimentary rocks from the Early Proterozoic of New Mexico and Colorado. The data are recalculated for the time of sedimentation, at ca. 1.7 Ga. Although there is a difference in the Sm/Nd ratio, all samples have ϵ_{Nd} consistent with a dominant provenance of 1.8 to 1.7 Ga with little or no older Archean crust. This indicates that substantial crust was generated in an active tectonic setting and subsequently stabilized, with deposition of a passive margin quartzite-shale sequence, over a very short period of time at 1.7 Ga. Unpublished data.

tho turbidite sequence from Cornwall, and major and trace element data are consistent with characteristics described above. Similarly, Dabard (1990) suggested a felsic volcanic provenance for Late Proterozoic turbidites from the Armorican Massif, with Eu/Eu* in the range 0.7 to 0.9.

Drawing a line between old and young differentiated crust is not necessarily straightforward. Under most circumstances during the Phanerozoic, the intracrustally differentiated component appears to be substantially older than new mantle-derived crust. The average Nd-model age of most Phanerozoic cratonic sediments is about 1.5 to 2.0 Ga, although there are examples where this is not the case. Early Proterozoic rocks of southwestern United States appear to record a major crust-forming episode at about 1.8 to 1.7 Ga (Nelson and DePaolo, 1985). The sedimentary sequence records a lower volcanogenic graywacke sequence, representing derivation from an arc terrane, and a younger quartzite-shale sequence, representing deposition on a stabilized passive margin (Soegaard and Eriksson, 1986). Trace element data from the fine grained sedimentary rocks are consistent with a dominantly young, undifferentiated arc terrane as a source for the lower graywackes and a reasonably well-mixed (i.e., uniform compositions) upper crustal source for the passive margin sediments (McLennan and Taylor, 1989; Fig. 13). Nd-isotopic data (Fig. 14), however, indicate that the dominant source of these sediments, including the quartzite-shale sequence, were derived from the mantle at about 1.7 to 1.8 Ga, just prior to sedimentation. This suggests an enormous volume of new crust was generated and stabilized over a very short period of time and thus places important constraints on the Proterozoic tectonic evolution of this region. Although a first-cycle origin for these quartzites is apparent from sedimentological data (Soegaard and Eriksson, 1989), combined geochemical and isotopic data permit quantitative constraints to be placed on the potential provenance components.

*5. **Exotic components.*** In some cases, identification of even trivial amounts of certain exotic components can have very important tectonic implications. For example, the first appearance of any components from a geochemically or isotopically distinct allochthonous terrane would provide important evidence regarding its proximity (e.g., Dickinson, 1988). Hiscott (1984) related variations in ferromagnesian trace elements in Taconic flysch to proximity of an obducted ophiolitic source terrane (Fig. 15). The lack of ophiolitic detritus in the southern Appalachians could be related to complexities in the nature of the Taconic orogeny, such as a change in regimes from more direct convergence in the north to either oblique subduction or strike-slip motion in the south (Hoscitt, 1984). In a different setting, McLennan et al. (1990) were able to identify an ocean crust component in modern forearc turbidites from the Marianas region, based on combined major and trace element geochemistry (e.g., low Ba/La, Th/U; high Mg, Cr) and Nd-Sr isotopic relations.

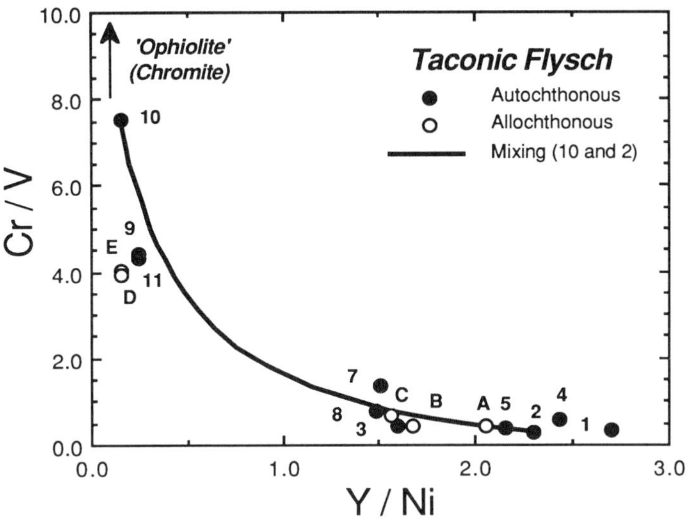

Figure 15. Plot of Cr/V versus Y/Ni for early Paleozoic Taconic flysch from eastern North America illustrating the greater importance of an ophiolitic provenance as one moves north. Higher numbers and letters indicate more northerly outcrops. Each data point represents the average of several analyses from one formation/area. Also shown is a mixing line generated from the two data points with the lowest and highest Cr abundances. The Cr/V ratio is an index of the enrichment of Cr over the other ferromagnesian trace elements, whereas Y/Ni monitors the general level of ferromagnesian trace elements (Ni) compared to a proxy for HREE (Y). Mafic-ultramafic sources tend to have high ferromagnesian abundances; such a provenance would result in a decrease in Y/Ni. A mineral such as chromite, important in ophiolitic sequences, tends to concentrate Cr preferentially over other ferromagnesian elements. Data from Hiscott (1984).

ARCHEAN PROVENANCE

Geochemical characteristics of Archean terranes are likely to be substantially different than in younger environments. There is abundant evidence to indicate that the composition of many plutonic and volcanic igneous rocks differed during the Archean compared to the Phanerozoic, most probably in response to differing thermal regimes (Taylor and McLennan, 1985; Goodwin, 1991). For example, granitic rocks formed during the Archean are more commonly Na- and plagioclase-rich granodiorites-tonalites-sanukitoids, whereas during the Phanerozoic they tended to be more K- and K-feldspar-rich granodiorites-monzonites-granites. Implications of this for petrographical and major element relationships have been discussed by McLennan (1984) and Taylor and McLennan (1985). Differences also appear to exist for trace elements (e.g., Fig. 16). Thus, Archean sedimentary rocks from volcanically active settings, when compared to Proterozoic and Phanerozoic sedimentary rocks from similar settings, tend to have (McLennan and Taylor, 1991): (1) less depletion in Eu, reflecting an absence of intracrustal differentiation in the source rock; (2) greater depletion in heavy REE (HREE; Gd-Yb) (i.e., high Gd/Yb ratios), reflecting igneous provenance with mantle sources having such characteristics or having had garnet (HREE enriched) as a fractionating phase during formation; and (3) higher Th/U ratios, perhaps indicating less depleted mantle sources for the igneous provenance or differing mechanisms of formation.

Accordingly, care must be taken when applying the above

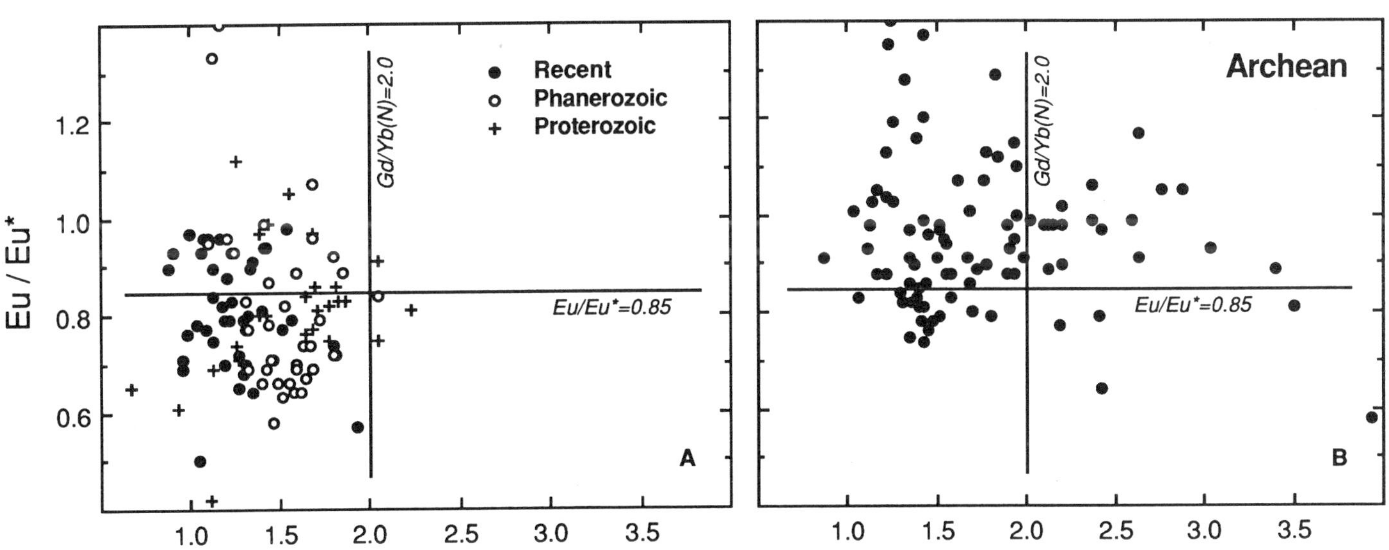

Figure 16. Plot of Eu/Eu* versus Gd_N/Yb_N for turbidites of various ages deposited in volcanically active tectonic settings. Archean turbidites tend to have higher Eu/Eu*, suggesting that the igneous/metamorphic provenance, on average, was less affected by intracrustal differentiation. They also commonly have $Gd_N/Yb_N > 2.0$, whereas such high ratios are virtually absent from post-Archean active margin sediments. Such differences indicate that caution is warranted in using post-Archean terrane types in interpreting Archean sedimentary rocks or sedimentary sequences derived primarily from Archean sources. From McLennan (1989a), where data sources are listed.

provenance concepts to Archean sedimentary rocks. By the end of the Archean, the upper crust appears to have stabilized to a state comparable to that seen in younger times. However, sedimentary rocks of any age, derived primarily from Archean terranes, especially from relatively localized sources, may be influenced by these differences. An example comes from the Early Proterozoic Chelmsford Formation derived from the Archean Superior Province (McDaniel and McLennan, 1991; McDaniel, 1992). Turbidites from the lowermost part of this sequence appear to be dominated by provenance components of stabilized upper continental crust, but most samples have REE patterns without negative Eu anomalies.

PB ISOTOPES AND COMPONENT ANALYSIS

The lead isotope system has great potential for evaluating both the origin and history of sedimentary rocks and their individual mineral constituents. U and Pb may be transported under a number of low-temperature geological processes relevant to sedimentary rocks. However, the occurrence of three independent, but geochemically related, Pb-isotope decay schemes, $^{238}U/^{206}Pb$, $^{235}U/^{207}Pb$, and $^{232}Th/^{208}Pb$, commonly allows for such processes to be evaluated and for considerable information regarding age and geochemical evolution of the rocks or minerals. It is not within the scope of this study to fully review the fundamentals of U-Th-Pb isotope systematics; for such a review, the reader is referred to Stacey and Kramers (1975), Doe and Zartman (1979), and Faure (1986).

For conglomeratic rocks it is possible to remove individual pebbles or cobbles and examine them in detail for geochemistry and age. Although of great use, such sedimentary rocks typically reflect only localized sources (e.g., Ingersoll, 1990). At the other extreme, single-grain components of siltstones and shales are difficult to separate and analyze. In this discussion, we concentrate on the evaluation of the provenance of sands and sandstones.

Dating components

U/Pb geochronology of zircons in sedimentary rocks as a technique for estimating the age of the provenance is well established (e.g., Tatsumoto and Patterson, 1964; Gaudette et al., 1981). Analyzing single zircon grains has been an especially attractive approach because it eliminates any ambiguity of mixed populations from different sources. Conventional mass spectrometry, using abraded zircon grains, and ion microprobe isotopic analyses of individual grains have both been used successfully. Although much of the work to date has been involved with identifying the earliest crustal components of the earth (e.g., Maas et al., 1992) some studies have also demonstrated the value of sedimentary zircon geochronology for tectonic problems (e.g., Ross et al., 1991; Ireland, 1992).

Zircons are of great value in understanding provenance history, not least due to their robustness and persistence in the sedimentary record. On the other hand, used in isolation for characterization of provenance, zircons have several shortcomings: (1) they represent only a very small fraction of any sedimentary rock, typically <0.1% (sandstones rarely have Zr>500 ppm and shales rarely have Zr>250 ppm); (2) they are more likely to be recycled in sedimentary rocks than are less resistant minerals (see above), and accordingly are likely to preferentially reflect old recycled components; and (3) they are absent in igneous rocks that form in magmatic environments that are undersaturated with respect to zircon and accordingly will not be represented in the sedimentary zircon record. Other trace minerals, including monazite and sphene, are also appropriate for U-Pb geochronological study but suffer from many of the same shortcomings as zircon, or they have additional complexities (e.g., sphene may form diagenetically).

Geochronological studies of major mineral phases provide an important new perspective on the age of sedimentary provenance. Feldspars, on their own, are not suitable for U/Pb dating due to extremely low U content, although K/Ar and Rb/Sr dating techniques hold promise (e.g., Harrison and Be, 1983). Clay minerals have been the center of much attention, especially for Rb/Sr geochronology, but these minerals are strongly affected by diagenesis and sedimentation ages or mixed ages are typically obtained (e.g., Clauer, 1982; Ohr et al., 1991). There is likely potential for K/Ar and Rb/Sr dating of detrital mica (Heller et al., 1985; Heller and Frost, 1988; Kelly and Bluck, 1989) and perhaps isotopic dating of individual sand-sized rock fragment grains; however, such approaches have yet to be investigated in any detail.

Quartz is the most abundant mineral in most sandstones. Recent advances in low-level analytical techniques have permitted U/Th/Pb isotopic data to be collected for very small quartz populations and under favorable circumstances for individual sand-sized quartz grains (Hemming et al., 1990, 1991). Early results suggest that U and Th are contained mainly in mineral and/or fluid inclusions and that, under a wide variety of geological conditions, quartz retains the original U/Th/Pb isotopic relationships. In addition, quartz commonly has high and variable U/Pb ratios, making the mineral suitable for Pb-Pb dating and, in cases of extremely high U/Pb ratios, suitable for U-Pb geochronology.

This approach is currently being investigated in Precambrian rocks from northern Minnesota and elsewhere (Hemming et al., 1990, 1991). In Figure 17, Pb-Pb isochrons are displayed for two Archean plutons from the Superior Province of northern Minnesota. In both cases, quartz—when combined with whole-rock and feldspar analyses—provides Pb-Pb ages in agreement with conventional isotopic approaches. In both cases, quartz is more radiogenic (higher $^{207}Pb/^{204}Pb$ and $^{206}Pb/^{204}Pb$) than the whole-rock analyses; however, the radiogenic nature of the quartz is highly variable. In Figure 18, the Pb isotopic composition of clear quartz from the early Proterozoic Pokegama Quartzite, representing the dominant quartz population, is plotted in terms of $^{207}Pb/^{204}Pb$ versus $^{206}Pb/^{204}Pb$. This formation is generally considered to be derived from the Archean Superior

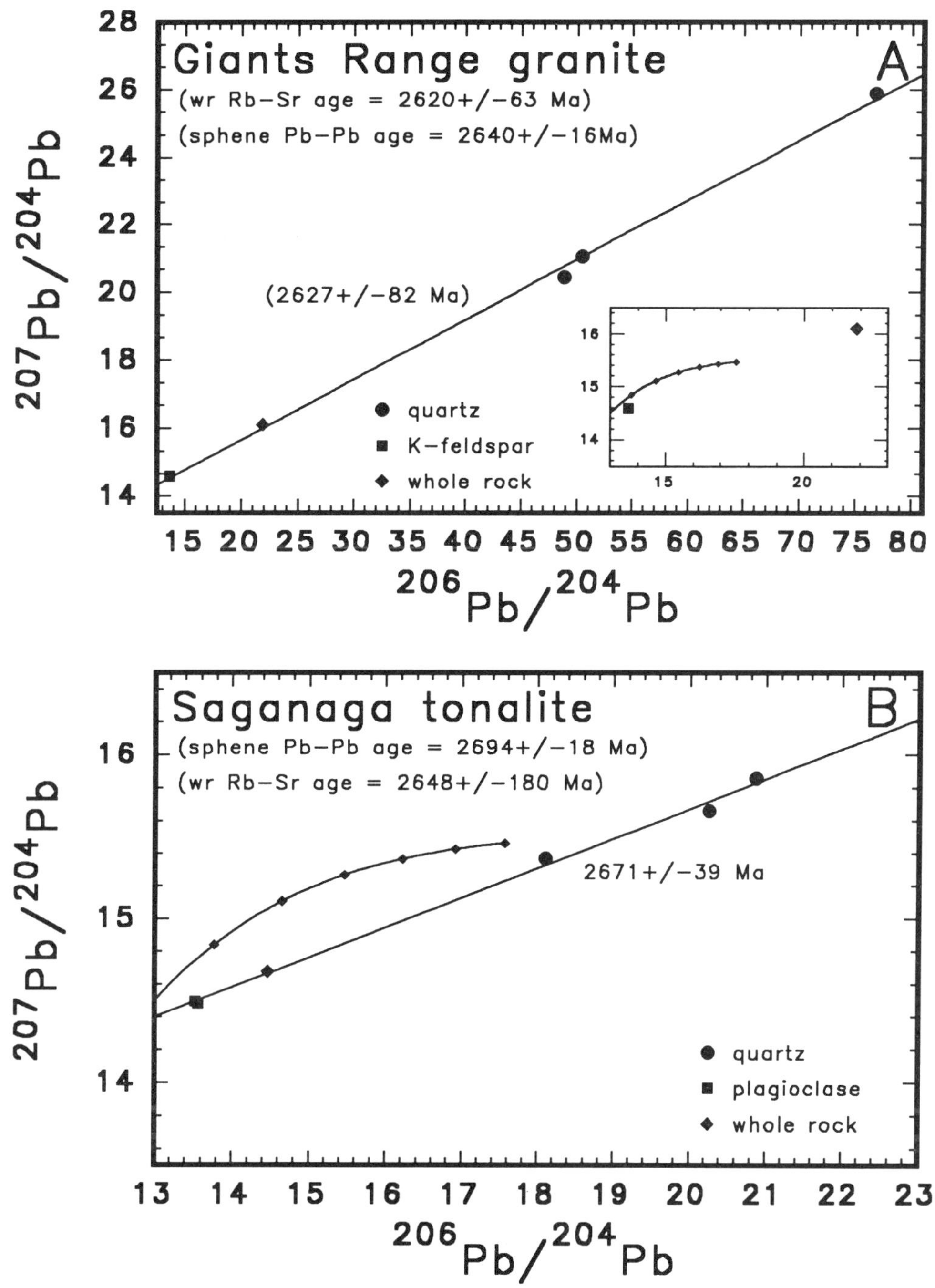

Figure 17. Plots of $^{207}Pb/^{204}Pb$ versus $^{206}Pb/^{204}Pb$ for samples from Archean plutons from Superior Province in northern Minnesota. Growth curves are for $\mu = 8.0$ with 500-Ma intervals. Note that quartz analyses are relatively radiogenic (i.e., high $^{206}Pb/^{204}Pb$) and are collinear with whole-rock and feldspar analyses. When combined, these data define ages consistent with ages derived from conventional radiometric techniques. The radiogenic nature of the quartz is variable between these plutons. This can be appreciated by noting that the inset for the Giants Range granite (A), which show the samples with lower $^{206}Pb/^{204}Pb$, has the same scale as the lower diagram, which shows data from the Saganaga tonalite (B). Unpublished data of S. Hemming. Conventional geochronological data compiled from several sources.

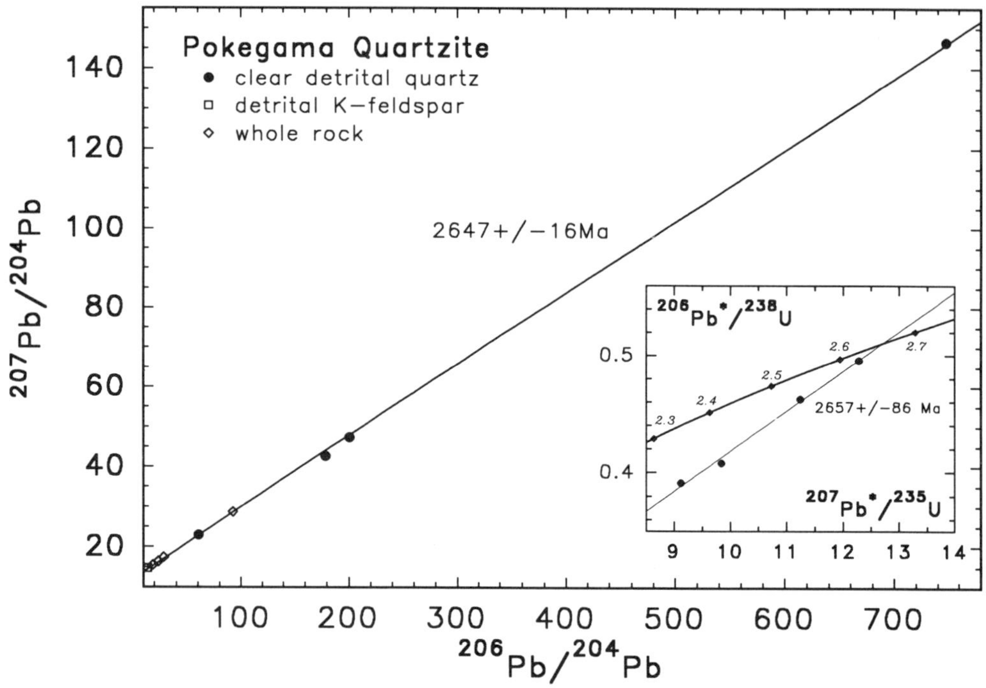

Figure 18. Plot of $^{207}Pb/^{204}Pb$ versus $^{206}Pb/^{204}Pb$ for samples from the Early Proterozoic Pokegama Quartzite from northern Minnesota. Whole rocks, feldspars, and quartz define a line of 2647 ± 16 Ma, consistent with a dominant provenance from the Superior Province. Inset, Concordia-discordia plot of radiogenic quartz samples from the Pokegama Quartzite. Increments are at 0.1-Ga intervals. These quartz samples display variable discordance and a range in $^{206}Pb/^{204}Pb$ (61 to 748) that cannot be correlated to the discordance. Together they define an upper intercept of 2657 ± 86 and lower intercept of 660 ± 661 Ma, with the scatter about the line being greater than would be expected simply from analytical uncertainty. Unpublished data of S. Hemming.

Province to the north. An age of 2647 ± 16 Ma is obtained, consistent with the age of the Superior Province and consistent with an age of 2673 ± 4 Ma for detrital zircons from the same unit (Hemming et al., 1991). In this case, the radiogenic character of the quartz permits U-Pb geochronology, and four clear quartz samples provide an age of 2657 ± 86 Ma (lower intercept 660 ± 661 Ma) with variable discordance (inset, Fig. 18).

Identification of terranes

The initial Pb isotopic composition of a rock provides important information about the origin and geochemical history because it is sensitive to the ratio of the parent-daughter isotopes of the source. Whole-rock Pb isotopic data, especially for shales (which efficiently mix provenance), may further constrain the timing of any U/Pb fractionation. For Pb, the initial U/Pb ratio, μ_i ($\mu = {}^{238}U/{}^{204}Pb$) may be determined either from the intersection of $^{207}Pb/^{204}Pb$-$^{206}Pb/^{204}Pb$ isochrons with an appropriate Pb isotopic growth curve or by analyzing minerals that contain little or no U, and thus preserve the Pb isotopic ratios at the time of formation. For sedimentary studies, both approaches may be useful; for example, the estimated μ_i for Pokegama Formation quartz data is about 7.9 (Fig. 18), which is entirely consistent with the K-feldspar data (see below) and with an origin from the Superior Province to the north (e.g., Tilton, 1983; Gariepy and Allegre, 1985).

Analyses of acid-leached detrital feldspar, particularly K-feldspar, also holds great promise for constraining tectonic history of sedimentary basins by identifying distinct provenance contributions. Feldspar is susceptible to alteration and recrystallization during diagenesis, and the potential for these processes to disturb the Pb isotope system within clastic sedimentary rocks has not been fully evaluated. It should be noted, however, that most low-temperature fluids have extremely low Pb content (e.g., seawater Pb < 0.1 ppb; river water Pb < 10 ppb; Taylor and McLennan, 1985), and considerable interaction is likely to be necessary for significant disturbance of the Pb isotopic system under such conditions. Furthermore, alteration that is restricted to the outer surfaces of the feldspar grains is removed during sample preparation. Where disturbance of the Pb isotopic system has occurred, it may be possible to evaluate the timing of diagenesis and the geochemical nature of the fluids.

A number of studies of igneous and metamorphic rocks have shown that differing tectonostratigraphic terranes are often characterized by distinct initial Pb isotopic compositions (e.g., Krogstad et al., 1989; Ayuso and Bevier, 1991). These differences relate both to differing mantle compositions (and ages of mantle depletion) and the mean age of the crust through which magmas pass. In many cases, it is the ratio $^{207}Pb/^{204}Pb$ that separates the Pb isotopic composition of many terranes. The ^{207}Pb parent isotope, ^{235}U, is relatively short-lived compared to ^{238}U (parent to ^{206}Pb) and was far more important in the earlier history of the earth. Accordingly, increases in U/Pb (e.g., during the formation of continental crust) that took place early in earth history result in elevated $^{207}Pb/^{204}Pb$ that persists through geological time. Increases in U/Pb that took place more recently result in elevated $^{206}Pb/^{204}Pb$ over time.

Two examples of using Pb isotopic compositions of feldspar to discriminate provenance, both from early Proterozoic sequences, are discussed below. Figure 19 shows the Pb isotopic composition of feldspars from the Proterozoic Chelmsford Formation, Sudbury Basin. Plotted for comparison are the fields for K-feldspar from the Archean Superior Province and the Proterozoic Penokean Orogen, both potential sources. The Chelmsford K-feldspars clearly have been derived almost exclusively from the Superior Province, thus providing an important constraint on the origin of these sedimentary rocks. Whole-rock data are also shown, and the trend between these samples and the least evolved feldspar coincides with isochrons defining Pb isotopic growth between 2.7 and 0.0 Ga, using the range of Superior Province K-feldspars as potential initial ratios. There is no evidence in these data for significant disturbance of the Pb isotope system at 2.0 to 1.8 Ga, which might be expected if Penokean-aged rocks were a significant component to the provenance.

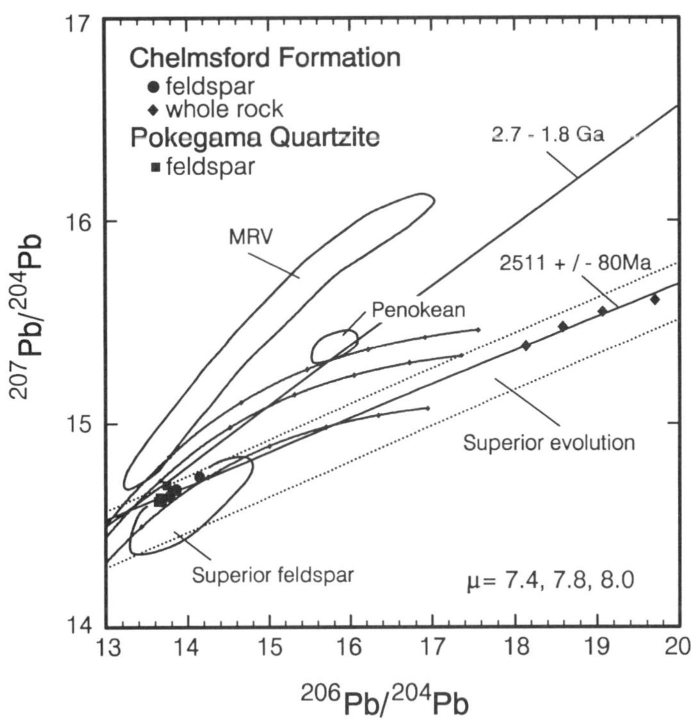

Figure 19. Plot of $^{207}Pb/^{204}Pb$ versus $^{206}Pb/^{204}Pb$ for leached feldspars from the Early Proterozoic Chelmsford Formation and Pokegama Quartzite and whole rocks from the Chelmsford Formation. Growth curves are for μ = 7.4, 7.8, and 8.0, with increments at 500 Ma. Also plotted are fields representing leached K-feldspars for Archean igneous rocks from the Superior Province and Minnesota River Valley and Proterozoic igneous rocks of Penokean affinity (compiled from many sources). Feldspars from the Chelmsford and Pokegama, both single grains and small populations, plot very close to the Superior Province. For the Chelmsford, this indicates very little or no involvement of Penokean-aged feldspars. For the Pokegama, neither the Penokean nor Minnesota River Valley gneisses are likely sources. The Chelmsford Formation whole rocks (sandstones and argillites) coincide with a reference line defining Pb growth from Superior Province K-feldspars, between 2.7 and 0.0 Ga (dotted lines); this suggests that, if Penokean-aged rocks represent more than a trivial component to the provenance, such a component did not contribute feldspar. Feldspars from the Chelmsford coincide with a line defining Pb growth between 2.7 and 1.8 Ga, and may indicate that the feldspars were influenced by Penokean-aged fluids, perhaps during diagenesis (plagioclase in these samples are albitized). Data from McDaniel (1992) and S. Hemming (unpublished).

There is a possibility that the feldspars form a linear array coinciding to an isochron defining a 2.7- to 1.8-Ga isochron, although such an interpretation is not demanded statistically. If such a trend exists, it could be interpreted as resulting from mixing of Superior Province Pb with 1.8-Ga fluids during diagenetic alteration of the feldspars (McDaniel, 1992).

K-feldspars from the Pokegama Quartzite (see Fig. 18) are also shown in Figure 19. In addition to the Superior Province and Penokean Orogen, plausible source rocks include the Minnesota River Valley (MRV) gneisses. The extended field for K-feldspars from MRV igneous rocks indicate a complex history and are also shown in Figure 19. The Pokegama K-feldspars plot within or very close to the field of the Superior province. Both the MRV and the Penokean Orogen can be excluded as significant sources for the detrital K-feldspars.

Applying such an approach to younger terranes in many cases is unlikely to be straightforward. Terranes that have undergone long and/or complex geological histories are likely to include some components of old crust. This will result in elevated $^{207}Pb/^{204}Pb$ ratios for K-feldspars that are derived directly from these older rocks and for K-feldspars formed during later melting of these terranes. In effect, the complexity of an extended geological history is likely to average out many of the very clear distinctions in Pb isotopic composition seen in older terranes. Nevertheless, where distinct Pb isotopic compositions do exist between terranes that are potential sources, this is likely to be reflected in the sedimentary record (e.g., Krogstad and Hanson, 1991). Accordingly, Pb isotopes provide an important additional tool in evaluating the tectonic history recorded in sedimentary rocks.

CONCLUSIONS

Our understanding of the controls on the chemical and isotopic composition of sedimentary rocks has advanced considerably. Major and trace element data provide constraints on both the provenance composition and the effects of sedimentary processes, such as weathering and sedimentary sorting. Nd isotopic data provide constraints on the mean age of the provenance. These techniques permit an evaluation of sediment history, which in turn affords insights into the prevailing tectonic conditions. To a large degree, the information gleaned from geochemistry complement those of petrography, and accordingly provide a different perspective on these problems.

Pb isotopic techniques have evolved to the stage where very small populations or single grains of most minerals can be analyzed for their Pb isotopic compsition. Interpretation of such analytical data may or may not have provenance-age significance, depending on questions such as diagenetic and metamorphic history, conditions of isotopic closure, and so forth. Interpretations depend on the geological context of the samples being analyzed, which is best done with the widest range of petrographic, geochemical, and isotopic data available.

ACKNOWLEDGMENTS

Many people have influenced our thinking about the application of geochemical and isotopic data to understanding the origin and history of sedimentary rocks, notably B. Bock, K. A. Eriksson, E. Krogstad, M. T. McCulloch, K. Mezger, H. W. Nesbitt, W. D. Sharp, S. R. Taylor, and G. M. Young. We are also grateful to Ross Taylor for commenting on an earlier draft, and to Mark Johnsson, Anne Linn, and Zell Peterman for helpful reviews.

This research was supported by National Science Foundation Grants EAR-8816386 and EAR-8957784 and by the Donors of the Petroleum Research Fund, administered by the American Chemical Society (to S.M.M.).

REFERENCES CITED

Andre, L., Deutsch, S., and Hertogen, J., 1986, Trace-element and Nd isotopes in shales as indexes of provenance and crustal growth: The Early Paleozoic from the Brabant Massif (Belgium): Chemical Geology, v. 57, p. 101–115.

Anwiller, D. N., and Mack, L. E., 1991, Diagenetic modification of Sm-Nd model ages in Tertiary sandstones and shales, Texas Gulf Coast: Geology, v. 19, p. 311–314.

Ayuso, R. A., and Bevier, M. L., 1991, Regional differences in Pb isotopic compositions of feldspars in plutonic rocks of the northern Appalachian Mountains, U.S.A., and Canada: A geochemical method of terrane correlation: Tectonics, v. 10, p. 191–212.

Banfield, J. F., and Eggleton, R. A., 1989, Apatite replacement and REE mobilization, fractionation, and fixation during weathering: Clays and Clay Minerals, v. 37, p. 113–127.

Barovich, K. M., Patchett, P. J., Peterman, Z. E., and Sims, P. K., 1989, Nd isotopes and the origin of 1.9–1.7 Ga Penokean continental crust of the Lake Superior region: Geological Society of America Bulletin, v. 101, p. 333–338.

Basu, A. R., Sharma, M., and DeCelles, P. G., 1990, Nd, Sr-isotopic provenance and trace element geochemistry of Amazonian foreland basin fluvial sands, Bolivia and Peru: Implications for ensialic Andean orogeny: Earth and Planetary Science Letters, v. 100, p. 1–17.

Ben Othman, D., White, W. M., and Patchett, J., 1989, The geochemistry of marine sediments, island arc magma genesis, and crust-mantle recycling: Earth and Planetary Science Letters, v. 94, p. 1–21.

Bhatia, M. R., 1981, Petrology, geochemistry and tectonic setting of some flysch deposits [Ph.D. thesis]: Canberra, Australian National University, 382 p.

Bhatia, M. R., 1983, Plate tectonics and geochemical composition of sandstones: Journal of Geology, v. 91, p. 611–627.

Bhatia, M. R., 1985, Rare earth element geochemistry of Australian Paleozoic graywackes and mudrocks: Provenance and tectonic control: Sedimentary Geology, v. 45, p. 97–113.

Bhatia, M. R., and Crook, K.A.W., 1986, Trace element characteristics of graywackes and tectonic setting discrimination of sedimentary basins: Contributions to Mineralogy and Petrology, v. 92, p. 181–193.

Blatt, H., 1967, Provenance determination and the recycling of sediments: Journal of Sedimentary Petrology, v. 37, p. 1031–1044.

Brimhall, C. H., and 9 others, 1991, Deformational mass transport and invasive processes in soil evolution: Science, v. 255, p. 695–702.

Chappell, B. W., 1968, Volcanic graywackes from the Upper Devonian Baldwin Formation, Tamworth-Barraba District, New South Wales: Journal of the Geological Society of Australia, v. 15, p. 87–102.

Chen, C.-H., Jahn, B.-J., Lee, T., Chem, C.-H., and Cornichet, J., 1990, Sm-Nd isotopic geochemistry of sediments from Taiwan and implications for the tectonic evolution of southeast China: Chemical Geology, v. 88, 317–332.

Clauer, N., 1982, The Rb-Sr method applied to sediments: Certitudes and uncertainties, *in* Odin, G. S., ed., Numerical dating methods in stratigraphy, 2 vols.: Chichester UK, John Wiley, p. 245–276.

Dabard, M. P., 1990, Lower Brioverian formations (Upper Proterozoic) of the Armorican Massif (France): Geodynamic evolution of source areas revealed by sandstone petrography and geochemistry: Sedimentary Geology, v. 69, p. 45–58.

Dickinson, W. R., 1970, Interpreting detrital modes of graywacke and arkose: Journal of Sedimentary Petrology, v. 40, p. 695–707.

Dickinson, W. R., 1985, Interpreting provenance relations from detrital modes of sandstones, *in* Zuffa, G. G., ed., Provenance of arenites: Proceeding of the NATO Advanced Study Institute on Reading Provenance from Arenites, Cosenza, Italy, p. 333–362.

Dickinson, W. R., 1988, Provenance and sediment dispersal in relation to paleotectonics and paleogeography of sedimentary basins, *in* Kleinspehn, K. L., and Paola, C., eds., New perspectives in basin analysis: New York Springer-Verlag, p. 3–25.

Dickinson, W. R., and Suczek, C. A., 1979, Plate tectonics and sandstone composition: American Association of Petroleum Geologists Bulletin, v. 63, p. 2164–2182.

Doe, B. R., and Zartman, R. E., 1979, Plumbotectonics, the Phanerozoic, *in* Barnes, H. L., ed., Geochemistry of hydrothermal ore deposits: New York, Wiley, p. 22–70.

Faure, G., 1986, Principles of isotope geology: New York, Wiley, 589 p.

Floyd, P. A., and Leveridge, B. E., 1987, Tectonic environment of the Devonian Gramscatho basin, south Cornwall: Framework mode and geochemical evidence from turbiditic sandstones: Journal of the Geological Society of London, v. 144, p. 531–542.

Folk, R. L., 1974, Petrology of sedimentary rocks: Austin, TX, Hemphill's, 170 p.

Frost, C. D., and Coombs, D. S., 1989, Nd isotope character of New Zealand sediments: Implications for terrane concepts and crustal evolution: American Journal of Science, v. 289, p. 744–770.

Gariepy, C., and Allegre, C. J., 1985, The lead isotope geochemistry and geochronology of late-kinematic intrusives from the Abitibi greenstone belt, and implications for late Archean crustal evolution: Geochimica et Cosmochimica Acta, v. 49, p. 2371–2383.

Gaudette, H. E., Vitrac-Michard, A., and Allegre, C. J., 1981, North American Precambrian history recorded in a single sample: High-resolution U-Pb systematics of the Potsdam sandstone detrital zircons, New York State: Earth and Planetary Science Letters, v. 54, p. 248–260.

Goodwin, A. M., 1991, Precambrian geology: The dynamic evolution of the continental crust: London, Academic Press, 666 p.

Harrison, T. M., and Be, K., 1983, $^{40}Ar/^{39}Ar$ age spectrum analysis of detrital microclines from the southern San Joachin Basin, California: An approach to determining the thermal evolution of sedimentary basins: Earth and Planetary Science Letters, v. 64, p. 244–256.

Heller, P. L., and Frost, C. D., 1988, Isotopic provenance of clastic deposits: Application of geochemistry to sedimentary provenance studies, *in* Kleinspehn, K. L., and Paola, C., eds., New perspectives in basin analysis: New York, Springer-Verlag, p. 27–42.

Heller, P. L., Peterman, Z. E., O'Neil, J. R., and Shafiqullah, M.,1985, Isotopic provenance of sandstones from the Eocene Tyee Formation, Oregon Coast Range: Geological Society of America Bulletin, v. 96, p. 770–780.

Helmhold, K. P., and van de Kamp, P. C., 1984, Diagenetic mineralogy and controls on albitization and laumonite formation in Paleogene arkoses, Santa Ynez Mountains, California, *in* McDonald, D. A., and Surdam, R. C., Clastic genesis: American Association of Petroleum Geologists Memoir 37, p. 239–276.

Hemming, S., McLennan, S. M., Hanson, G. N., Krogstad, E. J., and Mezger, K., 1990, Pb isotope systematics in quartz [abs.]: EOS Transaction of the American Geophysical Union, v. 71, p. 654.

Hemming, S., McLennan, S. M., Hanson, G. N., and Sharp, W. D., 1991, U-Pb isotope systematics in detrital quartz, feldspar and zircon: A provenance tool [abs.]: Geological Society of America Abstracts with Programs, v. 23(5), p. A108.

Hensel, H. D., McCulloch, M. T., and Chappell, B. W., 1985, The New England Batholith: Constraints on its derivation from Nd and Sr isotopic studies of granitoids and country rocks: Geochimica et Cosmochimica Acta, v. 49, p. 369–384.

Hiscott, R. N., 1984, Ophiolitic source rocks for Taconic-age flysch: Trace-element evidence: Geological Society of America Bulletin, v. 95, p. 1261–1267.

Hower, J., Eslinger, E. V., Hower, M. E., and Perry, E. A., 1976, Mechanisms of burial metamorphism of argillaceous sediment: 1. Mineralogical and chemical evidence: Geological Society of America Bulletin, v. 87, p. 725–737.

Ingersoll, R. V., 1988, Tectonics of sedimentary basins: Geological Society of America Bulletin, v. 100, p. 1704–1719.

Ingersoll, R. V., 1990, Actualistic sandstone petrofacies: Discriminating modern and ancient source rocks: Geology, v. 18, p. 733–736.

Ireland, T. R., 1992, Crustal evolution of New Zealand: Evidence from age distributions of detrital zircons in Western Province paragneisses and Torlesse graywacke: Geochimica et Cosmochimica Acta, v. 56, p. 911–920.

Johnsson, M. J., and Meade, R. H., 1990, Chemical weathering of fluvial sediments during alluvial storage: The Macuapanim Island point bar, Solimoes River, Brazil: Journal of Sedimentary Petrology, v. 60, p. 827–842.

Johnsson, M. J., Stallard, R. F., and Meade, R. H., 1988, First-cycle quartz arenites in the Orinoco River basin, Venezuela and Columbia: Journal of Geology, v. 96, p. 263–277.

Kelly, S., and Bluck, B. J., 1989, Detrital mineral ages from the Southern Uplands using $^{40}Ar/^{39}Ar$ laser probe: Journal of the Geological Society of London, v. 146, p. 401–403.

Krogstad, E. J., and Hanson, G. N., 1991, Pb isotope signatures in sediments in the northern Appalachians may record the accretion of Avalon during the Taconic Orogeny: Geological Society of America Abstracts with Programs, v. 23(1), p. 55.

Krogstad, E. J., Balakrishnan, S., Mukhopadhyay, D. K., Rajamani, V., and Hanson, G. N., 1989, Plate tectonics 2.5 billion years ago: Evidence at Kolar, South India: Science, v. 243, p. 1337–1340.

Land, L. S., 1984, Frio Sandstone diagenesis, Texas Gulf Coast: A regional isotopic study, *in* McDonald, D. A., and Surdam, R. C., Clastic genesis: American Association of Petroleum Geologists Memoir 37, p. 47–62.

Lee, D. E., and Bastron, H., 1967, Fractionation of rare-earth elements in allanite and monazite as related to geology of the Mt. Wheeler mine area, Nevada: Geochimica et Cosmochimica Acta, v. 31, p. 339–356.

Linn, A. M., DePaolo, D. J., and Ingersoll, R. V., 1991, Nd-Sr isotopic provenance analysis of Upper Cretaceous Great Valley fore-arc sandstones: Geology, v. 19, p. 803–806.

Maas, R., Kinny, P. D., Williams, I. S., Froude, D. O., and Compston, W., 1992, The Earth's oldest known crust: A geochronological and geochemical study of 3900–4200 Ma old detrital zircons from Mt. Narryer and Jack Hills, Western Australia: Geochimica et Cosmochimica Acta, v. 56, p. 1281–1300.

McCulloch, M. T., and Wasserburg, G. J., 1978, Sm-Nd and Rb-Sr chronology of continental crust formation: Science, v. 200, p. 1003–1011.

McDaniel, D. K., 1992, Geochemical and isotopic evidence for the origin of the Chelmsford Formation, Sudbury Basin, Canada [M.S. thesis]: Stony Brook, State University of New York.

McDaniel, D. K., and McLennan, S. M., 1991, Provenance characterization of an altered sandstone using combined petrographical and geochemical evidence: Geological Society of America Abstracts with Programs, v. 23(5), p. A108.

McLennan, S. M., 1984, Petrological characteristics of Archean graywackes: Journal of Sedimentary Petrology, v. 54, p. 889–898.

McLennan, S. M., 1988, Recycling of the continental crust: Pure and Applied Geophysics (PAGEOPH), v. 128, p. 683–724.

McLennan, S. M., 1989a, Rare earth elements in sedimentary rocks: Influence of provenance and sedimentary processes: Mineralogical Society of America Reviews in Mineralogy, v. 21, p. 169–200.

McLennan, S. M., 1989b, Archean sedimentary rocks and the Archean mantle, *in* Aswal, L. D., ed., Workshop on the Archean Mantle: LPI Technical Report 89-05, p. 57–59. Lunar and Planetary Institute, Houston.

McLennan, S. M., and Hemming, S., 1992, Samarium/neodymium elemental and isotopic systematics in sedimentary rocks: Geochimica et Cosmochimica Acta, v. 56, p. 887–898.

McLennan, S. M., and Taylor, S. R., 1980, Th and U in sedimentary rocks: Crustal evolution and sedimentary recycling: Nature, v. 285, p. 621–624.

McLennan, S. M., and Taylor, S. R., 1989, Proterozoic crustal evolution: Trace element evidence from sedimentary rocks of the ca. 1.7 Ga Ortega Group, New Mexico: 28th International Geological Congress Abstracts, v. 2, p. 404–405.

McLennan, S. M., and Taylor, S. R., 1991, Sedimentary rocks and crustal evolution: Tectonic setting and secular trends: Journal of Geology, v. 99, p. 1–21.

McLennan, S. M., Taylor, S. R., McCulloch, M. T., and Maynard, J. B., 1990, Geochemical and Nd-Sr isotopic composition of deep sea turbidites: Crustal evolution and plate tectonic association: Geochimica et Cosmochimica Acta, v. 54, p. 2015–2050.

Miller, R. G., O'Nions, R. K., Hamilton, P. J., and Welin, E., 1986, Crustal residence ages of clastic sediments, orogeny and crustal evolution: Chemical Geology, v. 57, p. 87–99.

Nance, W. B., and Taylor, S. R., 1976, Rare earth element patterns and crustal evolution—I. Australian post-Archean sedimentary rocks: Geochimica et Cosmochimica Acta, v. 40, p. 1539–1551.

Nance, W. B., and Taylor, S. R., 1977, Rare earth element patterns and crustal evolution—II. Archean sedimentary rocks from Kalgoorlie, Australia: Geochimica et Cosmochimica Acta, v. 41, p. 225–231.

Nathan, S., 1976, Geochemistry of the Greenland Group (Early Ordovician), New Zealand: New Zealand Journal of Geology and Geophysics, v. 19, p. 683–706.

Nelson, B. K., and DePaolo, D. J., 1985, Rapid production of continental crust 1.7–1.9 b.y. ago: Nd and Sr isotopic evidence from the basement of the North American midcontinent: Geological Society of America Bulletin, v. 96, p. 746–754.

Nelson, B. K., and DePaolo, D. J., 1988, Comparison of isotopic and petrographic provenance indicators in sediments from Tertiary continental basins in New Mexico: Journal of Sedimentary Petrology, v. 58, p. 348–357.

Nesbitt, H. W., 1979, Mobility and fractionation of rare earth elements during weathering of a granodiorite: Nature, v. 279, p. 206–210.

Nesbitt, H. W., and Young, G. M., 1982, Early Proterozoic climates and plate motions inferred from major element chemistry of lutites: Nature, v. 299, p. 715–717.

Nesbitt, H. W., and Young, G. M., 1984, Prediction of some weathering trends of plutonic and volcanic rocks based on thermodynamic and kinetic considerations: Geochimica et Cosmochimica Acta, v. 48, p. 1523–1534.

Nesbitt, H. W., and Young, G. M., 1989, Formation and diagenesis of weathering profiles: Journal of Geology, v. 97, p. 129–147.

Nesbitt, H. W., Markovics, G., and Price, R. C., 1980, Chemical processes affecting alkalis and alkaline earths during continental weathering: Geochimica et Cosmochimica Acta, v. 44, p. 1659–1666.

Newman, S., Macdougall, J. D., and Finkel, R. C., 1984, ^{230}Th-^{238}U disequilibrium in island arcs: Evidence from the Aleutians and the Marianas: Nature, v. 308, p. 268–270.

Ohr, M., Halliday, A. N., and Peacor, D. R., 1991, Sr and Nd isotopic evidence for punctuated clay diagenesis, Texas Gulf Coast: Earth and Planetary Science Letters, v. 105, p. 110–126.

Peterman, Z. E., Coleman, R. G., and Bunker, C. M., 1981, Provenance of Eocene graywackes of the Flournoy Formation near Agness, Oregon—A geochemical approach: Geology, v. 9, p. 81–86.

Potter, P. E., 1978, Significance and origin of big rivers: Journal of Geology, v. 86, p. 13–33.

Roser, B. P., and Korsch, R. J., 1986, Determination of tectonic setting of sandstone-mudstone suites using SiO_2 content and K_2O/Na_2O ratio: Journal of Geology, v. 94, p. 635–650.

Ross, G. M., Parrish, R. R., and Dudas, F. O., 1991, Provenace of the Bonner Formation (Belt Supergroup), Montana: Insights from U-Pb and Sm-Nd analyses of detrital minerals: Geology, v. 19, p. 340–343.

Schnetzler, C. C., and Philpotts, J. A., 1970, Partition coefficients of rare-earth elements between igneous matrix material and rock-forming mineral phenocrysts—II: Geochimica et Cosmochimica Acta, v. 34, p. 331–340.

Schwab, F. L., 1971, Geosynclinal compositions and the new global tectonics: Journal of Sedimentary Petrology, v. 41, p. 928–938.

Shirey, S. B., and Hanson, G. N., 1986, Mantle heterogeneity and crustal recycling in Archean granite-greenstone belts: Evidence from Nd isotopes and trace elements in the Rainy Lake area, Superior Province, Ontario, Canada: Geochimica et Cosmochimica Acta, v. 50, p. 2631–2651.

Soegaard, K., and Eriksson, K. A., 1986, Transition from arc volcanism to stable-shelf and subsequent convergent-margin sedimentation in northern New Mexico from 1.76 Ga: Journal of Geology, v. 994, p. 47–66.

Soegaard, K., and Eriksson, K. A., 1989, Origin of thick, first-cycle quartz arenite successions: Evidence from the 1.7 Ga Ortega Group, northern New Mexico: Precambrian Research, v. 43, p. 129–141.

Stacey, J. S., and Kramers, J. D., 1975, Approximation of terrestrial lead isotope evolution by a two-stage model: Earth and Planetary Science Letters, v. 26, p. 207–221.

Steidtmann, J. R., and Schmitt, J. G., 1988, Provenance and dispersal of tectogenic sediments in thin-skinned, thrusted terrains, *in* Kleinspehn, K. L., and Paola, C., eds., New perspectives in basin analysis: New York, Springer-Verlag, p. 353–366.

Sun, S.-s., 1980, Lead isotopic study of young volcanic rocks from mid-ocean ridges, ocean islands and island arcs. Philosophical Transactions of the Royal Society of London, v. A297, p. 275–311.

Tatsumoto, M., and Patterson, C., 1964, Age studies of zircon and feldspar concentrates from the Franconia sandstone: Journal of Geology, v. 72, p. 232–242.

Taylor, S. R., and McLennan, S. M., 1985, The continental crust: Its composition and evolution: Oxford, Blackwell's, 312 p.

Taylor, S. R., McLennan, S. M., and McCulloch, M. T., 1983, Geochemistry of loess, continental crustal composition and crustal model ages: Geochimica et Cosmochimica Acta, v. 47, p. 1897–1905.

Tilton, G. R., 1983, Evolution of depleted mantle: The lead perspective: Geochimica et Cosmochimica Acta, v. 47, p. 1191–1197.

Valloni, R., and Maynard, J. B., 1981, Detrital modes of recent deep-sea sands and their relation to tectonic setting: A first approximation: Sedimentology, v. 28, p. 75–83.

Veizer, J., and Jansen, S. L., 1979, Basement and sedimentary recycling and continental evolution: Journal of Geology, v. 87, 341–370.

Veizer, J., and Jansen, S. L., 1985, Basement and sedimentary recycling—2: Time dimension to global tectonics: Journal of Geology, v. 93, 625–643.

White, W. M., Dupre, B., and Vidal, P., 1985, Isotope and trace element geochemistry of sediments from the Barbados Ridge–Demerara Plain region, Atlantic Ocean: Geochimica et Cosmochimica Acta, v. 49, p. 1875–1886.

Manuscript Accepted by the Society December 24, 1992

Printed in U.S.A.

Geological Society of America
Special Paper 284
1993

Basaltic island sand provenance

Kathleen M. Marsaglia
Department of Geological Sciences, University of Texas at El Paso, El Paso, Texas 79968

ABSTRACT

The Hawaiian Islands are an ideal location to study basaltic sand provenance in that they are predominantly composed of basaltic flows and pyroclastic deposits that are exposed in arid to humid climates on progressively older islands. The major components of Hawaiian beach sands are calcareous bioclasts, volcanic lithic fragments, and monomineralic grains of dense minerals and plagioclase. A small to moderate percentage of volcanic lithic fragments are altered to iron oxides and/or clay minerals. The relative proportions of bioclastic, unaltered, and altered volcanic grains in these samples are highly variable, but are generally a function of island evolution, reflecting the degree of reef development and the intensity and duration of weathering and erosion.

Hawaiian volcanic lithic fractions consist of brown and black glassy volcanic fragments exhibiting vitric, microlitic, and lathwork textures and "other" holocrystalline volcanic lithic fragments. The proportions of these volcanic lithic types do not appear to vary systematically for sands with tholeiitic versus alkalic basalt provenance. Although these basaltic sands compositionally overlap sands from undissected magmatic arcs on traditional ternary plots (e.g., QFL, QmPK, LmLvLs), they can be compositionally and texturally distinguished from sands derived from island arc volcanic sources. In general, Hawaiian basaltic sands contain more olivine/pyroxene grains, fragments of tachylitic black glass, holocrystalline volcanic lithic fragments, and volcanic grains exhibiting lathwork textures.

INTRODUCTION

Basaltic sand provenance has received only minor attention in comparison to that of intermediate magmatic arcs (e.g., Dickinson and Suczek, 1979; Dickinson et al., 1983; Mack and Jerzykiewicz, 1989; Marsaglia and Ingersoll, 1992). The main goal of this study is to petrographically characterize the composition of sand with a basaltic provenance, focussing on textural aspects of volcanic grains that would allow the discrimination of basaltic from magmatic arc provenance. The Hawaiian Islands were chosen for this study because they are basaltic volcanoes that show a wide climatic range, from humid to arid, and display an evolutionary trend progressing from younger to older volcanoes along the chain. Beach sands rather than subaerial stream sands were sampled because the former are volumetrically more important, are more easily accessible, are often preserved during the formation of beach rock and eolianite, and are commonly redistributed into offshore environments. The composition of modern Hawaiian beach sands may be used as a guide for the interpretation of beach sediment transported into deeper water environments (Chamberlain, 1968; Moberly, 1968; Mathewson, 1970; Andrews and Bainbridge, 1972; Campbell and Hwang, 1982; Fornari and Campbell, 1987; Coulbourn et al., 1988; Moore et al., 1989), ancient equivalents associated with seamounts, or seamount-related sediments incorporated into accretionary prisms at subduction zones.

GEOLOGY OF THE HAWAIIAN ISLANDS

The Hawaiian Islands (Fig. 1) are a chain of basaltic domes or shield volcanoes composed of layered basalt flows and intercalated pyroclastic deposits. These islands lie at the southern extremity of a linear belt of islands and submerged seamounts, the Hawaiian-Emperor seamount chain, thought to have formed by sequential eruption during northwest passage of the Pacific plate over a "hot spot" source of magma. Cores and dredge

Marsaglia, K. M., 1993, Basaltic island sand provenance, *in* Johnsson, M. J., and Basu, A., eds., Processes Controlling the Composition of Clastic Sediments: Boulder, Colorado, Geological Society of America Special Paper 284.

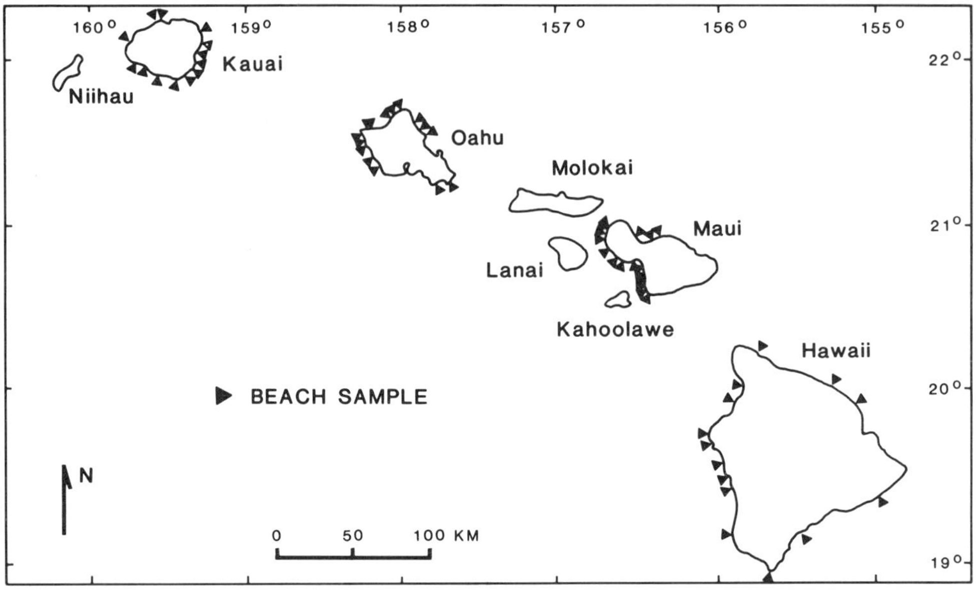

Figure 1. Generalized map of the Hawaiian Islands showing sample locations.

samples from seamounts northwest of the Hawaiian Islands show evidence for subaerial exposure and shallow marine deposition, implying that these were once volcanic islands that have since been submerged due to isostatic loading of the crust (McKenzie et al., 1980; Stearns, 1984). Correspondingly, the age of the volcanic rocks exposed on these islands and recovered from submarine seamounts increases northward up to 75 to 80 m.y. (Clague and Dalrymple, 1987; Clague and Hazlett, 1989). The Loihi seamount 30 km southeast of Hawaii is a submarine volcano described by Moore et al. (1982) as an "embryonic" Hawaiian Island.

Lavas exposed on the Hawaiian Islands range from Holocene to Pliocene in age (Bennison, 1974). Only the island of Hawaii is presently active, although all of the Hawaiian Islands except Lanai have had minor late Pleistocene eruptions (Stearns, 1984). The volcanic series on Hawaii, Kahoolawe, Maui, Lanai, and Molokai are Pleistocene and younger in age, whereas those on Oahu, Kauai, and Niihau are as old as Pliocene (Bennison, 1974).

The Hawaiian Islands are composed dominantly of basaltic lavas that have varied somewhat in composition during the developmental stages of each island: the early embryonic (preshield) submarine stage is alkalic in composition, the shield stage is tholeiitic, and the postshield stage is again alkalic (Clague and Hazlett, 1989). These compositional changes are reflected in the mineralogy of the basalts: tholeiitic basalts are olivine poor, whereas alkalic basalts usually contain olivine and small amounts of feldspathoids, alkali feldspar, titan-augite, or barkevikitic amphibole (Hyndman, 1972). The most abundant phenocryst minerals in these lavas are plagioclase feldspar (labradorite), olivine, augite, magnetite, and ilmenite (Moberly and Chamberlain, 1964; Macdonald and Abbott, 1970). Rare anorthoclase feldspar, nepheline, quartz, hypersthene, pigeonite, biotite, and melilite crystals are also present (Macdonald and Abbott, 1970).

Only limited outcrops of sedimentary, metamorphic, and igneous intrusive rocks are present on the Hawaiian Islands (Macdonald and Abbott, 1970; Bennison, 1974). Exposed sedimentary deposits include reef, eolian, lagoonal, and beach facies. Metavolcanic rocks occur as contact metamorphic rocks (hornfels) adjacent to intrusions, hydrothermally altered volcanic rocks containing talc, chlorite, and chalcedony, and rocks altered by fumarolic activity, forming siliceous residues, or reddish brown clay similar to laterite (Macdonald and Abbott, 1970). Mafic intrusive rocks occur as dikes, sills, laccoliths, plugs, and necks (Macdonald and Abbott, 1970).

EROSION AND WEATHERING PROCESSES ON THE HAWAIIAN ISLANDS

Both subaerial and marine erosion have been important in shaping the physiography of the Hawaiian Islands. High-energy wave conditions include the northeast trades, winter (north Pacific) swell, Kona storms, southern swell, and tropical storms and hurricanes (Inman and others, 1963; Moberly and Chamberlain, 1964; Moberly et al., 1965). Marine erosion is most important on the northeast coastlines, where steep sea cliffs have been formed by wave action; high waves associated with summer storms are responsible for some additional marine erosion on the southern extremes of these Islands (Stearns, 1984).

The Hawaiian Islands are in various stages of dissection, with the degree of erosion a function of their relative age (Morgan, 1983; Stearns, 1984). Stream development is initially hampered by the high porosity of the basaltic lava flows, but after a soil horizon forms, surface runoff is enhanced and surface streams develop (Moberly and Chamberlain, 1964; Stearns, 1984). The islands are characterized by radial drainage patterns with ephemeral streams in drier regions and perennial streams where rainfall is higher (Stearns, 1984).

Weathering and erosion may vary markedly across an island because of microclimates induced by the effects of volcano topography (Morgan, 1983; Stearns, 1984). Rainfall varies greatly (13 to 1,270 cm/yr) across individual islands (Sherman and Ikawa, 1968); precipitation is highest on the windward or northeast coasts, whereas the lee sides of islands are arid to semi-arid (Stearns, 1984; Coulbourn and others, 1988). Vegetation is a function of rainfall, ranging from tropical rain forest on the windward slopes to arid environments on the southern slopes (Inman et al., 1963). In general, chemical weathering is more important than mechanical disintegration (Macdonald and Abbott, 1970; Morgan, 1983; Stearns, 1984; Li, 1988), and it is most pronounced on the leeward sides of the islands, where rainfall, vegetation, and rates of carbonic acid production by microbial oxidation of organic matter are higher (Li, 1988). The lowest rates of chemical denudation are found on Hawaii, where the surface is characterized by relatively young basalt flows with little to no vegetation (Li, 1988).

Hawaiian soils are produced rapidly by chemical weathering of unstable minerals in basaltic parent rocks (Morgan, 1983; Wirthmann, 1986). Soils are red due to iron oxides formed by oxidation of magnetite (Stearns, 1984). Common insoluble weathering products of Hawaiian basalts are kaolinite, hematite, limonite, smectite, halloysite, and gibbsite (Macdonald and Abbott, 1970). When defoliated the soils can form laterites, which often exhibit a pelletal structure (Stearns, 1984). Oxide mineral nodules found in these soils are typically zoned with a darker colored, dense exterior layer composed of iron oxides, and a lighter colored, soft, friable interior zone composed of clay minerals (Sherman et al., 1969; Glasby et al., 1979). Gibbsite nodules also occur (Walker et al., 1969).

PREVIOUS STUDIES

Previous workers on Hawaiian beach sands have been primarily concerned with the littoral sand budget and its relationship to sand composition (Emery and Cox, 1956; Moberly and Chamberlain, 1964; Moberly et al., 1965; Chamberlain, 1968; Moberly, 1968; Coulbourn et al., 1988). Sand sources include coastal streams and rivers, coastal erosion, biological activity, and wind. The dominant source of sand is the biological activity of reef organisms; the dominant loss is to paralic sedimentation (Chamberlain, 1968). There is a general increase in volume of beach sand reservoir per kilometer of coastline from the younger (southern) to older (northern) Hawaiian Islands, which can be linked to greater biological production in better-developed reef tracts around the older islands (Moberly and Chamberlain, 1964; Chamberlain, 1968).

The calcareous component of Hawaiian beach sands has been emphasized by previous workers because it is the dominant component of the "important beaches of the State of Hawaii" (Moberly et al., 1965, p. 593). The major bioclasts present in Hawaiian beach sands are, in decreasing order of abundance, foraminifera (e.g., *Amphistegina* and *Heterostegina*), mollusk, calcareous red algae, echinoid, coral, calcareous green algae and bryozoan fragments (Inman et al., 1963; Moberly and Chamberlain, 1964; Moberly et al., 1965). Carbonate sand is transported to the beach from the reef flat and the edges of the fringing reefs (Inman and others, 1963; Moberly and Chamberlain, 1964; Moberly, 1968; Coulbourn and Resig, 1975); it may then be transported seaward through inlets formed where muddy stream waters inhibit reef growth (Inman et al., 1963). Inman et al. (1963) noted the dominance of reef-derived carbonate sand on the windward (northern) beaches, where there is little sand input from streams.

Although the proportions of various bioclasts and the controls on their distribution have been fairly well documented, the detrital/pyroclastic component of Hawaiian beach sands has been less studied. Moberly and Chamberlain (1964) and Moberly et al. (1965) recognized the following types of volcanic grains in these sands: fresh microcrystalline lava, fresh volcanic glass, and altered volcanic fragments. In general, they found that fresh basaltic lithic fragments are the most abundant component, except in glass-rich hyaloclastic beach sands produced by explosive reactions of basalt flows with water. Of the monomineralic grains, olivine is the most common, with lesser amounts of opaque magnetite-ilmenite, dark gray-green transparent grains of augite, and labradorite. In an independent study, Inman et al., (1963) found that the terrigenous components on Kauai beaches include olivine, opaque iron oxides, and feldspathic rock fragments.

Abrasion studies conducted by Moberly (1968) indicate that altered basalt grains are less resistant to abrasion than fresh basalt grains. He found that basalt grains abrade more rapidly than carbonate bioclasts, and that basalt abrades faster than olivine, which in turn abrades faster than volcanic glass.

Hawaiian beach sands range from coarse to fine grained, but generally are medium grained (Moberly and Chamberlain, 1964; Moberly et al., 1965). Beaches along the northeastern coasts are finer grained, except on Hawaii (Moberly et al., 1965). This contrast in grain size is a function of the increased rainfall/weathering and higher wave energy on the northeastern sides of the islands (Inman et al., 1963; Moberly et al., 1965).

PRODUCTION OF VOLCANICLASTIC SAND

Volcaniclastic sand-sized material in Hawaiian beach and stream sediments is produced either directly by volcanic eruptions (pyroclastic material) or by erosion of preexisting volcanic rocks and shallow intrusive rocks (epiclastic material). Only small quantities of vitric tuff can be attributed to subaerial lava fountains or phreatomagmatic eruptions (e.g., McPhie et al., 1990), whereas large amounts of fragmental material are produced as lava flows enter the ocean, or during the formation of tuff cones. Hawaii's famous black sand beaches are created by the littoral explosions that occur when lava flows enter the ocean (Macdonald and Abbott, 1979).

The type and the source of sand-sized volcaniclastic material produced on an intraoceanic, basaltic shield volcano are a function of its relative age and stage of development. Stearns (1984) outlined eight stages of Hawaiian volcano development. In the first, the explosive submarine stage, large amounts of ash and pumice are produced when magma comes in contact with sea water. When this fragmental cone builds above sea level during the second stage, it undergoes erosion until subsequent flows of basaltic lava drape over and protect the cone's surface. Stream erosion during this second stage is nonexistent; thus, the majority of sand produced in the first two stages is primary or slightly reworked pyroclastic debris, with minor epiclastic detritus probably produced by beach erosion. In the third stage, a caldera forms due to collapse of the volcano, and massive lava flows fill the caldera, eventually obliterating the caldera in stage four. Lithic and vitric tuff beds may form during stage three, and vitric tuff beds may form during stage four. Stage five is characterized by marine and stream erosion, resulting in high sea cliffs and deep valleys. In stage six, fringing reefs form around the island, which has been either eroded to sea level or submerged during a eustatic rise in sea level. The seventh stage is one of volcanic rejuvenation, with lava- and tuff-producing phreatomagmatic eruptions. Eventually an atoll may form (stage eight). According to Stearns (1984), all the Hawaiian Islands have passed through the first two stages. Mauna Loa and Kilauea volcanoes on Hawaii and Lanai are in the third stage, whereas Huallalai, Mauna Kea, and Kohala domes on Hawaii are in stage four. The remaining volcanoes on the islands of Maui, Oahu, Kauai, and Niihau are in the seventh phase: Haleakala, West Maui, East Molokai, Kahoolawe, Koolau, Waianae, Kauai, and Niihau.

METHODS

Loose sand samples were collected from the foreshore regions of various beaches on the islands of Hawaii, Maui, Oahu and Kauai (Fig. 1, Appendix 1). These islands are representative of the entire age range of volcanic series exposed along the chain. The number and the location of samples are predominantly a function of coastal accessibility, abundance of sand along the coast, and time constraints for field work on each island. Beaches along each of the major volcanic series for each island were sampled. Two samples were taken on beaches where sand color varied due to local segregation of dense minerals (marked A and B in Appendix 2). Note that beaches known to be artificially replenished with sand from other sources were not included in this study (e.g., Waikiki Beach: Coulbourn et al., 1988; Moberly, 1968); nevertheless, the possibility of artificial replenishment cannot be ruled out for some of the beaches sampled along more developed coastlines. A few stream samples were collected for comparison.

Collected samples were air dried and then sieved in order to concentrate the sand-sized (0.0625 to 2.0 mm) fraction. Thin sections of sand concentrates were prepared using the following process in order to maximize random grain orientation. First, sand is sprinkled onto an epoxied glass slide, slightly ground, and then a second slide is mounted to this ground surface. Then, the initial slide is ground away, and the epoxy/sand mixture is ground until grains are 30 μ thick. Thin sections were subsequently stained for both calcium-rich and potassium-rich feldspars to aid in the identification of possible trace amounts of quartz and potassium feldspar (see Marsaglia and Tazaki (1992) for staining technique). As many as 500 points were counted on each section using the Gazzi-Dickinson method (Ingersoll and others, 1984), thus minimizing dependence of composition on grain size. Grid spacings exceeded maximum grain size, but maximized coverage of thin sections. The total points counted (85 to 500; see Appendix 2) is a reflection of the grain size (fewer points for coarse sand), composition (fewer points for dominantly bioclastic samples), and the grain density/distribution across the thin section.

The Gazzi-Dickinson method divides volcanic lithic clasts into various textural types (Dickinson, 1970), which is especially significant in volcanic settings, where volcaniclastic components may constitute up to 100% of the sand-sized fraction. In addition, grains were separated according to glass color and alteration (i.e., orange, tan/brown and black; see Marsaglia, 1992). Counted grains were assigned to the monomineralic and polymineralic compositional categories listed in Table 1. Recalculated parameters, means, and standard deviations were calculated using a Lotus 123 spreadsheet program (Table 1; Appendices 2-5).

RESULTS

Grain types

The three major components of Hawaiian beach sands are calcareous bioclasts, monomineralic grains, and volcanic lithic fragments. Bioclastic debris is present in all but a few samples and the dominant component in 75% of the samples examined in this study (Appendix 3). The types of bioclasts are similar to those documented in previous studies (Coulbourn and Resig, 1975; Inman et al., 1963; Moberly and Chamberlain, 1964; Moberly, 1968; Moberly et al., 1965), including foraminifera, echinoid, calcareous red algae, pelecypod, and gastropod fragments (Fig. 2). Because the bioclastic component was not the focus of the present study, bioclasts were grouped into one category rather than subdivided (Appendix 2).

Monomineralic grains present are predominantly olivine with lesser pyroxene, plagioclase, opaque minerals, and rare amphibole (Fig. 3). The monomineralic totals listed in Appendix 2 (P and D categories) include both separate grains and sand-sized euhedral to subhedral crystals in volcanic lithic fragments (e.g., Figs. 3, 4). Some olivine crystals are partially to wholly altered to orange/brown iddingsite and others show resorption textures (Fig. 3). Minor feldspar alteration to clay minerals and feldspar dissolution is present in some altered lithic fragments.

TABLE 1. KEY TO COUNTED AND RECALCULATED PARAMETERS

P: Plagioclase feldspar
Lvo: Other volcanic lithic
Lvv: Vitric volcanic lithic
- brgl = brown glass
- orbg = orange-brown glass
- blgl = black glass
- algl = altered glass
- degl = devitrified glass

Lvmi: Volcanic lithic with microlitic texture
- brgl = brown glass
- orbg = orange-brown glass
- blgl = black glass
- algl = altered glass
- degl = defitrified glass

Lvl: Volcanic lithic with lathwork texture
- brgl = brown glass
- orbg = orange-brown glass
- blgl = black glass
- algl = altered glass
- degl = devitrified glass

D: Dense minerals
- opaq = opaque
- py/ol = pyroxene and/or olivine (and rare amphibole?)
- other = includes opaque and olivine intergrowths and olivine altered to iddingsite

Clay Alt = Altered volcanic glass or soil particle?
Glauc = Glauconite or celadonitized glass?
Bio = Predominantly carbonate bioclastic debris (may include some micritic beachrock cement)
Other = Organic material and beachrock
Total = Total points counted

Other Categories Included in Recalculated Parameters:
Q: Monocrystalline and polycrystalline quartz
clgl: Colorless glassy fragment
Lvf: Felsitic volcanic lithic

Recalculated Parameters:
Lvc = Lvf + Lvmi + Lvl
Lv = Lvv + Lvmi + Lvl (e.g., Lv[tot] = for algl + degl + blgl + brgl + orbg)
L = Lvo + Lvv + Lvmi + Lvl + (Clay alt)
F = P
%Bio = 100•Bio/Total
Lv%Alt = 100•(Lvv[orbg + alt] + Lvmi[orbg + alt] + Lvl[orbg + alt] + [Clay alt])/(Lv[tot] + Lvo)
CarbAltUnalt%Carb = 100•Bio/(Lv[tot] + Lvo)
CarbAltUnalt%Alt = 100•(Lv[algl + degl + orbg] + [Clay alt])/(Lv[tot] + Lvo + [Clay alt])
CarbAltUnalt%Unalt = 100•(Lv[blgl + brgl])/(Lv[tot] + Lvo + [Clay alt])

QFL%Q = 100•Q/(Q + F + L)
QFL%F = 100•F/(Q + F + L)
QFL%L = 100•L/(Q + F + L)

LvfLvmiLvl%Lvf = 100•Lvf/Lvc
LvfLvmiLvl%Lvl = 100•Lvl/Lvc
LvfLvmiLvl%Lvmi = 100•Lmi/Lvc

Volc%clgl = 100•Lv(clgl)/(Lv[brgl] + Lv[blgl] + Lv[clgl])
Volc%brgl = 100•Lv(brgl)/(Lv[brgl] + Lv[blgl] + Lv[clgl])
Volc%blgl = 100•Lv(blgl)/(Lv[brgl] + Lv[blgl] + Lv[clgl])

Volc%alt = 100•(Lv[degl + algl + orbg] + [Clay alt])/(Lv[tot] + [Clay alt])
Volc%brown = 100•Lv(brgl)/(Lv[tot] + [Clay alt])
Volc%black = 100•Lv(blgl)/(Lv[tot] + [Clay alt])

LvAccPlag%Lv = 100•(Lv[tot] + Lvo + [Clay alt])/(Lv[tot] + Lvo + [Clay alt] + D + P)
LvAccPlag%Acc = 100•D/(Lv[tot] + Lvo + [Clay alt] + D + P)
LvAccPlag%Plag = 100•P/(Lv[tot] + Lvo + [Clay alt] + D + P)

LvvLvmiLvl%Lvv = 100•Lvv/(Lvv + Lvmi + Lvl)
LvvLvmiLvl%Lvmi = 100•Lvmi/(Lvv + Lvmi + Lvl)
LvvLvmiLvl%Lvl = 100•Lvl/(Lvv + Lvmi + Lvl)
(LvvLvmiLvl percentages were calculated separately for brown and black glassy grains)

VitricCrysPlag%Vitric = 100•Lvv/(Lv[tot] + Lvo + P)
VitricCrysPlag%Crys = 100•(Lvmi + Lvl + Lvo)/(Lv[tot] + Lvo + P)
VitricCrysPlag%Plag = 100•P/(Lv[tot] + Lvo + P)

LvoLvmiLvl%Lvo = 100•Lvo/(Lvo + Lvl + Lvmi)
LvoLvmiLvl%Lvmi = 100•Lvmi/(Lvo + Lvl + Lvmi)
LvoLvmiLvl%Lvl = 100•Lvl/(Lvo + Lvl + Lvmi)

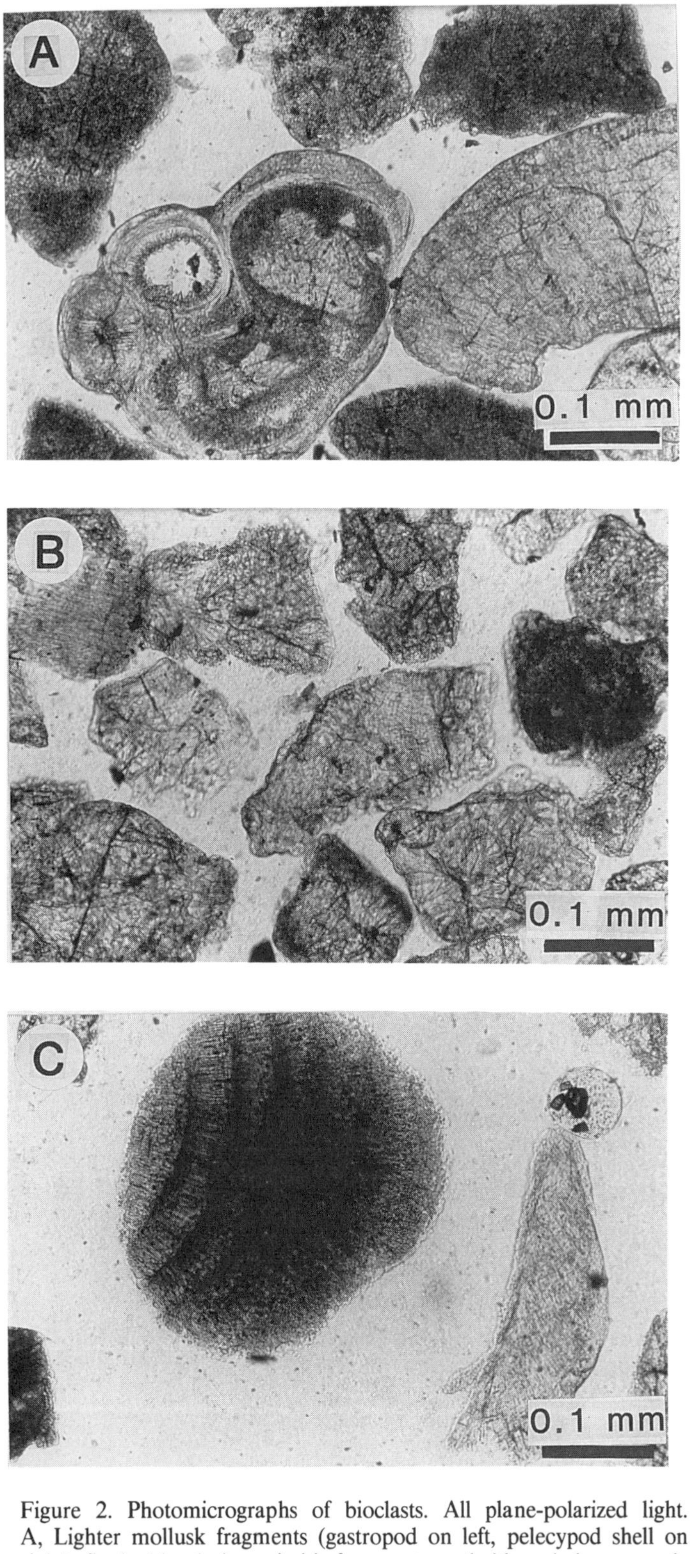

Figure 2. Photomicrographs of bioclasts. All plane-polarized light. A, Lighter mollusk fragments (gastropod on left, pelecypod shell on right), flanked by darker micritic fragments, probably red algae; sample KAU 10. B, Ultrastructures suggest that these are predominantly foraminifera fragments; sample OAU 8. C, Calcareous red algae (left) and mollusk fragment (right); sample HAW 20.

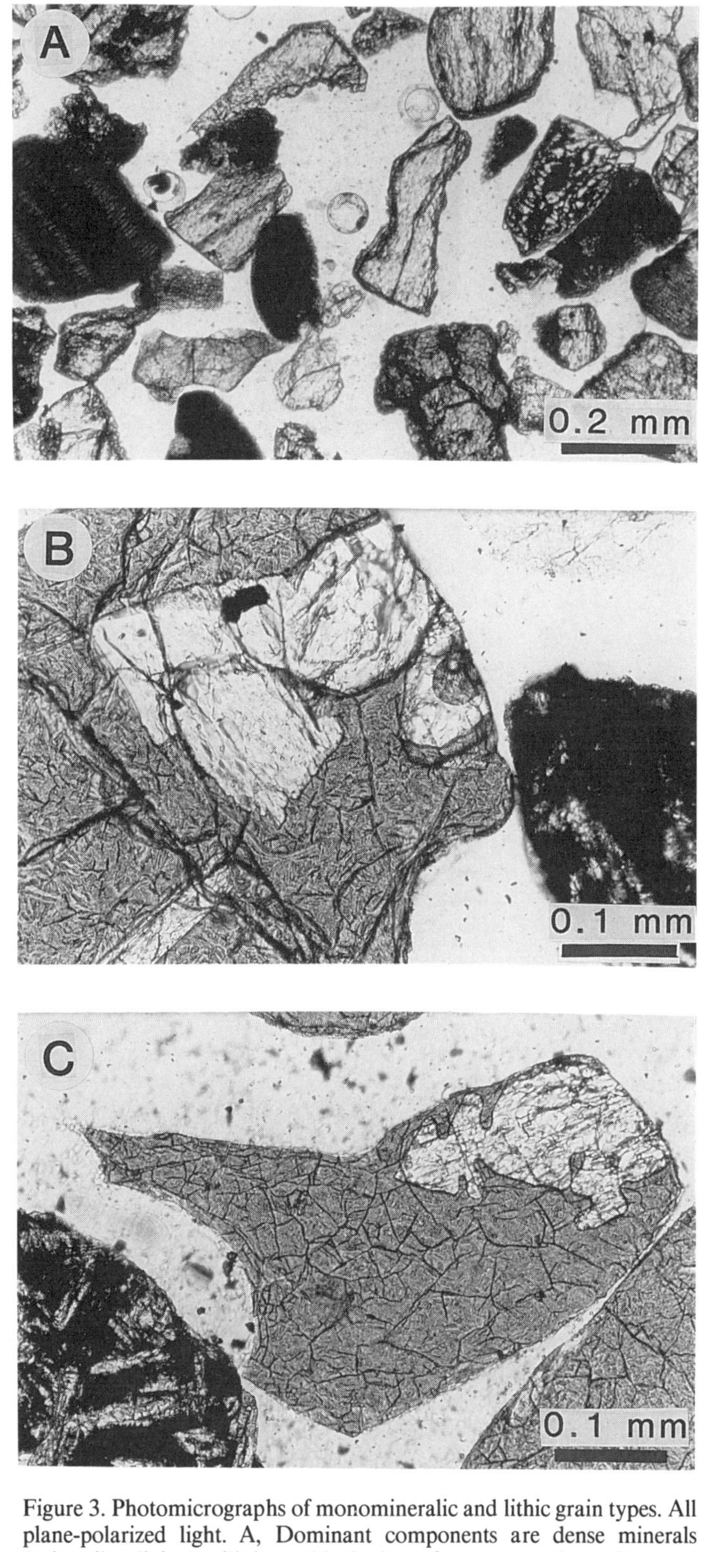

Figure 3. Photomicrographs of monomineralic and lithic grain types. All plane-polarized light. A, Dominant components are dense minerals (primarily olivine) with lesser black glassy fragments and one fragment of red algae; sample KAU 12. B, Brown lathwork glassy fragment with sand-sized phenocrysts of plagioclase, olivine, and pyroxene on left, and black microlitic fragment on right; sample HAW 37. C, Resorbed pyroxene crystal in brown glassy fragment on right and black lathwork fragment; sample HAW 6. Note that the irregular texture and dark color of brown glass in B and C are produced by etching and staining.

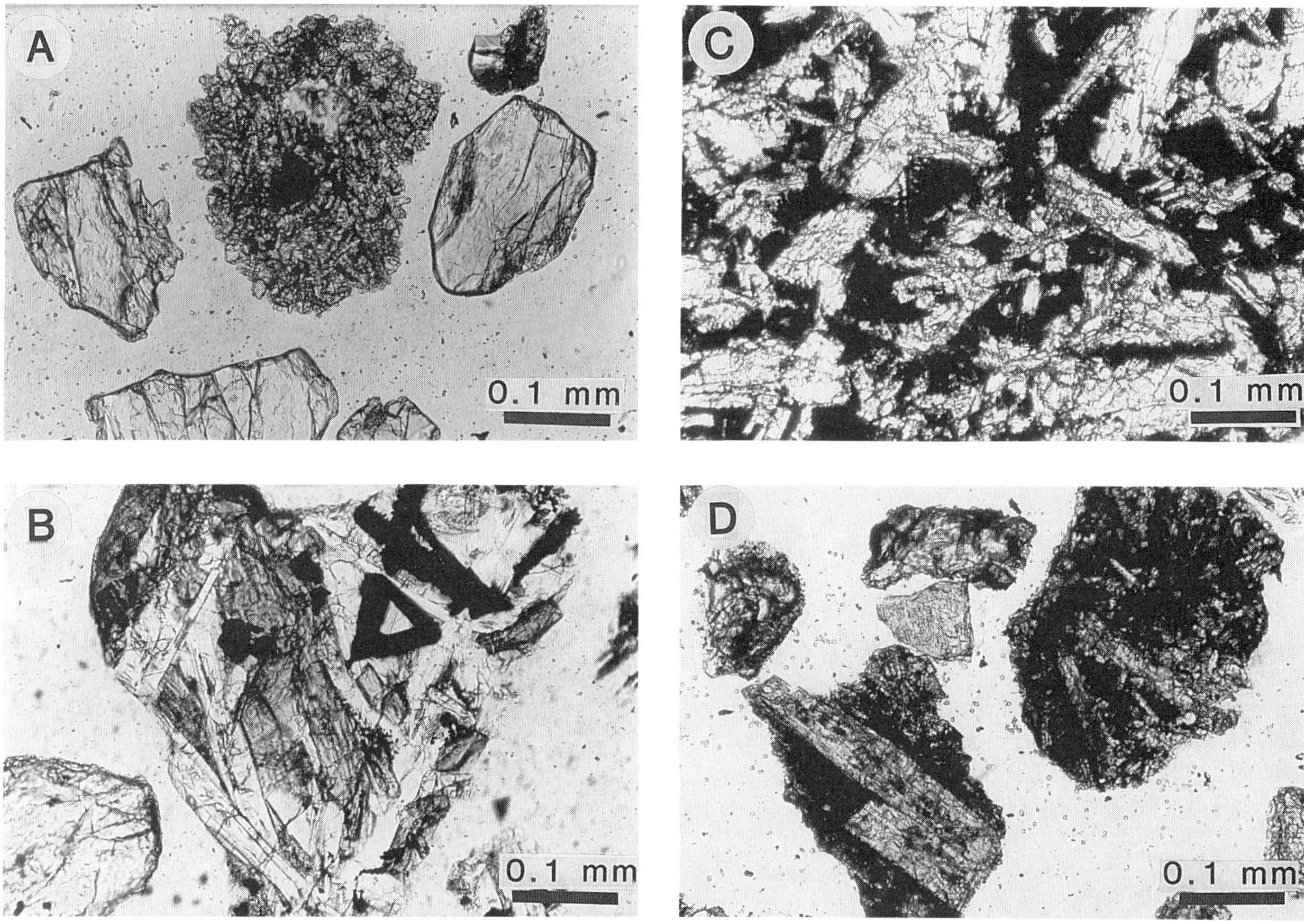

Figure 4. Photomicrographs of volcanic lithic fragments. All plane-polarized light. A, Microcrystalline "other" lithic fragment (olivine/pyroxene and opaques) and monocrystalline olivine grains; sample HAW 38. B, Holocrystalline "other" lithic fragment composed of plagioclase, pyroxene, and opaque crystals; sample HAW 7. C, Black lathwork fragment with plagioclase laths in an opaque groundmass; sample HAW 6. D, Plagioclase-rich sample with monomineralic plagioclase grain and black glassy lathwork to microlitic fragments; sample HAW 19.

Volcanic lithic fragments in Hawaiian beach sands range from holocrystalline to holohyaline with hypocrystalline varieties dominant. Glassy basalt fragments were subdivided into categories based on the color and texture of their glassy groundmass in plane-polarized light (tan/brown glass, black/gray glass, orange glass, glass altered to clay minerals, and devitrified glass; Table 1, Appendix 2). Tan/brown basaltic glass, sideromelane (Fig. 3), is produced when magma is rapidly quenched by water, whereas black/gray tachylitic glass (Figs. 3, 4) is formed in subaerial or subaqueous conditions where cooling rates are lower (Macdonald, 1972; Fisher and Schmincke, 1984; Cas and Wright, 1987). Tachylitic fragments are opaque due to the presence of Fe/Ti-rich microlites (Fisher and Schmincke, 1984), whereas the orange color of some glassy fragments may be a function of either oxidation during weathering or palagonitization. Tachylitic glassy fragments can be differentiated from opaque monomineralic grains by the presence of plagioclase microlites/phenocrysts and /or by their irregular texture.

Vitric fragments are characterized by their lack of microlites. Microlitic (Lvml) and lathwork (Lvl) volcanic lithic fragments are glassy fragments that contain variable amounts of plagioclase feldspar, olivine, pyroxene, and opaque crystals. Microlitic fragments contain silt-sized microlites (Figs. 3, 4) whereas lathwork fragments contain some sand-sized microlites/phenocrysts (Figs. 3, 4). A grain is counted as lathwork only if the crosshairs land on the glassy to microlitic groundmass adjacent to sand-sized phenocrysts (e.g., Ingersoll and others, 1984). When point counting coarse-grained samples where textures varied across a volcanic lithic fragment, the dominant texture in the vicinity of the crosshairs was counted.

The "other" volcanic category (Lvo: Table 1) includes holocrystalline aggregates of silt- to sand-sized crystals not covered by the Dickinson (1970) categories (e.g., Fig. 4). These are probably fragments of glomerocrysts or holocrystalline basalt to microgabbro. This category includes aggregates of plagioclase and opaque minerals; opaque minerals and olivine/pyroxene; olivine/pyroxene and plagioclase; and olivine/pyroxene, plagioclase, and opaque minerals. A grain is included in this category only if the crosshairs land on a silt-sized crystal during the point-counting process.

Other minor constituent grains include organic matter (e.g., plant fragments), beachrock fragments, a few grains of green clay (glauconite or celadonitized glass?), and rare clay alteration products (soil particles(?) discussed below). Sand-sized grains within beachrock fragments were placed in the appropriate monomineralic, volcanic, or bioclast category, whereas beachrock cements were placed in the "other" category. Isolated micritic cement fragments are not easily distinguished from micritic bioclasts, and so they may have been incorporated into the bioclastic category. Beachrock fragments are particularly common in Maui beach sands (Appendix 2).

A few Hawaiian stream samples were collected for petrographic study in order to determine the nature of sand-sized weathering products on the islands. Along with seemingly unaltered epiclastic volcanic detritus, these contain a high percentage of partially altered volcanic and monomineralic grains to wholly altered grains with and without relic pseudomorphic textures (Fig. 5; see Johnsson, 1990, for a discussion of alterites). These stream sands also contain rounded soil particles or pellets that resemble ferruginous silty claystone (Fig. 5), colorless gibbsite(?) grains, and lateritic pisolites. Many detrital grains exhibit pisolitic coatings.

Modal compositions

The total percentage of bioclastic debris and the percentage of volcanic grains that are altered are listed in Appendix 3, and the relative proportions of bioclastic debris and unaltered and altered volcanic debris in Appendix 4. These percentages are quite variable from island to island and across any given island (Figs. 6-8), with most samples being predominantly bioclastic. Recalculated parameters are given in Appendix 5 for those samples with a significant percentage of siliciclastic components (30% or more). When plotted on standard ternary compositional diagrams (e.g., QFL, QmPK, and LmLvLs diagrams; see Dickinson et al., 1983), Hawaiian sands are fairly uniform in composition. Although their ratio of feldspar to volcanic lithic fragments varies somewhat (Fig. 9), they would essentially all plot at the Lv apex of an LmLvLs diagram, and the P apex of a QmPK plot because their light mineralogy is dominated by volcanic lithic fragments with lesser amounts of plagioclase feldspar. Unaltered volcanic lithic fragments are composed of brown (sideromelane) and black (tachylite) glassy fragments exhibiting vitric, microlitic and lathwork textures (Figs. 10-12). There is much overlap in the textural attributes of the brown and black glassy fragments (i.e., vitric/microlitic/lathwork proportions; Fig. 11), confirming that the

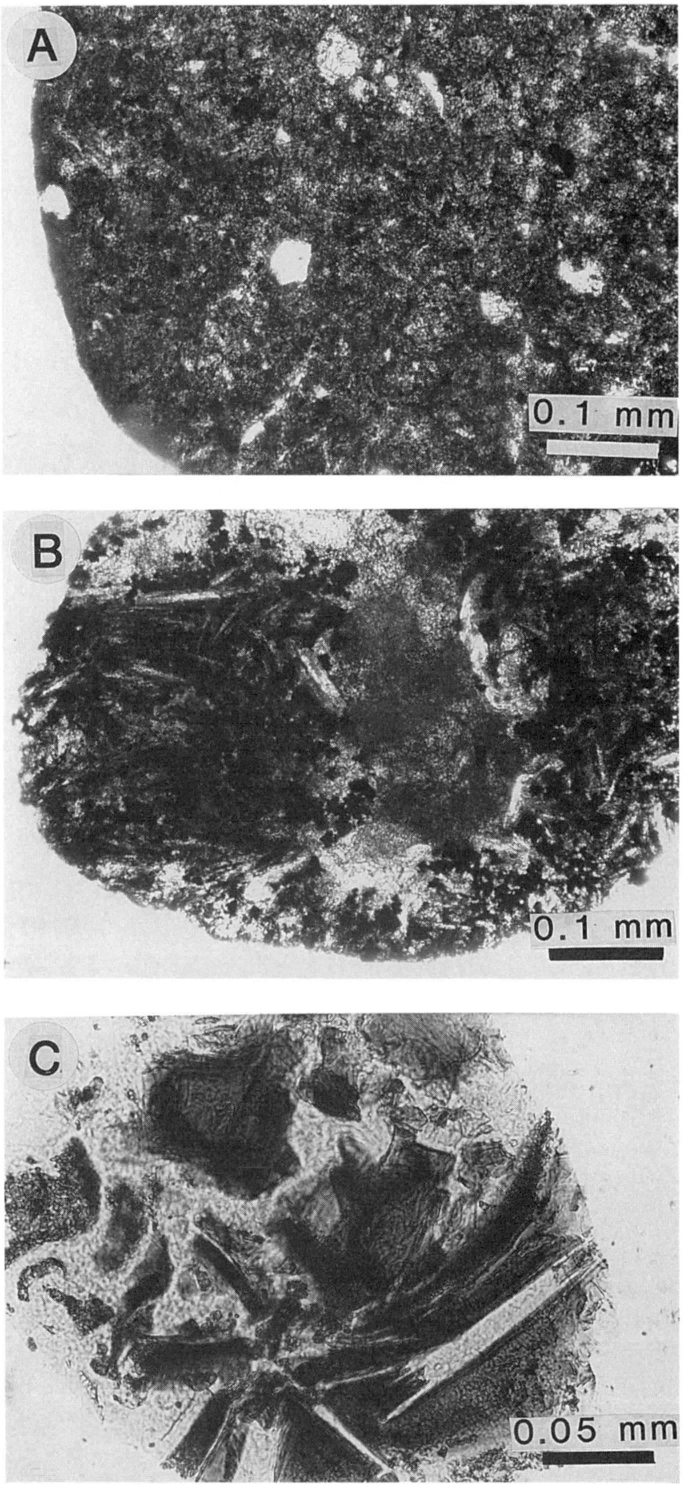

Figure 5. Photomicrographs of altered volcanic fragments. All plane-polarized light (each fragment isotropic under crossed Nicols; feldspar crystals completely altered/replaced or dissolved). A, Soil particle or alterite (Johnsson, 1990) from stream on the island of Hawaii. B, Altered microlitic fragment from stream on the island of Hawaii. C, Altered microlitic fragment from stream on the island of Maui.

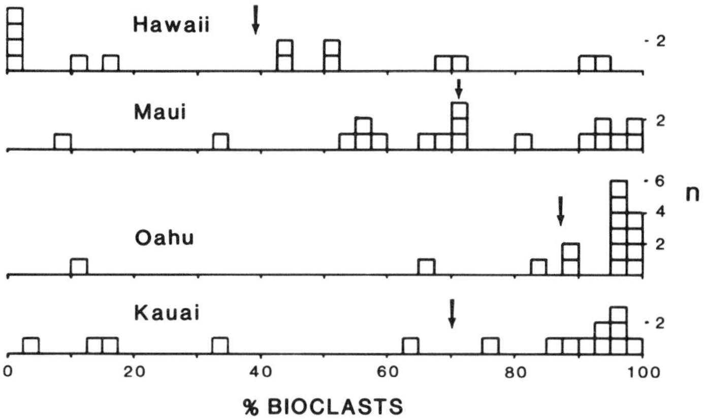

Figure 6. Histogram of total framework percentage of bioclasts (%Bioclasts: Table 1 and Appendices 2, 3) by island. Arrows mark mean values for individual islands.

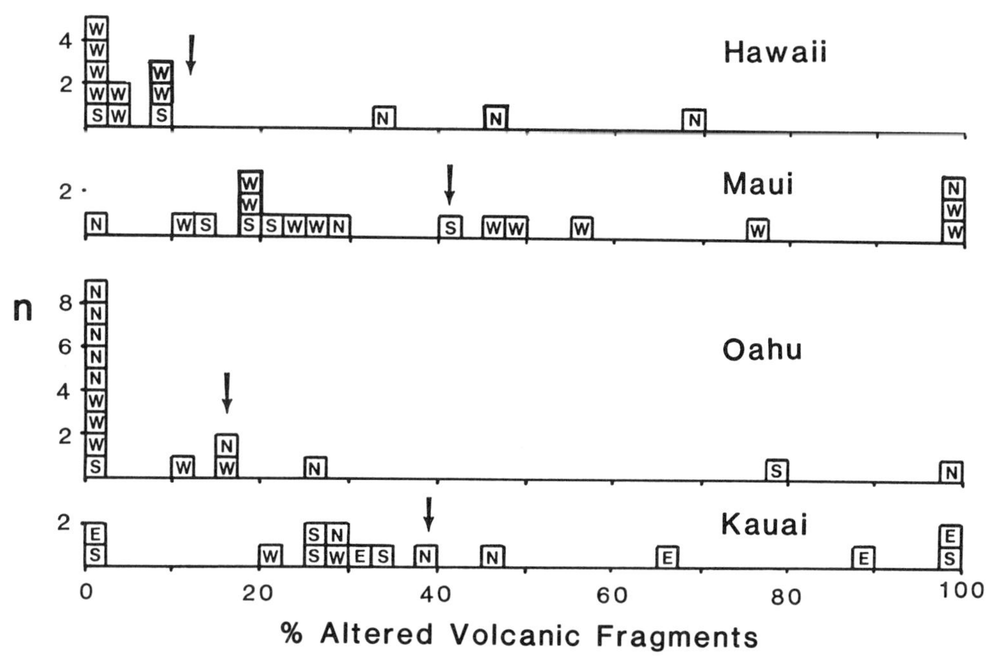

Figure 7. Histogram showing the distribution of all samples by island according to the total framework percentage of altered volcanic fragments including soil(?) particles (Lv%Alt: Table 1, Appendices 2, 3). Data have been keyed to geographic location (see Appendix 3), where E = east, S = south, W = west, and N = north. Arrows mark mean values for individual islands.

difference between tachylite and sideromelane is indeed not compositional, but a function of lava cooling rate. Note that some samples also contain significant amounts of holocrystalline lithic fragments (Figs. 13, 14), dense minerals (Fig. 15), and altered volcanic components (Fig. 16). The proportions of these components are discussed below.

DISCUSSION

Island evolution and sand composition

In general, the number of dominantly bioclastic samples per island increases with island age (Fig. 6; Appendix 3). On the youngest island, Hawaii, only 46% of the samples contain more than 50% bioclastic debris, whereas on the older islands a much greater proportion are predominantly bioclastic (Maui, 89%; Oahu, 93%; Kauai, 73%). This relationship appears to correlate with Stearns's (1984) model for island evolution: Hawaii has not yet developed significant fringing reefs, and so the supply of bioclastic debris is minimal, whereas the older islands are in more advanced stages of development, at which point fringing reefs are able to supply larger quantities of bioclastic material. This bioclastic debris probably dilutes the volcaniclastic contribution, or alternatively, the production of volcaniclastic debris may be less significant during the later stages of island development. The obvious exception to this trend is the island of Kauai, where late-stage erosion enhanced by higher rainfall may explain the relative decrease in bioclastic debris with respect to the volcanic component. In contrast to the other islands where ephemeral streams are common, Kauai is drained predominantly by a sys-

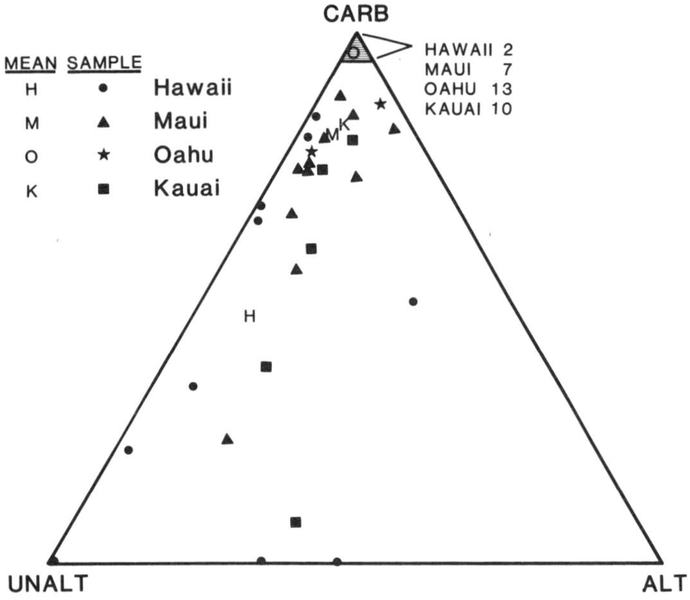

Figure 8. Ternary plot of percentage of bioclastic (CARB), unaltered volcaniclastic (UNALT), and altered volcaniclastic (ALT) debris by island (Appendix 4). Overlap of bioclastic samples at top of diagram resolved to the right; number of samples shown by island. Island means are as indicated.

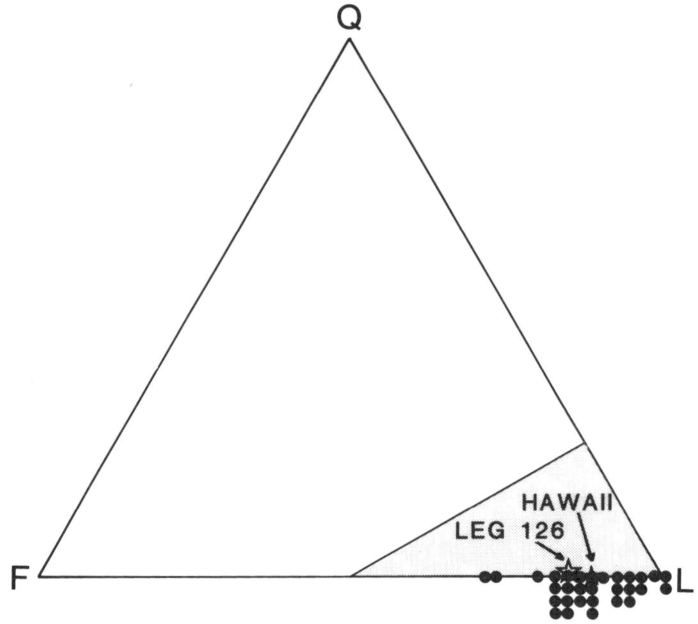

Figure 9. Triangular plot of QFL data for Hawaiian beach sands with mean composition indicated by filled star. Compositional field of magmatic arc provenance from Dickinson et al. (1983) is shaded. Overlap of data at base of diagram shown in histogram fashion. Mean value for Leg 126 (Izu-Bonin arc) samples (open star; n = 111) calculated from data presented by Marsaglia (1992).

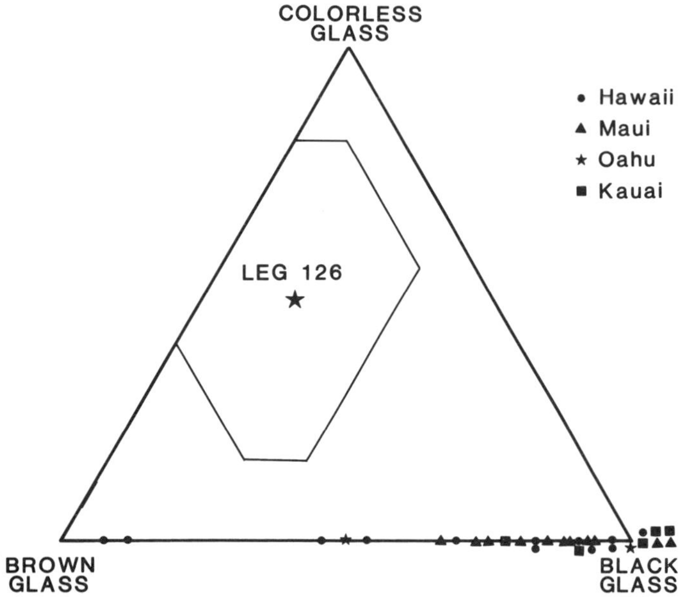

Figure 10. Triangular plot of colorless, brown (sideromelane) and black (tachylite) glassy fragments by island, where brown glass is Volc%brown and black glass is Volc%black (Table 1). Note that colorless glass was not observed in Hawaiian sand samples. Mean and field of variation (polygon defined by 1 standard deviation on either side of mean) for Leg 126, Izu-Bonin arc, samples calculated from data presented by Marsaglia (1992).

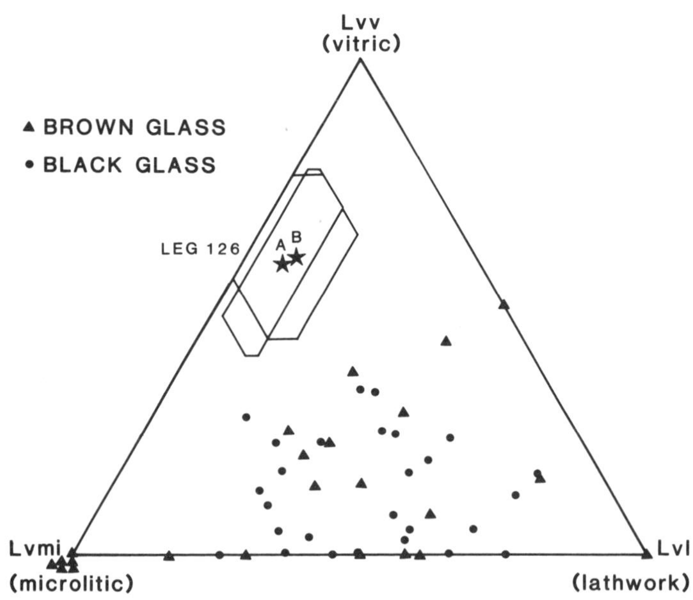

Figure 11. Triangular plot of vitric (Lvv), microlitic (Lvmi), and lathwork (Lvl) proportions for brown and black glassy fractions from Hawaiian beach sands. Mean and field of variation (polygon defined by 1 standard deviation on either side of mean) for Leg 126, Izu-Bonin arc, samples calculated from data presented by Marsaglia (1992); A = brown glass; B = black glass.

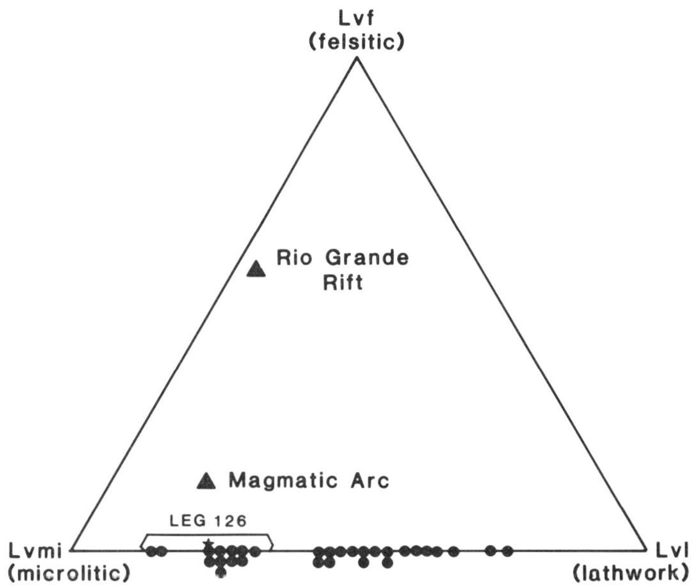

Figure 12. Triangular plot of felsitic (Lvf), microlitic (Lvmi) and lathwork (Lvl) fragments for Hawaiian beach sands (Table 1; Appendices 2, 5). Mean magmatic arc and Rio Grande rift values from Marsaglia (1991); mean (star) and field of variation (polygon defined by 1 standard deviation on either side of mean) for Leg 126, Izu-Bonin arc, samples calculated from data presented by Marsaglia (1992).

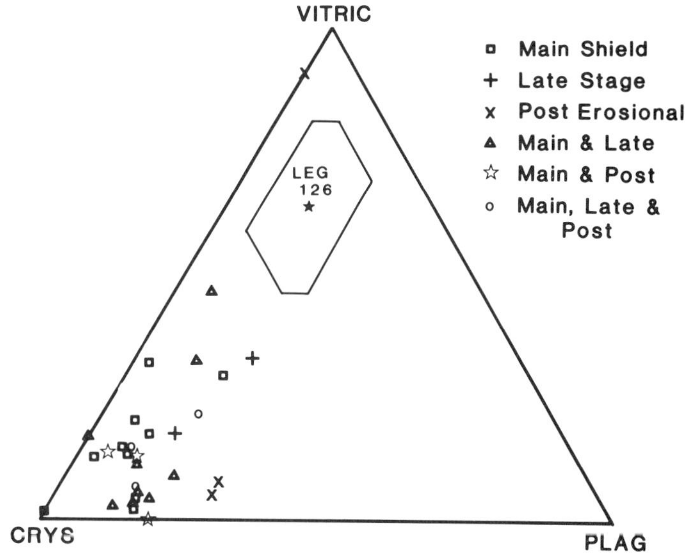

Figure 13. Ternary plot of relative proportions of vitric fragments (Vitric), crystalline (Crys: microlitic, lathwork and holocrystalline) fragments, and plagioclase (Plag) grains for Hawaiian beach sands (Table 1; Appendices 2, 5), identified by likely volcanic provenance (Appendix 1). Mean (filled star) and field of variation (polygon defined by 1 standard deviation on either side of mean) for Leg 126, Izu-Bonin arc, samples calculated from data presented by Marsaglia (1992).

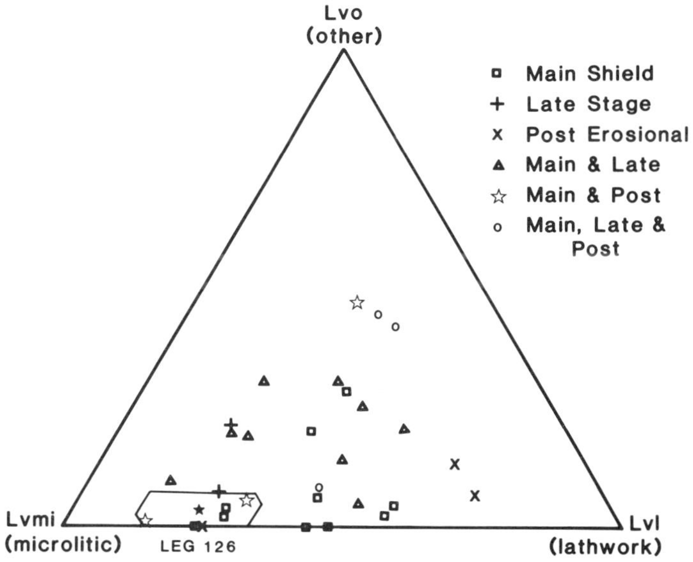

Figure 14. Ternary plot of relative proportions of holocrystalline (Lvo), microlitic (Lvmi) and lathwork (Lvl) volcanic fragments for Hawaiian beach sands (Table 1; Appendices 2, 5), by likely volcanic provenance (Appendix 1). Mean (filled star) and field of variation (polygon defined by 1 standard deviation on either side of mean) for Leg 126, Izu-Bonin arc, samples calculated from data presented by Marsaglia (1992).

tem of perennial streams and rivers. The samples with minimal bioclastic debris were collected from beaches at the mouths of the Waimea (KAU 11), Hanapepe (KAU 12), and Wainiha (KAU 7) Rivers and on the eastern edge of the Napali coast (KAU 6). Overall, these results are similar to the distribution of detrital versus carbonate fractions presented in Moberly et al. (1965).

Another component that might be expected to increase with island age and development is the proportion of weathered detritus, including soil particles and pedogenic oxide nodules liberated by soil erosion. The durability and the survivability of such grains in a high-energy beach environment are questionable, in that soil particles, pisolites, and pisolitic coatings were preferentially plucked from thin sections of stream samples; this suggests that they are much softer than other components, and therefore would be less likely to survive transport and wave reworking. Assuming that some beach sand is derived from stream input or soil erosion along beach zones, the lack of pisolitic or nodular fragments, and the paucity of identifiable soil particles in the beach samples examined in this study suggests that the other altered volcanic components are more resistant.

The relative proportion of weathered/altered particles shows little relationship to island age or stage of development, as shown in Figure 7. Hawaii and Oahu, and Maui and Kauai display similar distributions of the percentage of altered detritus, including soil particles and variably altered lithic fragments. On the island of Hawaii there is a marked increase in altered material in beach sands from the northern, wetter side of the island (Fig.

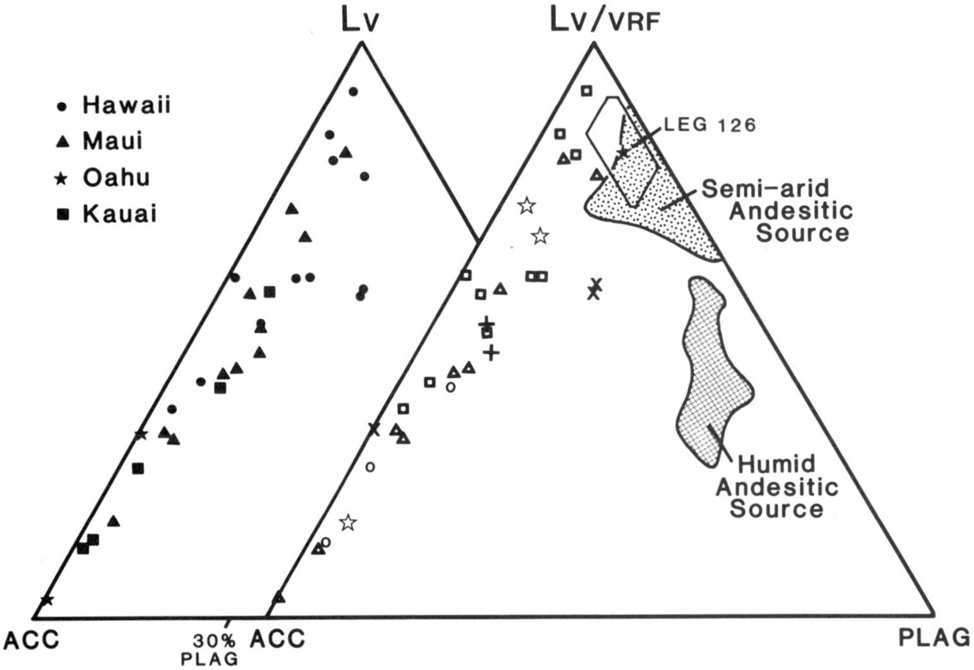

Figure 15. Ternary plots of relative proportions of volcanic rock fragments (Lv = Gazzi Dickinson method, Hawaii and Leg 126 samples; VRF = traditional method, Mack and Jerzykiewicz (1989)), accessory minerals (ACC) and monocrystalline plagioclase (PLAG) for Hawaiian sands (Table 1; Appendices 2, 5) and sands with an intermediate provenance. Semi-arid = New Mexico, Montana, and Colorado; Humid = Ranier, Fuego, and Pelee volcanoes. Leg 126 mean and field of variation recalculated from data presented by Marsaglia (1992). Hawaiian data subdivided by island on left and likely volcanic provenance on right (see Fig. 14 for legend).

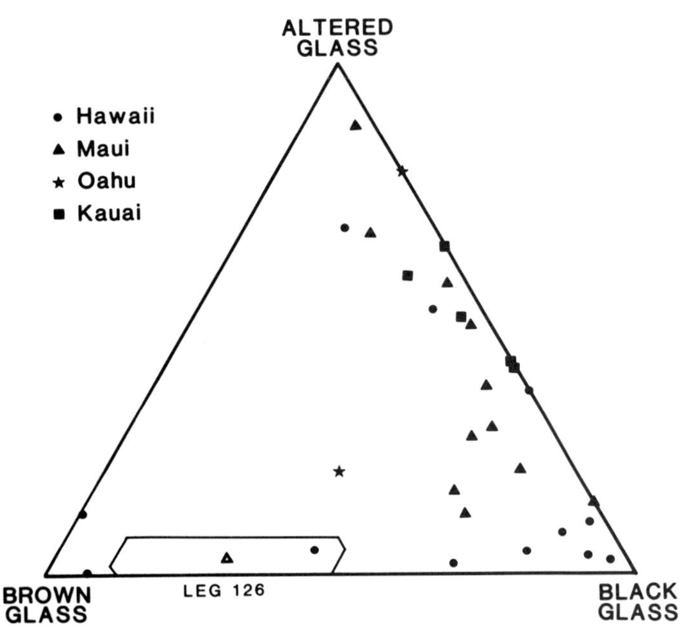

Figure 16. Triangular plot of relative proportions of altered (ALT), brown (BROWN), and black (BLACK) glassy volcanic fragments in Hawaiian beach sands by island (Table 1; Appendices 2, 5). Mean (open triangle) and field of variation (polygon defined by 1 standard deviation on either side of mean) for Leg 126, Izu-Bonin arc, samples calculated from data presented in Marsaglia (1992).

7). Enhanced soil erosion due to local agricultural and housing development may be contributing factors to higher proportions of altered volcanic detritus at some sites on the western side of Maui and the eastern side of Kauai (Fig. 7).

Theoretically, the evolutionary state of a Hawaiian island should be reflected in the relative proportions of reef-derived (carbonate) versus stream-derived (epiclastic: unaltered to completely altered) versus eruption-derived (hydroclastic: unaltered) material. In general, the overall proportion of bioclasts and unaltered and altered volcanic material in Hawaiian beach sands is variable across all the islands sampled except for Oahu, which is dominantly bioclastic (Fig. 8). The distribution of island means shows Oahu (older, reef dominated) and Hawaii (younger) as end member island types with Kauai and Maui as intermediate examples. Nascent basaltic islands should be dominated by unaltered volcanic material produced by submarine eruptions.

Evolutionary variations in Hawaiian basaltic lava composition, from alkalic to tholeiitic to alkalic, might also be reflected in beach sand compositions, assuming that the compositional differences between lavas are manifested in the types and proportions of phenocrysts and lithic textures. The probable source lavas (volcanic series) for each sample are listed in Appendix 1. Unfortunately, determination of discrete relationships between the volcanic series/lava composition and sand composition is hampered by the large number of samples with probable mixed provenance (Appendix 1). No distinct relationships have been identified between the proportions of various types of volcanic components and the likely source rocks for those samples (Figs. 13, 14). This also holds for relative proportions of monomineralic components (Fig. 15) and for relative proportions of vitric, microlitic, and lathwork textures in both brown and black glassy fragments (not pictured).

Possible effects due to relative age of the volcanic series were assessed by plotting these data by island. The only apparent trends are that beach sand samples from the island of Hawaii contain a higher proportion of brown (sideromelane) glass (Fig. 10) and less accessory minerals (Fig. 15). The enrichment of a few Hawaiian island samples in sideromelane can be linked to recent coastal volcanism, whereas the concentration of dense minerals in beach sands is probably a function of both the dense mineral content of source materials and the wave dynamics of the beach on which they were collected. The relative abrasion rates of detrital components of beach sands, as determined by Moberly (1968), indicate that olivine abrades more rapidly than volcanic glass but less rapidly than fresh basalt. Thus the proportions of dense minerals and glass fragments within Hawaiian beach sands may be influenced by wave intensity and sand residence time in the foreshore regions of a given beach environment.

Basaltic vs. andesitic provenance

Previous workers on volcaniclastic sands have primarily focussed on andesitic provenance (e.g., Dickinson and Suczek, 1979; Dickinson et al., 1983; Mack and Jerzykiewicz, 1989; Marsaglia and Ingersoll, 1992), with only minor attention to basaltic provenance (Ingersoll, 1984; Lundberg, 1991; Larue and Sampayo, 1990: Marsaglia, 1991). Ingersoll (1984) attributed the limited production of basaltic sand during weathering in subaerial continental settings to its rapid alteration/dissolution; as a result, the basaltic contribution is underrepresented, and is overwhelmed by other sources of siliciclastic detritus. This is less of a problem on the Hawaiian Islands where there are no other siliciclastic source rocks, only reef-derived bioclastic debris.

Hawaiian basaltic sands are similar in composition to magmatic arc-derived sands in terms of their relative QFL, QmPK, and LmLvLs percentages (e.g., Fig. 9; Marsaglia, 1989). These two volcanic provenances prove dissimilar, however, when viewed in terms of their volcanic lithic proportions. Only one previous study, of magmatic arc–derived sand by Marsaglia (1992), provides petrographic data similar to that presented in the present study, namely, volcanic glassy lithic fragments subdivided by glass color and crystallinity. Recalculated means and fields of variation for Izu-Bonin arc sands cored during Ocean Drilling Program Leg 126 (Marsaglia, 1992) are shown in Figures 9–11 and 12–16. There are compositional distinctions between these two provenances in that Hawaiian sands lack felsic components such as quartz, potassium feldspar, colorless glass (Fig. 10), and felsitic volcanic lithic fragments (Fig. 12), and are more enriched in lathwork volcanic lithic fragments (Figs. 11, 12, 14), "other" holocrystalline lithic fragments (Fig. 14), and dense accessory minerals (Fig. 15). Basaltic volcanic provenances are thus characterized by higher mean lathwork volcanic lithic proportions and can be distinguished from microlitic-rich intermediate (magmatic arc) and felsitic-rich felsic (e.g., Rio Grande Rift) volcanic settings (Fig. 12). These two data sets may be somewhat incompatible in that Leg 126 sands are predominantly composed of reworked pyroclasts, with lesser subaerial, island-derived material (Marsaglia, 1992), whereas the Hawaiian sands appear to have a large epiclastic component. Perhaps this disparity is reflected in the higher vitric component (Figs. 11, 13) and lower percentage of altered volcanic fragments (Fig. 16) in the Izu-Bonin arc samples.

In a study of andesitic stream sands from humid and semi-arid environments, Mack and Jerzykiewicz (1989) showed that epiclastic-andesitic sands are feldspatholithic, with total feldspar content controlled by weathering environment (Fig. 15). Leg 126 sands are most similar in composition to their semi-arid examples. In contrast, Hawaiian basaltic sands are relatively enriched in accessory, dense minerals. Direct comparison of these petrographic data sets is tentative, however, in that they were collected using different point-counting methods. Using the traditional method of Mack and Jerzykiewicz (1989), the percentage of volcanic lithic fragments would be somewhat higher in the Hawaiian and Leg 126 sands than pictured in Figure 15 (see discussion in Ingersoll and others, 1984). Nonetheless, Figure 15 illustrates that accessory/dense mineral content may help discriminate basaltic from andesitic sand provenance.

Previous workers have also pointed out the significance of microcrystalline volcanic lithic fragments in determining volcanic provenance (Dickinson, 1970; Lundberg, 1991; Marsaglia, 1991). Specifically, microlitic and lathwork volcanic lithic proportions have been used to distinguish intermediate from mafic sources; in at least one instance sands with basaltic provenance appear to be principally composed of microcrystalline volcanic grains exhibiting lathwork textures (Lundberg, 1991). Relative proportions of microlitic and lathwork textures determined for Hawaiian basaltic sands indicate a much broader range than that reported by Lundberg (1991), but they are significantly richer in lathwork grains than sands from magmatic arc volcanic settings (Figs. 11, 12). Hawaiian basaltic volcaniclastic lithic fragments (brown and black glass) are dominated by microlitic and lathwork textures, whereas intermediate to mafic brown and black glass fractions from magmatic arcs (see discussion in Marsaglia, 1992) are dominated by vitric and microlitic components (Fig. 11). This observation suggests that, even in similarly colored glass fractions, Hawaiian mafic volcanic clast populations can be distinguished from intermediate to mafic magmatic arc volcanic clast populations based on their textural attributes.

CONCLUSIONS

Major factors controlling Hawaiian beach sand composition include the degree of fringing reef development, the intensity of wave action along the coast, the source and nature of the volcanic detritus, and climate, which directly influences the degree of weathering. Beach sands on the Hawaiian Islands can be divided into bioclastic and volcaniclastic end members. The bioclastic sands appear to be more prevalent on the older islands, with better-developed fringing reefs, whereas the youngest island, Hawaii, has not yet developed a fringing reef system, and so shows the least bioclastic input. Abrasion studies by Moberly (1968) suggest that wave action may also alter Hawaiian beach sand composition, specifically increasing the relative proportions of bioclasts and olivine grains with respect to volcanic lithic fragments. Although the chemistry of basaltic source lavas does not appear to significantly affect sand composition, the mode of sand production, e.g., hydroclastic/pyroclastic versus epiclastic, affects the relative proportions of brown and black glassy fragments. The influence of climate on basaltic sand composition is best displayed on the island of Hawaii, where northern beaches are enriched in altered/weathered volcanic lithic fragments. Because most Hawaiian beach sands owe their composition to a combination of all of the above factors, further study is needed in order to determine the relative importance of each in shaping sand composition for a given beach. This study has been a regional examination of sand composition across the Hawaiian islands, which sets the stage for more detailed work.

Although Hawaiian beach sands are compositionally similar to sands derived from undissected magmatic arcs, they can be distinguished from these in that the Hawaiian sands lack a felsic volcanic component and are more enriched in tachylitic, lathwork, and holocrystalline volcanic lithic fragments. Preliminary results (Marsaglia, 1992) suggest that these textures are preserved even after severe diagenetic alteration. Thus, long after the chemical fingerprint of glass composition has been removed or altered, textural fingerprints of volcanic origin may remain and can be deciphered. Quantifying textural proportions of known end members, such as the basaltic Hawaiian Islands, is the first step in interpreting the provenance of ancient volcaniclastic deposits.

ACKNOWLEDGMENTS

This work was partially funded by a University of Texas at El Paso Research Initiation grant. Thanks to David Diamond for providing two sand samples (HAW 38 and 40), Michelle Pittenger for help in sample preparation, Robert Merrilees for assistance in the field, and Dale Crissman, Gary Girty, Mark Johnsson, and Neil Lundberg for their thoughtful reviews of the manuscript.

APPENDIX 1. HAWAIIAN SAMPLE LOCATIONS

Sample	Island	Beach (Volcano-Rock Unit[s]*)
HAW 1	Hawaii	Pu'uhonua Point (Mauna Loa–Ka'u Volcanic Series [V.S.†]—main shield)
HAW 2	Hawaii	Mioli'i Beach (Mauna Loa–Ka'u V.S.—main shield)
HAW 5	Hawaii	Ka Lae (South Point) (Mauna Loa–Kahuku V.S.—main shield)
HAW 6	Hawaii	Punaluu Beach (Mauna Loa–Ka'u V.S.—main shield)
HAW 7	Hawaii	Kahulu'u Beach (Hualahai-Hualahai V.S. —posterosional)
HAW 8	Hawaii	Honokohau Harbor (Hualahai-Hualahai V.S. —posterosional)
HAW 9	Hawaii	Keahole Point (Hualahai-Hualahai V.S.—posterosional)
HAW 10	Hawaii	Anaeho'omalu Bay (Mauna Loa–Ka'u V.S.—main shield)
HAW 11	Hawaii	Hapuna Beach (Mauna Kea–Hamakua V.S. —main shield)
HAW 14	Hawaii	Keokea Beach (Kohala—mixture of Pololu V.S. [main shield] and Hawi V.S. [late stage])
HAW 15	Hawaii	Laupahoehoe Point (Mauna Kea—mixture of Hamakua V.S. [main stage] and Laupahoehoe V.S. [late stage])
HAW 16	Hawaii	Kolekole Beach (Mauna Kea–Hamakua V.S.—main shield)
HAW 17	Maui	Waiehu Beach (West Maui—mixture of Wailuku V.S. [main shield] and Honolua V.S. [late stage])
HAW 20	Maui	H.P. Baldwin Beach (Haleakala-Kula V.S.—late stage)
HAW 21	Maui	Ho'okipa Beach (Haleakala-Kula V.S. [late stage] and Hana V.S. [posterosional])
HAW 22	Maui	D.T. Fleming Beach (West Maui–Honolua V.S. [late stage] and Wailuku V.S. [main shield])
HAW 23	Maui	Kapalua Beach (West Maui–Honolua V.S.—late stage)
HAW 24	Maui	Honokowa Beach (West Maui–Wailuku V.S.—main shield)
HAW 25	Maui	Hanaka'o'o Beach (West Maui–Wailuku V.S. [main shield] and Lahaina V.S. [post-erosional])
HAW 26	Maui	Puamana Beach (West Maui–Wailuku V.S.—main shield)
HAW 27	Maui	Ka'ili'ili Beach (West Maui–Wailuku V.S. [main shield] and Honolua V.S. [late stage])
HAW 28	Maui	Punahoa Beach (West Maui–Wailuku V.S. [main shield] and Honolua V.S. [late stage])
HAW 29	Maui	Beach at Palalua (West Maui–Wailuku V.S. [main shield] and Lahaina V.S. [post-erosional])
HAW 30	Maui	Mai Poina 'Oe La'u Beach (Haleakala-Kula V.S.—late stage)
HAW 31	Maui	Kama'ole Beach (Haleakala-Kula V.S.—late stage)
HAW 32	Maui	Wailea Beach (Haleakala-Kula V.S.—late stage)
HAW 33	Maui	Makena Bay (Haleakala-Honomanu V.S.—main shield)
HAW 34	Maui	La Pérouse Bay (Haleakala-Honomanu V.S. [main shield] and Hana V.S. [post-erosional])
HAW 35	Maui	Ahihi Bay (Haleakala-Honomanu V.S. [main shield] and Hana V.S. [posterosional])
HAW 36	Maui	Onleloa Beach (Haleakala-Honomanu V.S.—main shield)
HAW 38	Hawaii	Napoopoo Beach (Mauna Loa–Ka'u Volcanic Series [V.S.]—main shield
HAW 40	Hawaii	Kapaahu Beach (Kilauea-Puna V.S.—main shield)
OAH 1	Oahu	Wai'alae Beach (Koolau-Honolulu V.S. [post-erosional] and Koolau V.S. [main shield and minor late stage])
OAH 2	Oahu	Hanauma Bay (Koolau-Honolulu V.S.—post-erosional)
OAH 3	Oahu	Beach near Pali Alei (Waianae-Waianae V.S.—main shield and late stage)
OAH 4	Oahu	Beach near Pali Alei (Waianae-Waianae V.S.—main shield and late stage)
OAH 5	Oahu	Waimea Beach (Koolau-Koolau V.S.—main shield and late stage)
OAH 6	Oahu	Ehukai Beach (Banzai Pipeline) (Koolau-Koolau V.S.—main shield and late stage)
OAH 7	Oahu	Kuilima Cove (Koolau-Koolau V.S.—main shield and late stage)

APPENDIX 1. HAWAIIAN SAMPLE LOCATIONS (continued)

Sample	Island	Beach (Volcano-Rock Unit[s]*)
OAH 8	Oahu	La'ie Beach (Koolau-Koolau V.S.—main shield and late stage)
OAH 9	Oahu	Punalu'u Beach (Koolau-Koolau V.S.—main shield and late stage)
OAH 10	Oahu	Kalae'o'io Beach (Koolau-Koolau V.S.—main shield and late stage)
OAH 11	Oahu	Makua-ka'ena Beach (Waianae-Waianae V.S.—main shield and late stage)
OAH 12	Oahu	Makua-ka'ena Beach (Waianae-Waianae V.S.—main shield and late stage)
OAH 13	Oahu	Makaha Beach (Waianae-Waianae V.S.—main shield and late stage)
OAH 14	Oahu	Pu'u o Hulu Beach (Waianae-Waianae V.S.—main shield and late stage)
OAH 15	Oahu	Kahe Point Beach (Waianae-Waianae V.S.—main shield and late stage)
KAU 1	Kauai	Kamala Point (Koloa V.S.—posterosional)
KAU 2	Kauai	Hanaama'ulu Beach (Waimea Canyon V.S. [main shield and late stage] and Koloa V.S. [post-erosional])
KAU 3	Kauai	Wailua Beach (Waimea Canyon V.S. [main shield and late stage] and Koloa V.S. [post-erosional])
KAU 4	Kauai	Beach near Kealia (Waimea Canyon V.S. [main shield and late stage] and Koloa V.S. [posterosional])
KAU 5	Kauai	Anahola Beach (Waimea Canyon V.S. [main shield and late stage] and Koloa V.S. [post-erosional])
KAU 6	Kauai	Ke'e Beach (Waimea Canyon V.S.—main shield and late stage)
KAU 7	Kauai	Kolokolo Point (Waimea Canyon V.S. [main shield and late stage] and Koloa V.S. [post-erosional])
KAU 8	Kauai	Nawiliwili Bay (Kola V.S.—posterosional)
KAU 9	Kauai	Polihale Beach (Waimea Canyon V.S.—main shield and late stage)
KAU 10	Kauai	Kekaha Beach (Waimea Canyon V.S.—main shield and late stage)
KAU 11	Kauai	Lucy Wright Beach (Waimea Canyon V.S. [main shield and late stage] and Koloa V.S. [post-erosional])
KAU 12	Kauai	Hanapepe Bay (Waimea Canyon V.S. [main shield and late stage] and Koloa V.S. [post-erosional])
KAU 13	Kauai	Po'ipu Beach (Koloa V.S.—posterosional)

*Possible volcanic source of sediment indicated in parentheses after each location. Source based on proximity to outcrops including those in drainage basins of local streams, but does not take into account possible effects of longshore transport, and recycling of sediment from beach rock and Quaternary sedimentary deposits.

Approximate relative ages of volcanic series on the islands: Hawaii—Puna, Ka'u, Laupahoehoe, and Hualalai (0–50,000 yr) and Kahuka, Hamakua, and Pololu (50,000–<1,500,000 yr); Maui—Hana and Lahaina (<80,000 yr), and Kula, Honolua, Wailuku, and Honomanu (>100,000–1,500,000 yr); Oahu—Honolulu (<1 m.y.), Koolau (1.5–2.5 m.y.), and Waianae (2.5–<3.75 m.y.); Kauai—Koloa (>500,000–<1,500,000 yr), and Waimea Canyon (>3.25–<5.75 m.y.). Geological information derived from AAPG Geological Highway Map of Hawaii (Bennison, 1974).

†V.S. = volcanic series.

APPENDIX 2. RAW DATA

Sample	Island	P	Lvo	Lvv brgl	Lvv orbg	Lvv blgl	Lvv algl	Lvv degl	Lvmi brgl	Lvmi orbg	Lvmi blgl	Lvmi algl
KAU 1	Kauai	0	0	0	1	0	0	0	0	0	0	0
KAU 2	Kauai	0	0	0	4	0	2	0	0	0	0	0
KAU 3	Kauai	2	0	0	2	0	0	0	0	0	1	0
KAU 4	Kauai	0	0	0	0	0	0	0	1	1	0	0
KAU 5	Kauai	0	1	0	0	0	0	0	0	0	0	0
KAU 6A	Kauai	8	19	0	0	2	0	0	0	9	11	2
KAU 6B	Kauai	4	4	0	1	3	0	0	0	6	6	0
KAU 7	Kauai	4	2	0	2	0	0	0	1	4	7	0
KAU 8	Kauai	0	0	0	0	0	1	0	0	0	0	0
KAU 9A	Kauai	1	4	0	2	0	0	0	0	0	2	0
KAU 9B	Kauai	1	1	0	0	0	0	0	0	2	2	0
KAU 10	Kauai	0	0	0	0	0	0	0	0	0	2	0
KAU 11	Kauai	23	38	3	14	6	7	0	1	5	8	4
KAU 12	Kauai	6	24	0	5	6	0	0	0	1	6	4
KAU 13	Kauai	0	1	0	0	2	0	0	0	0	1	0
OAH 1	Oahu	0	2	0	0	1	0	0	1	0	1	0
OAH 2	Oahu	0	0	0	29	9	0	0	0	3	0	0
OAH 3	Oahu	0	3	0	0	1	0	0	0	0	3	0
OAH 4	Oahu	0	1	0	1	0	0	0	2	0	1	0
OAH 5	Oahu	1	0	0	0	0	0	0	0	0	0	0
OAH 6	Oahu	0	0	0	0	0	0	0	0	0	1	0
OAH 7	Oahu	0	0	0	0	0	0	0	0	0	0	0
OAH 8	Oahu	0	1	0	0	0	0	0	0	0	1	0
OAH 9	Oahu	1	3	0	1	0	0	0	0	0	0	0
OAH 10	Oahu	1	0	0	0	0	1	0	0	0	0	0
OAH 11	Oahu	1	0	0	1	0	0	0	0	0	2	0
OAH 12	Oahu	2	6	0	0	1	0	0	1	1	0	0
OAH 13	Oahu	1	1	0	0	0	0	0	0	0	4	0
OAH 14	Oahu	0	0	0	0	0	0	0	0	0	0	0
OAH 15	Oahu	1	0	0	0	0	0	0	0	0	0	0
HAW 1	Hawaii	6	2	1	0	1	0	0	0	0	16	1
HAW 2	Hawaii	31	4	0	0	5	0	0	2	1	67	0
HAW 5	Hawaii	3	1	0	1	4	0	0	1	1	6	0
HAW 6	Hawaii	13	0	135	0	0	0	1	122	0	27	0
HAW 7	Hawaii	49	17	5	0	4	0	0	4	2	22	0
HAW 8	Hawaii	30	5	1	0	7	0	0	4	2	10	0
HAW 9	Hawaii	3	2	0	0	0	0	0	0	0	2	0
HAW 10	Hawaii	1	0	0	0	0	0	0	28	3	23	0
HAW 11	Hawaii	1	0	0	0	1	0	0	1	0	8	0
HAW 14	Hawaii	13	15	0	2	4	0	0	5	12	3	4
HAW 15	Hawaii	16	6	0	68	49	2	0	0	4	47	1
HAW 16	Hawaii	30	27	2	33	11	5	0	4	13	15	1
HAW 17	Maui	29	34	0	2	3	2	0	3	9	24	9
HAW 20	Maui	0	0	0	0	0	0	0	0	0	0	0
HAW 21	Maui	0	0	0	0	0	0	0	0	0	0	0
HAW 22	Maui	4	2	0	9	1	0	2	1	14	1	0
HAW 22	Maui	2	0	0	3	0	0	1	0	2	2	0
HAW 24	Maui	0	0	0	2	0	0	0	0	3	0	0
HAW 25	Maui	7	3	0	3	5	0	0	5	3	24	0
HAW 26	Maui	7	15	2	0	10	1	0	2	6	20	5
HAW 27	Maui	14	15	0	3	2	1	0	7	10	26	3
HAW 28	Maui	13	15	1	2	0	0	0	3	4	14	0
HAW 29	Maui	4	8	0	0	0	0	0	0	0	2	2
HAW 30	Maui	26	13	0	9	10	23	0	3	9	20	4
HAW 31	Maui	12	4	0	6	3	5	0	2	11	19	5
HAW 32	Maui	1	0	0	0	2	0	0	1	4	10	0
HAW 33	Maui	4	1	2	2	3	0	0	5	4	17	2
HAW 34	Maui	6	1	0	0	16	1	0	0	9	73	3
HAW 35	Maui	2	0	0	0	0	0	0	0	1	4	0
HAW 36	Maui	0	0	0	0	0	0	0	0	1	0	0
HAW 38	Hawaii	11	5	1	1	29	2	0	2	7	70	4
HAW 40	Hawaii	5	0	21	0	0	0	0	76	0	3	0

APPENDIX 2. RAW DATA (continued)

Sample	Island	Lvmi dev	Lvl brgl	Lvl orbg	Lvl blgl	Lvl algl	Lvl dev	D opaq	D py/ol	D other	Clay alt
KAU 1	Kauai	0	0	0	0	0	0	0	1	0	0
KAU 2	Kauai	0	1	1	0	0	0	2	2	0	0
KAU 3	Kauai	0	0	0	3	0	0	0	18	0	0
KAU 4	Kauai	0	0	1	0	0	0	1	7	0	0
KAU 5	Kauai	0	0	0	2	0	0	1	1	0	0
KAU 6A	Kauai	0	0	7	14	0	0	3	37	0	0
KAU 6B	Kauai	0	0	3	9	0	0	6	211	0	0
KAU 7	Kauai	0	0	6	4	0	0	3	163	0	0
KAU 8	Kauai	0	0	0	0	0	0	5	3	0	0
KAU 9A	Kauai	0	0	0	2	0	0	0	10	0	0
KAU 9B	Kauai	0	0	0	2	0	0	2	37	0	0
KAU 10	Kauai	0	0	0	1	1	0	0	4	0	0
KAU 11	Kauai	0	3	10	11	4	1	3	145	0	0
KAU 12	Kauai	0	0	6	14	2	0	2	183	0	0
KAU 13	Kauai	0	0	0	2	0	0	1	2	0	0
OAH 1	Oahu	0	0	0	1	0	0	0	2	0	0
OAH 2	Oahu	0	0	1	0	0	0	0	89	0	0
OAH 3	Oahu	0	0	0	1	0	0	2	4	0	0
OAH 4	Oahu	0	0	0	1	0	0	144	29	0	0
OAH 5	Oahu	0	0	0	0	0	0	1	3	0	0
OAH 6	Oahu	0	0	0	0	0	0	0	2	0	0
OAH 7	Oahu	0	0	0	0	0	0	0	0	0	0
OAH 8	Oahu	0	0	0	0	0	0	1	7	0	0
OAH 9	Oahu	0	0	0	0	0	0	5	12	0	0
OAH 10	Oahu	0	0	0	0	0	0	1	6	0	0
OAH 11	Oahu	0	0	0	3	0	0	0	3	0	0
OAH 12	Oahu	0	0	0	1	0	0	1	19	0	0
OAH 13	Oahu	0	0	0	1	0	0	3	14	0	0
OAH 14	Oahu	0	0	0	0	0	0	0	0	0	0
OAH 15	Oahu	0	0	0	1	0	0	0	3	0	0
HAW 1	Hawaii	0	1	1	11	1	0	1	17	0	0
HAW 2	Hawaii	0	3	3	91	0	0	0	145	0	0
HAW 5	Hawaii	0	1	0	11	0	0	0	34	0	0
HAW 6	Hawaii	1	112	0	22	0	1	0	64	0	0
HAW 7	Hawaii	0	25	0	52	0	0	0	49	0	0
HAW 8	Hawaii	0	7	1	42	0	0	1	30	0	0
HAW 9	Hawaii	0	0	0	1	0	0	0	3	0	0
HAW 10	Hawaii	0	9	0	8	0	0	0	47	0	0
HAW 11	Hawaii	0	0	0	2	0	0	2	3	0	0
HAW 14	Hawaii	0	1	8	0	1	0	3	9	1	33
HAW 15	Hawaii	2	0	6	55	0	0	4	40	0	0
HAW 16	Hawaii	0	3	8	18	5	0	6	64	1	4
HAW 17	Maui	0	4	15	34	2	0	9	131	13	1
HAW 20	Maui	0	0	2	0	0	0	0	1	0	0
HAW 21	Maui	0	0	0	0	0	0	0	0	0	0
HAW 22	Maui	0	0	1	1	1	0	11	54	0	0
HAW 22	Maui	0	0	0	1	0	0	0	9	0	0
HAW 24	Maui	0	0	2	0	0	0	0	4	0	0
HAW 25	Maui	0	0	5	10	0	0	2	20	0	0
HAW 26	Maui	0	3	8	14	0	0	3	54	2	0
HAW 27	Maui	0	3	2	14	0	0	2	156	15	0
HAW 28	Maui	0	3	7	26	1	0	1	88	1	0
HAW 29	Maui	0	1	4	0	0	0	5	79	5	2
HAW 30	Maui	0	0	4	6	2	0	8	83	0	0
HAW 31	Maui	0	0	3	9	1	0	10	45	0	0
HAW 32	Maui	0	0	2	3	0	0	1	14	0	0
HAW 33	Maui	0	3	1	7	0	0	0	7	0	0
HAW 34	Maui	0	0	3	11	0	0	0	41	0	0
HAW 35	Maui	0	0	0	5	0	0	0	2	0	0
HAW 36	Maui	0	0	0	0	0	0	0	2	0	0
HAW 38	Hawaii	1	1	0	32	0	0	7	250	0	0
HAW 40	Hawaii	0	52	0	9	0	0	0	9	0	0

APPENDIX 2. RAW DATA (continued)

Sample	Island	Glauc	Bio	Other	Total
KAU 1	Kauai	0	198	0	200
KAU 2	Kauai	1	185	2	200
KAU 3	Kauai	0	174	0	200
KAU 4	Kauai	0	189	0	200
KAU 5	Kauai	0	195	0	200
KAU 6A	Kauai	0	188	0	300
KAU 6B	Kauai	0	47	0	300
KAU 7	Kauai	0	104	0	300
KAU 8	Kauai	0	189	2	200
KAU 9A	Kauai	0	179	0	200
KAU 9B	Kauai	0	153	0	200
KAU 10	Kauai	1	191	0	200
KAU 11	Kauai	0	10	4	300
KAU 12	Kauai	0	40	1	300
KAU 13	Kauai	0	191	0	200
OAH 1	Oahu	0	192	0	200
OAH 2	Oahu	0	269	0	400
OAH 3	Oahu	0	286	0	300
OAH 4	Oahu	0	21	0	200
OAH 5	Oahu	0	194	1	200
OAH 6	Oahu	0	197	0	200
OAH 7	Oahu	0	199	1	200
OAH 8	Oahu	0	190	0	200
OAH 9	Oahu	0	178	0	200
OAH 10	Oahu	0	191	0	200
OAH 11	Oahu	0	190	0	200
OAH 12	Oahu	0	167	1	200
OAH 13	Oahu	0	176	0	200
OAH 14	Oahu	0	85	0	85
OAH 15	Oahu	0	195	0	200
HAW 1	Hawaii	0	140	1	200
HAW 2	Hawaii	0	48	0	400
HAW 5	Hawaii	0	137	0	200
HAW 6	Hawaii	0	1	1	500
HAW 7	Hawaii	0	271	0	500
HAW 8	Hawaii	0	108	2	250
HAW 9	Hawaii	0	189	0	200
HAW 10	Hawaii	0	131	0	250
HAW 11	Hawaii	0	182	0	200
HAW 14	Hawaii	0	85	1	200
HAW 15	Hawaii	0	0	0	300
HAW 16	Hawaii	0	0	0	250
HAW 17	Maui	0	174	2	500
HAW 20	Maui	0	197	0	200
HAW 21	Maui	0	194	6	200
HAW 22	Maui	0	146	2	250
HAW 22	Maui	0	180	0	200
HAW 24	Maui	0	189	0	200
HAW 25	Maui	0	113	0	200
HAW 26	Maui	0	347	1	500
HAW 27	Maui	0	26	1	300
HAW 28	Maui	0	219	2	400
HAW 29	Maui	0	285	3	400
HAW 30	Maui	0	277	3	500
HAW 31	Maui	3	362	0	500
HAW 32	Maui	0	162	0	200
HAW 33	Maui	0	142	0	200
HAW 34	Maui	0	336	0	500
HAW 35	Maui	0	186	0	200
HAW 36	Maui	0	197	0	200
HAW 38	Hawaii	0	77	0	500
HAW 40	Hawaii	0	0	0	175

APPENDIX 3. RECALCULATED PARAMETERS*

Sample	Island	Side	Bio (%)	Alt (%)	Sample	Island	Side	Bio (%)	Alt (%)
KAU 1	Kauai	S	99.0	100.0	HAW 17	Maui	N	34.8	28.2
KAU 2	Kauai	E	92.5	87.5	HAW 20	Maui	N	98.5	100.0
KAU 3	Kauai	E	87.0	33.3	HAW 21	Maui	N	97.0	0.0
KAU 4	Kauai	E	94.5	66.7	HAW 22	Maui	W	58.4	75.7
KAU 5	Kauai	E	97.5	0.0	HAW 22	Maui	W	90.0	55.6
KAU 6A	Kauai	N	62.7	28.1	HAW 24	Maui	W	94.5	100.0
KAU 6B	Kauai	N	15.7	31.3	HAW 25	Maui	W	56.5	19.0
KAU 7	Kauai	N	34.7	46.2	HAW 26	Maui	W	69.4	23.3
KAU 8	Kauai	E	94.5	100.0	HAW 27	Maui	SW	8.7	22.1
KAU 9A	Kauai	W	89.5	20.0	HAW 28	Maui	SW	54.8	18.4
KAU 9B	Kauai	W	76.5	28.6	HAW 29	Maui	SW	71.3	42.1
KAU 10	Kauai	S	95.5	25.0	HAW 30	Maui	W	55.4	49.6
KAU 11	Kauai	S	3.3	38.2	HAW 31	Maui	W	72.4	45.6
KAU 12	Kauai	S	13.3	26.4	HAW 32	Maui	W	81.0	27.3
KAU 13	Kauai	S	95.5	0.0	HAW 33	Maui	W	71.0	19.1
		Average	70.1	42.1	HAW 34	Maui	S	67.2	13.7
		Standard Deviation	33.9	31.1	HAW 35	Maui	W	93.0	10.0
					HAW 36	Maui	W	98.5	100.0
OAH 1	Oahu	S	96.0	0.0			Average	70.7	41.4
OAH 2	Oahu	S	67.3	78.6			Standard Deviation	23.3	31.7
OAH 3	Oahu	N	95.3	0.0					
OAH 4	Oahu	N	10.5	16.7	HAW 38	Hawaii	W	15.4	9.1
OAH 5	Oahu	N	97.0	0.0	HAW 40	Hawaii	SE	0.0	0.0
OAH 6	Oahu	N	98.5	0.0	HAW 1	Hawaii	W	70.0	8.6
OAH 7	Oahu	N	99.5	0.0	HAW 2	Hawaii	W	12.0	2.3
OAH 8	Oahu	NE	95.0	0.0	HAW 5	Hawaii	S	68.5	7.7
OAH 9	Oahu	NE	89.0	25.0	HAW 6	Hawaii	SE	0.2	0.0
OAH 10	Oahu	NE	95.5	100.0	HAW 7	Hawaii	W	54.2	1.5
OAH 11	Oahu	W	95.0	16.7	HAW 8	Hawaii	W	43.2	3.8
OAH 12	Oahu	W	83.5	10.0	HAW 9	Hawaii	W	94.5	0.0
OAH 13	Oahu	W	88.0	0.0	HAW 10	Hawaii	W	52.4	4.2
OAH 14	Oahu	W	100.0	0.0	HAW 11	Hawaii	W	91.0	0.0
OAH 15	Oahu	W	97.5	0.0	HAW 14	Hawaii	NE	42.5	68.2
		Average	87.2	16.5	HAW 15	Hawaii	NE	0.0	33.8
		Standard Deviation	22.0	29.9	HAW 16	Hawaii	NE	0.0	46.3
							Average	38.9	12.5
							Standard Deviation	33.2	19.8

*All samples.

APPENDIX 4. RECALCULATED PARAMETERS*

		CarbAltUnalt%					CarbAltUnalt%		
Sample	Island	Carb	Alt	Unalt	Sample	Island	Carb	Alt	Unalt
KAU 1	Kauai	99.5	0.5	0.0	HAW 1	Hawaii	80.0	1.7	18.3
KAU 2	Kauai	95.9	3.6	0.5	HAW 2	Hawaii	21.4	1.8	76.8
KAU 3	Kauai	96.7	1.1	2.2	HAW 5	Hawaii	84.0	1.2	14.7
KAU 4	Kauai	98.4	1.0	0.5	HAW 6	Hawaii	0.2	0.7	99.1
KAU 5	Kauai	98.5	0.0	1.5	HAW 7	Hawaii	67.4	0.5	32.1
KAU 6A	Kauai	74.6	7.1	18.3	HAW 8	Hawaii	57.8	1.6	40.6
KAU 6B	Kauai	59.5	12.7	27.8	HAW 9	Hawaii	97.4	0.0	2.6
KAU 7	Kauai	80.0	9.2	10.8	HAW 10	Hawaii	64.9	1.5	33.7
KAU 8	Kauai	99.5	0.5	0.0	HAW 11	Hawaii	93.8	0.0	6.2
KAU 9A	Kauai	94.7	1.1	4.2	HAW 14	Hawaii	49.1	34.7	16.2
KAU 9B	Kauai	95.6	1.3	3.1	HAW 15	Hawaii	0.0	34.6	65.4
KAU 10	Kauai	97.9	0.5	1.5	HAW 16	Hawaii	0.0	46.3	53.7
KAU 11	Kauai	8.0	36.0	56.0	HAW 17	Maui	55.1	12.7	32.3
KAU 12	Kauai	37.0	16.7	46.3	HAW 20	Maui	99.0	1.0	0.0
KAU 13	Kauai	97.0	0.0	3.0	HAW 21	Maui	100.0	0.0	0.0
OAH 1	Oahu	97.0	0.0	3.0	HAW 22	Maui	81.6	15.1	3.4
OAH 2	Oahu	86.5	10.6	2.9	HAW 22	Maui	95.2	3.2	1.6
OAH 3	Oahu	97.3	0.0	2.7	HAW 24	Maui	96.4	3.6	0.0
OAH 4	Oahu	77.8	3.7	18.5	HAW 25	Maui	66.1	6.4	27.5
OAH 5	Oahu	100.0	0.0	0.0	HAW 26	Maui	80.1	4.6	15.2
OAH 6	Oahu	99.5	0.0	0.5	HAW 27	Maui	23.2	17.0	59.8
OAH 7	Oahu	100.0	0.0	0.0	HAW 28	Maui	74.2	4.7	21.0
OAH 8	Oahu	99.0	0.0	1.0	HAW 29	Maui	93.8	2.6	3.6
OAH 9	Oahu	97.8	0.5	1.6	HAW 30	Maui	72.9	13.4	13.7
OAH 10	Oahu	99.5	0.5	0.0	HAW 31	Maui	84.2	7.2	8.6
Oah 11	Oahu	96.9	0.5	2.6	HAW 32	Maui	88.0	3.3	8.7
OAH 12	Oahu	94.4	0.6	5.1	HAW 33	Maui	75.1	4.8	20.1
OAH 13	Oahu	96.7	0.0	3.3	HAW 34	Maui	74.2	3.5	22.3
OAH 14	Oahu	100.0	0.0	0.0	HAW 35	Maui	94.9	0.5	4.6
OAH 15	Oahu	99.5	0.0	0.5	HAW 36	Maui	99.5	0.5	0.0
					HAW 38	Hawaii	33.2	6.5	60.3
					HAW 40	Hawaii	0.0	0.0	100.0

*All samples.

APPENDIX 5. RECALCULATED PARAMETERS

Sample	Island	QFL%			LvfLvmiLvl%			Volc%			Volc%		
		Q	F	L	Lvf	Lvmi	Lvl	clgl	brgl	blgl	alt	brown	black
KAU 6A	Kauai	0	11.1	88.9	0	51.2	48.8	0	0.0	100.0	40.0	0.0	60.0
KAU 6B	Kauai	0	11.1	88.9	0	50.0	50.0	0	0.0	100.0	35.7	0.0	64.3
KAU 7	Kauai	0	13.3	86.7	0	54.5	45.5	0	8.3	91.7	50.0	4.2	45.8
KAU 11	Kauai	0	16.7	83.3	0	38.3	61.7	0	21.9	78.1	58.4	9.1	32.5
KAU 12	Kauai	0	8.1	91.9	0	33.3	66.7	0	0.0	100.0	40.9	0.0	59.1
OAH 2	Oahu	0	0.0	100.0	0	75.0	25.0	0	0.0	100.0	78.6	0.0	21.4
OAH 4	Oahu	0	0.0	100.0	0	75.0	25.0	0	50.0	50.0	20.0	40.0	40.0
HAW 1	Hawaii	0	14.6	85.4	0	54.8	45.2	0	6.7	93.3	9.1	6.1	84.8
HAW 2	Hawaii	0	15.0	85.0	0	41.9	58.1	0	3.0	97.0	2.3	2.9	94.8
HAW 5	Hawaii	0	10.3	89.7	0	40.0	60.0	0	8.7	91.3	8.0	8.0	84.0
HAW 6	Hawaii	0	3.0	97.0	0	52.6	47.4	0	88.3	11.7	0.7	87.6	11.6
HAW 7	Hawaii	0	27.2	72.8	0	26.7	73.3	0	30.4	69.6	1.8	29.8	68.4
HAW 8	Hawaii	0	27.5	72.5	0	24.2	75.8	0	16.9	83.1	4.1	16.2	79.7
HAW 10	Hawaii	0	1.4	98.6	0	76.1	23.9	0	54.4	45.6	4.2	52.1	43.7
HAW 14	Hawaii	0	12.9	87.1	0	70.6	29.4	0	46.2	53.8	67.5	15.0	17.5
HAW 15	Hawaii	0	6.3	93.8	0	47.0	53.0	0	0.0	100.0	35.5	0.0	64.5
HAW 16	Hawaii	0	16.8	83.2	0	49.3	50.7	0	17.0	83.0	55.1	7.6	37.3
HAW 17	Maui	0	17.0	83.0	0	45.0	55.0	0	10.3	89.7	36.4	6.5	57.0
HAW 22	Maui	0	10.8	89.2	0	84.2	15.8	0	25.0	75.0	87.1	3.2	9.7
HAW 25	Maui	0	10.8	89.2	0	68.1	31.9	0	11.4	88.6	20.0	9.1	70.9
HAW 26	Maui	0	7.5	92.5	0	56.9	43.1	0	13.7	86.3	28.2	9.9	62.0
HAW 27	Maui	0	14.0	86.0	0	70.8	29.2	0	19.2	80.8	26.8	14.1	59.2
HAW 28	Maui	0	14.6	85.4	0	36.2	63.8	0	14.9	85.1	23.0	11.5	65.6
HAW 29	Maui	0	17.4	82.6	0	44.4	55.6	0	33.3	66.7	66.7	11.1	22.2
HAW 30	Maui	0	20.2	79.8	0	75.0	25.0	0	7.7	92.3	56.7	3.3	40.0
HAW 31	Maui	0	15.0	85.0	0	74.0	26.0	0	6.1	93.9	48.4	3.1	48.4
HAW 33	Maui	0	7.8	92.2	0	71.8	28.2	0	27.0	73.0	19.6	21.7	58.7
HAW 34	Maui	0	4.9	95.1	0	85.9	14.1	0	0.0	100.0	13.8	0.0	86.2
HAW 38	Hawaii	0	6.6	93.4	0	71.8	28.2	0	3.0	97.0	10.0	2.7	87.3
HAW 40	Hawaii	0	3.0	97.0	0	56.4	43.6	0	92.5	7.5	0.0	92.5	7.5

APPENDIX 5. RECALCULATED PARAMETERS (continued)

Sample	Island	LvAccPlag%			Black LvvLvmiLvl%			Brown LvvLvmiLvl%		
		Lv	Acc	Plag	Lvv	Lvmi	Lvl	Lvv	Lvmi	Lvl
KAU 6A	Kauai	57.1	35.7	7.1	7.4	40.7	51.9			
KAU 6B	Kauai	12.6	85.8	1.6	16.7	33.3	50.0			
KAU 7	Kauai	13.3	84.7	2.0	0.0	63.6	36.4	0.0	100.0	0.0
KAU 11	Kauai	40.2	51.7	8.0	24.0	32.0	44.0	42.9	14.3	42.9
KAU 12	Kauai	26.3	71.4	2.3	23.1	23.1	53.8			
OAH 2	Oahu	32.1	67.9	0.0	100.0	0.0	0.0			
OAH 4	Oahu	3.4	96.6	0.0	0.0	50.0	50.0	0.0	100.0	0.0
HAW 1	Hawaii	59.3	30.5	10.2	3.6	57.1	39.3	50.0	0.0	50.0
HAW 2	Hawaii	50.0	41.2	8.8	3.1	41.1	55.8	0.0	40.0	60.0
HAW 5	Hawaii	41.3	54.0	4.8	19.0	28.6	52.4	0.0	50.0	50.0
HAW 6	Hawaii	84.5	12.9	2.6	0.0	55.1	44.9	36.6	33.1	30.4
HAW 7	Hawaii	57.2	21.4	21.4	5.1	28.2	66.7	14.7	11.8	73.5
HAW 8	Hawaii	56.4	22.1	21.4	11.9	16.9	71.2	8.3	33.3	58.3
HAW 10	Hawaii	59.7	39.5	0.8	0.0	74.2	25.8	0.0	75.7	24.3
HAW 14	Hawaii	77.2	11.4	11.4	57.1	42.9	0.0	0.0	83.3	16.7
HAW 15	Hawaii	80.0	14.7	5.3	32.5	31.1	36.4			
HAW 16	Hawaii	59.6	28.4	12.0	25.0	34.1	40.9	22.2	44.4	33.3
HAW 17	Maui	43.8	47.2	9.0	4.9	39.3	55.7	0.0	42.9	57.1
HAW 22	Maui	32.4	63.7	3.9	33.3	33.3	33.3	0.0	100.0	0.0
HAW 25	Maui	66.7	25.3	8.0	12.8	61.5	25.6	0.0	100.0	0.0
HAW 26	Maui	56.6	38.8	4.6	22.7	45.5	31.8	28.6	28.6	42.9
HAW 27	Maui	31.5	63.4	5.1	4.8	61.9	33.3	0.0	70.0	30.0
HAW 28	Maui	42.5	50.3	7.3	0.0	35.0	65.0	14.3	42.9	42.9
HAW 29	Maui	17.0	79.5	3.6	0.0	100.0	0.0	0.0	0.0	100.0
HAW 30	Maui	46.8	41.4	11.8	27.8	55.6	16.7	0.0	100.0	0.0
HAW 31	Maui	50.4	40.7	8.9	9.7	61.3	29.0	0.0	100.0	0.0
HAW 33	Maui	81.0	12.1	6.9	11.1	63.0	25.9	20.0	50.0	30.0
HAW 34	Maui	71.3	25.0	3.7	16.0	73.0	11.0			
HAW 38	Hawaii	36.6	60.8	2.6	22.1	53.4	24.4	25.0	50.0	25.0
HAW 40	Hawaii	92.0	5.1	2.9	0.0	25.0	75.0	14.1	51.0	34.9

APPENDIX 5. RECALCULATED PARAMETERS (continued)

Sample	Island	VitricCrysPlag%			LvoLvmiLvl%		
		Vitric	Crys	Plag	Lvo	Lvmi	Lvl
KAU 6A	Kauai	2.8	86.1	11.1	30.6	35.5	33.9
KAU 6B	Kauai	11.1	77.8	11.1	14.3	42.9	42.9
KAU 7	Kauai	6.7	80.0	13.3	8.3	50.0	41.7
KAU 11	Kauai	21.7	61.6	16.7	44.7	21.2	34.1
KAU 12	Kauai	14.9	77.0	8.1	42.1	19.3	38.6
OAH 2	Oahu	90.5	9.5	0.0	0.0	75.0	25.0
OAH 4	Oahu	16.7	83.3	0.0	20.0	60.0	20.0
HAW 1	Hawaii	4.9	80.5	14.6	6.1	51.5	42.4
HAW 2	Hawaii	2.4	82.6	15.0	2.3	40.9	56.7
HAW 5	Hawaii	17.2	72.4	10.3	4.8	38.1	57.1
HAW 6	Hawaii	31.3	65.7	3.0	0.0	52.6	47.4
HAW 7	Hawaii	5.0	67.8	27.2	13.9	23.0	63.1
HAW 8	Hawaii	7.3	65.1	27.5	7.0	22.5	70.4
HAW 10	Hawaii	0.0	98.6	1.4	0.0	76.1	23.9
HAW 14	Hawaii	8.8	72.1	19.1	30.6	49.0	20.4
HAW 15	Hawaii	46.5	47.3	6.3	5.0	44.6	50.4
HAW 16	Hawaii	29.1	53.7	17.1	28.7	35.1	36.2
HAW 17	Maui	4.1	78.8	17.1	25.4	33.6	41.0
HAW 22	Maui	32.4	56.8	10.8	9.5	76.2	14.3
HAW 25	Maui	12.3	76.9	10.8	6.0	64.0	30.0
HAW 26	Maui	14.0	78.5	7.5	20.5	45.2	34.2
HAW 27	Maui	6.0	80.0	14.0	18.8	57.5	23.8
HAW 28	Maui	3.4	82.0	14.6	20.5	28.8	50.7
HAW 29	Maui	0.0	81.0	19.0	47.1	23.5	29.4
HAW 30	Maui	32.6	47.3	20.2	21.3	59.0	19.7
HAW 31	Maui	17.5	67.5	15.0	7.4	68.5	24.1
HAW 33	Maui	13.7	78.4	7.8	2.5	70.0	27.5
HAW 34	Maui	13.8	81.3	4.9	1.0	85.0	14.0
HAW 38	Hawaii	19.9	73.5	6.6	4.1	68.9	27.0
HAW 40	Hawaii	12.7	84.3	3.0	0.0	56.4	43.6

REFERENCES CITED

Andrews, J. E., and Bainbridge, C., 1972, Submarine canyons off eastern Oahu: Pacific Science, v. 26, p. 108–113.

Bennison, A. P., 1974, Geological highway map of the state of Hawaii and the state of Alaska: American Association of Petroleum Geologists, United States Geological Highway Map Series 8, scale 1:500,000.

Campbell, J. F., and Hwang, D. J., 1982, Beach erosion at Waimea Bay, Oahu, Hawaii: Pacific Science, v. 36, p. 35–43.

Cas, R. A., and Wright, J. V., 1987, Volcanic successions: Modern and ancient: A geological approach to processes, products and successions: London, Allen & Unwin, 528 p.

Chamberlain, T., 1968, The littoral sand budget, Hawaiian Islands: Pacific Science, v. 22, p. 161–183.

Clague, D. A., and Dalrymple, G. B., 1987, The Hawaiian-Emperor volcanic chain, Part I, Geologic evolution, *in* Decker, R. W., Wright, T. L., and Stauffer, P. H., eds., Volcanism in Hawaii: U.S. Geological Survey Professional Paper 1350, p. 5–54.

Clague, D. A., and Hazlett, R. W., 1989, Geological field guide to the Hawaiian Islands: 28th International Geological Congress Field Trip Guidebook T188–304, 33 p.

Coulbourn, W. T., and Resig, J. M., 1975, On the use of benthic foraminifera as sediment tracers in a Hawaiian bay: Pacific Science, v. 29, p. 99–115.

Coulbourn, W. T., and 8 others, 1988, Sand deposits offshore Oahu, Hawaii: Pacific Science, v. 42, p. 267–299.

Dickinson, W. R., 1970, Interpreting detrital modes of graywacke and arkose: Journal of Sedimentary Petrology, v. 40, p. 695–707.

Dickinson, W. R., and Suczek, C. A., 1979, Plate tectonics and sandstone compositions: American Association of Petroleum Geologists Bulletin, v. 63, p. 2164–2182.

Dickinson, W. R., and 8 others, 1983, Provenance of North American Phanerozoic sandstones in relation to tectonic setting: Geological Society of America Bulletin, v. 94, p. 222–235.

Emery, K. O., and Cox, D. C., 1956, Beachrock in the Hawaiian Islands: Pacific Science, v. 10, p. 382–402.

Fisher, R. V., and Schmincke, H.-U., 1984, Pyroclastic rocks: New York, Springer-Verlag, 472 p.

Fornari, D. J., and Campbell, J. F., 1987, Submarine topography around the Hawaiian Islands, *in* Decker, R. W., Wright, T. L., and Stauffer, P. H., eds., Volcanism in Hawaii: U.S. Geological Survey Professional Paper 1350, p. 109–124.

Glasby, G. P., Rankin, P. C., and Meylan, M. A., 1979, Manganiferous soil concretions from Hawaii: Pacific Science, v. 33, p. 103–115.

Hyndman, D. W., 1972, Petrology of igneous and metamorphic rocks: New York, McGraw-Hill, 533 p.

Ingersoll, R. V., 1984, Felsic sand/sandstone within mafic volcanic provinces: Rift sand/sandstone shows no basaltic provenance: Geological Society of America Abstracts with Programs, v. 16, p. 548.

Ingersoll, R. V., Bullard, T. F., Ford, R. L., Grimm, J. P., Pickle, J. D., and Sares, S. W., 1984, The effect of grain size on detrital modes: A test of the Gazzi-Dickinson point-counting method: Journal of Sedimentary Petrology, v. 54, p. 103–116.

Inman, D. L., Gayman, W. R., and Cox, D. C., 1963, Littoral sedimentary processes on Kauai, a subtropical high island: Pacific Science, v. 17, p. 106–130.

Johnsson, M. J., 1990, Overlooked sedimentary particles from tropical weathering environments: Geology, v. 18, p. 107–110.

Larue, D. K., and Sampayo, M. M., 1990, Lithic-volcanic sandstones derived from oceanic crust in the Franciscan Complex of California: 'Sedimental memories' of source rock geochemistry: Sedimentology, v. 37, p. 879–889.

Li, Y.-H., 1988, Denudation rates of the Hawaiian Islands by rivers and groundwaters: Pacific Science, v. 42, p. 253–266.

Lundberg, N., 1991, Detrital record of the early Central American magmatic arc: Petrography of intraoceanic forearc sandstones, Nicoya Peninsula, Costa Rica: Bulletin of the Geological Society of America, v. 103, p. 905–915.

Macdonald, G. A., 1972, Volcanoes: Englewood Cliffs, New Jersey, Prentice-Hall: 492 p.

Macdonald, G. A., and Abbott, A. T., 1970, Volcanoes in the sea: Honolulu, University of Hawaii Press, 441 p.

Mack, G. H., and Jerzykiewicz, T., 1989, Detrital modes of sand and sandstone derived from andesitic rocks as a paleoclimatic indicator: Sedimentary Geology, v. 65, p. 35–44.

Marsaglia, K. M., 1989, Petrography, provenance, and diagenesis of arc-related sands and sandstones [Ph.D. thesis]: Los Angeles, University of California, 394 p.

Marsaglia, K. M., 1991, Provenance of sands and sandstones from the Gulf of California, a rifted continental arc: *in* Fisher, R. V., and Smith, G. A., eds., Sedimentation in volcanic settings, SEPM Special Publication 45, p. 237–248.

Marsaglia, K. M., 1992, Provenance and petrography of volcaniclastic sands recovered on ODP Leg 126, the Izu-Bonin arc: Proceedings of the Ocean Drilling Program, Scientific Results, v. 126, p. 139–154.

Marsaglia, K. M., and Ingersoll, R. V., 1992, A reassessment of magmatic-arc provenance: compositional trends in arc-related, deep-marine sand and sandstone: Geological Society of America Bulletin, v. 104, p. 1637–1649.

Marsaglia, K. M., and Tazaki, K., 1992, Diagenetic trends in ODP Leg 126 sandstones: Proceedings of the Ocean Drilling Program, Scientific Results, v. 126, p. 125–138.

Mathewson, C. C., 1970, Submarine canyons and the shelf along the north coast of Molokai Island, Hawaiian Ridge: Pacific Science, v. 24, p. 235–244.

McKenzie, J., Bernoulli, D., and Schlanger, S. O., 1980, Shallow-water carbonate sediments from the Emperor Seamounts: Their diagenesis and paleogeographic significance: Initial Reports of the Deep Sea Drilling Project, v. 55, p. 415–455.

McPhie, J., Walker, G. P., and Christiansen, R. L., 1990, Phreatomagmatic and phreatic fall and surge deposits from explosions at Kilauea volcano, Hawaii, 1970 A.D.: Keanakakoi Ash Member: Bulletin of Volcanology, v. 52, p. 334–354.

Moberly, R., Jr., 1968, Loss of Hawaiian littoral sand: Journal of Sedimentary Petrology, v. 38, p. 17–34.

Moberly, R., Jr., and Chamberlain, T., 1964, Hawaiian beach systems: Hawaii Institute of Geophysics Report HIG-64-2, 177 p.

Moberly, R., Jr., Baver, L. D., and Morrison, A., 1965, Source and variation of Hawaiian littoral sand: Journal of Sedimentary Petrology, v. 35, p. 589–598.

Moore, J. G., Clague, D. A., and Normark, W. R., 1982, Diverse basalt types from Loihi seamount, Hawaii: Geology, v. 10, p. 88–92.

Moore, J. G., Clague, D. A., Holcomb, R. T., Lipman, P. W., Normark, W. R., and Torresan, M. E., 1989, Prodigious submarine landslides on the Hawaiian Ridge: Journal of Geophysical Research, v. 94, p. 17465–17484.

Morgan, J., 1983, Hawaii, a geography: Boulder, CO, Westview Press, 293 p.

Sherman, G. D., and Ikawa, H., 1968, Soil sequences in the Hawaiian Islands: Pacific Science, v. 22, p. 458–464.

Sherman, G. D., Ikawa, H., and Matsusaka, Y., 1969, Aluminous-ferruginous oxide mineral nodules in tropical soils: Pacific Science, v. 23, p. 115–122.

Stearns, H. T., 1984, Geology of the State of Hawaii (second edition): Palo Alto, Pacific Books, 335 p.

Walker, J. L., Sherman, G. D., and Katsura, T., 1969, The iron and titanium minerals in the titaniferous ferruginous latosols of Hawaii: Pacific Science, v. 23, p. 291–304.

Wirthmann, A., 1986, Verwitterung und chemischer abtrag von basalt in den tropen: Geoökodynamik, v. 7, p. 113–128.

Manuscript Accepted by the Society December 24, 1992

Printed in U.S.A.

Geological Society of America
Special Paper 284
1993

Petrology of Mid-Cenozoic strike-slip basins in an accretionary orogen, St. Elias Mountains, Yukon Territory, Canada

Kenneth D. Ridgway* and Peter G. DeCelles*
Department of Geological Sciences, University of Rochester, Rochester, New York 14627

ABSTRACT

Eocene-Oligocene strike-slip faulting along the Denali fault system in Yukon Territory produced several local strike-slip basins that were filled by coarse-grained alluvial, fluvial, and lacustrine sediment referred to as the Amphitheatre Formation. The Amphitheatre Formation provides an excellent opportunity to study in detail the composition of siliciclastic sediment in a major, accretionary strike-slip orogen.

Petrographic analyses of sandstones and clast counts of conglomerates reveal a lithologically diverse provenance. On standard QFL and QmFLt diagrams, sandstones from the Burwash and Bates Lake basins are arkosic, overlapping the uplifted basement and dissected magmatic arc provenance fields of Dickinson and Suczek (1979). Conglomerates were derived from volcanic, plutonic, and medium- to low-grade metasedimentary rocks that currently are found in several accreted terranes associated with the Denali fault system. Wrangellia and the Alexander terrane provided voluminous volcanic, metavolcanic (greenstone), sedimentary, and low-grade metasedimentary material. The Yukon Crystalline terrane provided medium-grade (gneissic and schistose) material and plutonic material, and the Gravina-Nutzotin belt provided pelitic and metasedimentary material. These data support the contention that the Amphitheatre Formation was derived mainly from local, high-relief sources associated with major uplift in the eastern St. Elias Mountains, as well as the southern Yukon Crystalline terrane.

The Eocene-Oligocene climatic event apparently had no significant effect on Amphitheatre sandstone composition. This is attributed to the proximal character of the sandstones studied; steep local relief kept soil residence times to a minimum in spite of climatic change. In addition, the persistence of coal- and stream-dominated alluvial fan facies throughout the entire stratigraphic section indicates that climate remained humid in southwestern Yukon Territory, possibly owing to local orographic effects. It is proposed that, in general, climate should not have a major impact on detrital sand composition in large strike-slip orogens, where sediment accumulates within the orogen itself and is subject neither to long distances of transport nor to long-term residence in lowland soil profiles where most compositional modification is expected.

INTRODUCTION

Some of the world's largest mountain belts involve major transform fault systems (Sylvester, 1988; Sengor, 1990). A significant proportion of the vast amount of sediment generated in such settings ultimately may be trapped in local basins that develop along strike-slip faults within or marginal to the orogen. A key component of any basin analysis is a careful petrographic characterization of the basin fill, and numerous studies have demonstrated a close correspondence between various tectonic settings and detrital sand compositions (e.g., Ingersoll, 1978; Dickinson

*Present addresses: Ridgway, Department of Earth and Atmospheric Sciences, Purdue University, West Lafayette, Indiana 47907-1397; DeCelles, Department of Geosciences, University of Arizona, Tucson, Arizona 85721.

Ridgway, K. D., and DeCelles, P. G., 1993, Petrology of Mid-Cenozoic strike-slip basins in an accretionary orogen, St. Elias Mountains, Yukon Territory, Canada, *in* Johnsson, M. J., and Basu, A., eds., Processes Controlling the Composition of Clastic Sediments: Boulder, Colorado, Geological Society of America Special Paper 284.

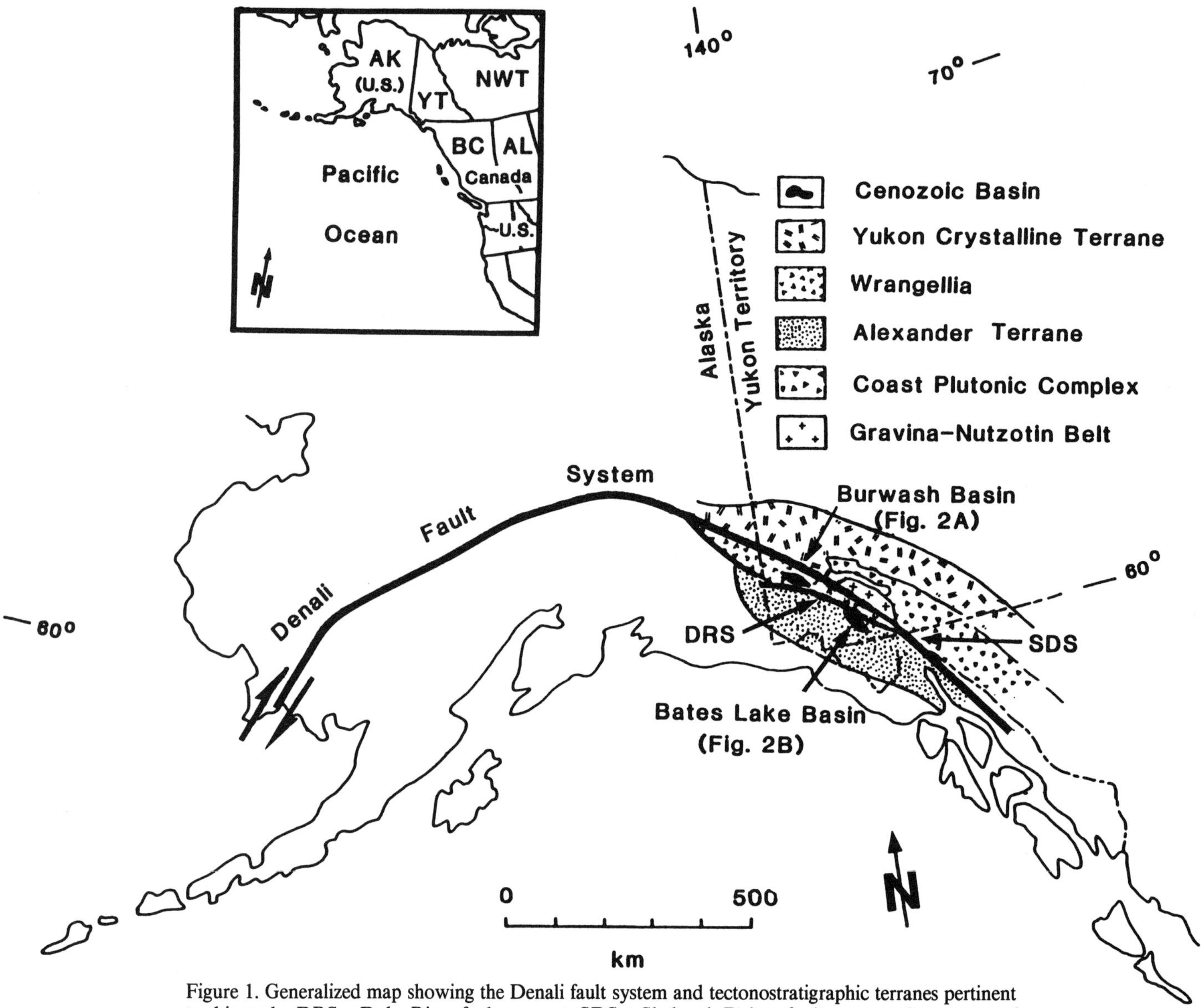

Figure 1. Generalized map showing the Denali fault system and tectonostratigraphic terranes pertinent to this study: DRS = Duke River fault segment; SDS = Shakwak-Dalton fault segment. Insert map of western Canada: AL = Alberta; BC = British Columbia; NWT = Northwest Territory; YT = Yukon Territory; AK = Alaska.

and Suczek, 1979; Dickinson et al., 1983; Valloni, 1985; Schwab, 1986; DeCelles and Hertel, 1989). Conspicuously underrepresented in the literature on tectonics and sandstone petrology are detailed petrologic studies of strike-slip basin fills. A notable exception is the Ridge basin, a well-known strike-slip basin in southern California (e.g., Link, 1982). Valloni (1985) reported data from oceanic settings in oblique (transform) accretionary margins, but aside from the Ridge basin studies, few data are available from nonmarine strike-slip basin fills.

In this study we present sandstone petrologic data from two Cenozoic nonmarine, strike-slip basins along the Denali fault in southwestern Yukon Territory, Canada. The Denali fault is a major continental transform fault system, with ~300 km of post–Late Cretaceous dextral offset (Eisbacher, 1976; Lanphere, 1978; Hickman et al., 1978). Subsidiary faults within the Denali fault system juxtaposed four distinct tectonostratigraphic terranes—Wrangellia, the Alexander terrane, the Yukon Crystalline terrane, and the Gravina-Nutzotin belt—during Late Mesozoic and early Cenozoic time (Fig. 1) (Read and Monger, 1976; Jones et al., 1977; Monger, 1984; Campbell and Dodds, 1982a,b). Coarse conglomerate and sandstone of the Amphitheatre Formation, derived in variable proportions from these four terranes, accumulated in several nonmarine strike-slip basins in Yukon Territory and British Columbia (Ridgway et al., 1992a; Ridgway and DeCelles, 1991a,b). Our data demonstrate the complex, hybrid character of sandstones and conglomerates derived from accretionary strike-slip orogens. Moreover, because these rocks were deposited during the well-known Eocene-Oligocene climatic deterioration (Ridgway et al., 1989, 1992a), we can assess the relative importance of climate and source-rock composition in controlling sand composition in strike-slip basins.

GEOLOGIC HISTORY AND PREVIOUS WORK

The sedimentary fills of all the mid-Cenozoic strike-slip basins along the Denali fault, Yukon Territory, have been collectively called the Amphitheatre Formation. The Amphitheatre Formation consists of 1,300 m of conglomerate, sandstone, shale, and coal (Eisbacher and Hopkins, 1977; Ridgway and DeCelles, 1991a). These rocks were deposited by stream-dominated alluvial fans, braided rivers, and lacustrine fan deltas (Ridgway and DeCelles, 1993). Palynological data indicate that the Amphitheatre Formation is mainly latest Eocene-Oligocene in age (Ridgway et al., 1992a,b); but the upper part intertongues with the Miocene Wrangell Lavas, which are at least 18 Ma in the northern part of the study area (Skulski, 1988; Skulski et al., 1992). Data presented herein are from two separate basins, the Burwash and Bates Lake basins (Figs. 1–3). Paleocurrent, isopach, and textural data indicate that the Burwash basin was filled mainly by a large, northwestward-prograding stream-dominated alluvial fan and by subsidiary sources along the northwestern flank of the basin (Fig. 2A). The Bates Lake basin was filled mostly by stream-dominated alluvial fans that prograded west-southwestward from the fault-bounded eastern basin margin (Fig. 2B). The sedimentology of the Amphitheatre Formation is addressed in separate papers (Ridgway et al., 1992a; Ridgway and DeCelles, 1993).

The Amphitheatre Formation is considered an overlap assemblage because it was deposited across several juxtaposed accreted terranes in the eastern Saint Elias Mountains (Fig. 1) (Campbell and Dodds, 1982a,b). Syndepositional faults, fault-wedge conglomerates, and the textural and compositional immaturity of the Amphitheatre Formation indicate that it contains the record of mid-Cenozoic displacement along the Denali fault system (Ridgway and DeCelles, 1991a,b; Ridgway, 1992).

The northeastern Saint Elias Mountains are located southwest of the Shakwak-Dalton segment of the Denali fault system and consist of two northwest-trending tectonostratigraphic accreted terranes (Figs. 1, 3) (Campbell and Dodds, 1982a,b). The terranes, the Alexander and the Wrangellia, are separated by the Duke River segment of the Denali fault system (Figs. 1, 3). Northeast of the Shakwak-Dalton segment, the Yukon Crystalline terrane, also referred to as the Yukon-Tanana terrane, merges to the southeast with the Coast Plutonic Complex (Figs. 1, 3) (Tempelman-Kluit, 1976; Mortensen and Jilson, 1985). In the southern part of the study area, the Gravina-Nutzotin belt abuts the Shakwak-Dalton segment (Figs. 1, 3). Wrangellia in the study area is about 30 km wide and underlies the Kluane Range of the Saint Elias Mountains (Read and Monger, 1976; Jones et al., 1977; Campbell and Dodds, 1982a,b). It is composed of probable upper Paleozoic (Carboniferous and (?)Lower Permian) submarine volcanic and pyroclastic rocks and minor sedimentary rocks, Middle Triassic pelite, Middle to Upper Triassic basalt (Nikolai Greenstone), and Upper Triassic carbonate, pelite, and lithic sandstone. Locally, Wrangellia is metamorphosed and intruded by Jurassic-Cretaceous and Tertiary(?) granitic rocks (Fig. 3) (Read and Monger, 1976; Campbell and Dodds, 1982a,b).

The Alexander terrane underlies much of the Icefield Ranges of the Saint Elias Mountains (Campbell and Dodds, 1982a,b; Gehrels and Saleeby, 1987). In southwestern Yukon Territory, this terrane consists of Cambrian or Lower Ordovician volcanic and sedimentary rocks, and Silurian(?), Devonian and Carboniferous(?) carbonate, pelitic, and volcanic rocks. These strata are locally complexly deformed, metamorphosed to greenschist facies, and intruded by granitic rocks of Permian and younger age (Fig. 3) (Read and Monger, 1976; Campbell and Dodds, 1982a,b). Sandstones of the Alexander terrane are lithic-rich (Campbell and Dodds, 1982a,b).

The Gravina-Nutzotin belt is located on the northeastern side of the Shakwak-Dalton fault in the southern part of the study area (Figs. 1, 3). It is composed of the Dezadeash Formation, an interbedded sequence of Oxfordian-Valanginian turbiditic sandstone and shale (Fig. 3) (Eisbacher, 1976). The Yukon Crystalline terrane is located on the northeast side of the Shakwak-Dalton fault in the northern part of the study area (Tempelman-Kluit, 1976). Much of the terrane consists of Paleozoic, high-grade metamorphic rocks that were intruded mainly by late Mesozoic and early Cenozoic granitic bodies such as the Ruby Range batholith (Tempelman-Kluit, 1979; Mortensen and Jilson, 1985). The Ruby Range batholith has a granodioritic composition and has been interpreted as the northern extension of the Coast Plutonic Complex (Fig. 3) (LeCouteur and Tempelman-Kluit, 1976). The Kluane schist is an extensive body of quartz-biotite-muscovite schist in fault contact with the Dezadeash Formation. Metamorphic ages from the Kluane schist cluster between 50 and 70 Ma (Fig. 3).

The terranes forming the northeastern Saint Elias Mountains are believed to have accreted to North America by mid-Cretaceous time (Jones et al., 1977; Coney et al., 1980; Monger et al., 1982; Monger, 1984). Subsequent northward migration from the Late Cretaceous through Paleogene occurred along major strike-slip faults such as those of the Denali fault system (Hickman et al., 1978; Lanphere, 1978; Nokleberg et al., 1985). Eisbacher (1976) documented a minimum right-lateral displacement of 300 km on the Shakwak-Dalton fault during Late Cretaceous to Eocene time in the Saint Elias Mountains.

The Amphitheatre Formation is preserved on the Alexander terrane (i.e., Bates Lake basin) in the southern part of the study area and on Wrangellia (i.e., Burwash basin) in the northern part of the study area (Figs. 1, 3). A key question in determining possible source rock control on Amphitheatre sandstone composition is the distance of tectonic transport of the Cenozoic strike-slip basins along the Duke River fault. Some constraints are provided by cross-cutting relationships between the Duke River fault, the Amphitheatre Formation, and the overlying Miocene Wrangell Lavas. The transition between the Amphitheatre For-

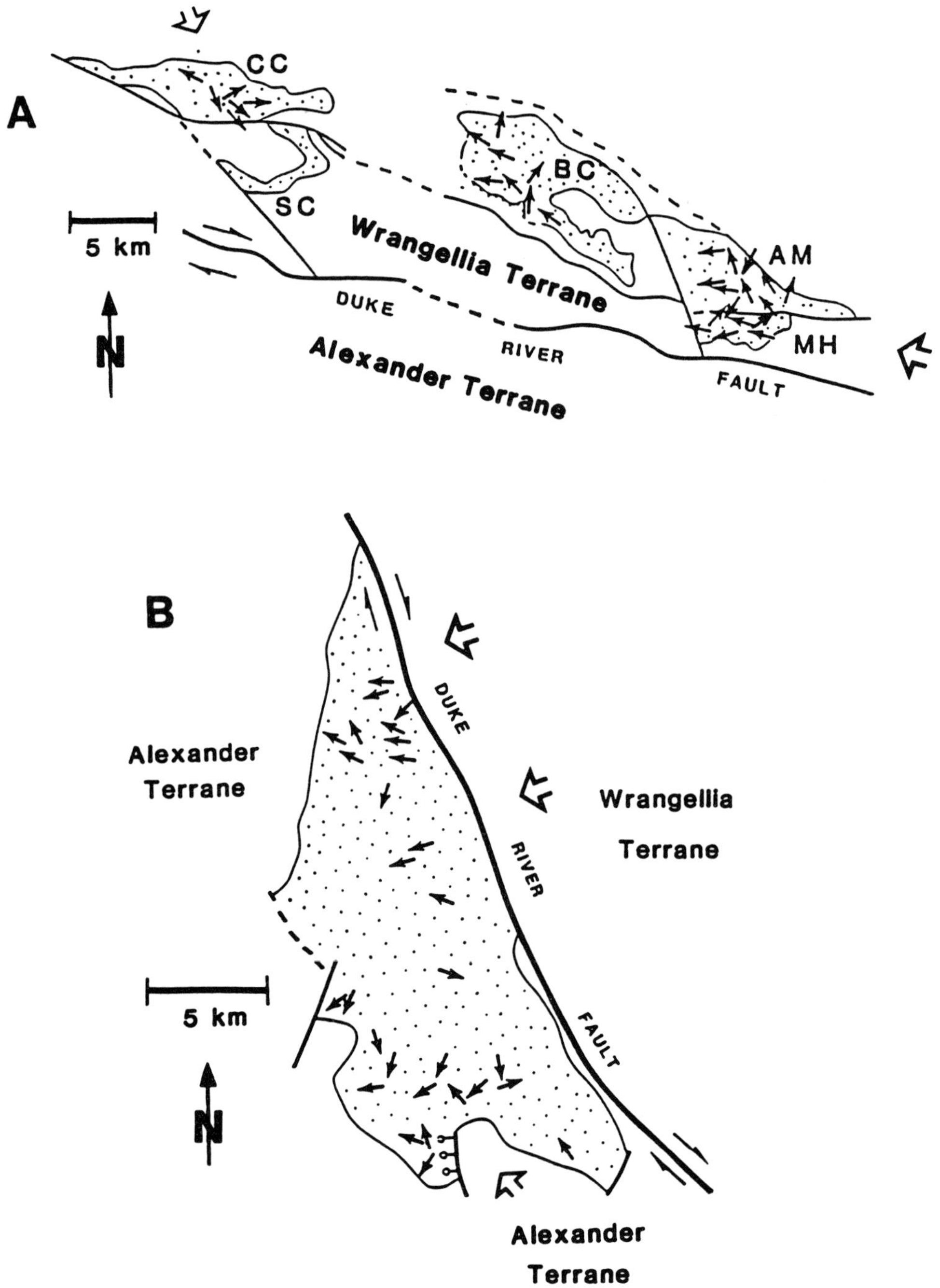

Figure 2. A, Generalized map of the Burwash basin, showing distribution of the Amphitheatre Formation (stippled), major faults (solid and dashed lines), and paleocurrent data (small black arrows) from imbricated conglomerate clasts. Individual arrows represent the average of paleocurrent data from each area. Large open black arrows represent the general sediment transport direction. The Duke River fault separates Wrangellia from the Alexander terrane in the Burwash and the Bates Lake basins. The Burwash basin is located on Wrangellia. Locations referred to in text and in Figures 5 through 9 are: CC = Cement Creek; SC = Steele Creek; BC = Burwash Creek; AM = Amphitheatre Mountain; MH = Mt. Hoge. B, Map of Bates Lake basin showing major faults and paleocurrent data. The basin is located on the Alexander terrane. A and B both modified from Ridgway and DeCelles (1993).

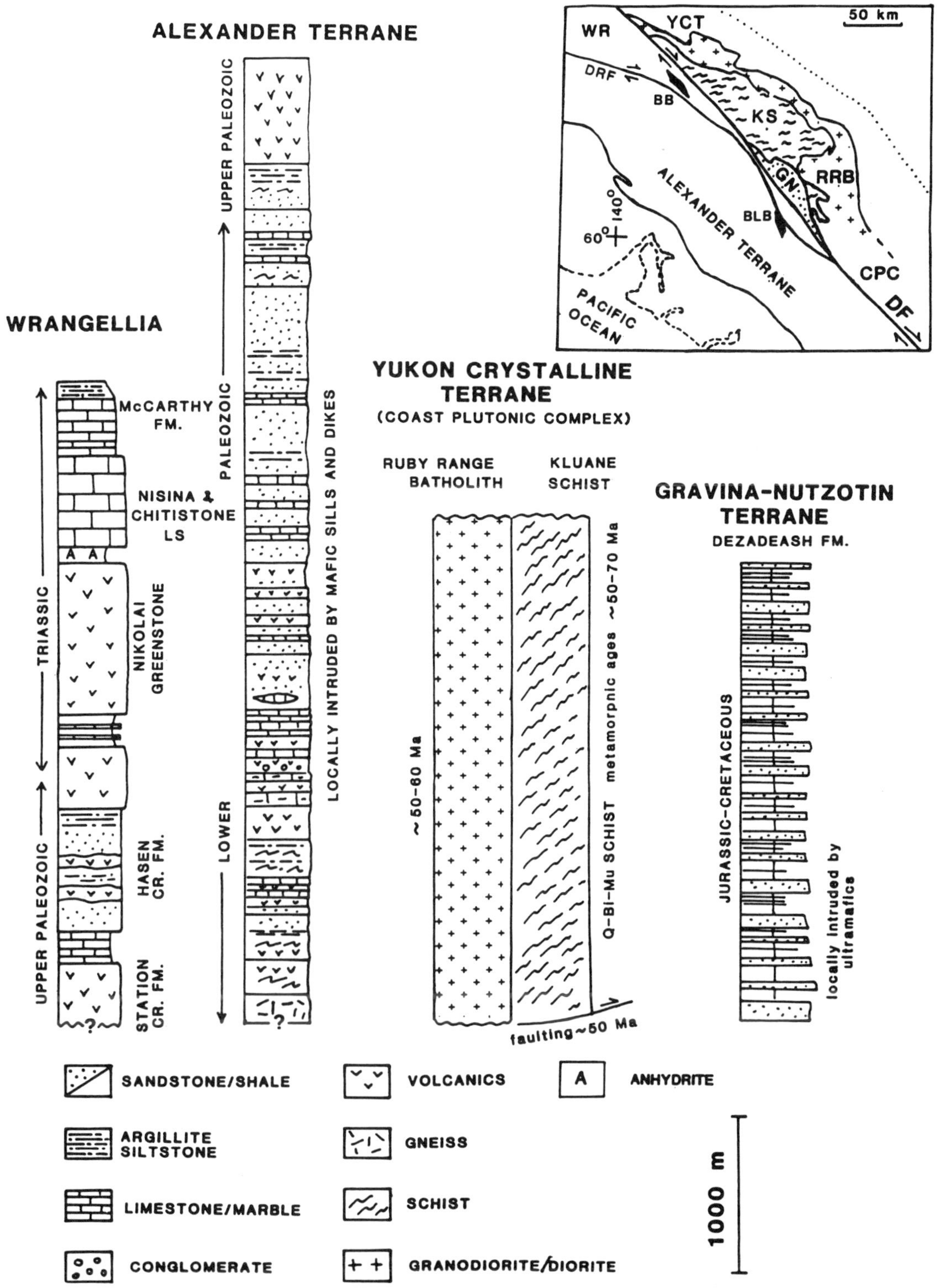

Figure 3. Schematic diagram illustrating general stratigraphy of the tectonostratigraphic terranes and belts that could have provided detritus to Amphitheatre strike-slip basins. Locally, these terranes have been metamorphosed and intruded by Mesozoic and Cenozoic plutons. Inset map shows aerial distribution of the terranes relative to the Burwash (BB) and Bates Lake (BLB) basins. WR = Wrangellia; GN = Gravina-Nutzotin belt; KS = Kluane Schist of the Yukon Crystalline terrane (YCT); RBB = Ruby Range Batholith, a northern extension of the Coast Plutonic Complex (CPC); DRF = Duke River fault; DF = Denali fault. Sources are Eisbacher (1976), Tempelman-Kluit (1976), Jones et al. (1977), Campbell and Dodds (1982a), and Gehrels and Saleeby (1987).

mation and Wrangell Lavas is gradational, exhibiting no evidence of a depositional hiatus or erosion (Souther and Stanciu, 1975; Skulski, 1988; Cole et al., 1989; Cole and Ridgway, 1993). The Duke River fault cuts only the lower part of the Wrangell Lavas, and is overlapped by the upper part of the sequence (Souther and Stanciu, 1975; Ridgway et al., 1992b). Thus, displacement of more than a few tens of kilometers between the Amphitheatre Formation and potential source terranes on either side of the Duke River fault is highly unlikely. Additional evidence of minimal postdepositional offset between the Amphitheatre Formation and eastern sources associated with the Duke River fault is provided by a very coarse granodioritic boulder conglomerate facies in the Burwash basin, which can be matched with a nearby granodioritic intrusion on the eastern side of the basin boundary fault (Eisbacher and Hopkins, 1977; Ridgway et al., 1992a). On the other hand, large amounts (hundreds of kilometers) of post-Oligocene dextral offset along the Shakwak-Dalton segment of the Denali fault are plausible. Potential complications of provenance are minimized, however, because the Coast Plutonic Complex (which would have been the major eastern source of detritus) extends at least 300 km to the southeast of the Amphitheatre basins.

TABLE. 1. RAW AND RECALCULATED POINT-COUNT PARAMETERS

Qm	Monocrystalline quartz
Qp	Polycrystalline quartz
P	Plagioclase feldspar
K	Potassium feldspar
C	Chert
Lml	Low-grade metamorphic rock fragments (slate)
Lmh	High-grade metamorphic rock fragments (mica-schist, gneiss)
Lv	Volcanic rock fragments
Lsi	Siltstone fragments
Lmud	Mudstone fragments
Lcarb	Carbonate rock fragments
Ch	Chlorite
S	Serpentine
M	Mica
D	Dense minerals (epidote, zircon, sphene, etc.)

Q = Qm + Qp + C
F = P + K
L = Lt - C
Lt = Lm + Ls + Lv
Ls = Lsi + Lmud + Larg + Lcarb + C
Lm = Lml + Lmh

METHODS

Petrographic analyses were conducted according to the methods of both the Gazzi-Dickinson (GD) and "traditional" schools (e.g., Ingersoll et al., 1984), with 450 counts per slide. All thin sections were stained for potassium and calcium feldspars, and only medium- to coarse-grained sandstones were analyzed. Table 1 lists the raw and recalculated point-count parameters used in this study. In order to facilitate comparisons with previous tectonic-petrologic studies, our data are recalculated and presented in terms of the GD method (Table 2). Most of the lithic grains counted are fine-grained, so the results of the two different point-counting methods are nearly identical. Appendix 1 contains the raw point-count data from the sandstones of the Amphitheatre Formation. Geographic and stratigraphic locations of samples designated in Table 2 and Appendix 1 can be found in Ridgway (1992, see Appendix 1, Plates 7-10).

Conglomerate compositions were determined in the field by lithologic identification of 200 clasts per sample locality. Petrographic thin sections of common clast types were examined in order to verify field identifications. Potential source areas were visited and sampled to facilitate petrographic comparison with Amphitheatre clast types.

RESULTS

Grain types

Amphitheatre Formation sandstones and conglomerates exhibit a wide variety of grain and clast types. Most of the grain types recognized in thin sections are of the standard varieties (e.g., Qm, K, P, chert). The abundance and the variety of lithic grain types require brief descriptions of the more important types.

Polycrystalline quartz (Qp). Much of the Qp in Amphitheatre sandstones is composed of aggregates of very fine (<0.05 mm), roughly equant crystals with minor mica. Many of the Qp grains are weakly foliated (Fig. 4A). We interpret these grains as sedimentary quartzite fragments, most likely derived from the Nasina Quartzite of the Yukon Crystalline terrane.

Volcanic lithic fragments (Lv). Several different types of volcanic lithic fragments occur in Amphitheatre sandstones. These include tachylitic basalt fragments with plagioclase laths (Fig. 4B); chloritized, mafic volcanic fragments, with altered feldspar crystals and fine-grained groundmass; and highly altered (mostly replaced by finely crystalline phyllosilicates) felsic(?) volcanic fragments. Plausible volcanic source rocks are widespread in both Wrangellia and the Alexander terrane. In Wrangellia, the most likely sources for these types of volcanic lithics are the Nikolai Greenstone, the Hasen Creek Formation, and the Station Creek Formation. A potential source of felsic volcanic detritus is the Mount Nansen volcanic suite of the Yukon Crystalline terrane.

High-grade metamorphic lithic fragments (Lmh). These grains are quartz-mica schist and garnet-quartz-mica schist fragments (Fig. 4C). The predominant mica phase is white mica, which forms along foliation surfaces between layers of polycrystalline quartz. Many of these grains exhibit a crenulated cleavage; some contain sheared and rotated plagioclase porphyroblasts. The relatively high-grade character of these lithic fragments requires a crustal source, probably in the Yukon Crystalline terrane.

Low-grade metamorphic lithic fragments (Lml). These grains are foliated, fine-grained phyllosilicate-rich metasedimen-

TABLE 2. RECALCULATED POINT-COUNT DATA FROM THE AMPHITHEATRE FORMATION

	QFL (%)			QmFLt (%)			QmPK (%)			QpLvLs (%)			LvLmLs (%)				
Location	Q	F	L	Qm	F	Lt	Qm	P	K	Qp	Lv	Ls	Lv	Lm	Ls	P/(P+K)	Qp/Lt
BURWASH BASIN																	
1. Mount Hoge Area																	
E1a-57	0.34	0.62	0.04	0.30	0.62	0.09	0.32	0.49	0.18	0.65	0.23	0.13	0.47	0.27	0.27	0.73	0.09
E1a-94	0.37	0.62	0.01	0.35	0.62	0.02	0.36	0.49	0.15	0.70	0.00	0.30	0.00	0.00	1.00	0.76	0.03
E1a-15	0.51	0.47	0.02	0.41	0.47	0.12	0.47	0.31	0.22	0.87	0.10	0.04	0.63	0.13	0.25	0.59	0.26
E1a-71	0.44	0.53	0.03	0.37	0.53	0.10	0.41	0.40	0.19	0.82	0.00	0.18	0.00	0.36	0.64	0.68	0.16
E1a-27	0.51	0.47	0.02	0.41	0.47	0.12	0.47	0.33	0.20	0.88	0.06	0.06	0.43	0.14	0.43	0.63	0.25
E1a-94	0.39	0.61	0.00	0.34	0.61	0.06	0.36	0.51	0.14	1.00	0.00	0.00	0.00	0.00	0.00	0.79	0.11
E1a-13	0.40	0.59	0.00	0.35	0.59	0.05	0.37	0.42	0.21	0.95	0.00	0.05	0.00	0.00	1.00	0.67	0.11
E1a-36	0.52	0.48	0.01	0.40	0.48	0.13	0.45	0.35	0.19	0.96	0.00	0.04	0.00	0.33	0.67	0.65	0.28
Average	0.44	0.55	0.01	0.37	0.55	0.09	0.40	0.41	0.19	0.85	0.05	0.10	0.22	0.18	0.61	0.67	0.17
Standard	0.06	0.06	0.01	0.04	0.06	0.04	0.05	0.07	0.03	0.12	0.08	0.09	0.26	0.14	0.29	0.06	0.19
2. Amphitheatre Mountain Area																	
Swc-25	0.46	0.37	0.17	0.35	0.37	0.28	0.48	0.28	0.24	0.46	0.35	0.19	0.50	0.22	0.28	0.54	0.23
Swc-87	0.56	0.27	0.17	0.47	0.27	0.27	0.64	0.21	0.15	0.43	0.29	0.29	0.38	0.23	0.38	0.56	0.23
Swc-142	0.51	0.31	0.18	0.42	0.31	0.27	0.58	0.24	0.18	0.44	0.38	0.17	0.41	0.40	0.19	0.51	0.20
Swc-192	0.49	0.36	0.15	0.36	0.36	0.28	0.50	0.30	0.20	0.54	0.39	0.08	0.66	0.21	0.13	0.49	0.27
Am-14	0.54	0.38	0.07	0.46	0.38	0.16	0.55	0.24	0.22	0.80	0.05	0.16	0.07	0.70	0.23	0.54	0.24
Am9-4	0.47	0.29	0.24	0.42	0.29	0.29	0.59	0.35	0.07	0.20	0.69	0.11	0.69	0.20	0.11	0.47	0.09
Am11-u	0.46	0.32	0.22	0.36	0.32	0.32	0.53	0.23	0.24	0.34	0.55	0.11	0.75	0.10	0.15	0.46	0.21
Am12-l	0.42	0.55	0.03	0.38	0.55	0.07	0.41	0.42	0.17	0.70	0.30	0.00	0.46	0.54	0.00	0.42	0.07
Am-15	0.47	0.41	0.12	0.42	0.41	0.17	0.51	0.24	0.25	0.30	0.55	0.15	0.72	0.07	0.20	0.47	0.13
Am9-3	0.44	0.33	0.23	0.39	0.33	0.28	0.55	0.33	0.12	0.26	0.60	0.15	0.49	0.39	0.12	0.44	0.09
Am2-121	0.44	0.33	0.24	0.35	0.33	0.32	0.52	0.38	0.11	0.32	0.48	0.20	0.51	0.27	0.21	0.44	0.14
Am9-6	0.46	0.27	0.27	0.36	0.27	0.37	0.57	0.31	0.12	0.36	0.52	0.12	0.51	0.37	0.21	0.46	0.17
Am3-15	0.54	0.45	0.01	0.43	0.45	0.13	0.49	0.33	0.19	0.93	0.00	0.07	0.00	0.00	1.00	0.54	0.29
Am2-15	0.43	0.54	0.03	0.34	0.54	0.12	0.39	0.40	0.21	0.78	0.00	0.22	0.00	0.17	0.83	0.43	0.18
Am9-2	0.37	0.63	0.00	0.34	0.63	0.03	0.35	0.44	0.20	1.00	0.00	0.00	0.00	0.00	0.00	0.37	0.06
Am-10-u	0.50	0.50	0.00	0.39	0.50	0.11	0.44	0.37	0.19	0.98	0.00	0.02	0.00	0.00	1.00	0.50	0.24
Am-12-m	0.52	0.45	0.03	0.40	0.45	0.14	0.47	0.38	0.15	0.83	0.02	0.15	0.09	0.09	0.82	0.52	0.25
Am-14	0.45	0.52	0.03	0.39	0.52	0.09	0.43	0.32	0.25	0.76	0.03	0.21	0.08	0.38	0.54	0.45	0.15
Am-6	0.45	0.47	0.08	0.40	0.47	0.12	0.46	0.42	0.12	0.38	0.11	0.51	0.18	0.00	0.82	0.45	0.10
Am-32	0.29	0.71	0.00	0.27	0.71	0.02	0.28	0.52	0.20	0.86	0.14	0.00	0.50	0.50	0.00	0.29	0.03
Am-12-u	0.36	0.63	0.01	0.34	0.63	0.04	0.35	0.49	0.16	0.69	0.13	0.19	0.40	0.00	0.60	0.36	0.05
Am-11-l	0.42	0.55	0.02	0.35	0.55	0.10	0.39	0.41	0.20	0.76	0.05	0.19	0.20	0.00	0.80	0.42	0.16
Am-5	0.41	0.53	0.06	0.34	0.53	0.13	0.39	0.49	0.12	0.55	0.08	0.37	0.16	0.12	0.72	0.41	0.12
Am9-5	0.32	0.60	0.08	0.30	0.60	0.10	0.33	0.53	0.14	0.19	0.81	0.00	1.00	0.00	0.00	0.32	0.03
Am3-47	0.47	0.53	0.00	0.41	0.53	0.06	0.44	0.37	0.19	1.00	0.00	0.00	0.00	0.00	0.00	0.47	0.15
Am-10-m	0.51	0.49	0.00	0.39	0.49	0.13	0.44	0.41	0.15	1.00	0.00	0.00	0.00	0.00	0.00	0.51	0.26
Swc-165	0.40	0.37	0.22	0.26	0.37	0.37	0.42	0.39	0.20	0.63	0.14	0.23	0.14	0.63	0.23	0.40	0.23
Swc-127	0.26	0.47	0.27	0.17	0.46	0.37	0.28	0.51	0.21	0.49	0.19	0.32	0.16	0.58	0.26	0.26	0.12
Swc-120	0.39	0.50	0.11	0.29	0.50	0.21	0.37	0.42	0.20	0.61	0.25	0.14	0.41	0.36	0.23	0.39	0.17
Swc-99	0.48	0.50	0.02	0.45	0.50	0.05	0.47	0.36	0.17	0.89	0.06	0.06	0.14	0.71	0.14	0.48	0.09

TABLE 2. RECALCULATED POINT-COUNT DATA FROM THE AMPHITHEATRE FORMATION (continued)

Location	QFL (%) Q	F	L	QmFLt (%) Qm	F	Lt	QmPK (%) Qm	P	K	QpLvLs (%) Qp	Lv	Ls	LvLmLs (%) Lv	Lm	Ls	P/(P+K)	Qp/Lt
Burwash Basin (continued)																	
2. Amphitheatre Mountain Area (continued)																	
Swc-74	0.40	0.53	0.07	0.35	0.53	0.12	0.40	0.41	0.19	0.49	0.26	0.26	0.33	0.33	0.33	0.40	0.10
Swc-17	0.42	0.47	0.11	0.33	0.47	0.20	0.42	0.38	0.20	0.64	0.16	0.20	0.20	0.57	0.24	0.42	0.16
Swc-57	0.28	0.65	0.06	0.26	0.65	0.09	0.29	0.51	0.21	0.35	0.09	0.57	0.09	0.35	0.57	0.28	0.04
Am-12-u	0.54	0.45	0.01	0.40	0.45	0.15	0.47	0.39	0.14	0.94	0.00	0.06	0.00	0.20	0.80	0.54	0.28
Am-13-u	0.36	0.64	0.00	0.31	0.64	0.04	0.33	0.51	0.16	1.00	0.00	0.00	0.00	0.00	0.00	0.36	0.08
Ua-259	0.46	0.48	0.06	0.40	0.48	0.13	0.45	0.40	0.14	0.51	0.00	0.49	0.00	0.00	1.00	0.46	0.13
Ua-309	0.25	0.67	0.08	0.22	0.67	0.11	0.25	0.60	0.15	0.27	0.24	0.49	0.32	0.03	0.65	0.25	0.04
Ua-248	0.35	0.64	0.01	0.30	0.64	0.06	0.32	0.65	0.04	0.95	0.00	0.05	0.00	0.75	0.25	0.35	0.07
Ua-263	0.49	0.43	0.08	0.47	0.43	0.10	0.52	0.32	0.16	0.30	0.38	0.32	0.40	0.26	0.34	0.49	0.05
Ua-282	0.35	0.27	0.38	0.29	0.27	0.44	0.51	0.36	0.13	0.15	0.79	0.06	0.83	0.11	0.06	0.35	0.09
Average	0.43	0.47	0.10	0.36	0.47	0.17	0.44	0.39	0.17	0.60	0.23	0.17	0.33	0.27	0.40	0.43	0.15
Standard	0.08	0.12	0.10	0.07	0.12	0.11	0.09	0.10	0.05	0.27	0.24	0.15	0.27	0.23	0.31	0.08	0.08
3. Burwash Creek Area																	
Bc3-10	0.45	0.41	0.13	0.35	0.41	0.24	0.46	0.49	0.05	0.55	0.32	0.13	0.45	0.38	0.18	0.90	0.17
Bc1-5	0.45	0.36	0.19	0.39	0.36	0.25	0.52	0.27	0.21	0.32	0.60	0.08	0.66	0.25	0.09	0.56	0.15
Bc2-8	0.44	0.55	0.01	0.34	0.55	0.11	0.38	0.49	0.13	0.91	0.00	0.09	0.00	0.00	1.00	0.79	0.18
Bc1-1	0.44	0.56	0.00	0.39	0.56	0.04	0.41	0.50	0.09	1.00	0.00	0.00	0.00	0.00	0.00	0.85	0.08
Bc2-12	0.36	0.61	0.02	0.28	0.61	0.10	0.31	0.57	0.12	0.78	0.04	0.18	0.20	0.00	0.80	0.83	0.13
Bc4-5	0.52	0.47	0.01	0.43	0.47	0.10	0.48	0.44	0.08	0.93	0.00	0.07	0.00	0.00	1.00	0.84	0.18
Bc1-42	0.45	0.53	0.02	0.34	0.53	0.13	0.39	0.47	0.14	0.89	0.00	0.11	0.00	0.14	0.86	0.77	0.21
Bc4-4	0.51	0.47	0.02	0.38	0.47	0.16	0.45	0.36	0.20	0.87	0.00	0.13	0.00	0.00	1.00	0.64	0.30
Bc2-9	0.42	0.58	0.00	0.34	0.58	0.08	0.37	0.48	0.15	0.94	0.00	0.06	0.00	0.00	1.00	0.76	0.15
Bc3-3	0.38	0.61	0.00	0.27	0.61	0.11	0.31	0.53	0.16	0.96	0.00	0.04	0.00	0.00	1.00	0.77	0.18
Bc2-18	0.41	0.57	0.02	0.31	0.57	0.12	0.35	0.52	0.12	0.86	0.00	0.14	0.00	0.00	1.00	0.81	0.18
Bc3-7	0.39	0.44	0.16	0.27	0.44	0.29	0.38	0.53	0.09	0.43	0.12	0.45	0.21	0.00	0.79	0.85	0.18
Bc4-7	0.43	0.56	0.01	0.36	0.56	0.08	0.39	0.45	0.16	0.84	0.00	0.16	0.00	0.17	0.83	0.74	0.13
Bc3-4	0.35	0.64	0.01	0.31	0.64	0.05	0.33	0.58	0.10	0.71	0.05	0.24	0.17	0.00	0.83	0.86	0.06
Bc3-9	0.49	0.49	0.02	0.40	0.49	0.11	0.45	0.39	0.16	0.83	0.00	0.17	0.00	0.00	1.00	0.71	0.20
Average	0.43	0.52	0.04	0.35	0.52	0.13	0.40	0.46	0.13	0.79	0.08	0.14	0.12	0.07	0.81	0.75	0.12
Standard	0.05	0.08	0.06	0.05	0.08	0.07	0.06	0.08	0.04	0.20	0.16	0.10	0.20	0.12	0.41	0.06	0.06
4. Steele Creek Area																	
Sc5-1	0.46	0.36	0.18	0.38	0.36	0.26	0.52	0.33	0.15	0.38	0.47	0.14	0.52	0.32	0.16	0.69	0.15
Sc2-10	0.33	0.53	0.15	0.30	0.53	0.18	0.36	0.43	0.21	0.18	0.23	0.59	0.26	0.08	0.67	0.68	0.05
Sc1-194	0.37	0.63	0.00	0.33	0.63	0.04	0.34	0.49	0.17	1.00	0.00	0.00	0.00	0.00	0.00	0.74	0.08
Sc2-100	0.50	0.44	0.06	0.41	0.44	0.15	0.48	0.34	0.18	0.60	0.06	0.34	0.14	0.04	0.82	0.66	0.21
Sc1-142	0.44	0.54	0.03	0.35	0.54	0.11	0.40	0.51	0.09	0.80	0.11	0.09	0.42	0.25	0.33	0.85	0.15
Sc1-190	0.43	0.55	0.02	0.35	0.55	0.10	0.38	0.49	0.12	0.83	0.04	0.13	0.25	0.00	0.75	0.80	0.16
Sc1-2	0.44	0.54	0.02	0.38	0.54	0.07	0.41	0.47	0.12	0.77	0.00	0.23	0.00	0.22	0.78	0.79	0.11
Sc5-3	0.42	0.58	0.00	0.38	0.58	0.05	0.39	0.43	0.18	1.00	0.00	0.00	0.00	0.00	0.00	0.70	0.11
Sc1-122	0.29	0.71	0.00	0.27	0.71	0.02	0.28	0.59	0.13	0.89	0.00	0.11	0.00	0.00	1.00	0.82	0.03

TABLE 2. RECALCULATED POINT-COUNT DATA FROM THE AMPHITHEATRE FORMATION (continued)

Location	QFL (%) Q	F	L	QmFLt (%) Qm	F	Lt	QmPK (%) Qm	P	K	QpLvLs (%) Qp	Lv	Ls	LvLmLs (%) Lv	Lm	Ls	P/(P+K)	Qp/Lt
BURWASH BASIN (continued)																	
4. Steele Creek Area (continued)																	
Sc1-160	0.38	0.61	0.00	0.27	0.61	0.12	0.30	0.55	0.15	1.00	0.00	0.00	0.00	1.00	0.00	0.78	0.20
Average	0.41	0.55	0.05	0.34	0.55	0.11	0.39	0.46	0.15	0.75	0.09	0.16	0.20	0.24	0.56	0.78	0.16
Standard	0.06	0.09	0.06	0.05	0.09	0.07	0.07	0.08	0.03	0.26	0.14	0.18	0.19	0.31	0.33	0.09	0.05
5. Cement Creek Area																	
Cc5-18	0.57	0.33	0.11	0.45	0.33	0.22	0.58	0.14	0.28	0.59	0.32	0.11	0.53	0.27	0.20	0.34	0.34
Cc1-7	0.57	0.25	0.18	0.41	0.25	0.34	0.62	0.11	0.26	0.59	0.30	0.14	0.43	0.36	0.21	0.30	0.39
Cc1-t-228	0.54	0.43	0.03	0.51	0.43	0.06	0.54	0.18	0.28	0.55	0.40	0.05	0.89	0.00	0.11	0.40	0.14
Cc1-t-181	0.31	0.29	0.40	0.23	0.29	0.49	0.44	0.44	0.12	0.30	0.61	0.19	0.38	0.48	0.14	0.79	0.12
Cc1-t-106	0.49	0.38	0.13	0.43	0.38	0.19	0.53	0.22	0.25	0.46	0.47	0.11	0.45	0.44	0.11	0.46	0.17
Cc1-t-6	0.49	0.38	0.12	0.43	0.38	0.18	0.53	0.22	0.25	0.49	0.37	0.16	0.32	0.53	0.15	0.46	0.16
Cc2-2	0.55	0.31	0.14	0.45	0.31	0.24	0.59	0.11	0.30	0.60	0.36	0.04	0.44	0.51	0.05	0.27	0.31
Cc1-t-88	0.44	0.40	0.16	0.41	0.40	0.19	0.50	0.21	0.28	0.32	0.57	0.14	0.35	0.57	0.09	0.43	0.09
Cc1-t-34	0.54	0.30	0.16	0.47	0.30	0.23	0.61	0.16	0.24	0.45	0.42	0.16	0.39	0.44	0.17	0.40	0.21
Cc5-f	0.35	0.38	0.27	0.24	0.38	0.38	0.39	0.42	0.20	0.30	0.61	0.15	0.73	0.07	0.20	0.68	0.17
Cc1-12	0.34	0.56	0.10	0.23	0.56	0.20	0.29	0.44	0.27	0.56	0.42	0.04	0.79	0.14	0.07	0.62	0.19
Cc1-282	0.57	0.29	0.14	0.44	0.29	0.26	0.60	0.18	0.21	0.48	0.51	0.02	0.90	0.06	0.03	0.46	0.31
Cc1-393	0.13	0.38	0.49	0.02	0.38	0.60	0.04	0.61	0.35	0.18	0.81	0.02	0.96	0.02	0.03	0.63	0.13
Cc1-372	0.53	0.43	0.04	0.33	0.43	0.24	0.44	0.31	0.26	0.90	0.01	0.09	0.06	0.38	0.56	0.54	0.43
Cc1-359	0.64	0.29	0.07	0.40	0.29	0.31	0.58	0.26	0.16	0.92	0.02	0.06	0.07	0.70	0.23	0.63	0.50
Cc1-186	0.56	0.40	0.04	0.33	0.40	0.27	0.46	0.43	0.11	0.87	0.01	0.12	0.05	0.21	0.74	0.80	0.38
Cc1-148	0.48	0.49	0.03	0.30	0.49	0.21	0.38	0.40	0.21	0.92	0.00	0.08	0.00	0.46	0.54	0.65	0.34
Cc1-85	0.40	0.55	0.05	0.24	0.55	0.21	0.31	0.53	0.16	0.84	0.05	0.12	0.19	0.33	0.48	0.76	0.26
Cc1-66	0.38	0.61	0.01	0.27	0.61	0.13	0.30	0.56	0.14	0.94	0.04	0.02	0.40	0.40	0.20	0.80	0.19
Cc1-34	0.30	0.63	0.07	0.16	0.63	0.21	0.20	0.67	0.13	0.75	0.16	0.11	0.40	0.30	0.30	0.84	0.19
Average	0.46	0.40	0.14	0.34	0.40	0.26	0.45	0.33	0.22	0.60	0.32	0.10	0.44	0.33	0.23	0.56	0.25
Standard	0.12	0.11	0.12	0.12	0.11	0.12	0.15	0.17	0.08	0.23	0.24	0.05	0.28	0.19	0.19	0.18	0.11
BATES LAKE BASIN																	
Bl2-108	0.36	0.51	0.12	0.05	0.51	0.44	0.09	0.90	0.01	0.82	0.00	0.18	0.00	0.45	0.55	0.99	0.33
Bl2-78	0.28	0.66	0.05	0.14	0.66	0.20	0.17	0.79	0.04	0.90	0.00	0.10	0.00	0.70	0.30	0.95	0.66
Bl2-29	0.34	0.64	0.02	0.20	0.64	0.16	0.24	0.72	0.04	0.97	0.00	0.03	0.00	0.78	0.22	0.95	0.64
Bl1-7	0.41	0.49	0.10	0.10	0.49	0.41	0.17	0.81	0.01	0.89	0.00	0.11	0.00	0.63	0.37	0.99	0.49
Bl2-2	0.37	0.58	0.05	0.20	0.58	0.22	0.25	0.72	0.03	0.95	0.00	0.05	0.00	0.79	0.21	0.96	0.58
Bl1-48	0.36	0.64	0.00	0.23	0.64	0.13	0.26	0.70	0.03	1.00	0.00	0.00	0.00	0.00	0.00	0.96	0.64
Bl-f	0.28	0.71	0.00	0.16	0.71	0.12	0.19	0.76	0.06	0.98	0.00	0.02	0.00	0.50	0.50	0.93	0.71
Bl-e-1	0.41	0.54	0.05	0.11	0.54	0.34	0.17	0.78	0.05	0.92	0.00	0.08	0.00	0.52	0.48	0.94	0.54
Bl2-133	0.34	0.63	0.03	0.17	0.63	0.19	0.22	0.76	0.02	0.89	0.00	0.11	0.00	0.31	0.69	0.97	0.63
Bl2-86	0.33	0.63	0.04	0.19	0.63	0.18	0.23	0.74	0.03	0.85	0.00	0.15	0.00	0.31	0.69	0.97	0.63
Bl5-mz	0.44	0.52	0.04	0.12	0.52	0.37	0.18	0.81	0.01	0.98	0.00	0.02	0.00	0.83	0.17	0.99	0.52
Bl2-125	0.45	0.50	0.05	0.17	0.50	0.33	0.26	0.71	0.03	0.97	0.00	0.03	0.00	0.80	0.20	0.96	0.50

TABLE 2. RECALCULATED POINT-COUNT DATA FROM THE AMPHITHEATRE FORMATION (continued)

Location	QFL (%)			QmFLt (%)			QmPK (%)			QpLvLs (%)			LvLmLs (%)			P/(P+K)	Qp/Lt
	Q	F	L	Qm	F	Lt	Qm	P	K	Qp	Lv	Ls	Lv	Lm	Ls		
Bates Lake Basin (continued)																	
Ml-a-1	0.39	0.58	0.03	0.09	0.58	0.33	0.13	0.85	0.02	0.93	0.00	0.07	0.00	0.17	0.83	0.98	0.58
Wc-231	0.22	0.77	0.00	0.15	0.77	0.08	0.16	0.75	0.09	0.97	0.00	0.03	0.00	0.50	0.50	0.89	0.77
Wc-11.5	0.36	0.57	0.07	0.08	0.57	0.35	0.13	0.85	0.03	0.90	0.00	0.10	0.00	0.58	0.42	0.97	0.57
Wc2-15	0.40	0.56	0.04	0.19	0.56	0.25	0.26	0.73	0.01	0.90	0.00	0.10	0.00	0.35	0.65	0.98	0.56
Wc2-95	0.40	0.41	0.19	0.15	0.41	0.44	0.26	0.73	0.01	0.90	0.00	0.10	0.00	0.85	0.15	0.99	0.41
Ml-c-1	0.29	0.54	0.17	0.14	0.54	0.31	0.21	0.67	0.12	0.53	0.04	0.43	0.07	0.23	0.70	0.84	0.54
Bl2-38	0.36	0.38	0.25	0.11	0.38	0.50	0.23	0.70	0.07	0.65	0.00	0.35	0.00	0.46	0.54	0.91	0.38
Bl2-7	0.28	0.44	0.27	0.05	0.44	0.51	0.10	0.87	0.03	0.82	0.00	0.18	0.00	0.81	0.19	0.97	0.44
Bl2-54	0.29	0.55	0.16	0.12	0.55	0.32	0.18	0.79	0.03	0.90	0.00	0.10	0.00	0.89	0.11	0.97	0.55
Bl2-156	0.28	0.59	0.13	0.10	0.59	0.31	0.14	0.81	0.04	0.96	0.00	0.04	0.00	0.94	0.06	0.95	0.59
Bl1-0.5	0.23	0.44	0.33	0.05	0.44	0.51	0.11	0.83	0.06	0.74	0.01	0.25	0.01	0.81	0.18	0.94	0.44
Bl1-91	0.38	0.37	0.26	0.12	0.37	0.51	0.25	0.72	0.04	0.85	0.01	0.15	0.01	0.82	0.17	0.95	0.37
Bl1-29	0.29	0.51	0.19	0.09	0.51	0.39	0.15	0.80	0.05	0.73	0.00	0.27	0.00	0.60	0.40	0.94	0.51
Bl-d-u	0.29	0.66	0.05	0.18	0.66	0.16	0.22	0.73	0.06	0.82	0.00	0.18	0.00	0.59	0.41	0.93	0.66
Wc2-65	0.39	0.57	0.04	0.18	0.57	0.25	0.24	0.76	0.00	0.95	0.00	0.05	0.00	0.69	0.31	1.00	0.57
Wc2-139	0.32	0.62	0.06	0.13	0.62	0.25	0.17	0.82	0.00	0.89	0.00	0.11	0.00	0.63	0.37	1.00	0.62
Olb1-6	0.49	0.11	0.40	0.08	0.11	0.80	0.43	0.50	0.08	0.70	0.26	0.03	0.38	0.58	0.05	0.87	0.11
Ol1-58	0.56	0.12	0.32	0.11	0.12	0.77	0.47	0.52	0.01	0.71	0.21	0.09	0.42	0.41	0.17	0.98	0.12
Ol1-5	0.76	0.04	0.20	0.04	0.04	0.92	0.51	0.49	0.00	0.86	0.06	0.08	0.24	0.41	0.34	1.00	0.04
Ola2-1	0.48	0.18	0.34	0.09	0.18	0.74	0.33	0.66	0.01	0.77	0.15	0.08	0.23	0.65	0.13	0.99	0.18
Average	0.37	0.50	0.13	0.13	0.50	0.37	0.22	0.74	0.03	0.86	0.02	0.12	0.04	0.60	0.36	0.96	0.50
Standard	0.10	0.17	0.11	0.05	0.17	0.20	0.10	0.10	0.03	0.11	0.06	0.10	0.11	0.20	0.21	0.04	0.18

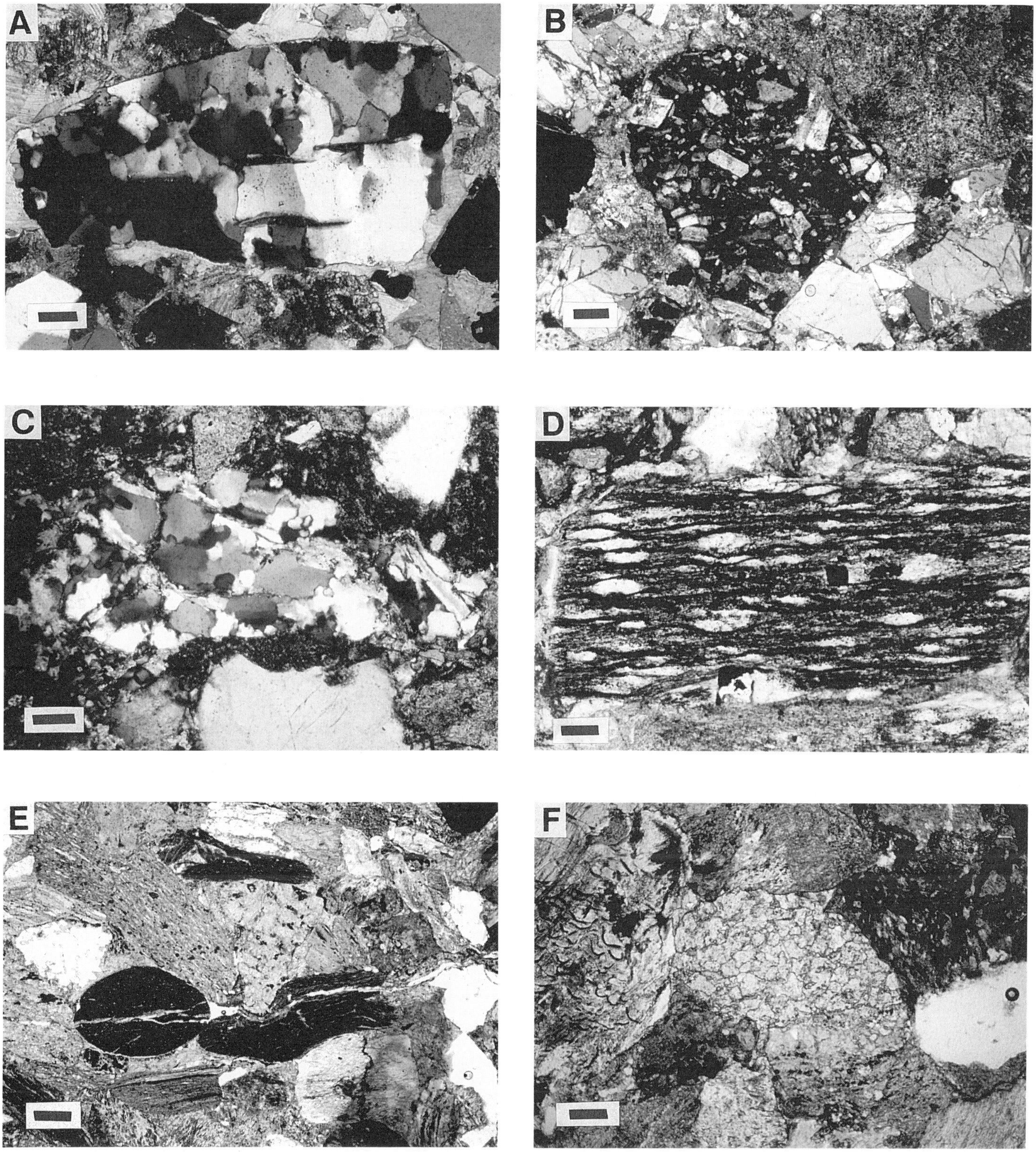

Figure 4. Photomicrographs of Amphitheatre Formation sandstones. A, Polycrystalline quartz (Qp). B, Volcanic lithic fragment (Lv), with plagioclase phenocrysts and glassy groundmass. C, Foliated quartz-mica schist (Lmh). D, Low-grade metasedimentary lithic fragment (Lml). E, Lithic-rich (mostly Lml and Ls) Amphitheatre Formation sandstone. F, Weakly foliated carbonate lithic fragment. Scale bars = 0.1 mm long in A and C; 0.24 mm long in B, D, and F; and 0.48 mm long in E.

tary lithic fragments. Most appear to be slate fragments (Fig. 4D). Potential sources of these grains include the pelitic rocks of Wrangellia and the Alexander terrane and low-grade metasedimentary rocks of the Gravina-Nutzotin belt.

Fine-grained sedimentary lithic fragments (Ls). Mudstone and siltstone grains occur in all Amphitheatre sandstones (Fig. 4E). These are generally pelitic rock fragments that could have been derived from Wrangellia, the Gravina-Nutzotin belt, or the Alexander terrane.

Carbonate lithic fragments. Some Amphitheatre sandstones contain minor amounts of coarse, sparry, calcitic lithic fragments. Many of these grains exhibit a weak foliation and may in fact be fragments of marble (Fig. 4F). The most likely sources of these grains are the Upper Triassic limestones of Wrangellia and middle to upper Paleozoic carbonate rocks of the Alexander terrane.

Accessory grains. Accessory grains occur in amounts usually less than 1%. These include various dense minerals, most important of which are epidote, sphene, zircon, and apatite. Sand-sized grains of serpentine comprise up to ~2% of some sandstones. These grains usually consist of coarse aggregates of serpentine crystals. The sources of the serpentine grains could be any of several highly altered (greenschist facies) ultramafic volcanic units of Wrangellia, the Alexander terrane, and locally within the Gravina-Nutzotin belt (Fig. 3).

Recalculated sandstone petrographic data are listed in Table 2 and plotted on various standard ternary diagrams (Figs. 5–10). Conglomerate data are shown in Figures 11 and 12. Sandstones from both the Burwash and Bates Lake basins are predominantly arkosic to subarkosic, although a few samples from Bates Lake basin are litharenites. Because the two basins exhibit contrasting petrology, the results from each are discussed separately.

Burwash basin

Sandstones from Burwash basin exhibit a broad range of compositions on QFL and QmFLt diagrams. The data (Figs. 5–9) are presented in terms of five subareas (Fig. 2A) that can be distinguished on the bases of sedimentological and structural criteria (Fig. 2A) (Ridgway et al., 1992a; Ridgway and DeCelles, 1993). The ratio Q:F ranges from ~1 to 2 and the feldspar population is dominated by calcic plagioclase with P/(P+K) = 0.4 to 0.96 (Table 2). The lithic fragment population consists of a heterogeneous mixture of sedimentary, volcanic, and metamorphic types. Particularly abundant are fragments of shale and slate; mica schist; weakly foliated, fine-grained quartzite (counted as Qp in Figs. 5–9); and various types of fine-grained volcanic fragments, including microlitic and felsic. Important dense minerals include zircon, tourmaline, sphene, and epidote, but in general these occur in amounts $<1\%$. Small amounts ($<1\%$) of serpentine grains were found in many samples.

Sandstones from the Cement Creek area of Burwash basin (Fig. 5) are slightly more quartzose- and lithic-rich than sandstones from the remainder of the basin. They also contain less plagioclase (average percent QmPK = 45,33,22). The lithic grain populations in the Burwash Creek and Steele Creek areas are dominated by quartzose (Qp) and sedimentary lithic grains, whereas lithic populations in the other areas of the Burwash basin are a mixture of volcanic and quartzose grains (compare Figs. 8 and 9 with Figs. 5–7). Conglomerates from Burwash basin are dominated by basic volcanic clasts (mainly tuffs and basalts) with subordinate argillite, and plutonic and acidic volcanic clasts (Fig. 11). Many of the volcanic clasts are at least partly chloritized and resemble greenstone. Medium-grade metamorphic clasts (mainly quartzites and quartz-mica schist) are relatively abundant in the Cement Creek area.

Bates Lake basin

Sandstones from the Bates Lake basin are arkosic to subarkosic but generally are more feldspathic and more plagioclase-rich than Burwash basin sandstones (Fig. 10). The average ratio P/(P + K) = 0.96 (Table 2). Polycrystalline quartz (Qp) is by far the most abundant type of rock fragment (Fig. 10). Lm populations consist mainly of schists and cataclasites with granulated plagioclase and minor quartz in a fine matrix of chlorite, epidote, and muscovite. Unlike the Burwash basin sandstones, those from Bates Lake basin contain lithic grain populations that are dominated by metamorphic and sedimentary rock fragments. Volcanic rock fragments are relatively rare in sandstones from Bates Lake basin. Bates Lake basin conglomerates are dominated by greenstone, basalt, siliceous argillite, and limestone clasts (Fig. 12). The greenstone clasts are predominantly diabases that have undergone low-grade metamorphism and consist of highly altered plagioclase crystals in a minor matrix of chlorite, sericite, and carbonates. Limestone clasts are particularly abundant in the southern part of the basin (Fig. 12).

INTERPRETATION

Burwash basin

Sandstone and conglomerate compositional data from Burwash basin indicate derivation primarily from volcanic and sedimentary (and low-grade metavolcanic and metasedimentary) rocks. The most likely source of most of this material is Wrangellia. The three stratigraphic units that crop out extensively in this part of Wrangellia are the Station Creek Formation, the Hasen Creek Formation, and the Nikolai Greenstone (Fig. 3). The Station Creek consists of moderately deformed, gray-green volcanic breccia and agglomerate, lithic and vitric tuff, and pale green to white siliceous tuffs; the Hasen Creek Formation is composed mostly of siliceous argillite; and the Nikolai Greenstone contains dark green and maroon amygdaloidal basalt and andesite, largely altered during greenschist metamorphism (Campbell and Dodds, 1982a,b). Conglomerate clasts representing all of these rock types are abundant in the Burwash basin (Fig. 11). The main sources of plagioclase grains in the sand fraction probably were the volcanic

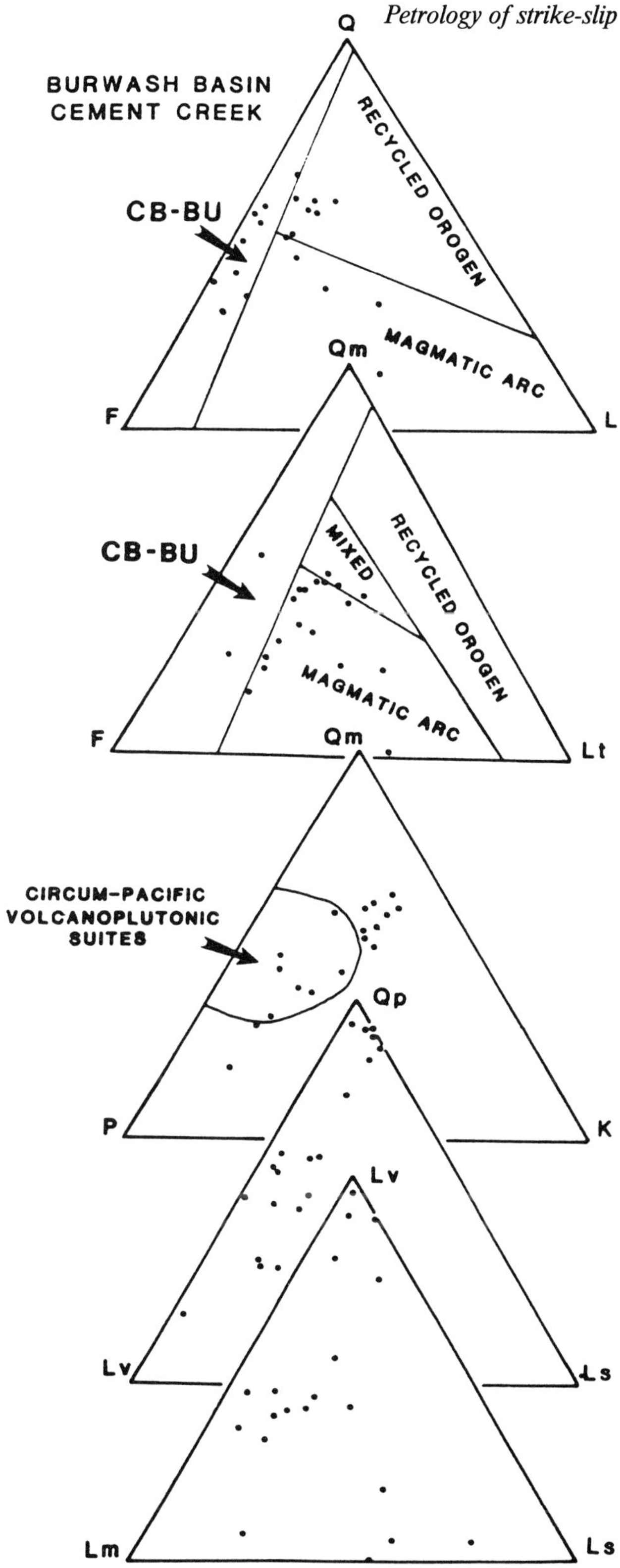

Figure 5. Ternary diagrams showing modal framework–grain compositions of Amphitheatre Formation sandstones from the Cement Creek area of the Burwash basin (see Fig. 2 for location). See Table 1 for grain parameters. Recycled Orogen, Magmatic Arc, and Continental Block–Basement Uplift (CB-BU) provenance fields are from Dickinson et al. (1983). The field for Circum-Pacific volcanoplutonic suites is from Dickinson (1982, 1985).

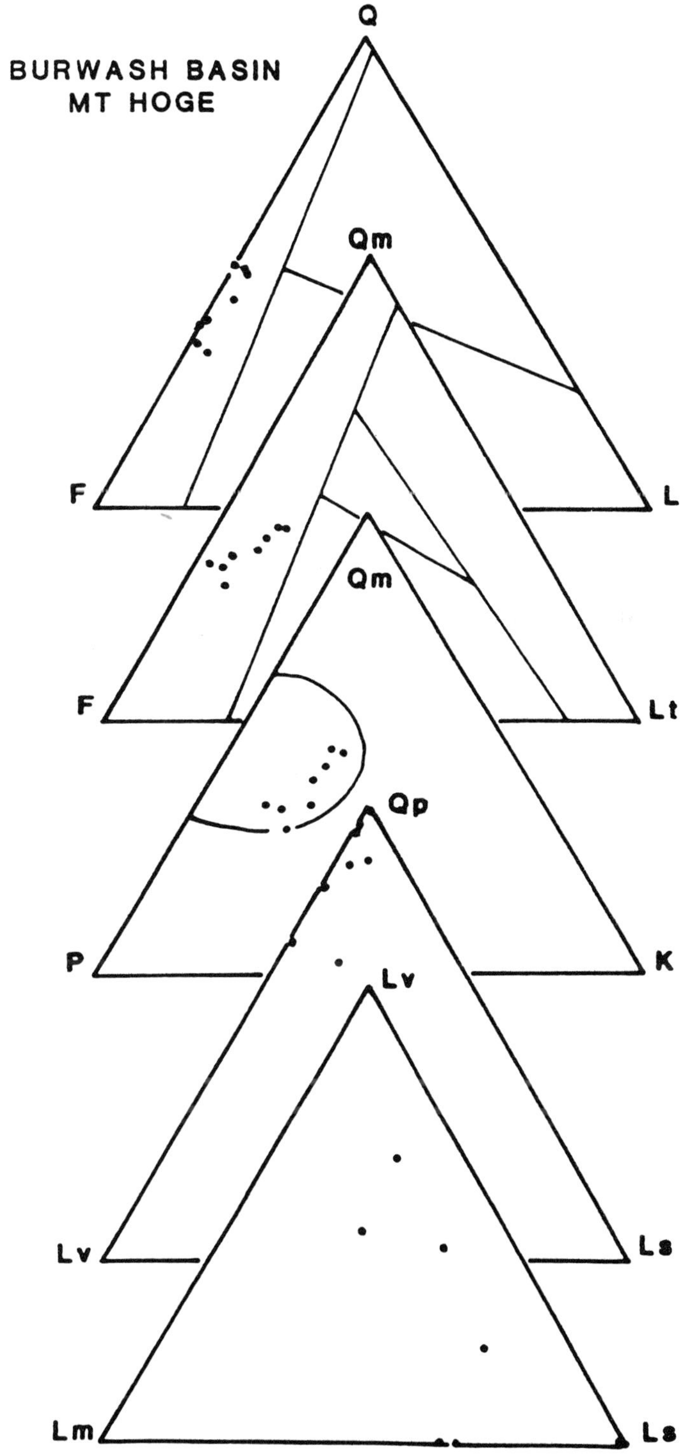

Figure 6. Modal framework–grain compositions of sandstones from the Mount Hoge area of the Burwash basin. See Figure 5 for explanation.

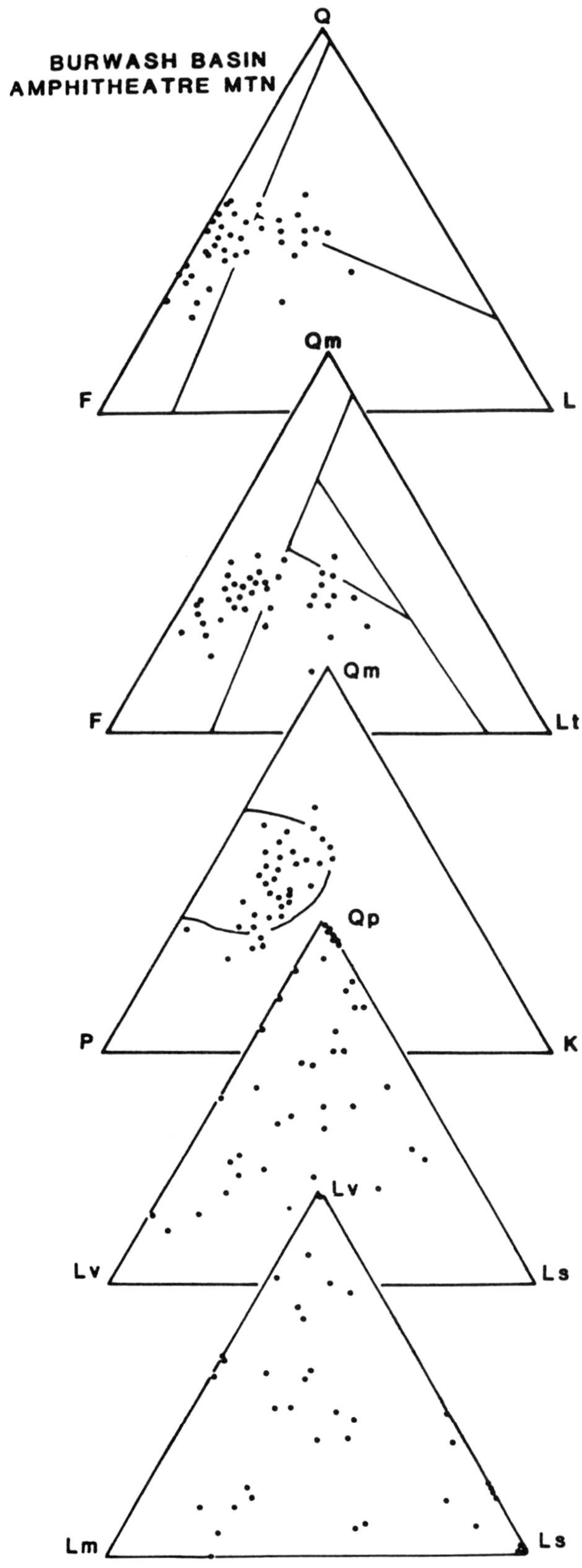

Figure 7. Modal framework–grain compositions of sandstones from the Amphitheatre Mountain area of the Burwash basin. See Figure 5 for explanation.

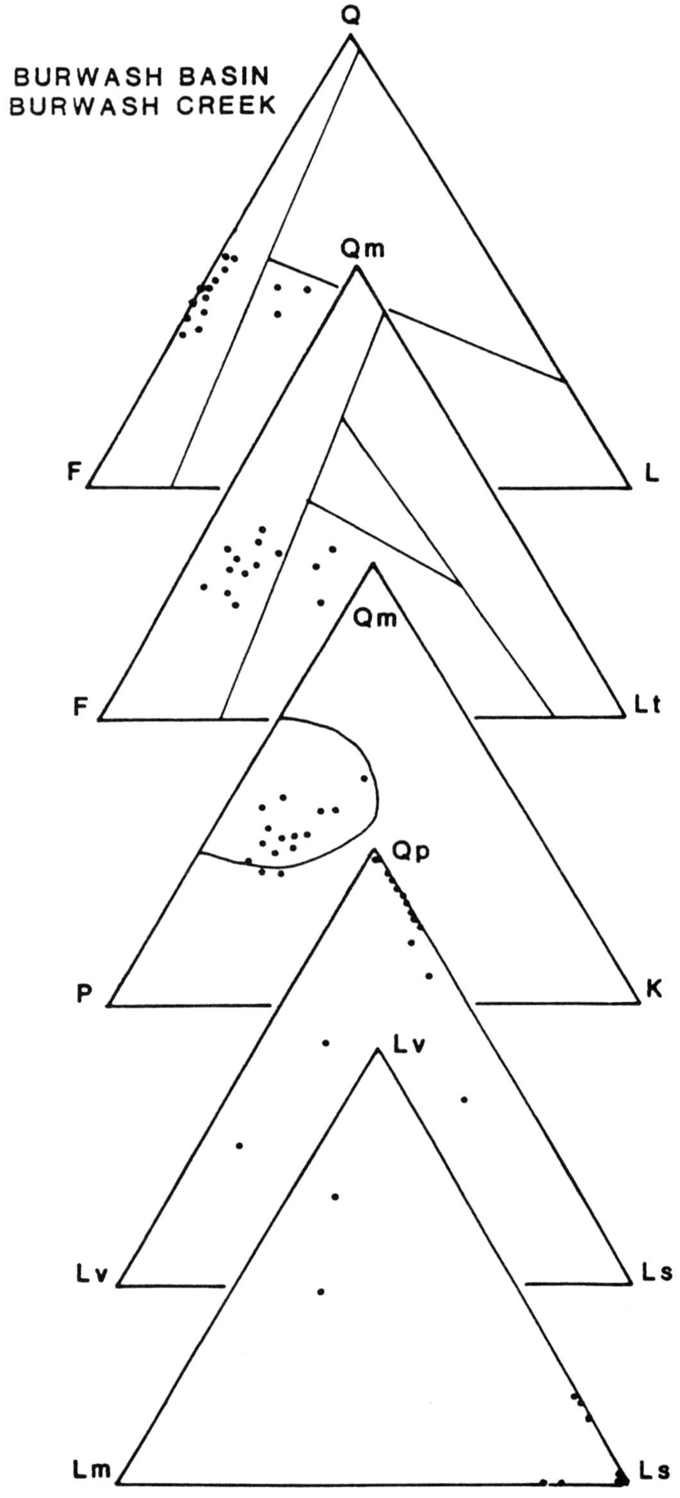

Figure 8. Modal framework–grain compositions of sandstones from the Burwash Creek area of the Burwash basin. See Figure 5 for explanation.

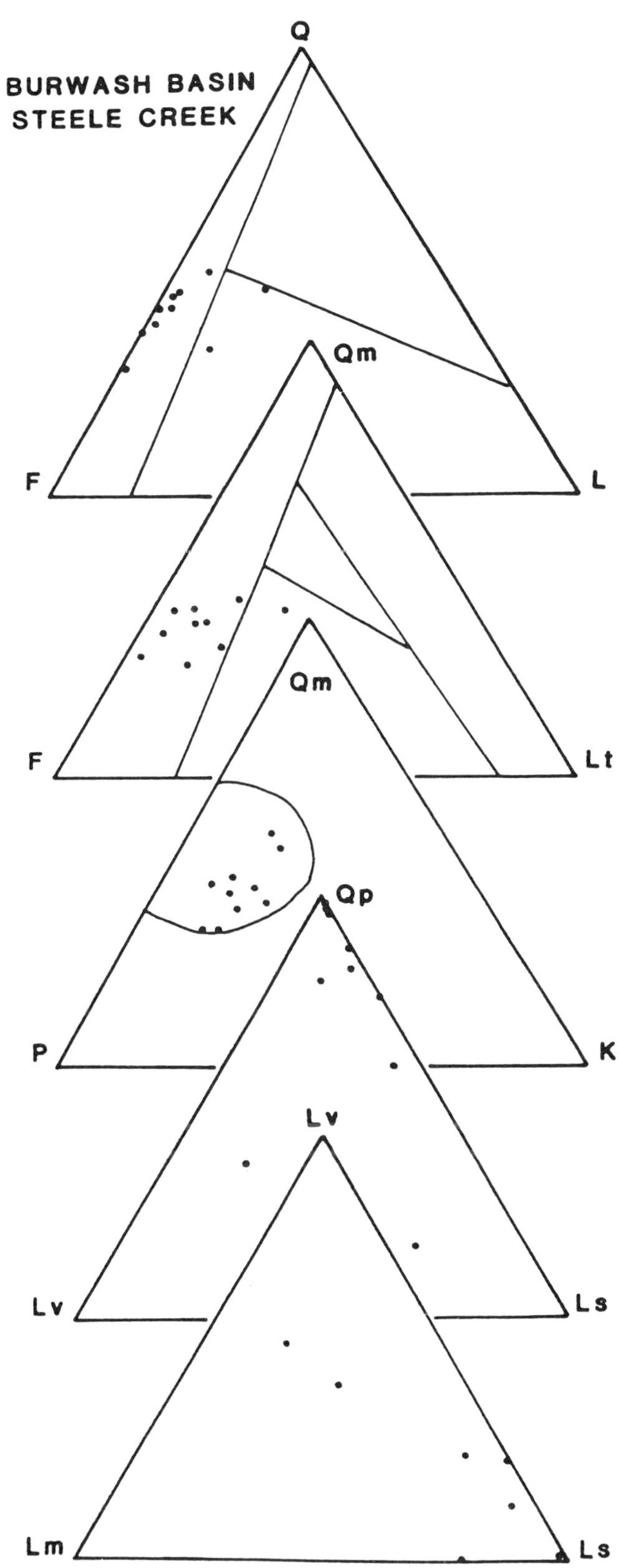

Figure 9. Modal framework–grain compositions of sandstones from the Steele Creek area of the Burwash basin. See Figure 5 for explanation.

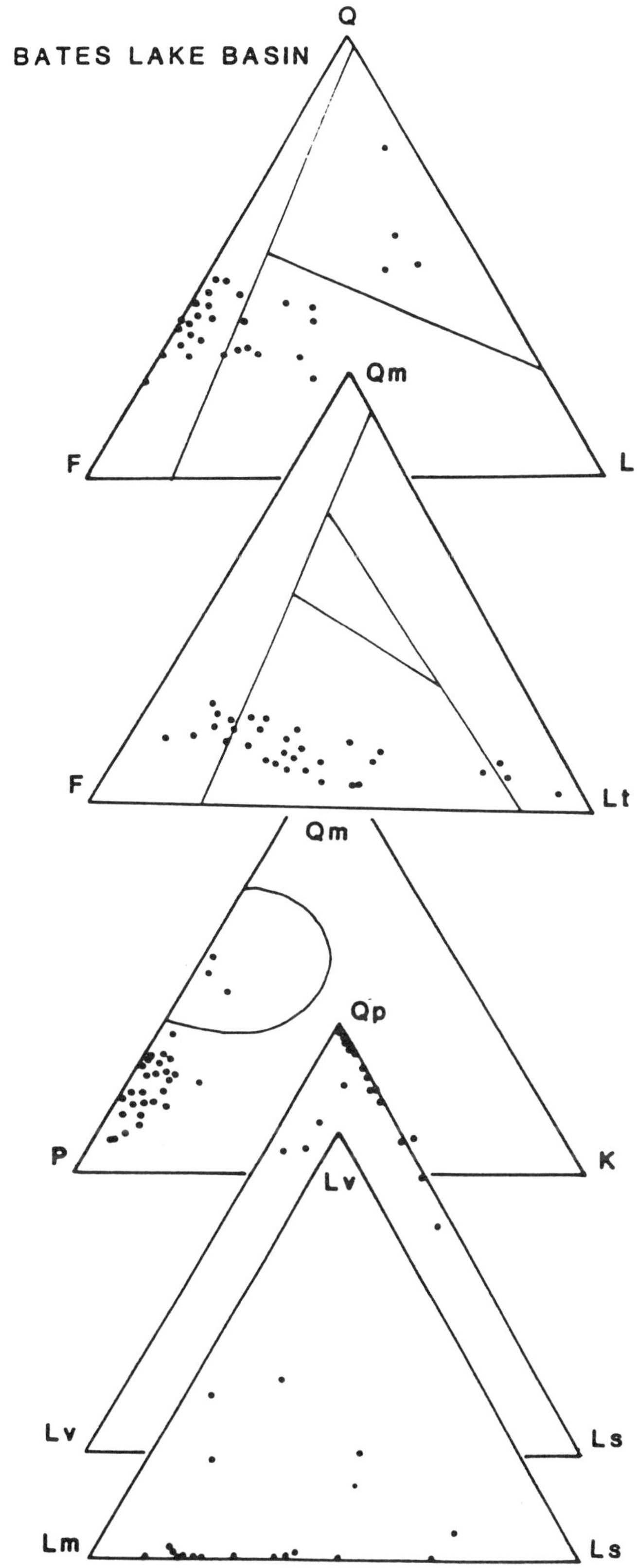

Figure 10. Modal framework-grain compositions of sandstones from the Bates Lake basin. See Figure 5 for explanation.

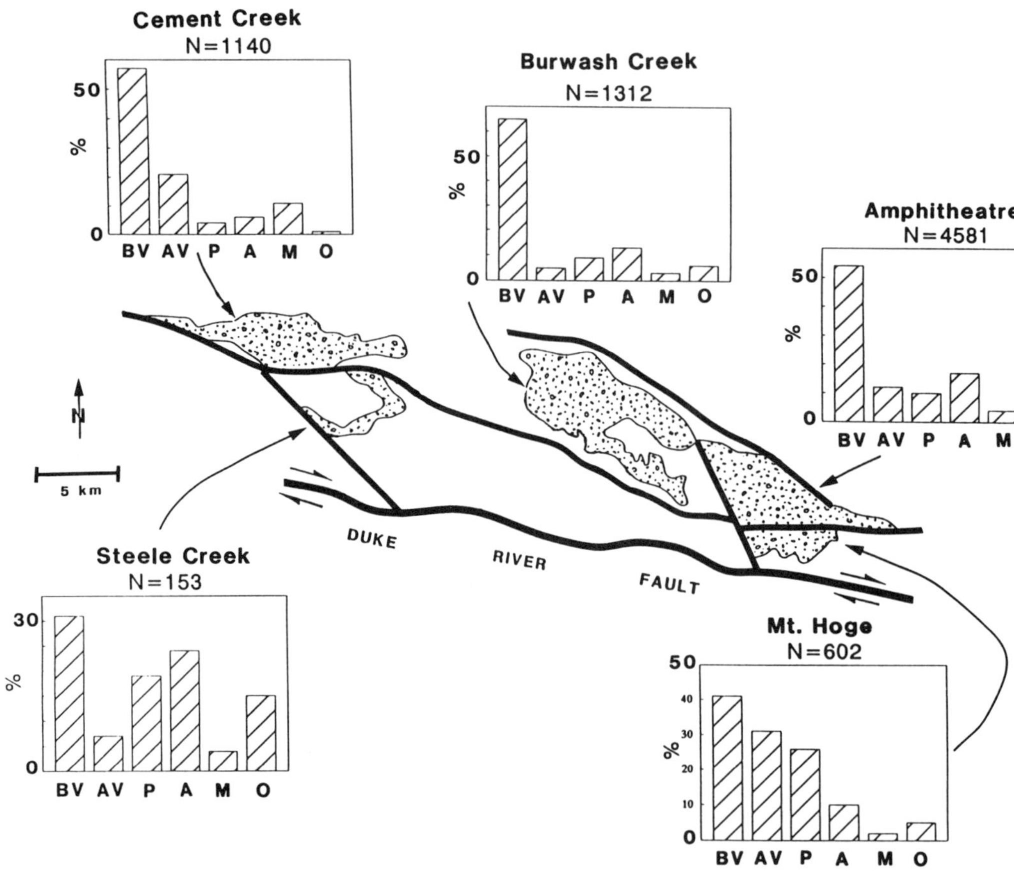

Figure 11. Map of the Burwash basin showing the distribution of conglomerate clast types in the Amphitheatre Formation. Histograms are based on total number of clasts counted in each area and are normalized to 100%. Areas represented by histograms correspond to respective areas in Figures 5 through 9 for sandstone compositions. Clast types: BV = basic volcanic; AV = acidic volcanic; P = plutonic; A = argillite and chert; M = high-grade metamorphic (e.g., quartzite, schist, gneiss); O = other types (mainly vein quartz with minor amounts of sandstone, siltstone, limestone, ultramafic). N = total number of clasts identified.

rocks of the Nikolai Greenstone and the Station Creek Formation. The abundant pelite and metapelite fragments probably were derived from the Hasen Creek Formation. The presence of medium- to high-grade metamorphic and plutonic detritus in some Burwash basin sandstones indicates that the Yukon Crystalline terrane also may have supplied material to the basin. This is corroborated by the presence of granite and banded quartzite clasts in the conglomerates, which probably were derived from the plutonic and metaplutonic roots of the Yukon Crystalline terrane and the Nasina Quartzite, respectively, which crop out extensively along the northeastern side of the Denali fault. The relative abundance of metamorphic clasts in conglomerates and quartz in the sandstones of the Cement Creek area suggest that a greater component of Yukon Crystalline terrane detritus was deposited in this part of the basin.

The majority of Burwash basin sandstones plot within the Continental Block (i.e., basement uplift) and dissected Magmatic Arc fields of Dickinson et al. (1983) (Figs. 5-9) and the "Plate Juncture Highland" provenance of Valloni (1985). Some samples from the Amphitheatre Mountain and Cement Creek areas plot within the Recycled Orogen field of Dickinson et al. (1983). On QmPK diagrams, Burwash basin sandstones completely overlap Dickinson's (1982) Circum-Pacific Volcanoplutonic suite (Figs. 5-9). The mixture of microlitic and felsic volcanic clasts suggests derivation from both andesitic and rhyolitic or rhyodacitic volcanic rocks. The small amounts of serpentine grains indicate that ultramafic subduction-related rocks (ophiolitic), possibly located in Wrangellia and/or the Dezadeash Formation, also contributed. The blend of sandstone compositions ranging between Magmatic Arc and Continental Block provenance fields, coupled with west-northwest sediment transport directions (Fig. 2A), is best interpreted as the result of mixing of volcanic rocks of Wrangellia and metaplutonic rocks of the Yukon Crystalline terrane.

Bates Lake basin

Sandstones and conglomerates in the Bates Lake basin are the erosional byproducts of volcanic and metasedimentary rocks

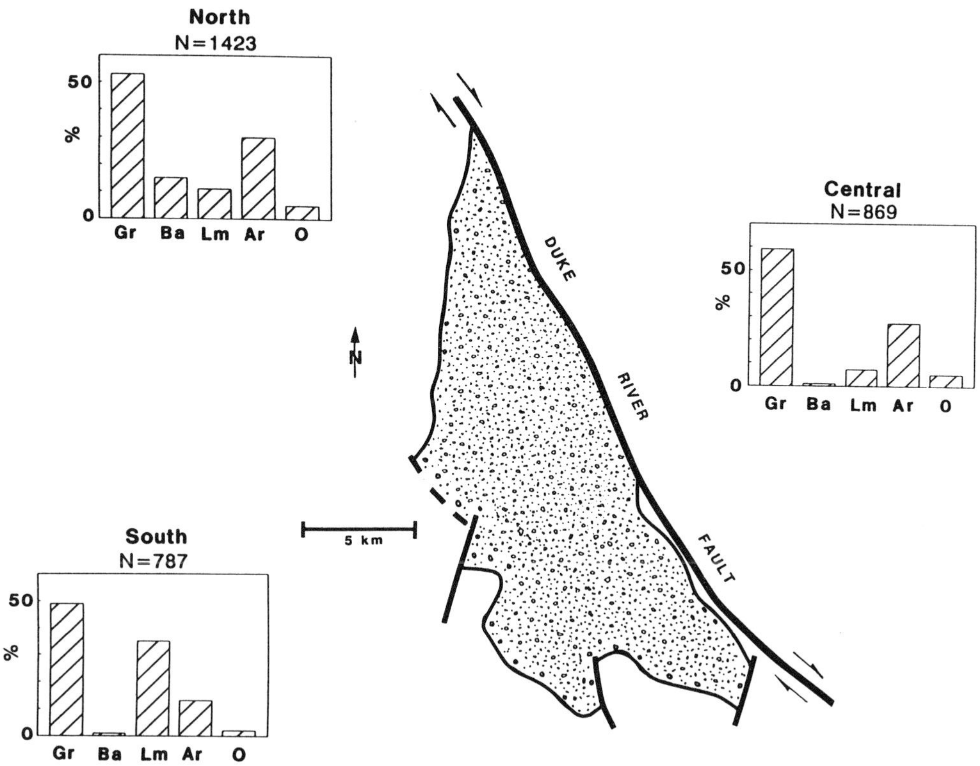

Figure 12. Map of the Bates Lake basin showing the distribution of conglomerate clast types in the Amphitheatre Formation. Histograms are based on total number of clasts counted in each area and are normalized to 100%. Clast types: Gr = greenstone; Ba = basalt; Lm = limestone; Ar = argillite and chert; O = other (mainly vein quartz and granodiorite with minor amounts of sandstone, siltstone, mica schist, quartzite, granite). N = total number of clasts identified.

of Wrangellia and igneous (greenstone) and sedimentary rocks of the Alexander terrane. Wrangellia in this area is dominated by volcanic and volcaniclastic rocks of the Station Creek Formation and the Nikolai Greenstone. The Alexander terrane in the vicinity of Bates Lake basin comprises an assemblage of graywacke, argillite, limestone, basaltic and andesitic flow rocks, and widespread gabbroic sills and dikes (Fig. 3). Virtually all of the clast types found in Amphitheatre sandstones and conglomerates in Bates Lake basin can be accounted for in these two terranes. However, the argillite clasts common in conglomerates of the central and northern part of the basin also could have been derived from the sedimentary rocks of the Gravina-Nutzotin belt located east of the basin (Fig. 3).

On standard QFL and QmFLt diagrams (Fig. 10), Bates Lake basin sandstones overlap the Basement Uplift and transitional to dissected Magmatic Arc fields of Dickinson et al. (1983), yet these sandstones contain relatively few volcanic rock fragments. Apparently, the volcanic rocks of Wrangellia and the Alexander terrane provided mainly plagioclase grains in the sand fraction, rather than lithic fragments. Based on thin sections of greenstone conglomerate clasts, the abundance of plagioclase (%QmPK = 22,74,3) and the lack of lithic volcanic fragments (%LvLmLs = 4,60,36) in the sandstone compositions may be due to the dominance of granular textures, consisting of euhedral plagioclase with very little matrix, in the source rocks. Unlike the Burwash basin sandstones, Bates Lake basin sandstones exhibit no overlap with Dickinson's (1982, 1985) Circum-Pacific Volcanoplutonic field, and the lithic component resembles what would be expected from a Recycled Orogenic terrane. Paleocurrent data from the Bates Lake basin (Fig. 2B) also suggest mixing of sediments derived from Wrangellia and the Alexander terrane. Dominant sediment transport was away from Wrangellia along the eastern strike-slip, faulted basin margin. Locally, sediment was transported away from the Alexander terrane along the normal-faulted southern margin of the basin.

In summary, sandstones and conglomerates from the Burwash and Bates Lake basins along the Denali fault system are extremely immature and reflect derivation from both local sources within Wrangellia and the Alexander terrane, as well as more distal sources within the Yukon Crystalline terrane and the southern part of the Gravina-Nutzotin belt. In comparison with other petrologic studies of transform settings, Amphitheatre sandstone compositions overlap almost completely the compositions of sandstones from the Ridge basin in California (Link, 1982) and are included within Valloni's (1985) general provenance field for plate juncture highlands (Fig. 13).

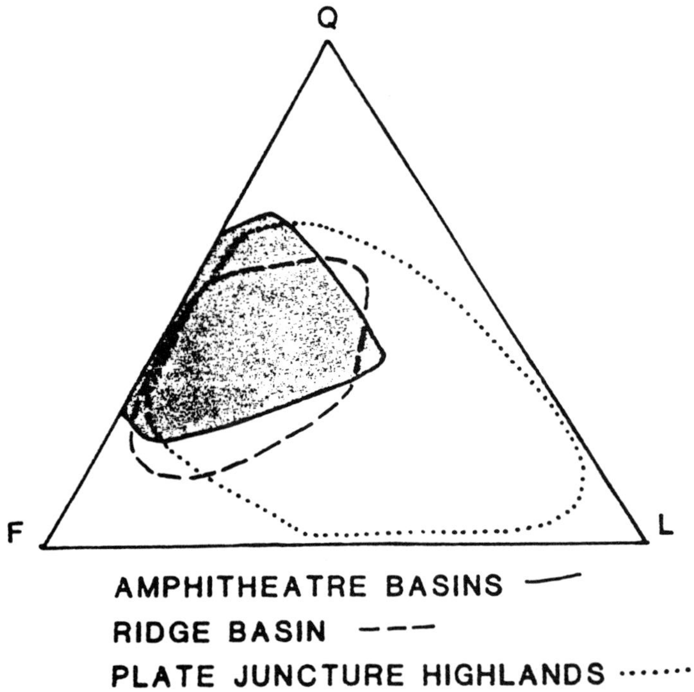

Figure 13. Distribution of detrital modes for sandstones from Cenozoic strike-slip basins of the Denali fault described in this study (shaded area) compared to those of the Ridge basin of southern California (from Link, 1982) and Valloni's (1985) data from plate junction highlands.

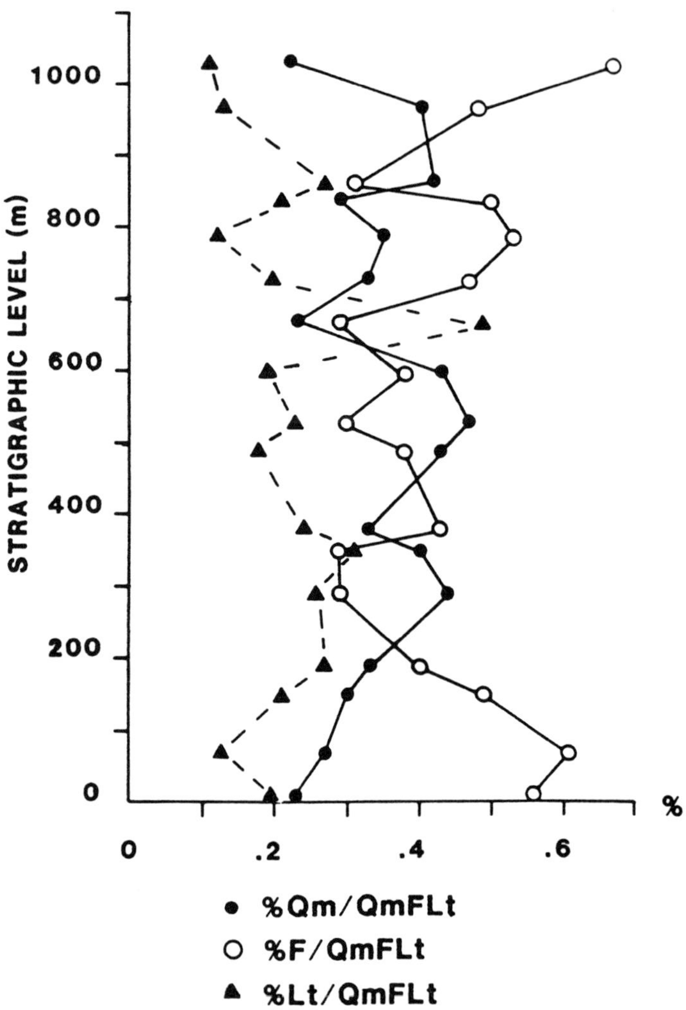

Figure 14. Stratigraphic distribution of the major, climate-sensitive framework modes in the Amphitheatre sandstones of the Burwash basin.

INFLUENCE OF CLIMATE

The abundance of coal and water-laid (mainly fluvial) facies in the Amphitheatre Formation indicates that Eocene-Oligocene paleoclimate was generally humid in southwestern Yukon Territory (Ridgway and DeCelles, 1993). Palynological and coal petrographic data, however, suggest that a change in forest character occurred during Amphitheatre deposition (Ridgway et al., 1989, 1992a). Palynomorphs and coal macerals from the lower part of the section are dominated by an angiosperm assemblage, derived from a temperate deciduous forest, whereas those from the upper part of the section consist mainly of a gymnosperm assemblage derived from a subtemperate, boreal coniferous forest. This change in forest character can be attributed to the terminal Eocene global cooling event. Such a change in paleoclimate raises the question of climatic control on detrital sand composition in the Amphitheatre strike-slip basins.

Figure 14 shows the stratigraphic distribution of the major, climate-sensitive framework modes in Amphitheatre sandstones of the Burwash basin, where we have the most thorough data base. No clearcut trends exist, but F content increases gradually upsection, mainly at the expense of Lt, in the upper half of the section. If this trend is meaningful, it may reflect slightly increasing feldspar stability in response to climatic deterioration. No reciprocal trend in Qm exists, however. We suspect that the apparent increase in F near the top of the section (Fig. 14) is related to factors other than climate.

Several factors probably contributed to the apparent inability of the terminal Eocene climatic event to leave an obvious signature in Amphitheatre sandstones. First, although the palynological data do indicate a decrease in paleotemperature, they provide no definitive information about rainfall. Sedimentological data indicate that Amphitheatre depositional systems were stream dominated during the entire period of deposition, and coal is evenly distributed throughout the stratigraphic section (Ridgway and DeCelles, 1993). Although mean annual temperature and temperature range are important climatic characteristics (Basu, 1985), temperature changes may have a relatively minor effect on sand composition if soil residence time is brief. Much more important is the amount of rainfall, which controls the density of vegetation and degree of soil development. In particular, extremely humid climate promotes destruction of feldspar and unstable lithic-grains (e.g., Basu, 1976; Suttner et al., 1981; Johnsson, 1990). The decline in temperature, which probably occurred during Amphitheatre deposition, apparently was not accompanied by a significant decline in rainfall, so a major decrease in sandstone maturity would not be expected at the Eocene-Oligocene boundary. Predicted rainfall patterns for the Cenozoic, based on global distributions of coal and evaporite (Parrish et al., 1982), also suggest that rainfall did not change appreciably at the Eocene-Oligocene boundary.

A second group of important and related factors that may have exerted control on sand composition includes the active tectonic setting, the high relief in the source terranes, and the resulting lack of short-term sediment storage capacity in the source drainage basins. Work by Basu (1976, 1985), Suttner et al. (1981), Johnsson et al. (1988), and Johnsson (1990) suggests that residence time of sands in soil profiles developed on interfluves (in both the source drainage basin and the depositional basin) is one of the most important variables in the process of sand maturation. Long residence times promote feldspar destruction and, in some cases, can produce first-cycle quartz arenities from originally lithic and feldspathic sands (Franzinelli and Potter, 1983; Johnsson et al., 1988). The textural immaturity of the Amphitheatre Formation and its proximity to what must have been high-relief source areas imply that erosion and transport of sediment were rapid. The steep interfluves in Wrangellia and the Alexander terrane would have inhibited soil development in spite of the humid climate. Sediment would have been rapidly conveyed into the depositional basin and undergone rapid burial, which promoted preservation of original petrologic compositions.

From a more general point of view, it would seem unlikely that climate would ever have a major impact on sand composition in basins associated with large strike-slip orogens. Regardless of rainfall, such settings are characterized by steep interfluves, high local relief, and rapid rates of uplift and subsidence; all of these factors favor preservation of immature sediments. Indeed, the Amphitheatre Formation is an excellent example of this general concept, having retained a faithful (and therefore immature) record of the lithologic composition of its provenance in spite of a humid climatic regime. Sandstones of the Ridge basin in southern California accumulated under somewhat drier climatic conditions (Crowell and Link, 1982) than those that prevailed during Amphitheatre deposition, but QFL compositions of Amphitheatre and Ridge basin sandstones are quite similar (Fig. 13).

CONCLUSIONS

The data reported herein clearly reflect the cosmopolitan character of sandstones and conglomerates in a complex accretionary orogen, where detritus derived from island arcs in various stages of dissection, ophiolites, and sedimentary and metasedimentary sources may mingle. The data also indicate that local uplifts in Wrangellia, and the Alexander, Gravina-Nutzotin, and Yukon Crystalline terranes provided detritus to the Amphitheatre basins. The petrographic data are corroborated by paleocurrent and lithofacies data which indicate that these basins were filled from several different directions.

On standard QFL and QmFLt provenance diagrams, these sandstones plot in the areas normally attributed to basement uplifts and magmatic arcs. This is not surprising because continental margin accretionary orogens usually involve island arcs and volcanoplutonic terranes in various stages of dissection. This suggests that, in the absence of independent criteria for strike-slip tectonics, sandstone petrologic data alone are not diagnostic of transform-orogenic settings.

The humid paleoclimate had little effect on Amphitheatre sand composition. Instead, the high relief and steep interfluve slopes of source drainage basins and rapid rates of sediment transfer and burial in the depositional basins preserved highly immature sands that accurately reflect source-terrane composition. The Eocene-Oligocene climatic deterioration also seems to have had little effect on Amphitheatre sandstones.

ACKNOWLEDGMENTS

This study was supported by a grant from the American Chemical Society, Petroleum Research Fund 23279-AC8-C (to P.G.D.). Additional funding (to K.D.R.) was provided by the American Association of Petroleum Geologists, the American Geological Institute Minority Participation Program, the American Indian Science and Engineering Society, the Geological Society of America, the Antoinette Lierman Medlin Scholarship Award (Geological Society of America Coal Division), Sigma Xi, and the Ford Foundation. Funding for initial reconnaissance work and logistical support throughout the study was provided by the Exploration and Geological Services Division (Whitehorse) of the Department of Indian Affairs and Northern Development (DIAND), Yukon. Trevor Bremner of DIAND is thanked for his help in showing us the geology and stratigraphy of Wrangellia in the Kluane Range. Comments by John Decker, Abhijit Basu, and an anonymous reviewer improved the quality of the manuscript.

APPENDIX 1. RAW POINT-COUNT DATA FROM THE AMPHITHEATRE FORMATION

Location	Qm	Qp	P	K	C	Lml	Lmh	Lv	Lsi	Larg	Lcarb	Lmud	Ch	S	M	D
							BURWASH BASIN									
1. Mount Hoge Area																
e1a-57	119	18	181	68	1	4	1	10	0	0	0	4	0	0	45	0
e1a-94	149	7	199	62	0	0	0	1	0	0	0	3	0	0	29	1
e1a-15	181	40	122	86	3	1	2	6	0	0	0	2	0	0	9	0
e1a-71	157	30	154	71	1	4	0	0	0	0	0	7	0	0	26	0
e1a-27	170	42	122	73	2	1	0	4	0	0	0	3	0	0	33	0
e1a-94	144	24	204	55	1	0	0	0	0	0	0	0	0	0	22	0
e1a-13	145	20	162	81	0	0	0	0	0	0	0	1	0	0	41	0
e1a-36	170	48	132	72	4	1	0	0	0	0	0	2	0	0	22	0
2. Amphitheatre Mountain Area																
swc-25	145	26	84	72	6	16	15	36	0	0	0	20	0	0	29	1
swc-87	202	30	67	49	9	17	3	28	0	0	0	28	0	0	17	0
swc-142	174	13	71	55	12	30	11	31	0	0	0	14	0	0	30	9
swc-192	151	24	93	61	19	13	14	41	0	0	0	8	0	0	18	8
am-14	190	16	82	76	5	21	14	2	0	0	2	5	16	8	28	12
am9-4	175	15	103	20	4	20	1	70	1	0	1	9	17	6	17	11
am11-u	151	27	65	70	13	9	2	68	2	0	9	3	0	7	29	0
am12-l	159	9	163	66	2	7	3	6	0	0	0	0	20	0	22	6
am-15	189	12	88	93	3	4	6	39	0	0	0	9	1	0	13	3
am9-3	171	12	104	38	6	39	3	49	4	0	2	6	7	3	16	21
am2-121	152	22	110	31	4	28	10	53	0	0	2	20	5	2	13	36
am9-6	151	27	81	33	10	42	4	59	5	0	4	5	4	0	17	22
am3-15	179	49	120	68	1	0	0	0	0	0	0	4	0	0	29	0
am2-15	143	32	148	76	4	2	1	0	0	0	0	10	0	0	33	1
am9-2	140	12	176	81	0	0	0	0	0	0	0	0	0	0	41	1
am-10-u	174	41	145	77	6	0	0	0	0	0	0	1	0	0	6	0
am-12-m	172	48	138	56	2	1	1	1	3	0	0	6	0	0	23	0
am-14	173	23	129	98	2	5	1	1	0	0	0	7	0	0	12	0
am-6	179	21	161	47	0	0	2	6	6	0	0	22	0	0	8	0
am-32	110	6	208	81	0	1	0	1	0	0	0	0	0	0	43	0
am-12-u	141	8	196	66	3	0	0	2	0	0	0	3	0	0	31	0
am-11-l	155	28	163	81	3	0	1	2	4	0	0	4	0	0	10	0
am-5	143	25	177	43	2	3	0	4	3	0	0	15	0	0	35	4
am9-5	134	9	214	55	0	0	2	0	0	0	0	0	0	0	38	0
am3-47	172	24	146	74	2	0	7	0	0	0	0	0	0	0	32	0
am-10-m	172	51	160	57	5	0	0	0	0	0	0	0	0	0	5	0
swc-165	110	36	103	52	23	59	0	13	2	0	0	22	0	0	31	0
swc-127	76	14	141	58	20	66	112	19	5	0	0	30	0	0	8	1
swc-120	123	28	141	68	10	16	6	18	0	0	0	10	3	0	23	5
swc-99	194	13	150	68	3	5	0	1	0	0	0	1	0	0	15	0
swc-74	144	15	147	67	4	10	0	12	3	0	0	9	1	0	36	1
swc-17	136	21	124	65	11	26	3	12	0	0	0	11	2	0	36	3
swc-57	95	7	168	69	1	8	0	3	0	0	0	13	0	0	85	1
am-12-u	179	52	149	53	8	1	0	0	1	0	0	3	0	0	4	0
am-13-u	138	17	214	69	2	0	0	0	0	0	0	0	0	0	10	0
ua-259	173	23	153	55	4	0	1	0	11	0	0	16	0	0	12	1
ua-309	94	11	227	58	0	1	2	11	4	0	0	18	0	0	23	1
ua-248	126	19	257	15	1	3	2	0	1	0	0	0	0	0	28	0
ua-263	207	8	128	62	2	9	2	14	2	0	5	5	7	2	23	1
ua-282	127	20	88	32	5	19	2	139	2	0	0	8	4	0	10	4
Burwash Creek Area																
bc3-10	147	15	155	17	2	21	26	25	2	0	0	8	10	0	11	19
bc1-5	165	17	85	68	3	20	8	52	0	0	2	5	4	0	13	9
bc1-1	170	18	206	37	1	0	0	0	0	0	0	0	0	0	18	4
bc2-8	141	39	179	47	2	0	0	0	0	0	0	4	0	0	38	1
bc2-12	121	33	219	45	2	0	2	2	0	0	0	8	0	0	20	0
bc4-5	191	39	178	33	1	0	1	0	0	0	0	3	0	0	5	3

APPENDIX 1. RAW POINT-COUNT DATA FROM THE AMPHITHEATRE FORMATION (continued)

Location	Qm	Qp	P	K	C	Lml	Lmh	Lv	Lsi	Larg	Lcarb	Lmud	Ch	S	M	D
								BURWASH BASIN (continued)								
3. Burwash Creek Area (continued)																
bc1-42	147	46	176	54	4	1	0	0	0	0	0	6	0	0	16	1
bc4-4	167	58	133	74	3	0	2	0	0	0	0	9	0	0	7	1
bc2-9	144	33	185	58	0	0	1	0	0	0	0	2	0	0	28	1
bc3-3	116	43	199	61	2	0	0	0	0	0	0	2	0	0	27	0
bc2-18	132	41	195	45	2	0	0	0	1	0	0	6	0	0	28	0
bc3-7	118	43	164	29	10	0	0	19	14	0	0	42	0	0	21	0
bc4-7	154	27	177	61	0	1	4	0	1	0	0	4	0	0	25	1
bc3-4	126	15	221	37	0	0	0	1	0	0	0	5	0	0	45	0
bc3-9	179	38	154	63	2	0	0	0	2	0	0	6	0	0	6	3
4. Steele Creek Area																
sc5-1	173	20	112	50	8	26	7	43	2	0	1	10	1	0	3	2
sc2-10	133	8	160	76	3	5	2	17	0	0	19	25	0	0	1	9
sc1-194	143	14	206	72	3	0	0	0	0	0	0	0	0	0	12	0
sc2-100	182	33	131	67	8	1	5	4	2	0	0	21	0	0	3	0
sc1-142	158	36	203	36	1	3	0	5	0	0	0	4	0	0	4	0
sc1-190	155	35	199	49	3	0	1	2	0	0	0	6	0	0	1	0
sc1-2	169	24	190	49	0	2	2	0	0	0	0	6	0	0	9	0
sc5-3	162	19	175	74	2	0	1	0	0	0	0	0	0	0	15	1
sc1-122	123	6	261	57	2	0	0	0	0	0	0	1	0	0	0	0
sc1-160	120	48	216	60	5	1	0	0	0	0	0	0	0	0	0	0
5. Cement Creek Area																
cc5-18	191	28	47	91	6	12	14	24	0	0	2	7	0	0	2	1
cc1-7	185	55	34	78	8	29	9	34	7	0	2	8	0	0	8	0
cc1-t-228	173	9	58	88	0	0	2	8	0	0	1	0	0	1	2	4
cc1-t-181	98	26	98	26	10	83	2	65	3	0	0	21	1	0	8	4
cc1-t-106	184	16	75	87	1	24	9	25	1	0	1	4	0	1	18	0
cc1-t-6	185	20	76	88	2	28	2	17	1	0	4	3	0	0	19	0
cc2-2	189	36	36	97	0	30	7	26	3	0	0	0	0	2	14	4
cc1-t-88	174	9	74	97	3	39	2	24	3	0	1	2	1	0	26	2
cc1-t-34	195	23	50	76	4	29	3	26	3	0	3	5	0	1	26	2
cc5-f	109	42	118	56	8	9	0	90	7	0	1	16	2	0	2	4
cc1-12	103	34	156	94	4	6	9	34	0	0	0	3	0	0	5	2
cc1-282	198	36	60	70	17	4	1	57	0	0	0	2	0	0	1	1
cc1-393	8	25	109	63	0	4	2	213	0	0	0	6	0	0	0	0
cc1-372	148	76	104	88	1	6	14	1	0	0	0	9	0	0	3	0
cc1-359	177	81	81	48	1	21	27	2	0	0	0	7	0	0	5	0
cc1-186	150	85	143	36	6	4	9	1	0	0	0	14	0	0	2	0
cc1-148	136	56	143	76	8	6	16	0	0	0	0	7	0	0	2	0
cc1-85	105	45	182	56	3	7	23	6	0	0	0	10	0	0	12	1
cc1-66	114	37	209	52	3	2	10	2	0	0	0	1	0	0	20	0
cc1-34	69	40	232	43	4	9	19	13	0	0	0	9	0	0	12	0
								BATES LAKE BASIN								
bl2-108	20	53	209	3	41	23	36	0	3	0	2	25	0	0	5	1
bl2-78	52	32	240	12	12	14	17	0	0	0	1	6	0	0	38	0
bl2-29	83	42	253	13	14	7	1	0	0	0	0	2	0	0	31	2
bl1-7	42	66	198	3	49	27	18	0	4	0	0	12	0	0	28	1
bl2-2	82	49	233	10	20	15	9	0	1	0	1	3	0	0	24	2
bl1-48	99	40	264	12	15	0	2	0	0	0	0	0	0	0	11	1
bl-f	68	38	278	21	9	1	4	0	0	0	0	1	0	0	29	1
bl-e-1	46	84	210	14	24	11	16	0	0	0	0	10	0	0	33	0
bl2-133	75	52	265	7	18	4	2	0	0	0	0	9	0	0	18	0
bl2-86	79	41	257	9	15	5	5	0	0	0	0	11	0	0	28	0
bl5-mz	47	87	210	2	28	15	24	0	2	0	0	8	0	0	26	0
bl2-125	70	93	194	8	12	16	13	0	0	0	0	4	0	0	36	4
ml-a-1	36	105	236	5	10	2	16	0	0	0	0	10	0	0	27	0

APPENDIX 1. RAW POINT-COUNT DATA FROM THE AMPHITHEATRE FORMATION (continued)

Location	Qm	Qp	P	K	C	Lml	Lmh	Lv	Lsi	Larg	Lcarb	Lmud	Ch	S	M	D
							BATES LAKE BASIN (continued)									
wc-231	64	19	299	36	14	1	1	0	0	0	0	1	0	0	12	0
wc-11.5	35	56	234	7	45	18	28	0	1	0	1	10	0	0	4	0
wc2-15	85	74	242	4	4	6	17	0	0	0	0	11	0	0	7	0
wc2-95	65	68	179	2	8	69	34	0	0	0	0	12	0	0	13	0
ml-c-1	63	55	201	37	3	17	5	5	1	0	0	51	0	0	13	0
bl2-38	50	68	155	16	14	52	30	0	1	0	0	60	0	0	14	2
bl2-7	29	64	244	7	16	127	52	0	1	0	0	28	0	0	9	0
bl2-54	63	65	276	9	3	72	17	0	1	0	0	8	0	0	8	0
bl2-156	48	63	276	15	6	58	22	0	0	0	0	4	0	0	16	0
bl1-0.5	33	75	249	17	10	164	21	2	5	0	0	31	0	0	8	0
bl1-91	68	113	198	10	12	119	19	1	0	0	0	25	0	0	9	0
bl1-29	49	89	255	15	11	61	6	0	2	0	0	38	0	0	15	1
bl-d-u	75	34	248	19	2	13	6	0	0	0	0	9	0	0	42	0
wc2-65	78	84	245	0	3	11	8	0	1	0	0	4	0	0	16	0
wc2-139	57	61	270	1	5	17	17	0	0	0	0	10	0	0	12	0
olb1-6	34	34	40	6	13	95	118	62	4	0	0	4	17	0	17	6
ol1-58	43	76	47	1	23	52	82	53	1	0	0	21	14	0	10	7
ol1-5	18	61	17	0	9	34	227	20	0	0	0	28	22	0	14	0
ola2-1	34	48	68	1	28	86	80	30	1	0	0	16	43	0	10	5

REFERENCES CITED

Basu, A., 1976, Petrology of Holocene fluvial sand derived from plutonic source rocks: Implications of paleoclimatic interpretation: Journal of Sedimentary Petrology, v. 46, p. 694–709.

Basu, A., 1985, Influence of climate and relief on compositions of sands released at source areas, *in* Zuffa, G. G., ed., Provenance of arenites: Dordrecht, D. Reidel, p. 1–18.

Campbell, R. B., and Dodds, C. J., 1982a, Geology of southwest Kluane Lake map area, Yukon Territory (115 G&F E1/2): Geological Survey of Canada Open File 829, scale 1:125,000.

Campbell, R. B., and Dodds, C. J., 1982b, Geology of the southwest Dezadeash map area, Yukon Territory (115A): Geological Survey of Canada Open File 831, scale 1:125,000.

Cole, R. B., and Ridgway, K. D., 1993, The influence of volcanism on fluvial depositional systems in a Cenozoic strike-slip basin, Denali fault system, Yukon Territory, Canada: Journal of Sedimentary Petrology, v. 63, p. 152–166.

Cole, R. B., Ridgway, K. D., and DeCelles, P. G., 1989, Late stage deposition within a Cenozoic strike-slip basin: Fluvial/basalt/pyroclastic transitions, St. Elias Mtns., Yukon Territory, Canada: Geological Society of America Abstracts with Programs, v. 21, p. A128.

Coney, P. J., Jones, D. L., and Monger, J.W.H., 1980, Cordilleran suspect terranes: Nature, v. 288, p. 329–333.

Crowell, J. C., and Link, M. H., 1982, Geologic history of Ridge Basin, eds., Society of Economic Paleontologists and Mineralogists, Pacific Section, Field Trip Guidebook, 304 p.

DeCelles, P. G., and Hertel, F., 1989, Petrology of fluvial sands from the Amazonian foreland basin, Peru and Bolivia: Geological Society of America Bulletin, v. 101, p. 1552–1562.

Dickinson, W. R., 1982, Compositions of sandstones in Circum-Pacific subduction complexes and fore-arc basins: American Association of Petroleum Geologists Bulletin, v. 66, p. 121–137.

Dickinson, W. R., 1985, Interpreting provenance relations from detrital modes of sandstones, *in* Zuffa, G. G., ed., Provenance of arenites: Dordrecht, D. Reidel, p. 333–361.

Dickinson, W. R., and Suczek, C. A., 1979, Plate tectonics and sandstone compositions: American Association of Petroleum Geologists Bulletin, v. 63, p. 2164–2182.

Dickinson, W. R., and 8 others, 1983, Provenance of North American Phanerozoic sandstones in relation to tectonic setting: Geological Society of America Bulletin, v. 94, p. 222–235.

Eisbacher, G. H., 1976, Sedimentology of the Dezadeash flysch and its implications for strike-slip faulting along the Denali Fault, Yukon Territory and Alaska: Canadian Journal of Earth Sciences, v. 13, p. 1495–1513.

Eisbacher, G. H., and Hopkins, S. L., 1977, Mid-Cenozoic paleogeomorphology and tectonic settings of the St. Elias Mountains, Yukon Territory: Geological Survey of Canada Paper 77-1B, p. 319–335.

Franzinelli, E., and Potter, P. E., 1983, Petrology, chemistry and texture of modern river sands, Amazon River System: Journal of Geology, v. 91, p. 23–39.

Gehrels, G. E., and Saleeby, J. B., 1987, Geologic framework, tectonic evolution, and displacement history of the Alexander terrane: Tectonics, v. 6, p. 151–174.

Hickman, R. G., Craddock, C., and Sherwood, K. W., 1978, The Denali Fault System and the tectonic development of southern Alaska: Tectonophysics, v. 47, p. 247–273.

Ingersoll, R. V., 1978, Petrofacies and petrologic evolution of the late Cretaceous fore-arc basin, northern and central California: Journal of Geology, v. 86, p. 335–352.

Ingersoll, R. V., Ballard, T. F., Ford, R. L., Grimm, J. P., Pickle, J. D., and Sares, S. W., 1984, The effect of grain size on detrital modes: A test of the Gazzi-Dickinson point-counting method: Journal of Sedimentary Petrology, v. 54, p. 103–116.

Johnsson, M. J., 1990, Tectonic versus chemical-weathering controls on the composition of fluvial sands in tropical environments: Sedimentology, v. 37, p. 713–726.

Johnsson, M. J., Stallard, R. F., and Meade, R. H., 1988, First-cycle quartz arenites in the Orinoco River basin, Venezuela and Colombia: Journal of Geology, v. 96, p. 263–277.

Jones, D. L., Silberling, N. J., and Hillhouse, J., 1977, Wrangellia—A displaced terrane in northwestern North America: Canadian Journal of Earth Sciences, v. 14, p. 2565–2577.

Lanphere, M. L., 1978, Displacement history of the Denali fault system, Alaska and Canada: Canadian Journal of Earth Sciences, v. 15, p. 815–822.

LeCouteur, P. C., and Tempelman-Kluit, D. J., 1976, Rb/Sr ages and a profile of initial Sr87/Sr86 ratios for plutonic rocks across the Yukon Crystalline

Terrane: Canadian Journal of Earth Sciences, v. 13, p. 319–330.

Link, M. H., 1982, Petrography and chemistry of sedimentary rocks, Ridge Basin, southern California, *in* Crowell, J. C., and Link, M. H., eds., Geologic history of Ridge Basin, southern California: Society of Economic Paleontologists and Mineralogists, Pacific Section, Field Trip Guidebook, p. 159–180.

Monger, J.W.H., 1984, Cordilleran tectonics: A Canadian perspective: Bulletin de la Societe Géologique de France, v. 7, p. 255–278.

Monger, J.W.H., Price, R. A., and Tempelman-Kluit, D. J., 1982, Tectonic accretion and the origin of the two major metamorphic and plutonic welts in the Canadian Cordillera: Geology, v. 10, p. 70–75.

Mortensen, J. K., and Jilson, G. A., 1985, Evolution of the Yukon-Tanana terrane: evidence from southeastern Yukon Territory: Geology, v. 13, p. 806–810.

Nokleberg, W. J., Jones, D. L., and Silberling, N. J., 1985, Origin and tectonic evidence of the Maclaren and Wrangellia terranes, eastern Alaska Range, Alaska: Geological Society of America Bulletin, v. 96, p. 1251–1270.

Parrish, J. T., Ziegler, A. M., and Scotese, C. R., 1982, Rainfall patterns and the distribution of coals and evaporites in the Mesozoic and Cenozoic: Palaeogeography, Palaeoclimatology, Palaeoecology, v. 40, p. 67–101.

Read, P. B., and Monger, J.W.H., 1976, Pre-Cenozoic volcanic assemblages of the Kluane and Alsek Ranges, southwestern Yukon Territory: Geological Survey of Canada Open File 381, 97 p.

Ridgway, K. D., 1992, Cenozoic tectonics of the Denali fault system, Saint Elias Mountains, Yukon Territory: synorogenic sedimentation, basin development, and deformation along a transform fault system [Ph.D. thesis]: Rochester, N.Y., University of Rochester, 508 p.

Ridgway, K. D., and DeCelles, P. G., 1991a, Stream-dominated alluvial-fan and lacustrine depositional systems in Cenozoic strike-slip basins, Denali fault system, Yukon Territory, Canada: Geological Society of America Abstracts with Programs, v. 23, p. A462.

Ridgway, K. D., and DeCelles, P. G., 1991b, Tectonic controls on depositional systems in non-marine strike-slip basins, Denali fault system, Yukon Territory: Geological Society of America Abstracts with Programs, v. 23, p. A92.

Ridgway, K. D. and DeCelles, P. G., 1993, Stream-dominated alluvial-fan and lacustrine depositional systems in Cenozoic strike-slip basins, Denali fault system, Yukon Territory, Canada: Sedimentology, v. 40, in press.

Ridgway, K. D., Cameron, A. R., Sweet, A. R., and DeCelles, P. G., 1989, Evolution of a strike-slip basin as defined by sedimentology, palynology, and organic petrology, St. Elias Mountains, Yukon Territory, Canada: Geological Society of America Abstracts with Programs, v. 22, p. 80.

Ridgway, K. D., DeCelles, P. G., Cameron, A. R., and Sweet, A. R., 1992a, Cenozoic syntectonic sedimentation and strike-slip basin development along the Denali fault system, Yukon Territory: Yukon Geology, v. 3, p. 1–26.

Ridgway, K. D., Skulski, T., and Sweet, A. R., 1992b, Cenozoic displacement along the Denali fault system, Yukon Territory [abs.]: EOS (American Geophysical Union Transactions), v. 73, p. 534.

Schwab, F. L., 1986, Sedimentary "signatures" of foreland basin assemblages: Real or counterfeit?, *in* Allen, P. A. and Homewood, P., eds., Foreland basins: International Association of Sedimentologists Special Publication 8, p. 395–410.

Sengor, A. M., 1990, Plate tectonics and orogenic research for 25 years: A Tethyan perspective: Earth Science Reviews, v. 27, p. 1–201.

Skulski, T., 1988, The origin and setting of anomalous arc magmatism in the Wrangell volcanic belt, southwest Yukon: Yukon Geology, v. 2, p. 88–89.

Skulski, T., Francis, D., and Ludden, J., 1992, Volcanism in an arc-transform transition zone: The stratigraphy of the St. Clare volcanic field, Wrangell volcanic belt, Yukon, Canada: Canadian Journal of Earth Science, v. 29, p. 446–461.

Souther, J. G., and Stanciu, C., 1975, Operation St. Elias, Yukon Territory: Tertiary volcanic rocks: Geological Survey of Canada Paper 75-1, part A, p. 63–70.

Suttner, L. J., Basu, A., and Mack, G. H., 1981, Climate and the origin of quartz arenites: Journal of Sedimentary Petrology, v. 51, p. 1235–1246.

Sylvester, A. G., 1988, Strike-slip faults: Geological Society of America Bulletin, v. 100, p. 1666–1703.

Tempelman-Kluit, D. J., 1976, The Yukon Crystalline Terrane: Enigma in the Canadian Cordillera: Geological Society of America Bulletin, v. 87, p. 1343–1357.

Tempelman-Kluit, D. J., 1979, Transported cataclasite, ophiolite, and granodiorite in Yukon: Evidence of arc-continent collision: Geological Survey of Canada Paper 79-14, 27 p.

Valloni, R., 1985, Reading provenance from modern marine sands, *in* Zuffa, G. G., ed., Provenance of arenites: Dordrecht, D. Reidel, p. 309–332.

Manuscript Accepted by the Society December 24, 1992

Geological Society of America
Special Paper 284
1993

Provenance evolution of Upper Paleozoic sandstones of north-central New Mexico

Kathleen A. Devaney* and Raymond V. Ingersoll
Department of Earth and Space Sciences, University of California, Los Angeles, California 90024-1567

ABSTRACT

Sandstone composition is a function of four complex and interrelated variables: provenance (itself a function of source rock, relief, and climate), transportation effects, depositional environment, and diagenesis. Documentation of source rock composition and comparison to derivative sandstone composition not only allow evaluation of sandstone provenance changes through time, they also provide evaluation of the relative importance of the variables controlling sandstone composition. Source rock composition is best determined by Gazzi-Dickinson point counts of the source rocks themselves, thus allowing direct comparisons between source rock and sandstone composition and taking into account source rock texture.

This study uses this technique combined with study of sandstone composition to pursue two main objectives: to investigate the relationship between tectonic history and sandstone composition, and to document sandstone compositions of the Pennsylvanian-Permian of north-central New Mexico. Upper Paleozoic strata of north-central New Mexico were derived from Precambrian crystalline rocks and were associated with the Ancestral Rocky Mountain orogeny. They display subtle changes in detrital modes with age that can be correlated with the area's Ancestral Rocky Mountain tectonic history. Periods of high tectonic activity resulted in deposition of sandstones with compositions closest to those of the Precambrian source rocks, so high tectonic activity allows source rock control to overwhelm other factors such as climate. During periods of low tectonic activity, other factors such as climate or transport/depositional environment were more important.

INTRODUCTION

Documentation of source rock composition is a crucial step in the study of sandstone provenance. Previous studies of crystalline source rocks have focused on specific attributes (e.g., Blatt and Christie [1963], Basu [1976, 1985] and other workers on quartz types), or have relied on regional studies by igneous and metamorphic petrologists, whose differing goals and techniques preclude direct comparisons between source rock composition and their derivative sandstones. Many workers bridge the gap between source rock and sandstone composition by collecting and studying modern sands (e.g., Ingersoll, 1990), and this is a valuable technique. We propose an additional technique: the direct application of the Gazzi-Dickinson point-counting method to the source rocks themselves.

This technique not only records the source rock composition in the same categories as used in the sandstones, it also takes into account source rock texture, whose importance is demonstrated by Heins (1992; this volume). Use of this technique allows direct comparison between crystalline source rock and derivative sandstone composition, allowing increased accuracy in evaluation of the relative importance of the complex factors affecting sandstone composition.

This study uses this technique combined with study of sandstone composition to pursue two main objectives: to investigate the relationship between tectonic history and sandstone composition, and to document sandstone compositions of the Pennsylvanian-Permian of north-central New Mexico. The Sangre de

*Present address: 4792 Sandia Drive, Los Alamos, New Mexico 87544.

Devaney, K. A., and Ingersoll, R. V., 1993, Provenance evolution of Upper Paleozoic sandstones of north-central New Mexico, *in* Johnsson, M. J., and Basu, A., eds., Processes Controlling the Composition of Clastic Sediments: Boulder, Colorado, Geological Society of America Special Paper 284.

Cristo, Sandia-Manzanita-Manzano (including Monte Largo), and Jemez-Nacimiento Mountains not only expose Pennsylvanian-Permian strata, they also expose Precambrian rocks representative of the Pennsylvanian-Permian source rocks, allowing direct comparison (Fig. 1).

Deposition of upper Paleozoic strata of north-central New Mexico was associated with the Ancestral Rocky Mountain orogeny. The Ancestral Rocky Mountains formed a series of uplifts and basins in what are now the Rocky Mountains of western North America. They developed during the Pennsylvanian-Permian, probably as an intraplate response to Ouachita-Marathon deformation, and their development exposed Proterozoic metamorphic and granitic terranes to erosion (Sutherland, 1963; Kluth and Coney, 1981; Kluth, 1986).

Paleozoic strata of the Sangre de Cristo Mountains were deposited in a major Ancestral Rocky Mountain basin, the Taos Trough (also called the Rowe-Mora basin); those of the Sandia-Manzanita-Manzano Mountains represent the edge of the Ancestral Estancia basin (Fig. 2). The Ancestral Estancia basin may have received sediment from the Pedernal uplift. The Paleozoic strata of the Jemez-Nacimiento Mountains are not associated with a major basin, but represent sediment shed off the opposite flank of one of the Taos Trough's source areas, the Uncompahgre uplift and from a smaller uplift, the Peñasco. Transport distances are less than 70 km for all areas (Dane and Bachman, 1965).

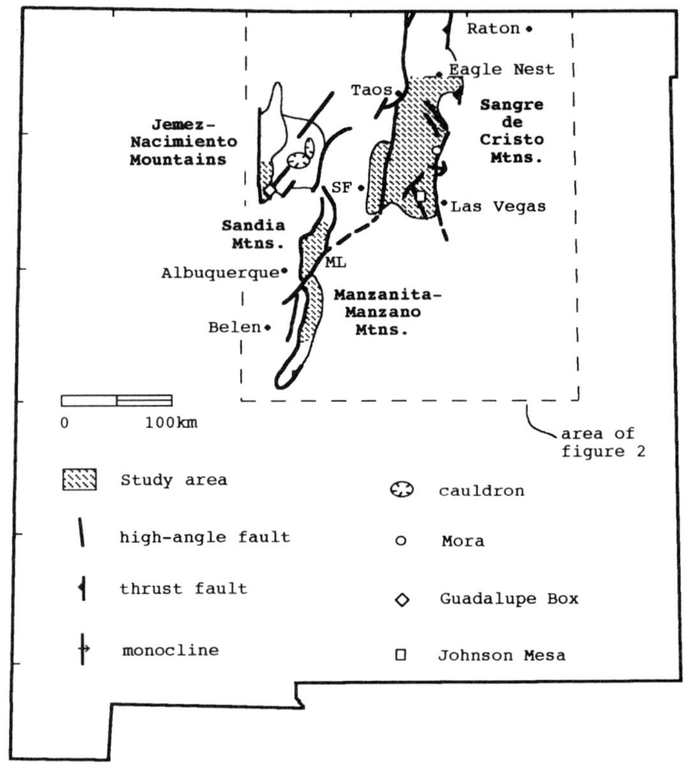

Figure 1. Tectonic map of north-central New Mexico showing the study areas and some locations discussed in the text. ML = Monte Largo; SF = Santa Fé. After Kelley, 1982.

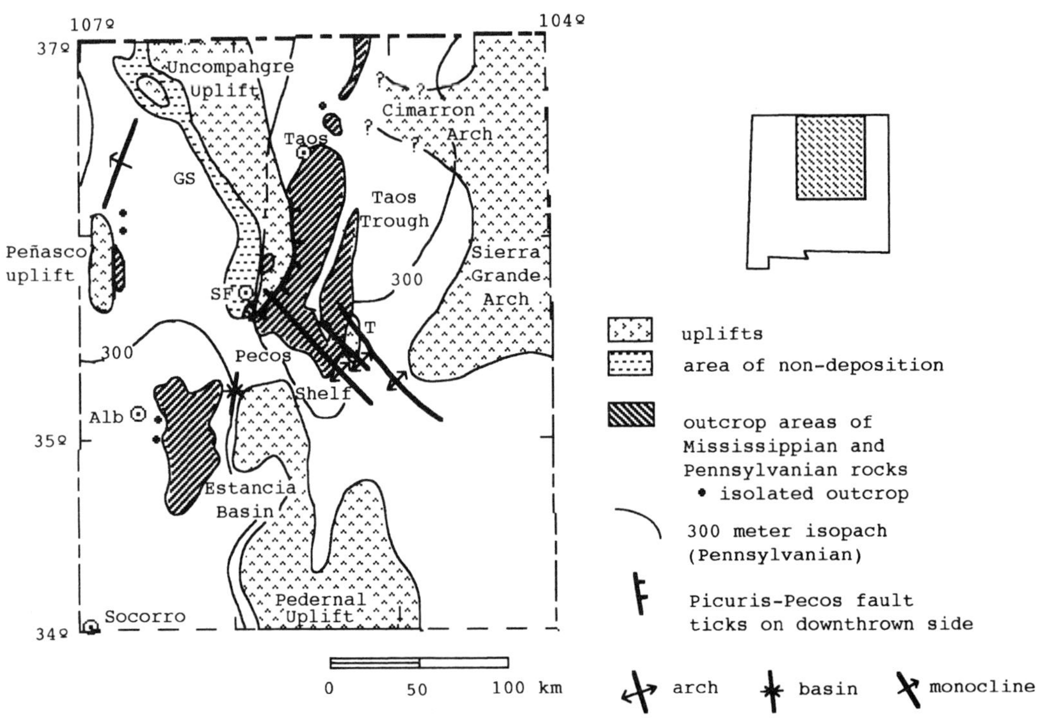

Figure 2. Map showing the Ancestral Rocky Mountain uplifts and basins of north-central New Mexico in relation to the present outcrops of Mississippian and Pennsylvanian rocks. GS = Gallina Shelf; SF = Santa Fé; T = Tecolote uplift. Compiled from Armstrong et al., 1979, and Dickerson, 1984.

STRATIGRAPHY AND DEPOSITIONAL ENVIRONMENTS

In the Sangre de Cristo and Sandia-Manzanita-Manzano Mountains, Proterozoic rocks are overlain in most places by the Mississippian Arroyo Peñasco Group (Armstrong, 1967; Baltz and Myers, 1984). In the Nacimiento Mountains, the Mississippian Arroyo Peñasco Group and Log Springs Formation are preserved only in small blocks downwarped prior to extensive erosion and deposition of Pennsylvanian strata (Woodward, 1987). Although the Mississippian section includes some siliciclastic sandstone, it is dominantly carbonate (Armstrong, 1967; Armstrong and Holcomb, 1989), and only a few sandstone samples from the Mississippian Arroyo Peñasco Group of the Sierra Nacimiento preserve provenance information well enough to be included in this study.

The Mississippian Arroyo Peñasco Group represents a time of transgression and regression on a carbonate platform (Armstrong and Mamet, 1974, 1977, 1979; Armstrong and Holcomb, 1989). The Log Springs Formation, in the Sandia and Jemez-Nacimineto area, was deposited in a continental-fluvial environment (Armstrong and Mamet, 1974; Armstrong and Holcomb, 1989).

Pennsylvanian nomenclature is notoriously complicated (Fig. 3). Lithofacies correlation has proven extremely difficult due to rapid facies changes (Sutherland, 1963). The lithostratigraphic classification used in the study area divides Pennsylvanian strata into the Sandia and Madera Formations (Read and others, 1944; Read and Wood, 1947). The Madera Formation is further subdivided into a lower limestone member and an upper arkosic limestone member (Read and others, 1944). Overlying nonmarine arkoses and red beds are excluded from the Sandia and Madera Formations; they are classified as the Sangre de Cristo Formation of Pennsylvanian-Permian age in the Sangre de Cristo Mountains and the Abo Formation in the Sandia-Manzanita-Manzano and Jemez-Nacimiento Mountains (Fig. 3).

Sangre de Cristo Mountains

Sedimentation in the Taos Trough was controlled by the tectonic activity of adjacent uplifts, resulting in deposition of deep-basin mud, shallow-shelf carbonate, coarse-grained sandstone, and proximal conglomerate; facies change rapidly both laterally and vertically (Sutherland, 1963). More than 2,300 m of sediment were deposited in the basin (Armstrong et al., 1979).

Baltz and Myers (1984) raised the Madera Formation, in the southeastern Sangre de Cristo Mountains, to group level and subdivided into the Porvenir Formation (new name) and the Alamitos Formation (originally defined by Sutherland, 1963) (Baltz and Myers, 1984). The Porvenir Formation is the lower limestone member of the Madera, dated as upper Desmoinesian using fusilinid data (Baltz and Myers, 1984).

The Morrowan section of the Sandia Formation represents predominantly shallow-marine nearshore sequences formed during periodic progradation of muddy strandplains into shallow-marine basins (Sutherland, 1963; Casey, 1980). The rest of the Sandia Formation in the Taos area records a period of increasing tectonic activity, resulting in deposition of coarse alluvial fan, braided stream, and fan delta complexes that prograded eastward into the basin (Casey, 1980). In the Pecos area, the Sandia Formation is primarily a shallow-marine nearshore sequence (Sutherland, 1963).

The Porvenir Formation is mainly marine, characteristic of deposition in the shallow sea of the Pecos Shelf (Baltz and Myers, 1984). Farther north, in the Taos area, sediment produced by uplift of the Cimarron arch in the earliest Desmoinesian, combined with sediment from the Uncompahgre and Sierra Grande uplifts, formed continental, alluvial fan, fluvial, and fan delta complexes that prograded onto the shallow shelf (Sutherland, 1963; Casey, 1980).

The lower Alamitos represents deposition of alluvial fans and deltas derived from the Uncompahgre uplift in the west alternating with periodic marine incursions (Sutherland, 1963). The upper Alamitos represents the final marine transgression into the southern Taos Trough, alternating with alluvial fan deposition (Sutherland, 1963).

The Sangre de Cristo Formation was deposited by extensive anastomosing alluvial fans. Oldest in the north, they prograded south and eventually filled the Taos Trough (Sutherland, 1963; Brown, 1984; Caldwell, 1987).

Sandia-Manzanita Mountains

The Pennsylvanian section of the Sandia-Manzanita-Manzano Mountains is considerably thinner than that exposed in the Sangre de Cristo Mountains (Kelley and Northrop, 1975; Myers, 1982). The Sandia Formation is dominated by siliciclastics (sandstone, shale and, to a lesser extent, conglomerate and siltstone), and represents a transgressive sequence deposited in shallow-marine and coastal environments as shallow seas covered the peneplained Precambrian surface (Read and Wood, 1947; Kottlowski, 1961; Ingersoll and Kelley, 1979).

Marine limestone dominates the Madera, with variable amounts of shale, sandstone, and conglomerate (Kelley and Northrop, 1975). Siliciclastic strata are more common in the upper part, the upper arkosic limestone member (Read et al., 1944; Myers, 1973, 1982). The lower member is dominated by limestone that caps Sandia Crest (Kelley and Northrop, 1975).

The lower Madera represents the time of maximum marine transgression (Read and Wood, 1947; Bachman, 1975; Casey, 1980). Depositional environments range from shallow-marine to nearshore to swamp and lagoonal to nonmarine (Ingersoll and Kelley, 1979). Alternation of marine and nonmarine deposition at many locations indicates cycles of transgression and regression (Ingersoll and Kelley, 1979). Wiberg and Smith (1991) noted that facies in the Sandia Mountains include delta plain terrigenous clastics, subtidal carbonates, and transgressively reworked sheet sandstones.

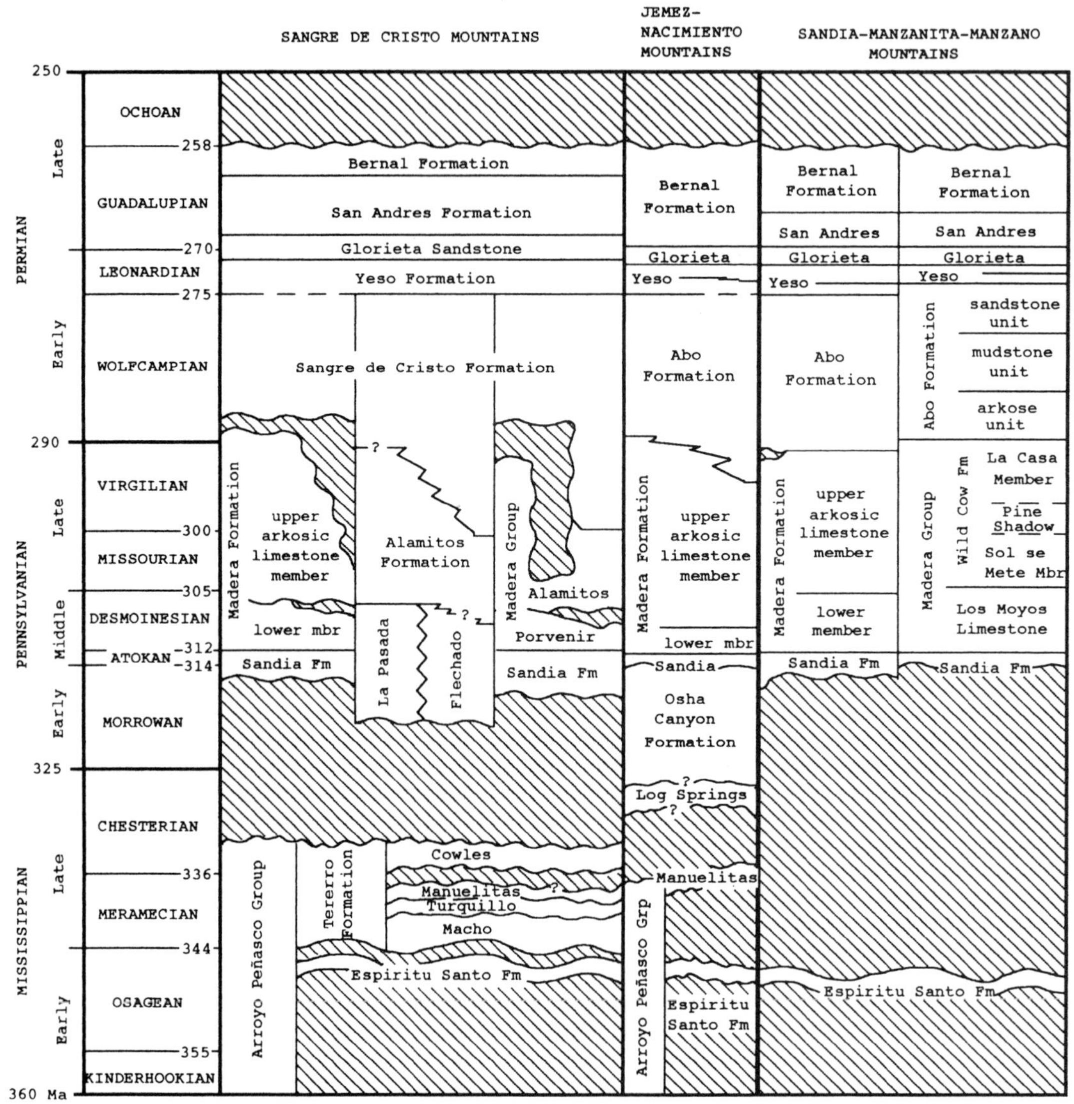

Figure 3. Upper Paleozoic stratigraphy of north-central New Mexico. The Log Springs Formation is present only as a thin paleosoil in the Sandia Mountains and is shown only for the Jemez-Nacimiento. The Sangre de Cristo Mountains columns in the Pennsylvanian represent the work of different authors: the first column is compiled from Read and Wood (1947) and Baltz (1972); the second column represents Sutherland's (1963) definitions; the third, the revisions by Baltz and Myers (1984). The three columns represent the transition from north to south, reading from left to right. In the Sandia-Manzanita-Manzano Mountains, the right column represents Myers's (1973) redefinition of the Pennsylvanian formations for the Manzano Mountains, and Hatchell and others' (1982) subdivision of the Abo Formation, also for the Manzano Mountains. Ages are in millions of years before present and are from Haq and Van Eysinga (1987). Also compiled from Armstrong and others (1979), Baars (1961, 1974), DuChene (1973), Kelley and Northrop (1975), Lucas and Hayden (1991), and Woodward (1987).

The Abo Formation consists of alternating sequences of dominantly reddish brown mudstone and sandstone with minor limestone and conglomerate, and represents deposition in fluvial-alluvial channels surrounded by floodplains (Kelley and Northrop, 1975; Ingersoll and Kelley, 1979).

Jemez-Nacimiento Mountains

The Sandia, Madera (both the upper and lower members), and Abo Formations were recognized in the Jemez-Nacimiento area by Wood and Northrop (1946). They split the Sandia into a lower limestone member and an upper clastic member. The lower limestone member and the lowermost part of the upper member have been reclassified as the Morrowan Osha Canyon Formation. The Sandia Formation has been restricted to the upper part of the upper clastic member (Woodward, 1987).

Pennsylvanian strata are thickest and best studied in the Guadalupe Box area of the Sierra Nacimiento (Fig. 1). The following is from DuChene (1973). The Sandia consists of coarse-grained, ledge-forming quartz sandstone overlain by interbedded shale, fine-grained silty sandstone, and highly fossiliferous limestone.

The lower limestone member of the Madera consists of fossiliferous, locally cherty limestone interbedded with thin beds of arkosic sandstone and fossiliferous shale. The lower member grades into the upper arkosic limestone member, which consists of arkosic limestone and arkose intercalated with highly fossiliferous calcareous shale. The amount of arkose increases upsection and the amount of carbonate decreases. The Abo Formation in the Guadalupe Box area consists of brownish red mudstone and shale interbedded with lenses of dark red, poorly sorted, coarse-grained arkosic sandstone (DuChene, 1973). The lower part of the formation includes fossiliferous limestone, with thinner and increasingly nodular limestone upsection (DuChene, 1973). The Abo represents predominantly alluvial plain/floodplain deposition with periodic marine incursions (Crabaugh and McPherson, 1985; Woodward, 1987).

Yeso Formation—All areas

In general, the Yeso consists of reddish orange sandstone, siltstone, and shale with some limestone in the Sandia-Manzanita-Manzano and Jemez-Nacimiento areas (Foster and others, 1972; DuChene, 1973; Martinez, 1974; Kelley and Northrop, 1975). In the Sandia-Manzanita-Manzano area, it contains gypsum locally (Kelley and Northrop, 1975). Salt deposits occur in the Yeso in the Las Vegas basin subsurface (Foster and others, 1972). The Yeso represents continental deposition transitional to the marine conditions of southern New Mexico, and so includes arid backreef-seaway or restricted lagoonal deposits, as well as some eolian deposits (Baars, 1961; Ingersoll and Kelley, 1979).

Glorieta Sandstone—All areas

The Yeso is overlain by the Glorieta Sandstone. In general, the Glorieta consists of fine- to medium-grained sandstone with minor shale (Foster and others, 1972; Martinez, 1974; Kelley and Northrop, 1975; Ingersoll and Kelley, 1979). Sandstones are generally well sorted; some contain frosted and polished quartz grains (Schowalter, 1969; Martinez, 1974). It represents high-energy littoral environment with local offshore-bar and eolian deposits (Baars, 1961).

SANDSTONE PROVENANCE

Sandstone composition is a function of four interrelated and variable factors: provenance, transportation, depositional environment, and diagenesis (Suttner, 1974). Sandstone provenance is strongly dependent on source rock composition, as well as weathering and relief in the source area (Suttner, 1974). "Tectonism exerts a dominating and overriding influence on sedimentation" (Suttner, 1974, p. 76), by influencing all four controlling factors to some extent, that of source rock most directly. By preferential removal of unstable grains, climate-controlled weathering may destroy source rock information; however, climatic destruction depends, in part, on residence time (Suttner, 1974; Basu, 1985). In areas of high relief (steep hill slopes), residence time may be low and grain destruction minor (Basu, 1985). Since diagenetic processes may alter or destroy unstable grains, it is vital to reconstruct the detrital composition of altered grains whenever possible and to reject those samples with evidence for abundant alteration (Helmold, 1985). Compositional variation caused by variation in depositional environment may be reduced by using the Gazzi-Dickinson method of point counting, which minimizes compositional dependence on grain size (Ingersoll et al., 1984).

All formations studied display the same types of diagenetic alteration to varying degrees. Feldspar alteration includes replacement by three general classes of minerals—clay minerals, carbonate minerals, and sericite—as well as by dissolution and/or development of vacuoles. Clay alteration is the most common and destructive feldspar alteration, the hardest to "look through" to the original grain composition; hence, sandstones with high percentages of interstitial clay (I) or clay alteration had to be excluded. Plagioclase is almost always more extensively altered and with a more complex diagenetic history than potassium feldspar (see also Fekete, 1988). Feldspar alteration is also present, although not as extensive, in Precambrian rocks and, as with upper Paleozoic strata, plagioclase is generally more altered than potassium feldspar.

Fine-grained metamorphic lithic fragments, especially those with dense minerals, are rarely found, partly because the metamorphic source areas are dominated by coarse-grained quartzite. In addition, diagenetic destruction as well as weathering of the source area may have been factors.

Sandia Formation sandstones have the highest number of rejected samples due to high I, which is consistent with their deeper burial depths; samples from the overlying Madera-Porvenir-Alamitos Formations do not contain high amounts of I, whereas a handful of samples from the Sangre de Cristo, Abo, and Yeso Formations contain sufficient I to warrant rejection.

If diagenesis was the main control of sandstone composition in north-central New Mexico, then the observed compositional trends with decreasing age would be the following: increase in P/F, QFL%F, LmLvLs%Lm, and framework% dense minerals. (This trend exactly parallels that of climate for north-central New Mexico.) This is not the case for sandstones with low I. The lowermost sandstones contain the highest QFLm%Lm and frm%D, and have similar P/F and QFLm%F to less deeply buried and younger sandstones.

METHODS

The main study area, and therefore the main sampling area, was the Sangre de Cristo Mountains (Fig. 1). The Sandia-Manzanita-Manzano area exposes a thinner upper Paleozoic section over a smaller geographic area than the Sangre de Cristo

Mountains. Samples were collected from throughout the Sandia Mountains, from Monte Largo, and from three locations in the Manzanita–northern Manzano Mountains. Samples from the Jemez-Nacimiento Mountains were collected from only a few areas: from the most complete upper Paleozoic section exposed at Guadalupe Box and from good exposures along Jemez Canyon. For sample location maps (as well as point-count data), see Devaney (1992).

All sandstones (of all sand-size ranges and mixed limestone-sandstone) and basement rocks in this study were point counted using the Gazzi-Dickinson method (Gazzi, 1966; Dickinson, 1970; Ingersoll et al., 1984). Basement counts are achieved by an extension of the Gazzi-Dickinson method of classifying lithic fragments: coarse components (>0.0625 mm) are counted as individual mineral grains and fine components (<0.0625 mm) are counted as lithic fragments. A minimum of 300 points were counted per slide; most were 400-point counts, and a few were 500-point counts. Point-count spacing varied with grain size; the largest spacing was used for the coarsest sand, to avoid counting a single grain more than once. The type of feldspar alteration and amount of interstitial material was noted, and thin sections with more than 20% interstitial material or grain alteration were rejected.

All thin sections were etched by immersion in concentrated hydrofluoric acid for 2 to 3 s and stained with sodium cobaltinitrite for potassium feldspar. Staining for plagioclase was not performed, since the acid etching serves to distinguish feldspar from quartz, and the potassium-feldspar stain differentiates feldspar types.

The point-counting categories and recalculated parameters are defined in Table 1. In the Gazzi-Dickinson method, granitic lithic fragments do not exist, only volcanic-hypabyssal, metamorphic, and sedimentary. In this study, only metamorphic and sedimentary lithic fragments were found. The metamorphic lithic fragments are derived from Proterozoic source rocks, but the sedimentary lithic fragments are generally different types of carbonate grains; most are likely intrabasinal (Sutherland, 1963). Except in sandy limestones, sedimentary lithic fragments are very minor components. Some were apparently derived from underlying limestone formations, but most show characteristics of intrabasinal provenance by Zuffa's definitions (1985), that is, they are rip-up clasts or are typically coarser than the terrigenous components, are generally bioclasts (especially echinoid debris) of roughly the same age as the formations in which they are found, and do not contain internal veins. No lower Paleozoic strata are found in north-central New Mexico due to the presence of the Transcontinental Arch (e.g., Dott and Batten, 1988). Upper Paleozoic strata were deposited directly on, and in many cases, apparently derived from, Precambrian crystalline rocks. Therefore, due to the probable lack of sedimentary source rocks, the intrabasinal character of many sedimentary lithic fragments, the low overall percentage of sedimentary lithic fragments, and little evidence of recycling (broken overgrowths are exceedingly rare), these sedimentary lithic fragments are not significant indicators of source rocks. The only lithic fragments that represent source rocks and so are used in the present study are metamorphic; therefore, QFL ternary plots become QFLm plots.

TABLE 1. CLASSIFICATION OF SYMBOLS OF GRAIN TYPES*

Quartzose grains (Q = Qm + Qp)
- Q = total quartz grains
- Qm = monocrystalline quartz (>0.0625 mm)
- Qp = polycrystalline quartz

Feldspar (F = P + K)
- F = total feldspar
- P = plagioclase
- K = total potassium feldspar
  - Kt = twinned potassium feldspar
  - Ku = untwinned potassium feldspar

Lithic grains (L = Lm + Ls)
- L = total lithic fragments
- Lm = total metamorphic lithic fragments
- Ls = total sedimentary lithic fragments

*After Dickinson, 1985.

Most of our ternary plots show data, means, and fields of variation around the mean. The "field of variation" is based on standard deviations, and although a useful representation of the variability of the data, is not a strictly defined statistical measure (Ingersoll and Suczek, 1979).

PETROLOGY

Precambrian source rocks

The Ancestral Rocky Mountain uplifts exposed Precambrian-cored highlands, including the Uncompahgre in Colorado and northern New Mexico and the Pedernal in central to southern New Mexico (Armstrong et al., 1979; Chronic, 1979). Precambrian rocks can be divided into two compositionally and geographically distinct groups: granitic and metamorphic rocks. The metamorphic rocks are a complex suite of metasedimentary and metavolcanic rocks, predominantly quartzite and greenstone in the Sangre de Cristo and Sandia-Manzanita-Manzano Mountains, and gneissic complexes in the Jemez-Nacimiento Mountains and Monte Largo (Montgomery, 1963; Kelley and Northrop, 1975; Condie, 1981; Cavin and others, 1982; Woodward, 1987). The approximate eastern extent of the Uncompahgre uplift is defined by the Picuris-Pecos Fault (Fig. 2). In the current exposures west of the Picuris-Pecos Fault, metamorphic rocks dominate to the north, in the Picuris Range, and granitic rocks dominate to the south, in the Santa Fé Range (Fig. 4). The Cimarron arch also contributed sediment to the Taos Trough (Casey, 1980). Dominantly granitic Precambrian rocks of the Cimarron arch are exposed in the Eagle Nest area (Callender et al., 1976; Condie, 1981).

Granitic rocks are too coarse, in general, to yield lithic fragments using the Gazzi-Dickinson method, and plot along the QF

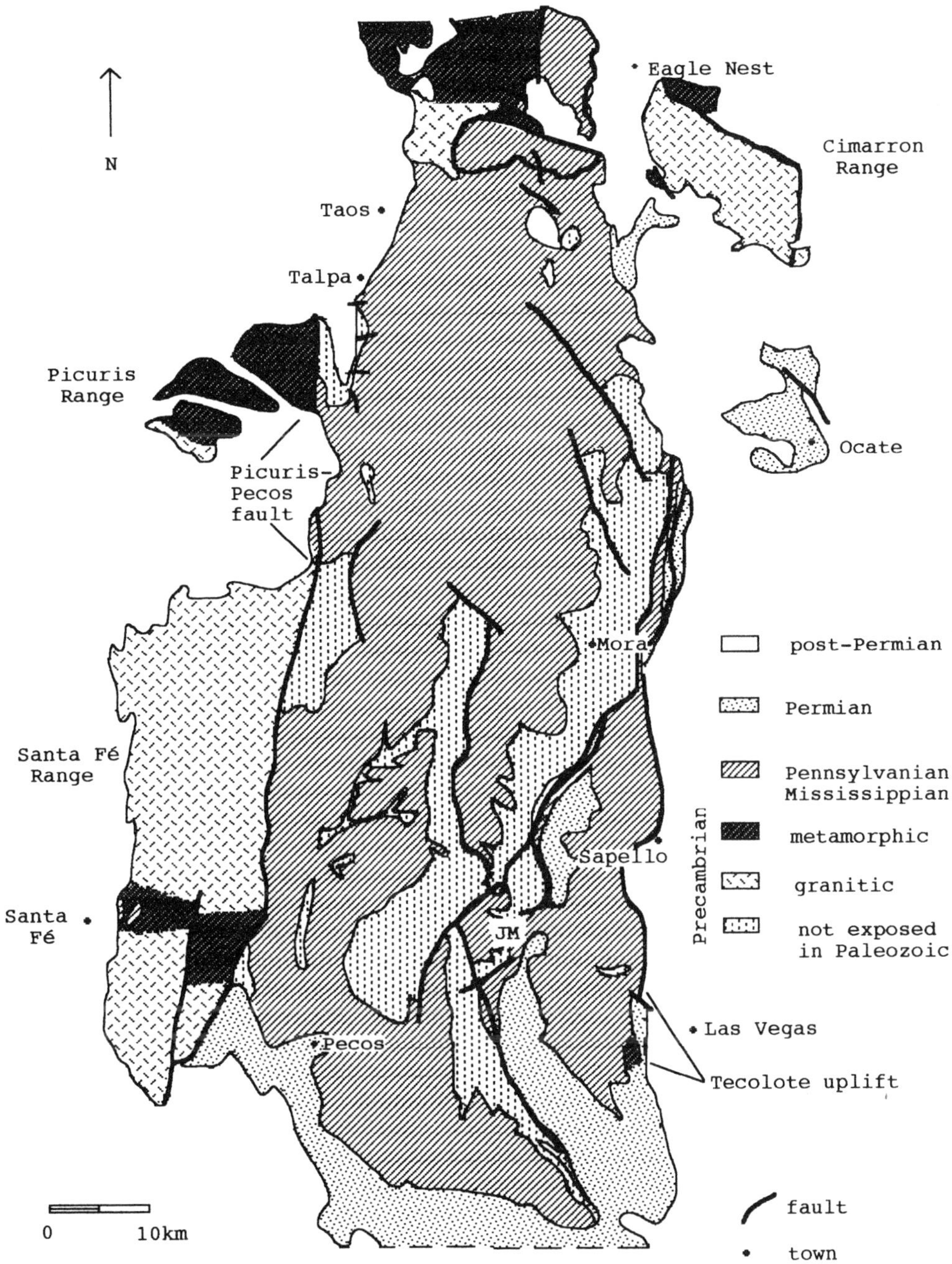

Figure 4. Generalized geological map of the southern Sangre de Cristo Mountains showing the present exposure of Precambrian source rocks and Paleozoic strata. JM = Johnson Mesa. Precambrian rocks are divided into three groups: rocks of dominantly granitic compositions exposed in the Santa Fé and Cimarron Ranges and northern Tecolote uplift; dominantly metamorphic rocks exposed in the Picuris Range and southern Tecolote uplift; both metamorphic and granitic rocks not exposed during Ancestral Rocky deformation. After Montgomery, 1963; Sutherland, 1963; Baltz, 1965; and Dane and Bachman, 1965.

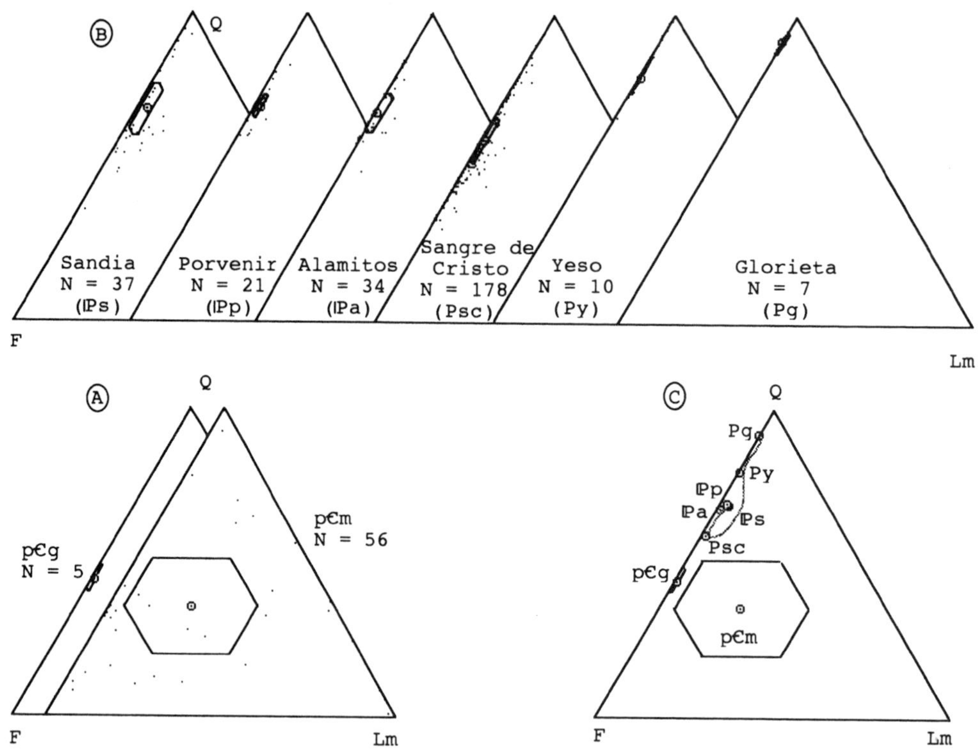

Figure 5. QFLm ternary plots for Precambrian rocks and Paleozoic sandstones of the Sangre de Cristo Mountains. See Table 1 for abbreviations. A, QFLm compositions of the Precambrian source rocks grouped by rock type, where p€g are Precambrian granitic rocks and p€m are Precambrian metamorphic rocks. B, QFLm compositions of Pennsylvanian and Permian sandstones. C, Paleozoic provenance evolution, showing the means for each formation and the Precambrian fields of variation for reference.

side of the QFLm ternary plot (Fig. 5A). Many metamorphic rocks contain material fine enough to be counted as lithic fragments; most of these contain some ferromagnesian minerals. These ferromagnesian minerals are especially sensitive to the effects of weathering and diagenesis. The proportion of feldspar in the metamorphic rocks is highly variable.

Modal QmKP composition of the two rock types are markedly different (Fig. 6A). Precambrian metamorphic rocks have considerably less potassium feldspar than do the granitic rocks, allowing differentiation of the influence of the two source areas. The differences in feldspar type of the two rock types is also shown in a PKtKu ternary plot comparing amount of plagioclase versus potassium feldspar twin type (Fig. 7A). Note that the metamorphic rocks are mostly plagioclase with some twinned potassium feldspar, whereas the granitic rocks contain significant amounts of untwinned potassium feldspar, and approximately equal amounts of plagioclase and twinned potassium feldspar.

Precambrian rocks in the Sandia-Manzanita-Manzano Mountains are cut by the Tijeras-Cañoncito Fault that also divides the two mountain ranges (Fig. 8). Precambrian rocks in the Sandia Mountains north of the fault are dominantly the Sandia Granite, except for the Juan Tabo metamorphic rocks in the northwest. Precambrian rocks in the Manzanita-Manzano Mountains, south of the fault, are mainly metamorphic. QFLm compositions of the Sandia Granite are high in feldspar, and include one fine-grained sample with 55 QFLm%Lm (Fig. 9A). Metamorphic rocks are lithic-rich. The fields of variation of the granitic and metamorphic rocks do not overlap. QmKP compositions of the Sandia Granite are similar to those of the Embudo Granite, although the former is lower in quartz (Figs. 6A, 10A). Metamorphic rocks contain no potassium feldspar (one sample rich in potassium feldspar is not plotted since it is so fine grained that the feldspars are counted as lithic fragments). Sandia Granite feldspars include all three types, with more plagioclase and twinned potassium feldspar than untwinned feldspar (Fig. 11A). Metamorphic rocks cluster at the plagioclase apex of the QmKP diagram.

Only two Precambrian samples were collected in the Nacimiento Mountains, both from the Joaquin Granite (Fig. 12). The QFLm compositions are similar to those of the Sandia Granite (Fig. 13A). QmKP compositions are higher in plagioclase than for either the Sandia or Embudo Granite (Fig. 14A). The samples have very few untwinned potassium feldspars (Fig. 15A). The Joaquin Granite comprises roughly the central third of the southern Nacimiento Mountains; the northern and southern sections are chiefly composed of gneiss (Fig. 12; Woodward, 1987). The northern Nacimiento Mountains are chiefly composed of quartz monzonite, tonalite, and granite. Both the northern and southern

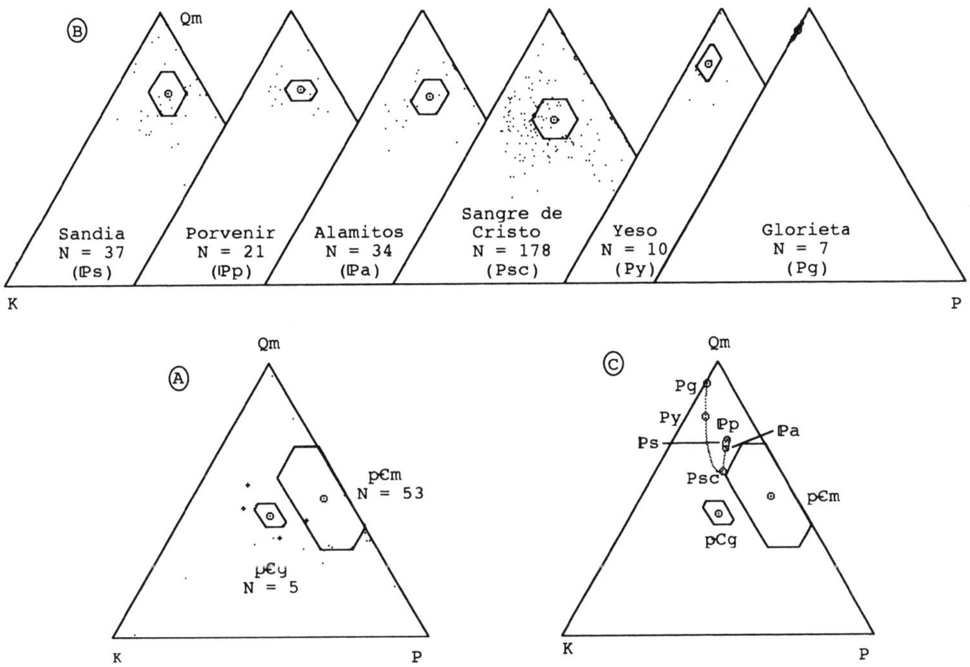

Figure 6. QmKP compositions of Precambrian and Paleozoic rocks studied in the Sangre de Cristo Mountains. A, QmKP compositions of the Precambrian rocks grouped by rock type. B, QmKP compositions of the Pennsylvanian and Permian strata. C, QmKP provenance evolution, showing the means for the Pennsylvanian and Permian in reference to the Precambrian fields of variation.

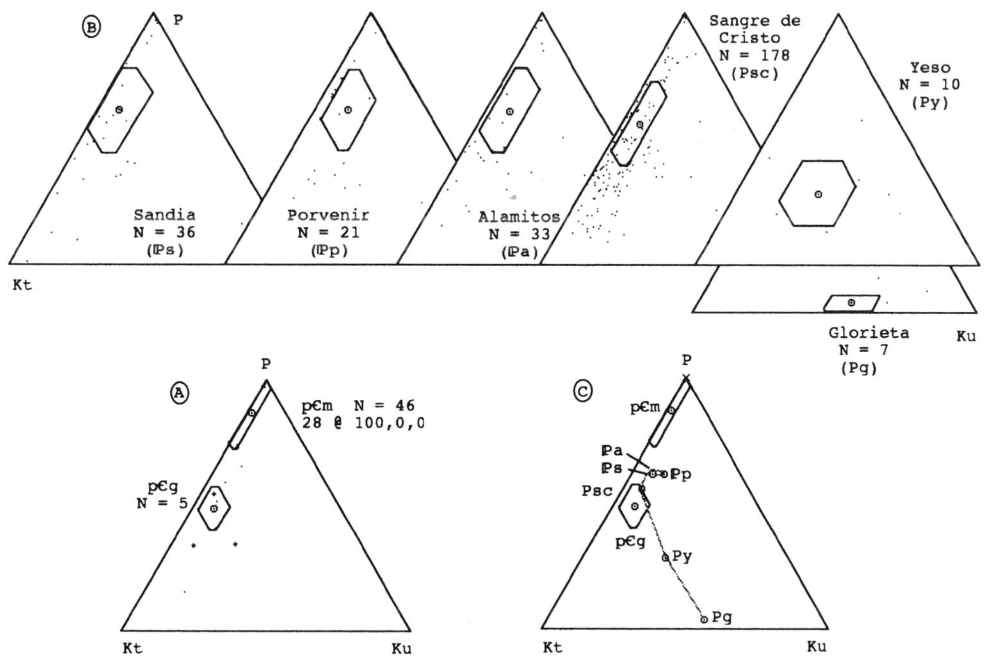

Figure 7. PKtKu compositions of Precambrian and Paleozoic rocks in the Sangre de Cristo Mountains. A, Precambrian feldspar compositions grouped by rock type. B, PKtKu compositions of the Pennsylvanian and Permian rocks. C, PKtKu provenance evolution, showing the means for the Paleozoic and their relations to compositions of source rocks.

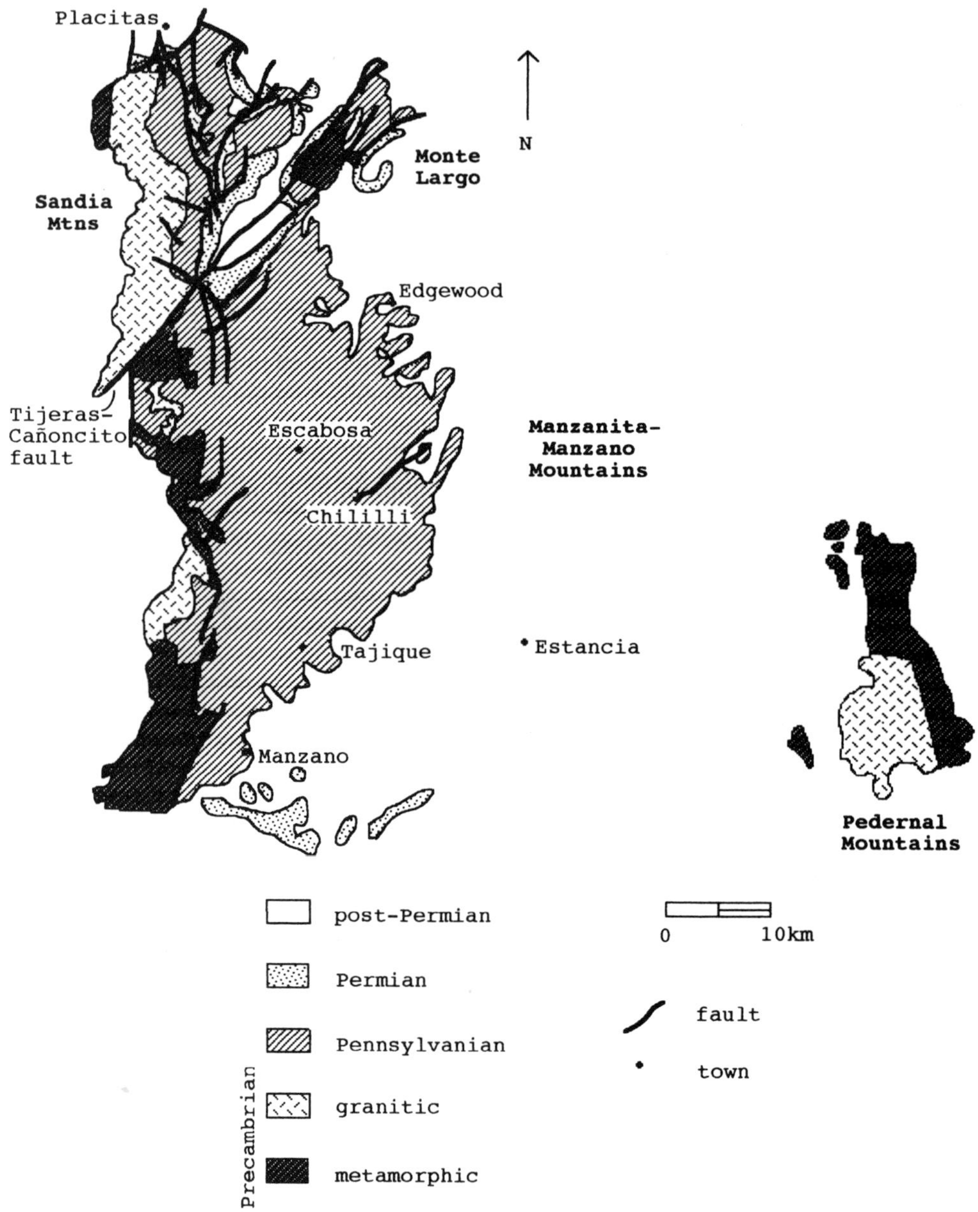

Figure 8. Generalized geological map of the study area in the Sandia-Manzanita-Manzano Mountains, the Monte Largo area and Pedernal Mountains showing the present exposure of Precambrian source rocks and Paleozoic strata. The Precambrian rocks are divided into two groups: dominantly granitic rocks exposed in the Sandias, portions of the Manzanita-Manzano Mountains and the central Pedernal, and dominantly metamorphic rocks exposed in the northern Sandias, Manzanita-Manzano Mountains, Monte Largo area and Pedernal Mountains. After Foster and Stipp, 1961; Dane and Bachman, 1965; Kelley and Northrop, 1975; Condie, 1981; and Cavin and others, 1982.

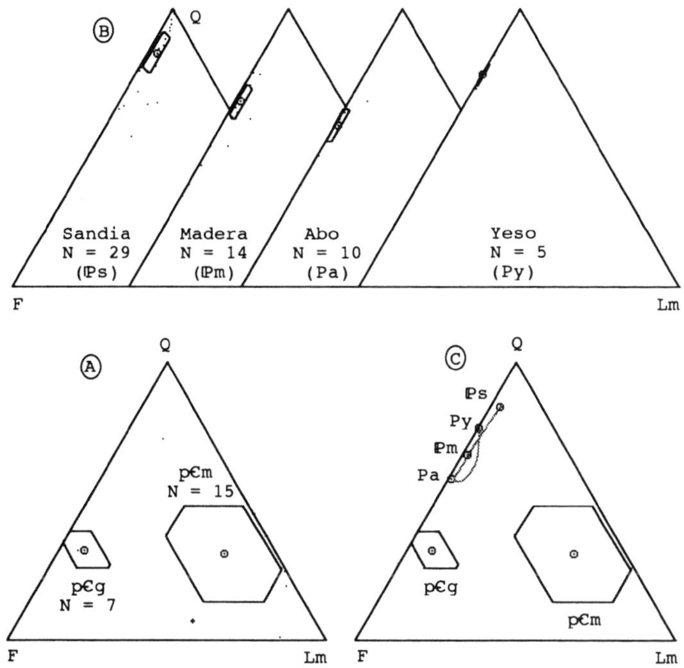

Figure 9. QFLm compositions of Precambrian and Paleozoic rocks from the Sandia-Manzanita-Manzano and Monte Largo Mountains (SMM). A, QFLm compositions of the metamorphic and granitic rocks of the SMM. The Sandia granites are higher in feldspar than those of the Embudo Granite of the Sangre de Cristo Mountains; one sample is fine-grained enough to yield lithics (plotted with a cross). B, QFLm compositions of Pennsylvanian and Permian rocks of the SMM. C, Paleozoic provenance evolution of the SMM; only means are shown with the fields of variation for the Precambrian rocks for reference.

Sierra Nacimiento contain subsidiary areas of metamorphic rocks (Woodward, 1987).

Paleozoic strata

Sangre de Cristo Mountains. Sandstones of the Sandia Formation of the Sangre de Cristo Mountains range from quartzarenitic to arkosic (by Folk's 1968 classification) and plot along the QF side of the QFLm ternary plot (Fig. 5B). Those of the Porvenir Formation are even less variable; they are arkosic to subarkosic and plot within the field of variation of the Sandia Formation. The Alamitos Formation also plots within the field of variation of the Sandia Formation. As with the older formations, QFLm compositional variation is small. Although the Sangre de Cristo Formation may contain up to 9 QFLm%Lm, most sandstones plot near the QF side. The Sangre de Cristo Formation's field of variation only partially overlaps that of the Sandia Formation, since the Sangre de Cristo generally contains more feldspar. The Yeso Formation contains very few lithic fragments (less than 1 QFLm%Lm), and contains more quartz than any of the older formations. The Glorieta Formation is the most quartzose of those studied and shows the least amount of QFLm variation. It has less than 1 QFLm%Lm.

Although most QFLm compositions overlap, there are differences worth noting. The Sandia Formation has more lithic fragments than any other unit, whereas the Porvenir has the fewest lithic fragments of any Pennsylvanian unit, and the Glorieta, of any Permian unit. QFLm means are slightly different; the Alamitos and Sangre de Cristo Formations average more feldspar than the Sandia and Porvenir, which in turn, have higher average feldspar than the Yeso and Glorieta.

Because Pennsylvanian-Permian sandstones are, in general, lithic-poor, the QFLm compositions are not dramatically different; QmKP compositions show greater variability (Fig. 6B). The Pennsylvanian formations appear essentially identical and are composed of a mix of plagioclase and potassium feldspars with higher average values of monocrystalline quartz than the averages of the Precambrian rocks and more plagioclase than potassium feldspar. However, the Porvenir Formation generally contains more potassium than plagioclase feldspar, except in Johnson Mesa area (Fig. 4). Sandstones in the Porvenir are not as common as those in bounding formations and most of the Porvenir sandstones were collected in Johnson Mesa; therefore, the data have been skewed toward the plagioclase apex. Johnson Mesa sandstones are plagioclase-rich throughout the upper Paleozoic section (Devaney, 1992). Sangre de Cristo sandstones contain the most feldspar of any of the formations studied and many plot within the field of variation of their Precambrian source rocks. The Yeso and Glorieta record a trend of decreasing feldspar, especially plagioclase, through the Permian. The Glorieta sandstones plot entirely along the QmK side, near the Qm apex.

Some differences in feldspar type shown on the PKtKu ternary plot are worth noting (Fig. 7B). Although the Pennsylvanian formations overlap, the Porvenir contains higher proportions of untwinned potassium feldspar than either the Sandia or the Alamitos. The Sangre de Cristo Formation sandstones generally contain all three types of feldspar. The Yeso Formation also contains all three feldspar types, but is dramatically different from the underlying Sangre de Cristo Formation. The Sangre de Cristo plots along the PKt side, overlapping the Precambrian fields of variation. The Yeso contains more potassium feldspar than plagioclase and many of the potassium feldspars are untwinned. The Glorieta contains very little plagioclase and a higher ratio of untwinned to twinned potassium feldspar. The Glorieta shows the greatest departure from the compositions of the Precambrian source rocks.

Sandia-Manzanita-Manzano Mountains. Sandstones of the Sandia Formation in the Sandia-Manzanita-Manzano Mountains range from arkosic to quartzarenitic, but are significantly higher in quartz than those of the Sangre de Cristo Mountains. As in the Sangre de Cristo Mountains, the Pennsylvanian-Permian has low lithic contents (Fig. 9B). The Sandia Formation contains as much as 10 QFLm%Lm. Madera Formation sandstones are strikingly higher in feldspar than those of the underlying Sandia Formation. Unlike the equivalent formation of the north, the

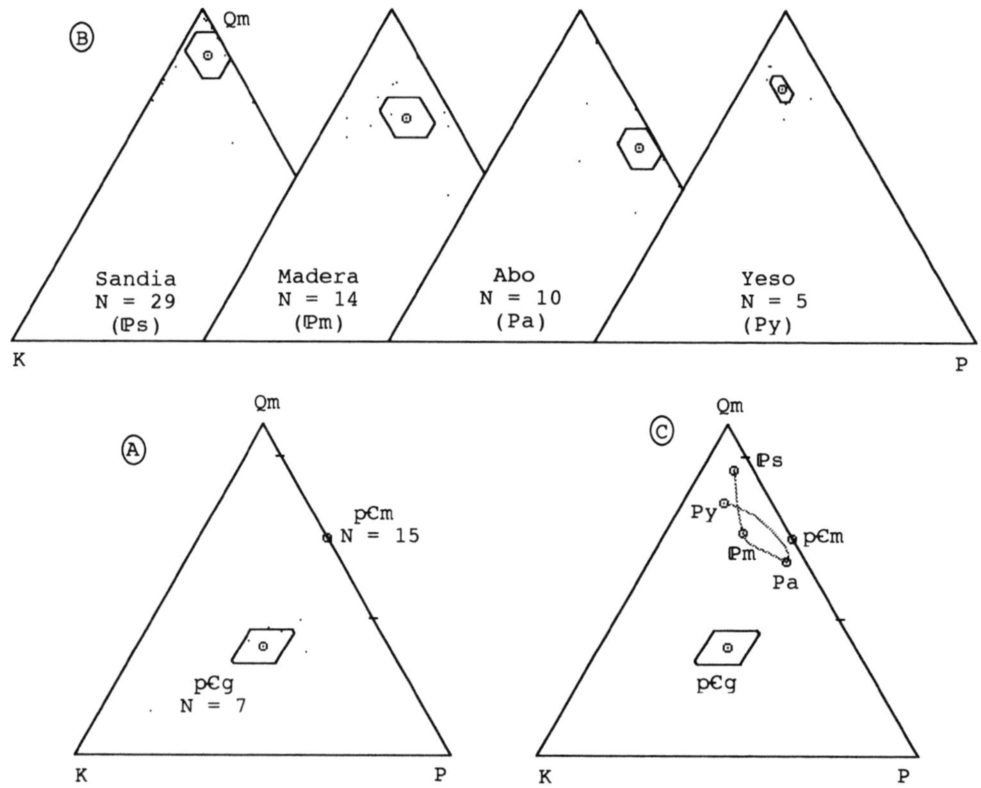

Figure 10. QmKP ternary plots for Precambrian and Paleozoic rocks of the Sandia-Manzanita-Manzano Mountains and Monte Largo. A, QmKP compositions for metamorphic and granitic rocks of the SMM. B, QmKP ternary plots for Pennsylvanian and Permian sandstones. C, Paleozoic provenance evolution.

fields of variation of the two formations do not overlap. The Madera contains a maximum of 8 QFLm%Lm. The overlying Abo Formation has 0 to 1 QFLm%Lm, except for one sample from the Manzanita Mountains with 5 QFLm%Lm. The Abo has more feldspar than either the Sandia or Madera; its field of variation only partially overlaps that of the Madera. The Yeso Formation has 0 to 1 QFLm%Lm, and so plots almost entirely on the QF side. It has a much lower feldspar content than either the Madera or Abo Formations.

The Sandia Formation QmKP compositions are separated from the overlying Paleozoic QmKP compositions by their high quartz content (Fig. 10B). The Madera QmKP field of variation is entirely separate from that of the Sandia since it is higher in feldspar and lower in P/F. The Abo Formation plots along the QmP side, its field of variation is separate from that of all other Paleozoic formations, and it overlaps the field of the Precambrian metamorphics. The Yeso contains a higher average percentage of potassium feldspar than the underlying Abo.

Comparison of feldspar type among the four Paleozoic formations in the Sandia-Manzanita-Manzano area shows that the Sandia contains some sandstones with high amounts of Kt and some Ku, but most Sandia Formation sandstones contain high amounts of plagioclase feldspar (Fig. 11B). The Madera contains

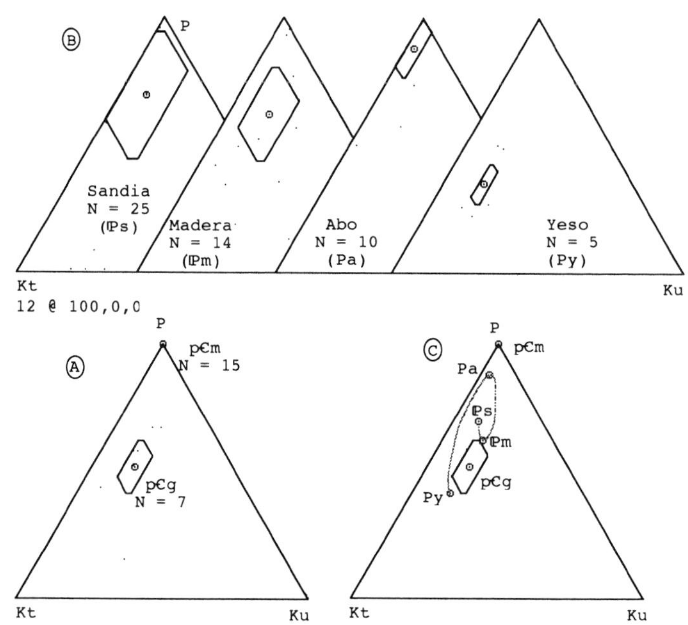

Figure 11. PKtKu compositions of Precambrian, Pennsylvanian, and Permian rocks of the Sandia-Manzanita-Manzano Mountains and Monte Largo. A, PKtKu compositions for Precambrian metamorphic and granitic rocks. B, PKtKu compositions of the Pennsylvanian and Permian. C, PKtKu ternary provenance evolution for the Sandia-Manzanita-Manzano and Monte Largo area.

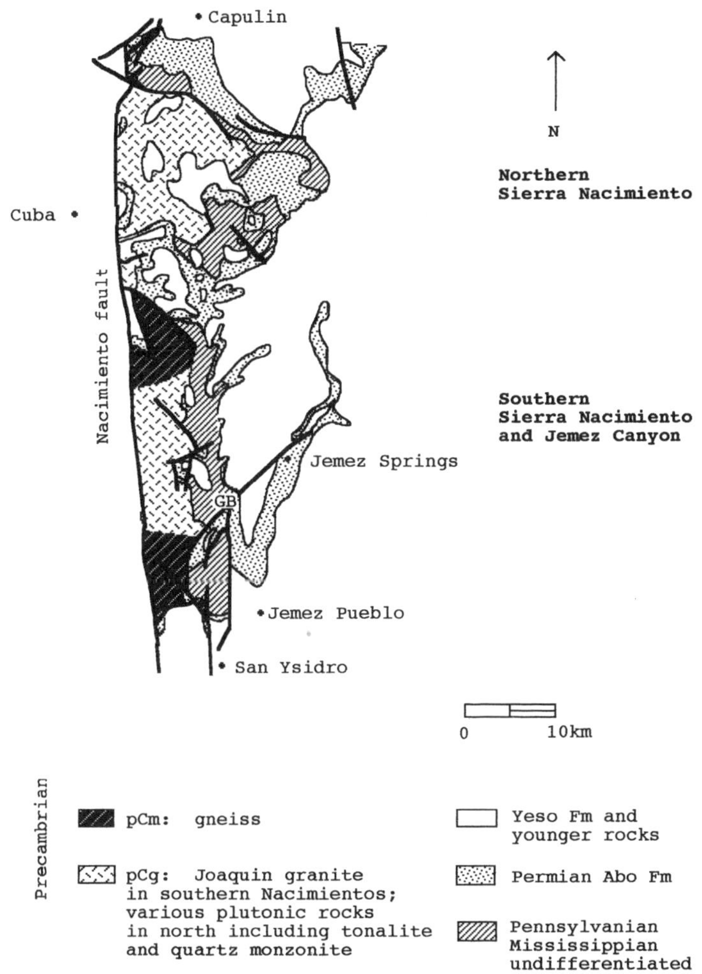

Figure 12. Generalized geological map of the study area in the Sierra Nacimiento and Jemez Canyon, showing present exposure of Precambrian source rocks and Paleozoic strata. Samples were collected from the southern end of the Joaquin granite and the adjacent area in the Sierra Nacimiento (GB = Guadalupe Box area) and from the Jemez Canyon area. The Precambrian rocks are divided into two groups: rocks of dominantly granitic composition exposed in the north and south-central Sierra Nacimiento; and dominantly metamorphic rocks exposed in the northern and southern sections of the southern Sierra Nacimiento. For explanation of symbols, see Figure 4. After Foster and Stipp, 1961; Dane and Bachman, 1965; and Woodward, 1987.

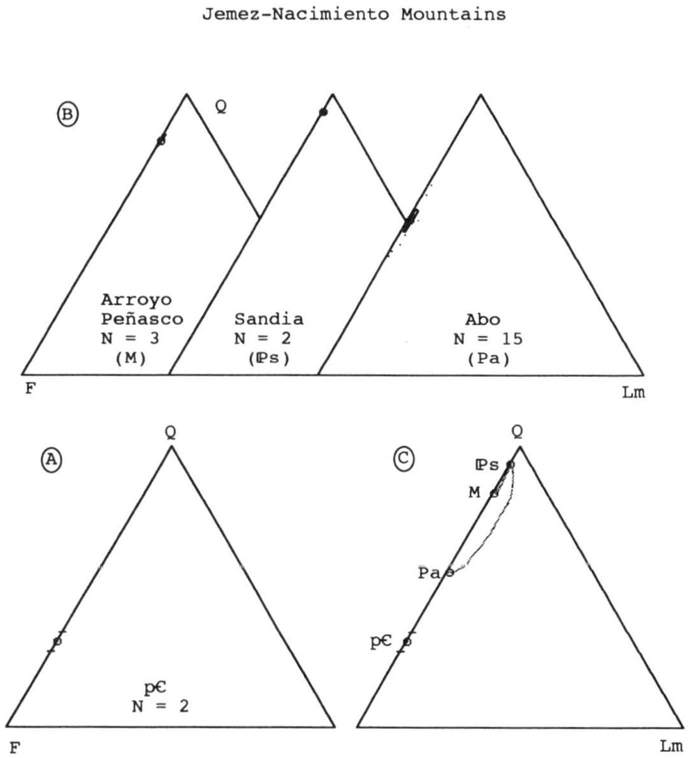

Figure 13. QFLm compositions for the Jemez-Nacimiento samples. A, QFLm compositions for two samples of the Joaquin granite. B, QFLm compositions of the three Mississippian Arroyo Peñasco samples, two Sandia and Abo Formation. No Madera Formation samples were available for the Jemez-Nacimiento area. C, Provenance evolution of the Jemez-Nacimiento strata.

all three feldspar types, including significant amounts of Ku. Its average PKtKu composition plots close to the Precambrian granite composition. The Abo Formation is very high in plagioclase feldspar, averaging close to the plagioclase feldspar apex and the Precambrian metamorphic composition. Although the Yeso Formation contains some plagioclase feldspar (as much as 50 PKtKu%P), its feldspar compositions cluster toward the KtKu side of the plot, containing higher average proportions of Ku than the Precambrian granitic rocks.

Jemez-Nacimiento Mountains. Study in the Jemez-Nacimiento area is preliminary; therefore, the number of samples is low. Three Mississippian sandy limestones of the Arroyo Peñasco Group were collected and studied from the Guadalupe Box area. Their QFLm compositions average 17 QFLm%F and less than 1 QFLm%Lm; this is 10 QFLm%F higher than the average for the overlying Sandia Formation (Fig. 13B). Only two Sandia Formation sandstones were collected, both from the Guadalupe Box area, and neither contains any Lm. No Madera Formation sandstones have been counted. The Abo Formation, from the Guadalupe Box and Jemez Canyon areas, contains a much higher percentage of feldspar, averaging 44 QFLm%F, and up to 3 QFLm%Lm.

The Mississippian Arroyo Peñasco contains potassium feldspar (mostly Kt with some Ku) with less than 1 QmKP%P (Figs. 14B, 15B). Since the samples are sandy limestones, the lack of plagioclase feldspar could be a result of diagenetic destruction, but no relict structures or other evidence of feldspar loss were observed in thin section. Mississippian samples from the other areas were rejected due to extensive diagenetic alteration. The Sandia Formation samples contain no potassium feldspar (Figs. 14B, 15B). The Abo Formation QmKP compositions are much closer to those of the Precambrian rocks and contain both potassium and plagioclase feldspar (averaging 23 QmKP%K and 22 QmKP%P), with more Kt than Ku.

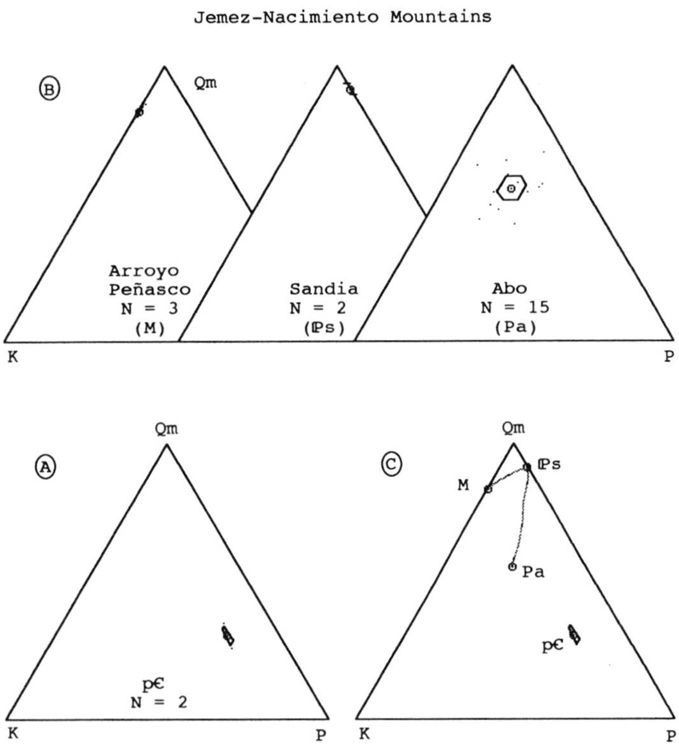

Figure 14. QmKP compositions of samples from the Jemez-Nacimiento area. A, QmKP compositions of the Precambrian Joaquin granite. B, QmKP compositions of Mississippian Arroyo Peñasco, Sandia Formation, and Abo Formation of the Jemez-Nacimiento area. C, Provenance evolution of the Jemez-Nacimiento area.

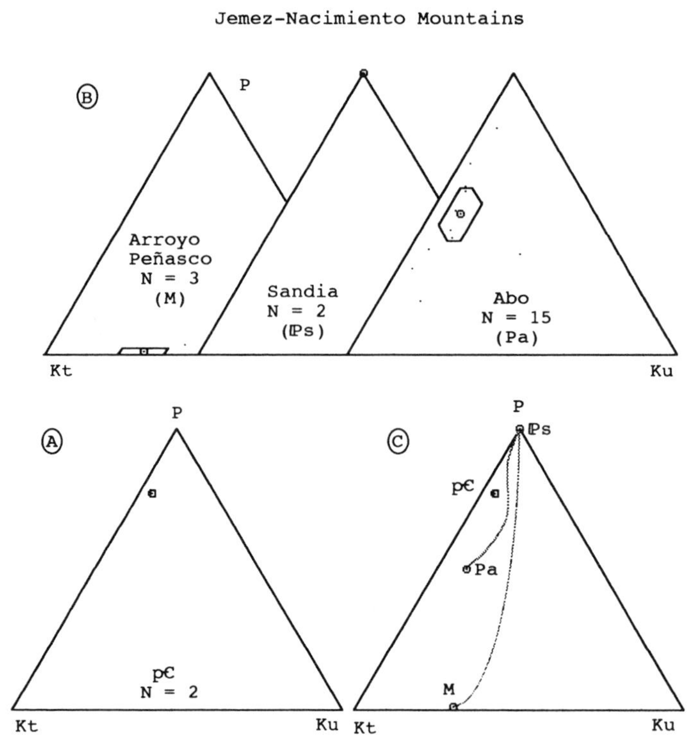

Figure 15. PKtKu compositions of the Jemez-Nacimiento area. A, Precambrian Joaquin granite PKtKu compositions. B, Mississippian Arroyo Peñasco, Sandia Formation, and Abo Formation feldspar compositions. C, Paleozoic provenance evolution of the Jemez-Nacimiento area.

PALEOCLIMATE

As discussed by Suttner (1974), climate is one of the factors controlling sandstone composition. Weathering in humid climates can selectively destroy labile minerals and lithic fragments; some workers record nearly complete destruction of labile minerals in tropical environments (e.g., Johnsson et al., 1991; Savage and Potter, 1991). On the other hand, Heins (1992; this volume) has reevaluated many earlier studies of climate and plutoniclastic sandstone compositions (including Suttner and Dutta, 1986) and states that source rock texture can probably explain most of the variation previously ascribed to climatic variation.

Suttner and Dutta (1986) examined upper Paleozoic strata (Fountain and Cutler Formations) of Colorado and Permian-Triassic strata (the Gondwana Supergroup) of India to try and delineate climatic effects on sandstone composition. Their data show a general trend of decreasing Q/(F+RF) and Qp/(F+RF) during a time of increasing aridity. Their study is relevant to the present study for two reasons (see Heins, 1992, and this volume, and Devaney, 1992, for a full discussion): (1) the Fountain and Cutler Formations were deposited under similar tectonic, depositional, and climatic conditions to those of the upper Paleozoic strata of north-central New Mexico; and (2) sandstone provenance studies must always consider climatic effects.

Seven samples representative of the Pennsylvanian-Permian section at Mora, New Mexico, were recounted using the Indiana method (i.e., Suttner and Dutta, 1986). It was not possible to select only medium–grain-size sandstones; most are coarse, but similar grain sizes were chosen for each age as much as possible. All samples have <15% interstitial material, and 300 points/slide were counted. The results are shown in Figure 16 and compared to results from Suttner and Dutta (1986), although exact comparisons are inappropriate due to differences in grain size.

North-central New Mexico, like Suttner and Dutta's area (1986), dried out during the Pennsylvanian-Permian (Habicht, 1979; Hunt, 1983; Rowley and others, 1985; Stanley, 1989), so if the Mora strata preserve a climate signature, then sandstone compositions should show a trend moving diagonally from the upper right of Figure 16 to the lower left with decreasing age (that is, Sandia–Porvenir–Alamitos–Sangre de Cristo–Yeso–Glorieta). If any trend is shown in the Mora section, it is the opposite to that expected; the Pennsylvanian sandstones represent an apparently drier climate than in the Permian. The Yeso and Glorieta samples are medium–coarse-grained sandstones, so their lower levels of RF could be due, in part, to smaller grain size. The Sandia Formation contains coal and so could not be the most arid. Of course, site means would be better than single samples/ formation; note variability of two samples from the Alamitos

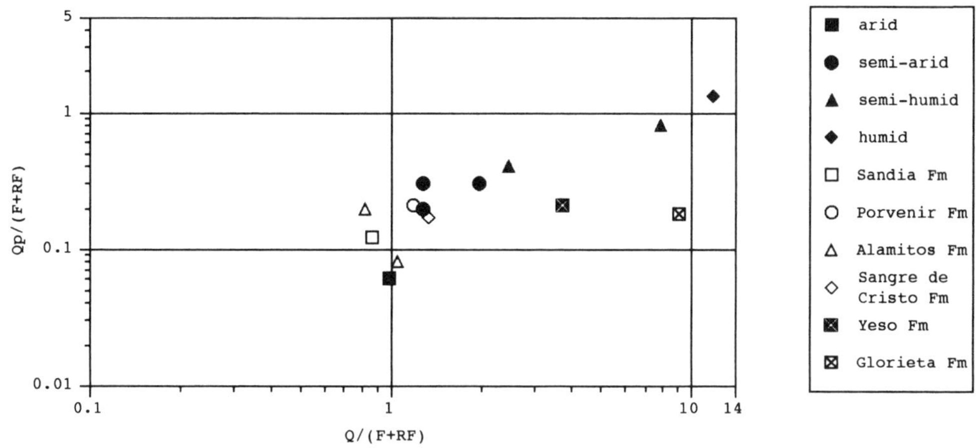

Figure 16. Log/log plot of Qp/F+RF versus Q/F+RF. Filled symbols are from Suttner and Dutta (1986) and have their climate assignments. The other points are from "Indiana" point counts of Mora section sandstones (Sangre de Cristo Mountains). From Devaney, 1992.

Formation. Paleoclimatic analysis of the Mora section suffers from similar problems as that of Colorado; it is likely that the "broad brush" (Suttner and Dutta, 1986, p. 342) of the tectonic signature overwhelms any trace of paleoclimatic change.

If climate is the major factor controlling sandstone composition, then upper Paleozoic strata of north-central New Mexico would show an overall decrease in quartz and kaolinite and a concomitant increase in QFLm%F, P/F, QFLm%Lm, mica, dense minerals, and Qp/Q as the area's climate stayed hot, but dried out. Instead, when diagenetically altered sandstones are removed, upper Paleozoic sandstone compositions deposited under the most humid paleoclimate conditions have the highest QFLm%Lm, mica, and dense minerals, as well as high P/F and higher QFLm%F than sandstones from the Yeso and Glorieta Fms (Figs. 5, 6, 9, 10, 13, 14) (Devaney, 1992). In addition, these compositions show variation with location consistent with source rock compositional control (Devaney, 1992). Although the Sangre de Cristo–Abo Formation has more feldspar than the formations of the Pennsylvanian, consistent with its drier climate, the overlying Yeso and Glorieta Formations, deposited in even drier climates, have less feldspar. Thus, although paleoclimate affects sandstone composition, tectonic control on source rock uplift overwhelms the "superimposed detail" (Suttner and Dutta, 1986, p. 342) of the climatic signature.

DISCUSSION

Provenance evolution of the Ancestral Rocky Mountains is reflected in detrital modes of upper Paleozoic sandstones, especially in the type and abundance of feldspar. Ancestral Rocky Mountain activity in the Sangre de Cristo area began in the Early Pennsylvanian with the development of the Uncompahgre and Sierra Grande uplifts and of the Taos Trough on the downthrown side of the Picuris-Pecos Fault (Sutherland, 1963). In the north, several alluvial fans built eastward into the Trough (the Sandia Formation) (Sutherland, 1963; Casey, 1980). In the south, the Pecos Shelf covered the area from Las Vegas to Santa Fé and was dominated by marine limestone deposition (Sutherland, 1963; Bachman, 1975). In the earliest Desmoinesian, the Cimarron arch was uplifted and fluvial–fan delta systems built southward (Casey, 1980). Later Desmoinesian deposition represents a major transgression and a decrease in uplift relief (Bachman, 1975; Casey, 1980).

Taos Trough sandstones associated with the first Ancestral Rocky Mountain tectonic pulse belong to the Sandia Formation and are quartzarenitic to arkosic (Fig. 5A). The Sandia Formation was apparently derived from a metamorphic or mixed metamorphic-granitic source area (Sutherland, 1963; and the present study). Later diminution of tectonic activity is reflected in the Porvenir Formation's sandstone by small reductions in the amount of feldspar and metamorphic lithic fragments (Figs. 5C and 6C) and a minor increase in the amount of untwinned potassium feldspar (Fig. 7C). Untwinned potassium feldspar is more resistant to mechanical weathering than is twinned potassium feldspar (Mack, 1978). Mack (1978) noted an increase in Ku/Kt due to mechanical weathering in high-energy littoral depositional environments. An increase in Ku/Kt could indicate an increasingly granitic provenance or an increase in the importance of mechanical weathering. Ku/Kt must, therefore, be analyzed in conjunction with other factors (such as depositional environment, distance of transport, and changes in other feldspar types and amounts) in order to differentiate between the two. The slight increase in the plagioclase/potassium feldspar ratio reflects a sampling problem (see Devaney, 1992). The small compositional changes in the Porvenir versus the Sandia may indicate that, although the tectonic lull was enough to allow a marine incursion, the uplifts remained as significant source areas for the Taos Trough.

Uplifts were rejuvenated in the late Desmoinesian, causing deposition of coarse fluvial-deltaic complexes in the northern Taos Trough; these complexes prograded southward (Sutherland, 1963; Casey, 1980). Sandstone composition of the Alamitos and

Sangre de Cristo Formations record this renewal of tectonic activity with an increase in feldspar (Figs. 5C, 6C) and a decrease in the proportion of untwinned to twinned potassium feldspar (Fig. 7C). The most dramatic increase in feldspar accompanies the marine regression caused by progradation of the alluvial fans of the Sangre de Cristo Formation (Fig. 6C) (Casey, 1980; Ross and Ross, 1986). The increase in feldspar is associated with a decrease in the QmKP%P, and an increase in the PKtKu%Ku, indicating a shift in source rock from that of metamorphic- to granitic-dominated, possibly because of erosion or because this tectonic pulse was more intense and extensive (Sutherland, 1963; Casey, 1980).

In contrast with the development of a major clastic wedge in the Taos Trough during the Early Pennsylvanian, the Sandia Formation of the Sandia Manzanita-Manzano and Jemez-Nacimiento areas represents a marine transgression (Read and Wood, 1947; Kottlowski, 1961; Ingersoll and Kelley, 1979). The Sandia Formation of these two areas has a relatively high QmKP%P and QFLm%Lm, indicating probable derivation from underlying Precambrian rocks (Figs. 9C and 10C) (see also Kelley and Northrop, 1975).

Tectonic activity in the south began in the Desmoinesian and continued into the Missourian, resulting in the development of the Estancia basin and the Pedernal arch (Bachman, 1975; Kluth, 1986; Ross and Ross, 1986). The Madera Formation from the Sandia-Manzanita-Manzano area records the increase of tectonic activity with a dramatic increase in QFLm%F over that of the Sandia (Fig. 9C). The fact that QFLm%F increases concurrently with a decrease in P/F and an increase in untwinned/twinned potassium feldspar ratio suggests that source area, rather than weathering factors, controlled sandstone composition. These changes are consistent with evolution from local metamorphic source areas on the peneplained Precambrian surface covered by the Sandia Formation to a more distant source area dominated by granitic rocks.

An increase in tectonic activity and a marine regression in the Permian is recorded by the arkosic Abo Formation in the Sandia-Manzanita-Manzano area. Like the Sangre de Cristo Formation, the Abo is the most arkosic of the nearby studied formations and the closest in composition to Precambrian source rocks. The Abo Formation of the Sandia-Manzanita-Manzano area contains more plagioclase feldspar than either the Sangre de Cristo Formation or the Abo of the Jemez-Nacimiento (Figs. 6, 10, 14), perhaps indicating uplift of a metamorphic source area near the Sandia-Manzanita-Manzano area (Kluth, 1986).

Preliminary data from the Jemez-Nacimiento show increase in the amount of feldspar in the Abo relative to Mississippian-Pennsylvanian sandstones (Fig. 13C), and decrease in P/F (Figs. 14C, 15C), probably reflecting the shift from the locally derived, plagioclase-rich source area to that of the increasingly active and granitic Uncompahgre uplift.

Tectonic activity decreased in north-central New Mexico during the Leonardian, and clastic sediments (the Yeso and Glorieta Formations) were deposited over nearly the entire area, eventually covering the Pedernal uplift (Kluth, 1986; Ross and Ross, 1986). The Yeso Formation has a lower proportion of feldspar and metamorphic lithic fragments (Figs. 5C, 9C), a lower P/F (Figs. 6C, 10C) and a higher Ku/Kt (Figs. 7C, 11C) than the underlying Sangre de Cristo and Abo Formations. The Glorieta Sandstone is even more stable than the Yeso (Figs. 5C, 6C, 7C).

CONCLUSIONS

Increasing Ancestral Rocky Mountain tectonism during the Pennsylvanian-Permian resulted in the deposition of more feldspathic sandstone (represented by the Sandia in the Taos Trough, the Madera in the Sandia-Manzanita-Manzano area, and the Sangre de Cristo and the Abo Formations in all three mountain ranges). The end of Ancestral Rocky Mountain activity in north-central New Mexico (represented by the Yeso and Glorieta Formations) resulted in a trend of decreasing feldspar, metamorphic lithic fragments, and P/F and increasing PKtKu%Ku. These trends all represent the influence of waning tectonism, erosion of residual highlands, and increasing stability of derived sandstones.

Although lithic fragment proportions are generally the best petrographic indicators of provenance (Ingersoll, 1990), feldspar is the most useful in sediments with low lithic percentages (Helmold, 1985). In north-central New Mexico, study of feldspar composition allows separation of metamorphic and igneous provenance, as well as documenting tectonic activity in the source area. For example, an increase in the proportion of untwinned potassium feldspar may indicate a decrease in tectonic activity, if coupled with a decrease in the overall amount of feldspar. High proportions of untwinned potassium feldspar coupled with a low P/F and a constant or increased overall amount of feldspar may indicate a granitic provenance.

Detailed knowledge of the composition of potential source rocks is helpful in provenance interpretations. Studies by metamorphic and igneous petrologists, however, do not allow direct compositional comparisons between source rock and sediment composition due to different goals and techniques. For a sedimentary provenance study, source rock composition is best documented using the Gazzi-Dickinson point-counting method on the source rocks themselves. Source rock and sediment compositions may then be directly compared and the relative importance of such modifying influences as weathering and diagenesis may then be evaluated.

ACKNOWLEDGMENTS

We gratefully acknowledge the Los Alamos National Laboratory for generously allowing one of us (K.D.) to use its facilities. We are also grateful to Andrea Kron for her assistance. Anne Linn, Bill Heins, and Gale Martin provided valuable discussions. Reviews by Timothy Lawton, Peter C. van de Kamp, and Mark Johnsson improved and clarified the manuscript. Finally, we extend sincere thanks to Joseph and Marjorie Devaney for their assistance in the field and for their patience and support during the course of the study.

REFERENCES CITED

Armstrong, A. K., 1967, Biostratigraphy and carbonate facies of the Mississippian Arroyo Peñasco Formation, north-central New Mexico: New Mexico Bureau of Mines and Mineral Resources Memoir 20, 80 p.

Armstrong, A. K., and Holcomb, L. D., 1989, Stratigraphy, facies and paleotectonic history of Mississippian rocks in the San Juan Basin of northwestern New Mexico and adjacent areas: New Mexico Geological Society, 40th Annual Field Conference Guidebook, p. 159–166.

Armstrong, A. K., and Mamet, B. L., 1974, Biostratigraphy of the Arroyo Peñasco Group, Lower Carboniferous (Mississippian), north-central New Mexico: New Mexico Geological Society, 25th Annual Field Conference Guidebook, p. 145–158.

Armstrong, A. K., and Mamet, B. L., 1977, Biostratigraphy and paleogeography of the Mississippian System in northern New Mexico and adjacent San Juan Mountains of southwestern Colorado: New Mexico Geological Society, 28th Annual Field Conference Guidebook, p. 111–127.

Armstrong, A. K., and Mamet, B. L., 1979, The Mississippian system of north-central New Mexico: New Mexico Geological Society, 30th Annual Field Conference, Guidebook p. 201–210.

Armstrong, A. K., and 6 others, 1979, The Mississippian and Pennsylvanian (Carboniferous) systems in the United States—New Mexico: U.S. Geological Survey Professional Paper 1110-W, 27 p.

Baars, D. L., 1961, Permian strata of central New Mexico: New Mexico Geological Society, 12th Annual Field Conference Guidebook, p. 113–120.

Baars, D. L., 1974, Permian rocks of north-central New Mexico: New Mexico Geological Society Guidebook 25, p. 167–169.

Bachman, G. O., 1975, Paleotectonic investigations of the Pennsylvanian system in the United States, Part I: Introduction and regional analyses of the Pennsylvanian system: U.S. Geological Survey Professional Paper 853-L, p. 233–243.

Baltz, E. H., 1965, Stratigraphy and history of Raton basin and notes on San Luis basin, Colorado–New Mexico: American Association of Petroleum Geologists Bulletin, v. 49, p. 2041–2075.

Baltz, E. H., 1972, Geologic map and cross-sections of the Gallinas Creek area, Sangre de Cristo Mountains, San Miguel County, New Mexico: U.S. Geological Survey Miscellaneous Investigations Map I-673, scale 1:24,000.

Baltz, E. H., and Myers, D. A., 1984, Porvenir Formation (new name)—and other revisions of nomenclature of Mississippian, Pennsylvanian and lower Permian rocks, southeastern Sangre de Cristo Mountains, New Mexico: U.S. Geological Society Bulletin 1537-B, 39 p.

Basu, A., 1976, Petrology of Holocene fluvial sand derived from plutonic source rocks: Implications to paleoclimatic interpretation: Journal of Sedimentary Petrology, v. 46, p. 694–709.

Basu, A., 1985, Influence of climate and relief on compositions of sands released at source areas, *in* Zuffa, G. G., ed., Provenance of arenites: Boston, D. Riedel, p. 1–18.

Blatt, H., and Christie, J. M., 1963, Undulatory extinction in quartz of igneous and metamorphic rocks and its significance in provenance studies of sedimentary rocks: Journal of Sedimentary Petrology, v. 33, p. 559–579.

Brown, C. E., 1984, Depositional environments of the Permo-Pennsylvanisn Sangre de Cirsto Formation, Coyote Creek District, Mora County, New Mexico: New Mexico Geological Society, 35th Annual Field Conference Guidebook, p. 115–122.

Caldwell, K. R., 1987, Depositional history and paleotectonic implications of the Pennsylvanian-Permian Sangre de Cristo Formation, northern New Mexico [M.S. thesis]: Richardson, University of Texas at Dallas, 150 p.

Callender, J. F., Robertson, J. M., and Brookins, D. G., 1976, Summary of Precambrian geology and geochronology of northeastern New Mexico: New Mexico Geological Society Guidebook 27, p. 129–135.

Casey, J. M., 1980, Depositional systems and basin evolution of the late Paleozoic Taos Trough, northern New Mexico [Ph.D. thesis]: Austin, University of Texas, 237 p.

Cavin, W. J., Connolly, J. R., Woodward, L. A., Edwards, D. L, and Parchman, M., 1982, Precambrian stratigraphy of Manzanita and north Manzano Mountains, New Mexico: New Mexico Geological Society, 33rd Annual Field Conference Guidebook, p. 191–196.

Chronic, J., 1979, The Mississippian and Pennsylvanian (Carboniferous) systems in the United States—Colorado: U.S. Geological Survey Professional Paper 1110-V, 25 p.

Condie, K. C., 1981, Precambrian rocks of the southwestern United States and adjacent areas of Mexico: New Mexico Bureau of Mines and Mineral Resources Resource Map 13, scale 1:1,562,500.

Crabaugh, J. P., and McPherson, J. G., 1985, Fluvial evolution in the Permian Abo and lower Yeso Formations, San Diego Canyon area, Sandoval County, New Mexico: New Mexico Geology, v. 7, p. 66.

Dane, C. H., and Bachman, G. O., 1965, Geologic map of New Mexico: U.S. Geological Survey State Geologic Map, scale 1:500,000.

Devaney, K. A., 1992, Provenance evolution of upper Paleozoic sandstones of north-central New Mexico [Ph.D. thesis]: Los Angeles, University of California, 330 p.

Dickerson, D. W., 1984, Structural controls on basin-margin sedimentation of the Pennsylvanian Taos Trough, New Mexico and contemporary Belize, central America: New Mexico Geological Society, 35th Annual Field Conference Guidebook, p. 101–105.

Dickinson, W. R., 1970, Interpreting detrital modes of graywacke and arkose: Journal of Sedimentary Petrology, v. 40, p. 695–707.

Dickinson, W. R., 1985, Interpreting provenance relations from detrital modes of sandstones, *in* Zuffa, G. G., ed., Provenance of arenites: Boston, D. Reidel, p. 333–361.

Dott, R. H., Jr., and Batten, R. L., 1988, Evolution of the earth (fourth edition): New York, McGraw-Hill, 643 p.

DuChene, H. R., 1973, Structure and stratigraphy of the Guadelupe Box and vicinity, Sandoval County, New Mexico [M.S. thesis]: Albuquerque, University of New Mexico, 100 p.

Fekete, S. R., 1988, Petrology and diagenesis of Pennsylvanian arkosic sandstones, Taos Trough, northern New Mexico [M.S. thesis]: Richardson, University of Texas at Dallas, 173 p.

Folk, R. L., 1968, Petrology of sedimentary rocks: Austin, Hemphill's, 170 p.

Foster, R. W., and Stipp, T. F., 1961, Preliminary geologic and relief map of the Precambrian rocks of New Mexico: New Mexico Bureau of Mines and Mineral Resources Circular 57, 37 p.

Foster, R. W., Frentress, R. M., and Riese, W. C., 1972, Subsurface geology of east-central New Mexico: New Mexico Geological Society Special Publication 4, 22 p.

Gazzi, P., 1966, Le arenarie del flysch sopracretaceo dell'Appennino modenese; correlazioni con il flysch di Monglhidoro; Mineralogia e Petrografica Acta, v. 12, p. 69–97.

Habicht, J.K.A., 1979, Paleoclimate, paleomagnetism and continental drift: American Association of Petroleum Geologists Studies in Geology, no. 9, 31 p.

Haq, B. U., and Van Eysinga, F.W.B., 1987, Geologic time scale: (A wall chart): Amsterdam, Elsevier.

Hatchell, W. O., Blagbrough, J. W., and Hill, J. M., 1982, Stratigraphy and copper deposits of the Abo Formation, Abo Canyon area, central New Mexico: New Mexico Geological Society, 33rd Annual Field Conference Guidebook, p. 249–260.

Heins, W., 1992, The effect of climate and topography on the composition of modern, plutonoclastic sand [Ph.D. thesis]: Los Angeles, University of California, 465 p.

Helmold, K. P., 1985, Provenance of feldspathic sandstones—The effect of diagenesis on provenance interpretations: A review, *in* Zuffa, G. G., ed., Provenance of arenites: Boston, D. Reidel, p. 139–163.

Hunt, A., 1983, Plant fossils and lithostratigraphy of the Abo Formation (Lower Permian) in the Socorro area and plant biostratigraphy of Abo red beds in New Mexico: New Mexico Geological Society, 34th Annual Field Conference Guidebook, p. 157–163.

Ingersoll, R. V., 1990, Actualistic sandstone petrofacies: Discriminating modern and ancient source rocks: Geology, v. 18, p. 733–736.

Ingersoll, R. V., and Kelley, V. C., 1979, Stratigraphy and paleoenvironments of the Hagan basin, north-central New Mexico: New Mexico Geological Society, 30th Annual Field Conference Guidebook, p. 197–200.

Ingersoll, R. V., and Suczek, C. A., 1979, Petrology and provenance of Neogene sand from Nicobar and Bengal Fans, DSDP sites 211 and 218: Journal of Sedimentary Petrology, v. 49, p. 1217–1228.

Ingersoll, R. V., Bullard, T. F., Ford, R. L., Grimm, J. P., Pickle, J. D., and Sares, S. W., 1984, The effect of grain size on detrital modes: A test of the Gazzi-Dickinson point-counting method: Journal of Sedimentary Petrology, v. 54, p. 103–116.

Johnsson, M. J., Stallard, R. F., and Lundberg, N., 1991, Controls on the composition of fluvial sands from a tropical weathering environment: Sands of the Orinoco River drainage basin, Venezuela and Colombia: Geological Society of America Bulletin, v. 103, p. 1622–1647.

Kelley, V. C., 1982, Tectonic map, published with the New Mexico Highway Geologic Map: New Mexico Geological Society, scale 1:3,500,000.

Kelley, V. C., and Northrop, S. A., 1975, Geology of Sandia Mountains and vicinity, New Mexico: New Mexico Bureau of Mines and Mineral Resources Memoir 29, 136 p.

Kluth, C. F., 1986, Plate tectonics of the Ancestral Rocky Mountains, *in* Peterson, J. A., ed., Paleotectonics and sedimentation in the Rocky Mountain region, United States: American Association of Petroleum Geologists Memoir 41, p. 353–369.

Kluth, C. F., and Coney, P. J., 1981, Plate tectonics of the Ancestral Rocky Mountains: Geology, v. 9, p. 10–15.

Kottlowski, F. E., 1961, Pennsylvanian rocks in north-central New Mexico: New Mexico Geological Society, 12th Annual Field Conference Guidebook, p. 97–104.

Lucas, S. G., and Hayden, S. N., 1991, Type section of the Permian Bernal Formation and the Permian-Triassic boundary in north-central New Mexico: New Mexico Geology, v. 13, p. 9–15.

Mack, G. H., 1978, The survivability of labile light-mineral grains in fluvial, aeolian and littoral marine environments: The Permian Cutler and Cedar Mesa Formations, Moab, Utah: Sedimentology, v. 25, p. 587–604.

Martinez, R., 1974, Geology of the Pajarito Peak area, Sandoval County, New Mexico [M.S. thesis]: Albuquerque, University of New Mexico, 72 p.

Montgomery, A., 1963, Precambrian rocks, *in* Miller, J. P., Montgomery, A., and Sutherland, P. K., Geology of part of the southern Sangre de Cristo Mountains, New Mexico: New Mexico Bureau of Mines and Mineral Resources Memoir 11, p. 7–21.

Myers, D. A., 1973, The upper Paleozoic Madera Group in the Manzano Mountains, New Mexico: U.S. Geological Survey Bulletin 1372-F, p. 1–13.

Myers, D. A., 1982, Stratigraphic summary of Pennsylvanian and Lower Permian rocks, Manzano Mountains, New Mexico: New Mexico Geological Society, 33rd Annual Field Conference Guidebook, p. 233–237.

Read, C. B., and Wood, G. H., 1947, Distribution and correlation of Pennsylvanian rocks in late Paleozoic sedimentary basins of northern New Mexico: Journal of Geology, v. 55, p. 220–236.

Read, C. B., Wilpolt, R. H., Andrew, D. A., Summerson, Ch. H., and Wood, G. H., Jr., 1944, Geologic map and stratigraphic sections of Permian and Pennsylvanian rocks of parts of San Miguel, Santa Fe, Sandoval, Bernalillo, Torrance, and Valencia counties, north-central New Mexico: U.S. Geological Survey Oil and Gas Investigations Preliminary Map OM-21, scale 1:190,000.

Ross, C. A., and Ross, J.R.P., 1986, Paleozoic paleotectonics and sedimentation in Arizona and New Mexico: American Association of Petroleum Geologists Memoir 41, p. 653–668.

Rowley, D. B., Raymond, A., Parrish, J. T., Lottes, A. L., Scotese, C. R., and Ziegler, A. M., 1985, Carboniferous paleogeographic, phytogeographic and paleoclimatic reconstructions: International Journal of Coal Geology, v. 5, p. 7–42.

Savage, K. M., and Potter, P. E., 1991, Petrology of modern sands of the Rios Guaviare and Inirida, southern Columbia: Tropical climate and sand composition: Journal of Geology, v. 99, p. 289–298.

Schowalter, T. T., 1969, Geology of part of the Creston Range, Mora County, New Mexico [M.S. thesis]: Albuquerque, University of New Mexico, 70 p.

Stanley, S. M., 1989, Earth and life through time (second edition): New York, W. H. Freeman, 689 p.

Sutherland, P. K., 1963, Paleozoic rocks, *in* Miller, J. P., Montgomery, A., and Sutherland, P. K., Geology of part of the southern Sangre de Cristo Mountains, New Mexico: New Mexico Bureau of Mines and Mineral Resources Memoir 11, p. 22–46.

Suttner, L. J., 1974, Sedimentary petrographic provinces: An evaluation: Society of Economic and Paleontologists and Mineralogists Special Publication 21, p. 75–84.

Suttner, L. J., and Dutta, P. K., 1986, Alluvial sandstone composition and paleoclimate, I. Framework mineralogy: Journal of Sedimentary Petrology, v. 56, p. 329–345.

Wiberg, T. L., and Smith, G. A., 1991, Cyclic sedimentology on a Middle Pennsylvanian (Desmoinesian) carbonate-clastic ramp, Sandia Mountains, north-central New Mexico: Geological Society of America Abstracts with Programs, v. 23, p. 105.

Wood, G. H., Jr., and Northrop, S. A., 1946, Geology of the Nacimiento Mountains, San Pedro Mountain and adjacent plateaus in parts of Sandoval and Rio Arriba Counties, New Mexico: U.S. Geological Survey Oil and Gas Investigations Map 57, scale 1:95,000.

Woodward, L. A., 1987, Geology and mineral resources of Sierra Nacimiento and vicinity, New Mexico: New Mexico Bureau of Mines and Mineral Resources Memoir 42, 84 p.

Zuffa, G. G., 1985, Optical analyses of arenites: influence of methodology on compositional results, *in* Zuffa, G. G., ed., Provenance of arenites: Boston, D. Reidel, p. 165–189.

Manuscript Accepted by the Society December 24, 1992

Printed in U.S.A.

Geological Society of America
Special Paper 284
1993

REE, Th, and Sc evidence for the depositional setting and source rock characteristics of the Quartz Hill chert, Sierra Nevada, California

Gary H. Girty and Robin W. Barber
Department of Geological Sciences, San Diego State University, San Diego, California 92720
Charles Knaack
Department of Geology, Washington State University, Pullman, Washington 99164

ABSTRACT

The post-Cambrian and pre–Upper Devonian Shoo Fly Complex represents the remnants of an ancient subduction system. The Quartz Hill chert, a thrust-fault–bounded, chlorite-grade greenschist facies unit, is included in the Shoo Fly Complex, and consists of about 25 m of interstratified radiolarite and argillite. Rare earth element (REE) data derived from 15 samples indicate that the Quartz Hill chert contains two chemically distinct groups of rock. Six Group 1 specimens display relatively flat REE/PAAS (post-Archean average Australian shale) distribution patterns, no Ce anomaly, and a variable positive Eu anomaly. In contrast, nine Group 2 specimens exhibit no Ce anomalies, a variable positive Eu anomaly, and display REE/PAAS values that increase from La to Eu, and then decrease from Eu to Lu. The REE, Th, and Sc characteristics of Group 1 samples are like those in Cretaceous, Tertiary, and Quaternary marine sediments containing particles derived from magmatic arcs. In contrast, the REE, Th, and Sc characteristics of Group 2 specimens are suggestive of a mixture of magmatic arc material and alkaline basaltic particulate matter derived from a seamount or ocean island.

Data presented here are consistent with the results from previous petrological and geochemical studies of rocks in the Shoo Fly Complex, and indicate that the Quartz Hill chert was deposited on the margin of an oceanic plate adjacent to a magmatic arc and a seamount(s) or ocean island(s). Thus, the data here and in the literature suggest that the REE, Th, and Sc characteristics of chert/argillite sequences deposited in or adjacent to active subduction systems are controlled primarily by source rocks in adjacent magmatic arcs, and in seamounts or ocean islands located within subducting plates.

INTRODUCTION

The pre–Upper Devonian Shoo Fly Complex, in the northern Sierra Nevada, California, has been interpreted to represent a subduction complex that developed along the margins of an early to middle Paleozoic island arc (Fig. 1) (Schweickert and Snyder, 1981; Girty et al., 1990, 1991). Remnants or debris shed from the island arc include the Sierra Buttes and Taylor Formations, and volcanogenic detritus within the Sierra City mélange, the structurally highest unit in the Shoo Fly Complex (Fig. 1). Although chert-argillite sequences are a common component of the Shoo Fly Complex, none have been studied in detail due to the difficulty of working with such fine-grained rocks.

Recent work suggests that Paleozoic, Mesozoic and Cenozoic chert and/or mudstone sequences deposited adjacent to active plate margins preserve in their rare earth element (REE), thorium (Th), and scandium (Sc) chemistry not only a record of their plate marginal setting, but also clues to source rock composi-

Girty, G. H., Barber, R. W., and Knaack, C., 1993, REE, Th, and Sc evidence for the depositional setting and source rock characteristics of the Quartz Hill chert, Sierra Nevada, California, *in* Johnsson, M. J., and Basu, A., eds., Processes Controlling the Composition of Clastic Sediments: Boulder, Colorado, Geological Society of America Special Paper 284.

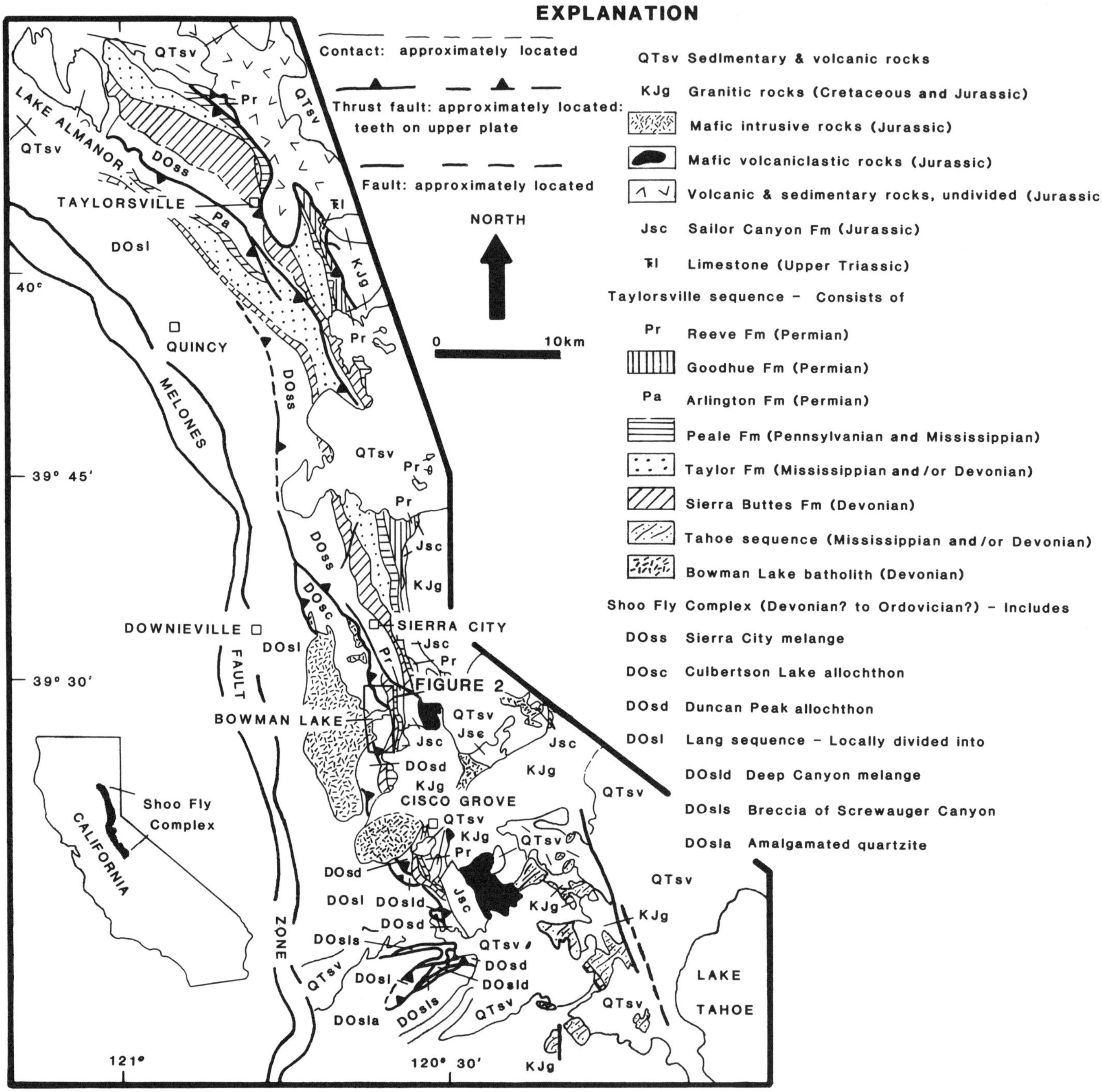

Figure 1. Simplified geological map showing location and geologic setting of the Shoo Fly Complex and Figure 2.

tions (Bhatia, 1985; Taylor and McLennan, 1985; Bhatia and Crook, 1986; McLennan, 1989; Musashino; 1990; Murray et al., 1990, 1991; McLennan et al., 1990; McLennan and Taylor, 1991; Girty and Barber, 1991; Barber, 1992). Although weathering, mass wasting, erosion, deposition, and diagenesis produce chemical changes in terrigenous sediments, some nuclides are transported, more or less, in unchanging proportions from their site of origin to their final site of deposition (Bhatia, 1985; Taylor and McLennan, 1985; Bhatia and Crook, 1986; McLennan, 1989; McLennan and others, 1990; McLennan and Taylor, 1991; Condie, 1991). Of the commonly transported nuclides, the REEs, Th, and Sc are the most useful for inferring plate depositional settings and provenance compositions because their distributions are not seriously affected by the above-mentioned processes and metamorphism up to and sometimes beyond amphibolite grade (Bhatia, 1985; Bhatia and Crook, 1986; McLennan, 1989; Mur-

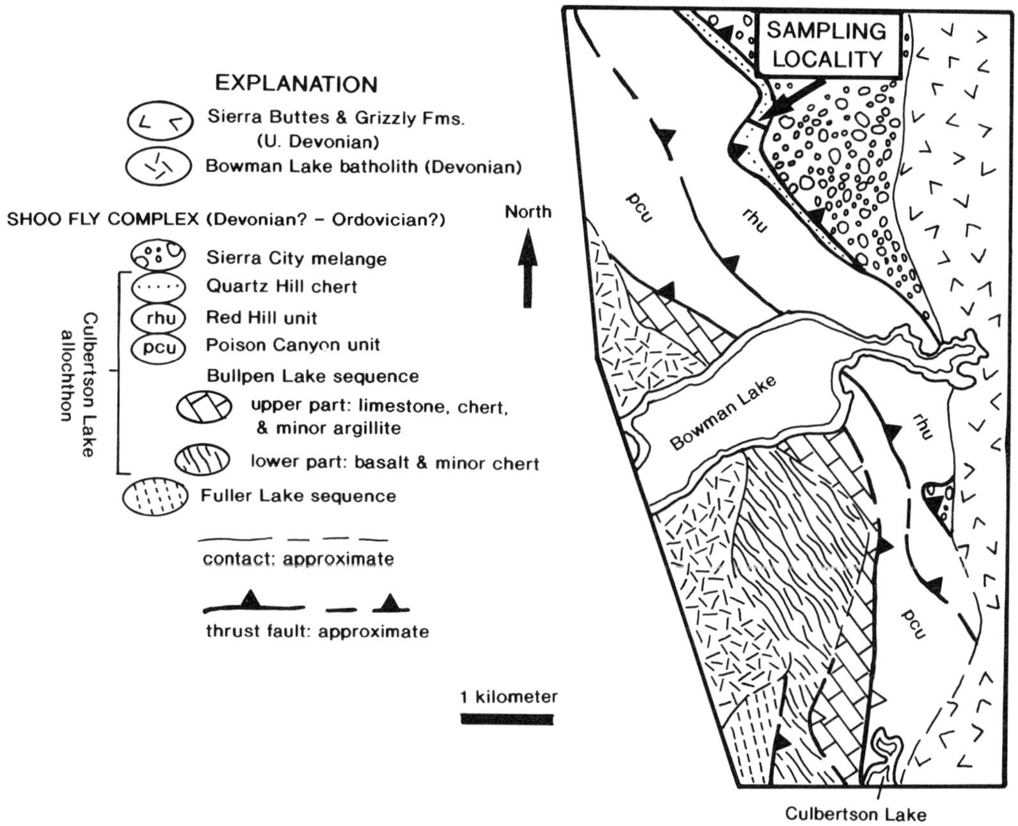

Figure 2. Geological map of a portion of the Shoo Fly Complex showing the location of the sampling traverse within the Quartz Hill chert. See Figure 1 for location of Figure 2.

ray and others, 1990, 1991; McLennan and Taylor, 1991; Condie, 1991).

In order to evaluate whether the general plate depositional setting and source rock characteristics of chert/argillite units in the Shoo Fly Complex could be determined by their REE, Th, and Sc chemistry, we collected 15 samples at 0.5- to 1.5-m intervals from an ~25-m-thick section of relatively undeformed chert/argillite (Fig. 2). We *informally* refer to the sampled chert/argillite unit as the Quartz Hill chert. In earlier publications, the Quartz Hill chert was referred to as the "chert-argillite 2" unit (e.g., Girty and Wardlaw, 1985; Girty and Pardini, 1987; Girty and others, 1990, 1991). We chose the section of Quartz Hill chert because it represents one of the least-deformed sections of chert and argillite mapped in the Shoo Fly Complex.

The Quartz Hill chert is the structurally highest unit in the Culbertson Lake allochthon, an imbricated succession of quartz-rich sandstones, argillites, and chert/argillites that structurally overlie the remnants of an alkalic seamount or oceanic island (i.e., the Bullpen Lake sequence) (Fig. 2) (Girty and others, 1990). Detrital zircon and sandstone petrological data indicate that the source of quartzose detritus in the Culbertson Lake allochthon included Precambrian crust, or material derived from such crust (Girty and Wardlaw, 1985).

The Sierra City mélange structurally overlies the Quartz Hill chert (Figs. 1, 2). It includes meter-scale blocks of gabbro, serpentinized peridotite, tholeiitic basalt, bedded radiolarian chert, and quartzose and feldspathic sandstone derived from subaqueous sediment gravity flows. The various blocks in the mélange are embedded in a fine-grained argillaceous matrix (Schweickert and others, 1984; Girty and Pardini, 1987; Legarre, 1988; Saleeby, 1990).

Detrital zircon and sandstone petrological data indicate that quartzose sandstones in the Sierra City mélange were derived from material shed from a craton-like source containing Precambrian rocks, or from material derived from such a source (Girty and Wardlaw, 1984; Girty and Pardini, 1987). In contrast, similar data suggest that feldspathic sandstones in the mélange were derived from debris eroded from a dissected, probably early Paleozoic, magmatic arc (Girty and Wardlaw, 1984; Girty and Pardini, 1987).

QUARTZ HILL CHERT

The Quartz Hill chert is metamorphosed to chlorite grade greenschist facies, and contains interstratified chert and argillite. Festoon ripple cross-laminations occur at several stratigraphic levels within the section that we studied (Fig. 2). They consistently indicate that the sampled section is stratigraphically upright.

Thin-section study, together with the results of dilute hydrofluoric acid digestions, revealed that chert beds in the Quartz Hill chert contain abundant radiolaria (Girty and Barber, 1991; Barber, 1992). Unfortunately, extracted radiolaria are not age-diagnostic, and vary in quality of preservation (B. Murchey, personal communication, 1991). Nevertheless, the presence of radiolaria is significant, because Cambrian radiolaria are uncommon (White, 1986). Thus, the Quartz Hill chert probably was deposited after the Cambrian and prior to the Late Devonian, the age of the Sierra Buttes and Grizzly Formations that unconformably overlie the Shoo Fly Complex (Fig. 2).

Chert and argillite beds within the Quartz Hill chert are 3 to 12 cm thick. Individual beds can be traced laterally for about 5 to 10 m before being covered by rock debris derived from mass wasting or vegetation.

Chert beds are easily distinguishable from argillaceous beds in the field by their white to gray color, extreme hardness, and tendency to break along conchoidal fractures when struck with a hammer. They commonly are composed of more than 80 to 90% spherical radiolarian tests, and are herein referred to as radiolarites. Radiolarian tests have average diameters of about 0.03 to 0.05 mm. Some radiolarites contain less than about 5 to 10% monaxon sponge spicules (B. Murchey, personal communication, 1991). Interstitial material is poorly resolved at ×400 magnification, but appears to be a mixture of chlorite, white mica, and quartz.

Bedding-parallel laminations (<2 cm thick) are prevalent in some beds in the Quartz Hill chert. They are produced by different textural and mineralogical types of argillite, and radiolarite similar to that described above. Grayish olive green to yellowish gray argillaceous beds, or argillaceous laminations in radiolarite, are composed either of a variable and poorly resolvable (at ×400 magnification) mixture of chlorite, white mica, and quartz, or of a variable mixture of mud- to silt-sized (0.10 to 0.01 mm) angular to subangular grains of quartz, angular to subangular slightly birefringent granular particles that appear opaque in plane light, subangular clots composed of multiple phases of microcrystalline white mica, and volumetrically minor angular to subangular feldspar. We refer to the former variety of argillite as type 1 and the latter as type 2. The grains and granular particles in type 2 argillites form a clastic framework assemblage that is enclosed in a matrix composed of variable proportions of chlorite and white mica, and minute particles that are poorly resolved at ×400 magnifications. Euhedral siderite(?) and pyrite are common secondary minerals in both types of argillite. No unequivocal radiolaria or sponge spicules were observed in any argillaceous layers.

The original material from which the mud- to silt-sized granular particles in type 2 argillites were derived is difficult to assess. The particles tend to be concentrated in yellowish gray laminae and cross-laminae (0.05 to 0.15 mm thick). In some of these laminations, the granular particles appear to be graded; in others they appear to be concentrated at the base of individual laminations. Under ×400 magnifications, the granular texture does not vary from particle to particle. In reflected light, the granular particles are cream colored, and do not display a metallic luster.

An 80 to 90% pure concentrate of the granular particles were studied by x-ray diffraction (XRD). The results of the XRD work indicated that the granular particles are composed largely of the metamorphic mineral clinochlore and minor quartz (Barber, 1992). The observation that the granular texture of the particles does not vary from particle to particle suggests that the original material now replaced by clinochlore and quartz was of relatively uniform composition and texture. In addition, the observation that the particles tend to be concentrated in the lower parts of individual laminations suggests that during deposition they were denser than surrounding phases. We speculate that the original material may have been of intermediate to basic volcanogenic origin. A volcanogenic origin for the particles is consistent with the above observations, and the REE, Th, and Sc characteristics of type 2 argillite described later in this chapter.

ANALYTICAL METHODS

Major, rare earth, and trace elements were analyzed at the GeoAnalytical Laboratory of Washington State University, Pullman, Washington. Eight of the 15 collected samples were analyzed for major elements following conventional x-ray fluorescence (XRF) techniques. Single 2:1 lithium tetraborate/rock-powder fused disks of each sample were analyzed for major and some trace elements on an automatic Rigaku 3370 spectrometer. Each element analysis was fully corrected for line interference and matrix effects. Loss-on-ignition was not determined. Resulting unnormalized data are listed in Table 1. Total iron is expressed as FeO. Uncertainties for major element data are estimated to be better than 1%. Uncertainties for trace element data analyzed by XRF probably are no better than about 5 to 20%, depending on the given element and its concentration.

Sample digestion techniques used in REE and trace element analysis at Washington State University are adapted from Crock and Lichte (1982). Each sample analyzed for REEs and trace elements was run on a Sciex Elan 250 inductively coupled plasma source mass spectrometer (ICP-MS), with cross-flow nebulizer, water-cooled torch and Brooks mass flow controllers. Oxide interference corrections were determined for each run using two solutions prepared from single element standards.

Sample intensities were corrected for isobaric oxide interferences (Lichte and others, 1987), and instrumental drift using a mass weighted average of the drift of In and Re (Doherty, 1989). Elemental intensities were calibrated against a curve constructed from three in-house standards: BCR-P, GMP-01, and MON-01. These standards were dissolved along with each batch of unknowns. The in-house standards have been characterized using a suite of international reference standards, and also have been analyzed by J. N. Walsh of Kings College, and by Allen Meier of the U.S. Geological Survey. Repeated analyses of BCR-P indicate that uncertainties for REE abundances are less than 5% (Table 2), and that uncertainties for abundance values listed in Table 3 are less than 10%. A more comprehensive discussion of analytical methods and petrological descriptions of each analyzed sample are provided by Barber (1992).

TABLE 1. X-RAY FLUORESCENCE–DETERMINED MAJOR AND TRACE ELEMENT ABUNDANCES LISTED FROM OLDEST TO YOUNGEST*

Sample/ Element	2†	8†	13†	21§	24§	27†	30†	32§
(wt.%)								
SiO_2	64.3	75.8	78.0	90.5	87.4	86.2	59.1	86.9
Al_2O_3	14.9	10.7	9.80	4.86	5.88	3.64	14.1	6.94
TiO_2	1.96	1.71	0.93	0.23	0.28	1.06	4.72	0.27
FeO^*	6.66	4.19	5.09	1.76	2.10	3.85	8.54	2.03
MnO	0.06	0.08	0.05	0.03	0.03	0.10	0.17	0.03
CaO	0.46	0.35	0.21	0.11	0.04	0.48	2.21	0.05
MgO	2.79	1.85	1.74	0.91	1.07	2.39	3.46	1.27
K_2O	4.20	2.99	2.36	1.21	1.40	0.11	2.74	1.90
Na_2O	0.34	0.17	0.30	0.19	0.24	.......**	0.70	0.12
P_2O_5	0.35	0.27	0.16	0.04	0.03	0.36	0.51	0.03
Total	96.0	98.1	98.5	99.8	98.5	98.2	96.3	99.5
MnO/TiO_2	0.03	0.05	0.05	0.13	0.11	0.09	0.04	0.11
(ppm)								
Ni	68	47	36	26	28	17	54	27
Cr	55	32	37	18	26	22	71	27
Sc	24	19	13	11	6	6	27	9
V	113	146	52	35	41	41	225	61
Sr	24	19	19	9	8	12	70	9
Zr	188	149	127	43	51	88	338	47
Ga	24	21	15	4	9	7	28	12
Cu	40	68	84	18	14	10	59	15
Zn	100	48	39	29	35	50	92	31

*LOI = Loss on ignition; not determined. Data are not normalized to 100%.
†Group 2 samples. Sample 30 is a laminated mixture of types 1 and 2 argillite. Sample 27 is a laminated mixture of type 2 argillite and radiolarite. All other Group 2 samples are type 1 argillite.
§Group 1 samples. Sample 21 is a radiolarite; all other Group 1 samples are type 1 argillites.
**None detected.

TABLE 2. RARE EARTH ELEMENT ABUNDANCES LISTED FROM OLDEST TO YOUNGEST

Sample Number	La	Ce	Pr	Nd	Sm	Eu	Gd	Tb	Dy	Ho	Er	Tm	Yb	Lu	Total
(ppm)															
1*	14.4	31.3	2.96	11.8	2.65	0.78	1.93	0.30	1.52	0.28	0.79	0.11	0.74	0.12	69.7
2†	28.2	65.1	7.49	33.4	8.53	2.07	6.76	1.05	5.19	0.95	2.49	0.33	2.01	0.31	164
5†	15.0	38.4	4.19	18.3	4.61	1.24	3.73	0.60	3.22	0.57	1.40	0.17	0.99	0.15	92.6
8†	19.4	46.2	5.64	26.1	6.78	2.05	5.38	0.86	4.55	0.82	2.09	0.26	1.51	0.22	122
11*	10.2	22.3	2.19	8.77	2.00	0.56	1.78	0.29	1.62	0.32	0.98	0.13	0.82	0.13	52.1
13†	19.2	45.6	4.71	20.0	5.08	1.25	4.10	0.68	3.74	0.71	1.86	0.25	1.56	0.23	109
15*	9.44	22.3	2.44	10.4	2.49	0.61	1.96	0.32	1.83	0.35	1.07	0.15	0.97	0.16	54.5
18†	28.8	64.3	7.50	32.1	8.12	2.01	6.72	1.07	5.68	1.03	2.58	0.33	2.00	0.29	163
21*	6.63	13.9	1.46	5.76	1.26	0.31	1.02	0.16	0.92	0.18	0.53	0.08	0.57	0.09	32.9
24*	11.3	22.0	2.31	8.66	1.70	0.38	1.27	0.20	1.13	0.23	0.69	0.10	0.68	0.11	50.8
25†	24.5	56.5	6.16	26.5	6.43	1.59	5.35	0.78	4.06	0.68	1.67	0.22	1.37	0.21	136
27†	24.4	50.6	6.85	31.4	8.22	1.99	6.52	0.98	4.80	0.0	2.03	0.25	1.56	0.25	141
30†	32.5	80.1	10.5	50.2	13.6	4.00	11.8	1.95	10.2	1.79	4.29	0.51	2.83	0.41	225
32*	8.17	17.2	1.67	6.32	1.21	0.32	1.16	0.16	0.94	0.21	0.59	0.09	0.60	0.09	38.7
34	5.34	11.1	1.20	4.87	1.03	0.19	0.67	0.11	0.52	0.10	0.27	0.03	0.21	0.03	25.7

*Group 1 samples.
†Group 2 samples.

TABLE 3. MASS SPECTROMETER–DETERMINED TRACE ELEMENT ABUNDANCES LISTED FROM OLDEST TO YOUNGEST

Sample Number	Ba	Th	Nb	Y	Hf	Ta	U	Pb	Rb	Cs
(ppm)										
1*	332	4.87	6.61	6.47	0.79	0.76	0.51	2.72	46.3	0.83
2†	926	6.07	25.3	21.9	2.83	1.34	1.60	18.1	129	1.56
5†	529	2.34	10.7	12.8	1.59	1.09	0.33	4.96	60.8	0.72
8†	723	2.68	16.2	18.6	2.14	1.04	0.49	4.25	103	1.12
11*	356	3.46	5.55	8.62	0.72	1.38	0.32	0.62	50.3	0.88
13†	532	4.88	15.8	15.4	2.28	1.04	0.54	1.77	70.5	0.98
15*	417	3.45	6.88	8.66	1.36	0.65	1.36	1.66	58.3	0.90
18†	736	4.24	20.0	24.6	2.81	1.05	0.68	3.27	102	1.19
21*	338	2.89	4.47	4.47	0.69	1.08	0.36	1.62	45.6	0.78
24*	343	3.95	5.33	5.54	0.82	1.03	0.49	1.66	44.9	0.99
25†	662	3.22	13.4	15.6	2.10	0.93	0.93	1.04	81.9	1.07
27†	23.0	2.60	15.0	19.3	1.63	1.50	1.34	0.64	2.74	0.10
30†	567	2.15	61.6	42.4	6.28	2.77	0.52	4.56	71.6	0.78
32*	486	3.93	5.87	4.91	1.02	0.77	0.34	0.81	75.4	0.91
34	80.0	0.84	4.78	2.53	0.37	2.91	0.29	2.73	8.76	0.17

Note: These determinations made on an inductively coupled plasma source mass spectrometer (ICP-MS).
*Group 1 samples. Sample 21 is a radiolarite; all other Group 1 samples are type 1 argillites.
†Group 2 samples. Sample 30 is a laminated mixture of types 1 and 2 argillite. Sample 27 is a laminated mixture of type 2 argillite and radiolarite. All other Group 2 specimens are type 1 argillite.

RESULTS AND DISCUSSION

As can be seen in Table 1, major element data are variable, but indicate that the analyzed samples are moderately to highly siliceous in character. For example, samples of type 1 argillite have SiO_2 values that range from 64.3 to 87.4 wt%. One laminated mixture of types 1 and 2 argillite has the lowest SiO_2 value (59.1%), whereas a laminated mixture of type 1 argillite and radiolarite has a value of 86.2 wt%. The one radiolarite specimen analyzed for major elements has an SiO_2 value of 90.5 wt% and is the most siliceous specimen analyzed.

Following the general procedures used by Murray and others (1991) and the recommendations of Condie (1991), we have normalized our REE data to a common well-characterized shale standard, i.e., post-Archean average Australian shale (PAAS) (see Table 2 in McLennan, 1989). In Figures 3 and 4, there are two types of REE/PAAS distribution patterns portrayed by our samples. We refer to those samples that: (1) display relatively flat REE/PAAS distribution patterns, (2) possess no Ce anomaly within analytical uncertainties, and (3) have a variable positive Eu anomaly as Group 1 specimens (i.e., samples 1, 11, 15, 21, 24, and 32) (Tables 1, 4; Figs. 3, 4). The remaining specimens make up Group 2 samples, which generally: (1) display REE/PAAS values that increase from La to Eu and then decrease from europium to Lu, (2) exhibit no Ce anomaly within analytical uncertainties, and (3) have a variable positive Eu anomaly (Tables 1, 4; Figs. 3, 4).

The Ce anomaly records the relative behavior of Ce with respect to its neighboring REEs, La and Pr. It is defined by the ratio $Ce/Ce^* = (Ce_{sample}/Ce_{shale})/Ce^*$ (e.g., Murray and others, 1991), where Ce^* is obtained by linear interpolation between shale (PAAS)-normalized La and Pr values (Murray and others, 1991). Ce/Ce^* values that are less than 1 are considered to be negative anomalies; those values that are greater than 1 are considered to be positive anomalies. No cerium anomaly is defined as $Ce/Ce^* = 1$.

The Eu anomaly (Eu/Eu^*) is defined by the ratio $Eu/Eu^* = (Eu_{sample}/Eu_{shale})/Eu^*$, where Eu^* is obtained by linear interpolation between PAAS-normalized Sm and Gd values. Eu/Eu^* values greater than 1 are considered to be positive anomalies.

Sample 34, a laminated radiolarite, displays an REE/PAAS distribution pattern that is broadly similar to Group 2 specimens, but contains only 25.7 ppm REE (Fig. 3, Table 2). In hand specimen, sample 34 is highly fractured, and is coated with a red iron-oxide stain. It occurs near the thrust fault that transects the top of the Quartz Hill chert (Fig. 2). The general appearance, and the stratigraphic/structural context of sample 34, suggest that it may have been altered by fluids generated during thrusting or by subsequent migration of fluids along the thrust fault that forms the upper boundary to the Quartz Hill chert. In the following paragraphs, we do not include sample 34 in our analysis.

Group 1 samples display no cerium anomaly and exhibit REE/PAAS distribution patterns that are similar to those determined for cherts and mudstones deposited adjacent to magmatic arcs (Figs. 3, 5). For example, some of the cherts and mudstones in the Cretaceous Franciscan Complex and Miocene Monterey Group were deposited adjacent to the Cordilleran continental margin magmatic arc. These well-characterized continental mar-

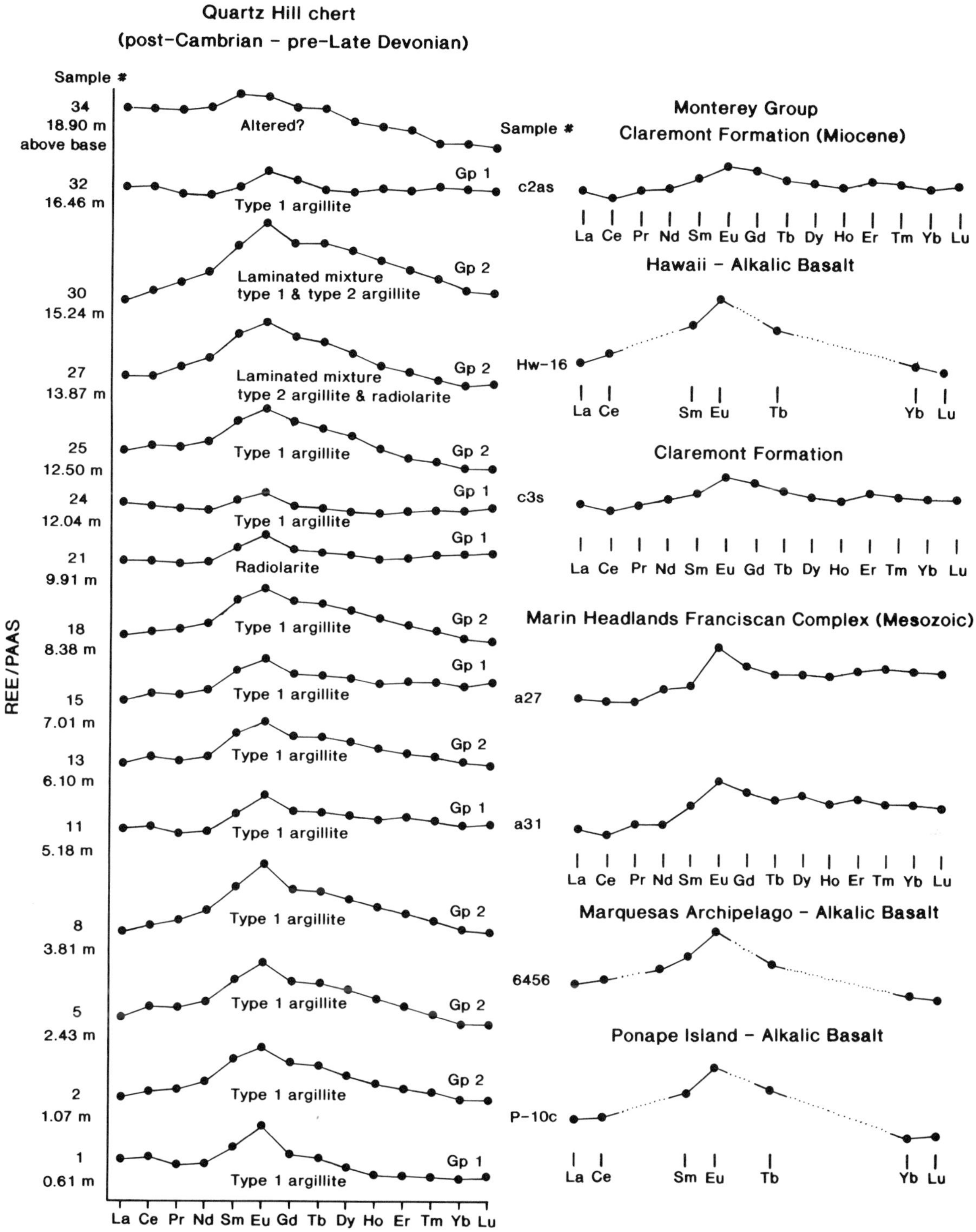

Figure 3. REE/PAAS distribution patterns for 15 samples of Quartz Hill chert. Shown on the right for comparison purposes are REE/PAAS distribution patterns for samples from the Monterey Group, Franciscan Complex, Ponape Island, Marquesas Archipelago, and Hawaiian Islands; data for these distribution patterns are from the Basaltic Volcanism Study Project (1981), Dixon and others (1984), Liotard and others (1986), and Murray and others (1991). Note how the REE distribution patterns for Group 1 samples (1, 11, 15, 21, 24, and 32) resemble the REE/PAAS distribution patterns of samples from the Monterey Group and the Franciscan Complex. The former samples are shales, the latter specimens are cherts. Note also how the REE/PAAS distribution patterns for Group 2 specimens resemble the REE/PAAS patterns derived from alkalic basalts from the Hawaiian Islands, Ponape Island, and the Marquesas Archipelago.

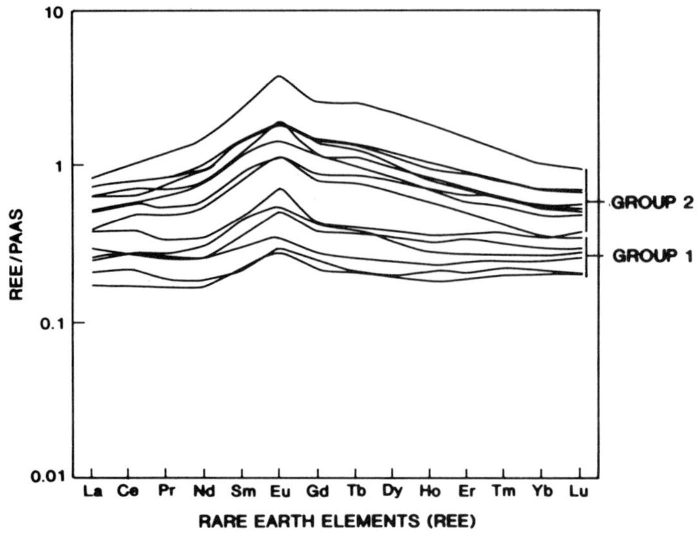

Figure 4. REE/PAAS plot comparing samples from Groups 1 and 2. Note the relatively flat REE/PAAS distribution patterns of Group 1 samples relative to Group 2 specimens.

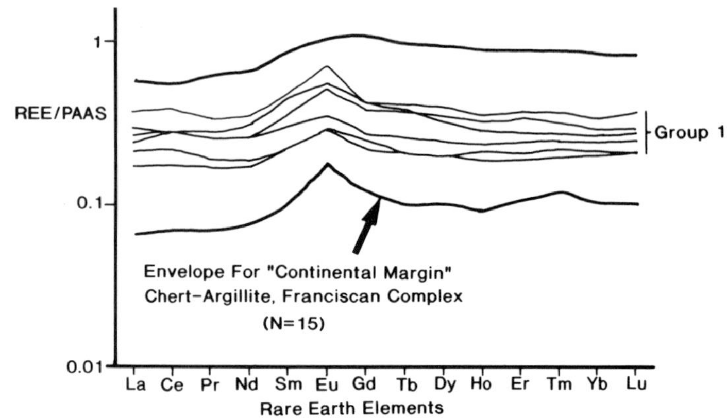

Figure 5. REE/PAAS plot comparing Group 1 samples with "continental margin" samples from the Franciscan Complex. Envelope for continental margin samples represents 1 standard deviation about the mean of data reported in Murray and others (1991). The samples from the Franciscan Complex probably contain particulate matter derived from the Mesozoic Cordilleran magmatic arc.

gin sediments display no or positive cerium anomalies and relatively flat REE/PAAS distribution patterns (Figs. 3, 5) (Murray and others, 1990, 1991). Judging from the paleogeographical/paleotectonic maps and/or data provided in a large number of published papers, the source of the REEs in the continental margin cherts and mudstones in the Franciscan Complex and Monterey Group probably included calc-alkaline magmatic material, and/or recycled particulate matter originally derived from such sources (e.g., Cole and Armentrout, 1979; Dickinson, 1979; Dickinson and others, 1982; Ingersoll, 1983).

On an La-Th-Sc ternary diagram, Group 1 samples plot within the field defined by Quaternary sand and mud deposited in magmatic arc–related basins along active plate margins (McLennan and others, 1990) (Fig. 6). This result is consistent with La/Th values for Group 1 samples that are like those typical of andesites, granodiorites, and granites (cf. McLennan and others, 1980; Gill, 1981) (Tables 4, 5). In addition, Th/Sc values for Group 1 samples are less than 0.90, suggesting a source rock composition more mafic than average granodioritic upper continental crust, e.g., PAAS (Tables 4, 5) (Taylor and McLennan, 1985). However, the spread of samples in the La-Th-Sc diagram from near the model andesite composition toward the PAAS and passive continental margin (or trailing edge) fields might suggest that Group 1 samples also contain a small component of the latter material (Fig. 6).

In the literature, Eu anomalies commonly are calculated from chondrite normalized data (e.g., McLennan and Taylor, 1991). For Group 1 samples Eu anomalies derived from chondrite normalized data vary from 0.79 to 1.06 (Table 4). These values are significantly greater than the average value of 0.65 for cratonally derived post-Archean sediment (e.g., PAAS), and are more like values obtained from sediment eroded from magmatic arcs (Taylor and McLennan, 1985; McLennan and others, 1990). These observations suggest that if a cratonal PAAS-like component is present in Group 1 specimens, it is present in only trace amounts.

Group 2 samples display REE/PAAS distribution patterns that are similar to the REE/PAAS patterns produced by alkalic basalt from modern-day ocean islands (Figs. 3, 7) (Basaltic Volcanism Study Project, 1981; Dixon and others, 1984; Liotard and others, 1986). This inference is supported by the relatively high TiO_2 values for most Group 2 samples, and MnO/TiO_2 values that are like modern-day ocean island alkalic basalts (Table 1) (cf. Dixon and others, 1984; Liotard and others, 1986). In addition, five Group 2 samples have La/Th values that are similar to those derived from modern-day ocean island alkalic basalts, whereas two samples have values suggestive of an andesitic source (cf. McLennan and others, 1980; Gill, 1981) (Tables 4, 5). Of the five samples for which Sc analyses are available, three Group 2 samples plot in the alkalic basalt field on the La-Th-Sc ternary diagram, whereas two specimens plot just outside the alkalic basalt field and within the field for magmatic arc–related sediment (Fig. 6). Th/Sc values for Group 2 samples are all less than 0.90, and are compatible with an intermediate to mafic source rock composition (Tables 4, 5).

The above observations suggest that alkaline basaltic material may have been mixed with mafic to intermediate magmatic arc material like that characteristic of Group 1 samples. In order to investigate this possibility, an REE mixing model was constructed using as end member compositions the average of Group 1 specimens and the average of alkalic basalt from Hawaii, Ponape Island, and the Marquesas Archipelago. The results

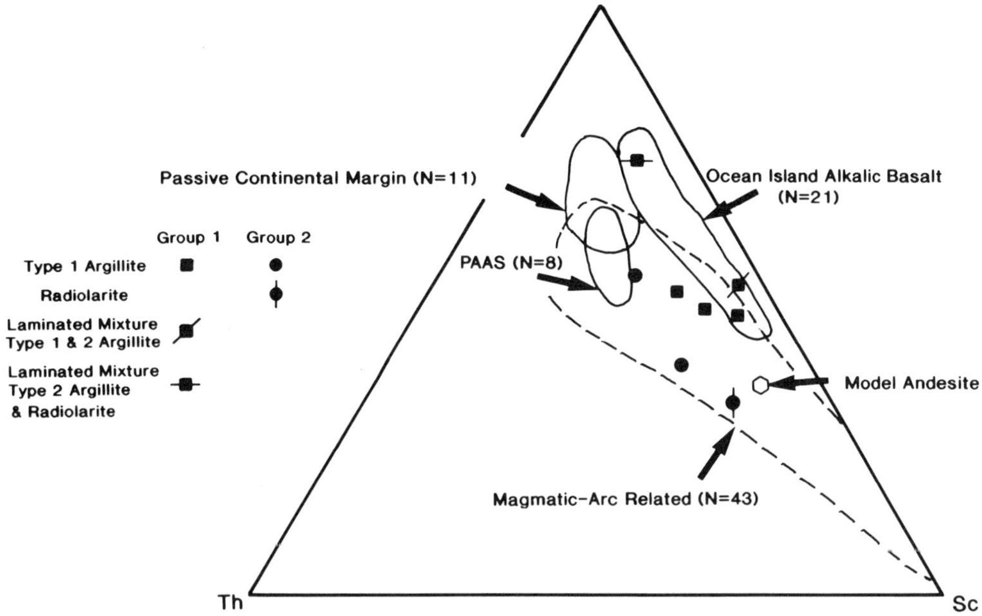

Figure 6. Samples from Groups 1 and 2 plotted on a La-Th-Sc ternary diagram. The magmatic arc–related field is derived from active-margin samples listed in McLennan and others (1990). Ocean island alkalic basalt field is derived from samples analyzed by Dixon and others (1984) and Liotard and others (1986). Passive continental margin (or trailing edge) field is derived from samples listed in McLennan and others (1990). The PAAS field is derived from samples listed in Taylor and McLennan (1985). Model andesite is from Taylor and McLennan (1985).

TABLE 4. CALCULATED ELEMENTAL RATIOS AND RARE EARTH ELEMENT ANOMALIES

Sample Number	La/Th	Th/Sc	Eu^2/Eu^*	$Ce^3/Ce^†$
1§	2.95	n.d.	1.06	1.11
2	4.65	0.25	0.83	1.03
5	6.41	n.d.	0.91	1.12
8	7.23	0.14	1.04	1.02
11§	2.94	n.d.	0.91	1.09
13	3.93	0.37	0.84	1.11
15§	2.74	n.d.	0.84	1.07
18	6.78	n.d.	0.83	1.01
21§	2.29	0.26	0.84	1.03
24§	2.86	0.15	0.79	0.99
25	7.61	n.d.	0.83	1.06
27	9.37	0.43	0.83	0.90
30	15.09	0.08	0.97	1.00
32§	2.08	0.44	0.83	1.07
34	13.20	n.d.	0.70	1.01

*Ratios determined from chondrite-normalized values using equation in McLennan and others (1990). Normalizing values are those given in McLennan (1989).

†Ratios determined from PAAS normalized values.

§Group 1 samples. All other samples are included in Group 2.

n.d. = no data available.

TABLE 5. La/Th and Th/Sc VALUES IN COMMON IGNEOUS ROCKS AND PAAS

Sample*	Lithology	La/Th	Th/Sc
PCC-1	Peridotite	15.0	0.001
BCR-1	Basalt	7.0	0.18
W-1	Diabase	5.7	0.07
ASV-1†	Andesite	7.0	0.48
GSP-1	Granodiorite	1.9	14.79
G-2	Granite	2.3	6.49
G-1	Granite	2.0	17.24
Ponape Island and Marquesas Archipelago§	Alkalic basalt	5.6 to 11.43	0.08 to 0.64
PAAS**	Shale	2.6	0.91

*Data sources are Flanagan (1973), Dixon and others (1984), Taylor and McLennan (1985), and Liotard and others (1986).

†See also Gill (1981), who suggested that La/Th values for island arc andesites vary from 2.0 to 7.0.

§Range in values is based on 20 analyses reported in Dixon and others (1984) and Liotard and others (1986). Note that Gill (1981) suggested that La/Th values for oceanic island basalt vary from 7.0 to 15.0.

**Post-Archean average Australian shale cited in Taylor and McLennan (1985). PAAS is derived from upper crustal cratonal sources and commonly is cited as closely approximating average granodioritic compositions (e.g., Taylor and McLennan, 1985).

obtained from the mixing model suggest that the average REE composition in Group 2 samples can be explained as an approximate 40:60 to 50:50 mixture of these two components (Fig. 8).

SUMMARY AND CONCLUSIONS

The Quartz Hill chert is composed of two chemically distinctive groups of marine sediment that we have designated as Group 1 and Group 2. Group 1 samples probably were deposited off shore of a magmatic arc, and are composed of magmatic arc–related particulate matter.

Group 2 specimens contain variable proportions of magmatic arc and alkalic basalt particulate matter, and are interstratified with Group 1 specimens (Fig. 3). The interstratified character of the two chemically distinctive groups of sediment is probably the result of variable rates of terrigeneous sedimentation. For example, during slack periods of terrigeneous input from the magmatic arc, or during periods of enhanced alkaline volcanic activity, more basaltic particulate matter would have been deposited, or vice versa (e.g., Dean and others, 1989). The above observations and inferences imply that the Quartz Hill chert was deposited on the margins of an oceanic plate adjacent to an ocean island(s) or seamount(s) and a magmatic arc.

Conclusions reached during this study are consistent with previous work that suggested that the Shoo Fly Complex contains the remnants of an alkalic seamount or ocean island and sand- to pebble-sized debris shed from a dissected magmatic arc (Girty and Wardlaw, 1984; Girty and Pardini, 1987; Girty and others, 1990). Thus, our data and the work of others, imply that the REE, Th, and Sc characteristics of chert/argillite sequences deposited in or adjacent to active subduction systems are controlled primarily by source rocks in adjacent magmatic arc systems and by seamount or ocean island volcanism located within the subducting plate. We hope that our work will stimulate more geologists to evaluate the depositional settings and source-rock compositions of other ancient chert/argillite sequences using their geochemical properties.

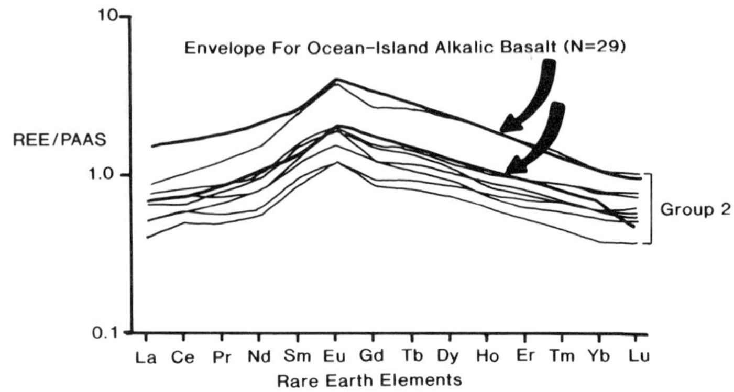

Figure 7. REE/PAAS plot of Group 2 samples comparing them to ocean island alkalic basalt. Envelope for ocean island alkalic basalt represents 1 standard deviation about the mean of data reported in the Basaltic Volcanism Study Project (1981), Dixon and others (1984), and Liotard and others (1986). Although some Group 2 samples contain slightly lower total REE abundances relative to modern-day ocean island alkalic basalt, they nevertheless display a REE/PAAS pattern similar in form to that of most alkalic basalt.

ACKNOWLEDGMENTS

Research reported here was funded by National Science Foundation Grant EAR-8902382 (to GHG). We thank R. W. Murray for preprints of his papers, and D. L. Jones, whose enthusiasm for cherts galvanized us into completing this project. R. W. Murray, P. K. Dutta, J. L. Graf, Jr., S. M. McLennan, and M. Johnsson read and provided constructive and thoughtful criticism of the manuscript.

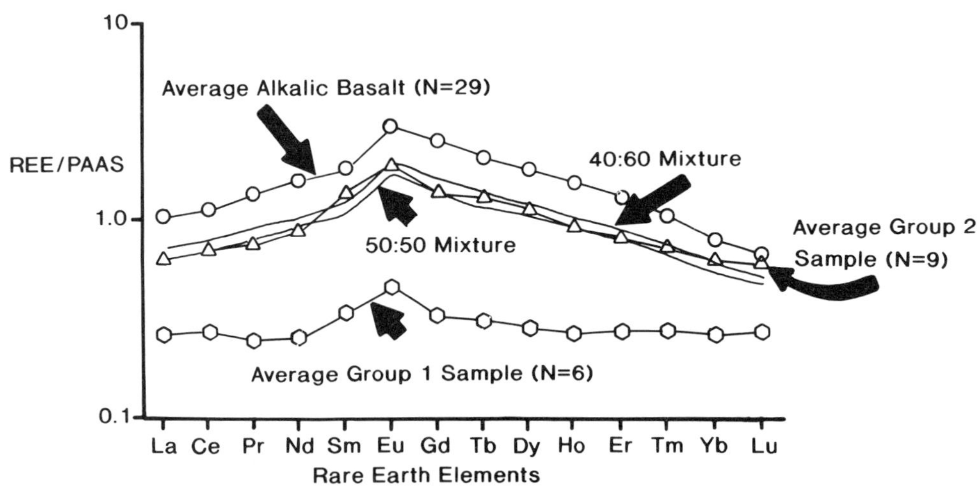

Figure 8. REE/PAAS plot summarizing the results of mixing average Group 1 REE data with average modern-day ocean island alkalic basalt. Note that a 50:50 to 40:60 mixture results in REE/PAAS curves that are grossly similar to the average REE/PAAS curve for Group 2 samples. Average alkalic basalt is derived from analyses listed in the Basaltic Volcanism Study Project (1981), Dixon and others (1984), and Liotard and others (1986).

REFERENCES CITED

Barber, R. W., 1992, Rare earth element analyses of cherts from the lower Paleozoic Shoo Fly Complex, northern Sierra Nevada, California [M.S. thesis]: San Diego, San Diego State University, 121 p.

Basaltic Volcanism Study Project, 1981, Basaltic volcanism on the terrestrial planets: New York, Pergamon Press, 1286 p.

Bhatia, M. R., 1985, Rare earth element geochemistry of Australian Paleozoic graywackes and mudrocks: Provenance and tectonic control: Sedimentary Geology, v. 45, p. 97–113.

Bhatia, M. R., and Crook, K.A.W., 1986, Trace element characteristics of graywackes and tectonic setting discrimination of sedimentary basins: Contributions to Mineralogy and Petrology, v. 92, p. 181–193.

Cole, M. R., and Armentrout, J. M., 1979, Neogene paleogeography of the western United States, *in* Armentrout, J. M., Cole, M. R., and TerBest, H., Jr., eds., Cenozoic paleogeography of the western United States: Society of Economic Paleontologists and Mineralogists, Pacific Section, Pacific Coast Paleogeography Symposium 3, p. 297–323.

Condie, K. C., 1991, Another look at rare earth elements in shales: Geochimica et Cosmochimica Acta, v. 55, p. 2527–2531.

Crock, J. G., and Lichte, F. E., 1982, Determination of rare earth elements in geologic materials by inductively coupled argon plasma/atomic emission spectrometry: Analytical Chemistry, v. 54, p. 1329–1332.

Dean, W. E., Gardner, J. V., and Parduhn, N. L., 1989, Influence of Shimada Seamount on sediment composition in the eastern tropical North Pacific: Geochimica et Cosmochimica Acta, v. 53, p. 1523–1536.

Dickinson, W. R., 1979, Cenozoic plate tectonic setting of the Cordilleran region in the United States, *in* Armentrout, J. M., Cole, M. R., and TerBest, H., Jr., eds., Cenozoic paleogeography of the western United States: Society of Economic Paleontologists and Mineralogists, Pacific Section, Pacific Coast Paleogeography Symposium 3, p. 1–13.

Dickinson, W. R., Ingersoll, R. V., Cowan, D. S., Helmold, K. P., and Suczek, C. A., 1982, Provenance of Franciscan graywackes in coastal California: Geological Society of America Bulletin, v. 93, p. 95–107.

Dixon, T. H., Batiza, R., Futa, K., and Martin, D., 1984, Petrochemistry, age and isotopic composition of alkali basalts from Ponape Island, western Pacific: Chemical Geology, v. 43, p. 1–28.

Doherty, W., 1989, An internal standardization procedure for the determination of yttrium and the rare earth elements in geological materials by inductively coupled plasma-mass spectrometry: Spectrochemica Acta, v. 44B, p. 263–280.

Flanagan, F. J., 1973, 1972 values for international geochemical reference samples: Geochimica et Cosmochimica Acta, v. 37, p. 1189–1200.

Gill, J., 1981, Orogenic andesites and plate tectonics: New York, Springer-Verlag, 390 p.

Girty, G. H., and Barber, R. W., 1991, A plate-marginal continental depositional setting for the Quartz Hill chert, Shoo Fly Complex, Sierra Nevada, California [abs.]: Geological Society of America Abstracts with Programs, v. 23, p. A429.

Girty, G. H., and Pardini, C. H., 1987, Provenance of sandstone inclusions in the Paleozoic Sierra City melange, Sierra Nevada, California: Geological Society of America Bulletin, v. 98, p. 176–181.

Girty, G. H., and Wardlaw, M. S., 1984, Was the Alexander terrane a source of feldspathic sandstones in the Shoo Fly Complex, Sierra Nevada, California?: Geology, v. 12, p. 339–342.

Girty, G. H., and Wardlaw, M. S., 1985, Petrology and provenance of pre–Late Devonian sandstones, Shoo Fly Complex, northern Sierra Nevada, California: Geological Society of America Bulletin, v. 96, p. 516–521.

Girty, G. H., Gester, K. C., and Turner, J. B., 1990, Pre–Late Devonian geochemical, stratigraphic, sedimentologic, and structural patterns, Shoo Fly Complex, northern Sierra Nevada, California, *in* Harwood, D. S., and Miller, M. M., eds., Paleozoic and early Mesozoic paleogeographic relations; Sierra Nevada, Klamath Mountains, and related terranes: Geological Society of America Special Paper 255, p. 43–56.

Girty, G. H., Gurrola, L. D., Taylor, G. W., Richards, M. J., and Wardlaw, M. S., 1991, The pre–Upper Devonian Lang and Black Oak Springs sequences, Shoo Fly Complex, northern Sierra Nevada, California: Trench deposits composed of continental detritus, *in* Cooper, J. D., and Stevens, C. H., eds., Paleozoic paleogeography of the western United States—II: Society of Economic Paleontologists and Mineralogists, Pacific Section, v. 2, p. 703–716.

Ingersoll, R. V., 1983, Petrofacies and provenance of late Mesozoic forearc basin, northern and central California: American Association of Petroleum Geologists Bulletin, v. 67, p. 1125–1142.

Legarre, H. A., 1988, Geochemistry of basaltic inclusions, Sierra City melange: Evidence for pre–Upper Devonian oceanic crust [Senior thesis]: San Diego, San Diego State University, 30 p.

Lichte, F. E., Meier, A. L., and Crock, J. G., 1987, Determination of rare earth elements in geological materials by inductively coupled mass spectrometry: Analytical Chemistry, v. 59, p. 1150–1157.

Liotard, J. M., Barsczus, H. G., Dupy, C., and Dostal, J., 1986, Geochemistry and origin of basaltic lavas from Marquesas Archipelago, French Polynesia: Contributions to Mineralogy and Petrology, v. 92, p. 260–268.

McLennan, S. M., 1989, Rare earth elements in sedimentary rocks: influence of provenance and sedimentary processes: Reviews in Mineralogy, v. 21, p. 169–200.

McLennan, S. M., and Taylor, S. R., 1991, Sedimentary rocks and crustal evolution: tectonic setting and secular trends: Journal of Geology, v. 99, p. 1–21.

McLennan, S. M., Nance, W. B., and Taylor, S. R., 1980, Rare earth element–thorium correlations in sedimentary rocks, and the composition of the continental crust: Geochimica et Cosmochimica Acta, v. 44, p. 1833–1839.

McLennan, S. M., Taylor, S. R., McCulloch, M. T., and Maynard, J. B., 1990, Geochemical and Nd-Sr isotopic composition of deep-sea turbidites: Crustal evolution and plate tectonic associations: Geochimica et Cosmochimica Acta, v. 54, p. 2015–2050.

Murray, R. W., Buchholtz ten Brink, M. R., Jones, D. L., Gerlach, D. C., and Russ, G. P., III, 1990, Rare earth elements as indicators of different marine depositional environments in chert and shale: Geology, v. 18, p. 268–271.

Murray, R. W., Buchholtz ten Brink, M. R., Gerlach, D. C., Russ, G. P., III, and Jones, D. L., 1991, Rare earth, major, and trace elements in chert from the Franciscan Complex and Monterey Group, California: Assessing REE sources in fine-grained marine sediments: Geochimica et Cosmochimica Acta, v. 52, p. 1875–1895.

Musashino, M., 1990, The Panthalassa—A cerium-rich Atlantic-type ocean: sedimentary environments of the Tamba Group, southwest Japan: Tectonophysics, v. 181, p. 165–177.

Saleeby, J. B., 1990, Geochronological and tectonostratigraphic framework of Sierran-Klamath ophiolitic assemblages, *in* Harwood, D. S., and Miller, M. M., eds., Paleozoic and early Mesozoic paleogeographic relations; Sierra Nevada, Klamath Mountains, and related terranes: Geological Society of America Special Paper 255, p. 93–114.

Schweickert, R. A., and Snyder, W. S., 1981, Paleozoic plate tectonics of the Sierra Nevada and adjacent regions, *in* Ernst, W. D., ed., The geotectonic development of California (Rubey Volume I): Englewood Cliffs, New Jersey, Prentice-Hall, p. 182–202.

Schweickert, R. A., Harwood, D. S., Girty, G. H., and Hanson, R. E., 1984, Tectonic development of the northern Sierra terrane; An accreted late Paleozoic island-arc and its basement, *in* Lintz, J., Jr., ed., Western geological excursions, v. 4: Reno, University of Nevada Mackay School of Mines, p. 1–65.

Taylor, S. R., and McLennan, S. M., 1985, The continental crust: Its composition and evolution: Oxford, Blackwell Scientific Publications, 312 p.

White, R. D., 1986, Cambrian radiolaria from Utah: Journal of Paleontology, v. 60, p. 778–780.

Manuscript Accepted by the Society December 24, 1992

Printed in U.S.A.

Geological Society of America
Special Paper 284
1993

Provenance controls on the Nd-Sr-O isotopic composition of sandstones: Example from Late Mesozoic Great Valley forearc basin, California

A. M. Linn* and D. J. DePaolo
Department of Geology and Geophysics, University of California, Berkeley, and Division of Earth Sciences, Lawrence Berkeley Laboratory, Berkeley, California 94720

ABSTRACT

Measurements of Nd, Sr, and O isotopes were made on whole-rock Great Valley sandstones in order to determine the extent to which isotopes preserve the composition and weathering history of tectonically active sources. Previous work has shown that the Nd isotopic composition of sedimentary rocks provides an accurate record of the source and can be used to evaluate the effect of weathering and diagenesis on Sr-O isotope systematics. Variations in the whole-rock Nd, Sr, and O isotopic compositions are similar within and between petrofacies; this suggests that the isotopic composition is controlled by provenance. The Nd-Sr isotopic compositions are sensitive to the large west-east variation in crust formation ages of the igneous and metasedimentary components. The sandstones decrease from +7 to –5 in ϵ_{Nd} and increase from 0.7045 to 0.7073 in $^{87}Sr/^{86}Sr$ with decreasing stratigraphic age. These values nearly encompass the range observed in the plutonic rocks and suggest that the Nd-Sr isotopic composition of the arc is preserved in the sandstones. In contrast, $\delta^{18}O$ decreases from +19 to +9 with decreasing stratigraphic age; these values are significantly higher than $\delta^{18}O$ values in the volcanic/plutonic arc source. Isotopic and petrographic variations are correlated; with increasing proportions of sedimentary and metasedimentary lithic fragments, values of ϵ_{Nd} decrease, and values of $^{87}Sr/^{86}Sr$ and $\delta^{18}O$ increase. The sedimentary and metasedimentary components decrease the whole-rock value by two to four ϵ_{Nd} units and by one to 10‰, depending on the fraction and lithology of the recycled component. The metasedimentary component has a significantly lower Sr concentration than the igneous component and therefore has a smaller effect on the whole-rock $^{87}Sr/^{86}Sr$ value. Subtraction of the sedimentary-metasedimentary component from the whole-rock sandstone compositions yields igneous isotopic compositions that are directly comparable to the composition of the arc source. Calculated igneous compositions of the sandstones decrease in ϵ_{Nd} from +7 to –5 with decreasing stratigraphic age and indicate that sediment sources were located at the eastward-migrating volcanic front. However, the calculated igneous components are +1 to +7 ‰ higher in $\delta^{18}O$ than the igneous source. The high $\delta^{18}O$ values are a measure of alteration of the source; increasing alteration in $\delta^{18}O$ is correlated with increasing values of CIA. This suggests that high mechanical erosion rates of tectonically active sources do not preclude significant chemical weathering.

*Present address: National Research Council, 2001 Wisconsin Avenue, N.W., Washington, D.C., 20007.

Linn, A. M., and DePaolo, D. J., 1993, Provenance controls on the Nd-Sr-O isotopic composition of sandstones: Example from Late Mesozoic Great Valley forearc basin, California, *in* Johnsson, M. J., and Basu, A., eds., Processes Controlling the Composition of Clastic Sediments: Boulder, Colorado, Geological Society of America Special Paper 284.

INTRODUCTION

In recent years, the isotopic and geochemical analysis of sedimentary rocks has provided new insights into the evolution of crustal, tectonic, and climatic processes through geologic time (e.g., Taylor and McLennan, 1985; Bowan, 1991). Nd isotopes are of particular interest in provenance studies because the Nd isotopic composition of the source rocks is preserved throughout the sedimentary cycle. Measurements of the Nd isotopic composition of siliciclastic rocks have been made to locate sediment source areas (e.g., Frost and Winston, 1987; Goldstein and Jacobsen, 1988; Nelson and DePaolo, 1988; Chen et al., 1990; McLennan et al., 1990; Linn et al., 1991, 1992), to reconstruct tectonic terranes (Miller and O'Nions, 1984; Clarke and Halliday, 1985), and to constrain models of crustal growth and sediment recycling (e.g., Hamilton et al., 1983; O'Nions et al., 1983; Allegre and Rousseau, 1984; Michard et al., 1985). Sr and O isotopes can also identify sedimentary sources (e.g., Clayton et al., 1978; Spiers et al., 1984; Yeh and Eslinger, 1986; Goldstein and Jacobsen, 1988; Nelson and DePaolo, 1988; Chen et al., 1990), but are more susceptible to modification by weathering, transport, and diagenetic processes. Nevertheless, Sr and O isotopes provide additional ways to locate source areas, and their analysis in association with Nd isotopes or other provenance indicators offers a means to test and refine provenance interpretations (see Heller et al., 1985, for an excellent example).

In this work, we investigate the relation between provenance and the Nd-Sr-O isotopic composition of Mesozoic Great Valley forearc sandstones of California. Great Valley sedimentary rocks are ideal for this purpose because the large west-east age and isotopic gradients in the Sierra Nevada arc source make it possible to do high-resolution provenance studies. Moreover, petrographic analysis indicates that a wide variety of lithologies are present in the sandstones (e.g., Dickinson and Rich, 1972; Ojakangas, 1968; Ingersoll, 1983); this provides constraints on the isotopic interpretations and allows us to test the relation between mineralogy and isotopic composition. Finally, the high erosion rates of active continental margin arcs (Dumitru, 1988; Linn et al., 1992) and the subsequent immaturity of the sandstones enhance the probability that the sediments will preserve the isotopic composition of the source. Linn et al. (1991, 1992) showed that the Nd isotopic and geochemical compositions of the Sierra Nevada arc are preserved in Great Valley sedimentary rocks and could be used to pinpoint source areas. In contrast, the Sr isotopic composition of the Upper Jurassic and Lower Cretaceous sandstones was modified by weathering, transport, or diagenetic processes and was less useful in determining provenance (Linn et al., 1992). The ability to use Nd isotopes as provenance indicators allows us to evaluate the extent to which the oxygen isotopic composition of arc sources is retained in sedimentary rocks. At the same time, the susceptibility of oxygen isotopes to weathering and diagenetic processes allows us to place constraints on the sedimentary processes that can modify the chemical and isotopic composition of siliciclastic rocks.

ISOTOPIC COMPOSITIONS OF SILICICLASTIC ROCKS

Nd isotopes

Sm and Nd are rare earth elements that have similar chemical properties, except that Nd has a slightly larger ionic radius. The major partitioning of Sm from Nd occurs during extraction of continental crust from the mantle (DePaolo and Wasserburg, 1976). Consequently the Nd isotopic composition of igneous sources varies between mantle and continental rocks and between crustal provinces of different ages. Isotopic variations in the source are most likely to be preserved in first-cycle sedimentary rocks; subsequent reworking of the sediments will yield an average age (denoted the crustal residence age) or isotopic composition of the source. Because the Sm/Nd value is similar in most igneous minerals and is preserved except under intense conditions of mechanical and chemical weathering of crustal rocks (e.g., Balashov et al., 1964; Fleet, 1984), the Nd isotopic composition of siliciclastic rocks is independent of grain size (Linn et al., 1992). Transport processes, however, may concentrate minerals or lithic fragments that differ in age and/or isotopic composition into different size fractions (e.g., Frost and Coombs, 1989; McLennan et al., 1989). Similarly, recrystallization of clay minerals or precipitation of cement during late-stage diagenesis may alter the Sm/Nd value of the rock (e.g., Balashov and Girin, 1969; Awwiller and Mack, 1991), although redistribution of the rare earth elements may be localized (Schieber, 1988). Finally, metamorphism can fractionate rare earth elements and form minerals that differ in Sm/Nd from the whole rock (Humphris, 1984; Nelson and DePaolo, 1988). This effect is significant only if the age difference between crust formation and metamorphism is large (Nelson and DePaolo, 1988), or if the rare earth elements are concentrated in the metamorphic minerals, as is the case in quartzose sediments (Cullers et al., 1979; Frost and Winston, 1987).

Sr isotopes

Fine-grained sedimentary rocks commonly fall on linear arrays in $^{87}Sr/^{86}Sr$-$^{87}Rb/^{86}Sr$ space; these "isochrons" yield the mineral, depositional or diagenetic age (e.g., Moorbath, 1969; Peterman et al., 1981; Heller et al., 1985; Faure, 1986). Isochron ages are valid only to the extent that the minerals or sedimentary rocks behave as a closed system with regard to Rb and Sr (Faure, 1986). However, the chemical properties of Rb, an alkali element, differ from Sr, an alkaline earth element. Sr is mobile relative to Rb, and chemical weathering, seawater exchange, and diagenesis can remove Sr from plagioclase and partition it into calcium carbonate cement. Rb extracted from micas and potassium feldspar during weathering is partitioned into clay minerals (e.g., Clauer, 1982), but tends to remain relatively inert during diagenesis (Perry and Turekian, 1974). Transport processes enhance the parent-daughter fractionation by mechanical sorting of fine-grained clays and micas that have high Rb/Sr values from coarser framework minerals that have lower Rb/Sr values. The magnitude of these fractionations can be difficult to quantify. Nevertheless, analysis of immature, relatively unaltered sandstones or mineral separates can yield geologically meaningful Sr isotopic compositions and provenance ages (e.g., Heller et al., 1985; Faure, 1986; Linn et al., 1992).

O isotopes

In contrast with the Nd-Sr isotope systems, the oxygen isotopic composition is primarily a function of the temperature, rather than the age, of rock formation. Igneous rocks typically crystallize at high temperatures and have relatively uniform $\delta^{18}O$ values of +6 to +11 (Hoefs, 1987). Sedimentary rocks are high in $\delta^{18}O$, ranging from about +10 to +40‰ (Hoefs, 1987), because they contain detrital or authigenic minerals that formed at low temperatures. For example, clay minerals formed during chemical weathering are 10 to 30‰ higher than the parent rock, depending on the temperature and isotopic composition of the meteoric waters (Savin and Epstein, 1970a,b). Rock fragments and coarser grained framework minerals are resistant to isotopic exchange during weathering (e.g., Lawrence and Taylor, 1972; Clayton and others, 1978), but may begin to equilibrate with formation fluids during diagenesis at temperatures exceeding 200° to 300°C (e.g., O'Neil and Kharaka, 1976; Yeh and Savin, 1977; Land and Milliken, 1981; Longstaffe, 1983; Eslinger and Yeh, 1986). The fine size fraction is more susceptible to modification by burial metamorphism, and may begin to equilibrate with pore fluids upon burial (Yeh and Savin, 1977). Whether diagenetic reactions increase or decrease, the $\delta^{18}O$ value of siliciclastic rocks depends on the origin and evolution of the formation fluids during the burial history of the sedimentary basin. For example, alteration of pelagic sediments by fluids derived from an underlying basalt may decrease the whole-rock $\delta^{18}O$ value (Smith, 1991), whereas recharge by meteoric fluids during uplift may increase the $\delta^{18}O$ value (e.g., Ayalon and Longstaffe, 1987; Tilley and Longstaffe, 1989). Because of the sensitivity of oxygen isotopes to alteration effects, only immature sandstones or quartz separates are likely to retain significant provenance information (Savin and Epstein, 1970a; Clayton et al., 1978; Eslinger and Yeh, 1986).

GEOLOGY OF SIERRA NEVADA ARC–GREAT VALLEY FOREARC

The Great Valley Group consists of nearly 15 km of Late Jurassic and Cretaceous sandstone, shale, and conglomerate that is exposed in a homocline along the western margin of the central valley of California (Fig. 1) (Dickinson and Rich, 1972). The forearc basin tectonically overlies the Franciscan accretionary complex to the west, extends below the surficial sediments of the Central Valley, and nonconformably overlies the eroded western flank of the Sierra Nevada arc and southern margin of the Klamath Mountains (Ingersoll, 1979). The Sierra Nevada arc is compositionally zoned with large west-east gradients in age, mineralogy, and geochemical and isotopic composition. Plutonic rocks in the western part of the arc are relatively mafic and are high in concentrations of FeO, MgO, and CaO, whereas plutonic rocks along the eastern margin are more felsic and are high in concentrations of K_2O, Rb, U, and Th (Dodge et al., 1982; Ague and Brimhall, 1988). From west to east, plutonic rocks decrease in ϵ_{Nd} from +7 to −6 (Fig. 1A) (Farmer and DePaolo, 1983; Linn et al., 1991) and increase in $^{87}Sr/^{86}Sr$ from 0.704 to 0.708 (Kistler and Peterman, 1973). The isotopic gradients are inferred to have been similar in the coeval volcanic cover (Linn et al., 1992). The gradients are largest in the southern part of the arc due to involvement of Cretaceous mantle-derived magma with Proterozoic continental shelf sediments (DePaolo, 1981) which are estimated to be −10 to −14 in ϵ_{Nd} and about 0.717 in $^{87}Sr/^{86}Sr$ (Linn et al., 1992). West-east crustal assimilation patterns in $\delta^{18}O$ were apparently overprinted by secondary processes, resulting in oxygen isotopic compositions which vary between +7 and +10‰ in the plutonic rocks (Fig. 1B) (Masi et al., 1981). Volcanic sources that supplied sediment to the early forearc basin were likely +6 to +7 in $\delta^{18}O$, similar to average island arcs (Hoefs, 1987).

The Sierra Nevada arc–Great Valley forearc system formed in Late Jurassic time following the Nevadan orogeny (Schweickert and Cowan, 1975; Saleeby, 1981). Late Jurassic and Early Cretaceous sediment derived from the Sierra Nevada arc and Klamath Mountains was deposited as submarine fans in a forearc slope setting that evolved into a deep forearc basin by Early Cretaceous time (Ingersoll, 1982). Throughout Cretaceous time, the forearc basin widened as a result of prograde westward accretion of the Franciscan complex and the eastward migration of arc magmatism (Dickinson, 1973; Ingersoll, 1979). As inactive parts of the Sierra Nevada arc and Klamath mountains cooled and subsided below sea level, the subaerial arc and its basement supplied an increasing proportion of detritus to the forearc basin and became the only sediment source in Late Cretaceous time (Ingersoll, 1979). Subduction ended in Neogene time with the development of a transform margin along the California coast (Atwater, 1970), and the Great Valley Group was tilted and uplifted above sea level (Ingersoll, 1979).

Variations in the relative mineral and lithic fragment abundances subdivide the Great Valley Group and record the evolution of sediment sources (e.g., Ojakangas, 1968; Dickinson and Rich, 1972; Ingersoll, 1983). The most comprehensive analysis of the sandstone petrology is reported in Ingersoll (1983) and is summarized in Table 1. The lower part of the stratigraphic section is exposed in the Sacramento Valley and is divided into the Stony Creek, Lodoga, and Boxer petrofacies. Upper Jurassic and lowermost Cretaceous Stony Creek sandstones are either composed of volcanic lithic fragments and plagioclase feldspar, or are high in chert and shale-argillite lithic fragments. Lower Cretaceous Lodoga sandstones are high in metamorphic lithic fragments, and middle Cretaceous Boxer sandstones are high in

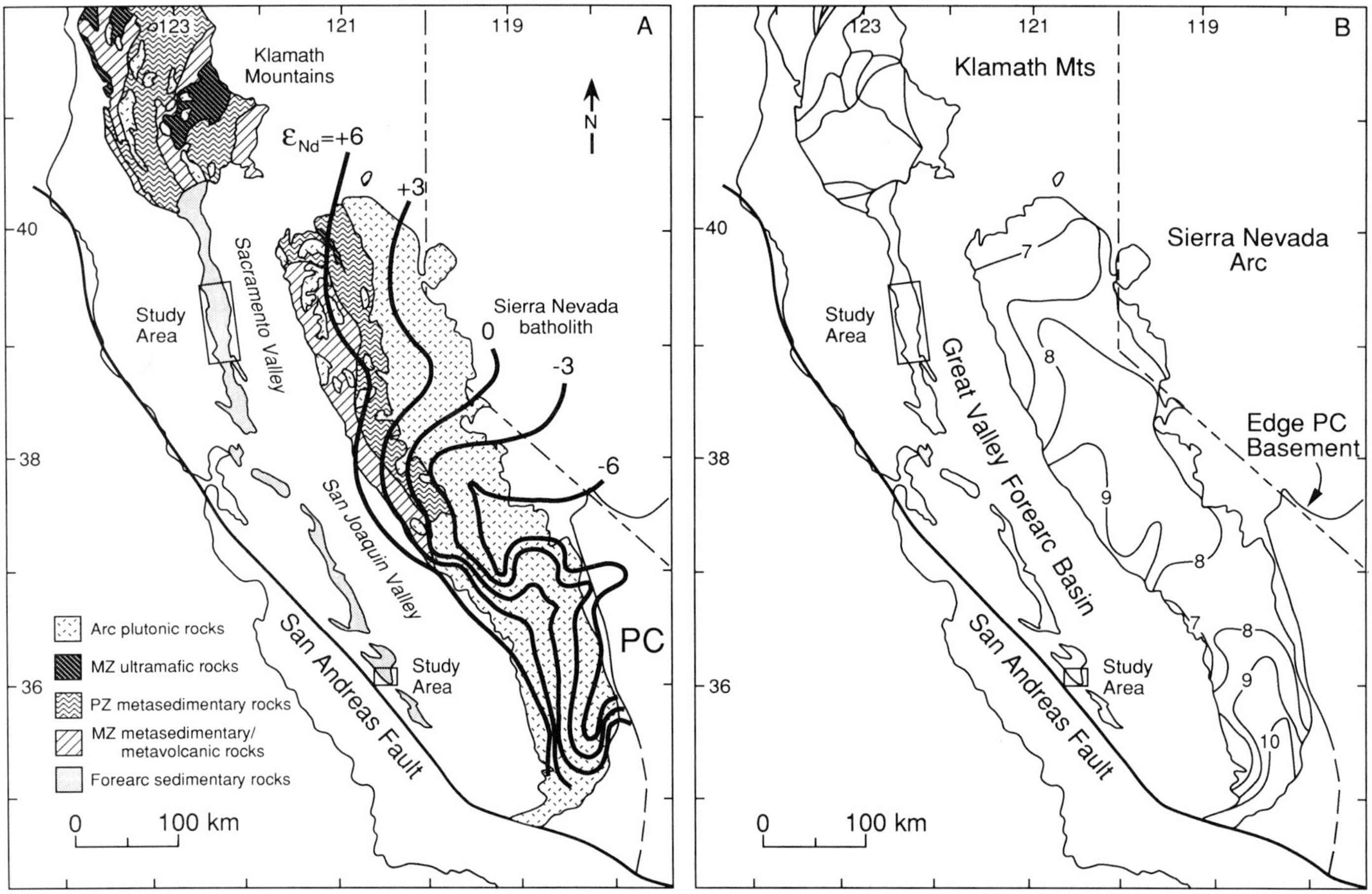

Figure 1. A, Present configuration of Sierra Nevada arc–Great Valley forearc system showing source lithologies and sandstone sample localities in the Sacramento and San Joaquin Valleys. Mesozoic and Paleozoic metasedimentary and metavolcanic rocks along the northwest margin of the Sierra Nevada batholith are the Foothills metamorphic belt. Heavy lines are contours of $^{143}Nd/^{144}Nd$ of Sierra Nevada plutonic rocks, expressed as ϵ_{Nd}, which decrease with increasing crustal assimilation toward the Proterozoic continental margin (PC). From Linn et al. (1992). B, Oxygen isotopic composition, expressed as contours of $\delta^{18}O$, of Mesozoic plutonic rocks in the Sierra Nevada batholith and Klamath Mountains. Modified from Masi et al. (1981).

TABLE 1. MEAN PETROGRAPHIC PARAMETERS OF GREAT VALLEY GROUP*

	QFL %			LmLvLs %					
Petrofacies	Q	F	L	Lm	Lv	Ls	P/F	Lv/L	Qp/Q
San Joaquin Valley									
Rumsey	40	36	24	31	60	9	0.55	0.60	0.03
Los Gatos	38	40	23	54	38	8	0.64	0.40	0.04
Grabast	30	30	41	57	37	6	0.68	0.37	0.09
Sacramento Valley									
Boxer	32	32	36	19	73	8	0.75	0.73	0.09
Lodoga	37	21	42	33	42	25	0.86	0.42	0.21
Stony Creek	25	22	53	12	60	28	0.92	0.60	0.37

*Modified from Ingersoll (1982), where Q = monocrystalline + polycrystalline (Qp) quartz; F = plagioclase (P) + potassium feldspar; L = metamorphic (Lm) + volcanic (Lv) + sedimentary (Ls) lithic fragments.

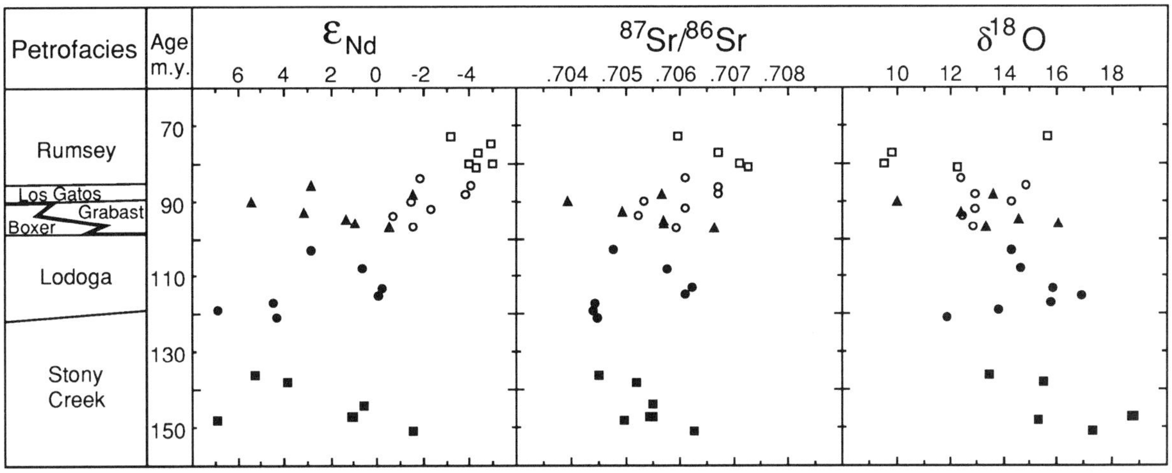

Figure 2. Correspondence in Nd-Sr-O isotopic composition with stratigraphic age in whole-rock Great Valley sandstones. San Joaquin Valley sandstones are divided into Rumsey (open boxes) and Los Gatos–Grabast petrofacies (open circles); Sacramento Valley sandstones are divided into Boxer (filled triangles), Lodoga (filled circles), and Stony Creek petrofacies (filled boxes). Time scale and stratigraphic relations after Ingersoll (1983) and Linn et al. (1992). Nd-Sr isotopic data from Linn et al. (1992).

volcanic lithic fragments. The upper part of the section is exposed in the San Joaquin Valley and is divided into the Grabast, Los Gatos, and Rumsey petrofacies. Middle Cretaceous Grabast and Los Gatos sandstones are high in metamorphic lithic fragments, and uppermost Cretaceous Rumsey sandstones are relatively quartzo-feldspathic with significant amounts of K-feldspar. The temporal change from lithic to quartzo-feldspathic sandstone has been interpreted as the unroofing sequence of the arc, and the relative increase in K-feldspar with decreasing stratigraphic age has been interpreted as indicating that source areas moved east during Cretaceous time (Ojakangas, 1968; Dickinson and Rich, 1972; Ingersoll, 1983).

PROVENANCE OF GREAT VALLEY GROUP

Isotopic stratigraphy

Stratigraphic variations in the whole-rock Nd-Sr-O isotopic compositions of Great Valley strata are shown in Figure 2 and are reported in Table 2. Isotopic variations in the sandstones are large but somewhat systematic among the petrofacies, and are strikingly similar in all the isotope systems. Because the Nd-Sr isotopic systems are best correlated, we use them to illustrate the stratigraphic variations in the sandstones, then compare the overall trends to the oxygen isotopic composition. Stony Creek sandstones are divided into two groups: one with ϵ_{Nd} values between +1.0 and −1.5 and one with ϵ_{Nd} values between +7 and +4. Although sandstones that are low in ϵ_{Nd} are high in $^{87}Sr/^{86}Sr$, there are no groups with distinct Sr isotopic compositions within the petrofacies. Rather, $^{87}Sr/^{86}Sr$ decreases from 0.7063 to 0.7045 in Upper Jurassic to Lower Cretaceous strata. Lodoga sandstones decrease in ϵ_{Nd} from +7 at the base to about 0 in the middle, then increase to +3 at the top of the petrofacies. The $^{87}Sr/^{86}Sr$ value also reverses in isotopic composition with decreasing stratigraphic age, and increases from 0.7044 to 0.7062, then decreases to 0.7048. Boxer sandstones are highly variable and encompass nearly the entire range of Nd and Sr isotopic compositions observed in the Sacramento Valley sandstones. Boxer sandstones vary in ϵ_{Nd} from +5.4 to −1.6, and in $^{87}Sr/^{86}Sr$ from 0.7039 to 0.7066. Nd-Sr isotopic compositions in San Joaquin Valley petrofacies are less variable than in Sacramento Valley petrofacies. Grabast and Los Gatos sandstones decrease in ϵ_{Nd} from −0.7 to −4 and increase in $^{87}Sr/^{86}Sr$ from 0.7052 to 0.7067 with decreasing stratigraphic age. Rumsey sandstones are low in ϵ_{Nd} and range from −4 to −5, although the uppermost sandstone is higher in ϵ_{Nd}. The apparent decrease in $^{87}Sr/^{86}Sr$ from 0.7073 to 0.7060 with decreasing stratigraphic age is likely a sampling artifact; two sandstones that were low in ϵ_{Nd}, and presumably high in $^{87}Sr/^{86}Sr$, were calcite-cemented and were not measured for Sr isotopes.

Variations in the oxygen isotopic composition of Great Valley sandstones are similar to variations observed in the Nd-Sr isotopes (Fig. 2). In general, sandstones that are high in ϵ_{Nd} and low in $^{87}Sr/^{86}Sr$ are high in $\delta^{18}O$. This observation is somewhat surprising, given the susceptibility of oxygen isotopes to modification by sedimentary processes. Stony Creek sandstones decrease in $\delta^{18}O$ from +19 to +13 with decreasing stratigraphic age, and, like Sr isotopes, are not divided into distinct groups. There is a reversal in oxygen isotopic composition in Lodoga sandstones, which is analogous to the reversal in Nd-Sr isotopic compositions. With decreasing stratigraphic age, $\delta^{18}O$ increases from +12 to +17, then decreases to +14. Boxer sandstones are again the most variable petrofacies, and range in $\delta^{18}O$ from +10 to +16. San Joaquin Valley sandstones are more variable in $\delta^{18}O$ than in ϵ_{Nd} and $^{87}Sr/^{86}Sr$, and the relation with Nd and Sr isotopes is less clear (Fig. 2). In contrast to the systematic stratigraphic

TABLE 2. ISOTOPIC AND PETROGRAPHIC COMPOSITION OF GREAT VALLEY SANDSTONES

Sample	Age (Myr)	Petrofacies	ε_{Nd}*	T_{DM}	$^{87}Sr/^{86}Sr$*	T_{UR}	$\delta^{18}O$	%Ls+ Lsm	CIA
San Joaquin Valley									
87GV-9	73	Rumsey	-3.2	1.09	0.70597	0.36	15.7		56
87GV-8	75	Rumsey	-4.9						
87GV-19	77	Rumsey	-4.3	1.19	0.70671	0.26	9.8	6	53
87GV-2	80	Rumsey	-5.0						
87GV-18	80	Rumsey	-4.0	1.16	0.70881	0.29	9.5	12	52
87GV-16	81	Rumsey	-4.3	1.19	0.70726	0.29	12.3	13	56
87GV-21	84	Grabast	-1.8	0.98	0.70703	0.39	12.4	22	53
87GV-15	86	Grabast	-4.0	1.17	0.70671	0.37	14.8		58
87GV-13	88	Grabast	-3.8	1.16	0.70795	0.37	13.0		55
87GV-10	90	Grabast	-1.4	0.90	0.70532	0.27	14.3	27	60
87GV-33	92	Grabast	-2.3	1.03	0.70715	0.39	12.9	22	54
87GV-31	94	Grabast	-0.7	0.89	0.70628	0.32	12.5	23	54
87GV-27	97	Grabast	-1.6	0.97	0.70938	0.32	12.9	27	59
Sacramento Valley									
89GV-48	86	Boxer	2.9						
89GV-44	88	Boxer	-1.6	0.96	0.70655	0.38	13.6	13	56
89GV-42	90	Boxer	5.4	0.35	0.70424	0.50	10.0	13	49
89GV-40	94	Boxer	3.2	0.55	0.70563	0.41	12.4	13	53
89GV-52	95	Boxer	1.4	0.71	0.70746	0.25	14.6		60
89GV-36	96	Boxer	1.0	0.75	0.70753	0.25	16.1		61
89GV-54	97	Boxer	-0.5	0.89	0.70805	0.37	13.3		58
89GV-53	103	Lodoga	2.9	0.59	0.70568	0.35	14.3		67
89GV-55	108	Lodoga	0.6	0.79	0.70675	0.46	14.7	24	59
89GV-56	113	Lodoga	-0.2	0.86	0.70770	0.39	15.9	33	63
89GV-57	115	Lodoga	0.0	0.85	0.70753	0.40	16.9		61
89GV-59	117	Lodoga	4.5	0.46	0.70560	0.29	15.8	33	63
89GV-58	119	Lodoga	6.9	0.25	0.70491	0.57	13.8		59
89GV-61	121	Lodoga	4.3	0.48	0.70499	0.56	11.9	10	56
89GV-64	136	Stony Creek	5.2	0.41	0.70514	0.54	13.5		58
89GV-60	138	Stony Creek	3.9	0.53	0.70597	0.62	15.5	13	58
89GV-65	144	Stony Creek	0.6	0.82	0.70668	0.50			
89GV-63	147	Stony Creek	1.1	0.78	0.70670	0.50	18.7	43	63
87GV-R8	147	Stony Creek	1.0	0.78	0.70651	0.53	18.8	31	63
87GV-R11	148	Stony Creek	6.9	0.27	0.70526	1.94	15.3	8	54
89GV-66	151	Stony Creek	-1.6	1.02	0.70873	0.37	17.3		64

*From Linn et al., 1992.

changes in the Nd-Sr isotopic composition of Grabast and Los Gatos sandstones, $\delta^{18}O$ varies between +12 and +15. Rumsey sandstones are low in $\delta^{18}O$ and range from +9 to +12, although the uppermost sandstone is again anomalous. The overall decrease in $\delta^{18}O$ with decreasing stratigraphic age in San Joaquin Valley sandstones is generally correlative with decreasing ϵ_{Nd} and increasing $^{87}Sr/^{86}Sr$. This trend is in the opposite sense as the Nd-Sr-O relation observed in Sacramento Valley sandstones.

The variation in isotopic composition of the petrofacies is correlative with the lithic fragment composition and thus reflects variations in provenance. Stony Creek sandstones, which are high in volcanic lithic fragments, are high in ϵ_{Nd} and low in $\delta^{18}O$, whereas sandstones, which are high in chert and shale-argillite lithic fragments, are low in ϵ_{Nd} and high in $\delta^{18}O$. There is also a correlation between high sedimentary lithic fragment content and high $^{87}Sr/^{86}Sr$, but the relation is less clear. Isotopic variations in the Lodoga, Grabast, and Los Gatos petrofacies are related to the metasedimentary lithic fragment content. For any stratigraphic age, the sandstone with the highest fraction of metasedimentary lithic fragments is generally lowest in ϵ_{Nd} and highest in $^{87}Sr/^{86}Sr$ and $\delta^{18}O$. The sensitivity of whole-rock Nd-Sr isotopes to the metasedimentary component reflects the large difference in crust formation ages of the volcanic/plutonic and metasedimentary sources. Sedimentary and metasedimentary lithic fragments were partly derived from recycled Proterozoic continental crust, whereas the arc source has Nd-Sr model ages that are intermediate between Cretaceous mantle and the Proterozoic sediments that it intruded and assimilated (Fig. 3). The Sr model ages of the batholith are young relative to the Nd model ages. This indicates that arc and crustal sources lost Sr during chemical weathering and recycling, and would lead to an underestimate of the age or fraction of the metasedimentary source.

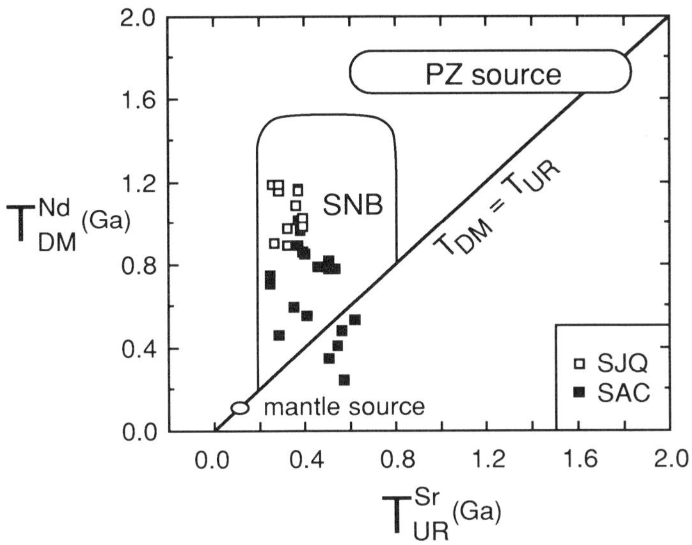

Figure 3. Comparison of Nd and Sr model ages of Great Valley sandstones and their sediment sources (data from Linn et al., 1992). Diagonal line represents equal Nd and Sr model ages; model ages to the right of the line indicate Rb loss; model ages to the left of the line indicate Sr loss (e.g., Nelson and DePaolo, 1988). Model ages of Sierra Nevada arc source (SNB) and Proterozoic crust (PZ) calculated by methods of DePaolo et al. (1991) and Farmer and DePaolo (1983), from Nd-Sr isotopic data reported in DePaolo (1981), Kistler et al. (1986) and Barbarin et al. (1989).

However, both Nd and Sr model ages of Great Valley sandstones are older than the depositional ages and indicate the presence of an older component (Table 2). Nd model ages (T_{DM}), which are defined as the age in which continental crust was extracted from a model depleted mantle, of the sandstones vary between 250 and 1190 m.y. This range is similar to the range in model ages of plutonic rocks in the Sierra Nevada batholith and suggests that the Nd isotopic compositions of the arc and its basement are preserved in the sedimentary rocks. In contrast, Sr model ages (denoted T_{UR} for crust extraction from a uniform mantle reservoir) of the sandstones range from 250 to 570 m.y. The Sr model ages of the sandstones are younger than both the Nd model ages of the sedimentary rocks and the Sr model ages of many of the plutonic rocks (Fig. 3). The old Sr model ages of some of the Sacramento Valley sandstones relative to model ages of the arc and the Upper Cretaceous sandstones is likely a result of Sr exchange by seawater interaction.

The whole-rock isotopic composition depends on the relative fraction, concentration, and isotopic composition of the sedimentary, metasedimentary, and igneous components. Assuming that the isootpic compositions and concentrations of the sedimentary and metasedimentary components are similar, the whole-rock Nd isotopic composition (ϵ_{WR}) of the sandstone is given by:

$$\epsilon_{WR} = \frac{(f.sed)\,[\mathrm{Nd}]_{sed}\,(\epsilon_{Nd})_{sed} + (1 - f.sed)\,[\mathrm{Nd}]_{ig}\,(\epsilon_{Ng})_{ig}}{(f.sed)\,[\mathrm{Nd}]_{sed} + (1 - f.sed)\,[\mathrm{Nd}]_{ig}},$$

where *f.* is the fraction of metasedimentary lithic fragments (*sed*), and *ig* (=igneous) is the volcanic/plutonic arc component determined by petrography. Linn et al. (1992) estimated that the Great Valley metasedimentary component has an Nd concentration of 20 ppm, similar to the plutonic rocks, and is –8 to –14 in ϵ_{Nd}, which is significantly lower than the igneous rocks. Subtraction of this component yields volcanic/plutonic compositions that are two to four ϵ_{Nd} units higher than the whole-rock value in average Lodoga, Grabast, and Los Gatos sandstones. This is a significant fraction of the overall isotopic variation and indicates that the Nd isotopic composition is a sensitive indicator of the pre-arc source. In contrast, the Sr concentration of the metasedimentary component is estimated to be about 100 ppm (Linn et al., 1992) which is only about 25% of the concentration of the igneous component due to Sr loss during sedimentary recycling. Consequently, the whole-rock $^{87}Sr/^{86}Sr$ value is controlled by the composition of the igneous source (Linn et al., 1992).

Location of source areas

In order to locate source areas, assumptions must be made regarding several factors: (1) the size and geometry of the drainage basin, (2) the erosion rate as a function of position within the drainage basin, and (3) the concentration and isotopic composition of sources contributing sediment to the forearc basin. Details of the first two parameters are described in Linn et al., (1992); the latter has been briefly discussed above. Subtraction of sedimentary and metasedimentary components from the whole-rock Nd-Sr isotopic compositions yields isotopic compositions that are representative of the arc source. The calculated igneous compositions are then compared with the isotopic gradient in the plutonic rocks, which varies only with geographic position, in order to locate the source area. Source locations are better constrained with Nd than Sr isotopes because the isotopic gradient in the plutonic rocks is larger and because Nd isotopes are less susceptible to modification by sedimentary processes.

With decreasing stratigraphic age, calculated igneous compositions of Great Valley sandstones decrease in ϵ_{Nd} from +7 to –5 and increase in $^{87}Sr/^{86}Sr$ from 0.704 to 0.707 (Linn et al., 1992). These values are the same as the range observed in the plutonic rocks of the Sierra Nevada arc and suggest that the Nd-Sr isotopic composition of the arc source is preserved in the sedimentary rocks. The decrease in ϵ_{Nd} and increase in $^{87}Sr/^{86}Sr$ with decreasing stratigraphic age is in the correct sense for the eastward-migration of source areas, which is in agreement with inferences drawn from sandstone petrography. Sediment source models show that San Joaquin Valley source areas moved from the central to the eastern part of the arc during Late Cretaceous time (Fig. 4A), and that Sacramento Valley source areas generally moved from the western to the central part of the arc during Early Cretaceous time (Fig. 4B). This conclusion does not depend on the values chosen for the metasedimentary component because the isotopic endmembers are Stony Creek and Rumsey sandstones which are low in metasedimentary lithic fragment and therefore have whole-rock isotopic compositions that are directly comparable to the composition of the arc. The large

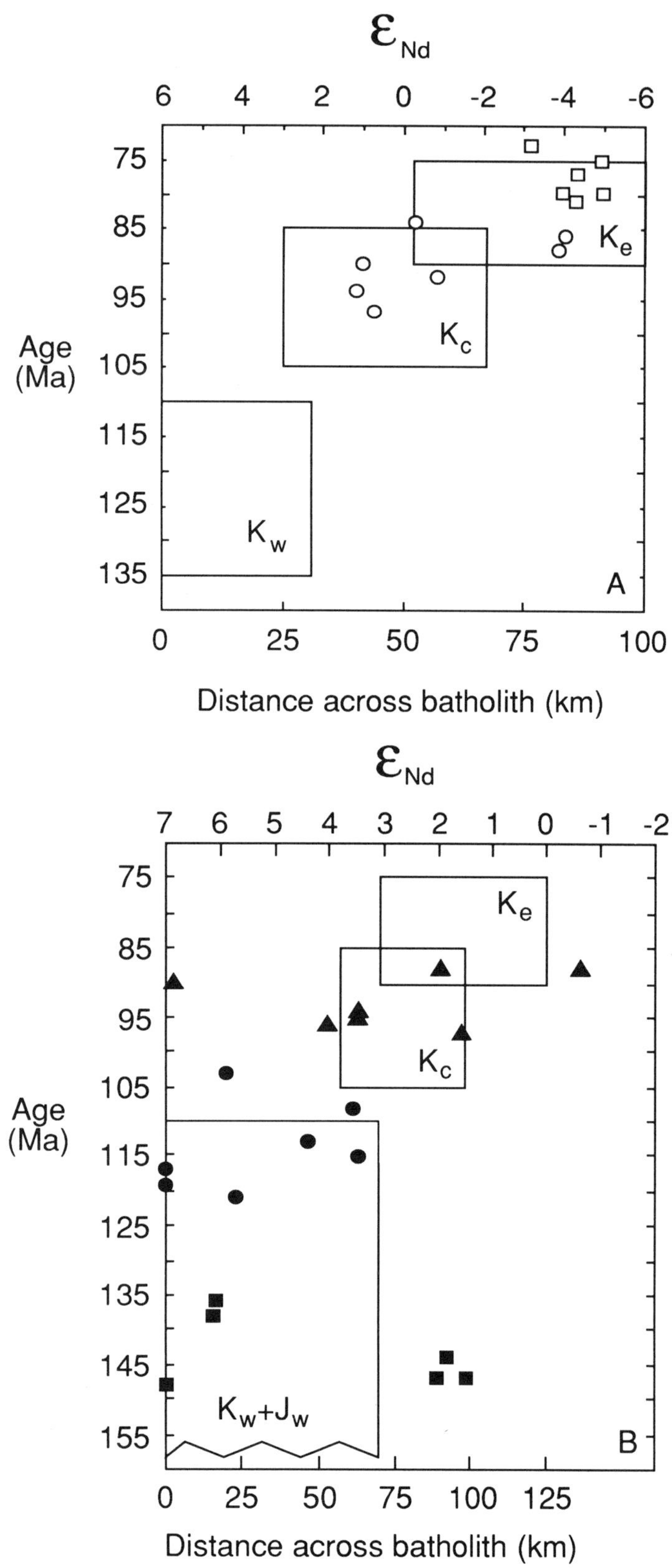

Figure 4. A, Depositional age and calculated igneous isotopic composition of San Joaquin Valley sandstones (open symbols), compared with crystallization age and isotopic composition of plutonic rocks (boxes) in an idealized west-east transect across the southern part of the Sierra Nevada batholith. K_w, K_c, and K_e are western, central, and eastern Cretaceous belts, respectively, from Saleeby (1981). B, Calculated Nd isotopic composition of the igneous component in Sacramento Valley sandstones (filled symbols) compared with age and geographical position of locus of magmatism (boxes) in west-east transect across the northern part of the arc. J_w is western Jurassic belt from Saleeby (1981). Symbols are the same as in Figure 2. From Linn et al. (1992).

range in ϵ_{Nd} suggests that source areas were localized at the migrating volcanic front; contributions of substantial amounts of sediment from previously active parts of the arc would yield a smaller range in ϵ_{Nd} because the values of the sources would be averaged. In general, the depositional age and calculated igneous isotopic composition of Great Valley sandstones are the same as the crystallization age and isotopic composition of the Sierra Nevada plutonic rocks. This indicates that the isotopic composition of the plutonic rocks is representative of the isotopic composition of the coeval volcanic cover and that erosion of volcanic sources was nearly contemporaneous with forearc deposition. Rapid erosion rates are also consistent with results from fission track dating (Dumitru, 1988), and with geomorphologic considerations which suggest that the volcanic front is a topographic high, which would quickly erode following cessation of magmatism (Linn et al., 1992).

Preservation of source $\delta^{18}O$

The oxygen isotopic composition of sedimentary rocks is a function of the composition of the source rocks and the amount of alteration in $\delta^{18}O$ produced during weathering and diagenesis. In contrast to Nd-Sr isotopes, the whole-rock $\delta^{18}O$ values of the sandstones are significantly higher than the arc source and are more similar to Proterozoic shelf sources (Fig. 5). This suggests that oxygen isotopes are particularly sensitive to the presence of recycled detritus. The relative contributions of sedimentary, metasedimentary, volcanic, and plutonic components to the whole-rock oxygen isotopic composition is determined in a manner analogous to the Nd mixing calculation above, except that the sources are: (1) sedimentary rocks with $\delta^{18}O$ between +20 in average shale-argillite (Eslinger and Savin, 1973; Savin and Epstein, 1970b; Yeh and Savin, 1977), and +28 in chert (Blatt, 1987); (2) metasedimentary rocks with $\delta^{18}O$ = +15, similar to the value of average metamorphic rocks (Hoefs, 1987); (3) western arc volcanic rocks with $\delta^{18}O$ = +6.5, similar to average island arcs (Hoefs, 1987); and (4) central and eastern arc plutonic rocks with $\delta^{18}O$ between +7 and +8 (Masi et al., 1981).

The sedimentary and metasedimentary fractions are determined by petrography, and the $\delta^{18}O$ values of the volcanic and plutonic sources are determined by inference with Nd isotopes (see above). The correlation between $\delta^{18}O$ and the lithic fragment content is illustrated in Figure 6. Stony Creek sandstones,

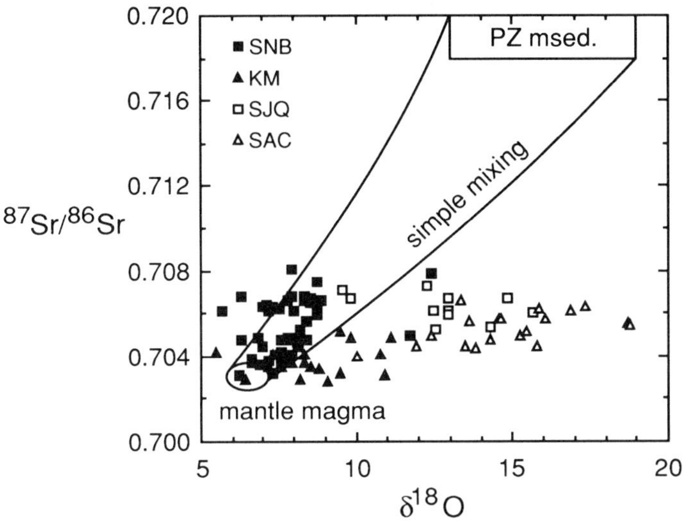

Figure 5. Sr-O isotopic systematics of Sierra Nevada (filled squares) and Klamath Mountain (filled triangles) plutonic rocks, modified from DePaolo (1981). Simple mixing of Cretaceous mantle magma with Proterozoic continental sediments produces the Sr-O isotopic compositions observed in most Sierra Nevada plutonic rocks. San Joaquin (open squares) and Sacramento Valley (open triangles) sandstones (this study) are similar in $^{87}Sr/^{86}Sr$ to the plutonic rocks, but are higher in $\delta^{18}O$. Isotopic data for the plutonic rocks from Masi et al. (1981).

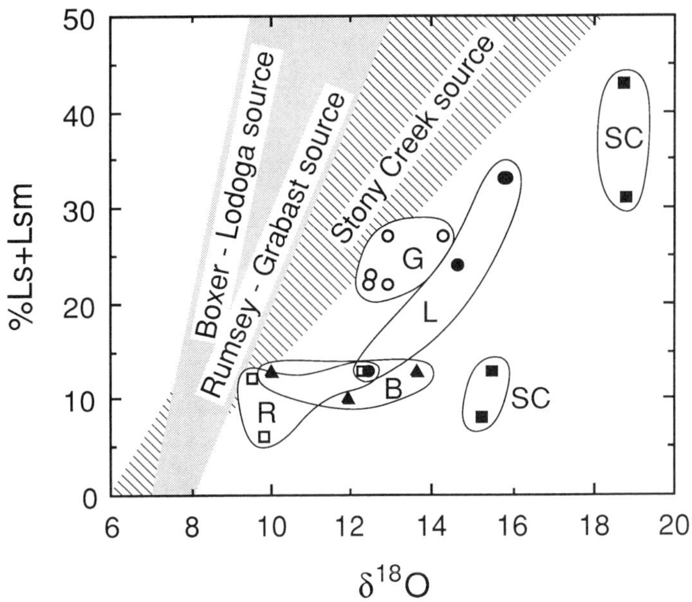

Figure 6. Correlation in whole-rock fraction of sedimentary and metasedimentary lithic fragments and $\delta^{18}O$. Fields encompass the petrofacies; Stony Creek sandstones are divided into volcaniclastic and high-chert varieties. Symbols are the same as in Figure 2.

which are high in chert and shale-argillite, have the highest $\delta^{18}O$ values of the Great Valley Group, whereas Boxer and Rumsey sandstones, which are high in volcanic or plutonic detritus, are lowest in $\delta^{18}O$. Lodoga, Grabast, and Los Gatos sandstones are high in metamorphic lithic fragments and have intermediate $\delta^{18}O$ values. The relation between petrography and $\delta^{18}O$ indicates that the oxygen isotopic composition of Great Valley sandstones is primarily controlled by provenance.

MODIFICATION OF SR AND O ISOTOPES

Although broad compositional variations in the sources are recorded in the Sr-O isotopic compositions of the sandstones, the isotopic compositions are shifted toward higher values than would be predicted for the source rocks. For example, $^{87}Sr/^{86}Sr$ values of the igneous component in Upper Cretaceous sandstones correspond to $^{87}Sr/^{86}Sr$ values of the plutonic rocks, but Lower Cretaceous and Upper Jurassic sandstones are significantly higher in $^{87}Sr/^{86}Sr$ than plutonic rocks of the same age (Linn et al., 1992). With regard to oxygen isotopes, Rumsey sandstones are 1 to 4‰ higher in $\delta^{18}O$ than the estimated compositions of the sources, and Grabast, Boxer, and Lodoga sandstones are up to 6‰ higher than their sources (Fig. 6). The volcaniclastic variety of the Stony Creek petrofacies ($\delta^{18}O \approx +15$) is nearly 9‰ higher than the source rocks. These observations suggest that the Sr-O isotopic compositions have been modified by sedimentary processes. In order to evaluate the extent to which the isotopes have been affected, we calculate the alteration factor, Δ, which is the difference between the measured isotopic composition ($\delta^{18}O_{sand}$) and the composition calculated for mixtures of the sedimentary, metasedimentary, and igneous components ($\delta^{18}O_{source}$), as determined by sandstone petrography. The oxygen isotopic values of the components are given above. The Sr isotopic composition of the igneous sources ($^{87}Sr/^{86}Sr_{calc.}$) are determined by analogy with Nd isotopes. Figure 7 shows that alteration of $\delta^{18}O$ varies between +1 and +5 in Upper Cretaceous sandstones, and varies between +3 and +8 in Lower Cretaceous and Upper Jurassic sandstones. The average shift in $\delta^{18}O$ of the sandstones is Δ = +4‰. The Sr alteration factor behaves similarly; it increases from Δ = 0 to 0.001 at the top of the section, to Δ = 0.002 at the base of the section (Fig. 7). The alteration factors indicate that, typically, 30 to 40% of the Sr and O in the sandstones is secondary.

Modification of both Sr and O isotopes increases with increasing stratigraphic age. However, it is difficult to determine when the alteration occurred because the sandstones that were the most deeply buried and diagenetically modified were derived from mafic sources, which are more susceptible to chemical weathering. The chemical index of alteration (CIA) provides a means of evaluating the effect of weathering on $\delta^{18}O$ and $^{87}Sr/^{86}Sr$. The CIA measures the degradation of feldspar to clay minerals and is determined from the abundances of K_2O, CaO, and Na_2O relative to Al_2O_3 (Nesbitt and Young, 1982; Taylor and McLennan, 1985). CIA varies from 50 in unweathered continental crust to 100 in clay minerals, and is generally less than 65 in Great Valley sandstones (Table 2). This value is consistent with the high mechanical erosion rates observed in tectonically active settings (e.g., Blatt et al., 1980) and with the CIA in modern continental margin arc sand (McLennan and others, 1990). CIA does not vary with stratigraphic age in the samples we analyzed;

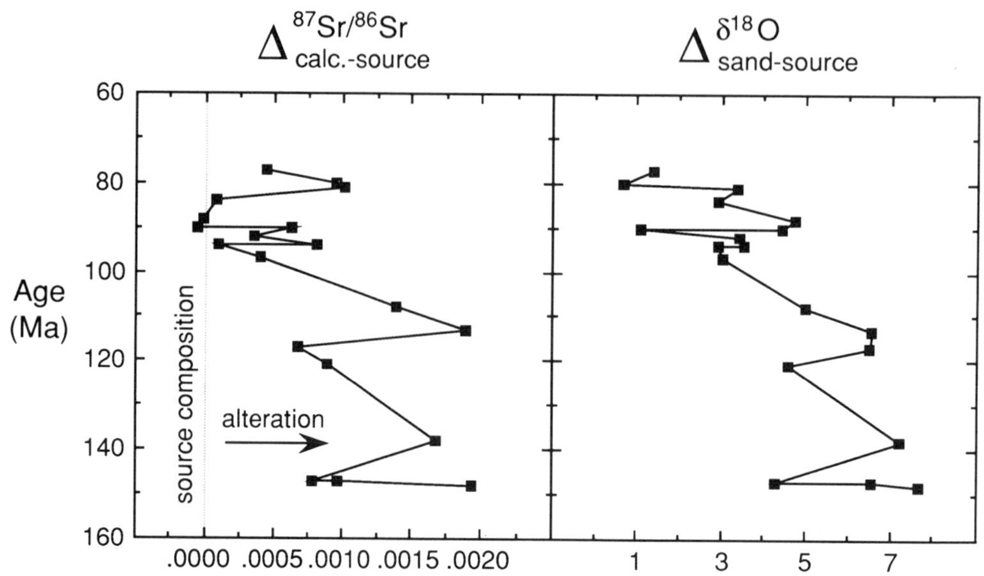

Figure 7. Degree of alteration (delta symbols) in Sr and O isotopic composition as a function of stratigraphic age. Alteration in Sr isotopic composition is the difference between the calculated value of the igneous fraction in the sandstones ($^{87}Sr/^{86}Sr_{source}$) and the value of the plutonic rocks in source areas inferred from Nd isotopes ($^{87}Sr/^{86}Sr_{calc.}$). Alteration in O isotopic composition is the difference between the measured whole-rock value ($\delta^{18}O_{sand}$) and the value calculated for mixtures of sedimentary, metamorphic, and igneous components ($\delta^{18}O_{source}$), as determined by sandstone petrography.

this suggests that diagenetic processes did not significantly modify the major element abundances. Figure 8 shows that increasing CIA is correlated with increasing $\delta^{18}O$ and with increasing alteration. The range in $\delta^{18}O$ is large compared with the range in CIA and suggests that the oxygen isotopic composition is a more sensitive indicator of chemical weathering than even mobile major elements. The Δ values indicate that the sandstones contain about 35% weathering products, whereas the CIA values are consistent with only about 15% alteration. However, the CIA measures only feldspar alteration, and may be expected to indicate lower amounts of alteration than whole-rock oxygen isotopes. The $\delta^{18}O$ data suggest that high erosion rates of tectonically active sources do not preclude chemical weathering.

CIA is not correlated with the Sr alteration factor; this suggests that weathering did not modify the $^{87}Sr/^{86}Sr$ values of the sandstones. The most likely way to increase the $^{87}Sr/^{86}Sr$ value, and therefore Δ, is to exchange Sr in plagioclase with Sr in seawater during deposition and/or submarine eruption of the earliest volcanic sources. $^{87}Sr/^{86}Sr$ values in seawater ranged from 0.7070 to 0.7075 during Early Cretaceous time (Burke et al., 1982), and were significantly higher than $^{87}Sr/^{86}Sr$ values of the arc.

Because alteration increases with increasing stratigraphic age, and therefore with burial depth, it is necessary to evaluate the effect of burial metamorphism on $\delta^{18}O$. We consider two end member models, which are illustrated in Figure 9. If the stratigraphic section behaves like an open system, an infinite reservoir of seawater will modify the isotopic composition of the sandstones toward higher $\delta^{18}O$ values at low temperatures and toward lower $\delta^{18}O$ values at high temperatures. The magnitude of

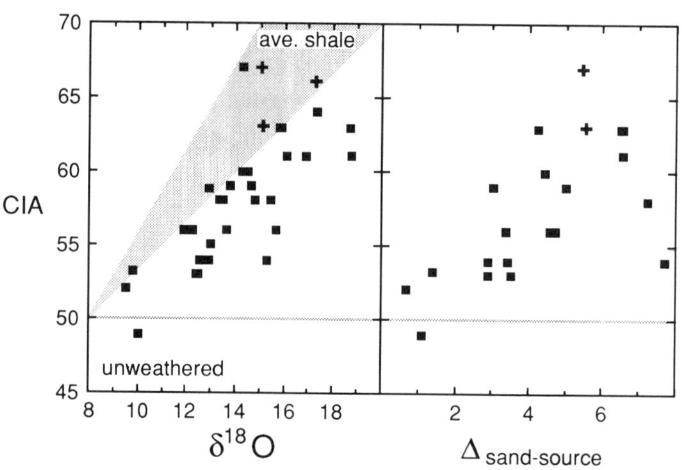

Figure 8. Correlation in $\delta^{18}O$ and chemical index of alteration in Great Valley sandstones. CIA defined in molecular proportions as $[Al_2O_3/(Al_2O_3+CaO+Na_2O+K_2O)] * 100$. Shaded region indicates range of compositions observed in unweathered to moderately weathered rocks.

the displacement depends on the fraction and mineralogy of the clay, and on the temperature gradient, because the clay-water fractionation decreases with increasing temperature. Great Valley sandstones which contain 30% authigenic chlorite should decrease from about +11 to +7‰ with increasing depth of burial. This is in the opposite sense as the trend observed in the sandstones (boxes), which *increase* in $\delta^{18}O$ with increasing depth of burial (Fig. 9). If the section behaves like a closed system, water/rock ratios are low and the isotopic compositions of the pore fluids are modified. As a result, increasing burial metamorphism could not significantly alter the whole-rock values. This is likely

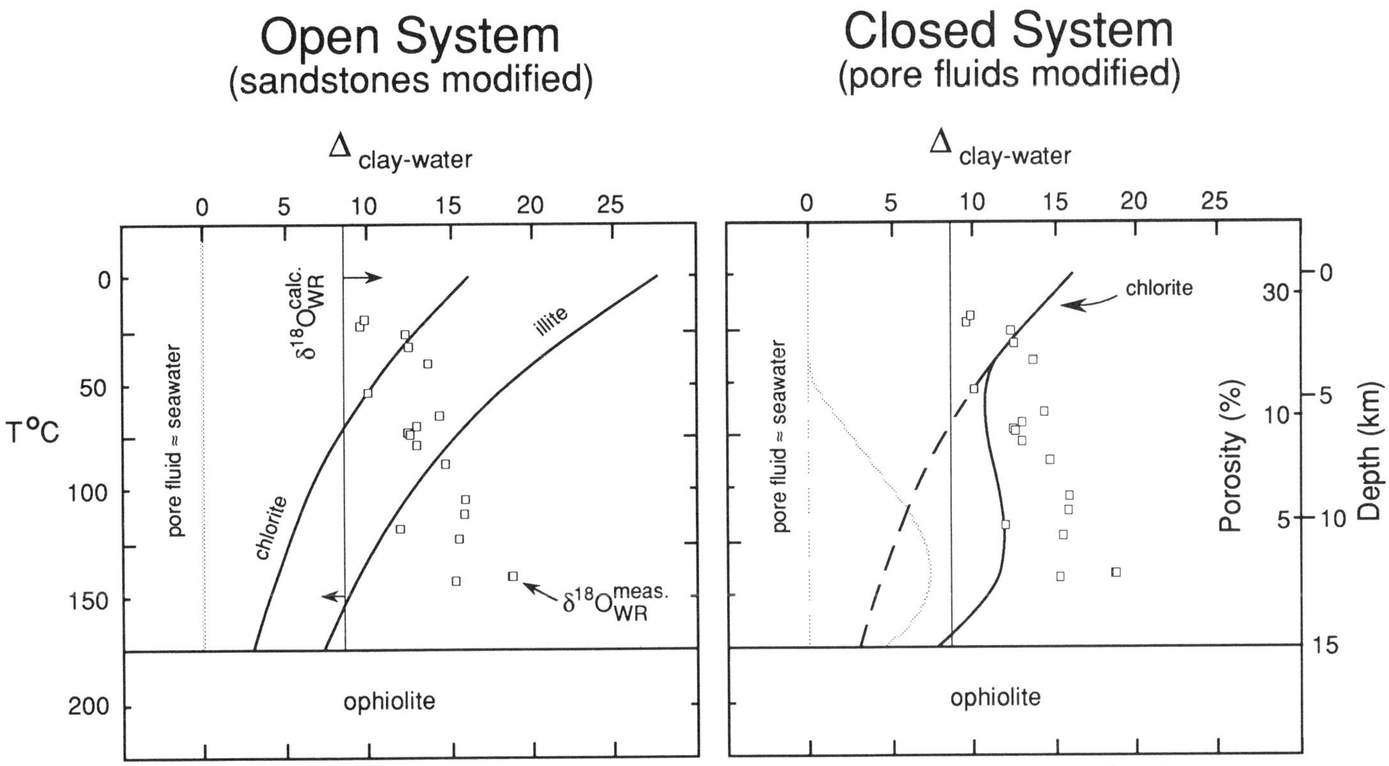

Figure 9. Model to evaluate the effect of burial metamorphism on the whole-rock oxygen isotopic composition of sandstones. $\Delta_{clay\text{-}water} = \delta^{18}O_{clay} - \delta^{18}O_{seawater}$. In an open system, the arrows denote the magnitude and direction of shift in calculated $\delta^{18}O$ values resulting from replacement of 30% of the bulk sandstone by chlorite. In a closed system, the calculated isotopic compositions are not modified by burial metamorphism. Chlorite and illite curves are calculated from equations in Longstaffe (1983) and Eslinger and Savin (1973). Stratigraphic depth and geothermal gradient from Dumitru (1988). Magnitude of pore fluid enrichment in closed system from Suchecki and Land (1983).

the case for the Great Valley Group because the section is inferred to have been buried nearly 15 km (Dumitru, 1988) which would result in very low porosity and permeability.

SUMMARY AND CONCLUSIONS

The whole-rock isotopic composition of sedimentary rocks is a function of: (1) the composition of the source rocks; and (2) shifts in the isotopic values produced by weatheirng, transport, and diagenetic processes. The Nd isotopic composition is generally preserved throughout the sedimentary cycle, and can be used to determine the provenance and to evaluate provenance interpretations of Sr and O isotopes that are more susceptible to weathering and diagenetic processes. Nd-Sr-O isotopic variations are similar in Great Valley sandstones and are correlated with the lithic fragment content. Metasedimentary sources decrease ϵ_{Nd} and increase $^{87}Sr/^{86}Sr$ in the sandstones. O isotopes are particularly sensitive to the sedimentary-metasedimentary source because recycled components contain clay minerals that form at low temperatures and that are consequently higher in $\delta^{18}O$ than the igneous source. This suggests that even the oxygen isotopic composition is mainly controlled by provenance. However, $^{87}Sr/^{86}Sr$ and $\delta^{18}O$ values are higher than would be expected for the sources determined by Nd isotopes, and are highest in those sandstones that were derived from relatively mafic volcanic sources. The high $^{87}Sr/^{86}Sr$ values are likely a result of Sr exchange with seawater either in the volcanic edifice or during deposition. However, alteration in $\delta^{18}O$ is correlated with CIA; this suggests that the oxygen isotopic composition was modified by low-temperature weathering processes, rather than diagenesis. This indicates that chemical weathering is important in tectonically active orogenic belts, particularly in those composed of mafic volcanic rocks. Modification of oxygen isotopes may also have been enhanced by the climatic conditions in the near-equatorial position of the Sierra Nevada arc during Cretaceous time. However, the intense chemical weathering regime associated with low latitudes apparently was not sufficient to disturb the Nd-Sr systematics or the major element ratios.

ACKNOWLEDGMENTS

Acknowledgment is made to the Donors of the Petroleum Research Fund, Grants PRF 19452-AC2 and PRF 22585-AC2-C, administered by the American Chemical Society, for partial support of this research. This work was also supported by the director, Office of Basic Energy Sciences, Engineering and Geosciences Division of the U.S. Department of Energy under Contract No. DE-AC03-76SF00098.

REFERENCES CITED

Ague, J. J., and Brimhall, G. H., 1988, Regional variations of major chemical constituents across the central Sierra Nevada batholith: Geological Society of America Bulletin, v. 100, p. 891–911.

Allegre, C. J., and Rousseau, D., 1984, The growth of the continent through geological time studied by Nd isotope analysis of shales: Earth and Planetary Science Letters, v. 67, p. 19–34.

Atwater, T., 1970, Implications of plate tectonics for the Cenozoic tectonic evolution of western North America: Geological Society of America Bulletin, v. 81, p. 3513–3535.

Awwiller, D. N., and Mack, L. E., 1991, Diagenetic modification of Sm-Nd model ages in Tertiary sandstones and shales, Texas Gulf Coast: Geology, v. 19, p. 311–314.

Ayalon, A., and Longstaffe, F. J., 1988, Oxygen isotope studies of diagenesis and pore-water evolution in the western Canada sedimentary basin: evidence from the Upper Cretaceous Basal Belly River sandstone, Alberta: Journal of Sedimentary Petrology, v. 58, p. 489–505.

Balashov, Y. A., and Girin, Y. P., 1969, On the reserve of mobile rare earth elements in sedimentary rocks: Geochemistry International, v. 6, p. 649–659.

Balashov, Y. A., Ronov, A. B., Migdisov, A. A., and Turanskaya, N. V., 1964, The effect of climate and facies environment on the fractionation of the rare earths during sedimentation: Geochemistry International, v. 1, p. 951–969.

Barbarin, B., Dodge, F.C.W., Kistler, R. W., and Bateman, P. C., 1989, Mafic inclusions, aggregates, and dikes in granitoid rocks, central Sierra Nevada batholith, California—Analytical data: U.S. Geological Survey Bulletin 1899, 28 p.

Blatt, H., 1987, Oxygen isotopes and the origin of quartz: Journal of Sedimentary Petrology, v. 57, p. 373–377.

Blatt, H., Middleton, G., and Murray, R., 1980, Origin of sedimentary rocks: Englewood Cliffs, NJ, Prentice-Hall, p. 17–42.

Bowan, R., 1991, Isotopes and climates: New York, Elsevier, 483 p.

Burke, W. H., Denison, R. E., Hetherington, E. A., Koepnick, R. B., Nelson, H. F., and Otto, J. B., 1982, Variation of seawater $^{87}Sr/^{86}Sr$ throughout Phanerozoic time: Geology, v. 10, p. 516–519.

Chen, C.-H., Jahn, B. M., Lee, T., Chen, C. H., and Cornichet, J., 1990, Sm-Nd isotopic geochemistry of sediments from Taiwan and implications for the tectonic evolution of SE China: Chemical Geology, v. 88, p. 317–332.

Clarke, D. B., and Halliday, A. N., 1985, Sm/Nd isotopic investigation of the age and origin of the Meguma Zone metasedimentary rocks: Canadian Journal of Earth Sciences, v. 22, p. 102–107.

Clauer, N., 1982, The rubidium-strontium method applied to sediments: Certitudes and uncertainties, *in* Odin, G. S., ed., Numerical dating in stratigraphy: New York, John Wiley & Sons, p. 245–276.

Clayton, R. N., Jackson, M. L., and Sridhar, K., 1978, Resistance of quartz silt to isotopic exchange under burial and intense weathering conditions: Geochimica et Cosmochimica Acta, v. 42, p. 1517–1522.

Cullers, R., Chaudhuri, S., Kilbane, N., and Koch, R., 1979, Rare-earths in size fractions and sedimentary rocks of Pennsylvanian-Permian age from the mid-continent of the USA: Geochimica et Cosmochimica Acta, v. 43, p. 1285–1301.

DePaolo, D. J., 1981, A neodymium and strontium isotopic study of the Mesozoic calc-alkalic granitic batholiths of the Sierra Nevada and Peninsular Ranges, California: Journal of Geophysical Research, v. 86, p. 10470–10488.

DePaolo, D. J., and Wasserburg, G. J., 1976, Nd isotopic variations and petrogenetic models: Geophysical Research Letters, v. 3, p. 249–252.

DePaolo, D. J., Linn, A. M., and Schubert, G., 1991, The continental crust age distribution: Methods for determining mantle separation ages from Sm-Nd isotopic data and application to the southwestern United States: Journal of Geophysical Research, v. 96, p. 2071–2088.

Dickinson, W. R., 1973, Widths of modern arc-trench gaps proportional to past duration of igneous activity in associated magmatic arcs: Journal of Geophysical Research, v. 78, p. 3376–3389.

Dickinson, W. R., and Rich, E. I.,1972, Petrologic intervals and petrofacies in the Great Valley sequence, Sacramento Valley, California: Geological Society of America Bulletin, v. 83, p. 3007–3024.

Dodge, F.C.W., Millard, H. T., Jr., and Elsheimer, H. N., 1982, Compositional variations and abundances of selected elements in granitoid rocks and constituent minerals, central Sierra Nevada batholith, California: U.S. Geological Survey Professional Paper 1248, 24 p.

Dumitru, T. A., 1988, Subnormal geothermal gradients in the Great Valley forearc basin, California, during Franciscan subduction: A fission track study: Tectonics, v. 7, p. 1201–1221.

Eslinger, E. V., and Savin, S. M., 1973, Oxygen isotope geothermometry of the burial metamorphic rocks of the Precambrian Belt Supergroup, Glacier National Park, Montana: Geological Society of America Bulletin, v. 84, p. 2549–2560.

Eslinger, E. V., and Yeh, H.-W., 1986, Oxygen and hydrogen isotope geochemistry of Cretaceous bentonites and shales from the Disturbed Belt, Montana: Geochimica et Cosmochimica Acta, v. 50, p. 59–68.

Farmer, G. L., and DePaolo, D. J., 1983, Origin of Mesozoic and Tertiary granite in the western United States and implications for pre-Mesozoic crustal structure, 1. Nd and Sr isotopic studies in the geocline of the northern Great Basin: Journal of Geophysical Research, v. 88, p. 3379–3401.

Faure, G., 1986, Principles of isotope geology (second edition): New York, John Wiley & Sons, 589 p.

Fleet, A. J., 1984, Aqueous and sedimentary geochemistry of the rare earth elements, *in* Henderson, P., ed., Rare earth element geochemistry: New York, Elsevier, p. 343–373.

Frost, C. D., and Coombs, D. S., 1989, Nd isotope character of New Zealand sediments: Implications for terrane concepts and crustal evolution: American Journal of Science, v. 289, p. 744–770.

Frost, C. D., and Winston, D., 1987, Nd isotope systematics of coarse- and fine-grained sediments: Examples from the middle Proterozoic Belt–Purcell Supergroup: Journal of Geology, v. 95, p. 309–327.

Goldstein, S. L., and Jacobsen, S. B., 1988, Nd and Sr isotopic systematics of river water suspended material: Implications for crustal evolution: Earth and Planetary Science Letters, v. 87, p. 249–265.

Hamilton, P. J., O'Nions, R. K., Bridgwater, D., and Nutman, A., 1983, Sm-Nd studies of Archaean metasediments and metavolcanics from west Greenland and their implications for the Earth's early history: Earth and Planetary Science Letters, v. 62, p. 263–272.

Heller, P. L., Peterman, Z. E., O'Neil, J. R., and Shafiquillah, M., 1985, Isotopic provenance of sandstones from the Eocene Tyee Formation, Oregon Coast Range: Geological Society of America Bulletin, v. 96, p. 770–780.

Hoefs, J., 1987, Stable isotope geochemistry (third edition): New York, Springer-Verlag, 241 p.

Humphris, S. E., 1984, The mobility of the rare earth elements in the crust, *in* Henderson, P., ed., Rare earth element geochemistry: New York, Elsevier, p. 317–342.

Ingersoll, R. V., 1979, Evolution of the Late Cretaceous forearc basin, northern and central California: Geological Society of America Bulletin, v. 90, p. 813–826.

Ingersoll, R. V., 1982, Initiation and evolution of the Great Valley forearc basin of northern and central California, U.S.A.: Geological Society of London Special Publication 10, p. 459–467.

Ingersoll, R. V., 1983, Petrofacies and provenance of Late Mesozoic forearc basin, northern and central California: American Association of Petroleum Geologists Bulletin, v. 67, p. 1125–1142.

Kistler, R. W., and Peterman, Z. E., 1973, Variations in Sr, Rb, K, Na, and initial Sr^{87}/Sr^{86} in Mesozoic granitic rocks and intruded wall rocks in central California: Geological Society of America Bulletin, v. 84, p. 3489–3512.

Kistler, R. W., Chappell, B. W., Peck, D. L., and Bateman, P. C., 1986, Isotopic variation in the Tuolumne intrusive suite, central Sierra Nevada, California: Contributions to Mineralogy and Petrology, v. 94, p. 205–220.

Land, L. S., and Milliken, K. L., 1981, Feldspar diagenesis in the Frio Formation, Brazoria County, Texas Gulf Coast: Geology, v. 9, p. 314–318.

Lawrence, J. R., and Taylor, H. P., Jr., 1972, Hydrogen and oxygen isotope systematics in weathering profiles: Geochimica et Cosmochimica Acta, v. 36, p. 1377–1393.

Linn, A. M., DePaolo, D. J., and Ingersoll, R. V., 1991, Nd-Sr isotopic provenance analysis of Upper Cretaceous Great Valley fore-arc sandstones: Geology, v. 19, p. 803–806.

Linn, A. M., DePaolo, D. J., and Ingersoll, R. V., 1992, Nd-Sr isotopic, geochemical and petrographic stratigraphy and paleotectonic analysis: Mesozoic Great Valley forearc sedimentary rocks of California: Geological Society of America Bulletin, v. 104, p. 1264–1279.

Longstaffe, F. J., 1983, Diagenesis 4. Stable isotope studies of diagenesis in clastic rocks: Geoscience Canada, v. 10, p. 43–58.

Masi, U., O'Neil, J. R., and Kistler, R. W., 1981, Stable isotope systematics in Mesozoic granites of central and northern California and southwestern Oregon: Contributions to Mineralogy and Petrology, v. 76, p. 116–126.

McLennan, S. M., McCulloch, M. T., Taylor, S. R., and Maynard, J. B., 1989, Effects of sedimentary sorting on neodymium isotopes in deep-sea turbidites: Nature, v. 337, p. 547–549.

McLennan, S. M., Taylor, S. R., McCulloch, M. T., and Maynard, J. B., 1990, Geochemical and Nd-Sr isotopic composition of deep-sea turbidites: Crustal evolution and plate tectonic associations: Geochimica et Cosmochimica Acta, v. 54, p. 2015–2050.

Michard, A., Gurriet, P., Soudant, M., and Albarede, F., 1985, Nd isotopes in French Phanerozoic shales: External vs. internal aspects of crustal evolution: Geochimica et Cosmochimica Acta, v. 49, p. 601–610.

Miller, R. G., and O'Nions, R. K., 1984, The provenance and crustal residence ages of British sediments in relation to paleogeographic reconstructions: Earth and Planetary Science Letters, v. 68, p. 459–470.

Moorbath, S., 1969, Evidence for the age of deposition of the Torridonian sediments of northwest Scotland: Journal of Geology, v. 5, p. 154–170.

Nelson, B. K., and DePaolo, D. J., 1988, Comparison of isotopic and petrographic provenance indicators in sediments from Tertiary continental basins of New Mexico: Journal of Sedimentary Petrology, v. 58, p. 348–357.

Nesbitt, H. W., and Young, G. M., 1982, Early Proterozoic climates and plate motions inferred from major element chemistry of lutites: Nature, v. 299, p. 715–717.

Ojakangas, R. W., 1968, Cretaceous sedimentation, Sacramento Valley, California: Geological Society of America Bulletin, v. 79, p. 973–1008.

O'Neil, J. R., and Kharaka, Y. K., 1976, Hydrogen and oxygen isotope exchange reactions between clay minerals and water: Geochimica et Cosmochimica Acta, v. 40, p. 241–246.

O'Nions, R. K., Hamilton, P. J., and Hooker, P. J., 1983, A Nd isotopic investigation of sediments related to crustal development in the British Isles: Earth and Planetary Science Letters, v. 63, p. 229–240.

Perry, E. A., and Turekian, K. K., 1974, The effect of diagenesis on the redistribution of strontium isotopes in shales: Geochimica et Cosmochimica Acta, v. 38, p. 929–935.

Peterman, Z. E., Coleman, R. G., and Bunker, C. M., 1981, Provenance of Eocene graywackes of the Flournoy Formation near Agness, Oregon—A geochemical approach: Geology, v. 9, p. 81–86.

Saleeby, J., 1981, Ocean floor accretion and volcanoplutonic arc evolution of the Mesozoic Sierra Nevada, *in* Ernst, W. G., ed., The geotectonic development of California: Englewood Cliffs, NJ., Prentice-Hall, p. 132–181.

Savin, S. M., and Epstein, S., 1970a, The oxygen and hydrogen isotope geochemistry of clay minerals: Geochimica et Cosmochimica Acta, v. 34, p. 25–42.

Savin, S. M., and Epstein, S., 1970b, The oxygen isotopic compositions of coarse grained sedimentary rocks and minerals: Geochimica et Cosmochimica Acta, v. 34, p. 323–329.

Schieber, J., 1988, Redistribution of rare-earth elements during diagenesis of carbonate rocks from the mid-Proterozoic Newland Formation, Montana, U.S.A.: Chemical Geology, v. 69, p. 111–126.

Schweickert, R. A., and Cowan, D. S., 1975, Early Mesozoic tectonic evolution of the western Sierra Nevada, California: Geological Society of America Bulletin, v. 86, p. 1329–1336.

Smith, B. M., 1991, Geologic implications of the oxygen isotope profile of the Toa Baja drill hole, Puerto Rico: Geophysical Research Letters, v. 18, p. 549–552.

Spiers, G. A., Dudas, M. J., Muehlenbachs, K., and Turchenek, L. W., 1984, Mineralogy and oxygen isotope geochemistry of clays from surficial deposits in the Athabasca Tar Sands area: Canadian Journal of Earth Sciences, v. 21, p. 53–60.

Suchecki, R. K., and Land, L. S., 1983, Isotopic geochemistry of burial-metamorphosed volcanogenic sediments, Great Valley sequence, northern California: Geochimica et Cosmochimica Acta, v. 47, p. 1487–1499.

Taylor, S. R., and McLennan, S. M., 1985, The continental crust: Its composition and evolution: Oxford, Blackwell Scientific, 312 p.

Tilley, B. J., and Longstaffe, F. J., 1989, Diagenesis and isotopic evolution of porewaters in the Alberta Deep Basin: The Falher Member and Cadomin Formation: Geochimica et Cosmochimica Acta, v. 53, p. 2529–2546.

Yeh, H.-W., and Eslinger, E. V., 1986, Oxygen isotopes and the extent of diagenesis of clay minerals during sedimentation and burial in the sea: Clays and Clay Minerals, v. 34, p. 403–406.

Yeh, H.-W., and Savin, S. M., 1977, Mechanism of burial metamorphism of argillaceous sediments: 3. O-isotope evidence: Geological Society of America Bulletin, v. 88, p. 1321–1330.

Manuscript Accepted by the Society Ddecember 24, 1992

Geological Society of America
Special Paper 284
1993

Source rock texture versus climate and topography as controls on the composition of modern, plutoniclastic sand

William A. Heins*
Department of Earth and Space Sciences, University of California, Los Angeles, California 90024-1567

ABSTRACT

The rock fragment content (measured by log of polycrystalline/monocrystalline sand grains) of sand derived from five nearly identical granodiorite plutons, located in strongly contrasting climatic zones of the United States and Mexico, shows a much stronger correlation with source rock texture than with any environmental factor or combination of factors. These factors include mean annual temperature (range, –4° to 28°C), mean annual precipitation range (range, 201 to 1,566 mm), mean annual surplus of precipitation over evapotranspiration (range, 0 to 1,138 mm), average drainage slope (range, 0.29 to 0.61), or relief ratio (range, 0.17 to 0.70). Source rock texture is measured by intercrystalline area/volume of source rock and has a range of 7.99 to 25.08 mm^{-1}. Stepwise regression analysis indicates that, in coarse sand, 71% of variation in rock fragment content is accounted for by variations in source rock texture versus 16% accounted for by variations in temperature. In medium sand, 96% of the variation in rock fragment content can be attributed to texture versus 1% for slope and <1% for temperature. In fine sand, 92% of the variation in rock fragment content stems from texture versus 3% from temperature. The same analysis does not document a statistically significant relationship between precipitation and rock fragment abundance. Many previous studies that appear to document a relationship between rock fragment abundance and precipitation in modern and ancient sand have not accounted for variations in source rock lithology that are more plausible explanations for observed variations in rock fragment abundance in derived sand.

INTRODUCTION

Previous studies of modern and ancient sand have apparently documented a strong correlation between precipitation and rock fragment abundance. Basu (1976) documented significantly more rock fragments in modern sand from the Rocky Mountains than in modern sand from the Appalachians; he attributed the difference to greater destruction of rock fragments by the greater precipitation in the Appalachians. Mack and Suttner (1977) used this result to conclude that the Pennsylvanian climate in the Colorado Front Range was more humid than the modern climate, because Pennsylvanian sand derived from similar granitic source rocks and transported under similar conditions contained fewer rock fragments than equivalent modern sand. Mack and Jerzykeiwicz (1989) documented a significant difference in the RFP index (rock fragments/[rock fragments + plagioclase]) of sand derived from andesite in arid versus humid settings; they attributed higher RFP in arid settings and lower RFP in humid settings to higher lability of rock fragments compared to plagioclase with increasing humidity. Grantham and Velbel (1988) documented a significant correlation between Cumulative Chemical Weathering Index (CCWI), which equals (runoff/relief ratio), and rock fragment abundance in sand derived from high-grade metamorphic rocks. They found that high CCWI correlates with less-abundant rock fragments. They argue that either high precipitation or low slope (leading to longer residence time), or both, will promote rock fragment destruction.

*Present address: Strategic Management Group, 3644 Market Street, Philadelphia, Pennsylvania 19104.

Heins, W. A., 1993, Source rock texture versus climate and topography as controls on the composition of modern, plutoniclastic sand, *in* Johnsson, M. J., and Basu, A., eds., Processes Controlling the Composition of Clastic Sediments: Boulder, Colorado, Geological Society of America Special Paper 284.

However, environmental interpretation based on a comparison of sand compositions requires that the characteristic compared (in this case, rock fragment abundance) varies as a result of environmental factors and not as a result of differences in the source rock. This study is designed to isolate the relative importance of environmental factors compared to source rock factors in determining rock fragment abundance by examining modern sand derived from locations where the environmental factors strongly contrast, but the composition of the source rock is constant. Environmental factors include the climatic variables of temperature and precipitation, the topographic variables of average slope or relief ratio, and the lithologic variable of source rock texture. Source rock texture is included in the analysis because it is an important determinant of the compressive strength of granite (e.g., Erkan, 1970) and has been linked to rock fragment abundance in modern sand derived from metamorphic rocks (Young, 1975).

TABLE 1. SOURCES OF GEOLOGICAL AND CLIMATIC DATA

Ocotito Granitoid Complex	
Location and Topo Map:	Sierra Madre del Sur, Guerrero (GR), Mexico (17.148 N, 099.541 W) Tierra Colorada 1:50,000 (E14C-48)
Geology:	de Cserna (1965), Negendank et al. (1987)
Climate:	Secretario de Programacion y Presupesto (1980, 1981), Universidad Nacional Autonom de Mexico (1970)
Granite Wash Mountains	
Location and Topo Map:	West-central Arizona (AZ), USA (033.708 N, 113,639 W) Hope NW 7.5' (T-sheet)
Geology:	Rehrig and Reynolds (1980), Reynolds et al. (1980), Lee (1984), Spencer et al. (1985)
Climate:	National Climatic Data Center (1988)
Kinney Lakes Granodiorite	
Location and Topo Map:	Sierra Nevada, California (CA), USA (038.568 N, 119.852 W) Ebbetts Pass 7.5'
Geology:	Armin et al. (1982, 1984), Huber (1983a, b)
Climate:	National Climatic Data Center (1988)
McCartney Mountain Stock	
Location and Topo Map:	Southwestern Montana (MT), USA (045.508 N, 112.647 W) Earl's Gulch 7.5'
Geology:	Friberg and Vitaliano (1981)
Climate:	National Climatic Data Center (1988)
Mt. Stuart Batholith	
Location and Topo Map:	Cascade Mountains, Washington (WA), USA (047.479 N, 120.805 W) Mt. Stuart 15'
Geology:	Erikson and Williams (1976), Erikson (1977), Tabor et al. (1982)
Climate:	National Weather Service (J. Holcombe, personal communication, 1989)

STUDY DESIGN

Sampling program

Sand was collected from 16 drainages of varying gradient and facing direction, developed on five granodiorite plutons in the United States and Mexico (Table 1). The sites were chosen to minimize source rock variation and maximize environmental variation among the sites. Sampling was restricted to 80 ± 10-m.y.-old, medium-grained, equigranul;ar, sub- to metaluminous (Shand, 1947) granodiorites (Streckeisen, 1976). This is a mineralogically and texturally simple rock type that is geographically widespread in the western United States (Heins, 1992, Appendix A). The five plutons were chosen to yield a hot/wet site (identified as GR), a hot/dry site (AZ), a cold/wet site (WA), a cold/dry site (MT), and an intermediate site (CA). These sites represent extremes only in the context of continental U.S./Mexico Cordilleran climate. Three or four drainages, representing a range of steepnesses and facing directions, were sampled on each pluton. All drainages have only one stream and are developed exclusively on the respective pluton. Detailed information about sample locations is reported in Heins (1992).

Climatic and topographic data for the drainages are summarized in Tables 2 and 3, respectively. Mean annual temperature ranges from –4° to 28°C; mean annual precipitation ranges from 201 to 1,566 mm; the surplus of precipitation over evapotranspiration ranges from 0 to 1,138 mm. The drainages are all

TABLE 2. SUMMARY OF CLIMATIC DATA

Drainage*	Temperature† (°C)	Precipitation§ (mm)	Surplus Water** (mm)
WAN, WAW	2.7	1,566	1,138
MTN, MTS, MTE, MTW	2.6	285	0
ORN, ORS, ORE	28.1	1,173	143
CAN, CAS	-2.0	995	743
CAE, CAW	-4.1	1,005	839
AZN, AZS, AZE, AZW	19.0	201	0

*Drainage codes are composed of the postal abbreviation for the state in which the pluton is located and the facing direction of the drainage.
†Temperatures are interpolated or extrapolated to the elevation of the sample sites from the nearest weather stations recorded by National Climatic Data Center (1988), assuming a constant local lapse rate.
§Precipitations are calculated from data of the same stations, accounting for elevation and prevailing local weather patterns, as determined by the staff meteorologist at the Department of Atmospheric Sciences, University California, Los Angeles.
**Surplus water was calculated with the program WATBUG (Willmott, 1977), as modified by its author, assuming 150-mm field capacity of the soil and evaporation curve G (e.g., Mather, 1978).
See also Heins (1992) for station data and calculation details.

TABLE 3. SUMMARY OF TOPOGRAPHICAL DATA

Drainage*	Elevation† (avg., m)	Area (km^2)	Aspect† (•)	Slope† (avg.)	Relief Ratio
WAN	2,155 ± 26	0.103	23 ± 77	0.48 ± 0.18	0.51
WAW	2,324 ± 83	0.191	291 ± 53	0.61 ± 0.18	0.70
MTW	1,949 ± 26	0.131	261 ± 54	0.35 ± 0.10	0.36
MTS	1,862 ± 52	0.102	142 ± 49	0.43 ± 0.11	0.36
MTN	1,847 ± 50	0.079	353 ± 23	0.55 ± 0.17	0.58
MTE	1,931 ± 27	0.023	84 ± 32	0.30 ± 0.08	0.29
ORS	271 ± 72	0.182	167 ± 52	0.40 ± 0.14	0.36
ORN	295 ± 32	0.038	358 ± 32	0.43 ± 0.12	0.44
ORE	333 ± 64	0.272	96 ± 46	0.29 ± 0.12	0.28
CAW	2,660 ± 55	0.231	262 ± 18	0.39 ± 0.15	0.39
CAS	2,399 ± 14	0.018	178 ± 52	0.30 ± 0.13	0.19
CAN	2,387 ± 24	0.030	11 ± 57	0.48 ± 0.20	0.32
CAE	2,616 ± 63	0.573	51 ± 44	0.27 ± 0.07	0.21
AZW	586 ± 35	0.108	255 ± 52	0.54 ± 0.13	0.35
AZS	595 ± 27	0.075	172 ± 50	0.42 ± 0.16	0.17
AZE	638 ± 34	0.119	148 ± 53	0.48 ± 0.15	0.44

*Drainage code is composed of the postal abbreviation for the state in which the pluton is located and the facing direction of the drainage.

†Mean values for elevation, aspect, and slope are calculated from ~100 random grid points per drainage (Strahler, 1956), plus or minus values are 1 standard deviation. Average aspect and standard deviation are calculated with the method of Batachelet (1981) as implemented by Krause and Geijer (1987).

mountainous and small (0.02 to 0.57 km^2), with average slope (Strahler, 1956) ranging from 0.29 to 0.61, or relief ratio (Schumm and Hadley, 1961; Carson and Kirkby, 1972) from 0.17 to 0.70. One to five samples each of source rock and derivative sand were collected from each drainage.

Petrographic analysis

Each sand sample was dry sieved into coarse (–1 to 1 ϕ), medium (1 to 2 ϕ), and fine (2 to 4 ϕ) fractions, then epoxy mounted and thin sectioned. Both sand and source rock sections were immersed for 2 to 3 s in concentrated hydrofluoric acid to clean and etch mineral surfaces (especially valuable for identifying untwinned plagioclase and grain boundaries), then stained for alkali feldspars with sodium cobaltinitrite.

Texture of the source rocks, described by the parameter s_v (surface/volume, e.g., Underwood, 1970), was measured in millimeters^{-1} (mm^2/mm^3) according to the method of Erkan (1970), using interfacial definitions from Heins (1992). The surface consists of the interfaces among mineral grains per volume of source rock; finer grained rocks have higher values of s_v. Values for the source rocks range from 7.99 to 25.08 mm^{-1} (Table 4).

The modal mineralogy of both source rock and derivative sand was determined by point counting. Points for both types of samples were counted according to the phase beneath the crosshairs, regardless of size, in the categories listed in Table 5. Five hundred points per slide were counted for source rock sections, 300 per slide for sand sections. A separate tally of rock fragments (Table 6) was kept during point counting of sand samples. This tally recorded, by category, how many points in the count were resident in sand grains composed of more than one crystal or phase. This is a broader definition of rock fragments than that used by many workers (e.g., Suttner and Basu, 1985, p. 616; Johnsson et al., 1988), but conforms to more "traditional" definitions (e.g., Franzinelli and Potter, 1983).

Monomineralic polycrystalline sand grains, which were counted as rock fragments, are extremely rare in these samples. The most common of this rare type of grain is polycrystalline plagioclase. Polycrystalline quartz is especially rare because of the nature of the source rocks. Quartz is usually an anhedral component interstitial to plagioclase and alkalai feldspar. Quartz-quartz contacts represent only 2 to 5%, on average, of the total area of intercrystalline boundaries in the source rock (Heins, 1992).

Modal mineralogy data, averaged for the source rock of each pluton and the sand of each drainage, can be found in Appendix 1. The individual analyses from which the averages were computed are reported in Heins (1992).

Statistical analysis

The strict limitation on source rock lithology and "experimental" selection of environmental conditions described above is necessary to bring the data set within the reach of even "quasi-statistical" (Rock, 1988) data analysis. I used single and stepwise multiple regression, implemented in the BMDP statistical packages 6D (Chasen, 1988) and 2R (Dixon and Jennrich, 1988) to explore the relationships among the dependent variable of rock fragment abundance and the independent variables of source rock texture, climate, and topography.

Rock fragment abundance is calculated as the log of the

TABLE 4. AREA OF MINERAL INTERFACES PER VOLUME OF SOURCE ROCK

State	Number of Samples	s_v* (mm^{-1})
WA	7	10.16 ± 1.59
MT	9	16.96 ± 2.70
OR	11	15.42 ± 3.79
CA (N, S)	7	9.74 ± 0.85
CA (E, W)	5	7.99 ± 2.21
AZ	15	25.08 ± 8.78

*Plus or minus values for s_v are 1 standard deviation of the mean for the source rock samples. This variation among samples is about two orders of magnitude more than the counting error in each sample.

TABLE 5. POINT-COUNTING CATEGORIES

P = Plagioclase feldspar (albite-anorthite)
- Twinned
- Untwinned

Q = Quartz
- Nonundulose (undulosity <5°)
- Undulose (undulosity ≥5°)
- Semicomposite ("modified igneous quartz" of Krynine, 1940)

K = Alkalai feldspar, except albite, as determined by staining and etching.

B = Biotite and phlogopite

O = Opaques

A = Accessories (all minerals not included in another category)

TABLE 6. ROCK FRAGMENTS/TOTAL SAND GRAINS, BY SIZE FRACTION AND DRAINAGE

Drainage	Number of Samples	Grain Size		
		Coarse	Medium	Fine
WAN	4	0.848	0.429	0.305
WAW	2	0.885	0.540	0.250
MTW	2	0.983	0.668	0.422
MTS	2	0.970	0.672	0.500
MTN	1	0.933	0.720	0.417
MTE	3	0.925	0.640	0.387
ORS	2	0.854	0.557	0.270
ORN	2	0.720	0.540	0.370
ORE	2	0.883	0.612	0.302
CAW	5	0.931	0.479	0.280
CAS	5	0.859	0.393	0.227
CAN	5	0.849	0.449	0.294
CAE	4	0.882	0.440	0.272
AZW	3	0.986	0.933	0.680
AZS	2	0.993	0.920	0.705
AZE	3	0.993	0.950	0.706

ratio of polycrystalline to monocrystalline sand grains. This definition rests on geological as well as statistical grounds.

Boundaries between minerals in crystalline igneous rocks are less stable than the grains themselves; as a result, boundaries between crystals are preferentially etched by acid (Wegner and Christie, 1985), and melt at lower temperatures (Büsch et al., 1974) than the grains themselves. I attribute this behavior to lower spatial density of intercrystalline versus intracrystalline chemical bonds. The area and kinds of grain boundaries within rock fragments derived from granitoids exhibit greater differences relative to the source rock than does the mineralogy of the rock fragments (Heins, 1992), which indicates that grain boundaries are more readily modified during disaggregation of the source rock than are the grains themselves. Even though different kinds of rock fragments certainly have varying durabilities, I exepct any granite-derived polycrystalline aggregate to be fundamentally less stable than any granite-derived monocrystalline grain, under the (very short) time and transport conditions relevant to this study. Although polycrystalline quartz probably defies this expectation, it is not a sufficiently abundant component to make a difference in the analysis.

A log ratio is more appropriate for statistical analysis than a unit-sum fraction like percent rock fragments (%R) or percent rock fragments among quartz, feldspar, and rock fragments (QFR%R), because it has an infinite range of variation and avoids spurious correlations introduced by the unit-sum constrain (Butler, 1979; Aitchison, 1986; Woronow, 1990; Woronow and Love, 1990).

The independent variable of source-rock texture is represented by s_v^2. The square is used because the number of grains per volume is a power function, not a linear function, of s_v. Climate is represented by mean annual temperature (°Kelvin, suitable for analysis of reaction kinetics), mean annual precipitation (millimeters), and mean annual surplus of precipitation over evapotranspiration (millimeters). Drainage morphology is represented by drainage area (kilometers^2), average slope, and relief ratio. Dependent and independent variable data used in the analysis are summarized in Table 7.

RESULTS

Bivariate results

The statistical relationships, as calculated with BMDP 6D, between rock fragment abundance in each size fraction and each environmental variable, are summarized in Table 8. The order of importance of environmental variables, for all grain sizes, seems to be parent rock texture (measured by s_v^2), moisture (measured by either precipitation or surplus water), and temperature. None of the other factors exhibits significant bivariate correlation with rock-fragment abundance.

Stepwise linear regression

Grantham and Velbel (1988) observed that physically meaningful combinations of environmental variables may corre-

TABLE 7. DATA USED FOR STATISTICAL ANALYSES

A*	B†	C†	D†	E§	F**	G**	H¶	I§§	J***	K†††
AZE	0.993	0.950	0.706	0.119	0.48	0.44	292.2	201	25.08	0.0
AZS	0.993	0.920	0.705	0.075	0.42	0.28	292.2	201	25.08	0.0
AZW	0.986	0.933	0.680	0.108	0.54	0.35	292.2	201	25.08	0.0
CAE	0.882	0.440	0.272	0.573	0.27	0.21	269.0	1,095	9.74	839.3
CAN	0.849	0.449	0.294	0.030	0.48	0.32	271.2	995	7.99	742.9
CAS	0.859	0.393	0.227	0.018	0.30	0.19	271.2	995	7.99	742.9
CAW	0.931	0.479	0.280	0.231	0.39	0.39	269.0	1,095	9.47	839.3
ORE	0.883	0.612	0.302	0.272	0.29	0.28	301.2	1,173	15.42	143.3
ORN	0.720	0.540	0.370	0.038	0.43	0.44	301.2	1,173	15.42	143.3
ORS	0.854	0.557	0.270	0.182	0.40	0.36	301.2	1,173	15.42	143.3
MTE	0.925	0.640	0.387	0.023	0.30	0.29	275.8	285	16.96	0.0
MTN	0.933	0.720	0.417	0.079	0.55	0.58	275.8	285	16.96	0.0
MTS	0.970	0.672	0.500	0.102	0.43	0.36	275.8	285	16.96	0.0
MTW	0.983	0.668	0.422	0.131	0.35	0.36	275.8	285	16.96	0.0
WAW	0.885	0.540	0.250	0.191	0.61	0.70	275.9	1,566	10.16	1,137.8
WAN	0.848	0.429	0.305	0.103	0.48	0.51	275.9	1,566	10.16	1,137.8

*Drainage code. See Tables 2 or 3 for explanation.
†Proportion of rock fragments/total sand grains (R) for coarse, medium, and fine fractions, respectively, log-ratio transformation was performed by BMDP as log (R/(1-R)).
§Drainage area in km^2.
**Slope and relief ratio.
‡Temperature, in °K.
§§Precipitation, in mm.
***s_v, in mm^{-1}. Transformation to s_v^2 was performed by BMDP.
†††Surplus of precipitation over calculated evapotranspiration, in mm.

late strongly with rock fragment abundance, even when the individual variables do not. To test that possibility with this data set, and to determine the relative importance of each provenance factor in explaining rock fragment variability, the data were examined with BMDP program 2R.

BMDP 2R steps independent variables into a linear-regression equation according to an F-value, which measures the power of the variable to minimize dependent-variable residuals. F-values for each variable not in the equation change after each step, according to the residuals left by variables already in the equation. The one variable with the highest F-value enters at each step, provided the value is above a predetermined significance level or F-to-enter. In this case, the F-to-enter was set at 4, which approximates $\alpha = 5\%$ for a traditional F-test. A complete explanation of the statistical theory and computational mechanics underlying the procedure are provided by Jennrich (1977), Hill (1987), and Dixon (1988, p. 565–567).

The independent variables chosen by BMDP for each grain size, and their associated regression parameters and correlations with the dependent variable, are summarized in Table 9. In all cases, source rock texture is the most important provenance variable by a wide margin; in medium sand, texture explains 55.3 times more of the variation in rock fragment abundance than the rest of the variables combined (~96.3% versus ~1.8%); for fine sand 28.6 times more (~92.3% versus ~3.2%), and for coarse sand 4.2 times more (~70.9% versus ~16.8%). Variations in abundance are associated with mean annual temperature, and, in the case of medium sand, with average slope, at statistically significant, but very low levels. Neither precipitation nor surplus water enters the stepwise regression.

DISCUSSION

Texture, temperature, and slope

The positive correlation between source rock texture and rock fragment abundance has an obvious geometric basis; the higher the number of crystals per volume (finer grained), and thus the higher the area of intercrystalline area per volume, in the source rock, the more likely it is that a sand grain of given volume will contain more than one crystal (e.g., Pettijohn, 1948; Young, 1975).

The negative correlation of temperature with rock fragment abundance might be attributed to kinetic control of mineral dissolution (Ollier, 1975; Lasaga, 1984), or control on biologic activity (Ollier,1975). Higher temperatures imply higher rate constants for mineral dissolution reactions and perhaps more intense biologic activity.

The positive correlation of slope with rock fragment abundance means that steeper slopes lead to more rock fragments, probably because both sand and water are removed more quickly from the weathering zone on steeper slopes (Elwell, 1984; Komura, 1984). Faster removal from the weathering zone reduces the time during which rock fragments can be destroyed. It is informative to note that slope can enter the multiple-regression model at a

TABLE 8. BIVARIATE REGRESSION RESULTS, ROCK FRAGMENT ABUNDANCE VS. ENVIRONMENTAL FACTORS

Grain Size/ Factor	b*	m†	r^2§	p**
Coarse				
Area	1.2458	-0.46094	0.014	0.666
Slope	0.8118	+0.87739	0.027	0.543
Relief ratio	1.2972	-0.30878	0.006	0.779
Temperature	0.2494	+0.00507	0.013	0.674
Precipitation	1.8418	-0.00084	0.646	<0.001
Surplus water	1.4184	-0.00065	0.292	0.031
s_v^2	0.5705	+0.00228	0.709	<0.001
Medium				
Area	0.3899	-0.69755	0.042	0.447
Slope	0.3471	+1.5187	0.108	0.214
Relief ratio	0.2385	+0.13785	0.001	0.885
Temperature	-4.6480	+0.01750	0.208	0.076
Precipitation	0.8202	-0.00067	0.552	<0.001
Surplus water	0.5395	-0.00068	0.424	0.006
S_v^2	0.3246	+0.00230	0.962	<0.001
Fine				
Area	0.1007	-0.61376	0.075	0.306
Slope	0.5610	+0.88815	0.085	0.273
Relief ratio	-0.1702	-0.04706	<0.001	0.940
Temperature	-2.8445	+0.00941	0.138	0.155
Precipitation	0.1862	-0.00048	0.637	<0.001
Surplus water	-0.0204	-0.00046	0.444	0.005
s_v^2	-0.9364	+0.04885	0.876	<0.001

*y-Intercept of the regression line.
†Slope of the regression line.
§Square of the correlation coefficient.
**$P(r^2 = 0)$ ("confidence level," α).

TABLE 9. STEPWISE REGRESSION RESULTS, ROCK FRAGMENT ABUNDANCE VERSUS ENVIRONMENTAL VARIABLES

Grain Size/ Variable	b*	m†	r^2§	Δr^2**	r^2sv/ r^2 other‡
Coarse					
S_v^2		0.00300	0.7087	0.7087	4.2
Temperature	6.5011	-0.02170	0.8771	0.1684	
Medium					
S_v^2		0.00238	0.9629	0.9629	55.3
Slope		0.44511	0.9732	0.0103	
Temperature	0.5623	-0.00388	0.9803	0.0071	
Fine					
S_v^2		0.00166	0.9225	0.9225	28.6
Temperature	0.8956	-0.00542	0.9548	0.0323	

*y-Intercept of the regression line for the equation that contains all the listed variables.
†Regression coefficient of each variable in the equation that contains all the listed variables.
§Square of the correlation coefficient for the regression equation that contains the listed variable and all previous variables. For medium sand, for example, r^2 for $S_v^2 = 0.9629$ is for the equation that contains only S_v^2 (see Table 8); r^2 for temperature = 0.9803 is for the equation that contains S_v^2, slope, and temperature.
**The improvement in r^2 that results from adding the listed variable to the equation. For medium sand, for example, the equation with S_v^2, slope and temperature has r^2 that is 0.0071 (= 0.9803 - 0.9732) higher than the equation with just S_v^2 and slope.
‡The ratio of the r^2 contribution of S_v^2 to the r^2 contribution of the environmental variable(s). This is a measure of the relative importance of source rock texture and environment in determining rock fragment abundance.

statistically significant level, even though it is not significant in the bivariate analysis, because it is highly correlated with the *residual* variability unaccounted for by the other variables.

Precipitation

The absence of either precipitation or surplus water in the model is surprising. The absence might be explained by a poor choice of moisture measurements, a pathologic data set, or limitations in the scope of environments sampled. Although all of these factors certainly play a role, as discussed below, there still is enough basis to accept source rock texture as a fundamenteal, or even the primary, control on rock fragment abundance and to downplay the importance of water.

Choice of water measurement. Mean annual precipitation or mean annual surplus water might not be the most relevant measure of environmental moisture for rock fragment destruction. The water input to each pluton varies not only in amount, but in form and timing as well (monsoonal rains, infrequent cloud bursts, snowmelt, etc.). Rock fragment destruction by water might depend more on the timing (especially relative to temperature fluctuations) and duration of water inputs than on the absolute amount. However, I do not think this is an important issue at this level of analysis. Interfaces of important individual mineral pairs, like biotite-quartz and quartz-alkalai feldspar, are significantly (in a rigorous statistical sense) depleted in rock fragments relative to the source rock according to functions that are linear with respect to both of the mean annual moisture measures (Heins, 1992), yet the mean annual moisture measures do not correlate as well with rock fragment abundance. From this relationship, I conclude that mean annual measures of water are adequate to describe the role of water in breaking up rock fragments, but that breaking up rock fragments is only a secondary modification imposed on a population whose size is primarily determined by source rock texture.

Pathologic data. A small and fortuitous correlation between source rock texture and precipitation may render this data set unsuitable for rigorously untangling the relative influence of the two factors. The correlation between s_v^2 and precipitation accounts for about 57% of the joint variability of the two factors, and the probability that the correlation is actually zero is only 7.3%, as determined by BMDP 6D. The fortuitous correlation arises because the finest grained source rocks are found in the driest area (AZ), and the source rocks from the wettest area (WA) are among the coarsest. Both finer grain size and drier conditions would be expected to produce more rock fragments, whereas coarser source rocks and wetter conditions would both be expected to produce fewer rock fragments.

Although the possibility that the stepwise regression model excludes precipitation solely because the fortuitous correlation destroys its F-to-enter after texture has already entered cannot be rigourously excluded, it seems unlikely. The Arizona and Montana source rocks, which experience nearly identical precipitation amount (Table 2) and timing (Heins, 1992), but have much different textures (Table 4), produce sand with much different rock-fragment abundances in every grain size (Table 6). The Washington drainages receive 50% more precipitation than the CAN and CAS drainages, but the source rocks have nearly the same texture and produce nearly the same rock-fragment abundance.

The conclusion that texture is more important than precipitation is also consistent with the much larger and more diverse (both geographically and in lithology of the source rocks) database of Basu (1976), even though that study came to the opposite conclusion. Basu found that sand samples from the colder, steeper, (drier) Rocky Mountains contains significantly more rock fragments in every sand fraction than did sand samples from the warmer, flatter, (wetter) Appalachians. However, nearly 80% of the Rocky Mountain sand samples were derived from fine-grained granitoids and silicic volcanics (Becraft, 1960a,b; Becraft and Pinkney, 1961; Pinkney and Becraft, 1961; Smedes, 1962; Smedes et al., 1962; Eggler, 1968; Eggler et al., 1969; Toth, 1983), whereas 90% of the Appalachian sand samples were derived from coarse-grained granitoids and high-grade metamorphics (Overstreet and Bell, 1965a,b; Stuckey, 1965; Jones and Walker, 1973; Whitney and Stormer, 1977; Stormer et al., 1980, Goldsmith et al., 1988; McSween et al., 1991). Among the Rocky Mountain localities, the finest grained source rocks (Boulder Batholith) produced the most abundant rock fragments, whereas the coarsest (Sherman Granite) produced the least abundant.

The data from Basu (1976) suffer an even stronger correlation among source rock texture, climate, and topography, and thus are not sufficiently well constrained to confirm or refute the results of this analysis, but the data are at least consistent with the conclusion of this study that the order of importance of provenance variables in determining rock fragment abundance is source rock texture >> temperature >> slope.

The only way to address the issue properly is to add drainges with coarser source rocks in drier climates and drainages with finer source rocks in wetter climates to the database to eliminate the fortuitous correlation. At present, it seems more reasonable to assume that the composition of short-traveled modern sand is more strongly controlled by source rock characteristics than by environment.

Environmental limitations. All of the locations sampled experience only weathering-limited erosion (Stallard, 1985; Johnsson, 1992) and only represent temperate climates. These weaknesses of the analysis can also only be addressed by the addition of more environments. It is possible, though logistically difficult, to substantially expand the environmental range of the study, since subaluminous granodiorites are very common everywhere in the world. They can be found in environments in which mean annual temperature ranges from about –30° to 40°C and precipitation from about 0 to >4 m (Heins, 1992). Because previous studies by Krynine (1935) and Ruxton (1970) suggest that even environments with mean annual temperature and precipitation of 22° and >4,000 mm and 30°C and 2,000 to 3,000 mm, respectively, produce "immature" plutoniclastic and volcaniclastic sediments where slopes are steep and erosion is rapid, it is also critical to examine a broad climatic range of transport-limited erosional environments. Unfortunately, the same level of source rock equivalence used in the current study is much more difficult to enforce in transport-limited settings.

Conclusion

Despite the difficulties inherent in analyzing this small, complex, natural data set, the conclusion that texture of the source rocks is the most important factor in determining rock fragment abundance in modern, short-traveled, plutoniclastic sand seems sound.

IMPLICATIONS FOR PREVIOUS AND FUTURE RESEARCH

If relatively small textural differences in a narrowly constrained class of source rocks (e.g., medium-grained, equigranular, subaluminous granodiorite) can lead to relatively large differences in the rock fragment content of derived sand, then it is critical to at least document in detail, if not explicitly account for, the effects of variations in source rock lithology when undertaking environmental analysis based on rock fragment abundance.

None of the studies cited in the introduction, which support environmental control on rock fragment abundance, adequately control for variations in source rock. The Rocky Mountain sand examined by Basu (1976) was derived primarily from fine-grained granitic and volcanic source rocks, whereas the Appalachian sand was derived primarily from coarse-grained granitic and metamorphic source rocks. This study suggests that the textural difference between the source rocks is a more likely cause of differences in sand composition than are variations in environment.

The source rocks for the Pennsylvanian sand examined by Mack and Suttner (1977) are much more texturally diverse and include coarser lithologies (Howard, 1966; Bryant et al., 1981) than the one or two types of source rocks for the modern sand (Scott et al., 1976). A test of quartz types (undulosity and polycrystallinity) is not sufficient to address the texture issue, or to demonstrate that the differences between the modern and ancient sand are not due only to textural differences among the source rocks.

The arid-setting source rocks examined by Mack and Jerzykeiwicz (1989) were all erupted in continental interiors, and thus are more potassic and less calcic ($Na_2O/K_2O \leqslant 0.8$; CaO = 1.4 to 3.8%) (Seager, 1981; Klepper et al., 1957; Larsen and Cross, 1956), whereas the humid-setting source rocks were erupted in a variety of marginal arcs, and are thus more sodic and more calcic ($Na_2O/K_2O \geqslant 2.2$; CaO = 5.5 to 8.1%) (Fiske et al., 1963; Chesner and Rose, 1984; Westercamp, 1976). The geochemical differences in the two kinds of source rocks might be a more plausible explanation than climate for the differences in RFP of the derived sands. RFP for the arid setting might be higher because there is less plagioclase in the source rocks, rather than because the arid climate less thoroughly destroys rock fragments. In addition, two of the three humid-area source rocks have significant, nonandesite, sources of plagioclase that might alone account for their higher RFP. Sand from Mt. Rainier might have extra plagiocalse derived from the Tatoosh Pluton (~10% of the source area; Fiske et al., 1963), and sand from Mt. Pelée might have extra plagioclase derived from plutonic xenoliths (Lacroix, 1904, Westercamp, 1976; Fichaut et al., 1989; Smith and Roobol, 1990).

The Coweeta Group source rocks investigated by Grantham and Velbel (1988) are represented by three formations, the Persimmon Creek Gneiss, the Coleman River Formation, and the Ridgepole Mountain Formation (Hatcher, 1979, 1980). In three of the four Coweeta Group watersheds examined (nos. 13, 27, 37), rock fragment abundance is also strongly correlated with the percentage of the watershed underlain by Coleman River Formation (Table 10). In the fourth watershed (no. 32), rock fragment abundance is lower than the area of Coleman River Formation would suggest, but that watershed clearly has long sediment residence time, documented by the extensive Quaternary cover. The results of multiple regression analysis (Table 11) of rock fragment abundance versus a measure of source rock lithology (percent Coleman River Formation in source area) and a residence time proxy (percent Quaternary cover in the source area), show that these factors offer an explanation that is equally as plausible as CCWI for explaining rock fragment abundance.

The evident correlation of rock fragment abundance with source rock type in the case of Grantham and Velbel's data also calls into question the use of quartz types to establish the lithologic equivalence of metamorphic source rocks. The formations of the Coweeta Group are different enough lithologically to be mapped in the field, and evidently different enough to produce varying amounts of rock fragments in derived sand, yet they yield identical quartz populations.

In the future, studies of the environmental information contained in the rock fragment content of modern or ancient sand should more thoroughly document and evaluate the details of source rock lithology before assessing environmental effects.

TABLE 10. PROPORTION OF SELECTED COWEETA HYDROLOGICAL LABORATORY WATERSHEDS UNDERLAIN BY COWEETA GROUP FORMATIONS AND OVERLAIN BY UNCONCOLIDATED QUATERNARY UNITS*

Watershed	Coleman River Formation	Persimmon Creek Gneiss	Ridgepole Mountain Formation	Quaternary Units
13	0.725	0.136	0.140	0.080
27	0.640	0.322	0.038	0.052
32	0.785	0.188	0.027	0.274
37	0.911	0.089	0.000	0.093

*Data derived by planimetric analysis of Hatcher (1980).

TABLE 11. MULTIVARIATE REGRESSION RESULTS, ROCKFRAGMENT ABUNDANCE VERSUS SOURCE ROCK LITHOLOGY AND RESIDENCE TIME, COWEETA GROUP WATERSHEDS

Grain Size	m_1*	m_2†	b§	r^2**
Coarse	0.7856	1.1331	-1.1715	0.938
Medium	0.4088	0.3592	-0.4249	0.951
Fine	0.1846	0.2442	-0.2573	0.832

*Regression coefficient for Coleman River Formation variable.
†Regression coefficient for Quaternary units variable.
§y-Intercept of the regression equation.
**Square of the correlation coefficient.

ACKNOWLEDGMENTS

This work was funded by National Science Foundation Grant EAR-9014225, a Calspace Grant (to R. V. Ingersoll), and a Calspace Graduate Fellowship (to W.A.H.). Fieldwork in Guerrero was supported by a grant from the Geological Society of America (to W.A.H.). Permission to collect samples in the Alpine Lakes Wilderness Area of the Wenatchee National Forest, Washington, was granted by Denny McMillen, resource assistant, Levenworth Ranger Station. Climate data for the Alpine Lakes area was provided by Jim Holcombe, National Weather Service, Wenatchee, Washington. James Murakami, Department of Atmospheric Sciences, University of California, Los Angeles (UCLA), assisted in projecting weather-station climate data to field sites. Surplus water calculations were made by Johannes Feddema, Department of Geography, UCLA, using the computer program WATBUG (Willmott, 1977), developed and modified at the University of Delaware. Eric Cox assisted with fieldwork in Washington and Montana. Thin sections were prepared by Ram Alkaly of the Department of Earth and Space Science, UCLA. Manuscript preparation was aided by facilities provided by the Geology Department, National Taiwan University, Taipei, and the Mineralogy and Applied Geology Departments, Eötvös University, Budapest, Hungary.

Many of the ideas presented in this work were stimulated or refined by discussions with J. J. Feddema (especially regarding the measurement of climate), R. V. Ingersoll, W. E. Reed, R. L. Shreve (especially regarding statistics), and S. W. Trimble. Preliminary versions of the manuscript profited from critical reviews by R. V. Ingersoll, and R. L. Shreve. The final version was greatly improved on the basis of reviews by S. A. Graham, M. J. Johnsson, and G. H. Mack.

APPENDIX 1. MODAL MINERALOGY SUMMARIES, BY DRAINAGE

	Parent (%)	Coarse (%)	Medium (%)	Fine (%)	All Sand (%)
AZ East					
Plagioclase	45.2	43.7	38.2	33.4	42.0
Quartz	16.0	17.4	18.4	26.9	18.4
Kspar	12.7	14.3	12.7	10.7	13.1
Biotite	6.9	6.3	11.2	6.9	7.1
Opaque	1.2	1.2	2.4	2.6	1.5
Accessories	18.0	17.9	17.0	19.6	17.9
P twinned	42.3	41.0	34.6	34.2	39.4
Q undulose	67.3	63.1	77.1	85.1	67.1
Q semicomposite	21.8	29.9	19.3	10.7	26.7
Q straight	10.9	7.0	3.6	4.1	6.2
AZ West					
Plagioclase	45.2	42.3	36.1	32.8	39.6
Quartz	16.0	18.8	21.0	20.6	19.6
Kspar	12.7	12.7	9.9	12.0	11.9
Biotite	6.9	5.4	9.4	7.9	6.8
Opaque	1.2	1.3	2.3	4.4	1.9
Accessories	18.0	19.4	21.2	22.3	20.2
P twinned	42.3	37.3	37.5	27.8	36.3
Q undulose	67.3	68.0	72.0	81.1	70.5
Q semicomposite	21.8	26.6	20.6	13.5	23.6
Q straight	10.9	5.3	7.4	5.4	5.9
AZ South					
Plagioclase	45.2	47.7	34.5	42.2	44.8
Quartz	16.0	17.5	18.8	19.5	17.9
Kspar	12.7	10.2	6.7	13.5	9.8
Biotite	6.9	6.7	11.7	8.7	7.8
Opaque	1.2	1.3	1.5	2.2	1.4
Accessories	18.0	16.7	26.8	14.0	18.4
P twinned	42.3	44.8	40.6	31.6	42.9
Q undulose	67.3	73.3	61.9	77.8	71.5
Q semicomposite	21.8	21.9	31.9	20.5	23.7
Q straight	10.9	4.8	6.2	1.7	4.8
CA East					
Plagioclase	39.2	32.7	50.1	48.3	37.6
Quartz	27.6	34.9	20.5	21.1	30.8
Kspar	25.3	24.4	17.1	17.4	22.3
Biotite	2.4	2.4	5.8	4.7	3.3
Opaque	0.5	0.7	2.7	5.0	1.5
Accessories	5.1	4.8	3.8	3.6	4.5
P twinned	40.5	33.5	27.5	16.1	30.5
Q undulose	31.7	22.0	39.0	53.8	28.6
Q semicomposite	29.5	30.0	14.8	8.7	24.9
Q straight	38.8	48.0	46.2	37.5	46.5
CA West					
Plagioclase	39.2	32.5	46.8	54.3	37.9
Quartz	27.6	29.9	17.2	14.9	25.2
Kspar	25.3	29.2	17.9	15.4	25.0
Biotite	2.4	2.9	12.3	10.1	6.3
Opaque	0.5	1.0	1.1	1.3	1.0
Accessories	5.1	4.6	4.6	4.0	4.6
P twinned	40.5	35.4	28.4	16.7	32.7
Q undulose	31.7	39.2	65.7	70.5	49.0
Q semicomposite	29.5	37.4	20.4	12..1	31.0
Q straight	38.8	23.4	13.9	17.4	20.0
CA North					
Plagioclase	40.6	48.9	54.5	49.3	49.7
Quartz	30.1	21.6	10.2	10.1	18.9
Kspar	22.5	17.8	10.7	8.5	15.9
Biotite	3.3	3.6	13.3	11.5	5.7
Opaque	0.9	1.5	1.7	3.1	1.7
Accessories	2.6	6.6	9.6	17.4	8.1
P twinned	62.7	35.9	20.7	15.8	31.8
Q undulose	60.5	28.9	45.9	40.4	32.3
Q semicomposite	29.5	34.6	11.7	5.8	28.7
Q straight	10.0	36.5	42.4	53.9	39.1
CA South					
Plagioclase	40.6	46.1	56.7	60.1	49.0
Quartz	30.1	25.6	11.4	7.7	21.9
Kspar	22.5	21.7	11.7	8.1	19.0
Biotite	3.3	2.2	9.0	5.9	3.5
Opaque	0.9	1.2	1.7	2.3	1.4
Accessories	2.6	3.1	9.6	15.9	5.2
P twinned	62.7	34.3	20.0	15.3	30.5
Q undulose	60.5	42.2	61.5	53.7	45.9
Q semicomposite	29.5	32.1	5.1	5.7	25.8
Q straight	10.0	25.8	33.3	40.6	28.3
GR East					
Plagioclase	40.4	22.2	24.5	37.3	27.4
Quartz	20.2	36.3	27.8	18.0	28.3
Kspar	15.7	22.3	17.3	10.5	17.3
Biotite	8.0	5.5	12.7	16.3	10.9
Opaque	1.0	1.3	3.5	2.5	2.3
Accessories	14.6	12.3	14.2	15.3	13.8
P twinned	54.7	48.9	33.3	30.4	38.7
Q undulose	55.7	53.7	68.9	79.6	65.9
Q semicomposite	26.3	28.9	18.6	9.3	20.0
Q straight	18.0	17.4	12.6	11.1	14.1
GR North					
Plagioclase	40.4	25.0	17.8	39.5	26.1
Quartz	20.2	31.0	29.5	17.0	28.1
Kspar	15.7	19.1	17.0	12.3	17.4
Biotite	8.0	4.9	12.2	11.2	7.7
Opaque	1.0	3.7	4.0	3.3	3.7
Accessories	14.6	16.2	19.5	16.7	17.0
P twinned	54.7	47.2	31.8	26.6	40.0
Q undulose	55.7	60.5	84.7	82.4	69.9
Q semicomposite	26.3	22.9	7.9	9.8	17.2
Q straight	18.0	16.6	7.3	7.8	12.9

APPENDIX 1. MODAL MINERALOGY SUMMARIES, BY DRAINAGE

	Parent (%)	Coarse (%)	Medium (%)	Fine (%)	All Sand (%)
GR South					
Plagioclase	40.4	29.9	19.8	34.8	28.2
Quartz	20.2	30.3	38.5	21.3	30.6
Kspar	15.7	21.4	19.0	12.0	18.6
Biotite	8.0	4.5	9.3	12.3	7.6
Opaque	1.0	1.7	1.2	4.8	2.3
Accessories	14.6	12.2	12.2	14.7	12.7
P twinned	54.7	49.4	37.8	27.8	41.3
Q undulose	55.7	61.5	65.8	82.8	67.5
Q semicomposite	26.3	28.2	18.2	8.6	20.9
Q sttraight	18.0	10.3	16.0	8.6	11.6
MT East					
Plagioclase	35.9	45.0	46.1	42.3	44.5
Quartz	19.6	28.3	12.7	15.2	20.6
Kspar	12.3	15.5	14.7	8.8	13.3
Biotite	3.8	5.0	20.0	22.0	13.6
Opaque	0.7	1.5	0.8	2.2	1.5
Accessories	5.5	4.7	5.8	9.4	6.3
P twinned	33.5	37.4	30.8	15.5	29.4
Q undulose	44.2	49.2	84.2	85.4	68.3
Q semicomposite	21.9	33.9	12.3	10.2	21.7
Q straight	11.7	16.9	3.5	4.4	10.0
MT North					
Plagioclase	35.9	51.0	39.7	38.0	43.9
Quartz	19.6	20.3	19.3	18.3	19.4
Kspar	12.3	12.7	11.0	12.3	12.1
Biotite	3.8	7.7	19.7	21.7	15.2
Opaque	0.7	1.0	1.0	1.7	1.2
Accessories	5.5	7.3	9.3	8.0	8.1
P twinned	33.5	46.4	44.5	23.7	38.8
Q undulose	44.2	41.0	58.6	78.2	57.3
Q semicomposite	21.9	45.9	31.0	9.1	30.4
Q straight	11.7	13.1	10.3	12.7	12.3
MT South					
Plagioclase	35.9	41.0	42.3	37.8	40.8
Quartz	19.6	29.8	10.3	12.5	22.2
Kspar	12.3	19.0	9.2	8.8	14.9
Biotite	3.8	4.2	30.7	31.7	15.2
Opaque	0.7	1.0	0.7	1.5	1.0
Accessories	5.5	5.0	6.6	7.7	5.9
P twinned	33.5	43.1	35.8	21.6	37.7
Q undulose	44.2	57.0	82.3	82.7	67.4
Q semicomposite	21.9	33.5	17.7	13.3	26.3
Q straight	11.7	9.5	0.0	4.0	6.3
MT West					
Plagioclase	35.9	43.3	48.5	51.3	46.5
Quartz	19.6	30.1	13.0	12.3	21.6
Kspar	12.3	16.2	11.8	10.0	13.6
Biotite	3.8	3.8	19.0	15.8	10.5
Opaque	0.7	0.8	0.5	1.8	1.0
Accessories	5.5	5.7	7.2	8.7	6.8
P twinned	33.5	39.4	23.4	23.1	31.5
Q undulose	44.2	52.2	83.3	86.5	68.2
Q semicomposite	21.9	36.7	11.5	4.1	22.6
Q straight	11.7	11.1	5.1	9.5	9.2
WA West					
Plagioclase	49.0	55.2	37.7	31.3	46.6
Quartz	22.5	22.8	21.3	19.7	21.9
Kspar	4.8	6.0	4.7	3.5	5.2
Biotite	10.8	8.8	32.0	36.7	19.3
Opaque	0.1	0.0	0.0	0.0	0.0
Accessories	12.8	7.2	4.3	8.8	7.0
P twinned	51.8	48.0	41.2	25.5	41.8
Q undulose	54.6	32.1	72.7	89.8	52.4
Q semicomposite	36.1	59.1	26.6	7.6	41.7
Q straight	9.3	8.8	0.8	2.5	5.9
WA North					
Plagioclase	49.0	46.7	33.6	56.3	47.6
Quartz	22.5	21.2	16.6	9.1	16.9
Kspar	4.8	4.6	3.4	1.5	3.5
Biotite	10.8	15.3	35.6	16.8	18.7
Opaque	0.1	0.1	0.3	1.8	0.6
Accessories	12.8	12.3	10.6	14.5	12.7
P twinned	51.8	52.3	40.9	24.9	42.5
Q undulose	54.6	49.2	68.8	89.0	63.9
Q semicomposite	36.1	38.6	21.6	7.3	26.8
Q straight	9.3	12.2	9.5	3.7	9.3

REFERENCES CITED

Aitchison, J., 1986, The statistical analysis of compositional data: London, Chapman and Hall, 416 p.

Armin, R. A., John, D. A., Moore, W. J., and Dohrenwend, J. C., 1982, Geologic map of the east part of the Raymond Peak Roadless Area, Alpine County, California, U.S. Geological Survey Map MF-1365-A, scale 1:62,500.

Armin, R. A., John, D. A., Moore, W. J., and Dohrenwend, J. C., 1984, Geologic map of the Markleeville 15-minute Quadrangle, Alpine County, California, U.S. Geological Survey Map I-1474, scale 1:62,500.

Basu, A., 1976, Petrology of Holocene fluvial sand derived from plutonic source rocks: Implications to paleoclimatic interpretation: Journal of Sedimentary Petrology, v. 46, p. 695–709.

Batschelet, E., 1981, Circular statistics in biology: London, Academic Press, 371 p.

Becraft, G. E., 1960a, Preliminary geologic map of the northern half of the Jefferson City Quadrangle, Jefferson and Lewis and Clark counties, Montana: U.S. Geological Survey Map MF-171, scale 1:24,000.

Becraft, G. E., 1960b, Preliminary geologic map of the southern half of the Jefferson City Quadrangle, Jefferson and Lewis and Clark counties, Montana: U.S. Geological Survey Map MF-172, scale 1:24,000.

Becraft, G. E., and Pinkney, D. M., 1961, Preliminary geologic map of the northwest quarter of the Boulder Quadrangle, Montana: U.S. Geologic Survey Map MF-183, scale 1:24,000.

Bryant, B., McGrew, L. W., and Wobus, R. A., 1981, Geologic map of the Denver 1° × 2° quadrangle: U.S. Geologic Survey Map I-1163, scale 1:250,000.

Büsch, W., Schneider, G., and Mehnert, K. R., 1974, Initial melting at grain boundaries, Part II: Melting in rocks of granodioritic, quartzdioritic and tonalitic composition: Neues Jahrbuch für Mineralogie, Monatscheft 8, p. 345–370.

Butler, J. C., 1979, Trends in ternary petrologic variation diagrams—Fact or fantasy?: American Mineralogist, v. 64, p. 115–121.

Carson, M. A., and Kirkby, M. J., 1972, Hillslope form and process: Cambridge, University Press, 475 p.

Chasen, S., 1988, Bivariate (scatter) plots (6D), *in* Dixon, W. J., ed., BMDP statistical software manual: Berkeley, University of California Press, v. 1, p. 357–388.

Chesner, C. A., and Rose, W. I., 1984, Geochemistry and evolution of the Fuego volcanic complex, Guatemala: Journal of Volcanology and Geothermal Research, v. 21, p. 25–44.

de Cserna, Z., 1965, Reconociemiento geológico en la Sierra Madre del Sur de México, entre Chilpancingo y Acapulco, Estado de Guerrero: Universidad Nacional Autónoma de México, Instituto de Geología Boletín 62, 77 p.

Dixon, W. J., ed., 1988, BMDP statistical software manual: Berkeley, University of California Press, v. 1, 619 p.

Dixon, W.J., and Jennrich, R., 1988, Stepwise regression (2R), *in* Dixon, W. J., ed., BMDP statistical software manual: Berkeley, University of California Press, v. 1, p. 357–388.

Eggler, D. H., 1968, Virginia Dale Precambrian ring-dike complex, Colorado-Wyoming: Geological Society of America Bulletin, v. 79, p. 1545–1564.

Eggler, D. H., Larson, E. E., and Bradley, W. C., 1969, Granites, grusses and the Sherman erosion surface, southern Laramie Range, Colorado-Wyoming: American Journal of Science, v. 267, p. 510–522.

Elwell, H. A., 1984, Soil loss estimation: A modelling technique, *in* Hadley, R. F., and Walling, D. E., eds., Erosion and sediment yields: Some methods of measurement and modelling: Norwich, UK, Geo Books, p. 15–36.

Erikson, E. H., 1977, Petrology and petrogenesis of the Mount Stuart Batholith—Plutonic equivalent of the high aluminum basalt association?: Contributions to Mineralogy and Petrology, v. 60, p. 183–207.

Erikson, E. H., and Williams, A. E., 1976, Implications of apatite fission track ages in the Mount Stuart Batholith, Cascade Mountains, Washington: Geological Society of America Abstracts with Programs, v. 8, p. 372.

Erkan, Y., 1970, Ein Versuch zur quantitative Erfassung der Festigkeitseigenschaften und zur quantitative Charakterisierung der Granite: Neues Jahrbuch für Mineralogie, Abhandlungen, v. 113, p. 91–109.

Fichaut, M., Marcelot, G., and Clocchiatti, R., 1989, Magmatology of Mt. Pelée (Martinique, F.W.I.) II: Petrology of gabbroic and dioritic cumulates: Journal of Volcanology and Geothermal Research, v. 38, p. 171–187.

Fiske, R. S., Hopson, C. A., and Waters, A. C., 1963, Geology of Mount Rainier National Park, Washington: U.S. Geological Survey Professional Paper 444, 93 p.

Franzinelli, E., and Potter, P. E., 1983, Petrology, chemistry and texture of modern river sands, Amazon river system: Journal of Geology, v. 91, p. 23–39.

Friberg, L. M., and Vitaliano, C. J., 1981, The petrology of the McCartney Mountain stock, southwest Montana: Northwest Geology, v. 10, p. 32–45.

Goldsmith, R., Milton, D. J., and Horton, J. W., 1988, Geologic map of the Charlotte 1° × 2° quadrangle, North Carolina and South Carolina: U.S. Geological Survey Map I-1251-E, scale 1:250,000.

Grantham, J. H., and Velbel, M. A., 1988, The influence of climate and topography on rock-fragment abundance in modern fluvial sands of the southern Blue Ridge Mountains, North Carolina: Journal of Sedimentary Petrology, v. 58, p. 219–227.

Hatcher, R. D., 1979, The Coweeta Group and Coweeta syncline: Major features of the North Carolina–Georgia Blue Ridge: Southeastern Geology, v. 21, p. 17–29.

Hatcher, R. D., 1980, Geologic map and mineral resources summary of the Prentiss Quadrangle, North Carolina (1:24,000), including geologic map of the Coweeta Laboratory (1:14,400): North Carolina Department of Natural Resources and Community Development, Division of Land Resources, Geological Survey Section, 2 sheets.

Heins, W. A., 1992, The effect of climate and topography on the composition of modern, plutoniclastic sand [Ph.D. thesis]: Los Angeles, University of California, 465 p.

Hill, M. A., 1987, Annotated computer output for regression analysis (BMDP Technical Report 48): Los Angeles, BMDP Statistical Software, 76 p.

Howard, J. D., 1966, Patterns of sediment dispersal in the Fountain Formation of Colorado: Mountain Geologist, v. 3, p. 147–153.

Huber, N. K., 1983a, Preliminary geologic map of the Dardanelles Cone quadrangle, central Sierra Nevada, California: U.S. Geological Survey Map MF-1436, scale 1:62,500.

Huber, N. K., 1983b, Preliminary geologic map of the Pinecrest quadrangle, central Sierra Nevada, California: U.S. Geological Survey Map MF-1437, scale 1:62,500.

Jennrich, R. I., 1977, Stepwise regression, *in* Enslein, K., Ralston, A. and Wilf, H.S., eds., Statistical methods for digital computers: New York, Wiley-Interscience, p. 58–75.

Johnsson, M. J., 1992, Chemical weathering controls on sand composition, *in* Nierenber, W. A., ed., Encyclopedia of earth system science: Orlando, FL, Academic Press, p. 455–466.

Johnsson, M. J., Stallard, R. F., and Meade, R. H., 1988, First-cycle quartz arenites in the Orinoco River basin, Venezuela and Columbia: Journal of Geology, v. 96, p. 263–267.

Jones, L. M., and Walker, R. L., 1973, Rb-Sr whole-rock age of the Siloam Granite, Georgia: A Permian intrusive in the southern Appalachians: Geological Society of America Bulletin, v. 84, p. 3653–3658.

Klepper, M. R., Weeks, R. A., and Ruppel, E. T., 1957, Geology of the southern Elkhorn Mountains, Jefferson and Broadwater counties, Montana: U.S. Geological Survey Professional Paper 292, 82 p.

Komura, S., 1984, A method for predicting sediment yields from watersheds: Natural Disaster Science, v. 6, p. 1–16.

Krause, R.G.F. and Geijer, T.A.M., 1987, An improved method for calculating the standard deviation and variance of paleocurrent data: Journal of Sedimentary Petrology, v. 57, p. 779–780.

Krynine, P. D., 1935, Arkose deposits in the humid tropics; A study of sedimentation in southern Mexico: American Journal of Science, v. 29, p. 353–363.

Krynine, P. D., 1940, Petrology and genesis of the Third Bradford Sand: Pennsylvania State College Mineral Industries Experiment Station Bulletin 29, 134 p.

Lacroix, A., 1904, La Montagne Pelée et ses éruptions: Paris, Masson, 622 p.

Larsen, E. S., and Cross, W., 1956, Geology and petrology of the San Juan region, southwestern Colorado: U.S. Geological Survey Professional Paper 258, 303 p.

Lasaga, A. C., 1984, Chemical kinetics of water-rock interactions: Journal of Geophysical Research, v. 89, p. 4009–4025.

Lee, D. E., 1984, Analytic data for a suite of granitoid rocks from the Basin and Range Province: U.S. Geological Survey Bulletin 1602, 54 p.

Mack, G. H., and Jerzykiewicz, T., 1989, Detrital modes of sand and sandstone derived from andesitic rocks as a paleoclimatic indicator: Sedimentary Geology, v. 65, p. 35–44.

Mack, G. H., and Suttner, L. J., 1977, Paleoclimate interpretation from a petrographic comparison of Holocene sands and the Fountain Formation (Pennsylvaniana) in the Colorado Front Range: Journal of Sedimentary Petrology, v. 47, p. 89–100.

Mather, J. R., 1978, The climatic water budget in environmental analysis: Lexington, MA, Lexington Books (D.C. Heath and Co.), 239 p.

McSween, H. Y., Speer, J. A., and Fullagar, P. D., 1991, Plutonic rocks, *in* Horton, J. W. and Zullo, V. A., eds., The geology of the Carolinas: Knoxville, University of Tennessee Press, p. 109–126.

National Climatic Data Center, 1988, Climate data TD3200 summary of the day—Cooperative observer network (West): Boulder, CO, EarthInfo, CD-ROM.

Negendank, J.F.W., Emmermann, R., Nun, N., and Schulz-Dobrick, B., Tobschall, H. and Drawczyk, R., 1987, The granitoid complexes of Acapulco, Xaltianguis and Ocotito (Sierra Madre del Sur, Mexico): Zentralblatt für Geologie and Paleontologie, Teil I, Heft 7/8, p. 705–718.

Ollier, C. D., 1975, Weathering: Edinburgh, Oliver & Boyd, 304 p.

Overstreet, W. C., and Bell, H., 1965a, The crystalline rocks of South Carolina: U.S. Geological Survey Bulletin 1183, 126 p.

Overstreet, W. C., and Bell, H., 1965b, Geologic map of the crystalline rocks of South Carolina: U.S. Geological Survey Map I-413, 1:250,000.

Pettijohn, F. J., 1948, Studies for students, A preface to the classification of the sedimentary rocks: Journal of Geology, v. 56, p. 112–117.

Pinkney, D. M., and Becraft, G. E., 1961, Preliminary geologic map of the southwest quarter of the Boulder quadrangle, Montana: U.S. Geological Survey Map MF-187, scale 1:24,000.

Rehrig, W. A., and Reynolds, S. J., 1980, Geologic and geochronologic reconnaissance of a northwest-trending zone of metamorphic core complexes in southern and western Arizona: Geological Society of America Memoir 153, p. 131–157.

Reynolds, S. J., Keith, S. B., and Coney, P. J., 1980, Stacked overthrusts of Precambrian crystalline basement and inverted Paleozoic sections emplaced over Mesozoic strata, west-central Arizona: Arizona Geological Society Digest, v. 12, p. 45–51.

Rock, N.M.S., 1988, Numerical geology: Berlin, Springer-Verlag, 427 p.

Ruxton, B. P., 1970, Labile quartz-poor sediments from young mountain ranges in northeast Papua: Journal of Sedimentary Petrology, v. 40, p. 1262–1270.

Schumm, S. A., and Hadley, R. F., 1961, Progress in the application of landform analysis in studies of semiarid erosion: U.S. Geological Survey Circular 437, 14 p.

Scott, G. R., Taylor, R. B., Epis, R. C., and Wobus, R. A., 1976, Geologic map of the Pueblo 1° × 2° quadrangle: U.S. Geological Survey Map MF-775, scale 1:187,000.

Seager, W. R., 1981, Geology of Organ Mountains and southern San Andres Mountains, New Mexico: New Mexico Bureau of Mines and Mineral Resources Memoir 36, 97 p.

Secretaría de Programaciao y Presupuesto, 1980, Carta de temperatura medias annuales 1:1,000,000: Coordinación General de los Servicios Nacionales de Estadística, Geografía E Información, México, D.F.

Secretaría de Programaciao y Presupuesto, 1981, Atlas nacional del medio físico: La Dirección General de Geografía del Territorio Nacional, México, D.R., 224 p.

Shand, S. J., 1947, Eruptive rocks, their genesis, composition, classification and their relation to ore deposits, with a chapter on meteorites (third edition): London, Thomas Murby, 488 p.

Smedes, H. W., 1962, Preliminary geologic map of the northern Elkhorn Mountains, Jefferson and Broadwater Counties, Montana: U.S. Geological Survey Map MF—243, scale 1:24,000.

Smedes, H. W., Klepper, M. R., Pinckney, D. M., Becraft, G. E., and Ruppel, E. T., 1962, Preliminary geologic map of the Elk Park quadrangle, Jefferson and Silver Bow Counties, Montana: U.S. Geological Survey Map MF-246, scale 1:48,000.

Smith, A. L., and Roobol, M. J., 1990, Mt. Pelée, Martinique; A study of an active island-arc volcano: Geological Society of America Memoir 175, 105 p.

Spencer, J. E., Richard, S. M., and Reynolds, S. J., 1985, Geologic map of the Little Harquahala Mountains, west-central Arizona: Arizona Bureau of Geology and Mineral Technology Open-File Report 85-9, scale 1:24,000.

Stallard, R. F., 1985, River chemistry, geology, geomorphology, and soils in the Amazon and Orinoco basins, *in* Drever, J. I., ed., The chemistry of weathering: Dordrecht, D. Reidel, p. 293–316.

Stormer, J. C., Whitney, J. A., and Hess, J. R., 1980, Petrology and geochemistry of the Elberton Granite, *in* Stormer, J. C., and Whitney, J. A., eds., Geological geochemical and geophysical studies of the Elberton batholith, eastern Georgia: Georgia Geological Society Fieldtrip Guidebook 19, p. 10–30.

Strahler, A. N., 1956, Quantitative slope analysis: Geological Society of America Bulletin, v. 67, p. 571–596.

Streckeisen, A. L., 1976, To each plutonic rock its proper name: Earth-Science Reviews, v. 12, p. 1–33.

Stuckey, J. L., 1965, North Carolina: Its geology and mineral resources: Raleigh, North Carolina Department of Conservation and Development, 550 p.

Suttner, L. J., and Basu, A., 1985, The effect of grain size on detrital modes: A test of the Gazzi-Dickinson point-counting method—Discussion: Journal of Sedimentary Petrology, v. 55, p. 616–617.

Tabor, R. W., Waitt, R. B., Frizzell, V. A., Swanson, D. A., and Byerly, G. R., 1982, Geologic map of the Wenatchee 1:100,000 quadrangle, Washington: U.S. Geological Survey Map I-1311.

Toth, M. I., 1983, Reconnaissance geologic map of the Selway-Bitterroot Wilderness, Idaho County, Idaho, and Missoula and Ravalli Counties, Montana: U.S. Geological Survey Map MF-1495-B, scale 1:250,000.

Underwood, E. E., 1970, Quantitative stereology: Reading, MA, Addison-Wesley, 274 p.

Universidad Nacional Autonoma de Mexico, 1970, Carta de Climas 1:500,000: Instituto de Geographia, La Direccion de Planeacion, Comision de Estudious del Territorio Nacional y Planeacion, Mexico, D.F.

Wegner, M. W., and Christie, J. M., 1985, General chemical etchants for microstructures and defects in silicates: Physics and Chemistry of Minerals, v. 12, p. 90–92.

Westercamp, D., 1976, Petrology of the volcanic rocks of Martinique, West Indies: Bulletin Volcanologique, v. 39, p. 175–200.

Whitney, J. A., and Stormer, J. C., 1977, Two-feldspar geothermometry, geobarometry in mesozonal granitic intrusions: Three examples from the Piedmont of Georgia: Contributions to Mineralogy and Petrology, v. 63, p. 51–64.

Willmott, C. J., 1977, WATBUG: A FORTRAN IV algorithm for calculating the climatic water budget: Publications in Climatology, v. 50, no. 1, 55 p.

Woronow, A., 1990, Methods for quantifying, statistically testing and graphically displaying shifts in compositional abundances across data suites: Computers in Geosciences, v. 16, p. 1209–1233.

Woronow, A., and Love, K. M., 1990, Quantifying and testing differences among means of compositional data suites: Mathematical Geology, v. 22, p. 837–854.

Young, S. W., 1975, Petrography of Holocene fluvial sands derived from regional metamorphic source rocks [Ph.D. thesis]: Bloomington, Indiana University, 145 p.

Manuscript Accepted by the Society December 24, 1992

Geological Society of America
Special Paper 284
1993

Intensity and duration of chemical weathering: An example from soil clays of the southeastern Koolau Mountains, Oahu, Hawaii

Mark J. Johnsson,* Stephen D. Ellen, and Mary Anne McKittrick
U.S. Geological Survey, 345 Middlefield Road, MS 999, Menlo Park, California 94025

ABSTRACT

Orographic precipitation on the southern flank of the southeastern Koolau Mountains produces a pronounced precipitation gradient. The corresponding gradient in the intensity of the chemical weathering environment provides an opportunity to address the effects of varying chemical weathering intensity on the composition of clay-size weathering products in soils developed on basalt. In addition, little-modified remnants of the constructional surface of the Koolau Volcano, isolated by stream dissection, remain as facets on the southern ends of the parallel ridges of the study area. By comparing clay mineralogy of soils developed on these older geomorphic surfaces with those developed on the younger sharp-crested ridges and steep side slopes, the effects of weathering duration on clay mineralogy can also be addressed.

Soil clays in this part of the Koolau Mountains are mineralogically complex; principal phases include smectite, kaolinite, and halloysite, but pure end member phases are uncommon. Rather, most phases contain some amount of mixed layering. Smectite may contain small (<5%) amounts of randomly interstratified halloysite. Similarly, kaolinite commonly contains a small proportion of halloysite interlayers. A complex halloysitic phase shows evidence of interstratification with both smectite and kaolinite. Nonphyllosilicates found in the clay fraction include gibbsite, goethite, rare quartz, and perhaps cristobalite.

The gradient in precipitation is reflected in soil clay mineralogy by varying proportions of dominantly smectitic, kaolinitic, and halloysitic phases. In regions of relatively low precipitation (<2,000 mm/yr), soils are dominated by the smectitic and halloysitic phases. With increased precipitation (as much as ~4,000 mm/yr), kaolinitic and halloysitic phases become the dominant clay minerals, and goethite and gibbsite become increasingly abundant.

Older soils developed on geomorphic surfaces representing the original constructional surface of Koolau Volcano are markedly more leached than those from younger landscapes in the same precipitation regime. Although smectite may be present, kaolinite is the dominant phase, and accumulations of Fe and Ti occur in the uppermost soil levels. Enrichment of Zr and Ti in these soils, as compared to concentrations in the original basaltic parent material, indicates that as much as 75% of the parent material has been lost. Thus weathering duration may affect soil clay composition in the same way as weathering intensity.

Because smectite and halloysite are expandable clay minerals, their presence in soils

*Present address: Department of Geology, Bryn Mawr College, Bryn Mawr, Pennsylvania 19010-2899.

Johnsson, M. J., Ellen, S. D., and McKittrick, M. A., 1993, Intensity and duration of chemical weathering: An example from soil clays of the southeastern Koolau Mountains, Oahu, Hawaii, *in* Johnsson, M. J., and Basu, A., eds., Processes Controlling the Composition of Clastic Sediments: Boulder, Colorado, Geological Society of America Special Paper 284.

may decrease slope stability and influence the nature of slope processes. Soil avalanches occur on steep slopes throughout the study area, whereas slow-moving landslides appear to be restricted to gentler slopes in drier parts of the study area where smectite is abundant. The clay mineralogy of soils thus appears to influence the nature of slope processes in the southeastern Koolau Mountains.

INTRODUCTION

Chemical weathering may have a profound effect on the composition of clastic materials. Although significant modification of sediment composition can occur through chemical weathering during transport and temporary storage of sediment (Rolfe and Hadley, 1964; Krook, 1969; Savage et al., 1988; Johnsson, 1990; Johnsson and Meade, 1990; Johnsson et al., 1991), the initial, and often principal, chemical weathering environment is the soil column: clastic sediments generally are not derived directly from source rock, but from the soils mantling them. Although sedimentary petrologists have focused much attention on chemical weathering controls on the composition of the coarser size fractions of detrital sediments, less attention has been directed at such controls on shale and mudstone composition. An understanding of the nature of secondary weathering products in soils provides insight into controls on the composition of fine-grained clastic rocks. In addition, an awareness of the nature of weathering processes in the pedogenic environment provides perspective on processes that shape the composition of coarser size fractions as well.

A "natural laboratory" for the evaluation of chemical weathering processes must allow factors believed to be important in determining the chemical weathering environment to vary sufficiently to allow observation of their effects. Equally important, these factors must be separable. The Hawaiian Islands provide an ideal setting in which to investigate weathering products of basalt. The degree to which chemical weathering alters the composition of bedrock is a function of both weathering intensity and duration. The prevailing northeasterly Trade Winds over the Hawaiian Islands create markedly orographic precipitation, resulting in a pronounced precipitation gradient. Because the intensity of chemical weathering is strongly controlled by the amount of water flushed through the weathering environment, this precipitation gradient presumably results in a corresponding gradient in chemical weathering intensity. Variations in the duration of chemical weathering may be evaluated either by examining islands of differing age or, as in this study, by examining soils on geomorphic surfaces of differing age. Finally, in Hawaii it is possible to choose a study area that is underlain by an extremely narrow range of rock types, lending confidence to the conclusion that variations in soil clay composition reflect variations in chemical weathering environment rather than differences in source rocks.

SETTING

The study area is located in the southeastern corner of the island of Oahu, in the Koolau Mountains north of Honolulu and its suburbs (Fig. 1). The surrounding region is heavily urbanized, but the study area itself, located in the uplands, has seen little human development. Deforestation may have occurred periodically in the 1,500 yr that the island has been inhabited (Kirch, 1982; Cuddihy and Stone, 1990), but most of our sample sites are in areas that have seen little perturbation during the past century. The study area provides an excellent natural laboratory for study of the effects of varying chemical weathering intensity and duration on soil clay mineralogy. An additional reason for its selection is that the city of Honolulu is affected by landslides that involve these weathered materials, and an understanding of chemical weathering processes might help identify areas at particular risk.

Geology

The island of Oahu is composed of two eroded shield volcanoes, represented by the Waianae Range on the western part of the island and the Koolau Range to the east. The Koolau Volcano is composed primarily of shield-stage tholeiites and olivine-tholeiites (Koolau Basalt) (Clague and Dalrymple, 1987; Langenheimand Clague, 1987); postshield alkalic rocks, which commonly cap shield-stage tholeiites on other Hawaiian volcanoes, are absent. The subaerial Koolau Volcano was active from 2.6 to 1.8 Ma (Doell and Dalrymple, 1973). The tholeiites are pierced by dozens of minor vents corresponding to the alkalic rejuvenated stage of Clague and Dalrymple (1987), which are responsible for isolated flows and air-fall deposits known as the Honolulu Volcanics (Stearns and Vaksvik, 1935; Wentworth and Winchell, 1947; Winchell, 1947; Clague and Dalrymple, 1987; Langenheim and Clague, 1987).

The study area is underlain by the Koolau Basalt (Fig. 2A), which ranges in age from 1.8 to 2.6 Ma (Doell and Dalrymple, 1973) and is composed of tholeiitic basalt, olivine tholeiitic basalt, and rare picritic tholeiitic basalt (Langenheim and Clague, 1987). In outcrop, dense basalt flows 1 to 10 m thick alternate with vesicular clinker zones of comparable thickness. Compositional differences between these distinct textures appear to be minimal. In thin section, the texture of the rocks varies little; generally they are intergranular-porphyritic, with groundmass varying between equigranular and felty textures according to the proportion of plagioclase microlites (Wentworth and Winchell, 1947). Phenocrysts, which constitute up to 50% of the rock, are typically olivine, orthopyroxene, or plagioclase. Magnetite and ilmenite are common accessory minerals, and generally occur as inclusions in olivine phenocrysts (Wentworth and Winchell, 1947). Bulk chemistry of six samples of unweathered Koolau Basalt is presented in Table 1.

The Koolau Basalt is locally overlain by alkalic basalts of

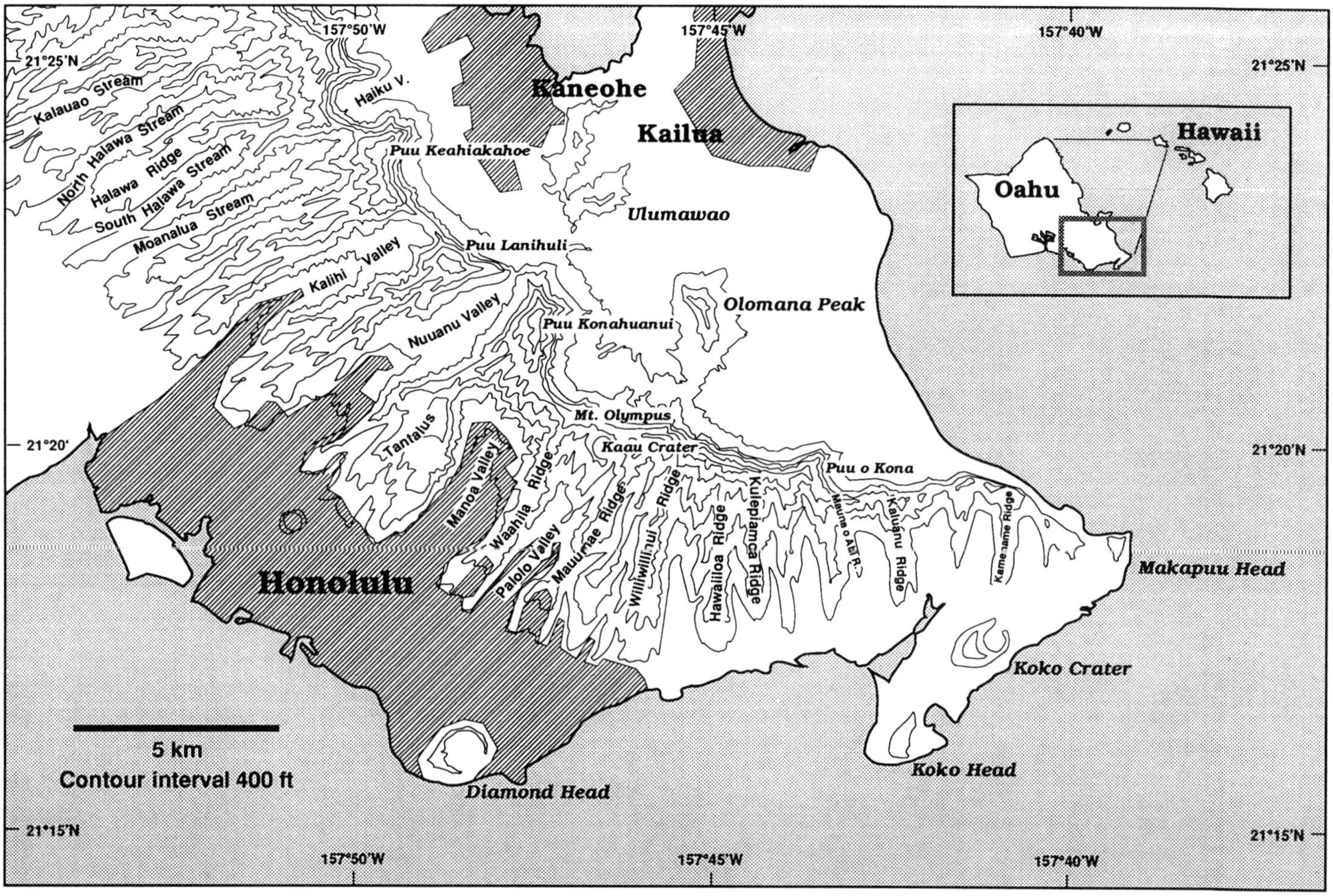

Figure 1. Location of study area showing major topographic features mentioned in text.

the Honolulu Volcanics (Fig. 2A), which range in age from 900 to 30 ka. This unit, probably equivalent to alkalic flood basalts on the sea floor north of Oahu (Clague et al., 1990), comprises alkalic basalts, basanites, nephelinites, and melilitites (Winchell, 1947; Langenheim and Clague, 1987). Because the Honolulu Volcanics are compositionally distinct from, and younger than, the Koolau Baslt, we avoided sampling soils developed on this unit.

Climate and vegetation

Oahu lies in the belt of the northern Trade Winds, and prevailing wind is from the northeast. Orographic precipitation results in extreme geographic variation in the distribution of rainfall (Stidd and Leopold, 1951; Taliaferro, 1959; Blumenstock and Price, 1967; Giambelluca et al., 1986). Within the study area, mean annual precipitation varies from approximately 500 mm near the coast to more than 4,000 mm near the crest of the Koolau Mountains (Giambelluca et al., 1986) (Fig. 2B). The rainfall maximum generally lies parallel to the crest of the Koolau Mountains and displaced about 1 km to the leeward side (south) (Mink, 1960, 1962). In general, the long ridges that characterize the Koolau Mountains in the study area lie perpendicular to isohyets. In distances as little as 5 km, precipitation may vary from 1,000 to 4,000 mm (Giambelluca et al., 1986).

Climate in the study area is somewhat seasonal. In Honolulu, typical of the drier parts of the study area, monthly average rainfall varies from about 95 mm in January to less than 10 mm in June (Blumenstock and Price, 1967). Mean monthly temperature varies slightly from a low of 22°C in January to a high of 26°C in August (Blumenstock and Price, 1967). Near the crest of the Koolau Mountains seasonality is much reduced, and precipitation varies little from a monthly mean of ~250 mm (Taliaferro, 1959).

Because of the strong correlation between precipitation and elevation, it might be anticipated that sea-level variations during the Pleistocene could have affected climate in the study area. Ruhe (1964) estimated past mean annual precipitation from established eustatic sea-level curves and modern isohyetal patterns, and found that precipitation at the Koolau summit would have varied only from 12% less than present to 20% more than present as a result of Pleistocene sea-level fluctuations, without altering overall precipitation patterns. Global secular change in climate is not addressed by this treatment, but a fairly uniform content of tropospherically deposited quartz in north-central Pacific pelagic

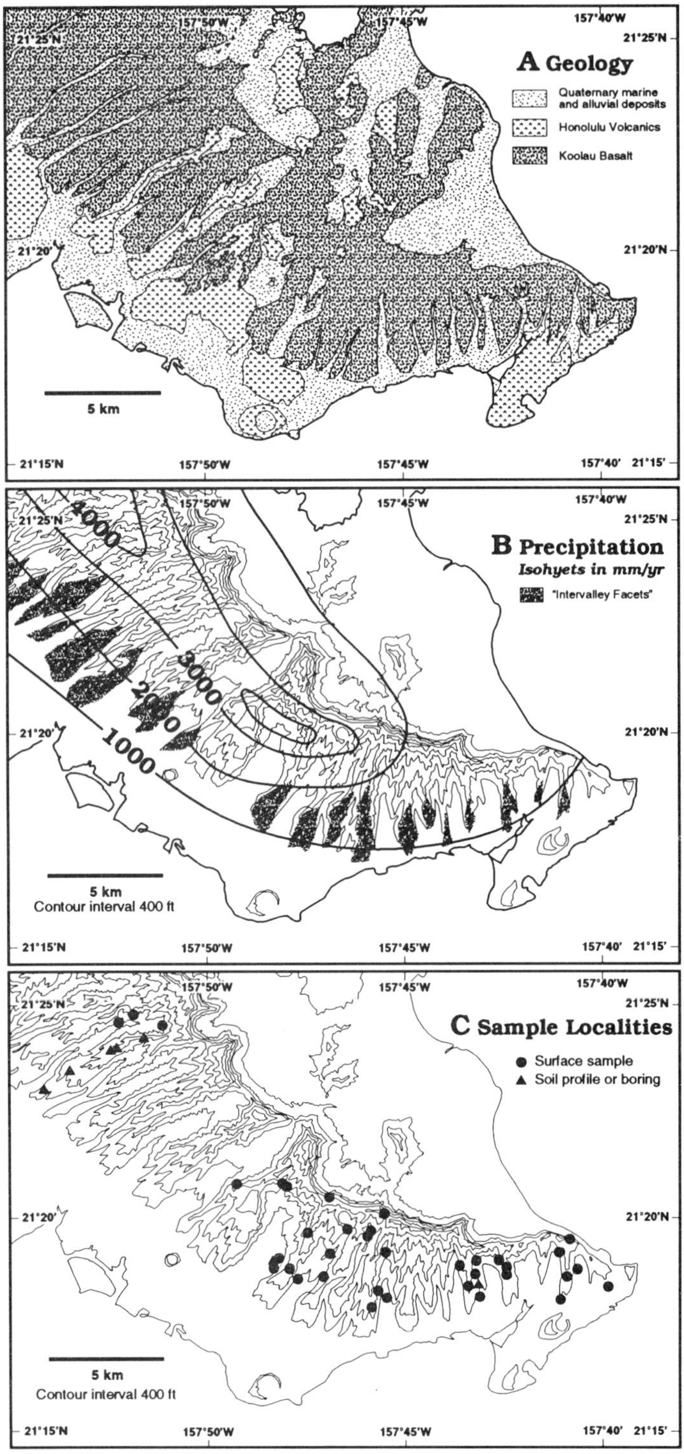

Figure 2. Maps of study area. A, Distribution of shield-phase (Koolau Basalt) and alkalic rejuvenated phase (Honolulu Volcanics) rocks. Adapted from Stearns (1939). B, Mean annual precipitation, in millimeters, adapted from Giambelluca et al. (1986). Also shown are locations of intervalley facets, little-modified remnants of the original constructional surface of Koolau Volcano. C, Sampling stations for this study. Surface samples (generally colluvium) were collected at all stations; soil profiles or multiple vertical samples in a boring were taken from stations indicated by triangles.

sediments suggests no appreciable change in jet-stream pattern for tens of millions of years (Jackson et al., 1971).

Vegetation in the study area is strongly correlated with precipitation (Ripperton and Hosaka, 1942; Cuddihy and Stone, 1990). Hillslopes in drier areas (mean annual precipitation, <2,000 mm) are generally grass covered. With increased precipitation grasses give way to shrubs, then to full-canopied trees. The wettest areas (mean annual precipitation, >3,000 mm), generally near the crest of the Koolau Mountains, harbor shrubs and ferns.

Geomorphology

Dissection of the Hawaiian shield volcanoes typically results in the creation of triangular- or diamond-shaped "intervalley facets" that represent remnants of a gently sloping erosion surface that only slightly modifies the original constructional surface of the volcano (Stearns and Vaksvik, 1935; Wentworth, 1943; Macdonald et al., 1983) (Fig. 3). The surfaces of these facets have not escaped modification by erosion; they appear in placed to truncate underlying flows (Stearns and Vaksvik, 1935), and soils are better developed on the facets than on other parts of the ridges. Nevertheless, these intervalley facets represent remnants, isolated by dissection, of a geomorphic surface that is older than the sharp-crested ridges and steep side slopes produced by such dissection. Intervalley facets increase in area toward the northwest (Fig. 2B) and converge to form the Schofield Plateau northwest of the study area. The study area is at an evolutionary stage midway between those shown in Figure 3B, C; intervalley facets remain at the lower (southern) ends of most ridges but are absent from the crest of the Koolau Mountains.

Mass-wasting processes in this steep terrain appear to be the dominant means by which solid weathering products are carried to local base level. Landslides take two principal forms: catastrophic soil avalanches (Wentworth, 1943), also called debris flows (Ellen et al., 1991), and slow-moving landslides, characterized by episodic downslope movement (Peck, 1967) that averages ~0.01 to 1.2 m/yr (R. L. Baum, written communication, 1992) for periods as long as 40 yr (De Silva, 1974). Slope failures associated with soil avalanches generally involve only the upper 0.5 to 1.0 m of weathered material and require a steep slope angle, whereas slow-moving landslides involve materials to depths as great as 20 m (Peck and Deere, 1960; De Silva, 1974) and can occur on slopes as gentle as 9° to 12° (Baum and Fleming, 1991). Soil avalanches are widespread but irregularly distributed throughout the study area. Interpretation of aerial photographs indicates that over the past 50 yr an average of about 35 mappable soil avalanches have occurred annually through the study area (D. M. Peterson, written communication, 1991). They occur most commonly during the rainly season (November–March), and great numbers may be triggered simultaneously by intense storms such as the New Year's Eve Storm of 1987–1988 (State of Hawaii, 1988; Dracup et al., 1991; Ellen et al., 1991). Slow-moving landslides have been detected only near the base of slopes near the southern ends of ridges, where they have caused

TABLE 1. BULK CHEMICAL COMPOSITION OF REPRESENTATIVE SAMPLES OF UNWEATHERED KOOLAU BASALT

Sample	SiO_2	Al_2O_3	Fe_2O_3*	MgO	CaO	Na_2O	K_2O	MnO	TiO_2	LOI†	Total
MJ91-L12	49.6	14.9	12.1	6.67	9.27	2.32	0.09	0.17	2.14	2.54	100.1
MJ91-L29A1	51.7	13.1	12.0	8.58	9.68	2.24	0.45	0.17	1.91	0.08	100.0
MJ91-L29A2	49.2	14.4	11.8	9.35	10.50	2.18	0.12	0.17	1.99	0.23	100.1
MJ91-L29A3	52.6	14.0	11.5	6.77	9.52	2.64	0.43	0.16	2.22	0.07	99.9
MJ91-L29A4	49.6	14.1	12.2	6.88	10.50	2.40	0.20	0.17	2.58	1.08	99.8
MJ91-L29A5	46.9	15.0	13.8	6.24	9.83	2.00	0.08	0.17	2.75	2.85	99.7
Mean	49.9	14.3	12.2	7.42	9.88	2.30	0.23	0.17	2.27	1.14	99.8

*All Fe reported as Fe^{+3}.
†Loss on ignition.
Note: All analyses reported in weight percent.

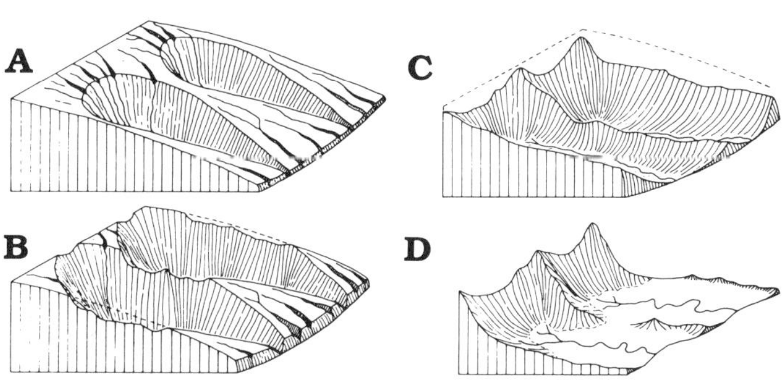

Figure 3. Model for geomorphic evolution of Hawaiian Islands. A, Original constructional surface of volcano is dissected by steep-walled "amphitheater-headed" valleys. B, Valley walls coalesce to form steep, knife-edge ridges cut below the constructional surface. Remnants of the constructional surface persist as intervalley facets at lower ends of ridges and upslope from heads of valleys. C, Continued erosion removes remnants of the constructional surface; slopes become gentler as colluvium accumulates at bases of side slopes, and alluvium fills valley bottoms. D, Continued denudation, accumulation of alluvium, and marine deposition during highstands lead to flat-floored valleys and coastal plains. Model originally from Stearns and Vaksvik (1935); figure from Macdonald et al. (1983).

considerable damage to human structures (Baum et al., 1989), but this apparent localization may be in part because they are subtle features that may escape notice unless they affect roads or structures.

Soil development

Soils in the study area consist of two principal types of material: residuum, consisting of more or less in situ weathered bedrock; and colluvium, consisting of weathered material that has been transported laterally by slope processes. Residuum is typically exposed along ridge crests and in stream cuts. Colluvium generally overlies residuum throughout the study area. Thickness of colluvium ranges from 10 cm or less on ridge crests to tens of meters where broad aprons of accumulation have formed at the base of steep slopes. Eolian deflation pits and dunes in the drier parts of the study area suggest that colluvium in places may contain a significant eolian component.

Soil development in these materials is generally weak throughout most of the study area. Soils developed in residuum generally show slightly greater horizon development than soils in colluvium. Intervalley facets, which occur only in the drier parts of the study area (mean annual precipitation, 1,000 to 2,000 mm) exhibit mature soils, locally with well-developed argillic horizons (Fig. 4A). These argillic horizons typically show moderate to strong prismatic or blocky structure, a red color, and have a relatively high clay content. Soils in the younger colluvium that typically overlies these older ridge-crest soils are weakly developed, showing weak blocky structure and little color or textural horizon development (Fig. 4B).

Soils formed on younger landscapes in the drier parts of the study area (mean annual precipitation, 1,000 to 2,000 mm) occur either in colluvial aprons, on benches between dense flows, or in colluvium-filled hollows on side slopes. These soils tend to show no horizon development, have a high clay content, high pH (6 to 7), and typically contain abundant basalt cobbles and boulders. Where mean annual precipitation exceeds ~2,000 mm, weathering is deeper and soils are highly leached, have low clay content, low pH (5 to 6), and show little horizon development (Fig. 4C). These soils typically have a thin cover of organic debris at the surface (O horizon).

SAMPLING STRATEGY

The alteration of basalt to a suite of weathering products, residual detritus, and ions in solution is affected by a complex suite of factors. The "state factors" of Jenny (1941, 1980)—climate, biology, topography, parent material, and time—provide a framework for the evaluation of weathering processes in soils. The study area provides an ideal natural laboratory in that these variables are both readily identified and isolated. We have concentrated on the effects of varying precipitation and duration of weathering. Vegetation is strongly correlated with precipitation, however, so we cannot separate the effects of vegetation from those of precipitation. Bedrock parent materials are fairly homogeneous throughout the study area. We have evaluated the effects of topography by comparing sample suites from valley bottoms, side slopes, and ridge crests in varying precipitation regimes.

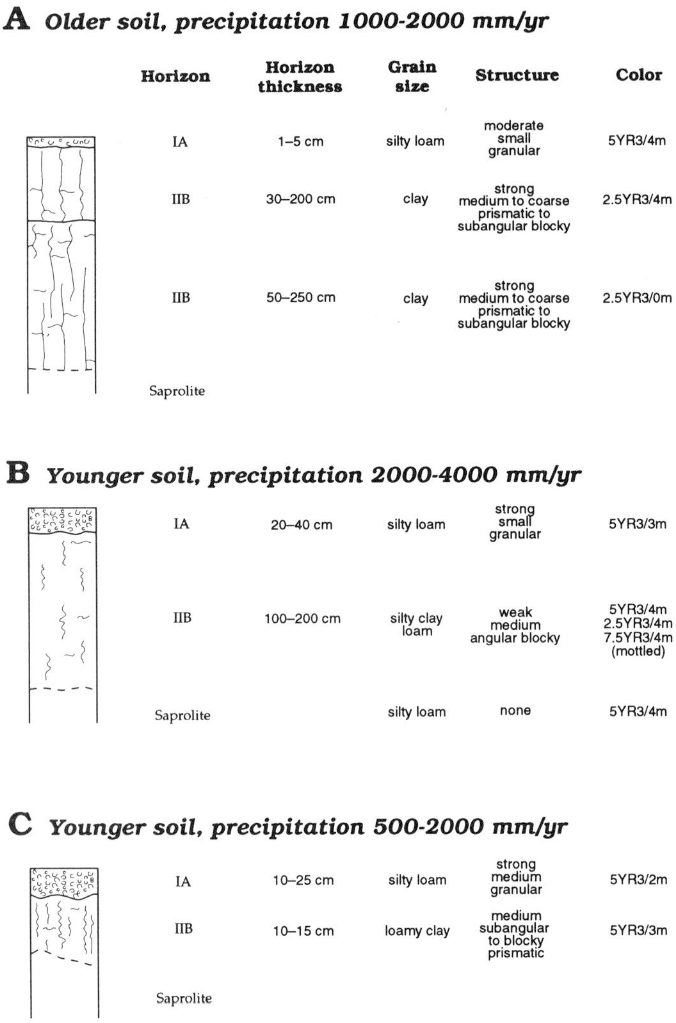

Figure 4. Schematic sketches and brief descriptions of typical soil profiles. A, Soils developed on intervalley facets. B, Younger soils in wetter portions of study area. C, Younger soils from drier portions of study area.

This report describes two sample suites. The first suite, consisting of 42 samples, was collected to determine the spatial distribution of clay minerals in the study area. Grab samples were taken from shallow depths (<50 cm) near the ground surface throughout the study area (Fig. 2C). These samples consist generally of colluvium, but where no colluvium was present, in situ residuum was sampled instead. These samples generally were found to represent the most weathered material at any given sampling station. The second sample suite is a set of six vertical profiles consisting of 4 to 10 samples each (Fig. 2C). Two profiles were collected from the intervalley facet on Halawa Ridge, near the northwestern corner of the study area, in areas receiving about 1,500 and 2,500 mm of precipitation. Three profiles were collected from younger landscapes higher on Halawa Ridge. Two of these profiles, receiving ~3,200 mm of precipitation annually, are developed on dense flow rock and on clinker, respectively, and were collected to evaluate the effects of different parent material textures. The third profile, developed on a clinker zone, receives over 4,000 mm of precipitation annually. An additional profile was collected in the "older alluvium" unit of Stearns (1939) in Kaalakei Valley (west of Mauna o Ahi Ridge; Fig. 1), in the eastern part of the study area (mean annual precipitation, ~1,500 mm). Samples were also collected from a drill core through a slow-moving landslide in Manoa Valley.

METHODS

Samples were sealed in plastic bags in the field to prevent dehydration of halloysitic clays, and were stored refrigerated as much as possible to retard organic metabolism. General sample preparation procedures are those of Moore and Reynolds (1989). Clays were released through ultrasonic disaggregation. Material passing through a 63-μm sieve was peptized with ~0.01-M sodium phosphate and centrifuged to isolate the <2-μm fraction. The settled material was resuspended, and centrifugation was repeated two more times to recover as much of the clay fraction as possible. Clay suspensions were concentrated by ultracentrifugation (to ~20,000 g), and oriented sample mounts were prepared by a filter-membrane peel technique similar to that of Derver (1973).

Qualitative identification of clay mineral phases was made by subjecting selected samples to the following chemical and heat treatments: (1) solvation by ethylene glycol (to expand interlayer sites in some expandable clays); (2) solvation by formamide (to expand interlayer sites in some expandable clays); (3) heat treatment to 110°C for 2 hr (to collapse interlayer sites in some expandable clays); (4) heat treatment to 350°C for 2 hr (to collapse interlayer sites in some expandable clays); (5) heat treatment to 500°C for 2 hr (to destroy kaolinite); and (6) untreated (dried in air at 25°C). These samples were examined both in their natural state and, following cation exchange, with only Mg in exchangeable sites. Ion exchange was performed by saturating the samples with Mg^{+2} by filtering 10 ml of 1 M $MgCl_2$ through a filter membrane containing a cake consisting of ~100 mg of sample (Moore and Reynolds, 1989). No oxidants were used to remove organic matter or amorphous iron because they were found to affect clay minerals as well. Because of the delicate nature of halloysitic clays, no heat was used in preparing the samples or in solvating them with organic molecules; the solvating compound was dripped directly on the x-ray diffraction mount and allowed to remain in contact with the clay for either 10 min (ethylene glycol) or 1 h (formamide) before being blotted away immediately prior to x-raying.

Quantitative analysis of the proportions of each sample composed of various clay species proved to be impossible for several reasons. First, as described below, most of the clay minerals proved to be complexly interstratified and compositionally far from stoichiometric end members, preventing the use of clay mineral standards and presenting challenges to computer modeling of predicted x-ray diffraction patterns. Second, a halloysitic phase makes up an important proportion of most samples. The spheroidal habit of halloysite is marked different from the platy habit of other clay minerals, and the resulting differences in preferred orientation between clay species make quantitative analysis of an oriented mount impossible. Although it would have been possible to perform quantitative analysis of a random powder, we found that the low diffraction intensities, coupled with difficulties in adequately solvating the clays with diagnostic organic materials, made this solution impractical. Accordingly, we report here only the relative peak areas of the 001 reflections of the principal clay minerals. These peak areas are proportional to clay mineral abundance, but not in a linear or simple manner. They may, nevertheless, be used to get a sense of *relative* differences in clay mineral abundances.

Samples for which peak areas are reported were solvated with formamide to allow discrimination between halloysite and kaolinite following the methods of Churchman et al. (1984). X-ray diffraction was performed on a Nicolet I2 diffractometer equipped with a 2.2-kW long fine-focus tube producing unfiltered Cu radiation at 35 kV and 40 mA. The machine was equipped with 1.2° Soller slits, a 3° beam slit, and a 1.5° goniometer slit. A graphite monochromator was installed to restrict analysis to CuK_α radiation and to eliminate fluorescence effects from amorphous Fe. Data were collected digitally for 4 s at 0.05° increments of 2θ. Raw data were smoothed by least squares by fitting a fifth-order polynomial to a running average of 11 data points. Background was then removed by manually identifying points on a baseline, splining a line through those points, and subtracting it from the data. Finally, a low-pass Fourier filter was applied to screen high-frequency noise. The areas (in counts per second) of the idealized peaks corresponding to 001 reflections were then determined.

Major element compositions of selected samples were determined through x-ray fluorescence spectroscopy by X-ray Assays Laboratories, Ltd[1]. Again, no oxidants were used to remove organic matter because they may have leached ions from the sample. Bulk samples (all size fractions) were ground to $<63\ \mu m$ in agate mills and fluxed with $LiBO_2$. Weight loss on ignition at 1,100°C was used for determination of volatile content. This component is reported here as H_2O, although some weight loss was undoubtedly the result of oxidation of organic matter and release of CO_2.

Clay texture was examined by scanning electron microscopy. Soil fragments were mounted on aluminum stubs with double-sided tape and coated with a 60:40 Au-Pd mixture to a thickness of ~150 Å.

RESULTS

Clay mineralogy

X-ray diffraction patterns from three representative samples, subjected to various chemical and heat treatments, are shown in Figure 5. Sample MJ91-L34c (Fig. 5A) is from the crest of the Koolau Range, in a region receiving between 1000 and 1,500 mm of precipitation annually. Sample MJ91-L42b (Fig. 5B) is from an intervalley facet at the southern end of Williwillinui Ridge (Fig. 1), an area that also received between 1,000 and 1,500 mm of rain annually. Sample MJ19-L31a3 (Fig. 5C) is colluvium from the lower side slopes of North Halawa Valley (Fig. 1), in one of the wettest parts of the study area; mean annual precipitation here exceeds 4,000 mm. Most of the phyllosilicate phases encountered in this study are represented among these samples.

The diffraction patterns of samples MF91-L34c and MJ91-L42b contain a prominent peak near 15 Å in the air-dried state that is little affected by Mg-saturation (Fig. 5a,b). On ethylene-glycol solvation this peak moves to near 17 Å. Less prominent peaks near 8.3, 5.6, 4.2, and 3.3 Å are multiples of the 17-Å reflection. In the air-dried state, the peak is unaffected by mild heat treatment (110°C), but collapses to 10 Å at 350 ° and 500 °C. This behavior is characteristic of minerals of the smectite group. The higher order peaks are slightly irrational; the coefficient of variation between the 001, 002, 003, and 005 in sample MJ91-L34c is 0.83%, and between the 001, 002, 003, 004, and 005 in sample MJ91-L42b it is 1.45%. According to the criteria of Bailey (1982), values greater than 0.75% define a structure sufficiently aperiodic to be classified as a mixed-layer mineral; the slight aperiodicity of these samples indicates interlayering with only a minor proportion of nonsmectic layers. Both kaolinite and halloysite are possible candidates for mixed layering; either could explain the slight discrepancies from rationality. Formamide solvation affects the 15-Å (air dried) peak by expanding it to 19 to 22 Å (Fig. 5a,b), the exact d-spacing varying between samples and depending on interlayer ion exchange (Mg saturation). Such variability suggests that the interlayering component in the mixed layer phase is halloysite, because a kaolinitic component should be unaffected by formamide-solvation and because the smectitic phase appears to be unaffected by Mg saturation unless it is formamide solvated. The $\Delta\ 2\theta$ test advocated by Moore and Reynolds (1989) indicates that the smectite component makes up more than 95% of the mixed layer phase in both samples.

Several samples were found to consist almost entirely of this dominantly smectite phase. Sample MJ91-L37b, from a side slope in the easternmost part of the study area, is nearly monomineralic smectite, also showing nearly rational peak behavior (coefficient of variation, 1.79%). The major element composition of this sample (Table 2) was used to arrive at the following structural formula following the procedure of Ross and Hendricks (1945) as outlined in Moore and Reynolds (1989):

$$(Na_{0.31}Ca_{0.08}K_{0.01})\ (Al_{0.93}Fe^{+3}_{0.76}Mg_{0.41})\ (Si_{3.65}Al_{0.35})\ O_{10}\ (OH)_2.$$

Ti and Mn were assigned to minor impurities not detectable by x-ray diffraction and were excluded from the smectite formula. All Fe is assigned to the trivalent state, a reasonable assumption

[1]Use of commercial names is for identification purposes only, and does not imply endorsement by the U.S. Geological Survey.

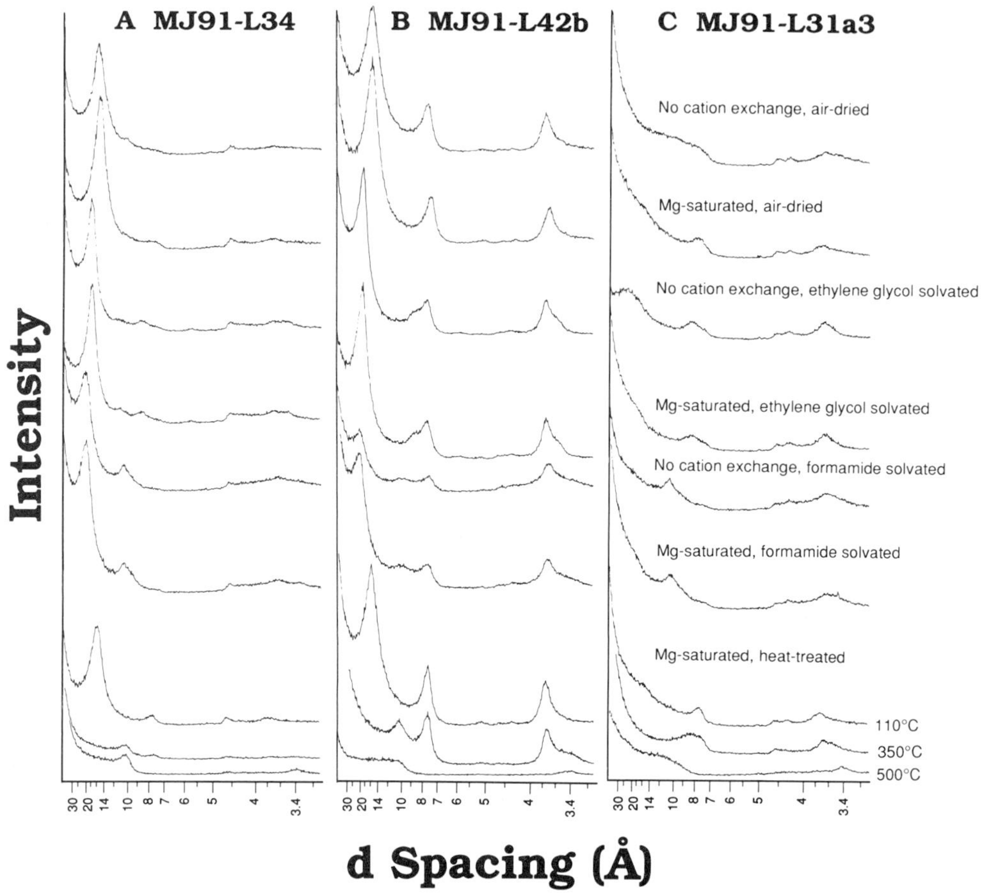

Figure 5. Sample x-ray diffraction patterns, corresponding to typical soils from drier (A, B) and wetter (C) parts of the study area. B is from an intervalley facet; A and C are from younger soils. Interpretation of clay mineralogy was through response of Mg-saturated and unexchanged clays to ethylene-glycol solvation, formamide-solvation, and heat treatment.

for a neoformed clay in a well-aerated soil. This formula represents a dioctahedral smectite of the montmorillonite-beidellite-nontronite series, containing tetrahedral Al intermediate between montmorillonite and beidellite, and octahedral Fe and Mg intermediate between montmorillonite-beidellite and nontronite. The preponderance of Na as the exchangeable cation in the interlayer sites probably reflects marine salt input to these soils; Ca greatly exceeds Na in the basaltic source rocks (Table 1).

The samples in Figure 5A,B also contain peaks near the positions expected for kaolinite. These reflections are far more prominent in sample MJ91-L42b, where they occur at 7.38 and 3.59 Å on the air-dried, unexchanged pattern (Fig. 5B). On Mg saturation, these d-spacings decrease slightly, to 7.20 and 3.55 Å. The peaks are little affected by ethylene-glycol solvation or formamide solvation. They sharpen somewhat with mild (110°C) heat treatment, show no further modification at 350°C, and disappear at 500°C. This behavior is roughly consistent with kaolinite, but the d-spacings are slightly larger than those of pure kaolinite and are slightly irrational, suggesting a small degree of mixed layering. A small proportion of either smectite or halloysite layers could produce the larger spacings observed in the air-dired patterns, but the sensitivity to Mg saturation suggests a halloysitic component, as the nearly pure smectite phase appears to be little affected by Mg saturation (no migration of the 17-Å peak on ethylene glycol-solvated patterns for samples MJ91-L34c and MJ91-L42b). The sharpening of both peaks with mild heat treatment (110°C) is also consistent with collapse of halloysite interlayers, as well as collapse of discrete halloysite in the sample. The $\Delta\ 2\theta$ test of Moore and Reynolds (1989) indicates that the kaolinite component makes up more than 95% of the mixed layer phase.

Discrete halloysite appears to be a minor component in the samples in Figure 5A,B. This phase is revealed only on the formamide-solvated patterns, where broad peaks appear roughly centered on 10 Å, regardless of Mg saturation (Fig. 5a,b). These peaks collapse with mild (110°C) heat treatment, contributing to the 7.5 to 7.3-Å peak, and disappear with that peak at 500°C. The generally poor preferred orientation in all samples, as expressed by the prominent *hkl* reflection present on all patterns at ~4.5 Å and by the low peak-to-background ratio, is also consistent with the presence of halloysite because of its nonplatey habit; indeed, preferred orientation appears to decrease with the propor-

TABLE 2. BULK CHEMICAL COMPOSITION OF REPRESENTATIVE SMECTITE FROM SOUTHEASTERN OAHU

Sample	SiO_2	Al_2O_3	Fe_2O_3*	MgO	CaO	Na_2O	K_2O	MnO	TiO_2	LOI†	Total
(wt.%)											
MJ91-L37b	49.2	14.6	12.7	3.72	0.97	2.17	0.09	0.23	1.80	13.7	99.18

*All Fe reported as Fe^{+3}.
†Loss on ignition.
Note: All analyses reported in weight percent.

tion of the 10 Å (formamide solvated) phase. The formamide-solvated pattern discriminates between discrete halloysite (10Å) and any kaolinitic mixed layer kaolinite/smectite (7.5-7.2 Å).

Sample MJ91-L31a3 is dominated by a quite different phase (Fig. 5C). With no cation-exchange, a broad bank appears on the air-dried pattern near 8 Å. This band can be deconvolved into two interfering peaks centered at roughly 10 and 7.7 Å. On Mg saturation, the 10-Å peak disappears and the 7.7-Å peak is enhanced; a small peak is also present near 15 Å. Ethylene-glycol solvation of the Mg-saturated sample results in a slight increase in the d-spacing of the larger peak, to ~8 Å, and an increase in the 15-Å peak to about 17 Å. If the sample is not Mg saturated, however, and is ethylene-glycol solvated, a prominent broad reflection occurs at 22 Å in addition to the 8-Å peak. On formamide solvation, both Mg-saturated and unexchanged samples show a strong 10-Å peak and weaker peaks near 21 and 7.2 Å. Mild heat treatment (110°C) results in a prominent peak at 7.5 Å, its irrational "multiple" at 3.6 Å, and a subdued reflection at 15 Å. Heat treatment to 350°C results in a broad peak at 8 Å and an irrational "multiple" at 3.55 Å. At 500°C, this peak is broadened and reduced in intensity to a broad band centered on ~10 Å. Such complex behavior is not distinctive of any pure mineral phase. This fact, together with the irrationality of apparent multiple reflections, indicates that the phase is mixed layer. The strong response to formamide solvation and Mg saturation, together with the lack of preferred orientation, indicates that the phase is primarily halloysitic.

Scanning electron microscopy shows that the bulk of this sample is composed of spheroids 2 to 3 μm in diameter (Fig. 6), a typical habit of halloysite. The spheroids have complexly crenulated surfaces (Fig. 6), as might be expected from repeated hydration and desiccation in a seasonally wet soil. The sample appears to contain both dehydrated (7 Å) and hydrated (10 Å) halloysite varieties, whose 001 reflections interfere on the air-dried diffraction profile to give the broad band centered near 8 Å; on formamide-solvation, both expand to 10 Å (Fig. 5C). The response of this peak to ethylene glycol solvation indicates that smectite is a component of the mixed layering; such a smectite component would also explain the persistence of the phase to 500°C. The prominent broad 22-Å peak that occurs when the sample is not Mg saturated and is ethylene-glycol solvated may indicate a partially ordered structure to the mixed layer halloysite/smectite. A 27-Å superstructure should exist in a perfectly ordered halloysite/smectite (ethylene-glycol solvated). If this structure is randomly interlayered with additional smectite layers, an interference should occur between the ordered halloysite/smectite peak and the smectite 001. Depending on the relative proportions of the ordered mixed layer halloysite/smectite and the smectite components, a peak near 22 Å might result. This partially ordered phase would collapse if the 10-Å halloysite collapsed to 7 Å through Mg saturation, destroying the superstructure reflection. Thus, our interpretation of this phase is that it

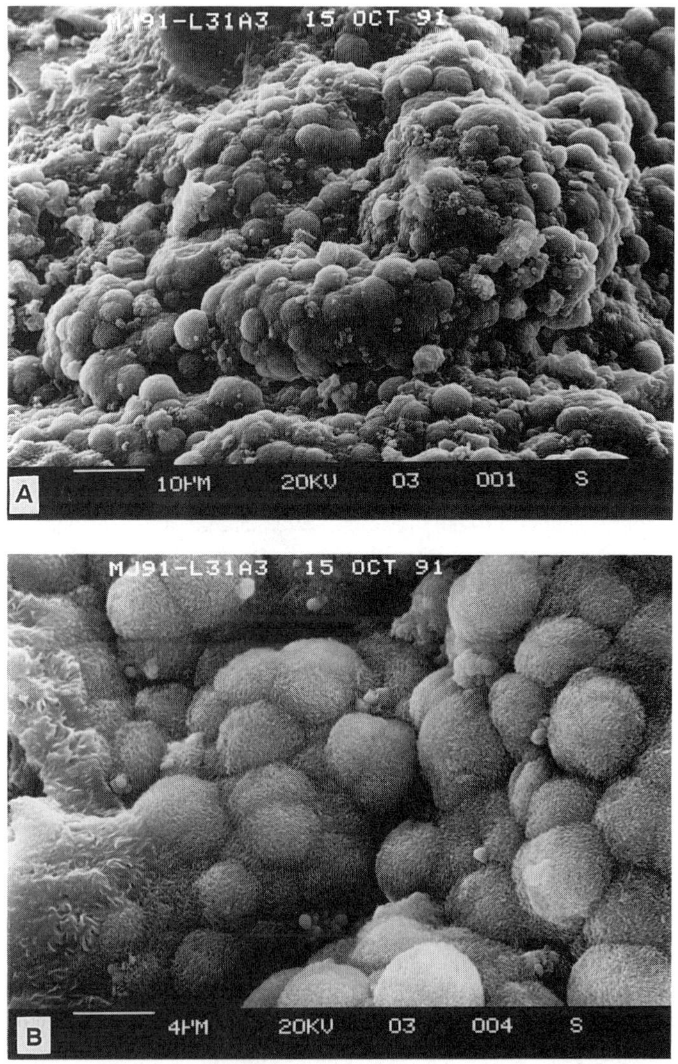

Figure 6. Scanning electron photomicrographs of sample MJ91-L31a3 (cf. Fig. 5C) from colluvium in North Halawa Valley (Fig. 1), showing halloysite in spheroidal habit. Note desiccation wrinkles on spheroid surfaces. Scale bar in micrometers.

is dominantly halloysitic, but contains a component of smectite layers, some of which are regularly ordered with halloysite layers. Weak peaks at 21 and 7.2 Å on the formamide-solvated patterns indicate the presence of minor amounts of discrete smectite and kaolinite as well.

The term "mixed layer" implies an alternation of unit cells corresponding to various discrete clay species along the *c*-axis of a clay crystallite many unit cells in thickness. Such a model for mixed layer clays, described by MacEwan and Wilson (1989) and adopted by Reynolds and other workers as a basis for the calculation of synthetic x-ray diffraction patterns (Reynolds and Hower, 1970; Reynolds, 1985) adequately explains observed x-ray diffraction patterns and provides a useful conceptual basis for mixed layering in clay minerals. Nadeau and Wilson (Nadeau et al., 1984; Wilson and Nadeau, 1985), however, have since shown high-resolution transmission electron microscopy that mixed layer clays generally do not consist of large crystallites, but of smaller "fundamental particles" only one or two unit cells in thickness, so that the layer sequence detected by x-ray diffraction is not fixed within a large crystallite but results instead from interparticle scattering

The bulk of the crystalline <2-μm fraction of all samples examined consists of varying porportions of the dominantly smectitic, kaolinitic, and halloysitic phases. Although complexly mixed layered, each appears to be near an end member phase in composition and structure; thus throughout the remainder of this paper the dominantly smectitic, halloysitic, and kaolinitic mixed layer phases will be referred to as smectite, halloysite, and kaolinite, respectively. It must be remembered, however, that few samples contain pure end members.

A few samples (not shown) contain a weak 14-Å peak that is stable on ethylene-glycol or formamide solvation, but collapses to 10 Å with heat treatment to 350°C. This behavior indicates a strong (vermiculite-like) residual lattice charge, and we identify this phase as vermiculite, although the "aluminum-chlorites" of Irion (1984) and hydroxy-interlayered vermiculites show similar behavior.

Many samples contain weak to prominent reflections near 4.82 and 4.18 Å (Fig. 5b,c). These d-spacings correspond to the most prominent lines from gibbsite and goethite, respectively, and probably represent those phases. Cristobalite also has a broad reflection near 4.18 Å, and may be present as well. Other lines that could correspond to any of these phases and confirm these interpretations were not observed, so positive identifications cannot be made. The low intensity of these peaks indicates that these minerals are present in relatively small concentrations in the <2-μm fraction, at least as highly crystalline phases.

Distribution of clay minerals

Surface samples. The dominantly smectitic, halloysitic, and kaolinitic phases are unevenly distributed throughout the study area. Although the data are noisy, several clear correlations can be made. Most significant is a marked variation in clay mineralogy with respect to precipitation (Table 3; Figs. 7-9). The dominant clay in the surface samples, operationally defined as that phase with the greatest 001 peak area, correlates to a large degree with mean annual precipitation (Figs. 7, 8); smectite is dominant only in those areas receiving less than ~2,000 mm of precipitation annually, whereas halloysite and kaolinite are dominant elsewhere. Kaolinite is generally more abundant than halloysite in the wettest areas, but this correlation is weaker. Gibbsite and goethite are present in most samples, but are present only in trace amounts in samples from drier areas; gibbsite is only abundant in areas where mean annual precipitation exceeds ~2,500 mm. A few samples are composed dominantly of gibbsite and goethite; these are all from areas receiving more than 3,000 mm of precipitation annually. A 14-Å phase, probably vermiculite, is found in only a few samples from the wettest areas. Quartz is not abundant in any sample, but is probably present in a few samples from the wettest areas. Where mean annual precipitation exceeds ~3,000 mm, no correlation exists between precipitation and the composition of the clay fraction; all samples from areas where mean annual precipitation exceeds 3,000 mm are dominated by kaolinite, halloysite, gibbsite, and goethite.

The second principal correlation is between clay mineralogy and soil age. Soils from the intervalley facets are compositionally distinct from soils from younger landscapes in the same rainfall regime. Because of their location at the lower ends of ridges, intervalley facets are found only in the drier parts of the study area (mean annual precipitation, $<2,000$ mm). Like other samples from drier areas, they contain smectite (Table 3). Unlike samples from younger parts of the landscape, however, they contain abundant kaolinite as well, and have high ratios of kaolinite to halloysite (Fig. 10). Kaolinite crystallinity, measured by x-ray diffraction peak width at half-maximum, is markedly greater in soils from the intervalley facets than in younger soils. Samples from younger landscapes that plot near the samples from older geomorphic surfaces in Figure 10 are, interestingly, all colluvium from lower side slopes of ridges. Their compositional similarity to the samples from the older geomorphic surfaces may reflect downslope movement of older soils. In few of these samples, however, is the kaolinite 001 peak the dominant reflection as it is in most of the soils from intervalley facets.

Although clay mineralogy correlates most strongly with precipitation and soil age, colluvium tends to be more leached than residuum at any particular location. Typically, colluvium is poorer in smectite and shows a higher ratio of kaolinite to halloysite than underlying in situ residuum. We saw no consistent variation in clay mineralogy among samples with respect to position in the landscape (ridge crest, side slope, or valley bottom; Table 3) nor with respect to parent material (dense flow rock or clinker).

Profile samples. The soil profiles examined in this study ranged from less than 0.5 to 4.0 m in exposed thickness (Table 4). Also examined were six samples that span an interval of 9 m cored from a borehole in a colluvial apron in Manoa Valley (Fig. 1) that is involved in a slow-moving landslide. None of these profiles extends to unweathered bedrock, so only the upper part

TABLE 3. MEAN CLAY MINERAL COMPOSITION OF UPPERMOST SOIL HORIZONS FROM SOUTHEASTERN OAHU

Precipitation (mm/yr)	Material	Topography	Number	Relative Areas of 001 Reflections Smectite (%)	Halloysite (%)	Kaolinite (%)	Vermiculite (%)
<1,000	Colluvium	All	3	88	7	5	0
<1,000	Colluvium	Ridge crest	1	82	7	11	0
<1,000	Colluvium	Side slope	2	88	7	5	0
1,000-2,000	Younger soils	All	15	64	18	18	0
1,000-2,000	Alluvium	Valley bottom	1	88	8	3	0
1,000-2,000	Colluvium	All	14	62	18	19	0
1,000-2,000	Colluvium	Side slope	11	64	17	18	0
1,000-2,000	Older soil	Intervalley facet	6	25	12	56	6
2,000-3,000	All	All	6	0	54	37	9
2,000-3,000	Colluvium	All	4	0	61	39	0
2,000-3,000	Colluvium	Ridge crest	1	0	84	16	0
2,000-3,000	Colluvium	Side slope	1	0	39	61	0
2,000-3,000	Colluvium	Valley bottom	2	0	61	39	0
2,000-3,000	Residuum	Ridge crest	2	0	41	33	27
3,000-4,000	All	All	9	3	59	38	0
3,000-4,000	Colluvium	All	4	8	50	43	0
3,000-4,000	Colluvium	Side slope	2	0	46	54	0
3,000-4,000	Colluvium	Valley bottom	2	15	54	31	0
3,000-4,000	Residuum	All	5	0	67	33	0
3,000-4,000	Residuum	Ridge crest	4	0	63	37	0
3,000-4,000	Residuum	Side slope	1	0	80	20	0
>4,000	All	All	3	0	54	46	0
>4,000	Colluvium	All	2	0	31	69	0
>4,000	Colluvium	Side slope	1	0	0	100	0
>4,000	Colluvium	Valley bottom	1	0	62	38	0
>4,000	Residuum	Ridge crest	1	0	100	0	0

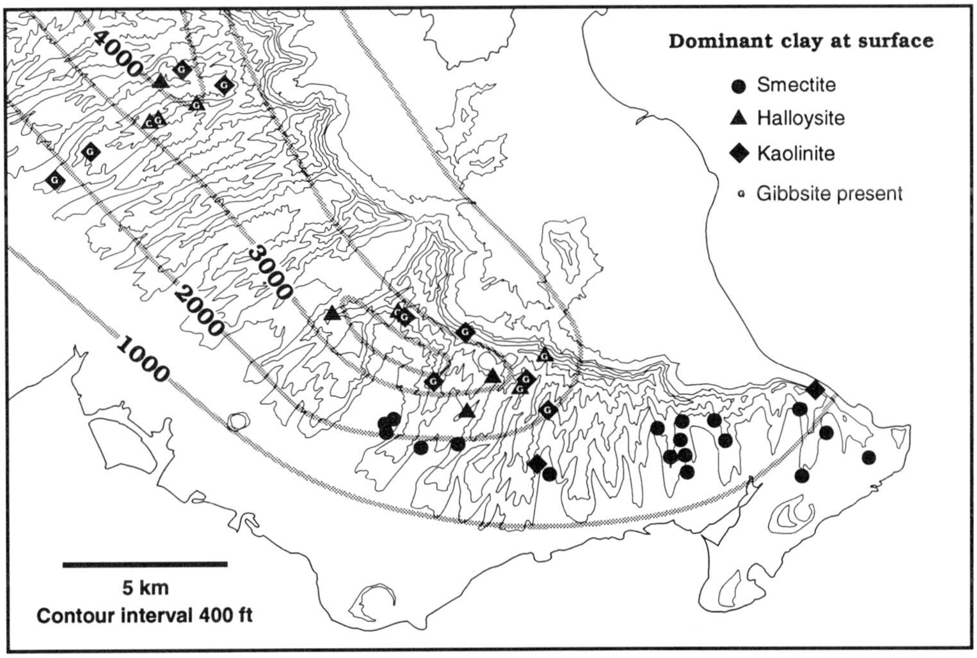

Figure 7. Map of study area showing geographical distribution of dominant clay minerals in uppermost soil level (generally colluvium). "Dominant" clay is operationally defined as that with the largest 001 peak area. Samples with abundant gibbsite are indicated. Isohyets, in millimeters per year, adapted from Giambelluca et al. (1986).

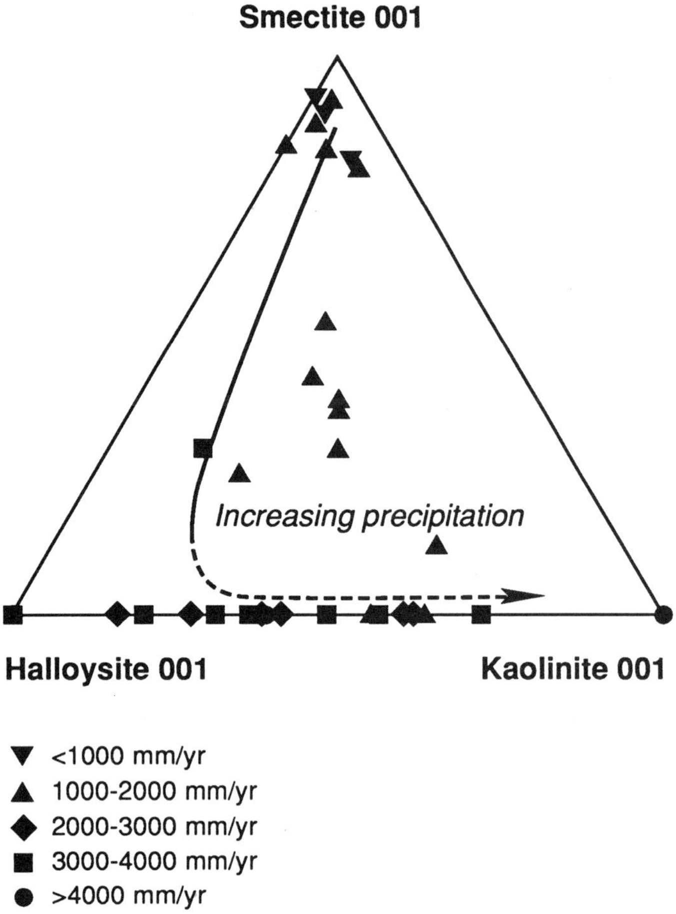

Figure 8. Ternary diagram showing relative 001 peak areas of smectite, halloysite, and kaolinite in samples from uppermost soil levels (generally colluvium). Symbols show mean annual precipitation to highlight correlation between smectite abundance and rainfall.

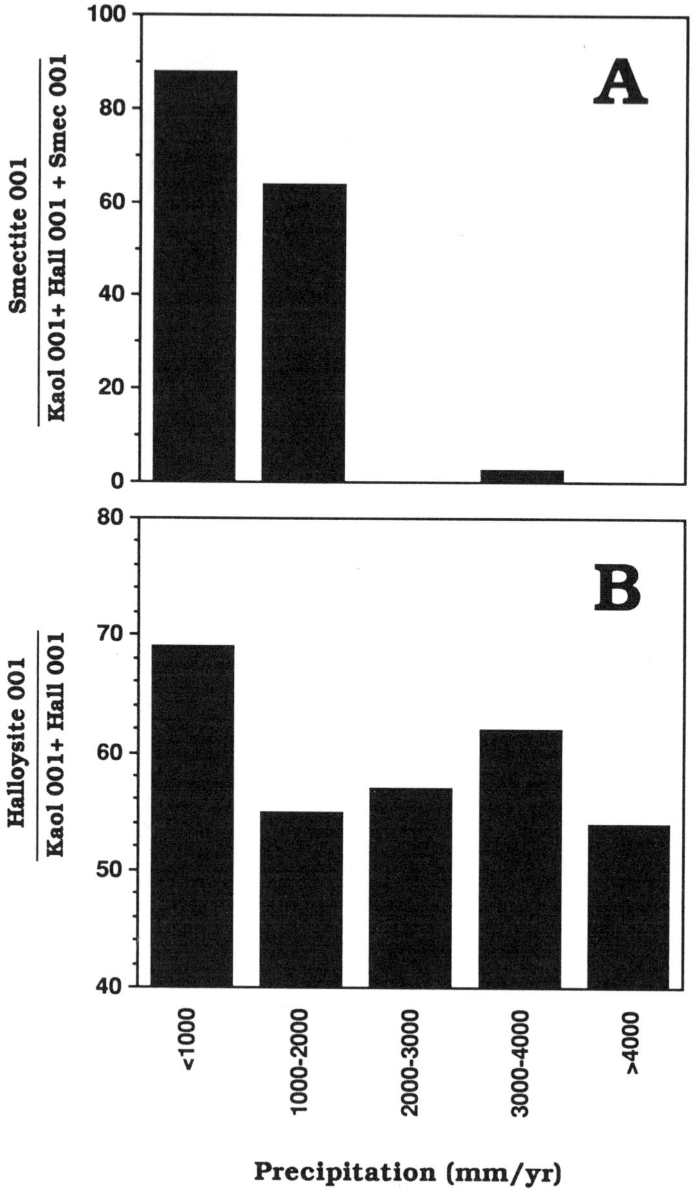

Figure 9. Bar graphs showing correlation of soil clay mineralogy with precipitation. A, Mean smectite 001 peak area relative to mean 001 areas of all clays. B, Mean ratio of halloysite 001 to halloysite 001 + kaolinite 001 peak areas.

of the weathering profile can be examined directly. Samples of unaltered bedrock (Table 1) do, however, give some indication of the composition of parent material.

Three weathering profiles were sampled near the crest of Halawa Ridge (Table 4; Figs. 1, 2C); the most complete of these is shown in Figure 11 (see also Fig. 4B). All three profiles are developed up the ridge from the intervalley facet at the southern end of the ridge, in a region where precipitation exceeds 3,000 mm annually. Clays in each profile consist dominantly of halloysite with generally subordinate kaolinite (Fig. 11A). In addition, gibbsite and goethite are important constituents of all samples, becoming more abundant near the top of the profile. Chemically, each profile is leached of alkali and other mobile cations; SiO_2, Al_2O_3, Fe_2O_3 and TiO_2 make up more than 98% of the nonvolatile component in every sample (Table 4; Fig. 11B). Loss on ignition, which increases markedly in the uppermost samples of each profile, varies from 15 to 36%. Loss on ignition is treated as H_2O in Figure 11 but also reflects abundant organic carbon. The strongest chemical trend in each profile is the increase in H_2O in the uppermost samples. This increase reflects organic material, including leaf and root debris, which is visually more abundant in the upper horizons. The profiles otherwise are quite homogenous.

Two profiles from similar topographic settings only ~300 m apart were sampled to evaluate the effects of bedrock texture on clay mineralogy. The profiles, one developed on dense flow rock and the other on clinker, show little vertical compositional variation and are mineralogically and chemically similar to the profile shown in Figure 11. Both are homogenous, intensely leached, and dominated by halloysite with subordinate kaolinite. No mineralogical or chemical distinction could be made between them.

Although oxide weight percent provides a useful measure of

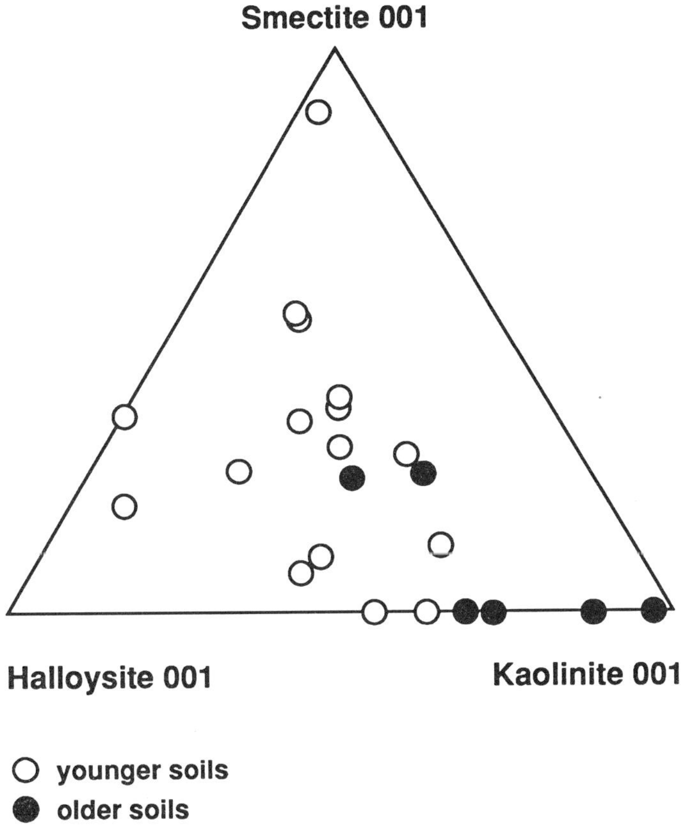

Figure 10. Ternary diagram showing relative 001 peak areas of smectite, halloysite, and kaolinite in old and young soils. Solid symbols, samples collected from intervalley facets; hollow symbols, samples from younger soils developed on ridge crests and side slopes. All samples from regions where mean annual precipitation is between 1,000 and 2,000 mm/yr.

the chemical composition of these samples, ions are not removed during weathering according to weight but according to the stoichiometry of weathering reactions. Accordingly, further insight into those reactions may be gained by replotting the data in Figure 11B in elemental moles/kg (Fig. 11C). Again, the largest variation is the increase in H in the upper fourth of the profile; this increase is largely absorbed by a decrease in Si. More apparent in this presentation, however, is the slight increase in Fe and Ti in the top few centimeters, mirrored by a decrease in Mn and the alkali elements. The composition of the phyllosilicate component of the <2-μm fraction, consisting only of halloysite and kaolinite (Fig. 11A), should contain Al and Si in a 1:1 ratio. The slight Al excess apparent in Figure 11c reflects the presence of the former element in gibbsite or amorphous phases in the bulk sample. K shows an increase at the top of the profile, likely reflecting input from plant debris. Finally, a slight increase in Zr, which can be assumed to behave conservatively (Carroll, 1953; Stallard, 1985), probably indicates progressive mass loss upward through the profile, suggesting that the profile represents a "normal" weathering gradient.

By assuming conservative behavior for Zr, the degree of enrichment of Zr in a sample relative to Zr in the parent material (Zr_o) can be used as a yardstick of weathering progression. Using the mean Zr concentrations of the Koolau lavas presented in Table 1 as representative of Zr_o, Zr/Zr_o generally increases toward the top of this profile (Fig. 11D). In practice, Zr was not analyzed to sufficient precision to be used as a consistent metric, and so a more reliable element must be sought. Ti generally behaves almost as conservatively as Zr in the weathering environment (Stallard, 1985), although the phases in which it is concentrated may change during weathering. Sherman (1952) showed, for example, that anatase is forming in some Hawaiian soils. Ti/Ti_o closely tracks Zr/Zr_o in this and other profiles from the study area (Figs. 11D, 12D, 13D), and the extent to which other components have been lost through weathering can thus be examined by plotting their concentration against Ti/Ti_o (Fig. 11E).

The results show that the profile in Figure 11 is, indeed, fairly uniform. The profile is quite homogenous and intensely leached compared to the source rock (Table 1; Fig. 11D), indicating that we are seeing only a small part of the weathering sequence in this 2-m profile. Disregarding the effects of abundant organic matter in the upper part of the profile, progressive weathering of the younger part of the landscape in a region receiving over 4,000 mm of precipitation annually is manifested primarily in an accumulation of Fe and Ti.

Two soil profiles were examined from the intervalley facet preserved at the base of Halawa Ridge. The profile in Figure 12 is from the northeastern tip of the intervalley facet, on the broad nose of Halawa Ridge in an area receiving just over 2,000 mm of precipitation annually. In contrast to the profiles described above, which formed on a younger, more active part of the landscape, the <2-μm fraction consists dominantly of kaolinite, with subordinate halloysite; area of the kaolinite 001 peak is two to four times the area of the halloysite 001 peak. Kaolinite crystallinity, measured by x-ray diffraction peak width at half-maximum, is markedly greater than that found in the younger and wetter profiles, and gibbsite and goethite are less abundant. Chemically, however, this profile is similar to that from the wetter, younger region higher on the ridge (Fig. 11). It is fairly homogenous, intensely leached (SiO_2, Al_2O_3, Fe_2O_3, and TiO_2 make up more than 99% of the nonvolatile component in every sample) and shows a surface enrichment in Fe and Ti. Less organic matter in the upper part of the profile reduces the abrupt change in the H/Si ratio at the top of the profile, although K still shows a strong increase in the upper meter. Al and Si are in 1:1 proportions throughout the profile (Fig. 12C, E), indicating that amorphous Al and gibbsite are not present. The enrichment in Fe and Ti in the upper 0.75 m is somewhat more pronounced than in the younger profile. Most dramatic, however, are the values of Zr/Zr_o and Ti/TR_o (Fig. 12D). Whereas these quantities indicated an enrichment in these conservative elements varying between 160 and 200% in the younger, wetter profile, they indicate enrichment varying between 200 and 320% in this older profile. Although compositionally similar, nearly twice as much mass has been lost from the older soil than from the younger.

TABLE 4. CHEMICAL COMPOSITION AND CLAY MINERAL COMPOSITION OF SOIL PROFILES FROM SOUTHEASTERN OAHU

Height Above Datum (m)	SiO_2	Al_2O_3	Fe_2O_3	MgO	CaO	Na_2O	K_2O	MnO	TiO_2	LOI	Zr*	Total	Relative Areas of 001 Reflections			
													Smectite (%)	Halloysite (%)	Kaolinite (%)	Vermiculite (%)
	(wt %)										(ppm)					
Profile MJ91-L9; Alluvium in Kaalakei Valley; Precipitation 1,250 mm/yr:																
0.00	41.6	21.1	16.7	1.82	0.33	0.27	<0.01	0.14	3.60	14.50	260	100.1	67	26	8	0
0.70	41.7	20.7	16.4	2.13	0.31	0.35	<0.01	0.14	3.49	14.70	260	99.9				
1.20	42.6	20.4	15.7	2.31	0.33	0.36	0.01	0.20	3.29	14.80	260	100.0	84	11	5	0
1.80	41.9	19.6	15.7	1.97	0.37	0.35	<0.01	0.27	3.27	16.90	260	100.3				
2.25	44.4	19.2	13.9	2.46	0.39	0.44	0.01	0.23	2.85	15.70	230	99.6	88	10	1	0
2.55	43.2	20.1	14.8	2.31	0.38	0.45	0.02	0.16	3.01	15.20	230	99.6				
2.85	43.4	18.0	15.2	2.00	1.87	0.73	0.14	0.25	3.09	14.60	230	99.3	88	8	3	0
3.00	44.2	19.1	15.7	1.94	1.09	0.66	0.10	0.29	3.07	13.80	230	100.0				
Profile MJ91-L11; Residuum developed on a clinker zone on Halawa Ridge; Precipitation >4,000 mm/yr:																
0.00	29.9	27.2	21.3	0.50	0.05	0.04	0.03	0.28	3.74	17.20	270	100.2	0	51	49	0
0.43	24.6	30.7	21.7	0.46	0.04	0.03	0.01	0.26	3.77	18.60	300	100.2				
0.86	29.9	26.8	31.4	0.52	0.04	0.04	0.01	0.32	3.77	16.90	270	99.7	0	76	24	0
1.27	21.4	32.3	21.8	0.95	0.03	0.02	0.01	0.25	3.75	19.50	270	100.0				
1.57	25.6	27.9	21.0	0.55	0.04	0.03	0.01	0.23	3.70	20.70	280	99.8	0	100	0	0
1.80	27.4	26.6	22.0	0.52	0.04	<0.01	<0.01	0.31	3.85	19.00	290	99.7				
1.93	23.5	28.9	20.8	0.52	0.03	0.02	0.02	0.25	3.63	22.70	300	100.4	0	100	0	0
2.21	14.2	27.4	23.9	0.44	0.03	0.03	0.02	0.29	3.81	29.90	330	100.0				
2.33	10.0	20.9	28.0	<0.01	0.09	0.03	0.07	0.16	4.53	36.00	320	99.8				
2.41	10.1	21.0	27.8	0.49	0.07	0.02	0.08	0.15	4.26	36.10	320	100.1				
Profile MJ91-L13; Residuum developed on flow rock on Halawa Ridge; Precipitation ~3,000 mm/yr:																
0.00	33.7	26.4	20.0	0.39	0.04	0.05	0.02	0.33	4.25	14.80	310	100.0	0	77	23	0
0.48	25.4	28.3	21.4	0.30	0.02	0.06	0.02	0.63	4.63	19.00	350	99.8				
0.75	33.2	26.2	20.4	0.38	0.03	0.04	<0.01	0.26	4.39	14.70	340	99.6	0	79	21	0
0.99	32.9	26.4	20.4	0.39	0.03	0.04	<0.01	0.36	4.34	15.40	340	100.3				
1.37	26.0	23.4	18.8	0.48	0.04	0.07	0.01	0.25	4.08	26.80	310	99.9	0	69	31	0
1.47	22.7	22.6	19.0	0.49	0.04	0.04	0.03	0.21	4.12	31.10	310	100.3				
Profile MJ91-L14; Residuum developed on a clinker zone on Halawa Ridge; Precipitation ~3,000 mm/yr:																
0.00	32.0	25.1	23.2	0.51	0.03	<0.01	<0.01	0.23	3.43	15.10	250	99.6	0	89	11	0
0.21	32.7	24.8	22.9	1.01	0.03	0.02	<0.01	0.43	3.35	14.50	240	99.7				
0.46	32.2	24.4	23.1	0.71	0.03	0.02	<0.01	0.44	3.41	15.40	230	99.7	0	100	0	0
0.70	30.4	27.2	21.4	0.55	0.03	0.03	0.01	0.27	4.04	16.10	310	100.0				
Profile MJ91-L15; Top of intervalley facet on Halawa Ridge; Precipitation ~2,000 mm/yr:																
0.00	33.3	27.0	19.5	0.62	0.03	0.05	<0.01	0.08	4.47	15.20	380	100.3	0	34	66	0
0.26	33.2	26.9	19.8	0.68	0.03	0.03	<0.01	0.08	4.53	14.90	390	100.2				
0.64	33.0	26.9	19.9	0.66	0.03	0.02	0.01	0.08	4.53	14.70	440	99.8	0	27	73	0
1.09	33.9	27.2	18.4	0.69	0.04	0.03	<0.01	0.06	4.14	15.70	400	100.2				
1.50	32.4	27.7	19.5	0.43	0.04	0.02	0.04	0.06	3.75	16.20	420	100.1	0	19	78	3
1.61	32.5	27.3	19.4	0.53	0.04	0.04	0.06	0.06	3.75	16.70	400	100.4				
178	30.0	26.1	20.0	0.38	0.03	0.06	0.10	0.49	3.96	18.40	350	99.5	0	17	74	9
2.09	25.7	23.5	24.2	0.45	0.02	0.06	0.19	0.07	4.78	20.80	370	99.8				
2.34	23.1	15.8	31.5	0.57	0.05	0.06	0.42	0.09	7.43	20.90	440	99.9	0	23	51	26
Profile Mj91-L43; Base of intervalley facet on Halawa Ridge; Precipitation ~2,000 mm/yr:																
0.00	37.0	25.7	18.3	0.88	0.03	0.02	0.01	0.04	3.90	14.20	310	100.1				
0.66	36.1	20.9	22.9	1.37	0.06	0.04	0.01	0.08	5.15	13.20	400	99.8	0	3	97	0
1.98	34.5	25.6	21.0	0.65	0.04	0.07	0.03	0.04	4.47	13.40	350	99.8				
3.30	321.2	23.5	26.5	0.42	0.05	0.03	0.10	0.03	5.12	12.90	420	99.9	0	29	71	0
4.00	30.5	23.5	26.0	0.41	0.05	0.06	0.17	0.03	5.16	14.10	390	100.0				

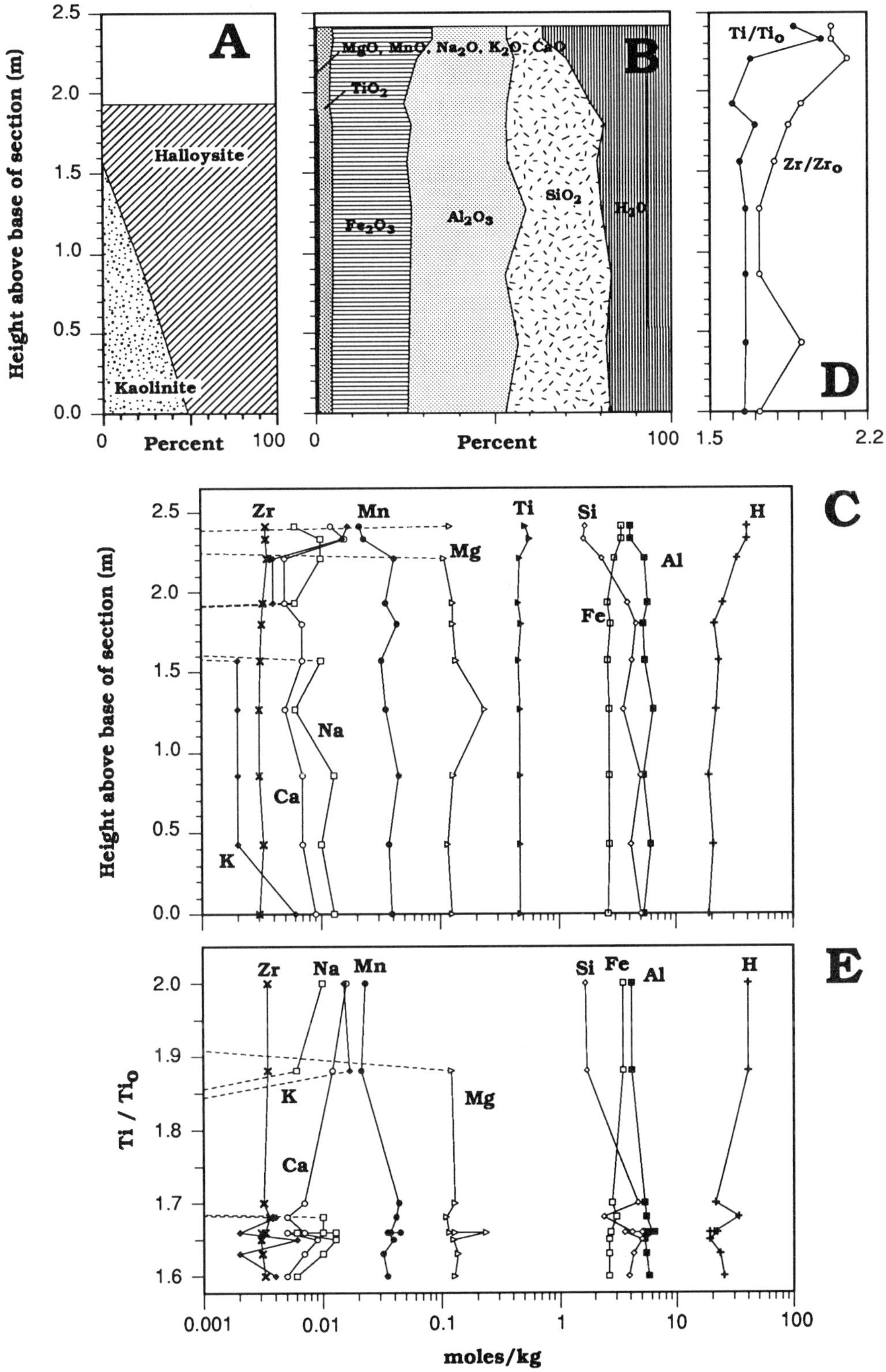

Figure 11. Clay mineralogy (A) and geochemistry (B-E) of soil profile MJ91-L11 (Table 4), from the crest of Halawa Ridge in a region receiving more than 4,000 mm precipitation annually. In A through D, quantities plotted against position in profile; in E, elemental concentration plotted against ratio of Ti in sample to Ti in unaltered Koolau Basalt (see Table 1).

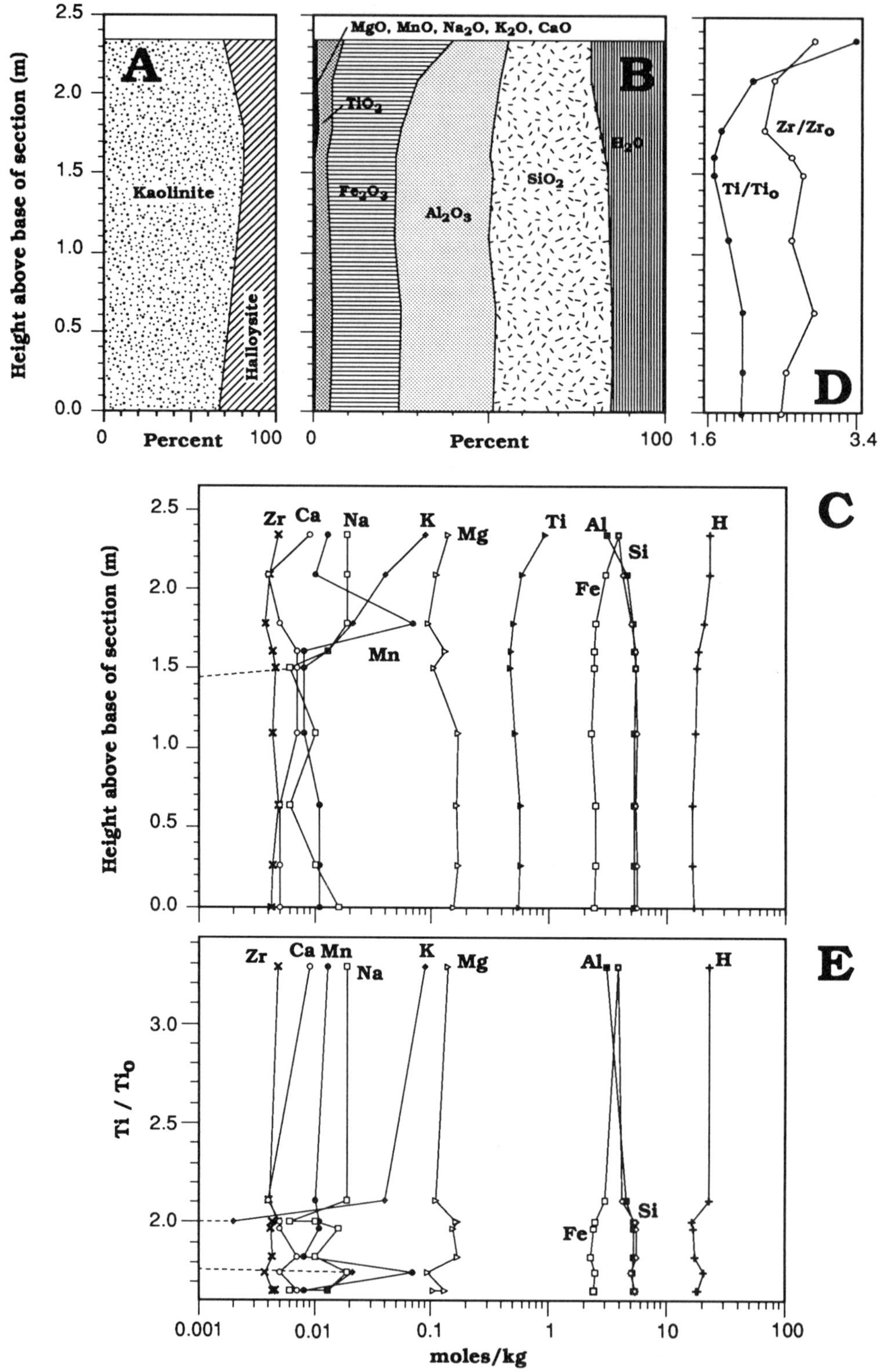

Figure 12. Clay mineralogy (A) and geochemistry (B-E) of soil profile MJ91-L15 (Table 4), from intervalley facet on lower part of Halawa Ridge (Figs. 1, 2) in a region receiving about 2,000 mm precipitation annually. In A through D, quantities plotted against position in profile; in E, elemental concentration plotted against ratio of Ti in sample to Ti in unaltered Koolau Basalt (see Table 1).

A profile in the "older alluvium" of Stearns (1939) was collected from Kaalakei Valley to investigate the effects of a weathering front superimposed on a stratigraphic alluvial sequence (Fig. 13). The profile, located in a stream cut at valley bottom, is in a region receiving about 1,250 mm of precipitation annually. The <2-μm fraction of this profile is dominated by smectite, although the combined 001 peak areas of halloysite and kaolinite are up to 33% of the area of the smectite 001 (Fig. 13A). Gibbsite and goethite are present, but only in trace amounts. The proportion of smectite increases upward in the profile, and there is a slight increase in the ratio of the halloysite to kaolinite 001 peak areas (Fig. 13A).

This profile is chemically distinct from those developed in wetter areas. Mg, Mn, and the alkali ions are more abundant, together making up 4 to 5% of most samples; of these ions, Mg is by far the most abundant. The concentration of these components, expressed either as oxide weight percent (Fig. 13B) are remarkably constant except for an enrichment in Ca and K in the upper fourth of the profile. The overall extent of weathering is greater at the base of the profile than at the top, as measured by Zr/Zr_0 and Ti/Ti_0 (Fig. 13D). Enrichment over bedrock concentrations of these conservative elements varies between about 130% at the top of the profile to nearly 160% at the base, indicating slightly greater weathering of older alluvium at the base of the profile than depositionally younger material at the top.

A slow-moving landslide in Manoa Valley occurs in a colluvial apron at the base of the western slope of Waahila Ridge in a region receiving about 1,750 mm of precipitation annually (Figs. 1, 2). A series of samples from a boring through this landslide shows that the <2-μm fraction of the colluvium composing the landslide is dominated by smectite and halloysite (Fig. 14). Smectite is not as abundant as in samples from drier areas, and increases irregularly with respect to halloysite with depth in the boring.

DISCUSSION

Intensity and duration of weathering

The strong correlation of clay composition with precipitation suggests that the intensity of the chemical weathering environment controls the composition of soil clays in the study area. Normalizing for the "state factors" of topography and parent material, we find precipitation, vegetation, and duration of weathering to be the principal determinants of soil clay composition in the study area. Precipitation and vegetation are strongly correlated and accordingly inseparable in the present study. Both contribute to intensity of chemical weathering; precipitation through facilitation of hydrolysis reactions, and vegetation both by introduction of organic acids and by anchoring soils (with their interstitial waters) to weathering bedrock surfaces. Duration of weathering is controlled by age of the soil and by the sample position with respect to a passing weathering front, and controls the extent to which weathering reactions are allowed to proceed.

Structurally complex three-layer clays rich in soluble cations are developed only in the drier parts of the study area; smectite is found only where mean annual precipitation is below 2,000 mm. Other researchers studying tropical weathering environments have noticed similar correlation of smectite with precipitation (Quantin et al., 1965; Barshad, 1966; Kantor and Schwertmann, 1974). Although smectite is generally believed to be favored in poorly drained soils, on Oahu as in Panama (Johnsson and Stallard, 1989), smectite appears to be the dominant clay mineral even in well-drained soils developed on steep slopes. In areas of greater rainfall, soils are extensively leached and cation-depleted phases such as gibbsite, kaolinite, and halloysite are formed. The correlation of leached, cation-depleted clay minerals to precipitation has been reported from elsewhere in Hawaii (Wentworth et al., 1940; Sherman, 1952; Tamura et al., 1953; Bates, 1962; Sherman et al., 1962, 1967; Hay and Jones, 1972; Wada et al., 1990) and in other tropical weathering environments (Barshad, 1966; Isphording, 1979; Singer, 1984; Curtis, 1990), but the distribution of clay minerals in soils of the Koolau Mountains is especially striking given their great variability over small distances and the relative homogeneity of the parent material.

Quartz is not present as a primary mineral in true basalts, and is exceedingly rare in any rocks in Hawaii (Macdonald et al., 1983). It is, however, deposited tropospherically in Hawaiian soils (Rex et al., 1969), and may be concentrated in soils in wetter areas (Jackson et al., 1971). In this study, we noticed the occurrence of clay-size quartz in soils. Clay-size quartz is not abundant in any sample, and is found only in samples from areas where precipitation exceeds ~3,000 mm annually. Jackson et al. (1971) attributed the correlation of tropospherically deposited quartz (in all size fractions) with precipitation in part to the removal of less stable phases through weathering, but also to greater flushing of dust from the atmosphere in regions of greater precipitation.

In any given rainfall regime, colluvium tends to be enriched in cation-poor phases and show a higher kaolinite/halloysite ratio than underlying residuum. Although this may reflect only greater leaching toward the ground surface, the abrupt nature of the compositional variation suggests that other processes may be at work. Colluvium generally contains more vegetal material than underlying residuum, and its depletion in soluble cations may reflect their removal through organic hydrolysis (Hem and Lind, 1974; Mast and Drever, 1987). In addition, colluvium may largely be made up of previously weathered material; in situ residuum, in contrast, is by definition freshly derived from unweathered bedrock.

The effects of weathering duration on soil clay composition can be evaluated by examining soils developed on geomorphic surfaces of different ages, by evaluation of soil profiles, or by evaluating stratigraphic trends in alluvial soils. Although smectite is abundant in samples from older geomorphic surfaces that occupy relatively dry parts of the study area, kaolinite is the dominant phase. This is in marked contrast to soils from younger parts of the landscape in the same rainfall regime, but also unlike the extremely leached soils in high rainfall regimes that similarly are

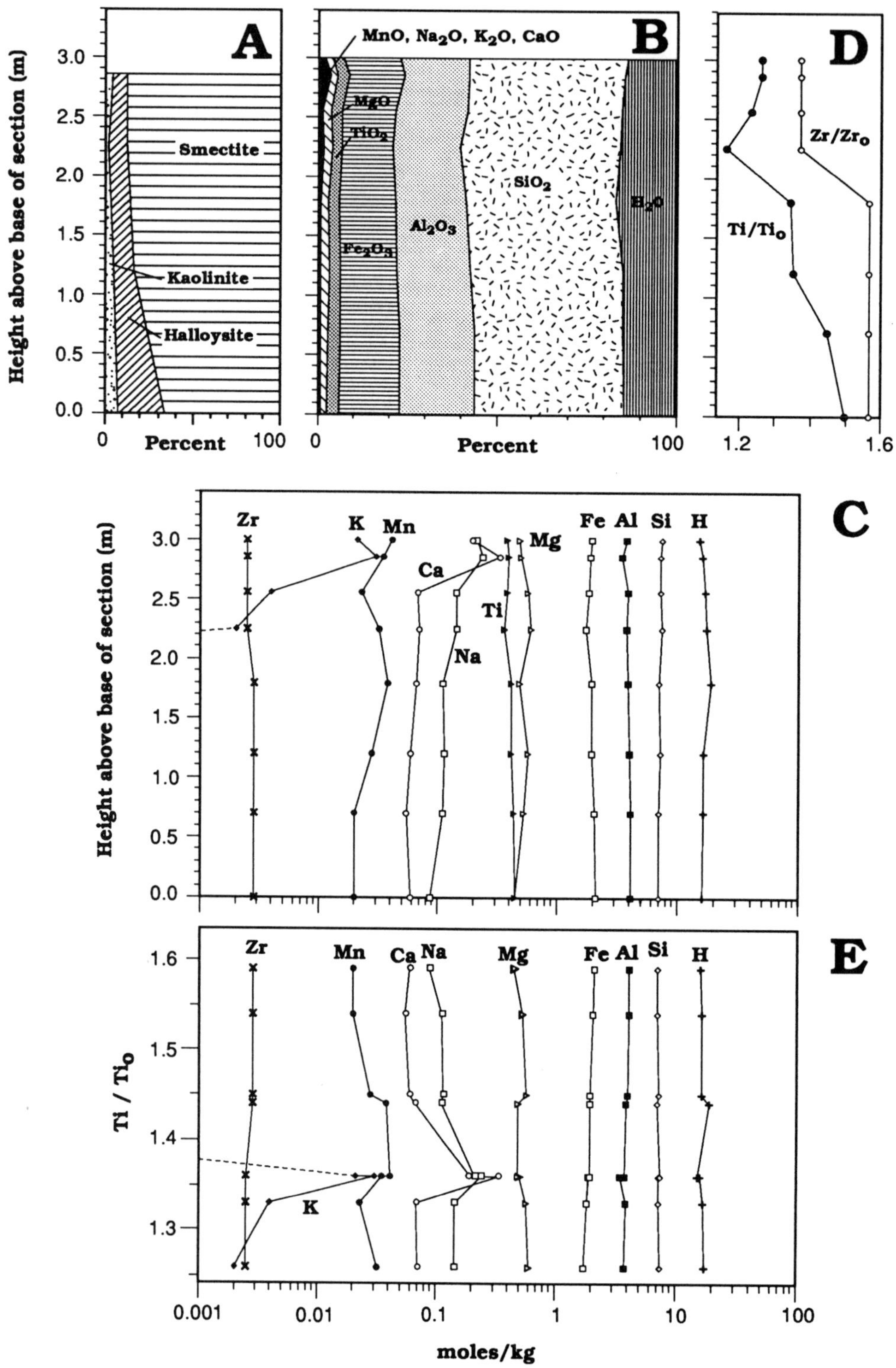

Figure 13. Clay mineralogy (A) and geochemistry (B-E) of soil profile MJ91-L9 (Table 4), from alluvial sequence in Kaalakei Valley (Fig. 2) in a region receiving about 1,250 mm of precipitation annually. In A through D, quantities plotted against position in profile; in E, elemental concentration plotted against ratio of Ti in sample to Ti in unaltered Koolau Basalt (see Table 1). Because this ratio increases slightly *down* the profile, trends shown in C are inverted.

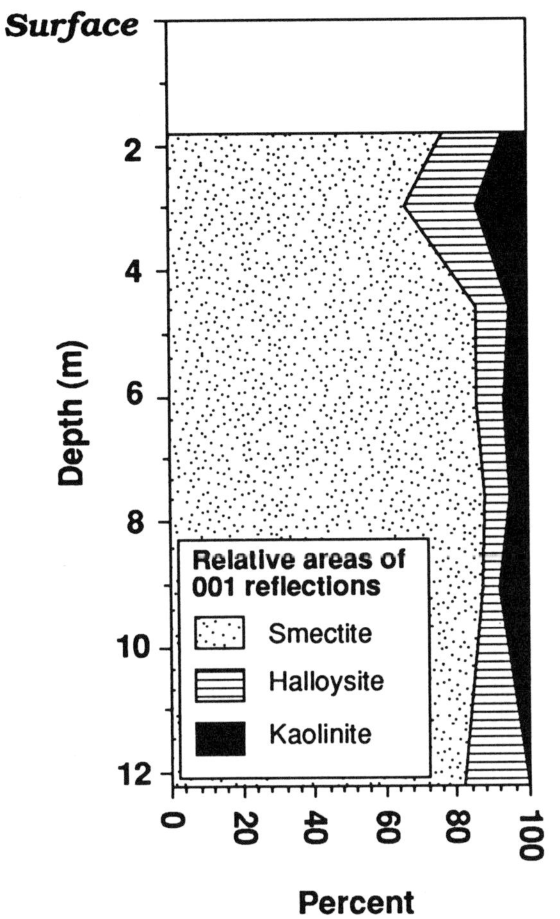

Figure 14. Relative 001 peak areas of clay minerals in colluvium from drill core through slow-moving landslide in Manoa Valley (Figs. 1, 2). Slight decrease in smectite abundance at top of sampled interval may represent a weathering front on the sliding colluvial deposit.

dominated by kaolinite but contain no smectite. Furthermore, kaolinite from older geomorphic surfaces tends to be more crystalline than that from younger landscapes. These observations suggest that kaolinite may be formed through progressive leaching of cation-rich clays such as smectite, but that smectite continues to be produced as fresh bedrock is weathered. Extensive pisolitic horizons or laterites, well-developed elsewhere in Hawaii (Sherman, 1949; Sherman et al., 1953, 1962, 1969; Sherman and Kanehiro, 1954; Walker et al., 1969), do not form on the intervalley facets in the study area, perhaps indicating that the extreme leaching necessary to form laterites requires either greater age or wetter conditions than are found on the intervalley facets.

Most of the weathering profiles examined, although recording a wide variation in overall weathering intensity as recorded by Zr/Zr_0 and Ti/Ti_0, show relatively little mineralogical or chemical variation with depth. It appears that the profiles we examined, although different from each other, record only a small part of the weathering front. An exception is the profile from alluvial deposits in Kaalakei Valley (Fig. 13), which shows a *decrease* in the extent of weathering toward the surface, as measured by Zr/Zr_0 and Ti/Ti_0, overall chemical composition, and clay mineralogy. The greater age of the material in the lower part of the depositional sequence, perhaps coupled with changes in deposition rate, has allowed time for weathering to progress further than it has in material in the upper part of the profile, resulting in an increase in cation-depleted clays in the lower part of the profile. A similar inversion of the weathering profile was observed in alluvial deposits of the Amazon by Johnsson and Meade (1990).

Weathering series

The relationship between precipitation and clay composition indicates that the intensity of chemical weathering is a prime control on the composition of secondary weathering products. What is not clear from those data alone, however, is whether bedrock alters directly to different products in different rainfall regimes, or whether chemical weathering proceeds at different rates but along the same reaction pathways in the different rainfall regimes, as suggested by Carroll and Hathaway (1963). Two lines of evidence suggest that, on Oahu, the latter is the case.

First, a structural continuum appears to exist between the clay species found in the Koolau soils; all phases are complexly mixed layers to varying degrees. Herbillon et al. (1981) found that halloysite or metahalloysite in Burundi was actually a mixed-layer kaolinite/smectite, and represented a transitional stage in the alteration of smectite to kaolinite. Other workers have reported mixed-layer kaolinite/smectites from similar weathering environments (Wilson and Cradwick, 1972; Herbillon et al., 1981; Norrish and Pikering, 1983; Robinson and Wright, 1987), and there is evidence that mixed-layer kaolinite/smectite may form through the alteration of montmorillonite under acidic weathering conditions (Altschuler et al., 1963; Srodon, 1980). Kaolinite has been shown to replace halloysite and be the dominant phase in basaltic weathering environments where there is no clear dry season (Quantin et al., 1965; Kantor and Schwertmann, 1974), as is the case in the high rainfall regions of our study area. Because the mixed layering observed by x-ray diffraction likely reflects only interparticle scattering, rather than interlayering of large numbers of silicate sheets (Nadeau et al., 1984; Wilson and Nadeau, 1985), the alteration of smectite to a mixed-layer halloysite/smectite may not involve the destruction of large silicate lattices, but simply the conversion of some proportion of smectite fundamental particles to halloysite. The same arguments apply to the halloysite-kaolinite transition.

Second, soils of different ages from the drier parts of the study area contain each of the principal clay phases (Table 3). Younger soils are dominated by smectite, but contain some halloysite and kaolinite; older soils consist of kaolinite with subordinate halloysite and smectite. The presence of smectite in even the oldest soils suggests that weathering of fresh material continues to produce smectite in these older soils.

Because we relied heavily on x-ray diffraction in this study, we did not observe amorphous phases that may be present in these soils. It is likely that the more crystalline weathering products are preceded or accompanied by precipitation of allophane and imogolite, as reported by Tamura et al. (1953) and Wada et al. (1990). These phases are quickly reorganized into smectites, which may initially be fairly pure phases, as are those from the driest parts of our study area. With continued weathering, the degree of mixed layering with a halloysitic component increases (Schmehl and Jackson, 1956; Tamura, 1956; Schultz et al., 1971; Yerima et al., 1985; Robinson and Wright, 1987). As soluble ions continue to be removed from the soil, a true halloysite may form, which is eventually reorganized into poorly crystalline kaolinite. This sequence is similar to that inferred from basaltic soils in Guam by Carroll and Hathaway (1963). On Oahu, longer weathering appears to result in an increase in kaolinite crystallinity.

Clay mineralogy and slope stability

Smectite and, to a lesser degree, halloysite, are expandable clays that are characterized by low frictional strength and associated low slope stability (Borchardt, 1977; Selby, 1982). Although soil avalanches have occurred on steep slopes throughout the study area, only in drier areas have landslides been observed on gentle slopes; these landslides are slow moving. The distribution of soil clays in relation to weathering environment provides a conceptual framework for evaluating susceptibility to slow-moving landslides. A line separating soils containing smectite from those without smectite (Fig. 15A) also serves to delineate the region where slow-moving landslides have been observed (Fig. 15B). This correlation suggests that smectite facilitates this kind of slope failure, and that the chemical weathering environment exerts a significant control on the distribution of these landslides. Furthermore, these results suggest that clay mineralogy can be used to identify parts of the study area where gentle slopes are susceptible to slow-moving landslides. If these relations are valid, then side slopes in areas receiving less than about 2,000 mm of precipitation annually are most susceptible to slow-moving landslides. Intervalley facets in the same rainfall regime, in contrast, should be less susceptible.

CONCLUSIONS

1. Soil clays in the southeastern Koolau Mountains are mineralogically complex. Pure smectite and kaolinite occur in the youngest and oldest soils, respectively, but most soils contain phases showing complex interstratification between smectite and halloysite, between kaolinite and halloysite, and between halloysite, smectite, and kaolinite. A structural continuum exists between smectite, halloysite, and kaolinite. Nevertheless, most phases encountered are stoichiometrically near one of these three end members.

2. The composition of soil clays in the southeastern Koolau Mountains correlates with annual precipitation. Precipitation would seem to affect the intensity of the chemical weathering environment because higher precipitation results in greater water-mineral contact time and greater flushing, thus increasing the intensity of chemical weathering. Smectite is the dominant soil clay in the driest parts of the study area, but becomes less abundant as rainfall increases, and is not present in soils receiving more than ~2,000 mm of precipitation annually. Halloysite and kaolinite are found throughout the study area but are most abundant in the wettest areas. The ratio of kaolinite to halloysite may increase slightly with precipitation. Gibbsite is abundant and tropospherically derived quartz is present in the $<2\text{-}\mu m$ fraction of soils where precipitation exceeds ~3,000 mm annually. Colluvium is generally more leached (higher in halloysite and kaolinite, higher kaolinite/halloysite ratio) than underlying in situ residuum at the same location.

3. The composition of soil clays in the southeastern Koolau Mountains correlates with the duration of chemical weathering. Normalized for precipitation, soils developed on older geomorphic surfaces are more leached than those from younger parts of the landscape. In parts of the study area where mean annual precipitation does not exceed ~2,000 mm, intervalley facets are dominated by kaolinite, with minor smectite and halloysite. In contrast, younger soils are dominated by smectite, with only minor halloysite and kaolinite.

4. Normalized for precipitation, age, and nature of the soil (colluvium versus residuum), soils in different topographic positions (ridge crest, side slopes, and valley bottoms) have similar compositions. Similarly normalized, no significant compositional differences were observed among soils developed on dense flow rock versus clinker.

5. The structural continuity among clay phases and the correlation of clay mineralogy with age suggest a weathering sequence in which bedrock alters first to smectite (possibly through some preliminary amorphous phase), through halloysite, to kaolinite and gibbsite. This sequence proceeds at different speeds depending on weathering intensity, and the 1.8 m.y. that have elapsed since the onset of weathering has allowed the sequence to reach different stages in different weathering regimes. Thus in dry areas only smectite is developed on younger landscapes, but kaolinite is dominant on older geomorphic surfaces. In wetter areas, leached soils dominated by kaolinite and gibbsite have formed in even the youngest soils.

6. Slow-moving landslides on gentle slopes have been documented only in parts of the study area where soils contain smectite. This relation suggests that smectite, which is characterized by low frictional strength, facilitates these landslides, and that its presence can be used to evaluate susceptibility of slopes to slow-moving landslides. Side slopes in areas receiving less than about 2,000 mm of precipitation annually are most susceptible to slow-moving landslides, intervalley facets in the same rainfall regime are less susceptible.

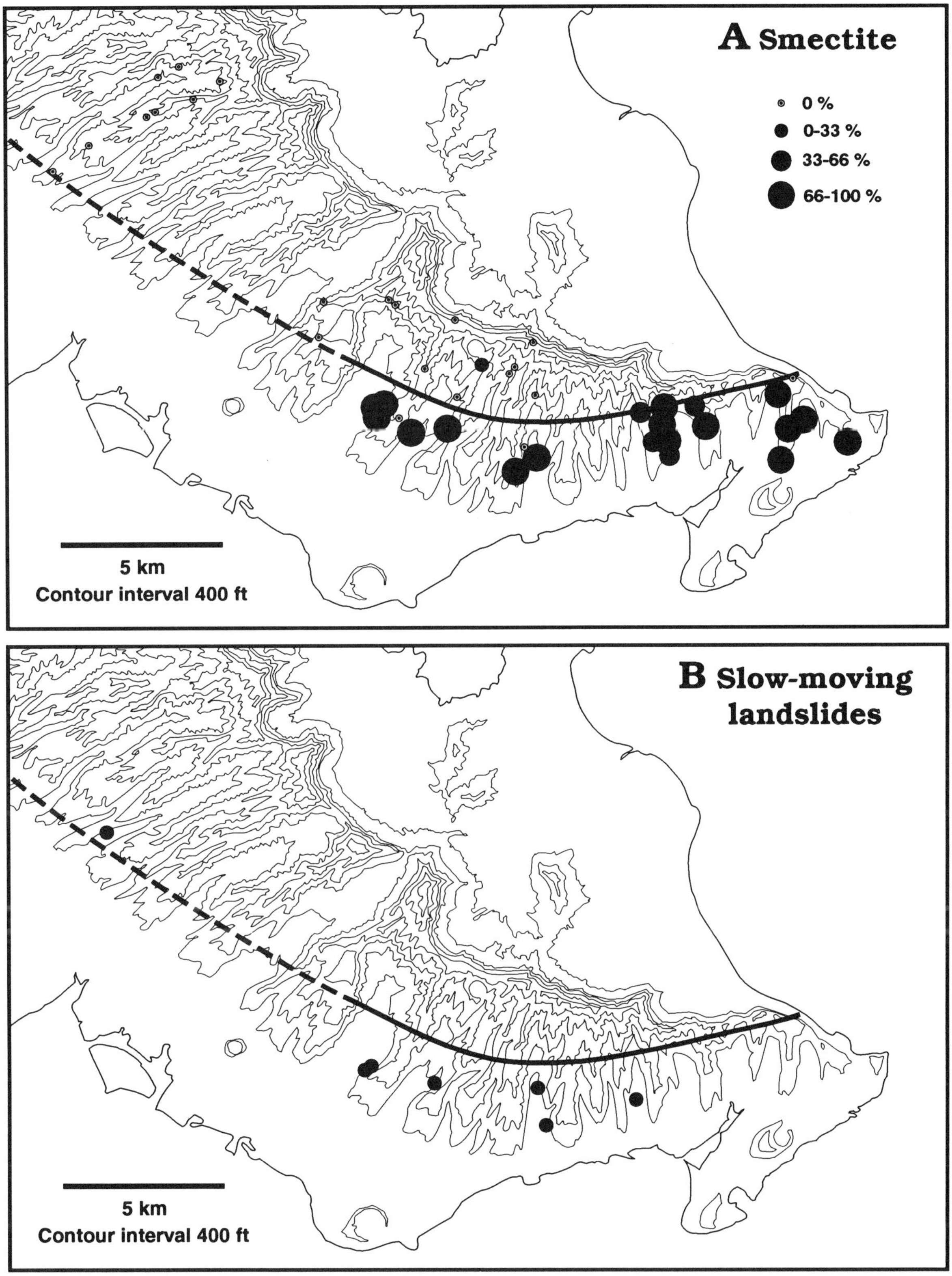

Figure 15. Maps of study area. A, Distribution of smectite in uppermost soil levels (generally colluvium), as expressed by area of smectite 001 relative to other 001 clay reflections. Thick line approximately delineates soils containing *any* smectite from soils without smectite. B, Distribution of known slow-moving landslides (dots) with respect to thick line from A.

ACKNOWLEDGMENTS

This research was supported in part by the City and County of Honolulu as part of a U.S. Geological Survey project on landslide and debris-flow hazards. We thank the Queen Emma Foundation, the Bishop Estates, and the Hawaiian Cement Company for allowing us access to their properties. B. R. Hill and M. F. Wong assisted in the field, and C. F. Hamilton assisted with sample preparation. We extend special thanks to R. H. Mariner, for access to his diffractometer and his generous donation of machine time, and to R. L. Oscarson, who contributed his expertise on the scanning electron microscope. Thanks to R. M. Pollastro, M. L. Cummings, and R. K. Gabor, who made available their results from earlier studies of soil clays on Hawaii, and who provided insights into the behavior of the halloysitic clays. Many of the ideas presented here grew out of discussions with R. W. Fleming, B. R. Hill, M. E. Reid, and H. J. Lee. The manuscript benefited from critical reviews by R. April, A. Basu, J. W. Harden, B. R. Hill, and D. W. Ming.

REFERENCES CITED

Altschuler, Z. S., Dwornik, E. J., and Kramer, H., 1963, Transformation of montmorillonite to kaolinite during weathering: Science, v. 141, p. 148–152.

Bailey, S. W., 1982, Nomenclature for regular interstratifications: American Mineralogist, v. 67, p. 394–398.

Barshad, I., 1966, The effect of a variation in precipitation on the nature of clay mineral formation in soils from acid and basic igneous rocks, *in* International Clay Conference: Jerusalem, Israel, p. 167–173.

Bates, T. F., 1962, Halloysite and gibbsite formation in Hawaii: Clays and Clay Minerals, v. 9, p. 315–328.

Baum, R. L., and Fleming, R. W., 1991, Use of longitudinal strain in identifying driving and resisting elements of landslides: Geological Society of America Bulletin, v. 103, p. 1121–1132.

Baum, R. L., Fleming, R. W., and Ellen, S. D., 1989, Maps showing landslide features and related ground deformation in the Woodlawn area of the Manoa Valley, City and County of Honolulu, Hawaii: U.S. Geological Survey Open-File Report 89-290, 16 p.

Blumenstock, D. I., and Price, S., 1967, Climates of the states: Hawaii: U.S. Department of Commerce, Environmental Science Services Administration, Environmental Data Service, Climatography of the United States, Report 60-51, 27 p.

Borchardt, G. A., 1977, Clay mineralogy and slope stability: California Division of Mines and Geology Special Report 133, 15 p.

Carroll, D., 1953, Weatherability of zircon: Journal of Sedimentary Petrology, v. 23, p. 106–116.

Carroll, D., and Hathaway, J. C., 1963, Mineralogy of selected soils from Guam: U.S. Geological Survey Professional Paper 403-F, 51 p.

Churchman, G. J., Whitton, J. S., Claridge, G.G.C., and Theng, B.K.G., 1984, Intercalation method using formamide for differentiating halloysite from kaolinite: Clays and Clay Minerals, v. 32, p. 241–248.

Clague, D. A., and Dalrymple, G. B., 1987, The Hawaiian-Emperor volcanic chain, Part I. Geologic Evolution, *in* Decker, R. W., Wright, T. L., and Stauffer, P. H., eds., Volcanism in Hawaii: U.S. Geological Survey Professional Paper 1350, p. 5–54.

Clague, D. A., Holcomb, R. T., Sinton, J. M., Detrick, R. S., and Torresan, M. E., 1990, Pliocene and Pleistocene alkalic flood basalts on the seafloor north of the Hawaiian Islands: Earth and Planetary Science Letters, v. 98, p. 175–191.

Cuddihy, L. W., and Stone, C. P., 1990, Alteration of native Hawaiian vegetation: Effects of humans, their activities and introductions: Honolulu, University of Hawaii Press, 138 p.

Curtis, C. D., 1990, Aspects of climatic influence on the clay mineralogy and geochemistry of soils, paleosols and clastic sedimentary rocks: Journal of the Geological Society of London, v. 147, p. 351–357.

De Silva, G.L.R., 1974, Slope stability problems induced by human modification of soil covered hill slopes of Oahu, Hawaii [Ph.D. thesis]: Honolulu, University of Hawaii, 471 p.

Doell, R. R., and Dalrymple, G. B., 1973, Potassium-argon and paleomagnetism of the Waianae and Koolau Volcanic Series, Oahu, Hawaii: Geological Society of America Bulletin, v. 84, p. 1217–1242.

Dracup, J. A., Cheng, E.D.H., Nigg, J. M., and Schroeder, T. A., 1991, The New Year's Eve flood on Oahu, Hawaii, December 31, 1987–January 1, 1988: National Research Council Natural Diaster Studies 1, 72 p.

Drever, J. I., 1973, The preparation of oriented clay mineral specimens for x-ray diffraction analysis by a filter-membrane peel technique: American Mineralogist, v. 58, p. 553–554.

Ellen, S. D., Iverson, R. M., and Pierson, T. C., 1991, Map showing the distribution of debris flows during the New Year's Eve storm of 1987–1988 in southeastern Oahu, Hawaii: U.S. Geological Survey Open-File Report 91-129, 1 sheet, scale 1:20,000.

Giambelluca, T. W., Nullet, M. A., and Schroeder, T. A., 1986, Rainfall Atlas of Hawaii: State of Hawaii, Department of Land and Natural Resources Report 76, 267 p.

Hay, R. L., and Jones, B. F., 1972, Weathering of basaltic tephra on the island of Hawaii: Geological Society of America Bulletin, v. 83, p. 317–332.

Hem, J. D., and Lind, C. J., 1974, Kaolinite synthesis at 25°C: Science, v. 184, p. 1171–1173.

Herbillon, A. J., Frankart, R., and Vielvoye, L., 1981, An occurrence of interstratified kaolinite-smectite minerals in a red-black soil toposequence: Clay Minerals, v. 16, p. 195–201.

Irion, G., 1984, Clay minerals of the Amazonian soils, *in* Sioli, H., ed., The Amazon: Limnology and landscape ecology of a mighty tropical river and its basin: Dordrecht, The Netherlands, W. Junk Publishers, p. 537–579.

Isphording, W. C., 1979, Chemical differentiation of temperate and tropical limestone derived clays: Transactions—Gulf Coast Association of Geological Societies, v. 29, p. 252–256.

Jackson, M. L., and 6 others, 1971, Geomorphological relationships of tropospherically derived quartz in the soils of the Hawaiian Islands: Soil Science Society of America Proceedings, v. 35, p. 515–525.

Jenny, H., 1941, Factors of soil formation: New York, McGraw-Hill, 281 p.

Jenny, H., 1980, The soil resource: New York, Springer-Verlag, 377 p.

Johnsson, M. J., 1990, Tectonic versus chemical-weathering controls on the composition of fluvial sands in tropical environments: Sedimentology, v. 37, p. 713–726.

Johnsson, M. J., and Meade, R. H., 1990, Chemical weathering of fluvial sediments during alluvial storage: The Macuapanim Island point bar, Solimões River, Brazil: Journal of Sedimentary Petrology, v. 60, p. 827–842.

Johnsson, M. J., and Stallard, R. F., 1989, Physiographic controls on the composition of sediments derived from volcanic and sedimentary terrains on Barro Colorado Island, Panama: Journal of Sedimentary Petrology, v. 59, p. 768–781.

Johnsson, M. J., Stallard, R. F., and Lundberg, N., 1991, Controls on the composition of fluvial sands from a tropical weathering environment: Sands of the

Orinoco River drainage basin, Venezuela and Colombia: Geological Society of America Bulletin, v. 103, p. 1622–1647.

Kantor, W., and Schwertmann, U., 1947, Mineralogy and genesis of clays in red-black soil topsequences on basic igneous rocks in Kenya: Journal of Soil Science, v. 25, p. 67–78.

Kirch, P. V., 1982, The impact of prehistoric Polynesians on the Hawaiian ecosystem: Pacific Science, v. 36, p. 1–14.

Krook, L., 1969, Investigations on the mineralogical composition of the Tertiary and Quaternary sands in northern Surinam: Verhandelingen van het Koninklijke Nederlandse Geologische Mijnbouwkundig Genootschap, v. 27, p. 89–100.

Langenheim, V.A.M., and Clague, D. A., 1987, The Hawaiian-Emperor volcanic chain, Part II. Stratigraphic framework of volcanic rocks of the Hawaiian Islands, *in* Decker, R. W., Wright, T. L., and Stauffer, P. H., eds., Volcanism in Hawaii: U.S. Geological Survey Professional Paper 1350, p. 55–84.

Macdonald, G. A., Abbot, A. T., and Peterson, F. L., 1983, Volcanoes in the sea: Honolulu, University of Hawaii Press, 517 p.

MacEwan, D.M.C., and Wilson, M. J., 1980, Interlayer and intercalation complexes of clay minerals, *in* Brindley, G. W., and Brown, G., eds., Crystal structure of clay minerals and their X-ray identification: London, Mineralogical Society, p. 197–248.

Mast, M. A., and Drever, J. I., 1987, The effect of oxalate on the dissolution rates of oligoclase and tremolite: Geochimica et Cosmochimica Acta, v. 51, p. 2559–2568.

Mink, J. F., 1960, Distribution pattern of rainfall in the leeward Koolau Mountains, Oahu, Hawaii: Journal of Geophysical Research, v. 65, p. 2869–2876.

Mink, J. F., 1962, Rainfall and runoff in the leeward Koolau Mountains, Oahu, Hawaii: Pacific Science, v. 16, p. 147–159.

Moore, D. M., and Reynolds, R. C., Jr., 1989, X-ray diffraction and the identification and analysis of clay minerals: Oxford, Oxford University Press, 332 p.

Nadeau, P. H., Wilson, M. J., McHardy, W. J., and Tait, J. M., 1984, Interstratified clays as fundamental particles: Science, v. 225, p. 923–925.

Norrish, K., and Pikering, J. G., 1983, Clay minerals, *in* Soils: An Australian perspective: London, Academic Press, p. 281–308.

Peck, R. B., 1967, Stability of natural slopes: Journal of the Soil Mechanics and Foundations Division, Proceedings of the American Society of Civil Engineers, v. 93, p. 403–417.

Peck, R. B., and Deere, D. U., 1960, Investigation of landslides for planning and remedial measures: AREA Bulletin, v. 556, p. 670–677.

Quantin, P., Badeaut-Trauth, D., and Weber, F., 1965, Mise en évidence de minéraux secondaires, argiles et hydroxydes, dans les andesols des Nouvelles-Hébrides, après la déferrification par la méthode de Endredy: Bulletin Group Français Argiles, v. 27, p. 51–67.

Rex, R. W., Syers, J. K., Jackson, M. L., and Clayton, R. N., 1969, Eolian origin of quartz in soils of Hawaiian Islands and in Pacific pelagic sediments: Science, v. 163, p. 277–279.

Reynolds, R. C., Jr., 1985, NEWMOD, A computer program for the calculation of one-dimensional diffraction patterns of mixed-layer clays: Hanover, NH, R. C. Reynolds.

Reynolds, R. C., Jr., and Hower, J., 1970, The nature of interlayering in mixed-layer illite-montmorillonites: Clays and Clay Minerals, v. 18, p. 25–36.

Ripperton, J. C., and Hosaka, E. Y., 1942, Vegetation zones of Hawaii: Hawaiian Agricultural Experiment Station Bulletin 89, 60 p.

Robinson, D., and Wright, V. P., 1987, Ordered illite-smectite and kaolinite-smectite: Pedogenic minerals in a Lower Carboniferous paleosol sequence, South Wales?: Clay Minerals, v. 22, p. 109–118.

Rolfe, B. N., and Hadley, R. F., 1964, Weathering and transport of sediment in the Cheyenne River basin, eastern Wyoming, *in* Bradley, W. F., ed., Clays and clay minerals: Proceedings of the 12th National Conference on Clays and Clay Minerals: New York, Macmillan, p. 649–670.

Ross, R. S., and Hendricks, S. B., 1945, Minerals of the montmorillonite group: U.S. Geological Survey Professional Paper 205-B, p. 23–79.

Ruhe, R. V., 1964, An estimate of paleoclimate in Oahu, Hawaii: American Journal of Science, v. 262, p. 1098–1115.

Savage, K. M., De Cesero, P., and Potter, P. E., 1988, Mineralogic maturity of modern sand along a high-energy tropical coast, Baixada de Jacarepaguá, Rio de Janeiro, Brazil: Journal of South American Earth Sciences, v. 1, p. 317–328.

Schmehl, W. R., and Jackson, M. L., 1956, Interstratification of layer silicates in two soil clays, *in* Clays and clay minerals, Proceedings of the 4th National Conference on Clays and Clay Minerals: University Park, Pennsylvania State University, Committee on Clay Minerals of the National Academy of Sciences, p. 423–428.

Schultz, L. G., Shepard, A. O., Blackmon, P. D., and Starkey, H. C., 1971, Mixed-layer kaolinite-montmorillonite from the Yucatan Peninsula, Mexico: Clays and Clay Minerals, v. 19, p. 137–150.

Selby, M. J., 1982, Hillslope materials and processes: Oxford, Oxford University Press, 264 p.

Sherman, G. D., 1949, Factors influencing the development of lateritic and laterite soils in the Hawaiian Islands: Pacific Science, p. 307–314.

Sherman, G. D., 1952, The titanium content of Hawaiian soils and its significance: Soil Science Society Proceedings, p. 15–18.

Sherman, G. D., and Kanehiro, Y., 1954, Origin and development of ferruginous concretions in Hawaiian latosols: Soil Science, v. 77, p. 1–8.

Sherman, G. D., Kanehiro, Y., and Matsusaka, Y., 1953, The role of dehydration in the development of laterite: Pacific Science, v. 7, p. 438–446.

Sherman, G. D., Ikawa, H., Uehara, G., and Okazaki, E., 1962, Types of occurrence of nontronite and nontronite-like minerals in soils: Pacific Science, v. 16, p. 57–62.

Sherman, G. D., Cady, J. G., Ikawa, H., and Blomberg, N. E., 1967, Genesis of the bauxitic Halii soils: Hawaii Agricultural Experimental Station Technical Bulletin 56, 46 p.

Sherman, G. D., Ikawa, H., and Matsusaka, Y., 1969, Aluminous-ferruginous oxide mineral nodules in tropical soils: Pacific Science, v. 23, p. 115–122.

Singer, A., 1984, The paleoclimatic interpretation of clay minerals in sediments—A review: Earth-Science Reviews,v. 21, p. 251–293.

Srodon, J., 1980, Precise identification of illite smectite interstratifications by X-ray powder diffraction: Clays and Clay Minerals, v. 28, p. 401–411.

Stallard, R. F., 1985, River chemistry, geology, geomorphology, and soils in the Amazon and Orinoco basins, *in* Drever, J. I., ed., The chemistry of weathering: Dordrecht, D. Reidel, p. 293–316.

State of Hawaii, 1988, Post flood report, New Year's Eve storm, December 31, 1987-January 1, 1988, windward and leeward east Oahu: State of Hawaii, Department of Land and Natural Resources Circular C119, 55 p.

Stearns, H. T., 1939, Geologic map and guide of the island of Oahu, Hawaii: Hawaii Division of Hydrography Bulletin 2, 75 p.

Stearns, H. T., and Vaksvik, K. N., 1935, Geology and ground-water resources of the island of Oahu, Hawaii: Hawaii Division of Hydrography Bulletin 1, 474 p.

Stidd, C. K., and Leopold, L. B., 1951, The geographic distribution of average monthly rainfall, Hawaii: Meteorological Monographs, v. 1, p. 24–33.

Taliaferro, W. J., 1959, Rainfall of the Hawaiian Islands: Honolulu, Hawaii Water Authority, 394 p.

Tamura, T., 1956, Weathering of mixed-layer clays in soils, *in* Clays and Clay Minerals, Proceedings of the 4th National Conference on Clays and Clay Minerals: University Park, Pennsylvania State University, Committee on Clay Minerals of the National Academy of Sciences, p. 413–422.

Tamura, T., Jackson, M. L., and Sherman, G. D., 1953, Mineral content of low humic, humic and hydrol humic latosols in Hawaii: Soil Science Society Proceedings, p. 343–346.

Wada, K., Kakuto, Y., and Ikawa, H., 1990, Clay minerals of two eutrandepts of Hawaii, having isohyperthermic temperature and ustic moisture regimes: Soil Science Society of America Journal, v. 54, p. 1173–1178.

Walker, J. L., Sherman, G. D., and Katsura, T., 1969, The iron and titanium minerals in the titaniferous ferruginous latosols of Hawaii: Pacific Science, v. 23, p. 291–304.

Wentworth, C. K., 1943, Soil avalanches on Oahu, Hawaii: Geological Society of America Bulletin, v. 54, p. 53–64.

Wentworth, C. K., and Winchell, H., 1947, Koolau basalt series, Oahu, Hawaii:

Geological Society of America Bulletin, v. 58, p. 49–78.

Wentworth, C. K., Wells, R. C., and Allen, V. T., 1940, Ceramic clay in Hawaii: American Mineralogist, v. 25, p. 2–33.

Wilson, M. J.,and Cradwick, P. D., 1972, Occurrence of interstratified kaolinite-montmorillonite in some Scottish soils: Clay Mineralogy, v. 9, p. 435–437.

Wilson, M. J., and Nadeau, P. H., 1985, Interstratified clay minerals and weathering processes, *in* Drever, J. I., ed., The chemistry of weathering: Dordrecht, D. Reidel, p. 97–118.

Winchell, H., 1947, Honolulu Series, Oahu, Hawaii: Geological Society of America Bulletin, v. 56, p. 1–48.

Yerima, B.P.K., Calhoun, F. G., Senkayi, A. L., and Dixon, J. B., 1985, Occurrence of interstratified kaolinite-smectite in El Salvador vertisols: Soil Science Society of America Journal, v. 49, p. 462–466.

MANUSCRIPT ACCEPTED BY THE SOCIETY DECEMBER 24, 1992

Printed in U.S.A.

Geological Society of America
Special Paper 284
1993

Factors influencing the composition of detrital heavy mineral suites in Holocene sands of the Apure River drainage basin, Venezuela

Andrew C. Morton
British Geological Survey, Keyworth, Nottingham NG12 5GG, United Kingdom
Mark J. Johnsson*
U.S. Geological Survey, 345 Middlefield Road, MS 999, Menlo Park, California 94025

ABSTRACT

Heavy mineral assemblages in rivers in the Apure River drainage basin of Venezuela and Colombia closely reflect the nature of the source regions, which lie in the Andean orogenic terranes to the west and northwest. The Caribbean Mountains, largely composed of greenschist-facies pelites, phyllites, carbonates, and metavolcanics, supply assemblages dominated by epidote and calcic amphibole. Minor amounts of the high-pressure index minerals glaucophane and lawsonite indicate the presence of blueschist-facies rocks, reflecting the origin of the Caribbean Mountains by subduction-related tectonism. The northern Mérida Andes, which comprise basement gneisses and granites overlain by unmetamorphosed to low-grade metamorphosed clastics, supply two types of assemblage reflecting these two lithological types: garnet-sillimanite-staurolite-amphibole suites from the basement rocks, and epidote-amphibole suites from the overlying cover sequence. The southern Mérida Andes supply stable heavy mineral suites reflecting recycling from the extensive unmetamorphosed sandstones that occur at outcrop. By considering suites from different physiographical provinces, the effects of short-term alluvial storage in the Llanos on heavy mineral assemblages have been evaluated. Weathering during alluvial storage appears to be effective in modifying the apatite-tourmaline ratio, which shows a steady, marked decline with distance from the mountain front, resulting from the removal of apatite during weathering. Clinopyroxene and garnet may also show evidence of loss through weathering, although the trends are poorly constrained statistically. Epidote and amphibole proportions remain essentially constant, possibly through a balance between loss through weathering and continual resupply from the breakdown of rock fragments. In general, the heavy mineral assemblages are less affected than the bulk mineralogy by alluvial storage on the Llanos.

INTRODUCTION

Although heavy minerals are accurate and sensitive indicators of the mineralogical composition of source regions that supply sand-sized detritus, the compositions of heavy mineral assemblages are affected by a variety of other processes that operate during the sedimentary cycle (Morton, 1985). These include weathering in the source area (prior to incorporation in the transport system), the effects of the transport process, the hydraulic conditions at the time of deposition, and the diagenetic processes that operate postdepositionally. These processes in combination may generate heavy mineral suites that are considerably less diverse compared with those present in the source rocks, causing difficulty in provenance reconstruction.

It is therefore important to understand the possible roles that each of these factors may play in generating the composition of heavy mineral assemblages. The effects of hydraulics have been extensively studied and, although complex, are now fairly well understood (see the review by Mange and Maurer, 1992). Recently there have been several well-constrained studies on the

*Present address: Department of Geology, Bryn Mawr College, Bryn Mawr, Pennsylvania 19010.

Morton, A. C., and Johnsson, M. J., 1993, Factors influencing the composition of detrital heavy mineral suites in Holocene sands of the Apure River drainage basin, Venezuela, *in* Johnsson, M. J., and Basu, A., eds., Processes Controlling the Composition of Clastic Sediments: Boulder, Colorado, Geological Society of America Special Paper 284.

effects of diagenesis, particularly on the effects of high-temperature pore fluids (Morton, 1984; Hansley, 1987; Milliken, 1988). The effects of weathering of source material on heavy mineral suites prior to their inclusion in the transport system (that is, in soil profiles) have been studied in some detail (see summary by Mitchell, 1975), with a number of heavy minerals showing marked depletion with increasing weathering. Inevitably, therefore, the transfer from source rock to transport system must involve a reduction in abundance of the more easily weathered species, particularly under conditions of intense weathering in low-relief areas, but this has yet to be quantified in any case study. However, in a gross sense the process does not appear to cause a marked decrease in mineral diversity, as present-day river sediments from all climatic regimes contain the same overall variety of heavy minerals as are found in the source rocks.

The effects of the factors operative during sediment transport are perhaps the least understood. Heavy mineral assemblages may be affected by three processes during transport: mechanical abrasion, hydraulic sorting, and chemical weathering. Abrasion during transport has the potential to destroy mechanically unstable heavy minerals, as shown by a number of experimental studies on the relative resistance of minerals to abrasion (Friese, 1931; Thiel, 1940, 1945; Dietz, 1973). However, studies of natural transport systems indicate that, in practice, although grains tend to become more rounded, individual mineral species are not entirely lost from assemblages by mechanical abrasion: there is no evidence, for example, for any downstream decrease in mineral diversity along major river systems such as the Mississippi (Russell, 1937), the Nile (Shukri, 1949) and the Rhine (van Andel, 1950). A recent study of the Cascade River in Westland, New Zealand, found no evidence for loss of olivine, experimentally determined to be one of the least mechanically resistant minerals, within the river system or in the high-energy beach zone (Morton and Smale, 1991). Therefore, the effects of abrasion on heavy mineral proportions during transport may be regarded as negligible. However, the possibility that abrasion may have a significant effect in exceptional circumstances should not be discounted. Although hydraulic processes control the absolute amount of heavy minerals carried by the transport system and affect the relative abundance of minerals with different hydraulic behavior, in general they cannot be held responsible for the absence of particular minerals from assemblages, and certainly not for their absence from the transport system as a whole.

Therefore, chemical weathering during fluvial transport and short-term storage has the greatest potential for reducing heavy mineral diversity within the transport system. Recent studies of fluvial systems in tropical areas have shown that this process has a marked effect on bulk sediment composition (Johnsson et al., 1988, 1991; Johnsson and Meade, 1990; Savage and Potter, 1991), causing immature lithic arenites to change into quartz arenites. This clearly introduces significant problems in the use of bulk sediment composition for provenance reconstruction, and offers an opportunity for examining the response of heavy mineral suites to the same process. This paper records the results of investigations into the effect of chemical weathering on heavy mineral suites in samples whose bulk mineralogy has been affected by this process.

The Llanos region of central Venezuela and Colombia (Fig. 1) provides an ideal setting to evaluate the effects of short-term alluvial storage and recycling on sediment composition. Johnsson et al. (1988, 1991) noted that storage of sediment on the broad alluvial plains of the region provides sufficient time for chemical weathering to modify significantly the composition of stored alluvium. Point-bar growth, overbank deposition, and reworking of the floodplain results in replacement of newly deposited detritus by older and compositionally more mature material. Rivers thus incorporate into their bed loads a progressively larger component of older, more chemically weathered alluvium as they cross the Llanos region. This process is reflected in the overall composition of bed load sands by the loss of unstable components with increasing distance from the mountain front. In this study, we investigate the effects of this process on the distribution of heavy minerals in sands of the Llanos region.

REGIONAL SETTING

The samples described here were collected primarily from the drainage basin of the Apure River (Fig. 1). The drainage basin comprises two principal physiographic provinces, the orogenic belt of the Mérida Andes and Caribbean Mountains, and the Llanos, or Andean foreland basin, a broad region of coalescing alluvial plains to the south and east of the orogenic belt. The Caribbean Mountains are the result of the accretion of a metapelitic subduction complex onto the northern part of South America, and are composed of greenschist-facies pelitic schists, phyllites, carbonates and minor metavolcanic rocks, blueschists, and eclogites (Harvey, 1972; Maresch, 1974; Santamaria and Schubert, 1974; Bonini and others, 1984). In contrast, the Mérida Andes consist of a core of granitic plutons and gneissic basement overlain by shallow marine shales, sandstones, and conglomerates and their low-grade metamorphic equivalents (Schubert, 1969; Shagam, 1972; Bellizzia and others, 1976). In the southernmost Mérida Andes, just north of the Colombian border, unmetamorphosed sandstone is the dominant rock type exposed at the surface, although granites are also found at outcrop (Bellizzia and others, 1976). The Llanos is a foreland basin receiving sediments shed from the rising orogen to the north and west. The foreland basin consists of a series of related subbasins separated by broad basement uplifts (Feo-Codecido and others, 1984). The portion of the Llanos within the Apure drainage basin is currently actively subsiding, and Holocene fluvial and lacustrine gravels, sands, and muds derived from the orogenic terranes are exposed at the surface (Bellizzia and others, 1976). This material is subject to continual periodic reworking by an unstable system of interconnecting river channels (Herrera and Heurtebise, 1974). Eolian activity also redistributes sand on the Llanos, and an extensive dune field occupies portions of the drainage basin.

Lowland regions of the Apure River drainage basin have a

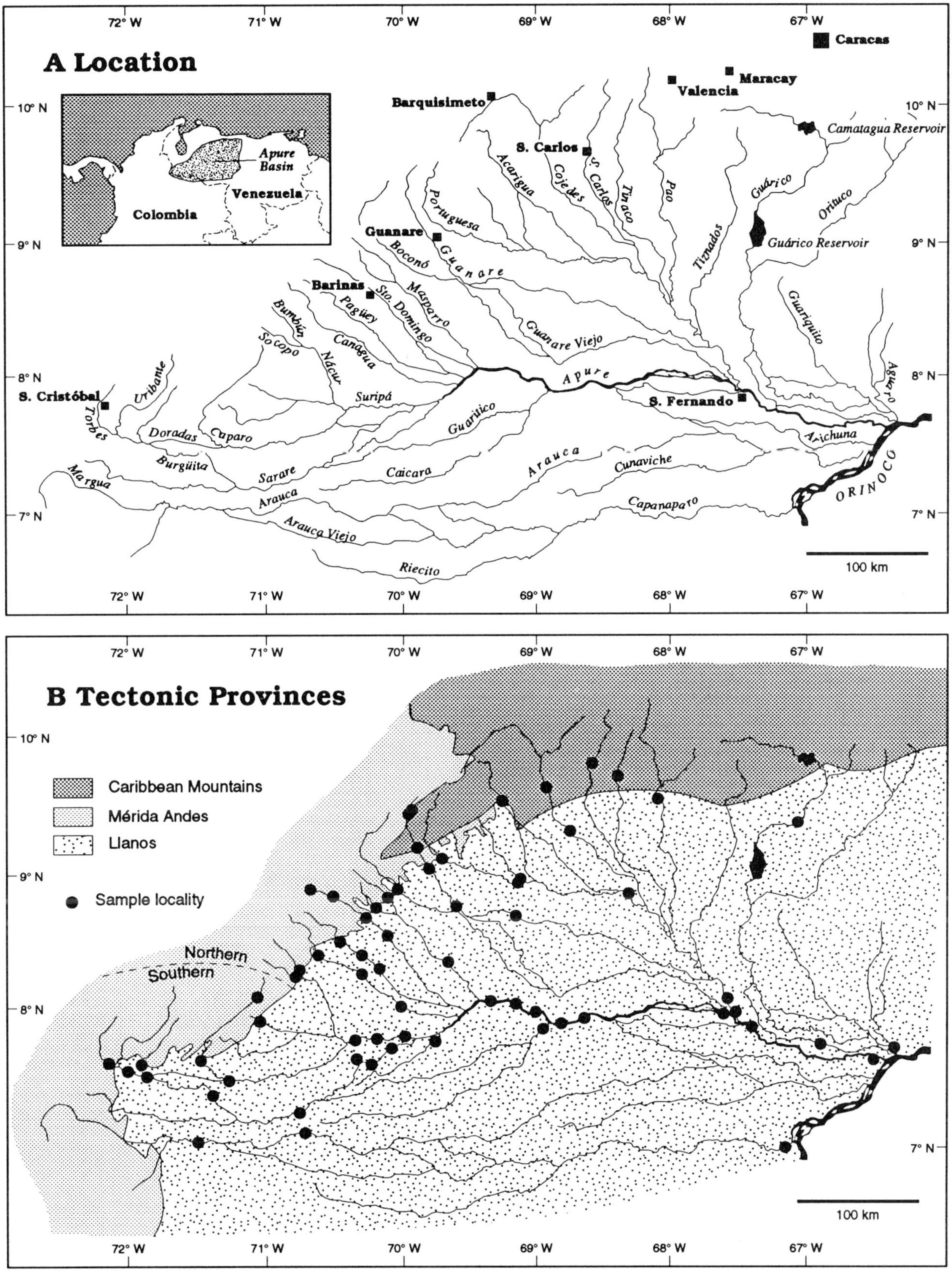

Figure 1. Maps of the Apure River drainage basin. A, Location with names of rivers (italics) and cities (boldface). B, Tectonic provinces. Adapted from Almeida, 1978. Sample locations indicated by dots.

humid tropical climate that is strongly seasonal, with a well-defined rainy season extending from late April to October. Climate is zoned by elevation and is generally temperate above 3,000 m (Ministerio del Ambiente y de los Recursos Naturales Renovables [Venezuela], 1983). Annual precipitation varies slightly across the basin from about 1,000 mm/yr in the northern Llanos to about 2,000 mm/yr near the Colombian border (Snow, 1976: Ministerio del Ambiente y de los Recursos Naturales Renovables [Venezuela], 1983). Palynological, ecological, and sedimentological evidence suggests that the climate of much of northern South America was considerably drier during the last glacial maximum (Van der Hammen, 1985). Vegetation is similarly zoned by elevation, tropical rain forests, and savannas in the foothills, giving way to cloud forests and páramo (sedge-dominated, stunted vegetation adapted to the cold) at higher elevations (United Nations Educational Scientific and Cultural Organization, 1980; Ministerio del Ambiente y de los Recursos Naturales Renovables [Venezuela], 1983). In Venezuela, the Llanos is primarily covered by seasonally dry savannas (United Nations Educational Scientific and Cultural Organization, 1980). Human activity has a great capacity to disturb erosion patterns, a problem of special concern to a study such as this. Until recently, erosional processes in the Apure River drainage basin were affected little by human activity, but deforestation and agricultural expansion are becoming increasingly widespread. This sample suite may represent a last opportunity to evaluate the effects of short-term alluvial recycling on sediment composition before human activities obscure the natural signal.

METHODS

Sand samples were collected from river beds using a pipe dredge or, where possible, from exposed parts of the bed. Care was taken to avoid samples that might have been contaminated by bank material. In larger rivers, the possibility of poor cross-channel mixing (Meade and others, 1983; Stallard, 1987) requires a check for cross-channel heterogeneity by examination of multiple samples across the channel. Except for regions immediately below major confluences, cross-channel heterogeneity proved to be minor in rivers of the Apure River drainage basin, and samples were generally collected near the center of the channel.

Samples were air dried at 100°C, then immersed in water and cleaned by ultrasonic probe in order to remove and disperse any clays or other grain coatings that may have adhered to grain surfaces. The sands were then wet sieved through 125- and 63-μm sieves. The resulting >125- and 63- to 125-μm fractions were dried in an oven at a temperature of 80°C. The 63- to 125-μm fractions were weighed prior to heavy mineral separation, which was achieved by gravity settling in bromoform (specific gravity, approximately 2.8 g cm^{-3}). The heavy mineral residues were weighed and split into two parts, with the majority made into permanent mounts for examination under the optical microscope. The optical mounts were prepared by sprinkling a representative amount on a glass slide, using Canada Balsam (refractive index, 1.537) as the mounting medium.

Heavy mineral proportions were estimated by counting 200 nonopaque detrital grains using the ribbon method described by Galehouse (1971). The remaining areas of each slide were scanned for rare minerals that may not have been detected earlier in the count. These data were used to calculate the ZTR (zircon + tourmaline + rutile) index (determined as the sum of the percentages of zircon, tourmaline, and rutile), regarded as an index of maturity because these three minerals have greater stability than all other heavy minerals (Hubert, 1962). In addition to the basic conventional analysis, the apatite-tourmaline ratio was determined by a separate count. Where possible, 100 grains of the mineral pair were counted, although small sample sizes precluded the achievement of this aim in some cases. The ratio is expressed as the apatite-tourmaline index (ATi), determined as (100 × apatite/apatite + tourmaline). This parameter was chosen because of its sensitivity to weathering: apatite is extremely unstable in acidic weathering conditions (Nickel, 1973; Grimm, 1973; Morton, 1984, 1986), whereas tourmaline is very stable. Furthermore, the two minerals have similar densities and grain shapes, and thus the ratio is not subject to fractionation through variable hydraulic conditions. Parameters such as the apatite-zircon ratio, for example, would vary not only with weathering but also with changing hydraulic conditions, because of the strong density contrast between the two minerals. Identification of problematic species was made using a Link Systems AN10/55S energy-dispersive x-ray analyzer attached to a Cambridge Instruments Microscan V electron microprobe; surface textures of heavy minerals were examined under a Cambridge Instruments Stereoscan Model 250 scanning electron microscope. Heavy mineral data are given in Table 1.

PROVENANCE

Many heavy mineral species were recorded in the study: allanite, andalusite, calcic amphibole ("hornblende"), cassiterite, clinopyroxene, chrome spinel, chloritoid, corundum, dumortierite, epidote (including zoisite and clinozoisite), garnet, gahnite (zinc spinel), kyanite, lawsonite, monazite, olivine, orthopyroxene, rutile (including anatase and brookite), sodic amphibole ("glaucophane"), staurolite, topaz, tourmaline, xenotime, and zircon. Many of these species are present in only trace amounts; major minerals (those forming more than 5% of some assemblages) are andalusite, calcic amphibole, clinopyroxene, chloritoid, epidote group, garnet, rutile, staurolite, tourmaline, and zircon. There is a great range in mineral assemblages, particularly between individual rivers flowing from the orogenic belt, indicating the strong local provenance control. This demonstrates the value of heavy mineral analysis as a sensitive indicator of local source rock mineralogy. For example, the River Bumbún (Fig. 1) clearly taps a geographically restricted outcrop of staurolite-rich schists in the Mérida Andes; staurolite forms 57% of the assemblage from this river, yet is scarce in the adjacent Socopo and

Canagua Rivers (Table 1). A similar pattern is shown by sillimanite, which is abundant in the Santo Domingo and Guanare Rivers but considerably less common in those adjacent (Table 1).

As well as the local source mineralogy control, variations in broader scale provenance are also reflected in the heavy mineral suite. The assemblages from the Caribbean Mountains are dominated by epidote and calcic amphibole, locally with abundant chloritoid or high ZTR indices. The northern Mérida Andes also supplies assemblages rich in epidote and calcic amphibole, but in addition there are several rivers that have abundant garnet, staurolite, and sillimanite. This reflects the presence of high-grade gneissic basement in the northern Mérida Andes and its absence in the dominantly greenschist-facies metapelitic rocks of the Caribbean Mountains. The distribution of glaucophane (sodic amphibole) and lawsonite, which are high-pressure index minerals (Mange and Maurer, 1992), is significant, in that these minerals only occur consistently in rivers draining the Caribbean Mountains (Table 1). This clearly reflects the origin of the Caribbean Mountains by subduction-related tectonism (Erlich and Barrett, 1990; Harvey, 1972; Maresch, 1974; Bellizzia, 1972).

Rivers draining the southern Mérida Andes are rich in zircon, giving rise to high ZTR values. These sediments also have low ATi values. This is consistent with the widespread outcrop of unmetamorphosed relatively mature sandstones in the hinterland. To the north, epidote becomes abundant, indicating the appearance of greenschist-facies rocks, which are common in the northern Mérida Andes and prevalent in the Caribbean Mountains.

The differences in the nature of the three source terrains, and the mixing relationships within the Llanos, are illustrated in the ternary plots that use ZTR, E (epidote), and GSS (garnet + sillimanite + staurolite) as poles (Fig. 2). Although previous studies have shown that epidote is less stable than the other minerals used in the ternary plot, it can be regarded as stable in the context of the Apure River drainage basin (see below). Samples from rivers draining the Caribbean Mountains generally plot in a small area close to the E pole, although two samples from the upper reaches of the River Portuguesa have high ZTR indices, evidently reflecting derivation from a small outcrop of unmetamorphosed mature sandstones. The rivers draining the northern Mérida Andes cover a relatively wide area, all with low ZTR values but with a great range in GSS values relative to E. The southern Mérida Andes samples all have high ZTR values. Mixing patterns are consistent with the individual source characteristics. Rivers that drain the Caribbean Mountains and the northern Mérida Andes fall within the overall field defined by the Caribbean Mountains and northern Mérida Andes samples, and lie slightly closer to the GSS pole than the samples derived solely from a Caribbean Mountains source. Rivers that drain a combined southern and northern Mérida Andes source plot further from the ZTR pole than those from the southern Mérida Andes. They tend to show mixing along a trend toward the E pole rather than the GSS pole. This is because of the geographic disposition of high GSS and E sources in the northern Mérida Andes, with high E material tending to occur in the west, adjacent to the southern Mérida Andes, and high GSS material in the east, geographically separated from the southern Mérida Andes. Finally, the samples that contain a combination of material derived from all three source regions fall toward the center of the ternary plot.

In some areas, there are marked downstream variations that do not appear to be related to introduction of fresh material from tributaries. Such instances are most marked in the upper reaches of the Apure and in the Suripá and Caparo Viejo Rivers. These variations are interpreted as resulting from local reworking of alluvium previously brought in from a different source to that presently being tapped. This is considered to be a result of the highly unstable nature of the river systems on the Llanos (Herrera and Heurtebise, 1974), and indicates that heavy mineral analysis could be used to help constrain earlier drainage patterns in the area.

A useful framework in which to evaluate the effects of chemical weathering on sediment composition is that of transport-limited and weathering-limited denudation regimes (Stallard and Edmond, 1983; Stallard, 1985; Stallard, 1988; Johnsson and others, 1991). If the transport processes removing weathered material from an area are potentially more rapid than the weathering processes generating the material, erosion is said to be weathering-limited. If, in contrast, the maximum weathering rate exceeds the ability of transport processes to remove material, then erosion is transport-limited. Under transport-limited conditions, weathering products have a longer time to react with soil and ground waters, strongly affecting the composition of detritus. Under weathering-limited conditions, detritus is quickly removed and is correspondingly less modified by chemical weathering.

Weathering-limited erosion is encountered on the steep slopes of the Mérida Andes and Caribbean Mountains. Sediment transport rates are high on such slopes, potentially exceeding rates at which weathering can generate loose material. Soils are thin, and bedrock is commonly exposed at the surface. Under these conditions, overall sand composition closely resembles source rock composition (Johnsson and others, 1991); the results of this study demonstrate that heavy mineral assemblages generated under such conditions are equally source-sensitive. Thus, even in tropical weathering environments, sediment composition is little modified under weathering-limited erosion regimes.

EFFECTS OF ALLUVIAL STORAGE

To evaluate the effects of alluvial storage on heavy mineral assemblages, the samples have been classified according to geographical and physiographical setting. Four categories have been used: (1) samples collected within the orogenic belt, (2) samples collected at proximal locations with the Llanos (<100 km from the mountain front), (3) samples collected at middle locations within the Llanos (100 to 200 km from the mountain front), and (4) samples collected at distal locations within the Llanos (>200 km from the mountain front). These categories are the same as those used by Johnsson and others (1991) in their evaluation of

TABLE 1. HEAVY MINERAL ASSEMBLAGES FROM RIVERS IN THE APURE RIVER DRAINAGE BASIN, IN FREQUENCY PERCENT

Sample Number	River	Area*	Source†	Minerals§									
				AN	AP	CA	CP	CR	CT	EP	GT	KY	LW
84-S3	Capanaparo	4	S	1.5	R	12.0	R			3.5	4.5		
84-S16	Pagüey	2	N	0.5	4.0	45.0	1.0			40.5	2.5		
84-S17	Canagua	2	N	R	1.0	61.0	1.0	R		23.5	5.0		
84-S18	Suripá	2	S, N	0.5	R	9.5	R		1.5	59.5	2.5		
84-S19	Arauca	2	S	R	0.5	7.0	R		R	3.0	11.0	0.5	
84-S20	Burguita	2	S	1.0	3.0	7.0				11.0	1.5	R	
85-S57	Apure	4	S, N, C	2.0	1.0	22.0	0.5	R	0.5	36.0	5.5	R	
85-S73	Portuguesa	3	N, C	5.5	1.5	14.0	4.0		8.0	42.0	4.0	0.5	
85-S74	Apure	3	S, N, C	1.0	1.5	16.5	0.5		0.5	23.0	15.0	0.5	
85-S77	Apure	3	S, N	0.5	0.5	29.5	R			25.0	9.5	0.5	
85-S78	Apure	3	S, N	1.0	2.5	27.0	R		1.0	29.0	7.0		
85-S79	Masparro	2	N	2.5	1.0	12.0	R			56.0	12.0	R	
85-S80	Guanare	1	N	1.0	1.0	21.0	0.5		3.0	10.5	46.0		
85-S81	Portuguesa	1	C	2.0	1.5					5.0	3.5		
85-S82	Acarigua	1	C	9.0	4.5	6.0	4.0		46.5	9.0	4.0		
85-S83	Cojedes	1	C	3.5	1.0	5.0	0.5		6.0	61.0	4.5		
86-S1	Orapa	1	C		3.0	14.5	0.5			76.5	R		
86-S3	Guanare	2	N	R	3.5	37.5	1.5		1.5	22.0	15.5		
86.S4	Portuguesa	2	C	8.5	1.5	12.5	3.0		42.5	19.5	3.5		
86-S6	Tinaco	1	C		2.5	42.5	1.5	1.5	R	38.5	3.0		
86-S8	Cojedes	2	C	3.5	0.5	2.0	R		39.0	43.0	0.5	R	
86-S11	Caño Guamai	2	C	1.0	1.0	18.5	17.0		1.5	55.5	1.5		
86-S12a	Portuguesa	2	C	0.5	1.0	1.0	1.0		R	4.5	1.5		
86-S14	Guanare	2	N	R	4.0	31.5	0.5		3.5	13.5	31.0	R	
86-S15	Guanare	2	N	R	2.0	41.5	1.0		5.0	19.5	15.5		
86-S17	Unnamed stream	1	N	0.5	3.0	40.5	0.5		3.0	20.0	23.0	R	
86-S18	Guanare	1	N	0.5	3.0	45.5	R		4.5	18.5	14.5	R	
86-S20	Canagua	2	N	R	0.5	55.5	R			30.5	3.5	R	
86-S21	Pagüey	2	N	0.5	3.5	51.0	R		R	26.0	5.5		
86-S22	Santo Domingo	2	N	0.5	3.0	40.0	0.5			29.5	12.0	R	
86-S23	Boconó	1	N	0.5	5.0	7.5	1.0	0.5	R	31.0	27.5		
86-S25	La Yuca	1	N	3.0	4.0	3.0				64.5	8.0	0.5	
86-S26	Santo Domingo	1	N	0.5	3.5	35.5	R			27.0	16.5		
86-S27	Santo Domingo	1	N	R	4.5	39.0				17.5	30.5		
86-S29	Santo Domingo	1	N		4.5	30.0	R			5.5	34.0		
86-S35b	Torbes	1	S	R	R				R	R	R		
86-S37	Uribante	1	S	4.0	2.5	2.0			R	24.5	1.5		
86-S38	Dorado	1	S	1.0	2.0	0.5				0.5		R	
86-S39	Caparo	1	S	R	1.0	6.5				35.0	5.0		
86-S41	Arauca	3	S	0.5	0.5	5.0	R			1.5	10.0		
86-S42	Caparo	2	5		0.5	6.5	R		0.5	25.0	2.0		
86-S43	Pedraza la Viela	1	N	2.5	0.5	3.0	0.5			63.0	11.0	R	
86-S44	Socopo	1	N	1.0	R	14.5	2.5		5.0	72.0	1.5		
86-S45	Bumbún	1	N		R	3.0				16.5	21.5		
86-S46	Canagua	1	N	1.0	0.5	71.0	0.5		1.0	18.5	1.5		
86-S47	Pagüey	1	N	3.0	2.0	61.5	1.0			23.0	0.5	1.0	
86-S48	Pao	1	C	R	0.5	10.0	1.0		3.0	72.5	4.0	R	
86B-S4	Apure	3	S, N	0.5	1.0	31.0			0.5	33.5	7.0		
86B-56	Apure	2	S, N	1.0	2.0	34.0	R		R	34.5	3.0		
86B-58	Apure	2	S	0.5	5.0	27.0	1.0			10.5	7.0		
86B-S13	Apure	2	S	R	6.0	22.5	2.0			12.5	10.0		
86B-S16	Sarare	2	S	R	1.5	9.5	R			6.0	9.0	0.5	
86B-S17	Apurito	4	C	0.5	1.5	26.5			3.0	44.5	4.5		
86B-S23	Caño Manati	4	S, N	R	R	16.5	R	R	0.5	36.0	5.5	R	
86B-S24	Apurito	4	C	R	0.5	20.0	R	R	1.0	29.0	11.0	R	
86B-S27	Portuguesa	3	N, C	4.0	0.5	24.5	1.5	R	2.5	39.5	9.0		
86B-S28	Apure	3	S, N	3.0	1.0	27.0	1.0		0.5	45.5	2.0		
86B-S31	Caño Guaritico	3	S, N	1.5	R	7.5	R		0.5	32.5	1.5		
86B-S35	Apure	3	S, N	1.0	1.5	30.0	R		R	22.5	7.0		

TABLE 1. HEAVY MINERAL ASSEMBLAGES (continued)

Sample Number	Minerals§ MO	OP	RU	SA	SL	SP	ST	TO	TP	XE	ZR	Other (rare)	ATi**
84-S3	R	R	8.5		3.0	1.5	1.5	2.5		R	61.5		2.3
84-S16	R		R		0.5	2.0	R	1.5	R		2.5		87.0
84-S17	R	R	0.5		1.0	2.5	1.0	R			3.5		67.1
84-S18			2.5		0.5	1.5	R	4.0		R	18.0		13.0
84-S19	1.0	R	5.0	R	1.0	0.5	R	1.0			69.5		19.7
84-S20	R		12.5		1.5		1.0	7.5			54.0		17.0
85-S57		R	5.0		2.5	4.0	0.5	4.5		R	16.0	Al	27.5
85-S73	R	0.5	2.5		7.0	0.5	0.5	5.0			4.5		22.6
85-S74	R		3.5	R	4.5	2.0	0.5	2.5	0.5	R	28.0	Al	37.0
85-S77		0.5	2.5		2.5	3.0	2.0	1.0			23.0	Al, Gh	40.0
85-S78		R	1.5		2.0	0.5	0.5	3.5			24.5		64.8
85-S79			0.5		4.0	2.0	5.0	2.0			3.0		58.3
85-S80			1.5		6.5	1.0	2.0	0.5			5.5		*88.9*
85-S81	R		11.5		0.5	0.5	0.5	5.5			69.5		24.0
85-S82		R	5.5		R			2.5			9.0		76.0
85-S83			4.0	2.0			1.0	2.5	4.0		5.0		17.4
86-S1	R					2.5		R			3.0		98.0
86-S3			R		14.5	2.0	0.5	0.5	R		1.0		79.0
86-S4			0.5	R	R	3.0		2.5	1.0		2.0		30.6
86-S6		2.0	0.5		R	5.5		R	0.5		2.0	Ol	97.0
86-S8			3.0	1.0		1.0	R	3.0	2.0		1.5		11.0
86-S11			R			1.5		0.5			2.0	Ol	52.1
86-S12a	R		8.5			0.5	R	3.0			78.5		19.0
86S-14	0.5	R	1.0		6.0	1.5	1.0	1.5			4.5	Ol	76.0
86-S15		R	0.5		6.0	1.5	4.0	1.0			2.5	Ol	75.0
86-S17			R		6.0	0.5	1.5	0.5			1.0	Gh	74.0
86-S18	R		1.0		8.5		1.5	1.0			1.5		77.0
86-S20		R	1.0		0.5	5.0	0.5	R			3.0	Al	67.0
86-S21		R	R		3.0	3.5	R	R			7.0	Al	97.0
86-S22		R	1.0		6.5	1.0		0.5	R		5.5	Ol	91.0
86-S23	0.5		3.0		1.5	3.0	2.5	2.0			14.5		75.0
86-S25			2.0		0.5	3.0	1.5	1.5			8.5		73.0
86-S26			2.5		7.5	4.0	R	0.5			2.5	Gh	85.0
86-S27			0.5		4.5	1.0		0.5			2.0		90.7
86-S29	0.5	R	0.5		18.0	2.0	1.5	1.0		R	2.5		80.0
86-S35b	1.0		4.0				R	2.5		R	92.5		11.6
86-S37			11.5		0.5	0.5	0.5	14.5			38.0		20.0
86-S38	1.5		12.5					18.0		1.5	62.5		6.0
86-S39	0.5		1.5				0.5	1.0			49.0		51.2
86-S41	0.5		5.0		2.5		R	5.5		R	69.0		14.0
86-S42	R		4.0		R	0.5	R	2.0		R	59.0		31.0
86-S43	R		10.0		0.5	0.5		2.0			6.5		30.0
86-S44			R		0.5	0.5	R	R			3.0		*54.5*
86-S45			R				57.0	R			2.0		*72.7*
86-S46		1.0	0.5		1.0	2.0	R	0.5			1.0	Ol	68.5
86-S47		R	1.5		2.0	2.0	0.5	R			2.0	Al	98.8
86-S48			0.5	R	R	3.0	0.5	1.0	0.5		3.5	Co, Du	23.0
86B-S4		0.5	3.5		3.0	3.0	2.5	0.5			13.5		*85.7*
86B-S6			2.5		1.5	1.5	3.5	3.5			13.0		26.5
86B-S8	1.0		5.0		4.5	2.0	R	7.0			29.5		38.0
86B-S13			6.5		3.0	2.0	0.5	8.0			27.0		46.0
86B-S16	R		8.5		3.0	1.0	R	5.0			56.0		41.0
86B-S17			2.5	R	2.0	5.5	2.0	1.5			6.0		34.0
86B-S23	R		2.0	R	3.5	1.5	2.0	2.0		R	30.5	Al	14.0
86B-S24	0.5	R	1.0	R	3.0	1.0	2.5	2.0		R	28.5	Cs	24.0
86B-S27	R	0.5	3.5	R	3.5	4.0	1.0	1.5			4.5		35.0
86B-S28		0.5	3.5		4.0	1.5	0.5	1.0			9.0		45.9
86B-S31	R		4.5		3.0	0.5	1.0	3.5		R	44.0		5.3
86B-S35	R	R	4.0		3.5	3.5	R	1.5		R	25.5		45.0

TABLE 1. HEAVY MINERAL ASSEMBLAGES (continued)

Sample Number	River	Area*	Source†	AN	AP	CA	CP	CR	CT	EP	GT	KY	LW
				Minerals§									
86B-S36	Apure	3	S, N	0.5	4.0	27.5	1.0		0.5	27.0	10.0		
86B-S39	Masparro	2	N	1.0	3.0	7.0	R		R	59.5	8.5	R	
86B-S42	Suripá	2	S, N	R	R	37.5	R		R	55.0	1.0		
86B-S43	Caño la Tigra	2	S, N	R	0.5	33.5	0.5		R	42.5	3.0		
86B-S46	Caparo Vieja	2	S	R	5.5	15.5	3.0			12.0	9.5		
88A-S22	Guárico	1	C			23.0	13.0			62.5			1.0
88D-S3	Caparo	1	S		R	3.0	1.0		2.0	35.5			

*1 = Within mountains, up to mountain front; 2 = proximal Llanos (<100 km from mountain front); 3 = middle Llanos (100–200 km from mountain front); 4 = distal Llanos (>200 km from mountain front).
†S = Southern Mérida Andes; N = northern Mérides Andes; C = Caribbean Mountains.
§Al = Allanite; AN = andalusite; AP = apatite; CA = calcic amphibole; CO = corundum; CP = clinopyroxene; CR = chrome spinel; (notes continued on facing page)

the changes in bulk sediment composition in the same sample set, and are adopted here for comparative purposes. The mean abundances of the main heavy minerals in these locations have been calculated, together with the mean ATi values (Table 2, Fig. 3). By considering mean values for each physiographical setting, the local distorting effects due to variations in provenance may be largely smoothed out. Uncertainties of ratio data, such as those reported in Table 2, are not statistically rigorous due to autocorrelation. Nevertheless, they do provide a useful framework with which to evaluate the confidence that can be placed on the estimate of the mean. The means of the suites reported in Table 2 are not well constrained, in part because of the small sample size, but more importantly because of the extreme compositional range of samples from each suite. As discussed above, this variability reflects not only the differences between sands from the different tectonic provinces within the orogenic belt, but also the marked

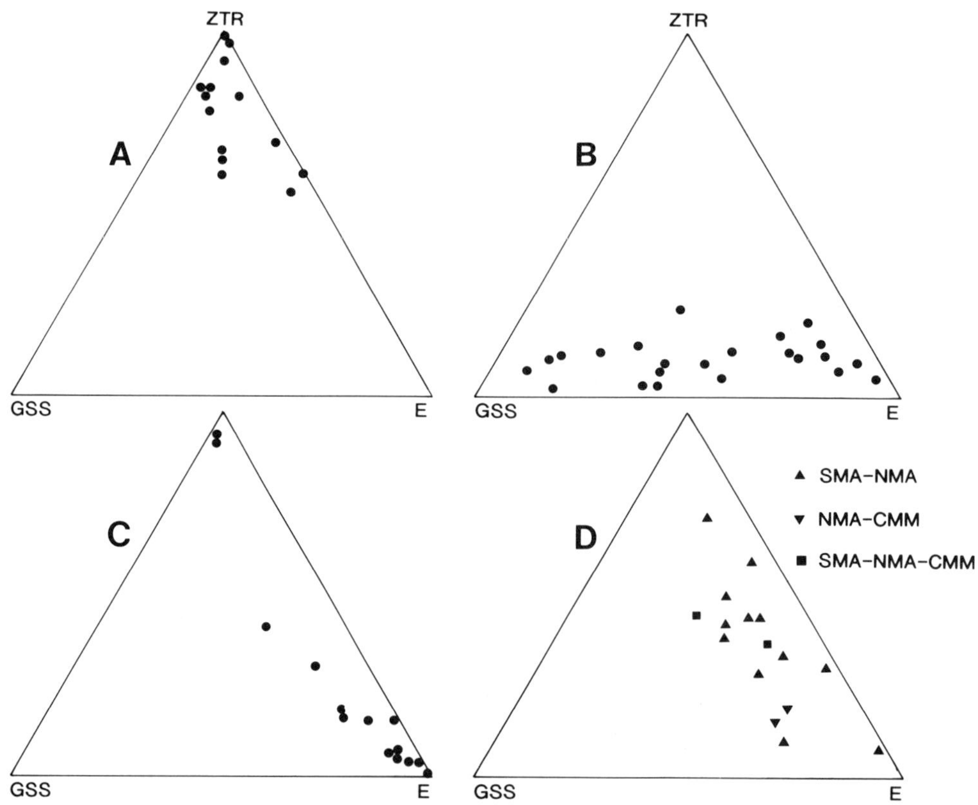

Figure 2. Ternary plots of the relative abundance of garnet + staurolite + sillimanite (GSS), epidote (E), and zircon + tourmaline + rutile (ZTR) in individual samples, showing the differences in provenance of river sediments draining four locations: A, the southern Mérida Andes; B, northern Mérida Andes; C, the Caribbean Mountains. D, Rivers draining combinations of the three terranes; SMA = southern Mérida Andes, NMA = northern Mérida Andes, CMM = Caribbean Mountains.

TABLE 1. HEAVY MINERAL ASSEMBLAGES (continued)

Sample Number	MO	OP	RU	SA	Minerals§ SL	SP	ST	TO	TP	XE	ZR	Other (rare)	ATi**
86B-S36			3.0		3.0	2.0	0.5	3.0			18.0	Gh	52.0
86B-S39			3.5		2.5	4.0	5.0	2.5			3.5		61.0
86B-S42			0.5		R	3.0	R	0.5			2.5	Al	61.5
86B-S43		0.5	1.0		0.5	3.0	10.5	2.0			2.5		43.6
86B-S46	R	0.5	9.0		3.0	2.0	R	4.5			35.5		50.0
88A-S22				0.5		R					R		-
88D-S3		R	3.5				R	5.5			49.5		14.0

CS = cassiterite; CT = chloritoid; DU = dumortierite; EP = epidote group; GT = garnet; GH = gahnite; KY = kyanite; LW = lawsonite; MO = monazite; OL = olivine; OP = orthopyroxene; RU = rutile, anatase, and brookite; SA = sodic amphibole; SL = sillimanite; SP = titanite; ST = staurolite; TO = tourmaline; TP = topaz; XE = xenotime; ZR = zircon.
**Apatite-tourmaline index. Values in italics refer to those determined on a very low total grain count (<20).

variation within each province. Thus, the large uncertainties in Table 2 are another manifestation of the fact that heavy mineral suites are sensitive indicators of provenance of sands from tectonically active regions where erosion is weathering limited.

As shown in Table 2 and Figure 3, the effect of alluvial storage on the abundances of most of the individual mineral species appears to be slight. Epidote contents remain virtually unchanged throughout, and zircon shows a tendency to increase (although variance on zircon values is high, particularly for the distal Llanos setting where the mean zircon is greatest). The ZTR value follows the zircon pattern, largely because zircon proportions have the greatest influence on the parameter in this study. Amphibole shows a rise in mean value between the mountain setting and the proximal Llanos, followed by a fall moving into the middle and distal parts of the Llanos. Garnet shows a slight overall fall. Of the less abundant minerals, only clinopyroxene shows an apparent trend, being virtually absent in the distal Llanos. However, the trends in mean values of epidote, amphibole, garnet, and clinopyroxene have no statistical significance because of the overlap in the errors around the mean of each mineral across the range of physiographical categories. Variations in chloritoid and staurolite are considered mainly to reflect strong local variations in their abundance from source area inhomogeneities. Sillimanite and rutile show no overall pattern, and titanite appears to show a slight increase.

TABLE 2. MEAN ABUNDANCES OF SELECTED HEAVY MINERALS, ZTR INDEX, AND APATITE-TOURMALINE RATIOS IN SAMPLES FROM DIFFERENT GEOGRAPHIC REGIONS IN THE APURE RIVER DRAINAGE BASIN*

	Within Orogenic Belt (Region 1)	Proximal Llanos (Region 2)	Middle Llanos (Region 3)	Distal Llanos (Region 4)
Calcic amphibole	19 ± 9	25 ± 7	23 ± 6	19 ± 8
Clinopyroxene	1 ± 1	5 ± 3	1 ± 1	0.1 ± 0.3
Chloritoid	3 ± 4	4 ± 5	1 ± 2	1 ± 2
Epidote	32 ±10	28 ± 8	32 ± 6	30 ±22
Garnet	11 ± 5	7 ± 3	7 ± 3	6 ± 4
Rutile	3 ± 4	5 ± 3	3 ± 1	4 ± 4
Sillimanite	2 ± 2	4 ± 2	4 ± 1	3 ± 1
Staurolite	3 ± 5	2 ± 2	1 ± 1	2 ± 1
Titanite	1 ± 1	2 ± 1	2 ± 1	3 ± 3
Zircon	17 ±11	20 ±10	17 ± 9	28 ±29
Number of samples	25	25	10	5
ZTR Index†	26 ±14	24 ±12	23 ±11	35 ±32
Apatite-tourmaline index (ATi)	57 ±13	51 ±11	43 ±17	20 ±17

*Data calculated from Table 1.
†ZTR index calculated as defined by Hubert (1962).
Uncertainties represent 95% confidence limits as determined by Student's t-distribution. Uncertainties of ratio data are not statistically rigorous.

The ATi value, however, shows a more significant trend. ATi declines markedly between the proximal and middle Llanos, and especially between the middle and distal Llanos; the difference in mean value between the two extreme physiographical categories is statistically significant at the 95% significance level (Fig. 3). The ATi parameter has been specifically determined because of its potential sensitivity to weathering in alluvial storage, as apatite is known to be unstable in acidic weathering environments typical of tropical, humid, well-vegetated settings, whereas tourmaline is essentially stable (Morton, 1984: Mange and Maurer, 1992). Because apatite and tourmaline have similar hydraulic behavior, owing to their similar density and grain shape, the ATi value is unlikely to be significantly affected by changing hydraulic conditions. Therefore, the overall reduction in ATi implies that apatite is being lost through weathering during short-term alluvial storage on the Llanos. Evidence for apatite dissolution is also given by the surface textures of apatite grains, many of which show corrosion (Fig. 4A). This substantiates the findings of the study of the framework minerals by Johnsson and others (1991), who showed that rock fragments and feldspars, particularly plagioclase, are lost through the same process.

Apart from the trend toward lower ATi values, however, the assemblages show little apparent change across the Llanos.

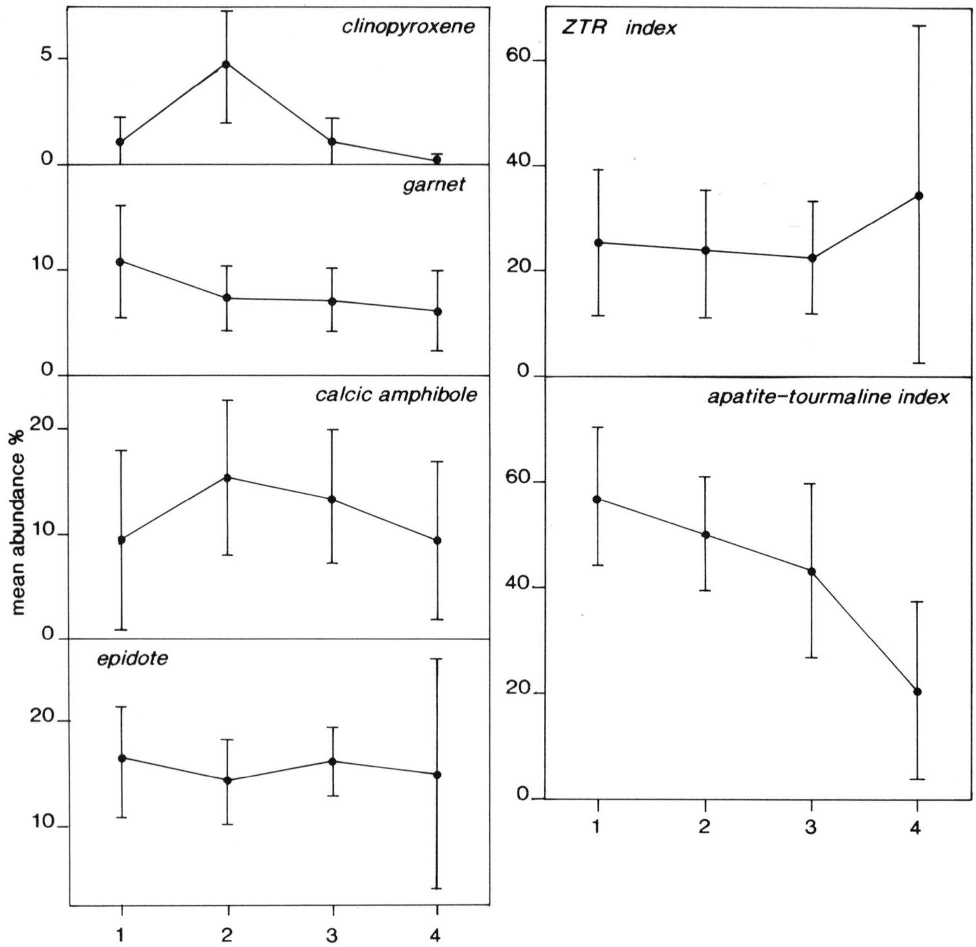

Figure 3. Variation of the major heavy mineral species, ZTR index and ATi value, across the four physiographic zones. 1 = orogenic belt, 2 = proximal Llanos, 3 = middle Llanos, and 4 = distal Llanos. Bars represent 95% confidence levels on the mean values.

Clinopyroxene and garnet, both of which are known to be susceptible to weathering (Morton, 1984), are apparently less abundant in the distal Llanos: this may be the result of dissolution during alluvial storage, but the trends are poorly constrained in statistical terms and no firm inferences can be made. In contrast, contents of amphibole and epidote, both of which are also susceptible to weathering (Morton, 1984), do not show any evidence for an overall decline. One possible explanation is that, although amphibole and epidote are partially destroyed by weathering during alluvial storage, they are continually resupplied to the very fine sand fraction by the progressive downstream destruction of lithic particles and feldspar grains. Johnsson and others (1991) showed that the most volumetrically important compositional change in the Apure River drainage basin is the progressive breakdown of rock fragments. The presence of internal subgrain boundaries within these particles cause them to have a high degree of mechanical instability that is accentuated when the grains are subjected to weathering. Breakdown of lithic clasts would release fresh, unweathered material previously insulated within the grains, thus maintaining levels of unstable heavy mineral and other framework species (such as feldspar) in the sediments. Release of feldspar through breakdown of volcanic lithic clasts under similar circumstances has been observed in point-bar sediments from the Amazon River (Johnsson and Meade, 1990). Both epidote and amphibole are common constituents of lithic clasts in the Apure River drainage basin, and in addition epidote is very common as an alteration product of feldspar (Fig. 5). Surface textures of these minerals in samples from the most distal parts of the Llanos are consistent with this hypothesis (Fig. 4B-D), in that etched grains coexist with highly angular, fresh material.

It therefore appears that heavy mineral suites can be affected by weathering during short-term alluvial storage in fluvial systems, with the apatite-tourmaline index providing the best indicator of the process. The sensitivity of the ATi value to weathering during alluvial storage implies that the parameter is not always an accurate guide to provenance. In many circumstances, notably in marine sediments and in fluvial and eolian sediments deposited in arid and semi-arid conditions, the ATi value is likely to be effective in discriminating sediments of differing provenance. How-

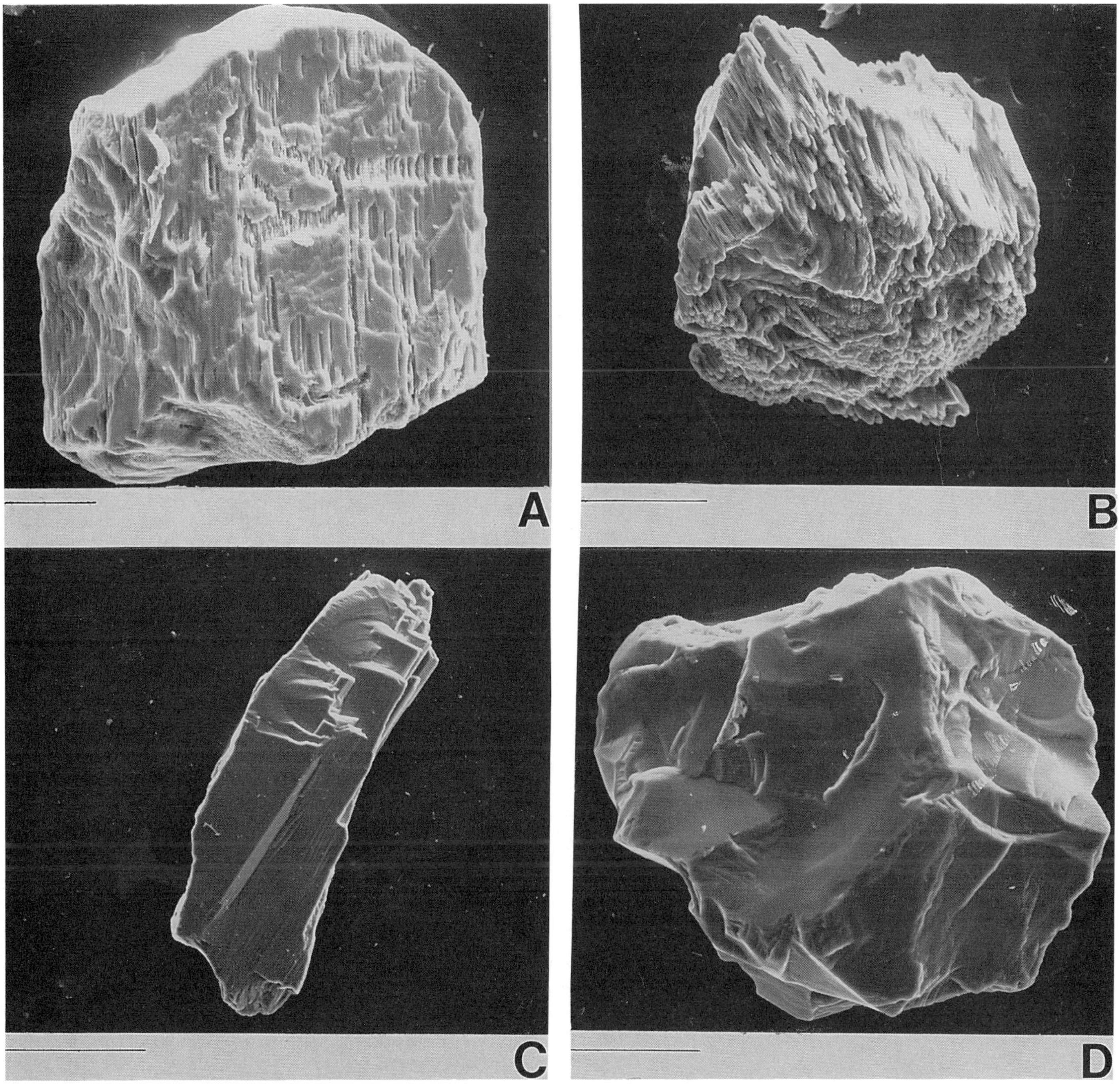

Figure 4. Scanning electron micrographs showing surface textures of heavy mineral grains from sample OR85-S57, taken from the Apure River in physiographic category 4 (distal Llanos). A, Apatite grain showing evidence of corrosion; scale bar = 20 μm. B, Corroded calcic amphibole grain; scale bar = 40 μm. C, Fresh, angular amphibole grain; scale bar = 20 μm. D, Fresh, angular epidote grain; scale bar = 40 μm.

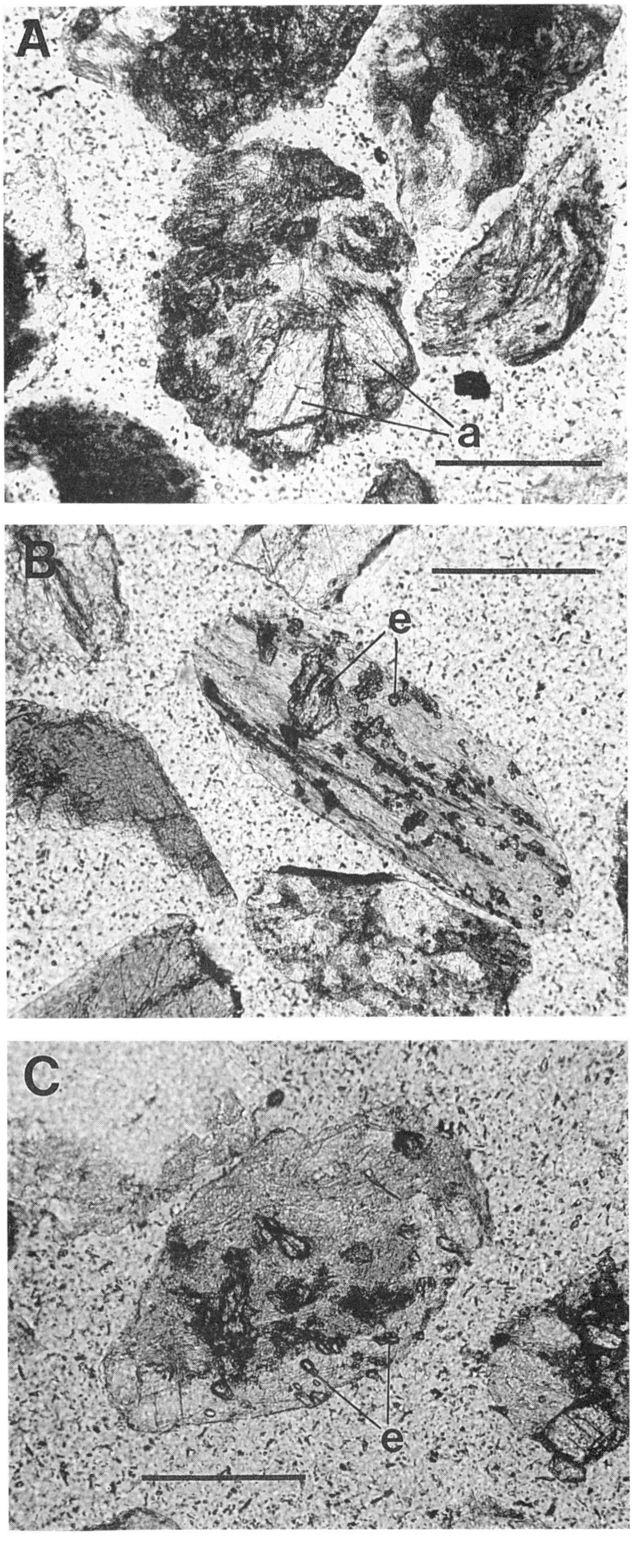

ever, in fluvio-deltaic sandstones deposited in humid tropical settings, such as Late Carboniferous sandstones in Britain (Anderton and others, 1979), variations in ATi are likely to be, at least in part, a function of weathering during alluvial storage, and should be used only as a guide to changes in provenance if there are other supporting data. Although there is an overall reduction in ATi across the Llanos, the findings of this study reveal that, over short distances, the ATi value is a useful provenance-sensitive parameter independent of hydraulic control. This is demonstrated by several of the tributaries flowing into the Apure, the best example being the Guanare. Six samples were studied from the Guanare system (Fig. 1), two within the orogenic terrane and four in the proximal Llanos. The heavy mineral suites in the Guanare samples are diverse (Table 1), dominated by amphibole, epidote, garnet, and sillimanite. The relative abundances of these species show considerable scatter and no downstream trend (Fig. 6). By contrast, the ATi value remains remarkably consistent (74.0 to 79.0, with the exception of one sample with a value of 88.9 from a low total count). The variations in relative abundance of the heavy mineral species are almost certainly hydraulically controlled; the lack of variation in the ATi value demonstrates that the parameter is not affected by this process and is effectively independent of it.

CONCLUSIONS

Heavy mineral assemblages from rivers in the Apure River drainage basin of Venezuela and Colombia accurately characterize the nature of the provenance, located in the Andean orogenic terranes to the north and west. The Caribbean Mountains supply assemblages dominated by epidote and calcic amphibole, derived from the greenschist-facies rocks that form the large part of the outcrop. Minor amounts of sodic amphibole (glaucophane) and lawsonite reflect the origin of the area by subduction-related tectonics. The northern Mérida Andes supply two assemblages, one dominated by garnet, sillimanite, and staurolite supplied by gneissic basement rocks, and the other by epidote and calcic amphibole supplied by the low-grade metapelitic cover sequences. The southern Mérida Andes supply assemblages dominated by zircon, reflecting the widespread outcrop of unmetamorphosed sandstones in the hinterland. Local source area heterogeneities are also picked out by heavy mineral assemblages, which identify, for

Figure 5. Photomicrographs of thin sections in plane-polarized light, showing polymineralic sand grains containing unstable heavy mineral subgrains. A, Metasedimentary lithic fragment with amphibole (labeled a) subgrains of very fine sand size, upper Guárico River, Caribbean Mountains. B, Foliated metasedimentary lithic fragment with epidote (labeled e) subgrains of very fine sand and silt size, upper Guárico River, Caribbean Mountains. C, Highly altered feldspar grain containing silt-size epidote (labeled e) subgrains, Orupe River (tributary to San Carlos River), Caribbean Mountains. Scale bar = 200 μm in all cases.

example, small outcrops of staurolite- and chloritoid-rich metamorphics. The accurate representation of source rock mineralogy by the heavy mineral assemblages is a result of the weathering-limited denudation regime prevalent within the orogenic terranes.

Previous studies of the Orinoco drainage basin demonstrated that weathering during short-term alluvial storage on the low-lying Llanos area has had marked effects on the framework mineralogy, causing progressive loss of rock fragments and feldspar (Johnsson and others, 1991). The heavy mineral assemblages are not as drastically affected by this process, with most heavy mineral species showing relatively little change with distance from the source terranes. Clinopyroxene and garnet show limited evidence for slight declines in abundance, but epidote and amphibole are virtually unchanged. However, alluvial storage has had a significant effect on the relative abundance of the readily weatherable mineral apatite compared with the stable and hydraulically equivalent mineral tourmaline. The apatite-tourmaline index (ATi) shows a clear and progressive decline between the proximal and distal Llanos, demonstrating loss of apatite through weathering. This ratio is therefore proposed as a useful guide to evaluating the effects of short-term alluvial storage in large-scale fluvial systems subject to a humid tropical climate. The lack of major variation in the abundance of epidote and calcic amphibole, which are known to be unstable under weathering conditions, may have resulted from continual resupply of fresh material to the very fine sand fraction through the progressive downstream disaggregation of rock fragments.

Over relatively short distances within individual tributaries to the Apure, the ATi value is remarkably constant compared with relative abundances of individual heavy mineral species. This shows the importance of determining parameters that are independent of hydraulic controls for provenance evaluation, and indicates that the apatite-tourmaline index is a useful guide to changes in provenance in many depositional settings, notably in marine sandstones and in fluvial or eolian sandstones deposited in arid or semi-arid conditions. In fluvial and fluvio-deltaic sandstones in which weathering during alluvial storage may have occurred, the parameter should not be used as a guide to changes in source unless supported by other evidence.

Finally, we consider that the inferences made herein regarding the effects of alluvial storage on heavy mineral suites need substantiation from other case studies. These should preferably involve situations where significant amounts of polycyclic material with low-diversity, ultrastable heavy mineral suites are not supplied to the drainage basin. In the Apure River study, the supply of polycyclic material from the southern Mérida Andes introduced considerable data scatter that may have masked some of the weathering-related trends. Case studies of the behavior of heavy minerals suites in transport-limited denudation regimes are also required to enable the continued development of the understanding of the processes controlling heavy mineral assemblages and their application to provenance studies.

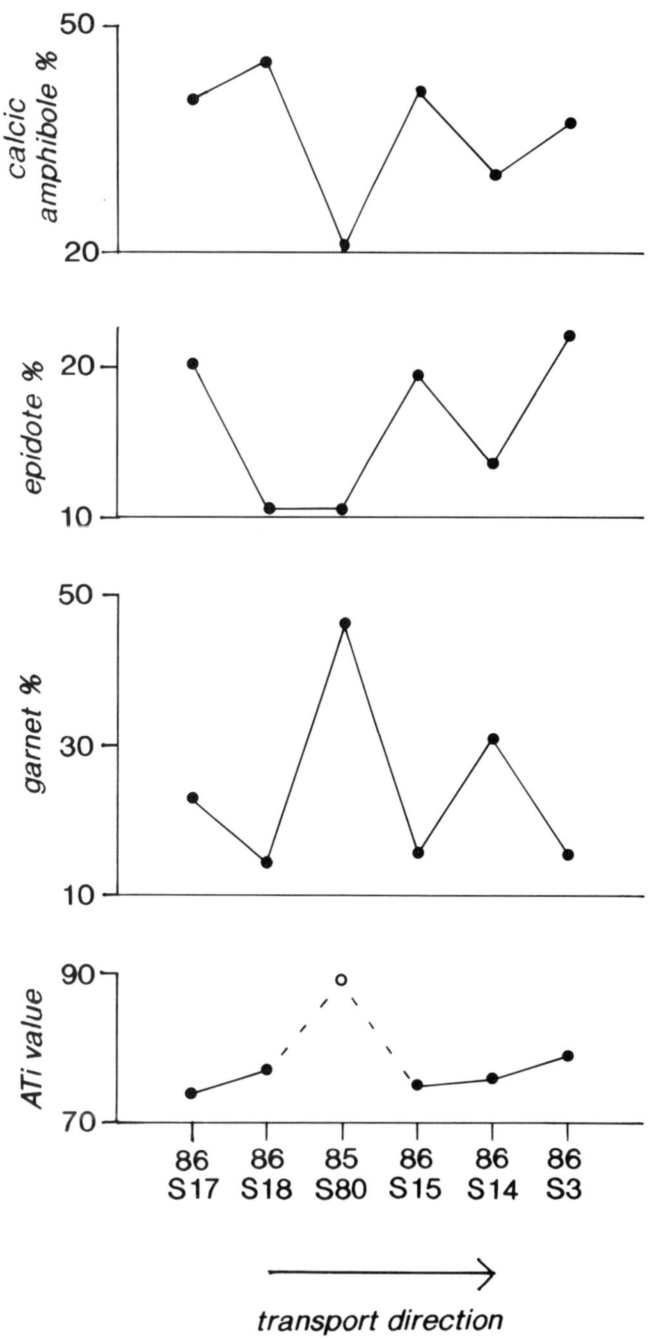

Figure 6. Downstream variations in selected heavy mineral abundances and ATi values in samples from the Guanare River system. Although heavy mineral abundances fluctuate markedly, ATi values remain virtually constant, confirming that this parameter is essentially unaffected by changes in hydraulic conditions. Sample 86-S17 is from an unnamed stream in the uppermost reaches of the Guanare drainage basin: other samples are from the Guanare River itself. The ATi value from OR85-S80 is included even though it is based on a very low grain count (only nine apatite plus tourmaline grains available). All other ATi values are based on a count of 100 grains.

ACKNOWLEDGMENTS

Samples were collected under the auspices of Proyecto Orinoco-Apure, a cooperative program between the Venezuelan Ministerio del Ambiente y los Recursos Naturales Renovables (MARNR) and the U.S. Geological Survey. Logistical support from MARNR was critical during sample collection; special thanks are due to José Manuel Saínz and David Pérez-Hernández. Robert H. Meade, Robert F. Stallard and Lois Koehnken made available many samples from their collections. Thanks to Omar Alvarado R., Luis H. Bravo, Clare C. Cranston, Steven L. Goldstein, Lois Koehnken, José Luis López, Neil Lundberg, Robert H. Meade, and Robert F. Stallard for assistance in the field and helpful discussion. Funding was provided by MARNR, the U.S. Geological Survey, Princeton University, and National Science Foundation Grants EAR-8407651 (to R. F. Stallard) and EAR-8616904 (to R. F. Stallard and Neil Lundberg). This manuscript has benefited greatly from the comments of Robert Knox, Claire Hallsworth, Emanuela Molinaroli, and an anonymous reviewer. It is published with the approval of the director of the British Geological Survey (NERC) and the director of the U.S. Geological Survey.

REFERENCES CITED

Almeida, M.F.F., 1978, Tectonic map of South America: Commission for the Geologic Map of the World, scale, 1:5,000,000, 2 sheets.

Anderton, R., Bridges, P. H., Leeder, M. R., and Sellwood, B. W., 1979, A dynamic stratigraphy of the British Isles: A study in crustal evolution: London, Allen & Unwin, 301 p.

Bellizzia, G. A., 1972, Is the entire Caribbean Mountain belt of northern Venezuela allochthonous? *in* Shagam, R., and 6 others, eds., Studies in earth and space sciences (Harry H. Hess memorial volume): Geological Society of America Memoir 132, p. 363–368.

Bellizzia, G. A., Pimentel, M. N., and Bajo, O. R., 1976, Mapa Geológico estructural de Venezuela: Caracas, Ministerio de Minas e Hidrocárburos de Venezuela, p. 30.

Bonini, W. B., Hargraves, R. B., and Shagam, R., 1984, The Caribbean–South America plate boundary and regional tectonics: Geological Society of America Memoir 162, 421 p.

Dietz, V., 1973, Experiments on the influence of transport on shape and roundness of heavy minerals: Contributions to Sedimentology, v. 1, p. 103–125.

Erlich, R. N., and Barrett, S. F., 1990, Cenozoic plate tectonic history of the northern Venezuela-Trinidad region: Tectonophysics, v. 9, p. 161–184.

Feo-Codecido, G., Smith, F. D., Jr., Abound, N., and de Di Giacomo, E., 1984, Basement and Paleozoic rocks of the Venezuelan Llanos basins, *in* Bonini, W. B., Hargraves, R. B., and Shagam, R., eds., The Caribbean–South America plate boundary and regional tectonics: Geological Society of America Memoir 162, p. 175–187.

Friese, F. W., 1931, Untersuchung von mineralen auf abnutzbarkeit bei verfractung im wasser: Tschermaks Mineralogisches und Petrographisches Mitteilungen, v. 41, p. 1–7.

Galehouse, J. S., 1971, Point-counting, *in* Carver, R. E., ed., Procedures in sedimentary petrology: New York, Wiley Interscience, p. 385–407.

Grimm, W. D., 1973, Stepwise heavy mineral weathering in the residual quartz gravel, Bavarian Molasse (Germany): Contributions to Sedimentology, v. 1, p. 103–125.

Hansley, P. L., 1987, Petrologic and experimental evidence for the etching of garnets by organic acids in the Upper Jurassic Morrison Formation, northwestern New Mexico: Journal of Sedimentary Petrology, v. 57, p. 666–681.

Harvey, S.R.M., 1972, Origin of the southern Caribbean Mountains, *in* Shagam, R., and 6 others, eds., Studies in earth and space sciences (Harry H. Hess memorial volume): Geological Society of America Memoir 132, p. 387–400.

Herrera, R., and Heurtebise, M., 1974, Neutron activation analysis of trace elements in quartz sands: Its possibilities in the assessment of provenance: Chemical Geology, v. 14, p. 81–93.

Hubert, J. F., 1962, A zircon-tourmaline-rutile maturity index and the interdependence of the composition of heavy mineral assemblages with the gross composition and texture of sandstones: Journal of Sedimentary Petrology, v. 32, p. 440–450.

Johnsson, M. J., and Meade, R. H., 1990, Chemical weathering of fluvial sediments during alluvial storage: The Macuapanim Island point bar, Solimões River: Journal of Sedimentary Petrology, v. 59, p. 768–718.

Johnsson, M. J., Stallard, R. F., and Meade, R. H., 1988, First-cycle quartz arenites in the Orinoco River basin, Venezuela and Colombia: Journal of Geology, v. 96, p. 103–227.

Johnsson, M. J., Stallard, R. F., and Lundberg, N., 1991, Controls on the composition of fluvial sands from a tropical weathering environment: Sands of the Orinoco drainage basin, Venezuela and Colombia: Geological Society of America Bulletin, v. 103, p. 1622–1647.

Mange, M. A., and Maurer, H.F.W., 1992, Heavy minerals in colour: London, Chapman and Hall, 147 p.

Maresch, W. V., 1974, Plate tectonics origin of the Caribbean Mountain System of northern South America: Discussion and proposal: Geological Society of America Bulletin, v. 85, p. 669–682.

Meade, R. H., Nordin, C. F., Jr., Pérez Herández, D., Mejía, B. A., and Pérez Godoy, J., 1983, Sediment and water discharge in Rio Orinoco, Venezuela and Colombia, *in* Proceedings of the Second International Symposium on River Sedimentation: Beijing, Water Resources and Electric Power Press, p. 1134–1144.

Milliken, K. L., 1988, Loss of provenance information through subsurface diagenesis in Plio-Pleistocene sediments, northern Gulf of Mexico: Journal of Sedimentary Petrology, v. 58, p. 992–1002.

Ministerio del Ambiente y de los Recursos Naturales Renovables (Venezuela), 1983, Venezuela en Mapas: Caracas, Ministerio del Ambiente y de los Recursos Naturales Renovables, 48 p.

Mitchell, W. A., 1975, Heavy minerals, *in* Gieseking, J. E., ed., Soil components. II. Inorganic: New York, Springer-Verlag, p. 449–480.

Morton, A. C., 1984, Stability of detrital heavy minerals in Tertiary sandstones of the North Sea basin: Clay Minerals, v. 19, p. 287–308.

Morton, A. C., 1985, Heavy minerals in provenance studies, *in* Zuffa, G. G., ed., Provenance of arenites: Dordrecht, Reidel, p. 249–277.

Morton, A. C., 1986, Dissolution of apatite in North Sea Jurassic sandstones: Implications for the generation of secondary porosity: Clay Minerals, v. 21, p. 711–733.

Morton, A. C., and Smale, D., 1991, The effects of transport and weathering on heavy minerals from the Cascade River, New Zealand: Sedimentary Geology, v. 68, p. 117–123.

Nickel, E., 1973, Experimental dissolution of light and heavy minerals in comparison with weathering and intrastratal solution: Contributions to Sedimentology, v. 1, p. 1–68.

Russell, R. D., 1937, Mineral composition of Mississippi River sands: Geological

Society of America Bulletin, v. 48, p. 1307–1348.

Santamaria, F., and Schubert, C., 1974, Geochemistry and geochronology of the southern Caribbean–northern Venezuelan plate boundary: Geological Society of America Bulletin, v. 85, p. 1085–1098.

Savage, K. M., and Potter, P. E., 1991, Petrology of modern sands of the Rios Guaviare and Inírada, southern Colombia: Journal of Geology, v. 99, p. 289–298.

Schubert, C., 1969, Geologic structure of a part of the Barinas Mountain Front, Venezuelan Andes: Geological Society of America Bulletin, v. 80, p. 443–458.

Shagam, R., 1972, Andean research project, Venezuela, *in* Shagam, R., and 6 others, eds., Studies in earth and space sciences (Harry H. Hess memorial volume): Geological Society of America Memoir 132, p. 449–463.

Shukri, N. M., 1949, The mineralogy of Nile sediments: Quarterly Journal of the Geological Society of London, v. 105, p. 511–529.

Snow, J. W., 1976, The climate of northern South America, *in* Schwertedfeger, W., ed., Climates of Central and South America: Amsterdam, Elsevier, p. 295–403.

Stallard, R. F., 1985, River chemistry, geology, geomorphology, and soils in the Amazon and Orinoco basins, *in* Drever, J. I., ed., The chemistry of weathering: Dordrecht, Reidel, p. 293–316.

Stallard, R. F., 1987, Cross-channel mixing and its effects of sedimentation in the Orinoco River: Water Resources Research, v. 23, p. 1977–1986.

Stallard, R. F., 1988, Weathering and erosion in the humid tropics, *in* Lerman, A., and Meybeck, M., eds., Physical and chemical weathering in geochemical cycles: Dordrecht, Kluwer, p. 225–246.

Stallard, R. F., and Edmond, J. M., 1983, Geochemistry of the Amazon. 2. The influence of geology and weathering on the dissolved load: Journal of Geophysical Research, v. 88, p. 9671–9688.

Thiel, G. A., 1940, The relative resistance to abrasion of mineral grains of sand size: Journal of Sedimentary Petrology, v. 10, p. 103–124.

Thiel, G. A., 1945, Mechanical effects of stream transportation in mineral grains of sand size: Geological Society of America Bulletin, v. 56, p. 1207.

United Nations Educational Scientific and Cultural Organization, 1980, Vegetation map of South America: 2 p.

van Andel, T. H., 1950, Provenance, transport and deposition of Rhine sediments: Wageningen, The Netherlands, Veenman en Zonen, 129 p.

Van der Hammen, T., 1985, The Plio-Pleistocene climatic record of the tropical Andes: Journal of the Geological Society of London, v. 142, p. 483–489.

MANUSCRIPT ACCEPTED BY THE SOCIETY DECEMBER 24, 1992

Printed in U.S.A.

Geological Society of America
Special Paper 284
1993

Climatic and tectonic control on sandstone composition in the Permo-Triassic Sydney foreland basin, eastern Australia

Prodip K. Dutta and Russel W. Wheat*
Department of Geography and Geology, Indiana State University, Terre Haute, Indiana 47809

ABSTRACT

The Permo-Triassic Sydney basin is located between a craton to the west and a volcanic source to the east. Systematic variations in sandstone composition, both in space and time, are characteristic patterns in the Sydney basin. A sharp change in sandstone composition is observed in the Late Permian from a quartzo-feldspathic subarkose to volcanic arenite. This change is attributed to the syndepositional development of a volcanic source along the eastern margin of the basin. Throughout the latter part of the Late Permian until Early Triassic time, the sediments were primarily derived from the volcanic source with a secondary mode from the craton. During this time, sandstone composition shows a gradual change in maturity from volcanic arenite to quartz arenite. This maturation was mostly related to chemical weathering in a temperate-humid climate in the source area. Petrographic data from the Sydney basin show that even temperate-humid climate can generate quartz arenite from a mixed cratonic-volcanic provenance and obliterate the plate tectonic signatures. In the Middle Triassic, the volcanic source was virtually shut off, and a quartz arenite unit was derived from the deeply weathered craton.

Spatial variations in sandstone composition within correlative units and cyclicity within the broad maturity trend are also observed. Sandstones rich in volcanic grains to the east and sandstones rich in cratonic quartzose grains to the west, are related to the proximity of the respective source terrain. The cyclicity in compositional trends is related to a subtle variation in the duration of chemical weathering and fluvial dynamics.

INTRODUCTION

Provenance interpretation can yield vital clues about plate tectonic setting, dispersal paths, location of source area, and climate. The nature of detrital grains leads us to determine the source rock composition and, in turn, plate tectonic setting. Paleocurrent studies help us to locate the source area and trace the dispersal pathway. Interpreting climate from framework grains, however, is either difficult or considered to be less significant because, until recently, climate has received little attention in most sandstone provenance studies. The effect of climate, however, is pervasive and may change sediment compositions through chemical weathering. Without a conscious effort, it is possible to overlook the climatic signature in sandstones, as has been the case in most sandstone provenance studies in the past. As a consequence, an important aspect of sand generation remains poorly understood.

One of the approaches in provenance studies is to consider sandstone composition in the context of a plate tectonic framework. The basic tenet in this approach is that the plate tectonic setting exerts first-order control on sand composition and overwhelms the effects of all other variables such as climate, environment of deposition, rigor of transport, etc. (Crook, 1974; Schwab, 1975; Dickinson and Suczek, 1979; Valloni and Maynard, 1981). First-order control of climate on Holocene sand composition has also been documented in big rivers (Potter, 1978; Franzinelli and Potter, 1983; Johnsson et al., 1988; Johnsson and Meade, 1990) as well as in low-order streams (Ruxton, 1970; Basu, 1976;

*Present address: Capital Environmental and Engineering Inc., Indianapolis, Indiana 46421.

Dutta, P. K., and Wheat, R. W., 1993, Climatic and tectonic control on sandstone composition in the Permo-Triassic Sydney foreland basin, eastern Australia, *in* Johnsson, M. J., and Basu, A., eds., Processes Controlling the Composition of Clastic Sediments: Boulder, Colorado, Geological Society of America Special Paper 284.

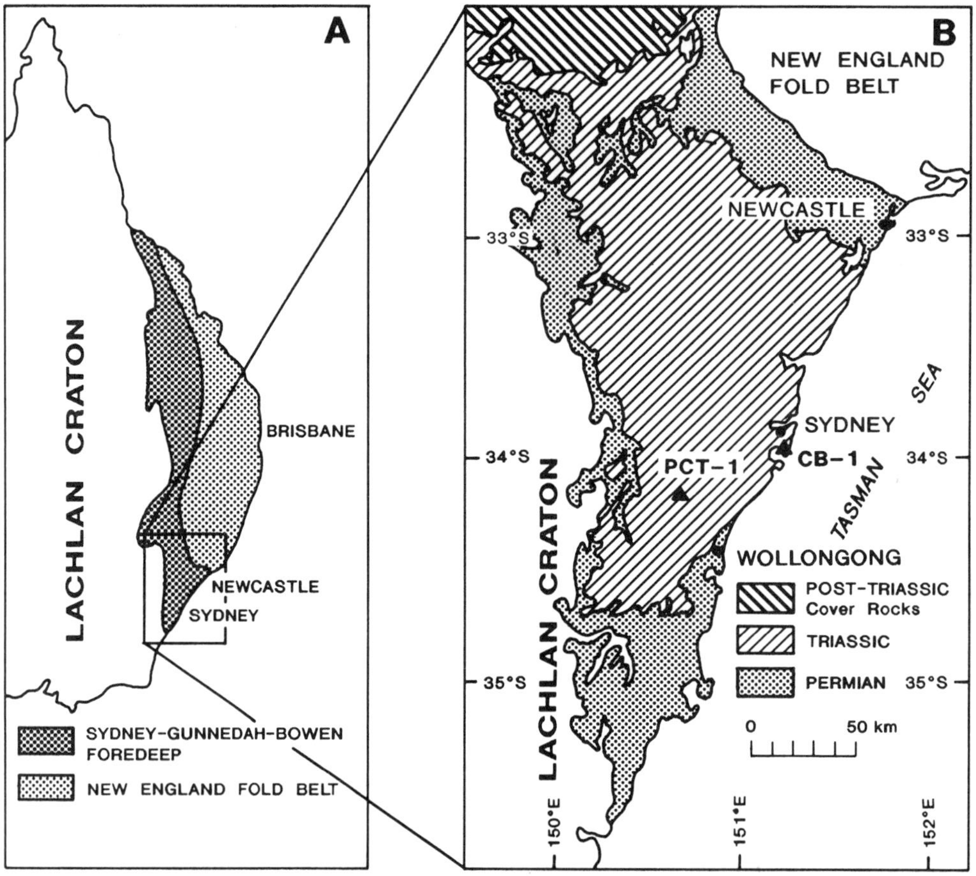

Figure 1. A, Map of the Sydney-Gunnedah-Bowen foredeep along the eastern seaboard of Australia (modified after Schiebner, 1985). B, Geological map of the Sydney basin showing borehole site locations in Picton (PCT-1) and Cape Banks (CB-1). Also shown on the map are the structural boundaries of the Sydney basin bounded by the New England Fold Belt (NEFB) in the northeast and the Lachlan craton in the west. After Herbert, 1980.

Young, 1976; Velbel, 1985; Grantham and Velbel, 1988; Johnsson and Stallard, 1989).

Climatic control on sand/sandstone composition has also been observed in the geological past. First-cycle Permo-Triassic sandstones in intracratonic Gondwana basins in India show a maturity trend, from arkose at the base to quartz arenite at the top of the sequence (Suttner and Dutta, 1986; Dutta, 1987). These authors attributed the genesis of quartz arenite to chemical destruction of unstable grains in a humid climate. Similar maturity trends are also observed in the Permo-Triassic sandstones of different Gondwana basins in eastern Australia (Conaghan et al., 1982; Conaghan, personal communication, 1989). Such similarities in sandstone compositional trends in the stratigraphically equivalent units of different Gondwana basins in eastern Australia lead to the question, has climate had any effect in modifying sandstone composition and in shaping the maturity trends? It is a particularly relevant question since the maturity trends have been interpreted to be related to the change in provenance from a volcanic to a cratonic source (Jensen, 1975; Conaghan et al., 1982). The purpose of this work is to address that question by studying the Permo-Triassic sandstones of the Sydney basin, the most-studied Gondwana basin in eastern Australia.

REGIONAL SETTING AND PREVIOUS WORK

The Sydney basin, the southernmost part of a north-south–trending foredeep in eastern Australia, is located between the Lachlan craton to the west and the New England Fold Belt (NEFB) to the northeast. The NEFB represents a sequence of basic to acidic volcanic rocks, moderately metamorphosed and deformed volcaniclastics, limestone, marine chert, and siltstone. The rocks in the NEFB are mostly of Paleozoic age. Structurally, the NEFB is a series of accretionary terranes, developed mostly during Late Cambrian through Permian (Scheibner, 1985), along the eastern margin of a westerly subducting convergent plate (Cawood, 1982). The nature of the eastern boundary of the basin has long been debated and still remains a matter of conjecture. The foredeep represents a retroarc foreland basin, stretching continuously for nearly 2,000 km along the seaboard of eastern Australia (Herbert, 1980). The Sydney basin represents the southernmost extension of the foredeep (Fig. 1) (Herbert, 1980;

Schiebner, 1985). The rock types in the craton range in age from Cambrian through Carboniferous and are typically granite, arenite, argillite, limestone, chert, and metasedimentary and metavolcanic rocks. A basic volcanic suite known as the Gerringong Volcanics of Late Permian age borders the southeastern part of the basin.

Sedimentation in the Sydney basin started during the Late Carboniferous. Fluvioglacial sediments were deposited within narrow, elongated depressions along the eastern margin of the Lachlan craton. In the Early Permian, the area subsided, creating a broad continental shelf where clastic sediments were deposited in shallow-marine environments. Also in the Early Permian, the subduction zone to the east became inactive, but the oceanic plate continued to push against the NEFB (Hunt and Hobday, 1984). The Sydney basin assumed its present foreland structural form during the mid-Permian Hunter orogeny (Herbert, 1980). Also during the mid-Permian time, deposition of the Gerringong Volcanics brought about a drastic reduction of open marine conditions. This allowed the development of extensive flood plains on which the fluvial Illawarra Coal Measures, Narrabeen Group, Hawkesbury Sandstone, and Wianamata Group were deposited (Fig. 2) (Herbert, 1980; Ward, 1980; Jones, 1990).

The Upper Permian–Middle Triassic sandstones of the Sydney basin show a compositional trend from volcanic arenite to quartz arenite (Herbert, 1980; Conaghan et al., 1982; Odins et al., 1990). This maturity trend generally has been interpreted as a result of change in source rock composition. The volcanic detritus was derived from a source to the northeast (NEFB), whereas the quartzose sediments came from the craton to the west (Herbert, 1980; Conaghan et al., 1982). Conaghan et al. (1982) evoked a dynamic fluvial system to explain the gradual change in sandstone composition. According to this model, the trunk river fed by transversely flowing tributaries received sediments from both volcanic and cratonic sources. The dynamic fluvial model, however, does not explain the compositional variations in space within correlative units or the maturity of some quartzose sandstones derived from the volcanic source, not does it address the minor cyclicity in compositional trends that we have observed in this work.

APPROACH AND METHODOLOGY

Sampling and field study

Sandstone samples were collected from two boreholes drilled by the Geological Survey of New South Wales (Fig. 1B). Borehole PCT-1, drilled near Picton, penetrated nearly 800 m of rocks including a part of the Shoalhaven Group, Illawarra Coal Measures, Narrabeen Group, Hawkesbury Sandstone, and the Wianamatta Group (Fig. 2). The second borehole, CB-1, located near Cape Banks, 50 km east-northeast of PCT-1, drilled more than 1,100 m. However, only 700 m of borehole core, covering the upper part of the Shoalhaven, Illawarra Coal Measures, and a part of the Narrabeen Group, were available for sampling. Fieldwork was carried out along the eastern margin of the basin from Wollongong to Newcastle (Fig. 1B). At certain sections along the eastern margin of the basin, pebbly, mineralogically immature braided stream deposits alternate with relatively finer and mature sandstone. The latter, products of a meandering trunk river system, exhibit a southerly paleocurrent pattern, while the braided stream deposits have westerly to southwesterly paleocurrent. Similar bimodal paleocurrent distribution was also observed by Ward (1980).

Modal analysis

The mineralogical composition of 60 sandstone samples from two boreholes was determined following the conventional petrographic method. Thin sections were stained for K-feldspar identification. More than 300 grains per thin section, representing a single sample, were counted. To avoid the problem of grain-size–dependent compositional variations, only the medium-grained sand-size fraction was counted (Suttner et al., 1981). Quartz grain types were identified following the methods of Basu et al. (1975) and Folk (1980). Because of their fine-grained nature, Wianamata sandstones were not included in this study. The grain types were identified according to the following scheme:

1. Coarse monocrystalline quartz of plutonic type (Q_m) was identified on the basis of undulosity.
2. Coarse polycrystalline quartz of plutonic and high-rank metamorphic varieties (Q_p) were counted separately.
3. Following Folk (1980), volcanic quartz (Q_v) was identified by its typical euhedral shape (Fig. 3A), embayments (Fig. 3B), and inclusion-free clear transparency (Fig. 3C).
4. Detritus with micro- and crypto-crystalline quartz aggregates was counted as quartzolithic grains (chert-like grains). A considerable number of such grains contain relicts of volcanic quartz (Fig. 4) and/or show relict volcanic texture.
5. Grains with quartzolithic appearance, but with relict plagioclase or any unstable mineral were grouped along with volcanic fragments (L_v; Fig. 5) because such grains represent chemically unstable volcanic constituents.
6. Lithic fragments of volcanic (L_v), plutonic (L_p), and sedimentary/metasedimentary (L_s) origins were differentiated.
7. A very few highly altered grains were observed in this study, but their presence was neglected in modal calculation.

The modal composition data are given in Appendix 1. Sharp and cyclical variations in sandstone composition are observed in both Picton and Cape Banks boreholes. This cyclical pattern necessitates the grouping of sandstones into different units (petrofacies) for convenience of discussion. Petrofacies classification has been based on QFL content in sandstones: seven petrofacies have been identified in CB-1 and six in PCT-1 (Appendix 1).

Microprobe analysis

In order to determine the composition of the volcanic source rock, the chemical compositions of feldspars in volcanic arenite were determined by electron microprobe analysis (Appendix 2). These feldspars occur as detrital grains and also as phenocrysts and microlites in volcanic lithic fragments (Fig. 6).

Oxygen isotopic composition of quartzolithic grains

Grains with microcrystalline and/or crypto-crystalline aggregates of quartz are common in sandstones from the Sydney basin. Such grains were interpreted to be of marine biogenic origin(Conaghan et al., 1982). Many of these grains also contain relicts of volcanic quartz and/or volcanic texture. However, such texture suggests that at least part of the chert-like quartzolithic grains in these sandstones is probably volcanogenic. Dutta (1990) determined the oxygen isotopic composition of 13 samples of chert-like quartzolithic grains from sandstones used in this study. The oxygen isotopic composition of these chert-like quartzolithic samples ranges between 12.7 and 20.8‰ (SMOW). This range of oxygen isotopic value indicates that the quartzolithics or their precursors were in isotopic equilibrium with meteoric water of

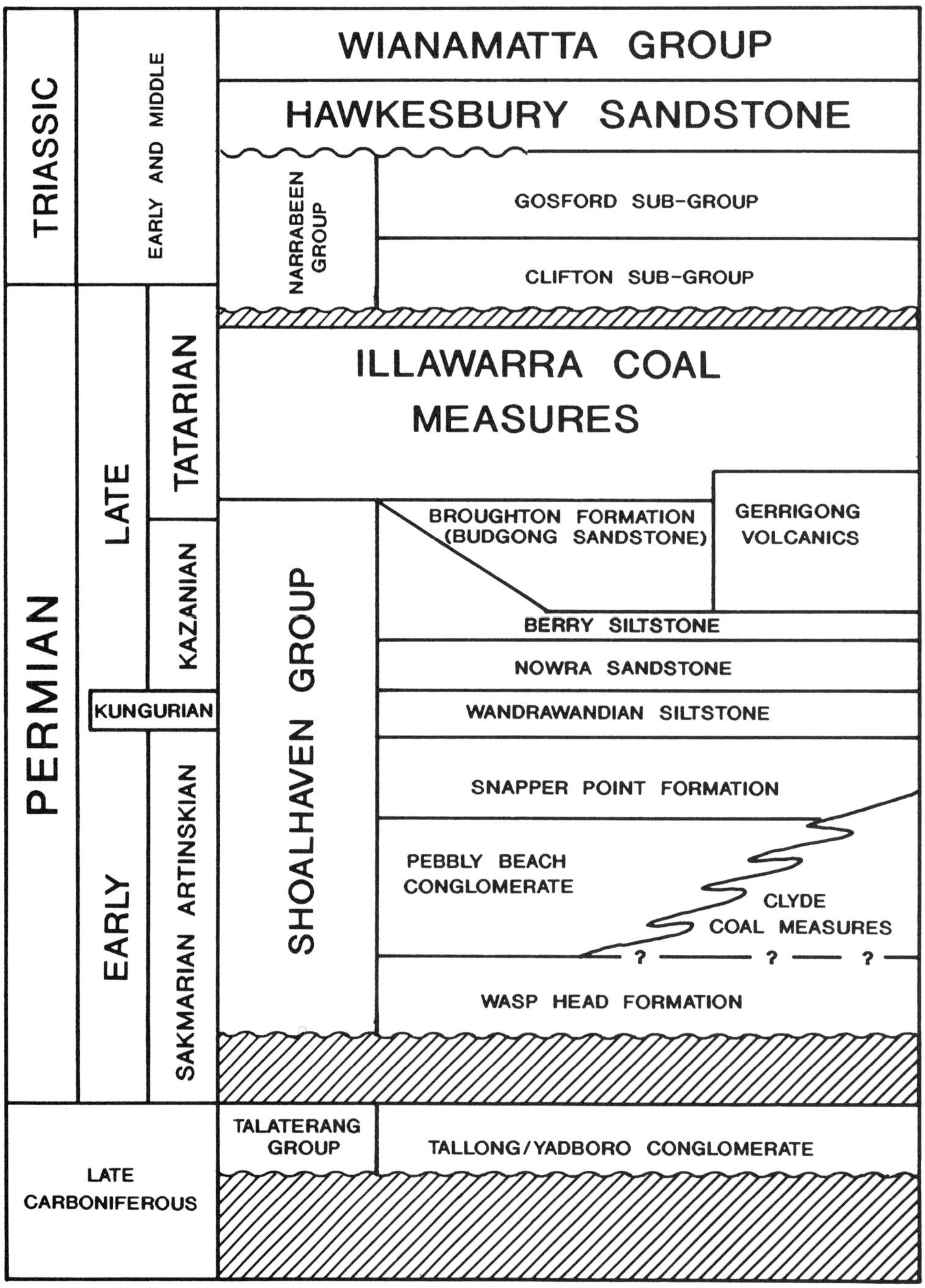

Figure 2. Stratigraphy of the southern Sydney basin. After Herbert, 1980, and Hutton et al., 1990.

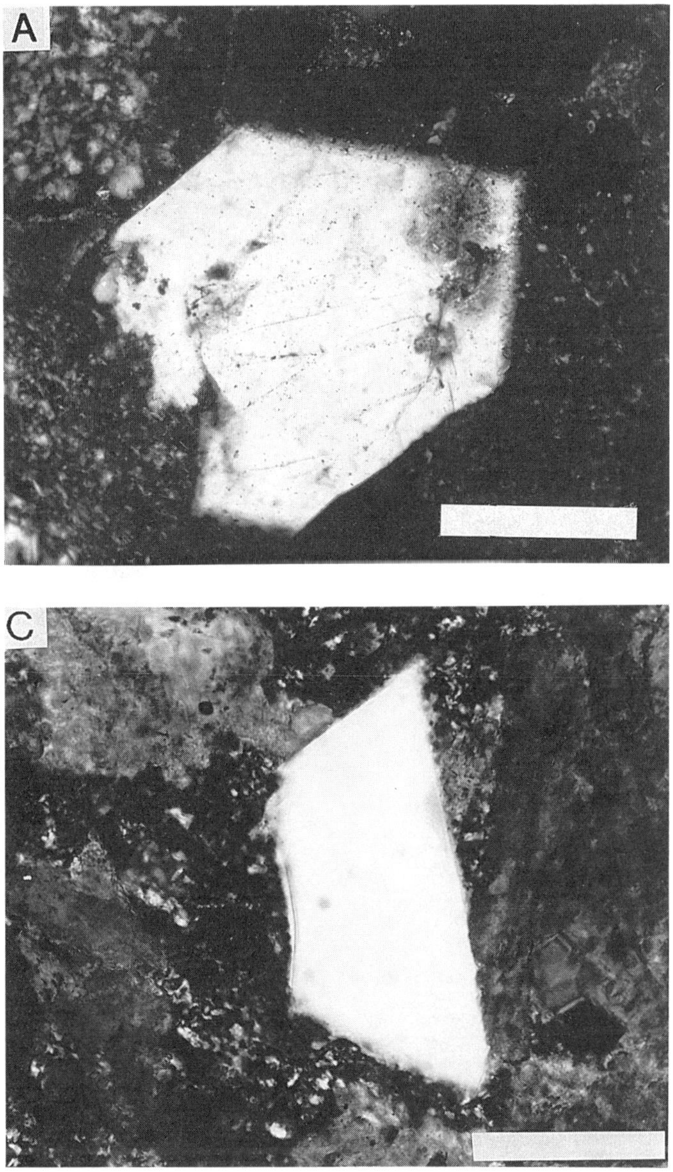

Figure 3. Photomicrographs of volcanic quartz showing typical euhedral face (A), embayments (B), and characteristic shape with inclusion-free clear transparency (C). Such volcanic quartz is present in all petrofacies except in petrofacies I and VI in the Picton area. Crossed Nicols. Scale bars in A and B = 0.25 mm; in C = 0.10 mm.

the Sydney basin during the Permo-Triassic sedimentation. Based on textural and oxygen isotopic evidence, Dutta (1990) interpreted the quartzolithic materials in the southern Sydney basin sandstones to have formed as a result of chemical alteration of volcanic rocks in a meteoric regime.

Classification of detrital grains

In this study, it has been possible to classify detrital grains on the basis of their source rock types, that is, granitic and volcanic. Most sandstones contain a mixture of both, but there are a few exceptions and some petrofacies were primarily derived from a single-source rock type. For example, petrofacies I and VI in the Picton borehole were derived entirely from the craton (Standard, 1969; Herbert, 1980). Quartz types in these sandstones are similar to what Basu et al. (1975) interpreted as plutonic/high-rank

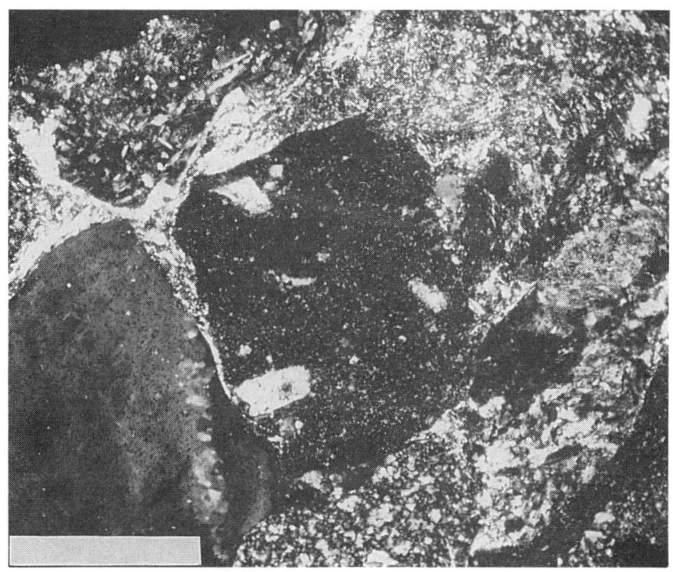

Figure 4. The quartzolithic grain at the center shows relicts of volcanic quartz in micro- and crypto-crystalline quartz aggregates. Crossed Nicols. Scale bar = 0.25 mm.

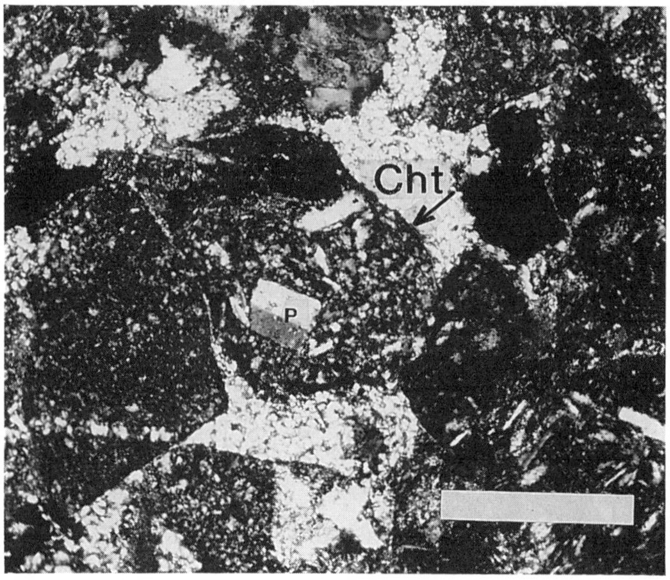

Figure 5. Photomicrograph of a detrital grain showing relict plagioclase (P) within a quartzolithic, chert-like grain (Cht). Volcanic origin of this chert-like grain is indicated by the presence of the euhedral plagioclase grain. Crossed Nicols. Scale bar = 0.62 mm.

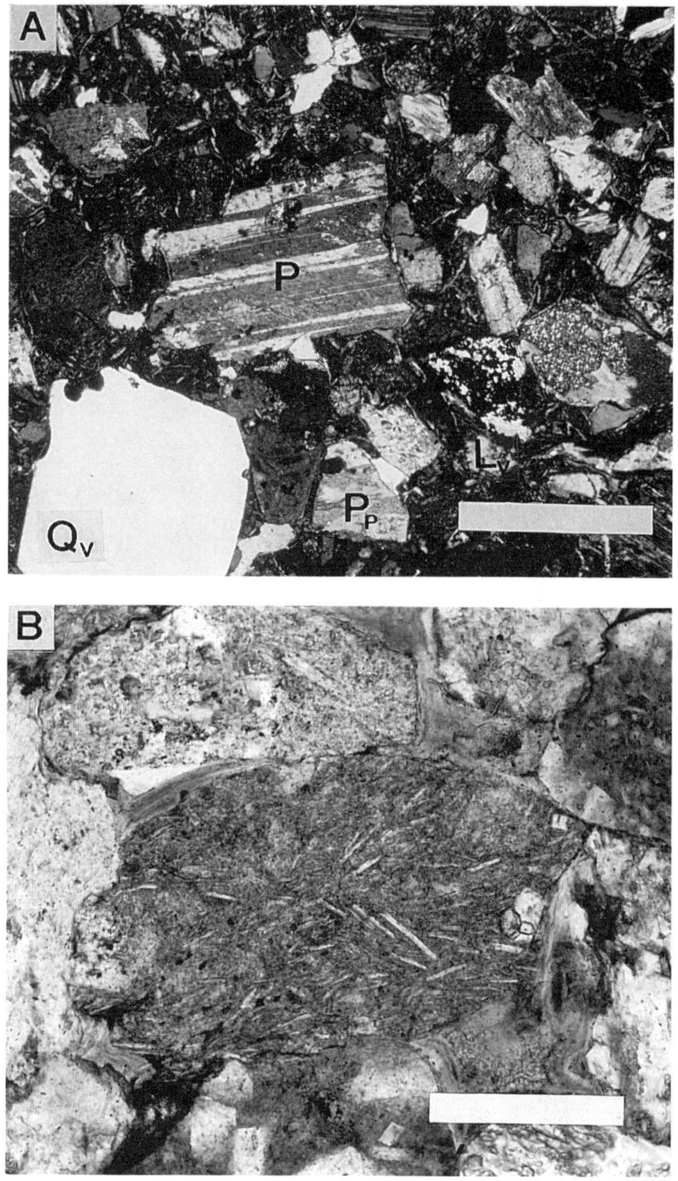

Figure 6. Photomicrographs of a volcanic arenite. A, Detrital plagioclase grain (P), detrital volcanic quartz (Q_v), and a volcanic lithic grain (L_v) with a plagioclase phenocryst (P_p); crossed Nicols. B, Laths of plagioclase in a volcanic lithic fragment; plain polarized light. Scale bars = 0.25 mm.

metamorphic variety. Plagioclase content is relatively low in petrofacies I (4%) but totally absent in petrofacies VI. Although the plagioclase grains in petrofacies I cannot be petrographically associated with any particular source rock type, their occurrence with microcline, orthoclase, plutonic/high-rank metamorphic quartz, and plutonic rock fragments suggests their granitic parentage. Therefore it is reasonable to infer that, in petrofacies I in the Picton borehole, the detritus derived from the craton contains 97% cratonic grains of granitic composition (Appendix 3).

Petrofacies I in the Cape Banks borehole, more proximal to the eastern volcanic source, contains 73% volcanic lithics and 1% volcanic quartz (Appendix 1). The origin of the quartzolithic (9%) and plagioclase (16%) grains in this petrofacies is difficult to ascertain.

Based on the following paleocurrent and paleogeographic evidence, it is possible to determine the origin of plagioclase and quartzolithic grains in the different petrofacies in the Picton and Cape Banks boreholes:

1. The contribution of plagioclase feldspar derived from the craton other than in petrofacies I, in the Picton borehole, is negligible. For instance, in petrofacies III, V, and VI in Picton borehole derived mostly from the granitic source, as indicated by their quartz types, plagioclase content is less than 1% (petrofacies II was derived from a volcanic source). On the other hand, plagioclase is common in volcanic arenite with abundant volcanic lithic and volcanic quartz grains. Therefore, plagioclase grains in sandstones, other than in petrofacies I in the Picton area, are all considered to be volcanogenic.

2. Contribution of quartzolithic grains, possibly of marine biogenic origin, from the craton in the east in petrofacies I and VI in the Picton area was negligible (Appendix 1).

3. Quartzolithic grains derived from the volcanic source in the west are essentially volcanogenic (Dutta, 1990).

4. Plutonic lithic fragments are rare, but a maximum abundance of 2% is observed in petrofacies I in the Picton area (Appendix 1). Their presence in any other petrofacies may be neglected.

5. Sedimentary and metasedimentary lithic fragments are

present in only minor amounts and are considered negligible (Appendix 1).

Based on the above rationale, the modal composition data in Appendix 1 is regrouped into source rock type categories as cratonic grains of granitic composition (CG) and volcanogenic grains (VG) (Appendix 3). Such an approach to classifying grain types is useful in differentiating between climatic control and plate tectonic control on sand composition. This is particularly important in mixed tectonic settings, as in the Sydney basin located between a craton and a volcanic terrain.

SOURCE ROCK COMPOSITION

The nature of detrital grains in Sydney basin sandstones suggests two major source rock types: an ultimate plutonic/high-rank metamorphic source of granitic composition and a volcanic source. The specific composition of the volcanic source is not well known. Microprobe analyses of detrital feldspars, feldspar phenocrysts, and fine-grained feldspar laths in volcanic clasts show albitic composition with the exception of two grains having low Na_2O content (Appendix 2).

It may be questioned whether the albitic composition, as observed in the Sydney basin, is diagenetic in origin; however, this is not likely. Albitization through replacement at depths greater than 2,500 m has been observed in many feldspar-rich sandstones (Land and Milliken, 1981; Boles, 1982; Walker, 1984; Morad et al., 1990). In the present study, the grains selected for microprobe analysis show no replacement texture. Moreover, feldspar's compositional bimodality in a single thin section; one group with Na_2O content varying between 0.1 and 3.65% and the other group with Na_2O content ranging from 10.46 to 11.17%, prompts us to conclude that the albitic composition is primary. The presence of sodic feldspars and volcanic quartz in the volcanic arenite suggests an acidic composition of the source rock.

SOURCE ROCK LOCATIONS

The paleogeographic (Herbert, 1980) and paleocurrent studies of earlier workers (Ward, 1980; Conaghan et al., 1982), as well as this research, suggest that the detritus of plutonic/high-rank metamorphic and volcanic parentages were derived from western and eastern sources, respectively. In addition, small amounts of sedimentary and metasedimentary lithic fragments were derived from both sources.

Although the granitic source to the east is well established by earlier workers and is confirmed in this study, the location and the age of the volcanic source, however, are uncertain. Regarding the source of volcanic detritus in Sydney basin sandstones, the NEFB is considered to be the only source by earlier workers (Herbert, 1980; Conaghan et al., 1982). However, several lines of evidence suggest a major source area other than the present location of the NEFB for the acid volcanic rocks:

1. Along the eastern margin of the basin from Newcastle in the north to Garie, south of Sydney, interbedded conglomerate, conglomeratic sandstones, and relatively finer grained sandstones of Illawara and Narrabeen groups show bimodal paleocurrent pattern. The coarser facies rich in volcanic detritus, deposited by braided streams, show westerly to southwesterly paleocurrent, whereas the finer sandstones representing meandering river deposits exhibit southerly paleocurrent direction (Fig. 7). Similar observations have also been made by Ward (1980) and Conaghan et al. (1982), indicating a volcanic source in the east.

2. An eastern source for the volcanics is also indicated by westerly alignments of tree stumps and fallen trunks within an ash bed in the Illawara Coal Measures exposed at the Swansea Heads, south of Newcastle (Diessel, 1983).

3. Magnetic anomalies indicate the presence of igneous rocks in the continental shelf within 3 km of the coast near Stanwell Park, south of Sydney (Fig. 7) (Ringis et al., 1970). The presence of igneous bodies has also been detected, in a geophysical study, a few kilometers offshore south of New Castle (C.F.K. Diessel, personal communication, 1990). Though the ages of these igneous bodies in the continental shelf are not known, their

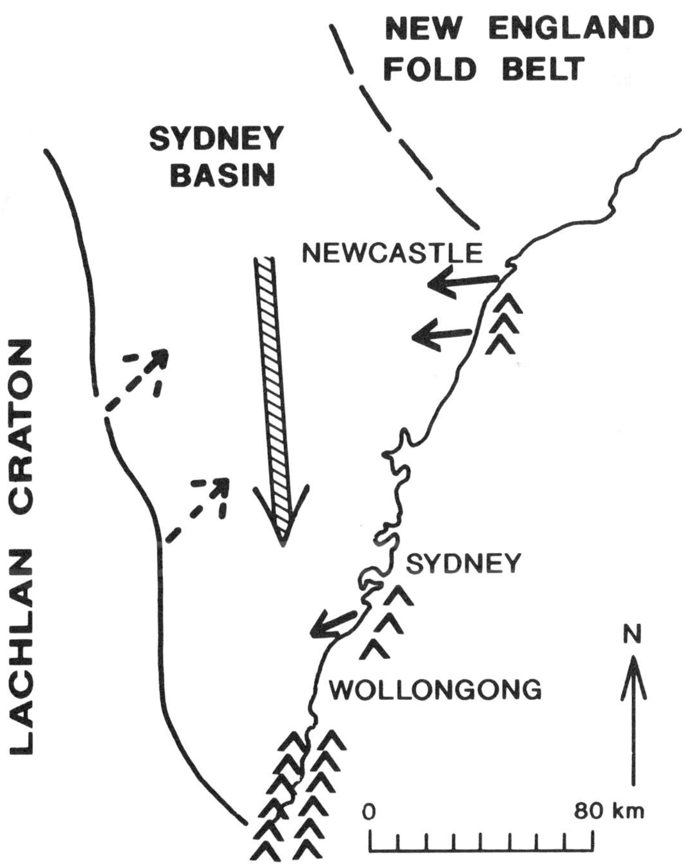

Figure 7. Paleocurrent pattern in the Sydney basin during Illawara and Narrabeen sedimentation. The primary southerly mode of the trunk river is shown by a long arrow; the westerly secondary mode is shown by small arrows. Another secondary northeasterly mode shown by dashed arrows was observed by Conaghan et al. (1982). Geophysical data locations in the offshore (south of Newcastle and south of Sydney), indicating the presence of igneous rocks and the Gerringong Volcanics outcrops (south of Wollongong) in the southern margin of the basin, are shown by inverted V symbols.

structural alignment with the Gerringong Volcanics (Fig. 7) possibly suggests their contemporaneity with the Permo-Triassic volcanism in the area.

4. Plagioclase feldspar in Carboniferous rocks from the NEFB shows an intermediate composition (Appendix 4) (Wilkinson, 1971). This is in sharp contrast with the composition of the volcanogenic feldspars in sandstones in the southern Sydney basin (Appendix 2). Feldspar composition indicates an alkalic composition for the volcanics. This find suggests that the NEFB may not be the only source of volcanic detritus in these sandstones.

5. Radiolarian cherts have been reported from the NEFB, but the oxygen isotopic evidence suggests that the chert-like quartzolithic grains in sandstones in this study were not of the radiolarian type. This again confirms that the NEFB was not the primary source area for the sediments in the southern Sydney basin.

Interbedded ashes and the petrified forest covered with ash beds within the Illawara Coal Measures, observed in the field, may indicate a proximal volcanic source during Permo-Triassic transition. The oxygen isotopic composition of volcanogenic quartzolithic grains supports their contemporaneity with this volcanism (Dutta, 1990). These observations suggest that the volcanogenic detritus was derived predominantly from a contemporaneous volcanic source, possibly a volcanic arc, developed along the eastern margin of the basin, close to the present coast (Fig. 7). This supports the interpretation, independently arrived at by Jones et al. (1987) that there was a landmass south of the NEFB, bordering the Sydney basin.

SANDSTONE COMPOSITION

In the Sydney basin, sandstones show wide variations in mineralogical composition. The Upper Permian through Middle Triassic sandstones in the Picton borehole exhibit major cyclicity in compositional variation with time: subarkose → volcanic arenite → quartz arenite. A similar maturation trend is observed in sandstones from Cape Banks. QFL diagrams (Fig. 8A, B) show the compositional trends in the two boreholes. Composition also is plotted along a stratigraphic profile, using a maturity parameter [quartzose (Q)/labile (L_a) grains] and a provenance parameter [cratonic (CG)/volcanogenic (VG) grains] (Fig. 9). Compositional variations are also observed in both space and time. Within the broad compositional trend, cyclicity is common.

Picton borehole

Petrofacies I is a subarkose with an average Q:F:L ratio of 85:11:4 (Appendix 1). A sharp change from subarkose to volcanic arenite is observed in petrofacies II, with an average Q:F:L proportion of 7:23:70. Petrofacies III, with Q:F:L values of 97:2:1 is a quartz arenite. Maturity reversal is again observed in petrofacies IV, a sublitharenite unit with a Q:F:L ratio of 77:10:13. Petrofacies V and VI are very similar with respective Q:F:L ratios of 98:1:1 and 100:0:0, but they differ significantly in grain types. Where petrofacies V contains 35% quartzolithic grains, petrofacies VI contains only 1%.

Cape Banks borehole

The Cape Banks section stratigraphically represents only a part of the Picton sequence. Both the basal (petrofacies I) and the upper sections (petrofacies VI and possibly the upper part of petrofacies V) of the Picton borehole are missing in Cape Banks (Fig. 9). The overall maturity trend is similar to that of the Picton area, but compositional variations between the two sections differ in scale. The sandstones in this borehole are mostly volcanic arenites. Petrofacies I has an average Q:F:L value of 10:16:74. A sharp change is observed in a single sample of petrofacies II (Q:F:L = 70:10:20). Petrofacies III swings back to immaturity with a Q:F:L ratio of 44:16:40. Petrofacies IV is a mature unit with Q:F:L ratio of 85:5:10. The overlying petrofacies V has a Q:F:L ratio of 47:10:43. A continuous gradual change in maturity is observed in petrofacies VI through VII with Q:F:L ratios of 72:10:18 and 87:7:6, respectively.

DISCUSSION

Four different kinds of mineralogical trends are observed in Sydney basin sandstones: (1) an overall mineralogical maturity trend through time (Appendix 1; Figs. 8, 9); (2) compositional variations in space and minor cyclical variations within broad maturity trends (Figs. 8, 9); (3) a sharp change in grain types from granitic (cratonic) to volcanic (from petrofacies I to II in the picton borehole) (Appendix 3; Fig. 9); and (4) a gradual change in grain types from volcanic to granitic (Appendix 3; Fig. 9).

In the Picton area, petrofacies I contains 100% craton-derived grains (CG) of granitic type. Petrofacies II contains 96% volcanogenic grains. In Cape Banks, petrofacies I, the correlative of petrofacies II in Picton, contains 100% volcanogenic grains. Such a sharp change in grain types in the Sydney basin reflects the rapid development of a volcanic arc along the eastern margin of the basin (Jones, 1990). This volcanic arc remained as an important source throughout Late Permian–early Early Triassic time and affected the development of all petrofacies in Cape Banks and petrofacies II through the lower part of petrofacies V in the Picton area. From the latter part of Early Triassic the influence of the volcanic arc started to decline, and eventually died out in the Middle Triassic during the deposition of petrofacies VI in the Picton area. The gradual evolution of the cratonic setting can be seen from the base of petrofacies IV with 12% CG to 100% CG in petrofacies VI in the Picton area.

There is a complete agreement between grain types, mineralogical composition, and plate tectonic interpretation in the case of petrofacies I (continental block), II (volcanic arc), and VI (continental block) in Picton and petrofacies I (volcanic arc) in Cape Banks. In each case, there is a direct relationship between sandstone composition and plate tectonic setting where the latter controlled sandstone composition (Dickinson and Suczek, 1979).

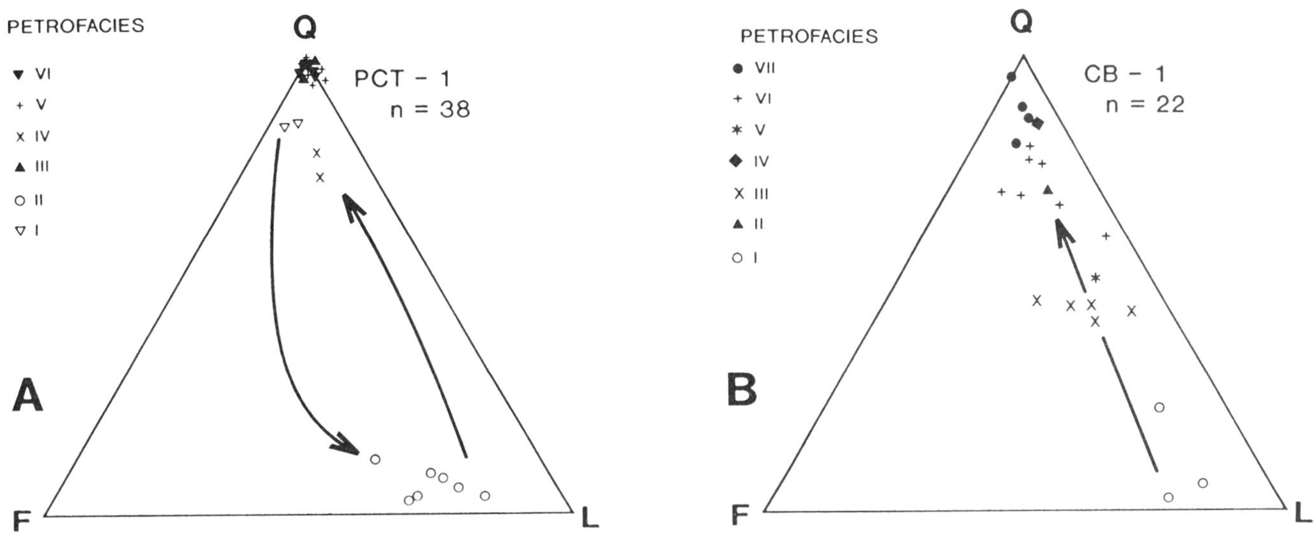

Figure 8. A, QFL compositional diagram for Picton borehole sandstones. B, QFL compositional diagram for Cape Banks borehole sandstones. Q = Coarse crystalline (Q_m + Q_p + Q_v) and chert-like quartzolithic (Q_l) grains; F = feldspars; L = lithic fragments. Roman numerals identify petrofacies (see Appendix 1). Sandstone compositional trends are shown by arrows.

On the other hand, the overall mineralogical maturity through time reflects climatic control on the sandstone composition. Petrofacies I in Cape Banks, derived from a volcanic source (100% VG) has a Q:F:L ratio of 10:16:74. This composition, as an approximation, may be considered as the initial soil-sand composition in the source area prior to sedimentation. This assumption is reasonable since the soil-sand in the source area is compositionally similar to fluvial sand (Suttner et al., 1981). Petrofacies I in the Picton area, entirely derived from the craton (100% CG), has a Q:F:L ratio of 85:13:2, which similarly can be considered as the sand composition of the soil horizon in the cratonic provenance.

Considering the initial composition of the precursor soil-sand in two respective provenances, the mineralogical maturity observed in the Permo-Triassic sandstones of the Sydney basin seems to be independent of the provenance factor. For example, the Cape Banks petrofacies I contains 100% VG (volcanic source) with 90% unstable grains, but petrofacies VI, with 89% VG, contains only 28% unstable grains. Similarly in Picton, petrofacies I contains 15% unstable grains, while petrofacies VI is entirely made up of quartzose grains. This elimination of unstable grains is independent of the source rock composition. This is shown in Figure 9, where the maturity parameter is independent of the provenance parameter.

The maturity of sandstones in both Picton and Cape Banks is related to chemical weathering. Feldspars and volcanic lithic fragments decayed chemically. Quartzose grains in the volcanic source were generated by the release of quartz phenocrysts from volcanic rocks, and by the formation of quartzolithic fragments as a result of chemical weathering/silicification of volcanic ash beds (Dutta, 1990). Independent evidence suggests that a humid climate prevailed during Late Permian and Early Triassic time. During the Late Permian, coal beds were deposited throughout the Sydney basin (Herbert, 1980; Jones, 1990). Chemical weathering of volcanic rocks resulted in thick and widespread kaolinitic claystone deposits during the Early Triassic (Ward, 1980). The supermaturity of petrofacies VI in the Picton area suggests a continuation of the humid climate, at least until the deposition of petrofacies VI in Middle Triassic.

The spatial variations between Cape Banks and Picton are easy to explain. The sandstones in Cape Banks contain more volcanic grains because of their proximity to the volcanic source, while the correlative sandstones closer to the craton contain more cratonic grains. However, the minor cyclical change within the broad maturity trend seems to be related to the combination of fluvial dynamics and climate.

The volcanic arc went through numerous quiescent and eruptive phases. During an eruptive phase after quiescence, the arc shed abundant detritus. This sudden surge in sediment supply caused the braided tributaries to extend and pushed the trunk river farther west. The immature sediments from the volcanic source were buried at a faster rate close to the arc, as indicated by the thinning of the sediment blanket toward the west (Fig. 3 in Conaghan et al., 1982). A large volume of volcanic detritus was laid down in the alluvial plain between the arc and the trunk river. The alluvial sediments, in storage, went through another phase of chemical weathering and matured. Maturity of sediments due to this additional phase of chemical weathering of detritus, while in storage, has also been observed in Holocene sediments (Johnsson et al., 1988; Johnsson and Meade, 1990).

During the next phase of quiescence the trunk river migrated to its preeruptive position in the east. Chemical weathering was active and produced relatively more mature detritus in the volcanic provenance. The trunk river transported and deposited the reworked alluvial sediments in storage, chemically weathered detritus from the arc and the craton over the earlier deposited, relatively immature sediments. This interaction of fluvial dynamics and climate explains the minor cyclicity in sandstone composition in the Sydney basin (Fig. 10).

The composition of Permo-Triassic sandstones in the Sydney basin seems to have been controlled both by plate tectonic setting and climate. A cyclical pattern in tectonic evolution of provenance: continental block → volcanic arc → continental block is clearly demonstrated through a sequential change in cratonic (CG), volcanic (VG), and cratonic detritus (Appendix 3). Conversely, climate evolved in an unidirectional manner. Sedimentation in the Sydney basin was initiated with a glacial climate in Late Carboniferous time (Herbert, 1980). Immaturity of petrofacies I in the Picton borehole reflects continuation of the cold spell where chemical weathering was hindered. In the Late

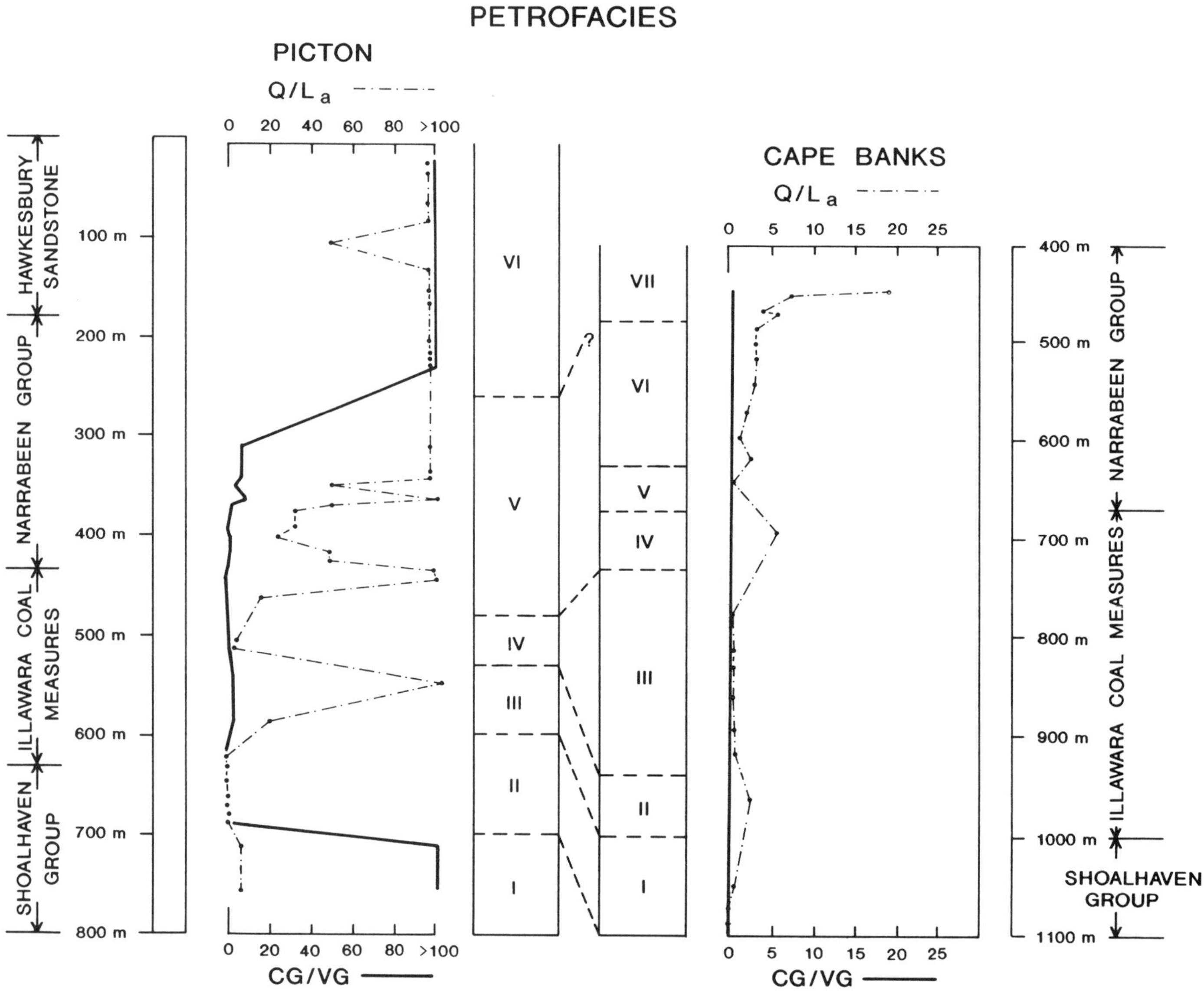

Figure 9. Stratigraphic variations in the framework composition of Picton and Cape Banks sandstones. Composition expressed in ratio of quartzose/labile grains (Q/L_a; maturity parameter) shows cyclical variations within broad overall maturity trends. Also shown in the diagram is a provenance parameter expressed as ratio of cratonic/volcanogenic grains (CG/VG). In the Picton area, maturity index shows wide variations without appreciable variations in provenance factor in petrofacies III, IV, and V. A similar relationship is also observed in the Cape Banks area within the stratigraphic intervals between petrofacies II and VI. This relationship between compositional variations and constancy of provenance demonstrates that the maturity of sandstone is independent of provenance. The petrofacies correlation between Cape Banks and Picton boreholes is based on trends in sandstone composition and the correlation of coal seams in the Illawara Coal Measures.

Permian, climate became sufficiently warm and humid to support lush vegetation and chemically destroy the unstable minerals. Considering the latitudinal location of the Sydney basin roughly around 65°S latitude (Smith et al., 1981; Veevers, 1984), the climate possibly was more temperate-humid rather than warm-humid. Petrographic data from the Sydney basin show that even temperate-humid climate can have a strong influence on sandstone composition and generate quartz arenite from a mixed cratonic-volcanic provenance and can obliterate the plate tectonic signature, unless the genetic imprints of the framework grains are known.

The climatic amelioration in the Sydney basin, from a glacial phase in Late Carboniferous time to a temperate-humid conditions during Permo-Triassic sedimentation, may be attributed to the overall global climatic change from a glacial icehouse state during the Late Paleozoic, to a warm-humid greenhouse state in the Mesozoic (Fischer, 1982). Initiation of sedimentation in the Sydney basin and elsewhere in other Gondwana basins of the Southern Hemisphere coincided with the beginning of this particular Late Paleozoic–Mesozoic phase of the 300-m.y. global climatic cycles (Fischer, 1982).

SUMMARY AND CONCLUSIONS

1. Two types of source rocks have been identified; plutonic/high-rank metamorphic and volcanic. Facies characteristics and petrographic and paleocurrent evidence suggest that the volcanic source was located along the eastern margin of the Sydney basin whereas the cratonic source lay toward the west.

2. Through identification of framework grain types, it has been possible to evaluate the dual role of tectonic setting and climate in controlling sand composition.

3. The overall maturity in sandstone composition in the Sydney basin is possibly related to the global climatic change from a cold glacial climate during the Late Paleozoic to a relatively warm, humid condition in Early Mesozoic time.

4. Even in a mixed cratonic-volcanic provenance, a temperate-humid climate may have a profound influence in controlling sand composition and the generation of quartz arenite. In such a case the climate can obliterate the plate tectonic signature, unless the genetic imprints of the framework grains are known.

5. Compositional variations in space, and minor cyclical change through time, are related to fluvial dynamics, induced by tectonics and modified by climate. As the trunk river shifts back and forth between a craton and a volcanic arc, it modifies the composition through repetition of the chemical weathering of detritus in a floodplain and mixing of detritus derived from volcanic and cratonic provenances.

6. The volcanic source was predominantly of an acid volcanic type to the east. Microprobe analysis of detrital feldspars, feldspar phenocrysts, and fine-grained feldspar laths in volcanic clasts in volcanic arenites show mostly an albitic composition. The predominance of volcanic quartz, along with sodic feldspar, suggests an acidic composition.

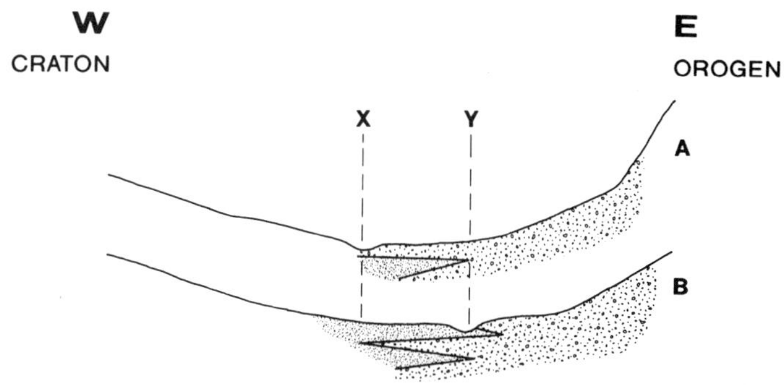

Figure 10. Schematic diagram showing the relative importance of volcanic and cratonic provenances through a single cycle of volcanic eruptions and quiescence. During volcanic quiescence, relief decreases and more intense chemical weathering takes place during which volcanogenic meteoric quartzolithic grains or its precursors are formed. Because of lowering relief, the river shifts to the east and reworks the braided river sediments from the volcanic source. During the height of volcanic activity (A), high-energy braided streams prograde toward basin axis, bring coarse clastics, and are deposited at a point Y (petrofacies I in Cape Banks). At point X the sediments are also volcaniclastic but have some contribution from the craton (petrofacies II in Picton borehole). During the waning stage of volcanism (B), the relief is lowered, the trunk river moves east (point Y), and reworks the earlier braided stream sediments, undergoing a second phase of chemical weathering (petrofacies II in Cape Banks). Petrofacies III in the Picton area (point X) was deposited at the same time; its maturity is related to both proximity of the craton and chemical destruction of unstable grains.

7. Chert-rich sandstones are commonly associated with folded mountain belts along convergent plate boundaries in conjunction with volcanic activity. In such a provenance the micro-crypto-crystalline quartz aggregates, usually referred to as chert, could be of marine biogenic origin derived from the folded mountains, or they could be volcanogenic from a volcanic source. Therefore, it is necessary to determine the origin of quartzolithic grains before any provenance interpretation is attempted.

ACKNOWLEDGMENTS

This research was supported by a grant from Indiana State University. Chris Herbert and Colin Ward reviewed an earlier version of this manuscript. The work in its present form was reviewed by Peter DeCelles, Shan de Silva, Peter Flemings and an anonymous reviewer. Norman Cooprider drafted the figures. The director of the New South Wales Geological Survey, Australia, kindly permitted us to collect the samples used in this study.

APPENDIX 1. MINERALOGICAL COMPOSITION DATA OF SANDSTONES FROM THE SOUTHERN SYDNEY BASIN, AUSTRALIA*

Sample Number	Q_m	Q_p	Q_v	Q_l	Q	K	P	F	L_v	L_p	L_s	L	L_a	Average Q:F:L for Each PF	Q/L_a	PF
PCT-1																
5	49	47	0	4	100	0	0	0	0	0	0	0	0		ID	VI
6	85	14	0	0	99	0	0	0	0	0	1	1	1		99.00	
7	85	12	0	3	100	0	0	0	0	0	0	0	0		ID	
8	87	12	0	0	99	1	0	1	0	0	0	0	1		99.00	
9	83	15	0	0	98	0	0	0	0	0	2	2	2		49.00	
10	84	14	0	2	100	0	0	0	0	0	0	0	0		ID	
11	88	12	0	0	100	0	0	0	0	0	0	0	0		ID	
12	81	18	0	1	100	0	0	0	0	0	0	0	0	100:0:0	ID	
13	71	25	0	3	99	0	0	0	0	0	1	1	1		99.00	
14	86	14	0	0	100	0	0	0	0	0	0	0	0		ID	
15	73	24	0	2	99	0	0	0	0	0	1	1	1		99.00	
16	83	15	0	2	100	0	0	0	0	0	0	0	0		ID	
22	69	18	0	13	100	0	0	0	0	0	0	0	0		ID	V
23	77	10	0	12	99	1	0	1	0	0	0	0	1		99.00	
24	67	11	0	20	98	2	0	2	0	0	0	0	2		49.00	
25	74	16	0	10	100	0	0	0	0	0	0	0	0		ID	
27	52	16	0	30	98	2	0	2	0	0	0	0	2		49.00	
28	55	9	0	33	97	2	0	2	0	0	1	1	3	98:1:1	32.33	
29	49	8	0	40	97	2	0	2	0	1	0	1	3		32.33	
30	32	9	0	53	94	0	2	2	1	0	3	4	6		15.67	
33	45	7	4	42	98	0	0	0	1	0	1	2	2		49.00	
35	54	6	5	33	98	0	2	2	0	0	0	0	2		49.00	
36	42	9	4	44	99	0	0	0	0	1	0	1	1		99.00	
37	17	5	21	56	99	0	0	0	0	0	1	1	1		99.00	
39	7	2	16	69	94	0	4	4	2	0	0	2	6		15.67	
43	13	3	2	61	79	0	9	9	12	0	0	12	21	77:10:13	3.67	IV
44	9	3	0	62	74	1	10	11	15	0	0	15	26		2.85	
45	55	20	0	25	100	0	0	0	0	0	0	0	0	97:2:1	ID	III
49	63	9	0	23	95	4	0	4	1	0	0	1	5		19.00	
50	3	1	0	4	8	0	20	20	71	0	1	72	92		0.09	II
52	2	1	0	6	9	0	22	22	69	0	0	69	91		0.10	
53	0	0	0	3	3	0	29	29	68	0	0	68	97		0.03	
54	2	0	0	2	4	0	14	14	82	0	0	82	96	7:23:70	0.04	
55	5	0	0	1	6	0	18	18	76	0	0	76	94		0.06	
56	4	0	0	0	4	0	27	27	69	0	0	69	96		0.04	
57	8	3	1	0	12	0	31	31	57	0	0	57	88		0.14	
58	77	5	0	3	85	8	4	12	0	1	2	3	15	85:11:4	5.66	I
59	78	4	0	4	86	6	4	10	0	2	2	4	14		6.14	
CB-1																
1	32	6	3	54	95	0	5	5	0	0	0	0	5		19.00	VII
3	34	7	8	39	88	0	6	6	5	0	1	6	12		7.33	
4	22	4	11	43	80	0	11	11	8	0	1	9	20	87:7:6	4.00	
6	17	8	9	51	85	0	6	6	9	0	0	9	15		5.67	
7	8	6	15	50	79	0	9	9	12	0	0	12	21		3.76	VI
8	5	2	19	44	70	0	19	19	9	2	0	11	30		2.33	
10	9	5	10	53	77	0	10	10	10	0	3	13	23		3.35	
12	8	3	8	57	76	0	8	8	14	0	2	16	24	72:10:18	3.17	
13	4	5	16	44	69	0	16	16	13	0	2	15	31		2.22	
14	3	0	24	33	60	0	4	4	35	0	1	36	40		1.50	
15	12	5	4	50	71	0	5	5	19	0	5	24	29		2.45	
17	6	7	6	28	47	0	10	10	39	0	4	43	53	47:10:43	0.89	V
19	6	5	17	57	85	0	5	5	8	0	2	10	15	85:5:10	5.67	IV

APPENDIX 1. MINERALOGICAL COMPOSITION DATA OF SANDSTONES* (continued)

Sample Number	Q_m	Q_p	Q_v	Q_l	Q	K	P	F	L_v	L_p	L_s	L	L_a	Average Q:F:L for Each PF	Q/L_a	PF
CB-1 (continued)																
21	2	1	18	18	39	0	15	15	45	0	1	46	61		0.64	III
22	6	3	16	20	45	0	18	18	36	0	1	37	55		0.82	
23	5	2	16	23	46	0	24	24	30	0	0	30	54	44:16:40	0.85	
26	2	1	11	30	44	0	7	7	47	0	2	49	56		0.79	
28	3	0	23	19	45	0	14	14	40	0	1	41	55		0.82	
30	9	0	23	38	70	0	10	10	19	0	1	20	30	70:10:20	2.33	II
33	0	0	3	20	23	0	12	12	63	0	2	65	77		0.30	I
34	0	0	0	6	6	0	12	12	82	0	0	82	94	10:16:74	0.06	
35	0	0	1	1	2	0	25	25	73	0	0	73	98		0.02	

*Key to Appendices 1 and 3:
Q_m = Monocrystalline quartz
Q_p = Polycrystalline quartz
Q_v = Volcanic quartz
Q_l = Chert-like quartzolithic grain
$Q = Q_m + Q_p + Q_v + Q_l$ = Total quartzose grain
K = Potassium feldspar
P = Plagioclase
F = Total feldspar grain
L_v = Volcanic lithic grain
L_p = Plutonic rock grain
L_s = Sedimentary and metasedimentary grain
$L = L_v + L_p + L_s$ = Total unstable lithic grain
$L_a = L + F$ = Total unstable grain
PF = Petrofacies
ID = Indeterminant
CG = Total cratonic grain
VG = Total volcanogenic grain

APPENDIX 2. MICROPROBE ANALYSIS OF FELDSPARS OCCURRING AS DETRITAL GRAIN AND AS PHENOCRYSTS AND GROUNDMASS LATHS IN VOLCANILITHIC GRAINS IN UPPER PERMIAN SANDSTONES IN THE SYDNEY BASIN

Sample		Chemical Analysis					
Grain Type and	Number	Na_2O	Al_2O_3	SiO_2	K_2O	CaO	Total
CB-1-25							
Detrital	1	11.35	20.46	68.43	0.02	0.00	100.26
Detrital	2	10.88	20.56	68.61	0.07	0.02	100.14
Detrital	3	10.36	21.73	66.41	0.89	0.35	99.74
Detrital	4	11.34	20.16	68.38	0.07	0.09	100.04
Detrital	5	10.62	21.04	68.26	0.21	0.37	100.50
Detrital	6	11.56	19.61	68.70	0.07	0.05	99.99
Lath	1	10.87	20.64	69.66	0.11	0.05	101.33
Lath	2	10.48	20.70	66.71	0.11	0.53	98.53
PCT-1-44							
Detrital	1	10.82	20.95	67.21	0.11	0.59	99.68
PCT-1-56							
Detrital	1	3.65	19.36	64.88	10.60	0.00	98.50
Detrital	2	0.10	18.71	64.93	15.21	0.00	98.95
Detrital	3	10.59	20.73	66.61	0.36	0.41	98.68
Detrital	4	11.15	20.21	69.01	0.04	0.02	100.44
Detrital	5	10.68	20.72	68.00	0.54	0.24	100.19
Phenocryst	1	11.14	20.70	68.91	0.25	0.15	101.16
Phenocryst	2	10.66	20.83	67.77	0.25	0.05	99.56
Phenocryst	3	10.80	20.47	68.08	0.18	0.08	99.61
Phenocryst	4	10.66	20.51	68.41	0.19	0.07	99.85
Phenocryst	5	11.17	20.05	69.20	0.09	0.07	100.59
Lath	1	10.55	20.15	67.07	1.36	0.22	99.36
Lath	2	10.72	20.65	68.32	0.15	0.15	100.01
Lath	3	10.55	20.25	67.07	1.36	0.22	99.36
Lath	4	10.46	20.49	68.27	0.46	0.25	99.94
Lath	5	11.02	20.04	68.75	0.13	0.03	99.97

APPENDIX 3. REGROUPING OF MINERALOGICAL COMPOSITION DATA FROM APPENDIX 2 INTO PROVENANCE CATEGORIES

Sample Number	Regrouped Data CG	Regrouped Data VG	Normalized to 100% CG	Normalized to 100% VG	Average CG:VG of Each PF	CG/VG	PF
PCT-1							
5	100	0	100	0		ID	VI
6	100	0	100	0		ID	
7	97	0	100	0		ID	
8	100	0	100	0		ID	
9	98	0	100	0		ID	
10	98	0	100	0		ID	
11	100	0	100	0	100:0	ID	
12	99	0	100	0		ID	
13	96	0	100	0		ID	
14	100	0	100	0		ID	
15	97	0	100	0		ID	
16	98	0	100	0		ID	
22	87	13	87	13		6.69	V
23	88	12	88	12		7.33	
24	80	20	80	20		4.00	
25	90	10	90	10		9.00	
27	70	30	70	30		2.33	
28	65	34	66	34		1.94	
29	59	40	60	40	60:40	1.50	
30	42	57	42	58		0.72	
33	52	47	53	47		1.13	
35	60	40	60	40		1.50	
36	51	48	52	48		1.08	
37	23	77	23	77		0.30	
39	9	91	9	91		0.10	
43	16	84	16	84	14:86	0.19	IV
44	12	87	12	88		0.14	
45	75	25	75	25	75:25	3.00	III
49	75	24	75	25		3.00	
50	4	96	4	96		0.04	II
52	3	97	3	97		0.03	
53	0	100	0	100		0.00	
54	2	98	2	98	4:96	0.02	
55	5	95	5	95		0.05	
56	4	96	4	96		0.04	
57	11	89	11	89		0.12	
58	97	0	100	0		ID	I
59	96	0	100	0	100:0	ID	
CB-1							
1	38	62	38	62		0.61	VII
3	41	58	41	59	32:68	0.69	
4	26	73	26	74		0.35	
6	25	75	25	75		0.33	
7	14	86	14	86		0.16	VI
8	9	91	9	91		0.10	
10	14	83	14	86	11:89	0.16	
12	11	87	11	89		0.12	
13	9	89	9	91		0.10	
14	3	96	3	97		0.03	
15	17	78	18	82		0.22	
17	13	83	14	86	14:86	0.16	V
19	11	87	11	89	11:89	0.12	IV

APPENDIX 3. REGROUPING OF MINERALOGICAL COMPOSITION DATA FROM APPENDIX 2 INTO PROVENANCE CATEGORIES (continued)

Sample Number	Regrouped Data CG	VG	Normalized to 100% CG	VG	Average CG:VG of Each PF	CG/VG	PF
CB-1 (continued)							
21	2	97	2	98		0.02	III
22	9	90	9	91		0.10	
23	7	93	7	93	5:95	0.07	
26	3	95	3	97		0.03	
28	3	96	3	97		0.03	
30	9	90	9	91	9:91	0.10	II
33	0	98	0	100		0.00	I
34	0	100	0	100	0:100	0.00	
35	0	100	0	100		0.00	

APPENDIX 4. CHEMICAL ANALYSIS OF FELDSPARS IN VOLCANICS OF THE CARBONIFEROUS OF NEW SOUTH WALES*

	1	2	3	4	5	6	7	8	9
SiO_2	54.19	55.25	55.16						
TiO_2	0.09	0.02	0.02						
Al_2O_3	29.02	28.42	28.63						
Fe_2O_3	0.15	0.14	0.18						
FeO	0.22	0.14	0.05						
MgO	0.14	0.08	0.06						
CaO	11.02	10.12	10.04	9.86	9.10	9.68	6.36	4.00	0.44
Na_2O	4.76	5.34	5.20	5.38	4.82	5.50	6.68	4.09	3.72
K_2O	0.53	0.49	0.83	0.64	0.63	0.45	0.80	0.62	10.73

*Southern New England Fold Belt; Wilkinson, 1971.
Key to table headings: 1 = Labradorite from hypersthene basalt; 2 = Labradorite from vitrophyric hypersthene-augite dacite; 3 = Labradorite from vitrophyric hypersthene-augite dacite; 4 = Labradorite from vitrophyric hypersthene-augite dacite; 5 = Labradorite from ignimbritic hypersthene-hornblende dacite; 6 = Andesine from ignimbritic hypersthene-hornblende dacite; 7 = Andesine from ignimbritic rhyodacite; 8 and 9 = Andesine and sanidine ignimbritic biotite rhyodacite.

REFERENCES CITED

Basu, A., 1976, Petrology of Holocene fluvial sand derived from plutonic source rocks: Journal of Sedimentary Petrology, v. 46, p. 694–709.

Basu, A., Young, S. W., Suttner, L. J., James, W. C., and Mack, G. H., 1975, Re-evaluation of the use of undulatory extinction and polycrystallinity in detrital quartz for provenance interpretation: Journal of Sedimentary Petrology, v. 45, p. 873–882.

Boles, J. R., 1982, Active albitization of plagioclase, Gulf Coast Tertiary: American Journal of Science, v. 282, p. 165–180.

Cawood, P. A., 1982, Structural relations in the subduction complex of the Paleozoic New England Fold Belt, eastern Australia: Journal of Geology, v. 90, p. 381–392.

Conaghan, P. J., Jones, J. G., McDonnell, K. L., and Royce, K., 1982, A dynamic fluvial model for the Sydney Basin: Geological Society of Australia Journal, v. 29, p. 55–65.

Crook, K.A.W., 1974, Lithogenesis and geotectonics: The significance of compositional variations in flysch arenites (graywackes), in modern and ancient geosynclinal sedimentation: Society of Economic Paleontologists and Mineralogists Special Publication No. 19, p. 304–310.

Dickinson, W. R., and Suczek, C. A., 1979, Plate tectonics and sandstone composition. Bulletin of the American Association of Petroleum Geologists, v. 63, p. 2164–2182.

Diessel, C.F.K., 1983, Excursion Synopsis: Seventeenth Newcastle Symposium: Newcastle, Australia, University of Newcastle, 6 p.

Dutta, P. K., 1987, Upper Kamthi: A riddle in the Gondwana stratigraphy of India, *in* McKenzie, G. D., ed., Gondwana Six: Stratigraphy, sedimentation, and paleontology: American Geophysical Union Geophysical Monograph 41, p. 299–238.

Dutta, P. K., 1990, Oxygen isotopic signature of detrital chert as a provenance indicator [abs.]: Thirteenth International Sedimentological Congress, Nottingham, p. 147–148.

Fischer, A. G., 1982, Long-term climatic oscillation recorded in stratigraphy, *in* Climate in earth history: Washington, D.C., National Academy Press, p. 97–104.

Folk, R. L., 1980, Petrology of sedimentary rocks: Hemphill, Austin, 182 p.

Franzinelli, E., and Potter, P. E., 1983, Petrology, chemistry, and textures of modern river sands, Amazon River system: Journal of Geology, v. 91, p. 23–39.

Grantham, J. H., and Velbel, M. A., 1988, The influence of climate and topography on rock-fragment abundance in modern fluvial sands of the southern Blue Ridge Mountains, North Carolina: Journal of Sedimentary Petrology, v. 58, p. 219–227.

Herbert, C., 1980, Depositional development of the Sydney Basin, *in* Herbert, C., and Helby, R., eds., A guide to the Sydney Basin: New South Wales Geological Survey Bulletin No. 26, p. 11–52.

Hunt, J. W., and Hobday, K. H., 1984, Petrographic composition and sulphur content of coals associated with alluvial fans in the Permian Sydney and Gunnedah basins, eastern Australia: Special Publications of the International Association of Sedimentologists 7, p. 43–60.

Hutton, A., Bamberry, J., and Jones, B., 1990, A revised stratigraphy for the southern Sydney Basin: Proceedings of Workshop on Southern and western Coalfields of the Sydney Basin: Geological Society of Australia Coal Geology Group, University of Wollongong, New South Wales, p. 43–58.

Jensen, A. R., 1975, Permo-Triassic stratigraphy and sedimentation in the Bowen basin, Queensland: Canberra, Australian Government Publication Service, Bureau of Mineral Resources Bulletin 154.

Johnsson, M. J., and Meade, R. H., 1990, Chemical weathering of fluvial sediments during alluvial storage: The Macuapanim Island point bar, Solimoes River, Brazil: Journal of Sedimentary Petrology, v. 60, p. 827–842.

Johnsson, M. J., and Stallard, R. F., 1989, Physiographic controls on the composition of sediments derived from volcanic and sedimentary terrains on Borro Colorado Island, Panama: Journal of Sedimentary Petrology, v. 59, p. 768–781.

Jones, B., 1990, The Shoalhaven Group: Implications for subsequent Coal Measure deposition in the southern Sydney Basin: Proceedings of Workshop on Southern and Western Coalfields of the Sydney Basin, Geological Society of Australia Coal Geology Group, University of Wollongong, New South Wales, p. 11–19.

Jones, J. G., Conaghan, P. J., and McDonnell, K. L., 1987, Coal measures of an orogenic recess: Late Permian Sydney Basin, Australia: Paleogeography, Paleoclimatology, Paleoecology, v. 58, p. 203–219.

Land, L. S., and Milliken, K. L., 1981, Feldspar diagenesis in the Frio Formation, Brazoria County, Texas Gulf Coast: Geology, v. 9, p. 314–318.

Morad, S., Bergan, M., Knarud, R., and Nystuen, P., 1990, Albitization of detrital plagioclase in Triassic reservoir sandstones from the Snorre field, Norwegian North Sea: Journal of Sedimentary Petrology, v. 60, p. 411–425.

Odins, P., Bamberry, J., Hutton, A., Jones, B., and Organ, M., 1990, Petrography of sandstones and other rocks, Illawarra Coal Measures: Proceedings of Workshop on Southern and Western Coalfields of the Sydney Basin, Geological Society of Australia Coal Geology Group, University of Wollongong, New South Wales, p. 59–69.

Potter, P. E., 1978, Petrology and chemistry of modern big river sands. Journal of Geology, v. 86, p. 423–449.

Ringis, J., Hawkins, L. V., and Seedsman, K., 1970, Offshore seismic and magnetic surveys of the southern coalfields off Stanwell Park: Australian Institute of Mining and Metallurgy Proceeding No. 234, p. 7–16.

Ruxton, P. B., 1970, Labile quartz-poor sediments from young mountain ranges in northwest Papua: Journal of Sedimentary Petrology, v. 40, p. 1262–1270.

Schiebner, E., 1985, Suspect terrains in the Tasman Fold Belt, eastern Australia, *in* David, H. G., ed., Tectonostratigraphic terrains of the Circum-Pacific regions. Earth Science Series 1: Houston, Circum-Pacific Council of Energy and Mineral Resources, p. 493–514.

Schwab, F., 1975, Framework mineralogy and chemical composition of continental margin-type sandstone: Geology, v. 3, p. 487–490.

Smith, A. G., Hurley, A. M., and Briden, J. C., 1981, Phanerozoic paleocontinental world maps: Cambridge, Cambridge University Press.

Standard, J. C., 1969, Hawkesbury Sandstone, Sydney Basin, Australia: Journal of the Geological Society of Australia, v. 16, p. 407–417.

Suttner, L. J., and Dutta, P. K., 1986, Alluvial sandstone composition and paleoclimate, I. Framework mineralogy: Journal of Sedimentary Petrology, v. 56, p. 329–345.

Suttner, L. J., Basu, A., and Mack, G. H., 1981, Climate and the origin of quartz arenite: Journal of Sedimentary Petrology, v. 51, p. 1235–1246.

Valloni, R., and Maynard, J. B., 1981, Detrital modes of Recent deep-sea sands and their relation to tectonic setting: A first approximation: Sedimentology, v. 31, p. 75–83.

Veevers, J. J., 1984, Phanerozoic earth history of Australia: Oxford, Claredon Press, 418 p.

Velbel, M. A., 1985, Geochemical mass balances and weathering rates in forested watersheds of the southern Blue Ridge: American Journal of Science, v. 285, p. 904–930.

Walker, T. R., 1984, Diagenetic albitization of potassium feldspar in arkosic sandstones: Journal of Sedimentary Petrology, v. 54, p. 3–16.

Ward, C. R., 1980, Notes on the Bulgo Sandstone and Bald Hill Claystone: New South Wales Geological Survey Bulletin, v. 26, p. 176–186.

Wilkinson, J.F.G., 1971, The petrology of some vitrophpyric calc-alkaline volcanics from the carboniferous of New South Wales: Journal of Petrology, v. 12, p. 587–619.

Young, S., 1976, Petrology of fluvial sand derived from regionally metamorphosed source rocks [Ph.D. thesis]: Bloomington, Indiana University, 144 p.

Manuscript Accepted by the Society December 24, 1992

Printed in U.S.A.

Geological Society of America
Special Paper 284
1993

A theoretical study of mineralogical maturation of eolian sand

Prodip K. Dutta
Department of Geography and Geology, Indiana State University, Terre Haute, Indiana 47809
Ziye Zhou
Department of Physics, Indiana State University, Terre Haute, Indiana 47809
Paulo R. dos Santos
Instituto de Geociencias, Universidade de Sao Paulo, CEP 01498, Sao Paulo, Brazil

ABSTRACT

Eolian sands tend to be fine grained and quartz rich. This mineralogical maturation may be attributed to mechanical breakdown of sand-size feldspar grains and their subsequent winnowing from dune fields. During eolian transport, ballistic impacts between saltating grains and stationary grains on the ground are very common. If the impact is sufficiently strong, the collision may be inelastic and part of the kinetic energy of the incident grain may be dissipated through inelastic deformation and subsequent breakdown of the grains involved in the collision. The theoretical calculation presented in this chapter indicate that the mechanical breakdown of feldspar sand grains through ballistic impact can be achieved if wind speed exceeds 10 m·s^{-1}. Therefore, quartz-rich sand can result from ballistic impacts between sand grains in deserts.

INTRODUCTION

Eolian sands tend to be fine grained and quartz rich (McKee, 1983; Ahlbrandt, 1979). Through inductive logic, geologists have inferred that mineralogical maturity in eolian sand is accomplished through mechanical disintegration of feldspar sand into finer silt and clay fractions. Subsequent winnowing of the finer fractions by wind leaves behind quartz-rich sand in deserts.

Size reduction on a limited scale and rounding of sand particles because of abrasion have been documented in laboratory experiments. Abrasion results in the chipping off of corners of angular grains and generating cracks (Greeley and Iversen, 1985; Kuenen, 1960; Marshland and Woodruff, 1937). Disintegration of sand grains into silt fractions in deserts is also accomplished through salt crystallization (Goodie et al., 1979; Evans, 1970). However, such processes are possibly important only locally (Cooke and Warren, 1973; Cooke and Smalley, 1968; Wellman and Wilson, 1965) and may not be significant in the maturation of sand.

Extensive experimental works on eolian transport of sand grains have been conducted since Bagnold (1941) tried to quantify the mechanics of sand movement in the desert environment (Werner, 1990; Anderson, 1987; Willetts and Rice, 1986; Rumpel, 1985), but the question of sand maturation by eolian process has not been addressed on a quantitative level. In this chapter, we focus on the mineralogical maturation of eolian sand, first by performing a theoretical calculation on ballistic impact of sand-size feldspar grains to show whether such grains would break into a finer fraction, and then observing whether they really break through a wind tunnel experiment.

In the wind tunnel experiment, the breakage of the coarse sand grains is clearly observed. However, because the sand used in the experiment is a mixture of quartz and feldspar sand grains, the experiment does not distinguish between the comminution behaviors of feldspar and quartz sand grains. Our theoretical calculations, which are presented here and in a subsequent paper that is currently in preparation, show that feldspar grains break more easily than quartz grains through ballistic impact.

THEORETICAL MODEL OF BALLISTIC IMPACT BETWEEN SAND GRAINS

Background

Sand grains in deserts are generally quartzofeldspathic in composition. The crystal structure of a grain depends on its chem-

Dutta, P. K., Zhou, Z., and dos Santos, P. R., 1993, A theoretical study of mineralogical maturation of eolian sand, *in* Johnsson, M. J., and Basu, A., eds., Processes Controlling the Composition of Clastic Sediments: Boulder, Colorado, Geological Society of America Special Paper 284.

ical composition and on the conditions under which the grain is formed. The electrical attraction between cations and anions holds the crystal together and gives rise to the bond energy of the crystal. The strength of the bond energy is usually different across different cleavage planes. When a grain experiences a low-speed impact, it tends to break apart along the weakest cleavage plane, across which the bond energy is the lowest.

In deserts, wind-blown sand moves by three processes—suspension, saltation, and creep—of which saltation is particularly important. Grains in saltation receive momentum from the wind and transfer it to the bed at a rapid rate. This keeps the bed surface in a disturbed state, prompting further saltation, stimulating creep, and releasing fine material into suspension (Willetts and Rice, 1986). Saltation transport is initiated by aerodynamic entrainment and drag force of the wind. After a sand grain rises above the ground, it moves in the direction of the wind and falls toward the ground. As the grain falls, its velocity increases both vertically, due to gravity, and horizontally, due to the drag force in the direction of the wind. The grain's kinetic energy and momentum also increases as the velocity increases. When the grain strikes the ground, it collides with the stationary grain(s) on the bed. Because of the wind drag, sand grains in eolian transport usually strike the bed with more energy than that with which they left the bed. A typical collision usually results in one ricocheted grain and a large number of emergent grains (Willetts and Rice, 1986). Therefore, the initial entrainment of grains leads to a chain reaction with an increasing number of grains mobilized in saltation. The process reaches a steady state when neither the number nor the velocity distribution of ejected grains changes with time any more. This can be achieved through the reduction of wind speed near the bed due to the drag exerted by the grains on the air (the reaction to the force that accelerates grains in the direction of the wind).

In a collision, the kinetic energy of the incident grain can be distributed among three processes (Anderson, 1987): rebound of the original grain from the bed, ejection of a number of grains originally stationary on the bed, and dissipation by inelastic deformation and frictional rearrangement of sand grains on the ground. If the impact is weak, sand grains will not deform or break on collision. Such collisions can be considered elastic, and the energy can be lost only through frictional rearrangement of sand grains. However, if the impact is so strong that the energy involved in the collision is considerably larger than the bond energy across a cleavage plane, the colliding grains will absorb part of the impact energy and break into pieces.

In a collision, the kinetic energy of the incident grain is proportional to mass and the square of the velocity of the grain. The mass of a grain is equal to the density multiplied by the volume of the grain. Therefore, for a given velocity, the impact is proportional to the cube of the linear dimension of the grain. Alternatively, the bond energy is proportional to the cross-sectional area of the cleavage plane, which is proportional to the square of the linear dimension of the grain. Therefore, the smaller a grain is, the harder it is to break.

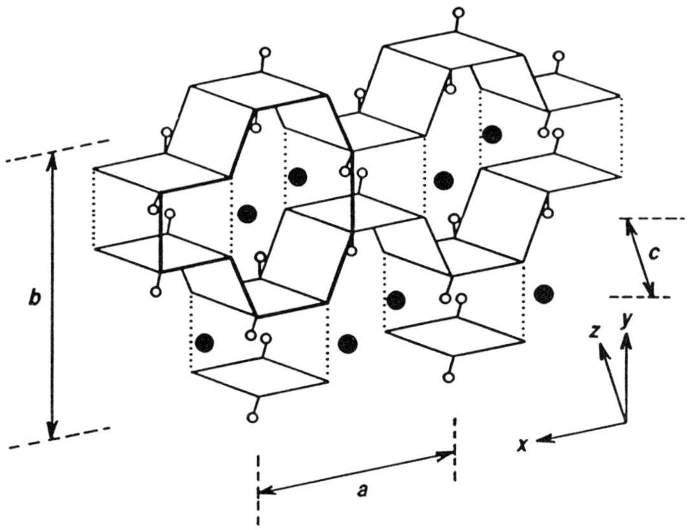

Figure 1. Feldspar, perspective diagram of framework, showing "crankshafts" reflected by the mirror plane (Megaw, 1973). Solid lines join T cations (Si or Al). The potassium ions in the interstices of the structure are represented by solid circles. The T_1-O bonds across the (001) cleavage plane are represented by the lines joining open circles and the slab. The slab repeats itself translationally in the z direction. A unit cell, outlined by bold solid lines, contains four $KAlSi_3O_8$ molecules.

To study eolian transport and saltation, an assumption is usually made that sand grains are spherical, uniform in size and mass, rigid, and perfectly elastic (Rumpel, 1985). Therefore, the grains are considered unbreakable on ballistic impact. In this chapter, sand grains are assumed to be spherical, uniform in size and mass, and rigid, but not perfectly elastic. This assumption is based on the evidence that wind velocity in deserts can be sufficiently high and can generate enough energy in saltating grains so that, on collision, they behave inelastically and break into pieces.

Bond energy of feldspar

Feldspar is a tectosilicate mineral with a three-dimensional framework configuration. The lattice structure of potassium feldspar, a common feldspar in eolian sand, is monoclnic C-face-centered (Megaw, 1973), with

$$a = 8.56\ \text{Å} \qquad \alpha = 90.0^\circ$$
$$b = 13.03\ \text{Å} \qquad \beta = 116.0^\circ$$
$$c = 7.17\ \text{Å} \qquad \gamma = 90.0^\circ.$$

A perspective view of the linkage scheme of the feldspar framework is shown in Figure 1. A unit cell is outlined by the bold solid lines connecting cations (either Si or Al). Each unit cell contains four molecules, represented by the chemical formula $KAlSi_3O_8$.

There are three cleavage planes in the feldspar structure: (100), (010), and (001). Among the three cleavage planes, the (001) plane is the most prominent, across which the bond energy is the lowest. The cross section area of the (001) cleavage plane is

$$A = ab\sin\gamma = 111.537\ \text{Å}^2. \quad (1)$$

The bond energy, the minimum energy required to break a crystal across a cleavage plane, is equal to the total energy required to break all chemical bonds across that cleavage plane. There are four such bonds across the (001) plane in each unit cell, all of which are T_1—$O_A(1)$ bonds (Megaw, 1973). The T_1 position in feldspar is occupied by either an Si atom or an Al atom. The bond energies for the Si-O and Al-O covalence bonds (Kerr, 1990) are:

$$E\ (\text{Si-O})=(799.6 \pm 13.4)\ \text{kJ/mole}=(1.328 \pm 0.022)\times 10^{-11}\ \text{erg/bond}; \quad (2)$$
$$E\ (\text{Al-O}) = (512.1 \pm 9.2)\ \text{kJ/mole} = (8.504 \pm 0.153) \times 10^{-12}\ \text{erg/bond}. \quad (3)$$

In the microcline structure, the most ordered potassium feldspar, the Al ions occupy the $T_1(O)$ positions (Megaw, 1973). When the crystal is broken along the (001) plane, there are two Si-O bonds and two Al-O bonds to be broken in each unit cell. The energy required to break these four bonds in a unit cell is:

$$E_{cell}\ (microcline) = 2E\ (\text{Si-O}) + 2E\ (\text{Al-O}). \quad (4)$$

The energy required to break a microcline crystal of unit cross section area along the (001) plane is:

$$E\ (\text{microcline}) = \frac{2E\ (\text{Si-O}) + 2E\ (\text{Al-O})}{ab\sin\gamma} = (39.06 \pm 0.67)\ \text{erg/mm}^2. \quad (5)$$

The bond energy of a spherical microcline grain of radius R across a (001) cleavage plane through the center of the grain is:

$$E_B\ (\text{microcline}) = E\ (\text{microcline}) \cdot \pi R^2 = \frac{2E\ (\text{Si-O}) + 2E\ (\text{Al-O})}{ab\sin\gamma}\pi R^2. \quad (6)$$

In the sanidine structure, the most disordered potassium feldspar, the replacement of Si ions by Al ions is random. Statistically, each cation in the sanidine is 75% Si and 25% Al. The energy required to break four bonds in a unit cell is:

$$E_{\text{cell}}\ (\text{sanidine}) = 4[0.75E(\text{Si-O}) + 0.25E(\text{Al-O})]. \quad (7)$$

The energy required to break a sanidine crystal of unit cross section area along the (001) plane is:

$$E\ (\text{sanidine}) = \frac{4[0.75E\ (\text{Si-O}) + 0.25E\ (\text{Al-O})]}{ab\sin\gamma} \quad (8)$$
$$= (43.34 \pm 0.74)\ \text{erg/mm}^2.$$

The bond energy of a spherical sanidine grain of radius R across a (001) cleavage plane through the center of the grain is

$$E_B\ (\text{sanidine}) = E\ (\text{sanidine}) \cdot \pi R^2 = \frac{4[0.75E\ (\text{Si-O}) + 0.25E\ (\text{Al-O})]}{ab\sin\gamma}\pi R^2. \quad (9)$$

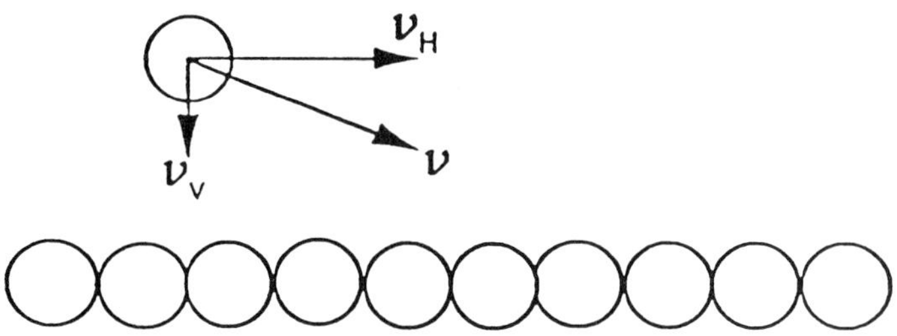

Figure 2. A grain falls toward the ground. The velocity of the grain, v, is the vector sum of the horizontal wind speed, v_H, and the vertical falling speed, v_V, of the grain.

Collision process

Consider a spherical sand grain of radius R falling toward the ground with velocity, v, which is determined by the velocity of wind and the terminal falling velocity of the grain in the air (Fig. 2), and colliding with a stationary grain on the ground. The kinetic energy of the grain is

$$E_m = \frac{1}{2}mv^2 = \frac{1}{2}\rho\left(\frac{4\pi}{3}R^3\right)v^2. \quad (10)$$

where $m = \rho(4\pi/3)R^3$ is the mass of the grain and $\rho \approx 2.57$ g/cm^3 is the density of feldspar. Because of the conservation of momentum, when a grain collides with other grains, only part of the kinetic energy of the incident grain can dissipate through inelastic deformation and frictional rearrangement of sand grains. The impact energy, the part of the energy that can dissipate during a collision, is equal to the energy of relative motion of the two grains involved in the collision and is determined by the velocity of the incident grain and the masses of the incident and target grains.

When a grain of mass, m, collides with a stationary grain on the ground with effective mass M (Figs. 3, 4), the velocity of the incident grain can be resolved into two components, $v_{\parallel}$ and $v_{\perp}$, parallel and perpendicular, respectively, to the line connecting the centers of the incident and target grains. In an elastic collision, if the two grains have the same mass, the parallel component of momentum of the incident grain, $mv_{\parallel}$, is transferred to the stationary target grain and to the sand bed and the perpendicular component of the momentum of the incident grain, $mv_{\perp}$, is retained as the initial momentum for the rebound (Fig. 3). If m is smaller than M in an elastic collision, the incident grain will bounce backward after colliding with the stationary grain on the ground. However, if the parallel component of the momentum, $mv_{\parallel}$, is sufficiently large, the collision will be inelastic and the grains involved in the collision may be broken into several pieces.

In an inelastic collision, only the kinetic energy associated with the parallel component of the momentum can be partially absorbed by the colliding grains. The kinetic energy associated with the perpendicular component of the momentum is retained as part of the initial kinetic energy for the rebound in the saltation process and/or is dissipated through frictional rearrangement of

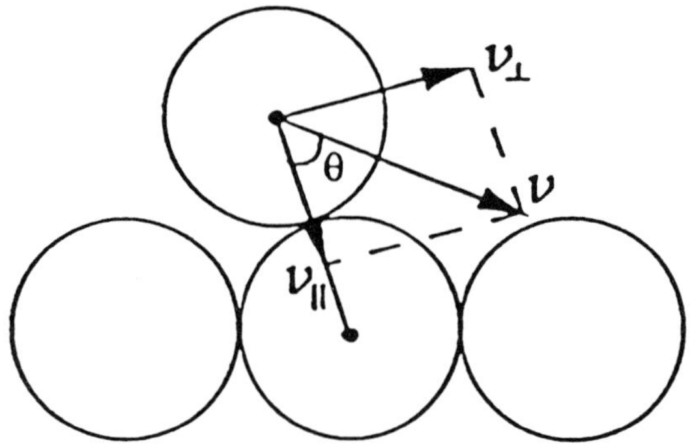

Figure 3. An incident grain collides with a stationary grain on the ground. The velocity of the incident grain, v, can be resolved into two components, $v_{\|}$ and $v_{\perp}$. Only the parallel component of the momentum of the incident grain, $mv_{\|}$, can be partially absorbed in a collision. The perpendicular component, $mv_{\perp}$, is retained as the initial momentum for the rebound of the grain.

sand grains on the ground. The kinetic energy associated with the parallel component of the momentum (Fig. 3) is:

$$E_K = \frac{1}{2} mv_{\|}^2 = \frac{1}{2} m\,(v \cos \theta)^2. \tag{11}$$

The parallel component of the velocity of the center of mass of the system (Resnick et al., 1992) is:

$$v_{\mathrm{CM}} = \frac{m}{m+M}\, v \cos \theta. \tag{12}$$

The parallel components of the velocities of the incident and target grains relative to the center of mass are:

$$V_m = v_{\|} - v_{CM} = \frac{M}{m+M}\, v \cos \theta \tag{13}$$

and

$$V_M = v_M - v_{CM} = \frac{m}{m+M}\, v \cos \theta, \tag{14}$$

respectively. The kinetic energy of relative motion of the system associated with the parallel component of the momentum is:

$$E_{rel} = \frac{1}{2} mV_m^2 + \frac{1}{2} MV_M^2 = \frac{1}{2}\frac{mM}{m+M}(v \cos \theta)^2. \tag{15}$$

In an inelastic collision, only the kinetic energy of relative motion, E_{rel}, can dissipate through inelastic deformation, breakage, and frictional rearrangement of sand grains. This part of kinetic energy is also called the energy of impact, E_I. The remaining energy, the kinetic energy of the center of mass, $E_{CM} = (1/2)(m+M)v^2{}_{CM}$, does not change during a collision due to the conservation of momentum. A parameter, $f = E_{rel}/E_K = M/(m+M)$, which is equal to the ratio between the kinetic energy of relative motion and the total kinetic energy of the system, is introduced here to indicate the efficiency of collision in such a way that the energy of impact can be expressed as:

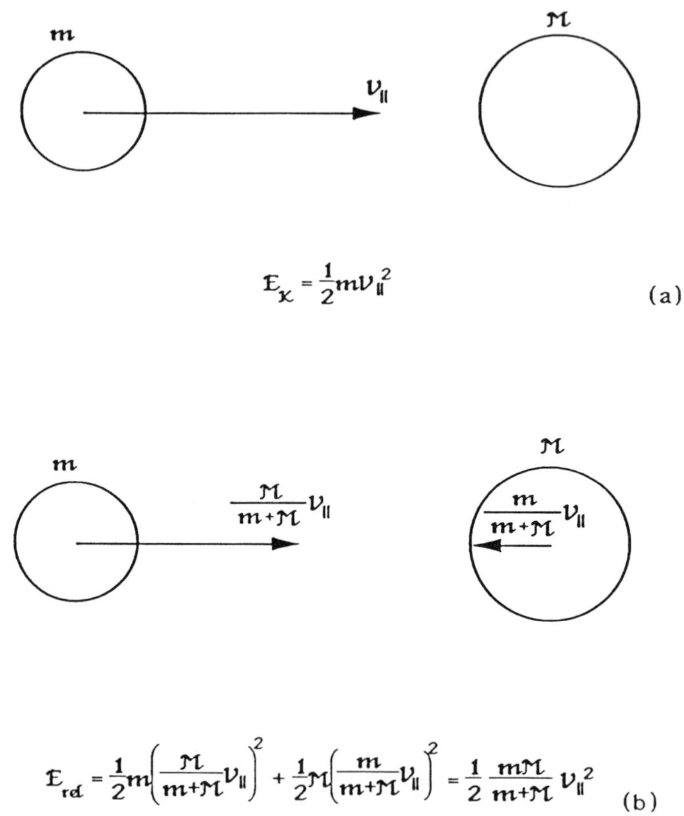

Figure 4. A grain of mass, m, collides with a stationary grain of effective mass, M. The effective mass of the grain on the ground, usually different from the real mass of the grain, is determined by the mass of the grain, as well as by how closely packed grains are on the ground and by the friction between the grain and its neighbors. (a) In a reference frame at rest relative to the ground, the grain on the ground is stationary, and the incident grain approaches the stationary grain with velocity $v_{\|}$. The kinetic energy of the system is equal to that of the incident grain. (b) In a reference frame at rest relative to the center of mass of the two grains involved in the collision, both grains are moving toward the center of mass of the system where the collision takes place. Only the kinetic energy of the system in such a reference frame, E_{rel}, can be absorbed by the two grains involved in an inelastic collision.

$$E_I = f \cdot E_K = \frac{1}{2} f\rho\left(\frac{4\pi}{3} R^3\right)(v \cos \theta)^2. \tag{16}$$

A necessary condition for a grain to be broken on collision is that the impact energy is larger than the bond energy of the grain, $E_I > E_B$. This condition can also be expressed as:

$$\frac{1}{2} f\rho\left(\frac{4\pi}{3} R^3\right)(v \cos \theta)^2 > \frac{E_{cell}}{ab \sin \gamma}\pi R^2. \tag{17}$$

Therefore, the minimum striking velocity, $v_{\min}$, required to break a spherical grain of diameter $D = 2R$ is:

$$v_{\min} = \frac{1}{\cos \theta}\left(\frac{3E_{cell}}{fD\rho ab \sin \gamma}\right)^{1/2}. \tag{18}$$

TABLE 1. MINIMUM STRIKING VELOCITY REQUIRED TO BREAK A MICROCLINE SAND GRAIN

θ (°)	*f* (%)	v_{min} (m·s⁻¹) *D* = 500 μm	*D* = 250 μm	*D* = 125 μm	*D* = 62.5 μm
	75	3.49	4.93	6.97	9.86
0	50	4.27	6.04	8.54	12.08
	25	6.04	8.54	12.08	17.08
	75	3.61	5.11	7.22	10.21
15	50	4.42	6.25	8.84	12.51
	25	6.25	8.84	12.51	17.69
	75	4.03	5.69	8.05	11.39
30	50	4.93	6.97	9.86	13.95
	25	6.97	9.86	13.95	19.73
	75	4.93	6.97	9.86	13.95
45	50	6.04	8.54	12.08	17.08
	25	8.54	12.08	17.08	24.16
	75	6.97	9.86	13.95	19.73
60	50	8.54	12.08	17.08	24.16
	25	12.08	17.08	24.16	34.17
	75	13.47	19.05	26.95	38.11
75	50	16.50	23.34	33.00	46.67
	25	23.34	33.00	46.67	66.00
90				∞	

TABLE 2. MINIMUM STRIKING VELOCITY REQUIRED TO BREAK A SANIDINE SAND GRAIN

θ (°)	*f* (%)	v_{min} (m·s⁻¹) *D* = 500 μm	*D* = 250 μm	*D* = 125 μm	*D* = 62.5 μm
	75	3.67	5.19	7.35	10.39
0	50	4.50	6.36	9.00	12.72
	25	6.36	9.00	12.72	17.99
	75	3.80	5.38	7.61	10.76
15	50	4.66	6.57	9.31	13.17
	25	6.57	9.31	13.17	18.63
	75	4.24	6.00	8.48	12.00
30	50	5.19	7.35	10.39	14.69
	25	7.35	10.39	14.69	20.78
	75	5.19	7.35	10.39	14.69
45	50	6.36	9.00	12.72	17.99
	25	9.00	12.72	17.99	25.45
	75	7.35	10.39	14.69	20.78
60	50	9.00	12.72	17.99	25.45
	25	12.72	17.99	25.45	35.99
	75	14.10	20.07	28.38	40.14
75	50	17.38	24.58	34.76	49.16
	25	24.58	34.76	49.16	69.53
90				∞	

For microcline sand grains, the minimum striking velocity is:

$$v_{min} = \frac{1}{\cos\theta}\left(\frac{3[2E\,(\text{Si}-\text{O}) + 2E\,(\text{Al}-\text{O})]}{fD\rho ab \sin\gamma}\right)^{1/2}. \qquad (19)$$

For sanidine sand grains, the minimum striking velocity is:

$$v_{min}\,\frac{1}{\cos\theta}\left(\frac{3[4(0.75E\,(\text{Si}-\text{O}) + 0.25E\,(\text{Al}-\text{O}))]}{fD\rho ab \sin\gamma}\right)^{1/2}. \qquad (20)$$

The theoretical values of minimum striking velocities required to break microcline and sanidine grains have been calculated by using equations (19) and (20), respectively. The results are tabulated in Tables 1 and 2.

Because the speed of wind in nature often exceeds 10 m · s^{-1}, the data in Tables 1 and 2 show that the feldspar sand grains will break in small-angle ballistic impacts during eolian processes.

WIND TUNNEL EXPERIMENT

In order to verify the theoretical results derived in the previous section, a wind tunnel experiment was conducted to demonstrate whether feldspar sand grains will break through ballistic impacts.

Figure 5 shows the setup of the experiment. The wind tunnel was constructed of a wooden bed with two side walls. A squirrel cage air blower was mounted to the near end of the tunnel, an inclined board was mounted to the far end of the tunnel to trap sand and reduce air turbulence, a plastic film was stapled to the top of the side walls on the near end, and a filter made of synthetic fabrics was stapled on the top of the side walls on the far end.

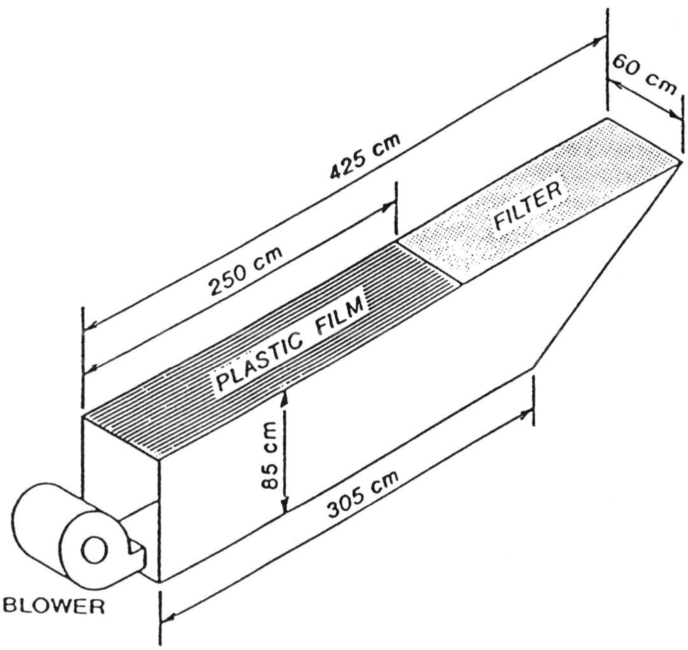

Figure 5. The schematic diagram of the wind tunnel. The wind tunnel experiment demonstrated that coarse feldspar sand grains do break into finer grains as a result of ballistic impacts.

The sand used in the experiment is feldspar rich, collected from first-order streams draining coarse-grained Butler Hill Gran-

TABLE 3. SAND SIZE DISTRIBUTION IN SAMPLES COLLECTED BEFORE WIND TUNNEL EXPERIMENT

Grain Size (μm)	Size Distribution of Sand Grains — Sample 1 (g) (%)	2 (g) (%)	3 (g) (%)	4 (g) (%)	5 (g) (%)	Average (%)
500 ~ 1,680	**98.82** 92.46	**85.10** 86.78	**84.84** 87.08	**86.51** 87.92	**86.96** 88.34	88.52
250 ~ 500	**7.65** 7.54	**12.96** 13.22	**12.59** 12.92	**11.89** 12.08	**11.48** 11.66	11.48

TABLE 4. SAND SIZE DISTRIBUTION IN SAMPLES COLLECTED AFTER WIND TUNNEL EXPERIMENT

Grain Size (μm)	Size Distribution of Sand Grains — Sample 1 (g) (%)	2 (g) (%)	3 (g) (%)	4 (g) (%)	5 (g) (%)	Average (%)
500 ~ 1,680	**114.43** 77.95	**101.07** 82.59	**108.32** 87.40	**94.51** 82.66	**99.85** 86.73	83.47
250 ~ 500	**32.36** 22.05	**21.31** 17.41	**15.61** 12.60	**19.82** 17.34	**15.28** 13.27	16.18

ite in the St. Francois Mountains, Missouri. About 120 kg of sand was used in the experiment. The sand was first sifted through a sieve of mesh size 1.68 μm. After the sand was mixed thoroughly, five samples were taken and analyzed for their size distribution (Table 3). Because the porous texture of the filter is permeable to grains smaller than 250 μm, only the sand grains larger than or equal to 250 μm in dimension were trapped in the wind tunnel. Therefore the data of the size distribution analysis include only two fractions, 0.5~1.68 and 0.25~0.5 mm.

The sand was put on the bed of the wind tunnel. In order to increase the probability and efficiency of collisions and to accelerate the process of mechanical breakdown, several granite blocks were put on the tunnel bed. The air blower was then turned on. The speed of air flow at the nozzle of the blower was approximately 18 m · s^{-1}. Due to the rough surface of the bed, the large cross-sectional area of the tunnel, and the air resistance of the filter, the speed of air flow decreased as the air flowed away from the blower into the tunnel. The average speed of air flow in the tunnel was estimated to be approximately 10 m · s^{-1}. The wind was blown intermittently for about 5 h. Each time the blower was turned off, the sand was redistributed mechanically throughout the bed in the tunnel. After the experiment, the sand was thoroughly mixed, and five samples were taken and similar size analyses were made. The result is shown in Table 4.

With respect to sand grains of size ranging from 0.25 to 1.68 mm, the wind tunnel may be considered as a closed system throughout the experiment. Unlike the desert environment, the experimental result was not affected by the winnowing process. The change in the relative concentration of different fractions in the samples taken before and after the wind tunnel experiment was purely caused by the breakage of sand grains.

By comparing the data in Tables 3 and 4, it is evident that, in the wind tunnel experiment, the coarse sand grains larger than 500 μm do break into finer grains due to ballistic impact. The whole process could be observed through the open end of the wind tunnel and the transparent plastic film on the top of the tunnel. The saltation of the sand grains was clearly observed.

CONCLUSION

In deserts, collision velocities attained by saltating sand grains are often higher than the theoretical values of minimum striking velocities, obtained in the section on Theoretical Model of Ballistic Impact above, required to break a feldspar grain. In many areas, the wind velocity is higher than 10 m · s^{-1}, the average speed of wind in the wind tunnel experiment. Elimination of feldspar sand-size fraction through eolian process has been observed in active desert environments. The theoretical analysis supports the inductive logic that explains why eolian sands tend to be fine grained and quartz rich. Although it does not distinguish between the comminution behaviors of quartz and feldspar grains, the experimental result suggests that sand grains do break in the eolian process.

ACKNOWLEDGMENTS

We thank Mark Cintala and Gary Kocurek for their critical comments on the manuscript. We are indebted to George Palmer for his advice on the wind tunnel experiment and for usage of the squirrel cage blower, both of which were essential to the experiment, and to Dale Manwaring for building the wind tunnel. We also thank Ryan Corman, David MacKenzie, and Mose Nasser for their valuable help throughout the wind tunnel experiment and Norman Cooprider for drafting the diagrams.

REFERENCES CITED

Ahlbrandt, T. S., 1979, Textural parameter of eolian deposits, *in* McKee, E. D., ed., A study of global sand seas: U.S. Geological Survey Paper 1052, Chapter B, p. 21–51.

Anderson, R. S., 1987, A theoretical model for eolian impact ripples: Sedimentology, v. 34, p. 934–946.

Bagnold, R. A., 1941, The physics of blown sand and desert dunes: London, Methuen, 265 p.

Cooke, R. U., and Smalley, I. J., 1968, Salt weathering in deserts: Nature, v. 220, p. 1226–1227.

Cooke, R. U., and Warren, A., 1973, Geomorphology in deserts: Berkeley, University of California Press, 374 p.

Evans, I. S., 1970, Salt crystallization and rock weathering: A review: Revue de Geomorphologie Dynamique, v. 19, p. 153–177.

Goodie, A. S., Cooke, R. U., and Doornkamp, J. C., 1979, The formation of silt from quartz dune sand by salt weathering processes in deserts: Journal of Arid Environments. v. 2, p. 105–112.

Greeley, R., and Iversen, J. D., 1985, Wind as a geological process on Earth, Mars, Venus and Titan: Cambridge, Cambridge University Press, 333 p.

Kerr, J. A., 1990, Strength of chemical bonds, *in* Lide, D. R., ed., Handbook of chemistry and physics (71st edition) Section 9: Boca Raton, FL, CRC Press, p. 86–94.

Kuenen, Ph. H., 1960, Experimental abrasion: 4. Eolian action: Journal of Sedimentary Petrology, v. 40, p. 427–449.

Marshland, P. S., and Woodruff, J. G., 1937, A study of the effects of wind transportation on grains of several minerals: Journal of Sedimentary Petrology, v. 7, p. 18–30.

McKee, E. D., 1983, Eolian sand bodies of the world, *in* Brookfield, M. E. and Ahlbrandt, T. S., eds., Eolian sediments and process: Development in Sedimentology, v. 38, p. 1–25.

Megaw, H. D., 1973, Crystal structures: A working approach: Philadelphia, W. B. Saunders Company, 563 p.

Resnick, R., Halliday, D., and Krane, K. S., 1992, Physics (fourth edition): New York, John Wiley & Sons, v. 1, 592 p.

Rumpel, D. A., 1985, Successive aeolian saltation: Studies of idealized collisions: Sedimentology, v. 32, p. 267–280.

Wellman, W. H., and Wilson, A. T., 1965, Salt weathering, a neglected geological erosive agent in coastal and arid environments: Nature, v. 205, p. 1097.

Werner, B. T., 1990, A steady-state model of wind-blown sand transport: Journal of Geology, v. 98, p. 1–17.

Willetts, B. B., and Rice, M. A., 1986, Collision in eolian transport: The saltation/creep link, *in* Nickling, W. G., ed., Aeolian geomorphology: London, Allen & Unwin, p. 1 17.

MANUSCRIPT ACCEPTED BY THE SOCIETY DECEMBER 24, 1992

Printed in U.S.A.

Geological Society of America
Special Paper 284
1993

Application of a tumbler experiment using granodioritic grus to examine the character of quartz-grain fracture in high-gradient streams

Robert H. Osborne, Edwin J. Bomer III,* Yeon-Chang Wang, and Yi Lu
Department of Geological Sciences, University of Southern California, Los Angeles, California 90089-0740

ABSTRACT

A tumbler experiment using the 0.0625- to 4-mm fraction of four granodioritic grus samples was performed to investigate the nature of fracturing in the production of siliciclastic sand. Petrographical modal analysis was performed on 240 subsamples representing four major lithological constituents (quartz, potassium feldspar, plagioclase, and polymineralic rock fragments), four grain-size fractions (0.50 to 0.70, 0.35 to 0.49, 0.25 to 0.34, and 0.177 to 0.24 mm), and four time periods (untumbled grus; 2, 4, and 8 days). Each tumbling day represents a maximum of 25 km of transport in water. Tests of sample means and analysis of variance suggest that all subsamples are statistically homogeneous with respect to the number of quartz, potassium feldspar, and plagioclase grains among these size fractions. There is a highly significant difference in the number of rock fragments in the 0.50- to 0.70-mm fraction as compared to finer grained fractions, but no such differences occur within the medium- and fine-grained sand fractions. These results are inconsistent with the concept that lithic composition of the detrital light-mineral fraction is strongly size dependent within the 0.177- to 0.50-mm range. Where such compositional differences do occur, they may reflect sedimentological processes other than the comminution of grains from a single protolith.

Grains from three grus samples examined for size changes displayed relatively rapid rates of disintegration during the first few days of tumbling, and then continued to fracture at reduced rates from days 4 through 8. From 8.9 to 16.2% of the relatively coarser grained fractions experienced comminution during 8 days (≈200 km) of transport. Of this amount, the sand fraction increased from 1.7 to 3.4%, whereas the silt plus clay fraction increased from 6.4 to 14.5%. For each sample, the weight percent of the fine- and/or medium-sand fraction remained nearly constant.

Shape-ratio values (length of short axis/length of long axis of the maximum projection grain outline) as well as amplitude-ratio values for harmonics 2 and 3, as measured by Fourier analysis, were used to examine the character of quartz-grain fracture as evidenced in the 0.25- to 0.50-mm size fraction. Although some across-grain fracture occurred during the first day of tumbling, an overall increase in the frequency of smaller shape-ratio values indicated the dominance of grain-parallel fracture in all three samples. Scanning electron microscopy indicated that abrasion was not obvious even after 8 days (≈200 km) of transport.

---

*Present address: Exxon Company, U.S.A., P.O. Box 2528, Corpus Christi, Texas 78403-2528.

Osborne, R. H., Bomer III, E. J., Wang, Y.-C., and Lu, Y., 1993, Application of a tumbler experiment using granodioritic grus to examine the character of quartz-grain fracture in high-gradient streams, *in* Johnsson, M. J., and Basu, A., eds., Processes Controlling the Composition of Clastic Sediments: Boulder, Colorado, Geological Society of America Special Paper 284.

Average shape-ratio values were computed for at least 200 quartz grains in each of 49 samples from four high-gradient streams in southern California. Overall, the average shape-ratio values are remarkably consistent within each of the four streams, and closely resemble the value characteristic of the associated grus sample. Although significant decreases in average shape-ratio values occur along some local, very steep stream gradients due to high rates of grain-parallel fracture and/or shape sorting, the values reequilibrate after only 7 to 10 km of transport along lower gradient stream segments. Reequilibration probably reflects the dilution of more elongate by less elongate grains along lower stream gradients. No significant changes in the average shape-ratio values occur where small tributaries draining the same protolith enter the trunk streams, from sidewall erosion of the same source rock or across a contact with younger granitic source rock.

Simple shearing is known to produce self-similar fracturing. Such fracturing is suggested by the lithological analysis as well as highly elongate quartz grains as observed in this and other studies. Self-similar fracturing may be an important mechanism in explaining the compositional and textural aspects of nascent siliciclastic sand derived from plutons subjected to emplacement and/or postcrystallization tectonism.

INTRODUCTION

The fundamental assumption for grain-shape studies of detrital quartz for source determination is that the shape population from a given protolith or local source reflects the genetic history of that source and is distinct from the quartz grain-shape populations of other sources contributing to a given sediment dispersal system. Using Fourier grain-shape analysis and scanning electron microscopy (SEM), Mazzullo and Magenheimer (1987) concluded that quartz grains from 125 to 180 μ in diameter derived from crystalline rocks have highly nonspherical, angular outlines and numerous crystalline nodes, grain embayments, and fractures that reflect crystallization and deformation at elevated temperatures and pressures. Although Mazzullo and Magenheimer (1987) showed that quartz grain shapes from crystalline as well as wind-transported and quartz-cemented sedimentary rocks are distinct, a great deal has yet to be learned regarding the origin of grain-shape populations from crystalline rocks, particularly those of samples used as end members in Fourier grain-shape analysis. Although the general mechanisms affecting quartz grain-shape distributions of acid plutonic rocks are known, specific knowledge concerning a number of fundamental controls of grain shape is lacking. For example, what are the size and shape characteristics of quartz crystals within a given pluton? What are the orientation and spacing of fracture sets within a given pluton? How much shape variation occurs within closely associated plutonic bodies of slightly different composition and texture that sedimentary petrologists might consider a single source terrane? What is the nature of change in grain-shape populations as protolith weathers to grus under various climatic regimes? What degree of change in the grain-shape population occurs during transport in high-gradient streams? Although it is not necessary to answer such questions to use Fourier grain-shape analysis for many sourcing and sediment-transport analyses, answers to these questions, among others, are essential to understand the origin of detrital quartz grain-shape populations within a given sedimentary system.

Nascent detritus derived from crystalline rocks has been little used in experimentation. Kuenen's (1959) experiments concerning the modification of quartz sand were performed using small cubes cut from large quartz crystals. Rogers et al. (1963) used doubly crushed milky quartz. Moss (1966) reported that some varieties of quartz may be much stronger than ordinary acid-plutonic or metamorphic quartz. Therefore, experimental results using quartz crystal or only the most resistant grains derived from a given protolith may be misleading. Using field and laboratory experiments, Moss (1966, 1972) and Moss et al. (1973) demonstrated the dominance of fracturing in determining the character of nascent quartz grains derived from an acid-plutonic rock. A high degree of fracturing occurs because nascent quartz is coarse grained, brittle, and criss-crossed by partially healed cracks, which were acquired before the protolith was exposed to weathering. Their research indicated that grain abrasion was negligible during 4-h ($\simeq$14.4 km) tumbling runs. The breakage of quartz, particularly in the presence of pebbles, was thought to be due largely to reopening along more poorly bonded, dominantly across-grain fractures. Progressive fragmentation of detrital quartz grains eventually must reach a stage wherein grains consisting of a single fracture-bounded element are produced. Such grains, which lack significant internal weakness, would strongly resist further breakage. Moss et al. (1973) have suggested that these fracture-bounded elements in the material they examined may be produced at a grain diameter slightly less than 20 μ (medium silt).

The experimental detritus used by Moss and others (1973) for their tumbler experiments was derived from the deeply weathered Shannon's Flat Granodiorite. In this material, from 54 to 78% of the grains contain more than 90% quartz for sieve sizes from 0.50 to 4.76 mm, whereas feldspars comprise only 6 to 8% of the discrete grains. Although suitable for the objectives of their study, the mineral composition and perhaps grain-size characteristics are atypical when compared to detritus supplied to the headwaters of many mountain streams.

The purpose of this work is to examine changes in the mineralogy, grain-size, and quartz grain-shape composition of more typical granodioritic grus samples during as much as 200 km of transport in a tumbler. The grus samples were collected in the Lake Tahoe basin of California and Nevada, and formed under alpine conditions where mechanical weathering predominates. As such, the mineral composition and grain-size characteristics are more similar to detritus supplied to the headwaters of many streams than the material studied by Moss and others (1973). Attention is directed to the sand fraction (0.0625 to 2.00 mm) and particularly to the 0.25- to 0.50-mm size fraction due to the suitability of this size range for identifying source rocks using Fourier grain-shape analysis. Of particular concern is the general character of grain fracture during transport, and its effect on resultant mineralogical composition, grain-size distributions, and grain-shape characteristics. Concepts derived from the grain-shape analysis associated with the tumbling experiment were then applied to four high-gradient streams in southern California to better understand the relative importance of fracture and shape sorting in natural settings.

EXPERIMENTAL MATERIALS

Granodiorite associated with the Sierra Nevada batholith is the most abundant plutonic rock exposed in the Lake Tahoe basin. Diorite, quartz diorite, and gabbro also occur as small bodies commonly adjacent to metamorphic rocks (Burnett, 1968). Bateman and Wahrhaftig (1966) explained the occurrence of the more mafic rock bodies as wall-rock contamination of parent magma. Inasmuch as the work herein concerns changes in quartz grain-shape populations, attention was directed to sampling grus derived from the granodiorite.

Each of four grus samples from the localities described in part A of Appendix 1 consisted of approximately 5 kg of material scraped from the surface of the protolith with a small trowel. Modal and normative petrochemical analyses indicate that all of the protolith samples used in this study are hornblende granodiorites according to the plutonic rock classification of Streckeisen (1976). Mineralogically, the protolith samples consist of 27 to 40% quartz, 38 to 39% plagioclase feldspar, 11 to 17% potassium feldspar, 10 to 17% hornblende, and minor amounts of opaque and nonopaque accessory minerals.

The four grus samples were wet sieved to obtain the −2.0 to 4 ϕ (0.0625 to 4 mm) size fraction. For each sample, this fraction was divided into five equal parts. One portion was retained as the untumbled grus (0-day) sample, whereas each of the other four parts was placed in a rotating drum with water for a period of 1, 2, 4, or 8 days, respectively. The inside circumference of the tumbling drum is 60 cm, and the rotation rate is 30 cycles/min.

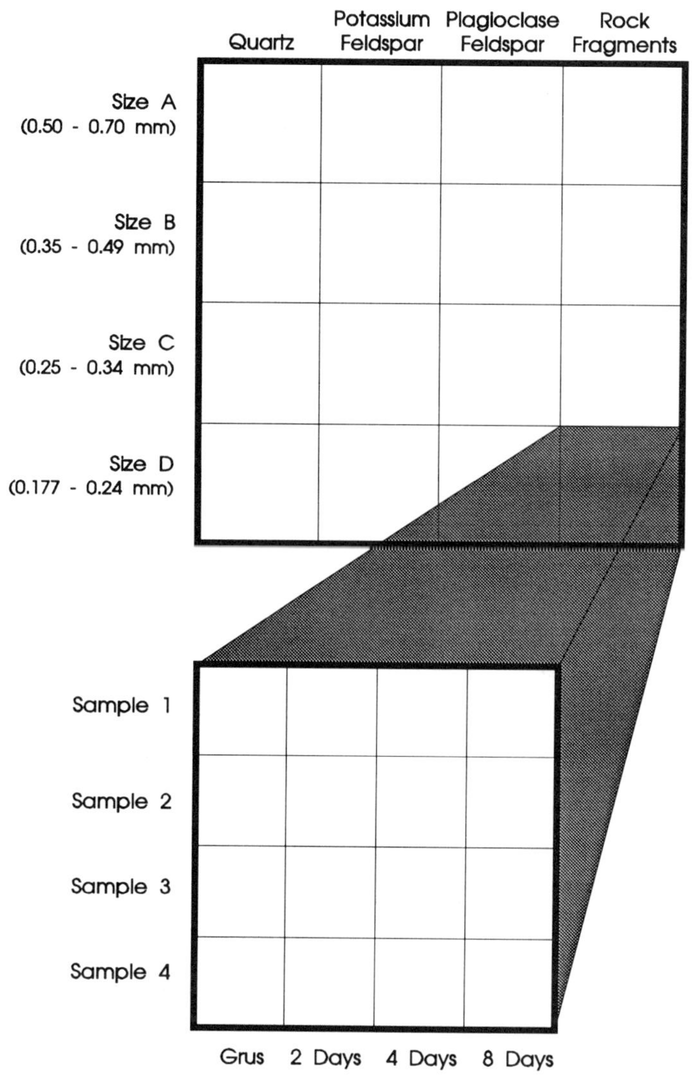

Figure 1. Nested analysis of variance design for the reduction of the obtained lithological data.

Therefore, each sample experienced a maximum of 25 km of transport/day. After tumbling, petrographical, grain-size and grain-shape analyses were performed on the untumbled grus and various size fractions of samples retrieved from the drum after 1, 2, 4, and 8 days of transport. The details of the procedures used for each type of analysis are described in the appropriate section of this study.

LITHOLOGICAL ANALYSIS

Petrographical modal analysis was performed on 240 subsamples derived from the four grus samples from the Lake Tahoe basin. The experimental design employed is shown schematically in Figure 1. Sixty subsamples were point counted using the area method (Galehouse, 1971) for each of the four principal lithic constituents, namely, quartz, potassium feldspar, plagioclase, and polymineralic rock fragments (Table 1). For each lithic constitu-

TABLE 1. RESULTS OF PETROGRAPHIC MODAL ANALYSIS

Sample Number*	Quartz	Potassium Feldspar	Plagioclase Feldspar	Rock Fragments	Number of Counts†
A. 0.50- to 0.70-mm Grain-Size Fraction					
1.0	2	4	252	43	302
1-2	10	10	242	55	320
1-4	7	37	237	19	302
1-8	4	30	238	26	301
2-0	90	20	153	29	306
2-2	107	0	155	25	307
2-4	87	3	169	28	301
2-8	96	3	156	19	300
3-0	72	5	183	14	300
3-2	83	12	152	24	300
3-4	95	23	149	19	300
3-8	112	10	146	23	304
4-0	154	101	24	20	300
4-2	170	73	44	9	301
4-4	143	120	28	8	304
4-8	132	91	60	11	302
B. 0.35- to 0.49-mm Grain-Size Fraction					
2-0	117	0	173	5	304
2-2	99	1	169	11	300
2-4	90	11	172	7	304
2-8	118	10	144	12	303
3-0	92	0	183	10	300
3-2	103	2	165	9	302
3-4	78	3	175	5	301
3-8	85	8	184	12	300
4-0	125	24	130	15	305
4-2	138	73	66	28	308
4-4	131	77	80	10	307
4-8	148	49	92	12	304
C. 0.25- to 0.34-mm Grain-Size Fraction					
1.0	28	20	254	9	312
1-2	35	19	230	14	301
1-4	6	23	248	15	302
1-8	20	17	251	11	302
C. 0.25- to 0.34-mm Grain-Size Fraction (continued)					
2-0	118	4	157	8	301
2-2	104	2	172	8	306
2-4	121	14	147	8	300
2-8	134	2	142	14	302
3-0	137	17	104	22	304
3-2	128	22	142	6	311
3-4	122	25	153	3	320
3-8	122	14	120	30	302
4-0	121	9	168	10	326
4-2	108	6	166	14	300
4-4	97	18	171	16	302
4-8	113	10	172	9	305
D. 0.177- to 0.24-mm Grain-Size Fraction					
1-0	60	20	207	23	314
1-2	36	19	229	19	304
1-4	37	28	230	5	300
1-8	34	24	226	16	304
2-0	101	2	182	12	310
2-2	139	0	104	12	303
2-4	108	1	156	11	301
2-8	107	1	183	5	301
3-0	116	43	134	3	302
3-2	133	0	128	4	300
3-4	136	0	108	12	300
3-8	142	1	138	7	311
4-0	88	11	193	14	313
4-2	69	21	208	7	307
4-4	126	15	157	6	306
4-8	117	14	154	8	303

*For each sample, the first number is the grus sample number; the second number represents untumbled grus (0) or the day (2, 4, or 8) that this subsample was retrieved from the tumbler.

†The number of counts includes opaque and nonopaque heavy minerals that are not considered in this analysis.

ent, four grain-size fractions (0.50 to 0.70, 0.35 to 0.49, 0.25 to 0.34, and 0.177 to 0.24 mm), as well as four temporal subsamples (untumbled grus; 2, 4, 8 days) were analyzed. Inasmuch as randomized areas with widths greatly exceeding the maximum grain diameters of each size fraction were used, the point-counting data may be considered number percentages (Galehouse, 1971). At least 300 grains were counted for each subsample, which provides an average accuracy of approximately ±5.4% for quartz, ±2.9% for potassium feldspar, ± 5.7% for plagioclase feldspar, and ±2.5% for rock fragments at the 95% confidence level (Van der Plas and Tobi, 1965).

The abundance of each lithic constituent as a function of grain size and tumbling time display considerable scatter (Fig. 2), but possible trends are discernable. For grus samples 1, 2 and 3, the number of quartz grains increases with decreasing grain size. However, sample 4 generally shows a decrease in the abundance of quartz with decreasing grain size. These trends appear to be compensated by a decrease in plagioclase with decreasing grain size for samples 1, 2, and 3, and an increase in plagioclase with decreasing grain size for sample 4. Sample 4 also displays a general decrease in the abundance of potassium feldspar with decreasing grain size, whereas trends for potassium feldspar are obscure for samples 1, 2, and 3. The number of rock fragments generally decreases from the 0.50- to 0.70-mm fraction to the 0.35- to 0.49-mm fraction, and appears to be relatively consistent within and among the finer grain sizes for samples 1, 2, and 3. The number of rock fragments for sample 4 is much the same in all the subsamples.

Untransformed or "raw" point counts were employed for significance testing of the lithological data. Although the use of raw numbers does eliminate the ratio aspect, the data do remain nearly constant sum. The use of constant-sum data may produce

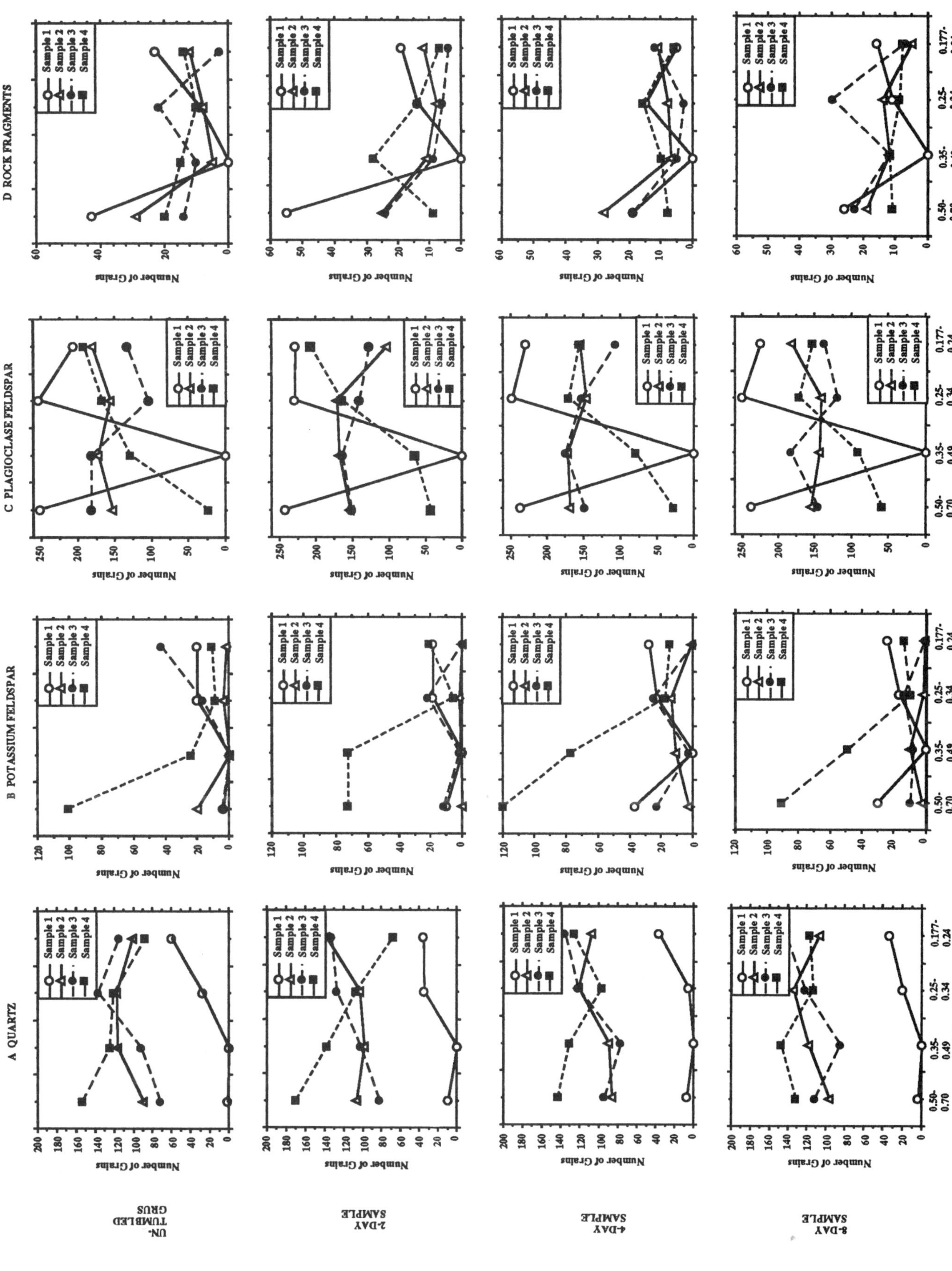

Figure 2. Plots showing the compositional changes for quartz, potassium feldspar, plagioclase feldspar, and polymineralic rock fragments for samples 1 through 4 as a function of grain size and time of exposure to tumbling (untumbled grus; 2, 4, 8 days).

spurious statistical results, but at present there is no completely satisfactory way to measure the strength or correct for such possible effects (Davis, 1986). As such, concern regarding the use of constant-sum data largely remains in the realm of theoretical rather than applied statistics. Although the use of such data may not produce mathematically exact probabilities, it does seem that useful, if not perfect, sedimentological information may be obtained by significance testing. If the use of nearly constant-sum data is shown to introduce substantive error, the compatibility of the statistical results combined with those obtained by other lines of inquiry becomes more important in decision making.

For each lithic constituent, a two-tailed Student t-test was performed for pairs of mean values for the four grain-size fractions of untumbled and tumbled grus to measure the strength of the previously described trends. No statistically significant differences between grain-size fractions were obtained for quartz or plagioclase at the 5% level. Differences between mean values for potassium feldspar are barely significant between the 0.50- to 0.70- and 0.177- to 0.24-mm fractions at the 5% level ($|t| = 2.06 > 2.04$ for 30 d.f.), and are not significant at the 1% level ($|t| = 2.06 < 2.75$ for 30 d.f.). No other significant differences occur with respect to the number of feldspar grains. Rock fragments show highly significant differences between the 0.50 and 0.70 mm and each of the three finer grained fractions; however, the three finer grained fractions are statistically homogeneous with respect to mean values. Due to the somewhat different behavior suggested by the previously described trends for sample 4, Student t-tests were computed without this sample. The results are the same for quartz, plagioclase, and rock fragments; however, no significant differences occurred with respect to potassium feldspar at the 5% level. With the exception of rock fragments, comparison of the mean values suggests a high degree of uniformity with respect to each lithic constituent within the four grain-size fractions examined.

An analysis of variance was carried out to measure the grain size and temporal relationships among the 60 subsamples for each of the four major lithic constituents. Inasmuch as sample 1 contains no grains in the 0.35- to 0.49-mm size fraction (a missing cell in statistical terms), it was assumed that samples are a random effect, and therefore may be nested within grain size. Data for quartz, potassium feldspar, plagioclase, and rock fragments (Table 1) were treated separately to avoid the problem of pooled variance estimation due to large differences in variance among them. The arcsine square root data transformation (Alder and Roessler, 1977) alleviated heterogeneity concerns associated with the potassium feldspar and rock fragment data. For each lithic constituent, the sample by grain-size interaction effect was pooled with the model error term to form a combined error term.

Tables 2 through 5 show that there are no statistically significant differences between or among the subsamples representing the four grain-size fractions, the subsamples representing the days, or the grain size by days interaction effect for quartz, potassium feldspar, and plagioclase; in short, the 180 subsamples are homogeneous with respect to these three lithic constituents. Table 5 shows that there is a highly significant difference for rock fragments with respect to grain size; however, the days and grain size by days interaction effect is not statistically significant. Examination of Table 6 indicates that significant to highly significant differences in the number of rock fragments occur where the coarse sand fraction (0.50 to 0.70 mm) is compared to the two medium- and one fine-sand fractions; however, no significant differences occur among the medium- and fine-sand fractions. The observed difference in the number of identifiable rock fragments in the coarse-sand fraction versus the medium- and fine-grained fractions most likely represents the disintegration of rock fragments into discrete mineral grains at or near the boundary between coarse- and medium-grained sand (0.50 mm). Of interest is the consistency in the abundance of rock fragments throughout the size range 0.177 to 10.49 mm. Thus, except for rock fragments, the results of the analysis of variance do not

TABLE 2. RESULTS OF ANALYSIS OF VARIANCE FOR QUARTZ

Source of Variation	Sum of Squares	Degrees of Freedom	Mean Square	Calculated F-Value	$F_{0.95}$
Grain size	0.005	3	0.018	0.66	2.84*
Days	0.006	3	0.002	0.07	2.84*
Grain size x days	0.018	9	0.002	0.07	2.12*
Error term	1.216	44	0.028		
Total	1.294	59			

*Nonsignificant result at 5% level.

TABLE 3. RESULTS OF ANALYSIS OF VARIANCE FOR POTASSIUM FELDSPAR

Source of Variation	Sum of Squares	Degrees of Freedom	Mean Square	Calculated F-Value	$F_{0.95}$
Grain size	0.128	3	0.042	1.60	2.84*
Days	0.038	3	0.013	0.48	2.84*
Grain size x days	0.076	9	0.008	0.32	2.12*
Error term	1.168	44	0.027		
Total	1.411	59			

*Nonsignificant result at 5% level.

TABLE 4. RESULTS OF ANALYSIS OF VARIANCE FOR PLAGIOCLASE

Source of Variation	Sum of Squares	Degrees of Freedom	Mean Square	Calculated F-value	$F_{0.95}$
Grain size	0.089	3	0.030	0.74	2.84*
Days	0.004	3	0.001	0.03	2.84*
Grain size x days	0.019	9	0.002	0.05	2.12*
Error term	1.766	44	0.040		
Total	1.877	59			

*Nonsignificant result at 5% level.

TABLE 5. RESULTS OF ANALYSIS OF VARIANCE FOR ROCK FRAGMENTS

Source of Variation	Sum of Squares	Degrees of Freedom	Mean Square	Calculated F-Value	$F_{0.95}$	$F_{0.99}$
Grain size	0.089	3	0.030	8.28	2.84§	4.31†
Days	0.011	3	0.004	0.99	2.84*	
Grain size x days	0.014	9	0.002	0.42	2.12*	
Error term	0.159	44	0.004			
Total	0.273	59				

*Nonsignificant result at 5% level.
§Significant result at 5% level.
†Highly significant result at 1% level.

TABLE 6. STUDENT'S PROBABILITIES THAT THE SAMPLE MEANS FOR ROCK FRAGMENTS IN THE FOUR GRAIN-SIZE CLASSES ARE EQUAL

	A 0.50–0.70 mm	B 0.35–0.49 mm	C 0.25–0.34 mm	D 0.177–0.24 mm
A				
B	0.016§			
C	0.016§	0.852*		
D	0.005†	0.698*	0.537*	

*Nonsignificant result at 5% level.
§Significant result at 5% level.
†Significant result at 1% level.

support the credibility of the possible lithic trends discussed earlier in this section.

GRAIN-SIZE ANALYSIS

Grus samples 1, 2, and 3 were randomly selected to carry out the grain size and quartz grain-shape analysis for the tumbler experiments. For each of the three grus samples, grain-size analysis was performed by mechanical sieving (Ingram, 1971) at 0.5 ϕ intervals for the –2.0 to 4.0 ϕ (0.625 to 4.0 mm) size range for the untumbled grus as well as the 1-, 2-, 4-, and 8-day samples retrieved from the tumbling drum. The Udden-Wentworth scale is used to describe grain-size data (Wentworth, 1922). The results of these 15 analyses are presented in Figures 3 through 6. Figure 6 shows that all three samples displayed relatively rapid rates of disintegration during the first few days of tumbling, and then continued to disintegrate at reduced rates from days 4 through 8. The coarser grained fractions in samples 1 were reduced by 5.2% by the end of the first day, an additional 1.4% disintegrated during the second day, and a total 2.3% disintegrated during the final 6 days. The average rate of grain reduction for the 6-day period is 0.4%/day. The breakage of the coarser fractions in sample 2 apparently was linear for the first 2 days, with an average reduction rate of 4.5%/day. The breakage continued from 8.9 to 12.2% during the next 6 days, which produced an average reduction rate of 0.5%/day during this period. Sample 3 experienced the most comminution: during the first day, a 10.9% reduction occurred, and for the next 3 days, an additional 4.0% disintegrated, yielding an average rate of 1.3%/day. An additional 1.4% was comminuted during the last 4 days, which yields an average reduction rate of 0.4%/day. These results show that a total of 8.9 to 16.2% of the relatively coarser grained fractions experienced comminution during 8 days, or 200 km of transport, in the tumbler. Of this amount, the sand fraction only increased from 1.7 to 3.4%, whereas the silt plus clay fraction increased from 6.4 to 14.5% during the same period.

The rate of production of silt and clay for each of the samples is shown in Figure 6. For sample 1, this fine-grained fraction rapidly increased from 0.0 to 3.9% after the first day of tumbling, increased to 4.6% by the end of the second day, and then gradually increased to a total of 6.4% after tumbling for 8 days. The fine-grained fraction in sample 2 increased to 4.7% after 1 day, continued to increase at almost the same rate to 7.4% after 2 days, and then gradually increased to a total of 8.8 percent after tumbling for 8 days. For sample 3, the fine fraction increased to 8.9% after only 1 day of tumbling, and then more gradually increased to a total of 14.5% after 8 days of tumbling. The production of silt and clay occurs mostly during the first day of tumbling, as indicated by the results of the regression analysis listed in Table 7.

The grain-size data shown in Figures 3 through 6 show systematic decreases in the coarse-grained size fraction of grus samples and corresponding increases in the finer grained fractions above and below a rather persistent fine- and/or medium-sand size fraction for each sample. For sample 1, the relatively stable medial fraction ranges from about 1.75 to 2.25 ϕ (0.21 to 0.30 mm); for sample 2, from about 1.25 to 1.75 ϕ (0.30 to 0.42 mm); and for sample 3, from about 2.25 to 2.75 ϕ (0.149 to 0.21 mm).

There is an 8.9 to 16.2% decrease in the coarser grained fractions of the three grus samples and a corresponding increase in the finer sand and silt plus clay fractions during tumbling. Considerable fragmentation of relatively coarser grained fractions occurred, particularly during the first 2 days of tumbling, but the obtained results do not prove from which of the coarser size fractions the material in any finer grained fraction was derived. The curves for all three samples continue to rise from days 4 through 8, indicating that equilibrium conditions with respect to grain breakage were not achieved during as much as 100 to 200 km of transport in the tumbling drum.

FOURIER GRAIN-SHAPE ANALYSIS

Introduction

Since its introduction by Schwarcz and Shane (1969) and Ehrlich and Weinberg (1970), Fourier analysis of detrital quartz grain shapes has been applied to a wide variety of natural tracer and petrofacies problems. The interested reader is referred to the following papers: Ehrlich et al. (1974), Brown et al. (1980), Clark and Osborne (1982), Mazzullo (1986), and Osborne and Yeh (1991).

Ehrlich and Weinberg (1970) describe a closed form Fourier method to analyze the observed variation of two-dimensional, maximum projection, grain-shape area. Grain shape may be estimated by an expansion of the periphery radius as a function of angle about the grain's center of gravity by a Fourier series. In Fourier analysis, a series of sine and cosine curves with periods equal to fundamental harmonics are fit to the observed

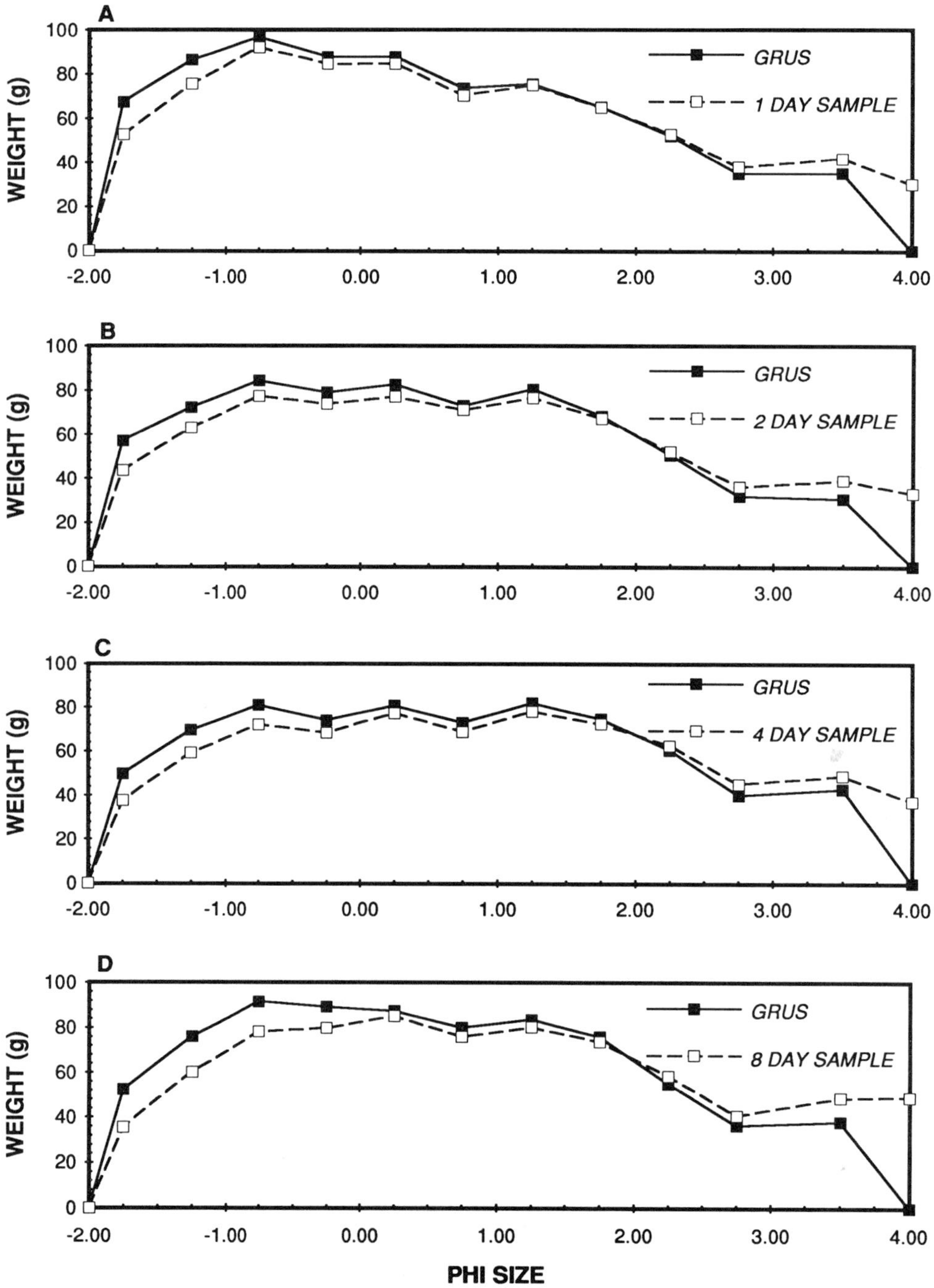

Figure 3. Weight (g) for the grain-size fractions of untumbled grus sample 1 (solid square) and associated samples retrieved after tumbling (1-, 2-, 4-, 8-day intervals).

data by a least squares technique. Fundamental harmonics are the prime fractions (1/3, 1/4, . . . 1/N), where N equals half the number of digitized points used to define the periphery of a grain. As the number of fundamental harmonics is increased, the computed curve converges with the observed data. The highest frequency that can be estimated is known as the Nyquist frequency, which is equal to twice the distance between successive observations (Davis, 1986). If the Nyquist frequency is exceeded, error may be introduced by the incorporation of irresolvable high frequencies into lower frequencies (aliasing). The radius of the grain is given by:

$$R(\theta) = Ro + \sum_{n=1}^{\infty} Rn \cos(n\theta - \phi n),$$

where θ is the polar angle measured from an arbitrary reference line. The first term in the series, Ro, is equivalent to the average radius of the grain in the maximum projection orientation. For the remainder of the terms, n is the harmonic order or number, Rn is the harmonic amplitude, and ϕn is the phase angle. The nth

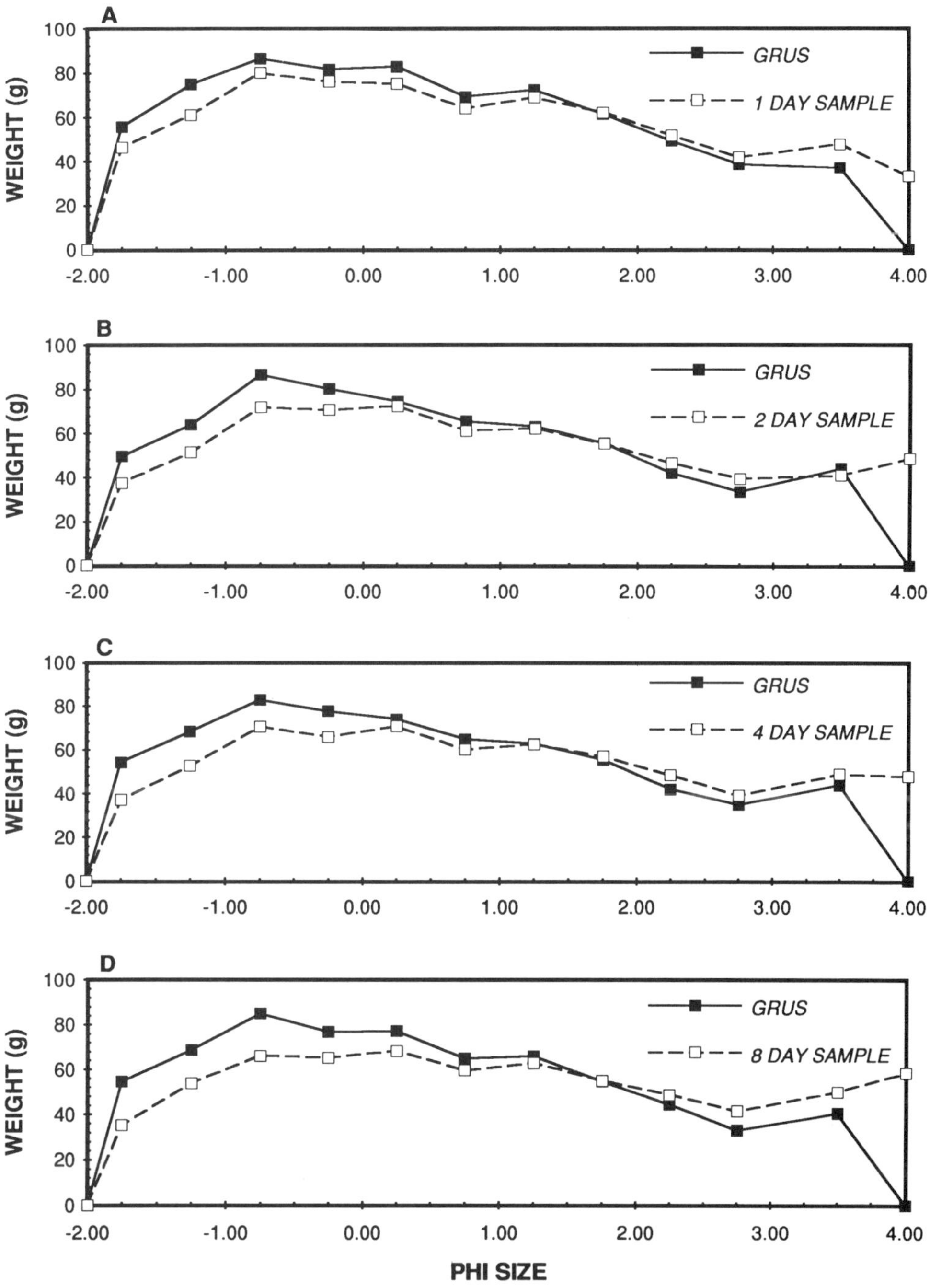

Figure 4. Weight (g) for the grain-size fractions of untumbled grus sample 2 (solid square) and associated samples retrieved after tumbling (1-, 2-, 4-, 8-day intervals).

harmonic contributes to the explanation of the observed shape variation as a figure with n "bumps." For example, the "zeroth" harmonic is a centered circle with an area equal to that of the maximum projection; the first harmonic is an off-centered circle; the second is a figure eight; the third is a trefoil, and so forth. The amplitude of each harmonic represents the contribution of each basic shape component to the overall shape of the maximum projection outline for each grain. The phase angle represents the orientation of each component. With given combinations of amplitudes and phase angles, a regularly shaped two-dimensional maximum projection outline can be partitioned into its fundamental shape components. The center of gravity of the maximum projection shape is used as the origin of the radius expansion to simplify interpretation of the Fourier series (Full et al., 1984). Coordinates of points along the periphery of the maximum projection outline are required for the Fourier expansion. At least twice the number of such points must be known as the number of the highest desired harmonic. The initial origin of the periphery

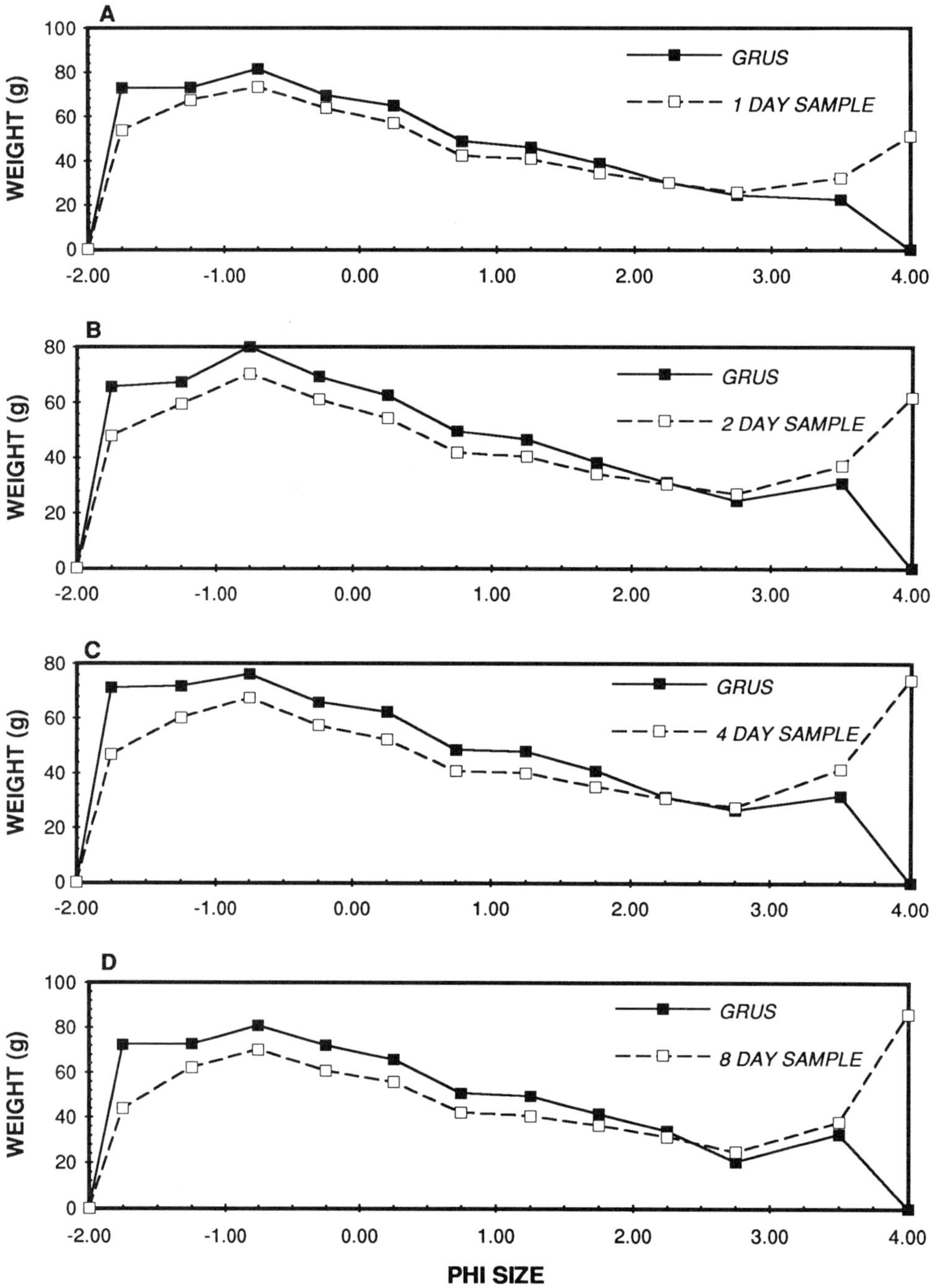

Figure 5. Weight (g) for the grain-size fractions of untumbled grus sample 3 (solid square) and associated samples retrieved after tumbling (1-, 2-, 4- and 8-day intervals).

points may be arbitrary, because a later transformation places the origin at the center of gravity of the maximum grain-projection area. If a harmonic or periodic function exists within the data, the amplitude of the sine and cosine curves with periods close to the natural harmonic will be considerably larger than the amplitudes of other harmonics in the sequence.

Shape information is contained in both amplitude and phase angle values. Amplitude is invariant with respect to grain orientation, whereas phase angle is not. As the orientation procedure is time consuming and spectra of harmonic amplitude values normally serve to distinguish between and/or among constituent grain-shape populations, consideration of the phase angle is usually not considered in Fourier grain-shape analysis. Inasmuch as the present study deals with elongate grus grains and attempts to evaluate potentially small changes in quartz grain shape resulting from fracture, the phase angle was considered by alignment with respect to the longest apparent radius of the maximum projection outline for each grain. As the phase angles are oriented with respect to an objective and unique criterion, both the harmonic amplitude and phase angle values are unique.

The shape of the nascent quartz grain in grus is due principally to the size and shape of the original crystal from which it

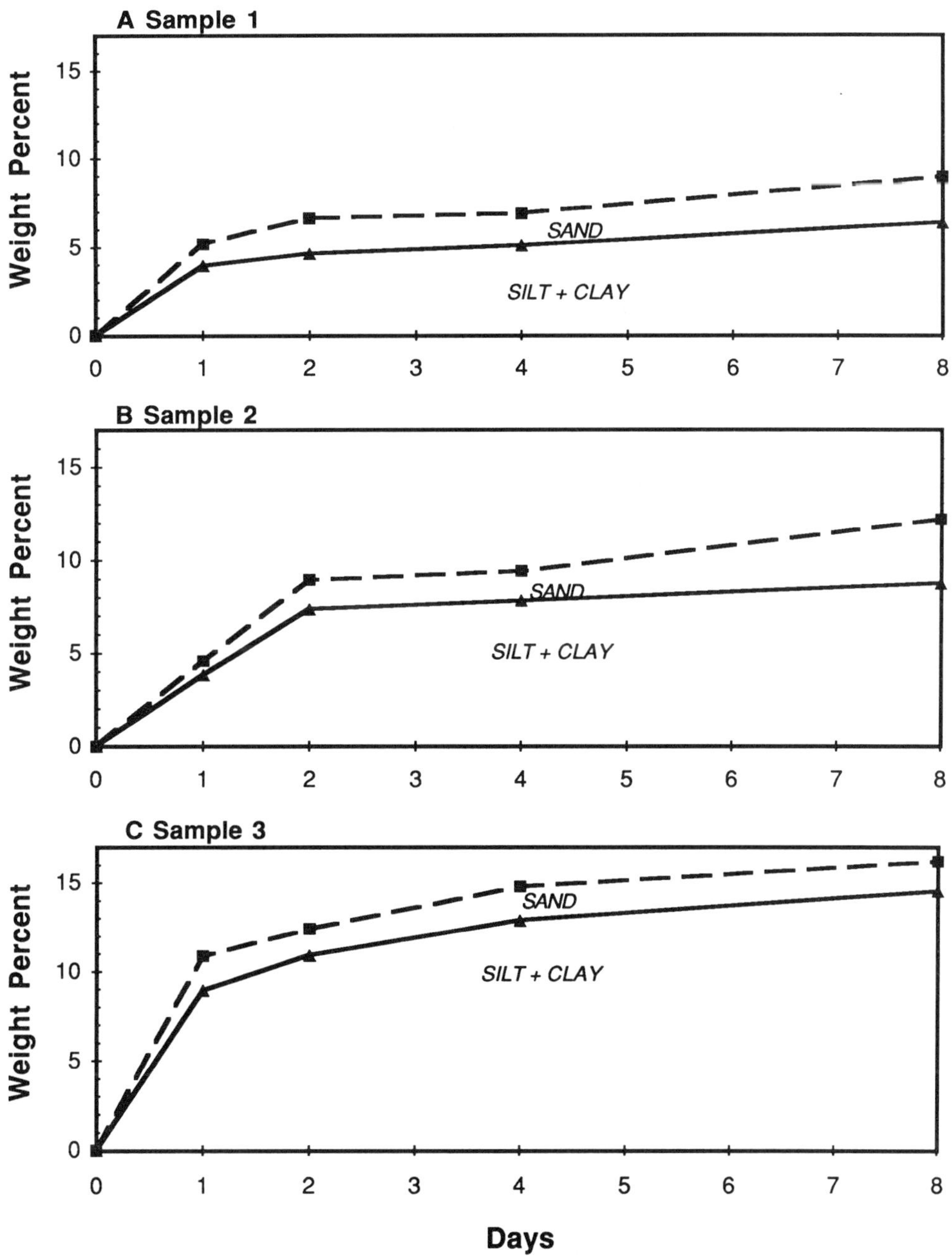

Figure 6. Rate of change in grain size for grus samples 1, 2, and 3 during 8 days (≈200 km) of transport in tumbling drum.

TABLE 7. RESULTS OF TESTING RELATIONSHIP BETWEEN INCREASE IN PERCENT OF SILT PLUS CLAY PRODUCED DURING 1-DAY TUMBLING PERIOD

Variable	Degrees of Freedom	Parameter Estimates	Standard Error	t for Null Hypothesis: Parameter=0	Probability >|t|
Intercept	1	0	1.10	0.00	1.00*
Tumbling (1 day)	1	5.86	1.56	3.76	0.02§

*Nonsignificant result at 5% level.
§Significant result at 5% level.

was derived and to breakage along dominant and subordinate microfracture sets present in the protolith at the time of weathering. Inasmuch as nascent quartz is typically elongate, there are at least two fracture sets oriented more or less parallel to the elongation direction as well as one or more other sets which may trend at high angles to the direction of elongation. In terms of understanding quartz grain-shape composition as an indicator of source rock, it is important to know whether breakage during transport dominantly occurs along fracture sets parallel to or at high angles to the direction of elongation. In the first case, elongate grains would continue to form during a transport, and the grain-shape distributions for a given size fraction of a transported sand would closely match that of the grus from which it was derived. In this case, the grain-shape composition of a single-source sand sample would be an excellent indicator of the associated protolith. In the second case, there would be a tendency for elongate grus grains to break across the long dimension during transport, and therefore become more equant. In this case, it would be more difficult to assign detrital quartz grains to a unique protolith using Fourier grain-shape analysis. Analysis of the distributions of shape-ratio values (length of short axis/length of long axis of the maximum projection grain outline), as well as an analysis of the amplitude spectra for harmonics 2 and 3, was employed to examine the character of change in fundamental quartz-grain shape from grus through as much as 200-km of transport in a rotating drum. The results of this experiment were then used to examine changes in the grain-shape of four high-gradient streams in southern California.

Although conceptually similar to the closed form methodology described by Ehrlich and Weinberg (1970), the methodology employed in this report makes use of the newer and more widely used Fast Fourier Transform described by Brigham (1974). This procedure involves the calculation of many values of the line spectrum using the Fast Fourier Transform computer algorithm to produce a smoothed estimate of the continuous spectrum. As its name implies, this algorithm is extremely rapid and requires only $n \cdot \log_2 n$ arithmetic operations rather than the n^2 operations as do alternate methods.

Sample preparation

Approximately 215 quartz grains from the 0.25 to 0.50-mm (1.0-2.0 ϕ) fraction from each sample are used for the Fourier grain-shape analysis. Samples are first wet sieved to obtain the desired size fraction to minimize attrition of the quartz grains during sieving. Sieving is followed by drying in convection ovens at 40°C. All strongly magnetic minerals are removed from the samples using a hand magnet. The samples are boiled in a solution of 10% hydrochloric acid with stannous chloride crystals for 20 min to remove iron-oxide coatings and carbonate cement (Ingram, 1971). The hydrochloric acid residue was removed by rinsing the samples in deionized water. The quartz grains are etched clean by washing them in a hydrofluoric acid bath for 1 min. Such etching does not significantly alter the shape of the grains (Schultz, 1980). All samples are then rinsed in deionized water and dried thoroughly.

Samples are examined under a petrographical microscope using a reflecting light source. Quartz grains are picked by using a very small brush. These grains are dry mounted on a glass slide, which enhances the contrast between the grain and its background (Ehrlich and Weinberg, 1970).

Fourier methodology

Digitizing of the individual quartz grains is accomplished using an optical digitizer and the EDGE Fortran program written by Tim Fogarty in 1985. In this process, a black and white camera projects the maximum two-dimensional image of the quartz grain from a standard binocular petrographical microscope to a video screen. The image of the grain appears dark in relation to the surrounding background since the microscope is lighted from below. A maximum contrast between grain edge and background allows the computer to scan the pixels of the video image and record all boundary points of the grain as Cartesian (X-Y) coordinates. A pixel is determined as a boundary point if it is dark and one of its neighbors is light. While the Cartesian coordinates are computed, the centroid of the projected grain also is computed.

The next step in the digitizing process is to prepare the data for the Fourier transform. The Cartesian coordinates are converted to polar coordinates. The radius is the distance from a boundary point to the centroid; the angle used is the angle from the horizontal line bisecting the centroid to the boundary point. In order to examine the entire wavenumber amplitude spectrum (each wavenumber is analogous to one harmonic) of a grain during statistical analysis, the Fast Fourier Transform (Brigham, 1974) is used. One requirement of the Fast Fourier Transform is that the data be evenly spaced. Because the boundary points of a grain are not so spaced, the data must be linearly interpolated. The boundary points are interpolated to 128 evenly spaced points. The Fast Fourier Transform also requires that the data set be centered around zero. To center the data, the mean radius is subtracted from the radius versus angle data. This radius is later placed in the final amplitude spectrum set.

On completion of the Fast Fourier Transform, the amplitude of each wavenumber is calculated by the following equation (Brigham, 1974):

$$\sqrt{(\text{Real})^2+(\text{Imaginary})^2}\ .$$

Because the input data are real rather than imaginary, half of the wavenumbers are reflections and are not used in the statistical analyses. This procedure potentially leaves 64 wavenumbers available for analysis. Each wavenumber is represented by a vector and the occurrence of 64 wavenumbers results in the analysis of 64 dimensions in space. Although 64 wave numbers are available for analysis, only the first 24 were used due to increased dispersion associated with higher wavenumbers.

Results

Figure 7 shows the distribution of values for the shape ratio (length of short axis/length of long axis of the maximum projection grain outline) for each of 215 quartz grains from grus samples 1, 2, and 3 as a function of time. An increase in shape-ratio values signifies that the grain shape became more equant, with across-grain fracture being the dominant process, whereas a decrease in such values indicates that grain shape became more elongate, with grain-parallel fracture being the dominant process. Accordingly, changes in the distribution of shape-ratio values may be used to determine the relative importance of grain-parallel versus across-grain fracture.

For sample 1, there is a loss of the most elongate portion of the shape-value distribution (0.3 to 0.5) and an increase in values from 0.5 to 0.7 after tumbling for 1 day. This sudden decrease in the most elongate portion of the distribution appears to be evidence for the dominance of across-grain fracture. After 8 days in the tumbler, the proportion of grains in the 0.3 to 0.5 range are much the same as in the grus sample, which suggests the resurgence of grain-parallel fracture. Overall, there is a shift from a modal value of approximately 0.7 to 0.6 after 1 day of tumbling, which appears to persist through day 8. On the other hand, a comparison of mean values (Table 8) shows no measurable change from grus through day 1 and a decrease from mean values of 0.67 to 0.65 from days 1 through 8. A Student t-test for differences in mean values between the 1 and 8-day samples is significant at the 10% level for a two-tailed test ($|t| = 1.66$).

Although it appears that across-grain breakage occurred during the first day of tumbling, the effects of grain-parallel fracture appear to dominate after tumbling for 8 days. For sample 2, there is a substantive shift toward smaller ratio values from grus through day 1, which persists through day 8. This rapid and persistent shift toward the more elongate portion of the shape-ratio distribution reflects the pervasive effects of grain-parallel fracturing during this period. A Student t-test between the means for grus and the one-day sample is significant at the 1 percent level ($|t| = 3.29$). There is a modest shift toward smaller ratio values for sample 3, which also suggests the dominance of grain-parallel fracture. A Student t-test for grus and the 1-day tumbler sample is not statistically significant; however, there is a difference between grus and the 8-day sample at a 10% significance level ($|t| = 1.70$).

Although there is some evidence for initial across-grain fracture in sample 1, the overall increase in smaller shape-ratio values demonstrates the dominance of grain-parallel fracture in all three samples. The *mean* shape-ratio values for grus and the 1-day tumbler samples show a highly significant difference for sample 2, but nonsignificant differences for samples 1 and 3. For samples 1 and 3, significant differences in the shape-ratio means occur between the 1- and 8-day tumbler samples at a 10% level. Thus, depending on the choice of central tendency measurement, one could argue on the basis of a general decrease of shape-ratio values that important changes in the distributions occur during the first day of tumbling.

However, if one relies on the comparison of distribution means, only sample 2 showed a significant decrease after 1 day of tumbling, whereas samples 1 and 3 showed significant but smaller decreases only after 8 days of tumbling. Standard deviation values (Table 8) consistently decrease at the end of the first tumbling day, and then increase from days 1 through 8 to nearly equal or exceed that of the associated grus sample. Linear regression shows that on average, 1 day of tumbling has a significant effect on reducing the standard deviation of the shape-ratio values at a 7% level (Table 9). An additional 7 days of tumbling has the effect of increasing the standard deviation of the shape-ratio values at the 1% significance level (Table 10). These results confirm that the shape-ratio distributions display systematic decreases followed by increases in standard deviation values in a closed particulate system. It therefore seems likely that relatively high rates of grain breakage occur during the first day of tumbling ($\simeq$25 km of transport) to approach a quasi-stable distribution, after which continued grain elongation, dominantly due to grain-parallel fracture, increases the standard deviation values.

For samples 1 through 3, the mean amplitude values for harmonics 2 and 3 are listed in Table 11. These harmonics are of interest, because they contain information concerning fundamental shapes that are likely to arise from grain-parallel or across-grain fracturing. Inasmuch as harmonic 2 is the dominant measure of grain elongation, changes in associated amplitude values may indicate the degree of grain-parallel fracture. Harmonic 3 is the principal measure of the contribution of the triangular shape component, which may reflect the degree of across-grain breakage. The third harmonic was selected as a measure of across-grain breakage, because cross-fracture sets are unlikely to be oriented exactly normal to the long-grain dimension. Therefore, grains with a triangular aspect are much more likely to be produced than those with a square or rectangular aspect (harmonic 4). Preliminary measurements of fracture orientations in quartz grains in protolith samples appear to support this assertion.

Of 12 tests of significance among the grus, 1- and 8-day samples using Duncan's New Multiple Range Test (Alder and Roessler, 1977), only three significant results were obtained at the 5% level. No significant differences were obtained for harmonic 3, which indicates that the contribution of triangularity to the grain-shape composition remained unchanged during the tumbler experiment. This is interpreted to mean that the production of triangular shaped grains by across-grain breakage was insignificant. In sample 1, the contribution of harmonic 2 decreased significantly after 1 day of tumbling, and then increased from the 1- to 8-day samples. For sample 2, harmonic 2 increased significantly in the 1-day sample, and remained unchanged from the 1- to 8-day samples. No significant differences for harmonic 2 were obtained for sample 3. Values for the amplitude ratio of harmonic 2/harmonic 3 (Table 11) increase for all three samples as a function of transport distance, which argues for the dominance of grain-parallel fracture. These results are much the same as those obtained using the shape-ratio distributions.

Inasmuch as it appears that breakage of detrital quartz

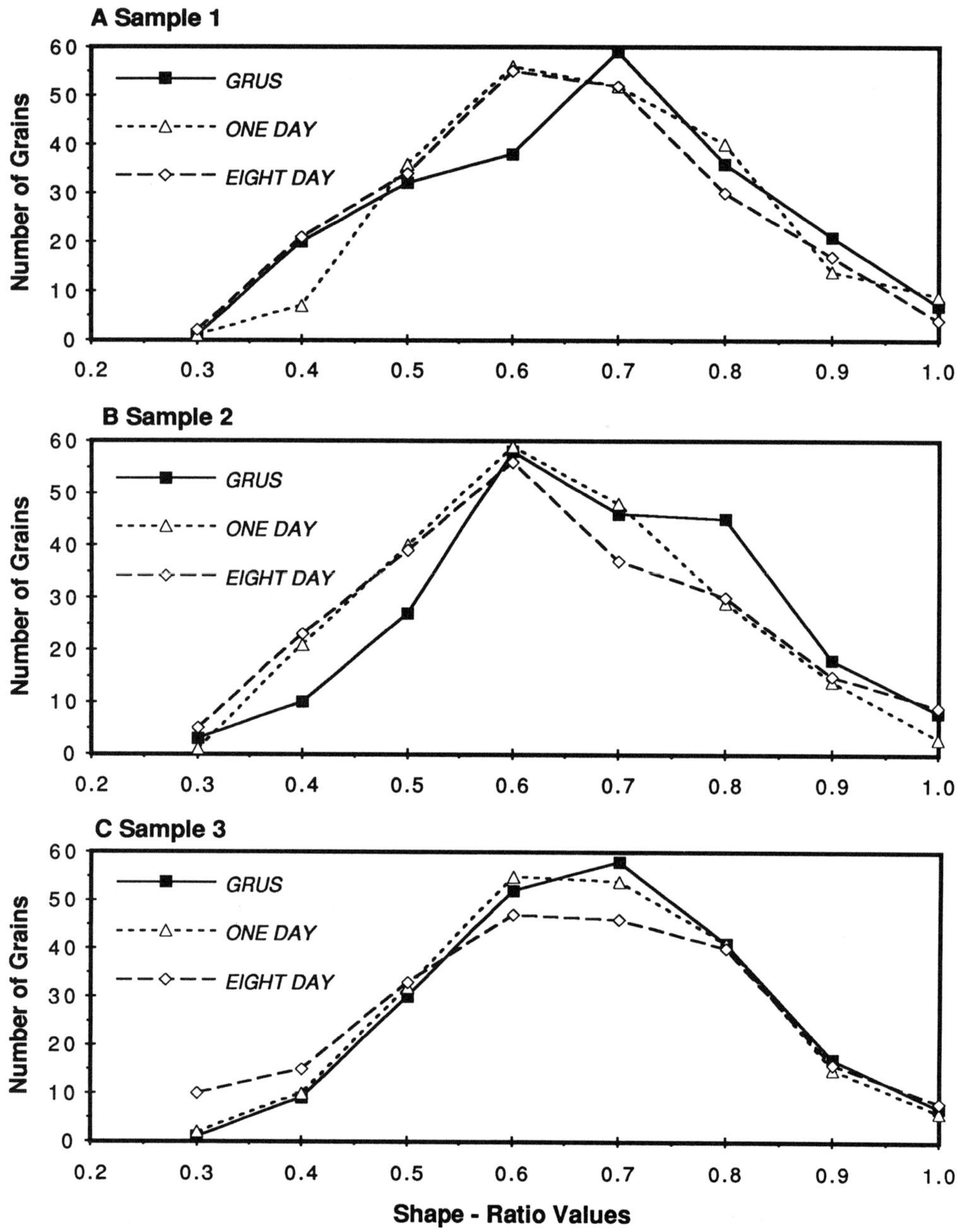

Figure 7. Temporal changes in the distributions of shape-ratio values for untumbled (solid square) and tumbled grus samples 1, 2, 3.

TABLE 8. MEANS AND STANDARD DEVIATIONS OF SHAPE-RATIO VALUES

		Grus Sample	1-day Sample	8-day Sample
Sample 1	Mean	0.67	0.67	0.65
	Standard Deviation	0.16	0.14	0.15
Sample 2	Mean	0.68	0.63	0.64
	Standard Deviation	0.15	0.14	0.16
Sample 3	Mean	0.68	0.67	0.65
	Standard Deviation	0.14	0.14	0.16

TABLE 9. REGRESSION ANALYSIS TESTING RELATIONSHIP BETWEEN STANDARD DEVIATION VALUES OF SHAPE-RATIO DISTRIBUTIONS AND 1 DAY OF TUMBLING

Variable	Degrees of Freedom	Parameter Estimate	Standard Error	t for Null Hypothesis: Parameter=0	Probability >ltl
Intercept	1	0.15	0.00	51.62	0.00
Tumbling (1 day)	1	-0.01	0.00	-2.45	0.07

TABLE 10. REGRESSION ANALYSIS TESTING RELATIONSHIP BETWEEN STANDARD DEVIATION VALUES OF SHAPE-RATIO DISTRIBUTIONS AND 8 DAYS OF TUMBLING

Variable	Degrees of Freedom	Parameter Estimate	Standard Error	t for Null Hypothesis: Parameter=0	Probability >ltl
Intercept	1	0.14	0.00	37.62	0.00
Tumbling (8 days)	1	+0.00	0.00	4.35	0.01

TABLE 11. MEAN AMPLITUDE AND RATIO VALUES FOR HARMONICS 2 AND 3

	Untumbled Grus	One-day Sample	Eight-day Sample
A. Sample 1			
Harmonic 2	10.76	10.43	11.66
Harmonic 3	4.00	3.84	3.69
Harmonic 2/Harmonic 3	2.69	2.72	3.15
B. Sample 2			
Harmonic 2	10.43	11.99	12.25
Harmonic 3	4.17	4.02	3.77
Harmonic 2/Harmonic 3	2.50	2.98	3.25
C. Sample 3			
Harmonic 2	10.50	10.61	11.49
Harmonic 3	4.32	4.15	4.05
Harmonic 2/Harmonic 3	2.43	2.55	2.84

grains during transport occurs dominantly along the same fracture sets that produced the shape distributions found in associated grus samples, Fourier grain-shape analysis may be used to identify unique plutonic source rocks provided the shape characteristics of the grus from each potential protolith are known. These results strictly apply to only the 0.25- to 0.50-mm size fraction, but it is likely that this relationship is true for at least fine- and medium-grained sand. SEM photomicrographs (Fig 8) show little difference between representative quartz grains from grus and associated 8-day tumbler samples.

APPLICATION OF THE SHAPE RATIO TO SAND FROM HIGH-GRADIENT STREAMS

Each rotating drum experiment described herein consisted of a closed particulate system involving grus from a single protolith; therefore, complications arising from a number of variables associated with sediment transport in natural fluvial systems were avoided. Changes in stream gradients, channel dimensions, channel shapes, sediment sizes, bed roughness, and hydraulic factors all are important in determining the character of fluvial sediment discharge at a given point. Carefully controlled field and/or experimental data are required to determine the primary and interaction effects of such variables (Schumm et al., 1987). Stream gradient is the only geomorphological variable considered in the present study.

The obtained experimental results, as well as those of Moss (1972), suggest that most grain-shape modification occurs during the early stages of transport. Therefore the upper reaches of four high-gradient streams (part B of Appendix I; Fig. 9) were selected to evaluate any potential downstream trend in grain-shape composition for the 0.25- to 0.50-mm (1.0 to 2.0 ϕ) size fraction, which, from the experimental results, might be expected to be dominated by grain-parallel breakage. However, any such trend might be perturbed by the influx of grains eroded from the adjacent canyon walls consisting of the same source rock, additions from minor tributaries draining the same protolith, the addition of new grain shapes where a stream crosses contacts separating different protoliths, and hydraulic sorting by shape. A grus sample was taken near the head of each stream, and 1-kg samples were collected at 0.8-km intervals for the sampled section of each stream. The average shape ratio was computed by analyzing at least 200 grains per sample.

The sampled portion of Culp Canyon Creek flows through Mesozoic tonalite, and has an average gradient of 71 m/km. Boden Canyon Creek flows through Mesozoic tonalite, and was sampled for a distance of 11.5 km from its head. Mill Creek dissects Permo-Triassic granite and Precambrian anorthosite, and was sampled for a distance of 10 km. It has an average gradient of 64 m/km along this 10-km length. Mill Creek crosses a contact with a younger granite at the 10.0 km station (Fig. 10). The sampled section of Big Tujunga Creek cuts across gneissose granite for a distance of 10 km, and has an average gradient of 41.7 m/km. Samples were collected from Culp and Boden Canyon

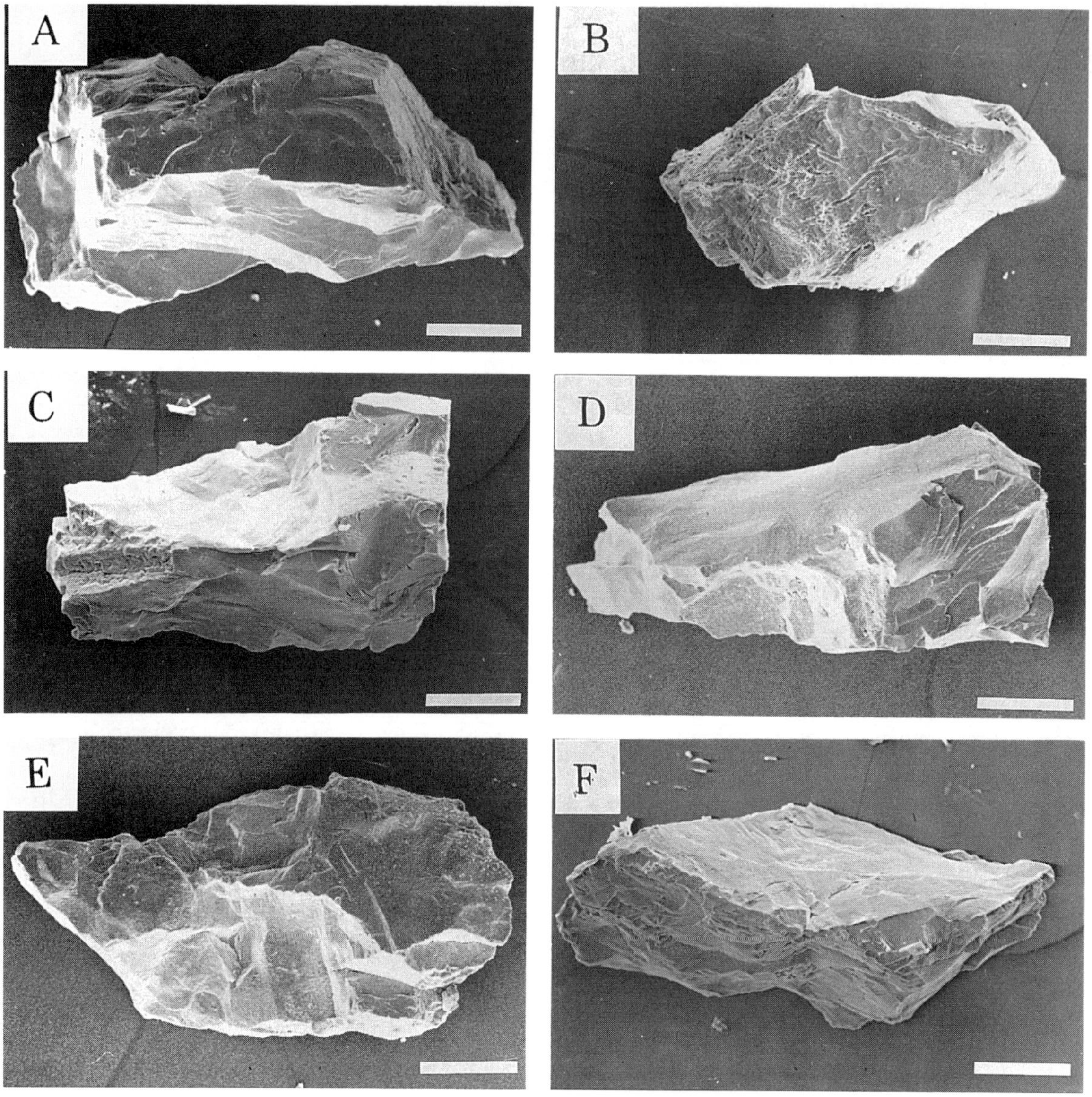

Figure 8. SEM photomicrographs showing planar fracture sets and angularity of elongate quartz grains from untumbled grus (A, C, E) and 8-day (B, D, F) tumbler samples. A and B, Grains from Grus sample 1. C and D, Grus sample 2. E and F, Grus sample 3. Scale bars = 200 μ.

Creeks during June 1984, and from Mill and Big Tujunga Creeks during May 1985. Each stream was divided into two or more segments based on differences in average gradient (Fig. 10, Table 12). The average shape-ratio values for each sample are indicated in Figures 10 and 11, and are grouped by stream segment in Table 12.

Table 12 shows that, except for Segment A in Boden Canyon Creek, the range of differences in shape-ratio values within each stream segment is less than 0.04. Differences in shape-ratio values less than 0.05 generally are not statistically significant at the 5% level; therefore, only differences equal to or exceeding 0.05 will be considered sedimentologically important. Accordingly, these data indicate very little change in the shape-ratio values within assigned stream segments. For Segment A in Boden Canyon Creek, the value for the shape ratio increases from 0.70 to 0.79 during the first 0.8 km of transport, and then decreases to 0.71 at the 1.6-km station. The observed increase, then decrease, in shape values indicate an initial increase in the proportion of more equant grains followed by an increase in the proportion of more elongate grains. These changes in proportion may reflect an increase in the rate of production of more equant grains by across-grain fracturing, which is followed by either the selective transport of more elongate grains and/or the renewed dominance of grain-parallel fracture, producing a higher proportion of elon-

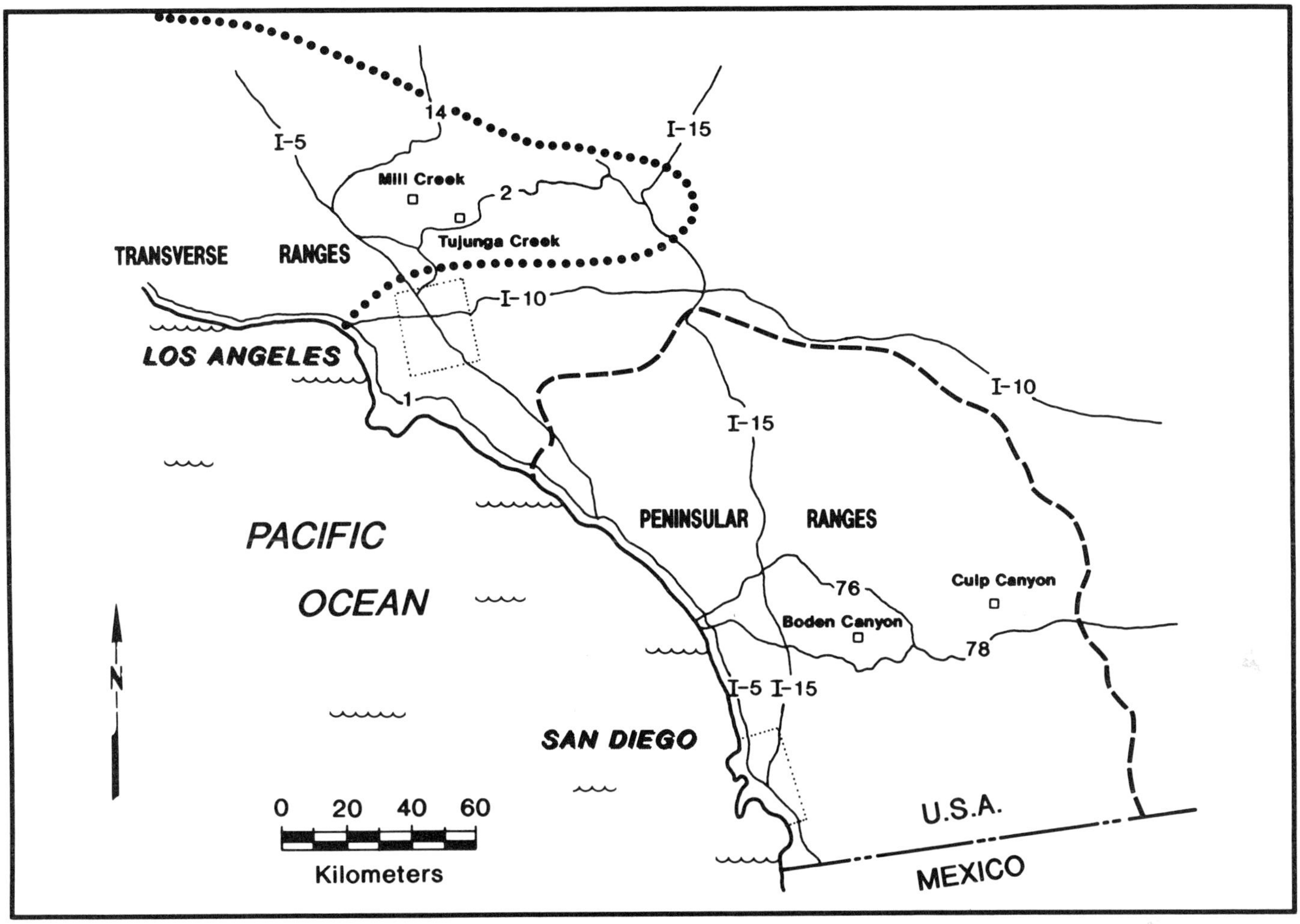

Figure 9. Map showing the location of the headwaters of Culp Canyon Creek, Boden Canyon Creek, Mill Creek, and Big Tujunga Creek. The Transverse and Peninsular Ranges are outlined by dotted and dashed lines, respectively.

gate grains. Other factors being equal, if selective transport of more elongate grains were the dominant process, shape-ratio values might be expected to consistently increase downstream. The shape-ratio values remain much of the same along the 9.6-km-long channel assigned to Segment B, therefore it appears more likely that elongate quartz grain shapes were added preferentially to the 0.25- to 0.50-mm size fraction by means of grain-parallel fracture of coarser grained quartz. At any given point, the production of "new" quartz grains operated at a slightly higher rate than grains originally in the 0.25- to 0.50-mm fraction were being fractured to produce smaller, but probably similarly shaped, particles. In short, a dynamic near-equilibrium condition apparently was achieved with respect to quartz grain-shape composition in the 0.25- to 0.50-mm size fraction.

It also is important to examine the obtained shape-ratio values between adjacent stream segments for differences related to changes in stream gradient (Figs. 10, 11). Again, there is a difference of 0.08 (0.79 to 0.71) to 0.10 (0.79 to 0.69) from the 0.8-km station along Boden Canyon Creek to the 1.6-km and 2.4-km stations, respectively. Otherwise, there are no substantive differences across the boundaries in the stream-gradient profiles. The observed decrease in shape-ratio values for Boden Creek indicates an increase in the proportion of more elongate grains. In Segments C and D of Culp Canyon Creek, there also is a decrease of 0.05 from the 8.0-km station to the 11.2-km station, which reflects an increase in the proportion of more elongate grains through Segment C into Segment D. These increases in the proportions of more elongate grains may signify either selective transport and/or a relative increase in the rate of grain-parallel fracture along the highest gradient portions of Boden and Culp Canyon Creeks (Table 12). Figures 10 and 11 show that no other significant change in shape-ratio values occurs across the other stream segment boundaries.

A comparison of shape-ratio values for grus, stream-sample average and the most downstream sample for each of the four streams examined is provided in Table 13. For each stream, the

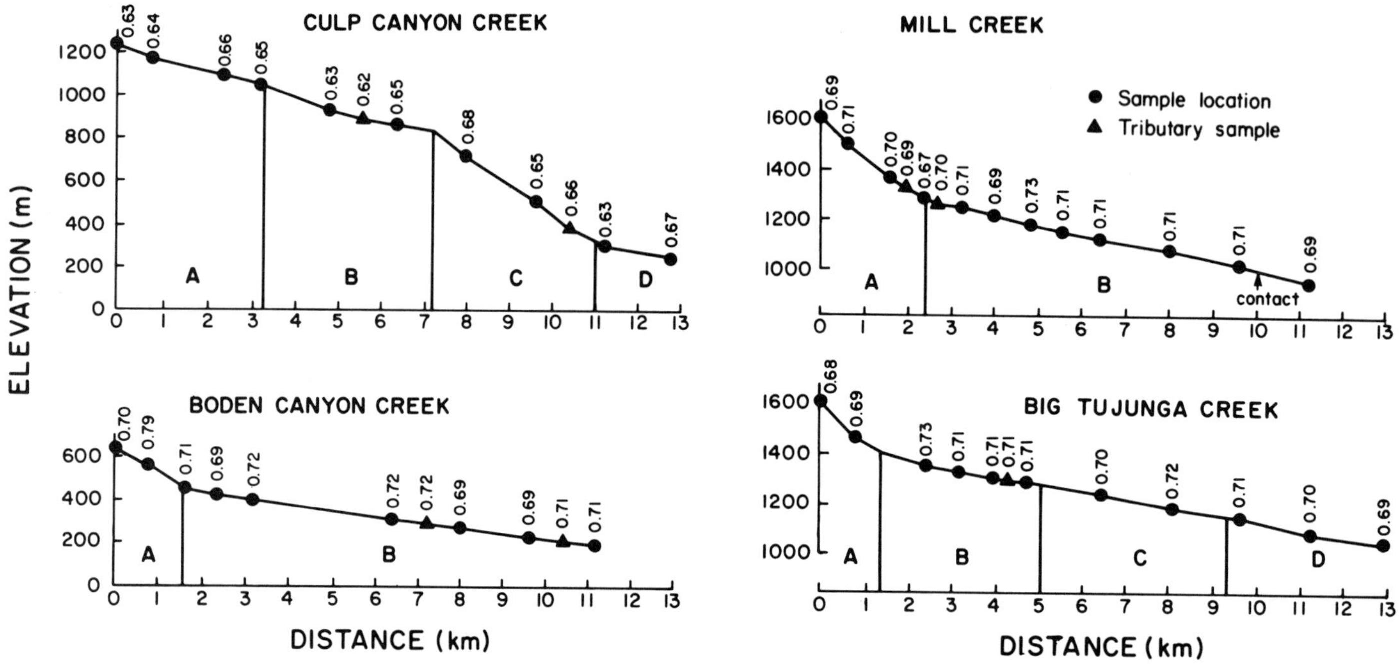

Figure 10. Topographic profiles and shape-ratio values for Culp Canyon Creek, Boden Canyon Creek, Mill Creek, and Big Tujunga Creek. For each stream, the shape-ratio value for grus is plotted at 0 km.

TABLE 12. RANGE OF SHAPE-RATIO VALUES AND AVERAGE GRADIENTS WITHIN SEGMENTS OF THE HEADWATERS OF FOUR HIGH-GRADIENT STREAMS IN SOUTHERN CALIFORNA

Stream Segment	Distance along Channel from Source (km)	Number of Samples	Range of Shape-Ratio Values	Average Gradient (m/km)
A. Culp Canyon Creek				
A	0.0–3.2	4	0.63–0.66	62
B	3.2–7.2	3	0.63–0.65	79
C	7.2–11.0	3	0.65–0.68	196
D	11.0–12.8	2	0.63–0.67	42
B. Boden Canyon Creek				
A	0.0–1.6	3	0.70–0.79	125
B	1.6–11.2	8	0.69–0.72	56
C. Mill Creek				
A	0.0–2.4	5	0.67–0.71	146
B	2.4–11.2	9	0.69–0.73	34
D. Big Tujunga Creek				
A	0.0–1.6	2	0.68–0.69	125
B	1.6–5.0	5	0.71–0.73	29
C	5.0–9.3	2	0.70–0.72	33
D	9.3–12.8	3	0.69–0.71	36

differences among these shape-ratio values is no more than 0.04, which indicates a high degree of shape-compositional homogeneity. This uniformity appears to be due to rather consistent breakage along dominant grain-parallel fracture sets (Fig. 12).

Figures 10 and 11 also indicate that no important change in shape-ratio values is associated with the small tributaries that enter the sampled portions of the four streams or is associated with the change in granitic source rock at the 10.0-km station along Mill Creek, which has its highest gradient in Segment A. The shape-ratio value at the most downstream station for each stream is equal or nearly equal to that of the grus from which the sample was derived. It appears that although significant variations in shape-ratio values may sometime occur along very steep stream gradients, the grain-shape composition reequilibrates to become equal or very nearly equal to that of the grus from which it was derived after only 7 to 10 km of transport along lower gradient stream segments.

DISCUSSION

Lithology

Petrographical modal analysis of untumbled grus, 2-, 4-, and 8-day tumbler samples from four granodioritic source localities, indicates compositional homogeneity with respect to quartz, plagioclase feldspar, and potassium feldspar for the 0.50- to 0.70-, 0.35- to 0.49-, 0.25- to 0.34-, and 0.177- to 0.24-mm size fractions. Although polymineralic rock fragments are more frequent in the coarse sand fraction (0.50 to 0.70 mm), no statistically significant differences in the frequency of these grains occur among the fine- and medium-sand fractions examined (0.177 to 0.49 mm). This relationship clearly reflects the disintegration of coarser grained rock fragments into discrete mineral grains at or near the boundary between medium and coarse sand (0.50 mm).

Our results are inconsistent with the concept that the lithological composition of the detrital light-mineral fraction is strongly size dependent within the 0.177- to 0.70-mm range (Blatt et al., 1980). As predicted by Blatt et al. (1980, Fig. 8–21), a significant decrease in polymineralic rock fragments occurs between the 0.50- to 0.70- and 0.35- to 0.49-mm size fractions. Blatt et al. (1980, Fig. 8-21) have also predicted a continued decrease in the percentage of rock fragments within the range from 0.49 to 0.177 mm; however, our results indicate that the number of polymineralic rock fragments remains much the same. Figure 8-21 also predicts increases in the percentages of total feldspar and total quartz from 0.70 to 0.177 mm; however, our results indicate that there is no statistically significant change in

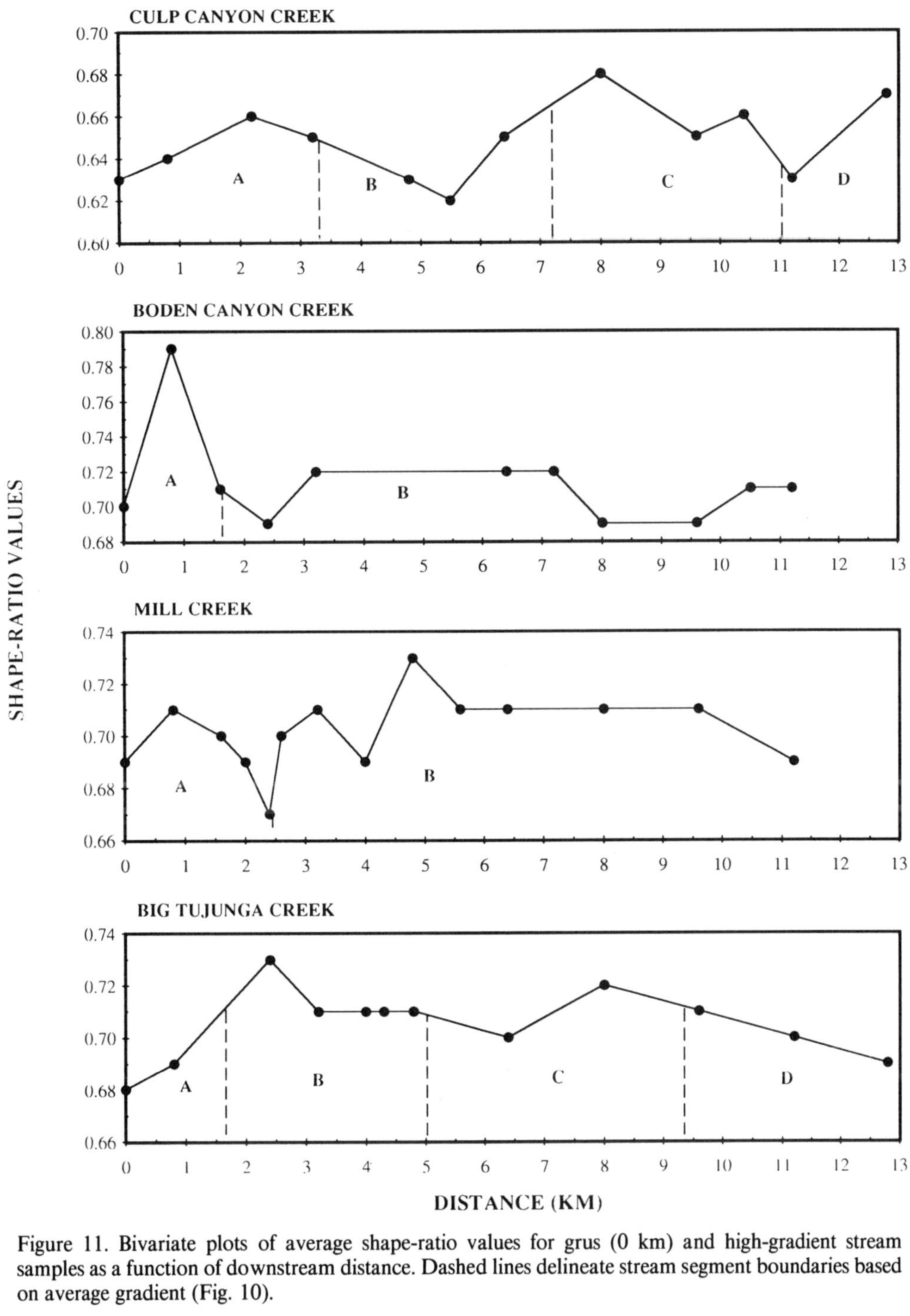

Figure 11. Bivariate plots of average shape-ratio values for grus (0 km) and high-gradient stream samples as a function of downstream distance. Dashed lines delineate stream segment boundaries based on average gradient (Fig. 10).

TABLE 13. SHAPE-RATIO VALUES FOR GRUS, STREAM-SAMPLE AVERAGE, AND MOST DOWNSTREAM SAMPLE ASSOCIATED WITH HEADWATERS OF FOUR HIGH-GRADIENT STREAMS IN SOUTHERN CALIFORNIA

Stream	Grus	Average Value for Stream Samples	(n)	Value for Most Downstream Sample	(km)
Culp Canyon Creek	0.63	0.65	(11)	0.67	(12.8)
Boden Canyon Creek	0.70	0.72	(10)	0.71	(11.2)
Mill Creek	0.69	0.70*	(12)	0.71	(9.6)
Big Tujunga Creek	0.68	0.71	(11)	0.69	(12.8)

*The value for the 11.2-km sample was omitted in this calculation due to a change in source rock at the 10-km station.

the number of potassium feldspar, plagioclase feldspar, or quartz grains within this size interval. Systematic differences in lithological composition do occur within "natural" sample sets as a function of grain size; however, such compositional variation most likely involves processes other than the comminution of grains from a single protolith.

The fact that the mineral composition is independent of grain size within the observed range indicates that crystal breakage associated with the production of grus, as well as breakage during grain transport, is controlled by fracture that crossed mineral boundaries. As pointed out by Moss et al. (1973), coeval and postemplacement deformation, as well as sheeting during the uplift of plutons, produces fracture sets along which grain breakage potentially could occur during weathering and transport.

In a similar experiment using nascent natural detritus, Moss et al. (1973) reported that quartz was the dominant fragmentation product in all size fractions of the sample they studied down to a size of 2 μ. They also noted that when pebbles were mixed with sand in their rotating drum experiments, the breakage of quartz grains was increased greatly. Their results appear to contradict those of the present study; however, the experimental materials employed were quite different, even though the protolith was granodiorite in both cases. Moss et al. (1973) used a large gravel sample collected from the headwaters of Gibraltar Creek, which is located 35 km southwest of Canberra, Australia. Gibraltar Creek drains the Shannon's Flat Granodiorite, which is, in fact, an adamellite whose modal constituents average 34.1% quartz, 26.1% plagioclase feldspar, 25.7% potassium feldspar, and 7.1% biotite. Petrographical modal analysis of sediment derived from this highly weathered pluton indicates that from 54 to 78% of the grains studied contain more than 90% quartz for sieve sizes from 0.50 to 4.76 mm, whereas feldspars comprise only 6 to 8% of the discrete grains. Thus the fact that quartz was the dominant fragmentation product in their study simply reflects its overwhelming dominance in the gravel sample used for their experiments. Of considerable importance to the present study is the fact that Moss et al. (1973, p. 498) pointed out that the Gibraltar Creek detritus showed no major compositional differentiation when experimentally fragmented.

Bomer (1985) performed petrographical modal analysis on quartz grains from grus and the 12.8-km sample from Culp Canyon Creek, as well as from grus and the 9.6-km sample from Mill Creek, to determine whether any significant changes in quartz type occurred as a function of transport. The quartz grain classification of Basu and others (1975) was employed to examine the 0.25- to 0.50-mm size fraction of these samples. For Culp Canyon Creek, the differences for monocrystalline quartz (undulose plus nonundulose), polycrystalline quartz with two to three subunits, and polycrystalline quartz with more than three subunits are 6.8, 8.6, and 1.8%, respectively. For Mill Creek, these differences are 1.1, 1.2, and 0.1%, respectively. These observed differences are within the limits of point-counting error at the 95% confidence interval. Therefore, fracturing along quartz crystal boundaries does not appear to change in the 0.25- to 0.50-mm size fraction, at least during initial stages of transport.

Sammis et al. (1987) reported that the particle-size distribution in fault gouge is fractal (self-similar). They explain this self-similarity using a model in which the probability of fracturing a given particle depends only on the relative sizes of neighboring grains rather than the absolute size or mineralogy of the grain itself, or the mineralogy of neighboring grains. Biegel et al. (1989) verified this model in the laboratory by means of a simple shear experiment. Inasmuch as simple shearing of granitic protolith has been demonstrated to produce self-similar fracturing, this model may be important in explaining the mineral composition, as well as the grain-size and grain-shape distributions of nascent siliciclastic sand.

Grain size

All three grus samples examined displayed relatively rapid rates of disintegration in the absence of pebbles during the first few days of tumbling and then continued to disintegrate at reduced rates from days 4 through 8. No equilibrium condition was attained with respect to grain size during an exposure of 4 to 8 days ($\approx$100 to 200 km) of transport in the tumbler; therefore, grain breakage may be expected to continue at relatively low rates with additional transport. A total of 8.9 to 16.2% of the relatively coarse-grained fractions experienced comminution during 8 days ($\approx$200 km) of transport. Of this amount, the sand fraction only increased from 1.7 to 3.4%, whereas the silt plus clay fraction increased from 6.4 to 14.5% during the same period. These data indicate that the grain-size distribution after as much as 200 km of transport is almost unchanged for the sand fraction and only moderately changed for the silt plus clay fraction when compared to the grus from which each sample was derived.

Moss et al. (1973) reported larger size reductions in a tumbler experiment using Gibraltar Creek detritus. In their standard 4-h (≈14.4 km) tumbler runs, weight losses ranged from 9.5% for the finest fraction used (0.500 to 0.595 mm) to 33.5% for the coarsest fraction (4.00 to 4.76 mm). The breakage products showed a continuous size spectrum from the initial size down to, and probably less than, 0.06 micron, which was the technical lower limit of their size analyses. In our study experimental losses to the sand fractions were greater than those found in Gibraltar Creek (Moss, 1972) for all except the coarsest size fraction. However, Moss (1972) terminated his stream study only 50 km downstream of the source, and thus natural breakage probably was not completed. Moss et al. (1973) concluded that the degree of breakage produced experimentally was of the same order of magnitude as that produced in natural streams. Unfortunately, it is not possible to directly compare our results with those reported by Moss et al. (1973) because of the differences in grain size and mineralogy of the experimental materials used.

Grain shape

The shape of individual sedimentary grains during transport may be modified by breakage, abrasion, solution, and/or precipi-

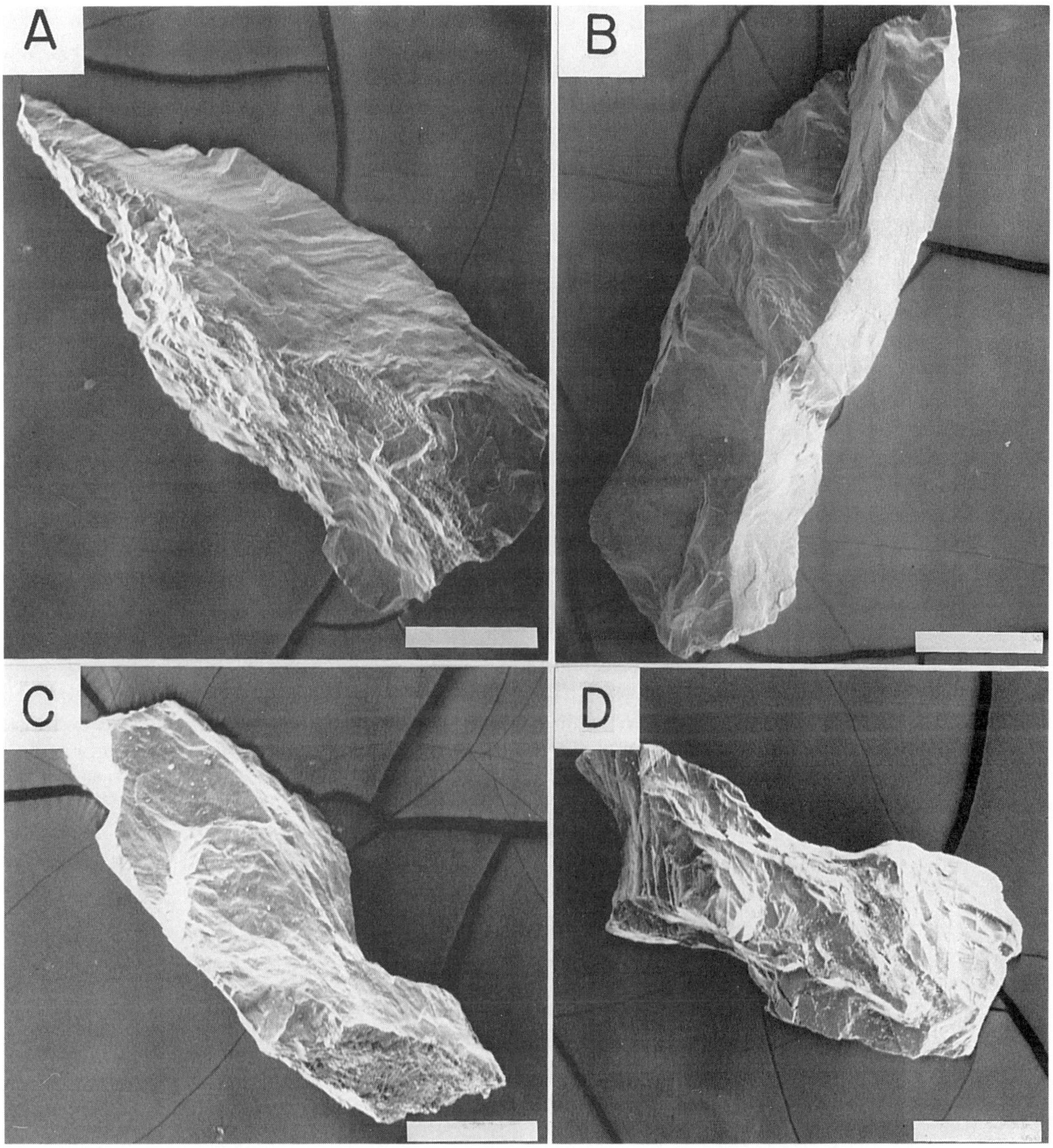

Figure 12. SEM photomicrographs showing planar fracture sets and angularity of elongate, detrital quartz grains from Culp Canyon Creek. A and B, Grus sample. C, The 0.8-km sample. D, The 6.4-kkm sample. Scale bars = 200 μ.

tation. Of the shape modification processes, chemical solution and precipitation can be eliminated from consideration for quartz used in the tumbling experiment due to the short time intervals involved. Furthermore, hydraulic sorting by grain size, specific gravity, and shape, as well as mixing of grains derived from different sources, may alter the grain-shape composition of a given sample without shape modifications of individual grains (Kennedy and Ehrlich, 1985). Inasmuch as only the 0.25- to 0.50-mm fraction for quartz was examined, hydraulic sorting by grain size and density can be eliminated from further consideration. Even so, a great deal remains to be learned concerning the primary and interaction effects of the remaining variables under different climatic regimes and at differential temporal and spatial scales.

Our study shows that generally smaller shape-ratio values occur after 1 day of tumbling (≈25 km of transport). However, if one relies on the comparison of distribution means, only grus sample 2 displayed an increase in the proportion of more elongate grains after tumbling for 1 day, whereas samples 1 and 3 showed significant but smaller changes only after 8 days (≈200 km) of transport. Standard deviation values consistently decreased after tumbling for 1 day, and then increased from days 1 through 8 to nearly equal or exceed that of the associated grus sample. The change in standard deviation values suggest a relatively high rate of grain breakage during the first day of tumbling to approach a quasi-stable distribution, after which more elongate grains were produced by a higher rate of grain-parallel fracture to increase the standard deviation values. Analysis of amplitude-ratio values for the second and third harmonics indicates that grus samples 1, 2 and 3 appear to have been dominated by grain-parallel fracture during transport.

For the 0.25- to 0.50-mm size fraction, these results indicate that grain-parallel breakage along those fracture sets that produced the elongate quartz grains in the associated grus sample continued to produce similarly shaped grains during as much as 200 km of transport in a tumbler. Although there is evidence that across-grain breakage may initially occur in some cases, grain-parallel fracture appears to predominate. SEM has been used to demonstrate that detrital quartz grains commonly possess dominant and subordinate sets of microfractures spaced as little as 1 to 10 μ apart (Blatt et al., 1980). Although there generally is more than one set of identifiable fracture sets present, one set appears to be dominant so that the observed structure is referred to as sheeting. Moss and Green (1976) have shown that the orientation of fracture-bounded sheets bears little relation to crystallographic properties of quartz. The occurrence of long, parallel fracture sets bounding individual particles also has been observed in fault zones, where tensile fractures tend to form preferentially in the direction of the largest principal stress. The mechanics responsible for the generation of such material is described in detail by Ashby and Sammis (1990). Since plutonic rocks are subjected to considerable shear during emplacement and postemplacement deformation, the production of fracture-bounded, elongate quartz grains may be expected.

Shape-ratio values within each of the four high-gradient streams examined appear to be remarkably consistent within and among the assigned stream segments. Where significant differences occur, they indicate an increase in the proportion of elongate grains due either to an increased rate of grain-parallel fracture and/or shape sorting. No important change in shape-ratio values occurred where small tributaries draining the same source rock enter the trunk streams, from sidewall erosion of the same protolith or across a contact with younger granitic source rock along Mill Creek. Using a more conventional Fourier grain-shape analysis in which amplitude values for higher harmonics were used but phase angle was not considered, Bomer (1985) reported no significant progressive downstream change in the grain-shape composition of Boden Canyon Creek and Big Tujunga Creek, a minor downstream increase in more elongate grains along Mill Creek, and a minor downstream increase in more equant grains in Culp Canyon Creek. Thus, except for perhaps Culp Canyon Creek, there is good agreement between the results obtained by Bomer (1985) and the present study.

Although significant change in shape-ratio values occurs along some local, very steep stream gradients, the grain-shape composition as measured by such values reequilibrates to become very nearly equal to that of the grus from which the grains were derived after only 7 to 10 km of transport along lower gradient stream segments. This reequilibration probably reflects the dilution of the modest increase in more elongate grained associated with some steeper stream segments by the preponderant, less elongate grains along lower gradient segments. In contrast, the tumbler experiments indicate that the rate of grain-parallel breakage had not equilibrated with respect to the medium sand fraction after as much as 200 km of transport. The results from the stream samples examined agree with those of Peterson and Mazzullo (1987), who reported no significant downstream change in the shape composition of bedload for a 500-km length of the low-gradient Mississippi River, except where the Mississippi River sediment was diluted by the sediment load of its tributaries. They further reported that the SEM analysis revealed no noticeable abrasion of source-inherited crystal faces or fracture surfaces on the quartz grains.

Kennedy and Ehrlich (1985) and Peterson and Mazzullo (1987) determined that the major cause of change in the shape composition of fluvial sediment is due to the introduction of new shape elements from different sources. It is clear from our results that sediment derived from small tributaries draining much the same protolith had no major effect on the shape composition of the trunk stream. We attribute our results to a high degree of similarity in the orientation and spacing of fracture sets in the protoliths from which the quartz grains were derived. Kennedy and Ehrlich (1985) reported that provenance determinations using Fourier grain-shape analysis improves with increasing grain size from coarse silt to fine sand. This observation may reflect the increased importance of across-grain breakage within finer grain-size fractions, which would render the resultant grain-shape composition less like that of the associated protolith.

Most of the shape-ratio values from the three granodioritic grus samples from the Lake Tahoe basin range from approximately 0.55 to 0.75, and those from the grus samples associated with the four high-gradient streams range from 0.63 to 0.70. Although the similarity in the range of these values may be coincidental, such similarly may reflect rather common petrogenetic, deformational, and weathering histories of the sampled plutons. In any case, it seems clear that we must examine the crystallization history and the mechanical properties of acid-plutonic source rocks to understand the origin of quartz grain shapes. A step in this process is to relate the grain-shape composition of grus with the orientation and spacing of fracture sets within associated protolith.

CONCLUSIONS

With the exception of rock fragments in the 0.50- to 0.70-mm (coarse sand) fraction, there is a high degree of compositional uniformity within fine- to coarse-sand fractions for quartz, potassium feldspar, plagioclase, and rock fragments after as much as 8 days (≃200 km) of transport in a rotating drum. Sedimentological processes other than the comminution of grains from a single protolith may be operative when the lithological composition of the light-mineral fraction is strongly size dependent within the 0.177- to 0.70-mm size range.

Only 8.9 to 16.2% of the relatively coarser grained fractions experienced comminution during 8 days (≃200 km) of transport. Of this amount, the sand fraction increased from 1.7 to 3.4%, the silt plus clay fraction increased from 6.4 to 14.5%, and the weight percent of the fine- and/or medium sand fraction remained almost constant.

Examination of changes in the distribution of shape-ratio values for quartz grains in the 0.25- to 0.50-mm fraction demonstrated the dominance of grain-parallel as opposed to across-grain fracture. This condition reflects persistent breakage along those fracture sets that produced the highly elongate grains in associated grus samples. SEM analysis indicates that abrasion is not an important factor during as much as 200 km of transport. Inasmuch as the fundamental shapes of the transported grains are quite similar to those of the grus from which they were derived, Fourier grain-shape analysis appears to be a reliable technique for the identification of a given protolith. This assertion assumes that the distribution of quartz crystal sizes and shapes, as well as the orientation and spacing of fracture sets, is consistent within and unique to a given source rock.

Average shape-ratio values within sand for four high-gradient streams in southern California are remarkably consistent, and closely resemble the value characteristic of the associated grus sample. Significant decreases (increased elongation) in average shape-ratio values occur along some localized, very steep stream gradients due to high rates of grain-parallel fracture and/or shape sorting; however, such values reequilibrate after only 7 to 10 km of transport along lower gradient stream segments. Such reequilibration probably is due to the dilution of more elongate by less elongate grains along lower stream gradients.

Simple shearing is known to produce self-similar fracturing, as suggested by the lithological analysis and highly elongate grain shapes observed in this and other studies. Consequently, this mechanism may be important in explaining compositional and textural aspects of nascent siliciclastic sand derived from plutons subjected to emplacement and/or postcrystallization tectronism.

ACKNOWLEDGMENTS

We thank Randall Moory for his assistance in obtaining the four grus samples at Lake Tahoe. Discussions with Bruce Mackey and Charlie Sammis were most helpful concerning the selection of the most appropriate ANOVA design to reduce the lithological data and possible mechanisms to explain the production of the observed quartz grain shapes, respectively. Modal analysis of the protolith samples from Lake Tahoe were performed by Arthur Lee. We also thank Sam Boggs, Mark Johnsson, Stephen Kennedy, Jinyou Liu, Gretchen Luepke, and Jim Mazzullo for their cogent reviews. The final line drawings were prepared by Janet Dodds, Arthur Lee, Jinyou Liu, and Rory Robinson. The manuscript was typed by Sue Turnbow.

APPENDIX 1. SAMPLE LOCALITIES

Part A. Grus Samples from Lake Tahoe Basin, California and Nevada

1. Collected at an elevation of about 1,913 m along Lakeshore Drive in Incline Village, Nevada, immediately south of Mill Creek (39°14′N latitude, 119°56′W longitude
2. Collected at an elevation of 1,921 m along Nevada State Route 28 about 2.4 km south of Mill Creek (39°13′N latitude, 119°56.6′W longitude)
3. Collected at an elevation of 1,982 m along California State Highway 89 immediately south of Rubicon Point (38°59.6′N latitude, 120°05.6′W longitude)
4. Collected at an elevation of 1,921 m along Nevada State Route 28 about 0.5 km west of First Creek (39°15′N latitude, 119°59′W longitude)

Part B. Stream Samples from Southern California

1. Culp Canyon Creek is located in the Anza Borrego Desert about 5 km southwest of the City of Borrego Springs (33°12′N latitude and 116°25′W longitude)
2. Boden Canyon Creek is located north of the town of Ramona (33°08′N latitude and 116°53′W longitude)
3. Mill Creek is situated in the Angeles National Forest north of Los Angeles, California, and heads at Mill Creek Summit (34°24′N latitude, 118°06′W longitude)
4. Big Tujunga Creek also is located in the Angeles National Forest (34°17′N latitude, 118°0′W longitude)

REFERENCES CITED

Alder, H. L., and Roessler, E. B., 1977, Introduction to probability and statistics: New York, W. H. Freeman, 426 p.

Ashby, M. F., and Sammis, C. G., 1990, The damage mechanics of brittle solids in compression: Pure and Applied Geophysics, v. 133, p. 489–521.

Basu, A., Young, S. W., Suttner, L. J., James, W. C.,and Mack, G. H., 1975, Re-evaluation of the use of undulatory extinction and polycrystallinity in detrital quartz for provenance interpretation: Journal of Sedimentary Petrology, v. 45, p. 875–882.

Bateman, P. E., and Wahrhaftig, C., 1966, Geology of the Sierra Nevada, *in* Bailey, E. H., ed., Geology of northern California: California Division of Mines and Geology Bulletin 190, p. 107–172.

Biegel, R. L., Sammis, C. G., and Dieterich, J. H., 1989, The frictional properties of a simulated gouge having a fractal particle distribution: Journal of Structural Geology, v. 11, p. 827–846.

Blatt, H., Middleton, G., and Murray, R., 1980, Origin of sedimentary rocks: Englewood Cliffs, New Jersey, Prentice-Hall, 782 p.

Bomer, E. J., 1985, Quartz grain-shape modification in high-gradient fluvial environments: Fourier grain-shape analysis [M.S. thesis]: Los Angeles, University of Southern California, 139 p.

Brigham, E. O., 1974, The fast Fourier transform: Englewood Cliffs, New Jersey, Prentice-Hall, 252 p.

Brown, P. J., Ehrlich, R., and Colquhoun, D. J., 1980, Origin of patterns of quartz sand types on the southwestern United States continental shelf and implications on contemporary shelf sedimentation—Fourier grain shape analysis: Journal of Sedimentary Petrology, v. 50, p. 1095–1100.

Burnett, J. L., 1968, Geology of the Lake Tahoe basin, *in* Evans, J. R., and Matthews, R. A., eds., Geological studies in the Lake Tahoe area, California and Nevada: Sacramento, Geological Society of Sacramento, Annual Field Trip Guidebook, 99 p.

Clark, R. A., and Osborne, R. H., 1982, Contribution of Salinas River sand to the beaches of Monterey Bay, California, during the 1978 flood period: Fourier grain-shape analysis: Journal of Sedimentary Petrology, v. 52, p. 807–822.

Davis, J. C., 1986, Statistics and data analysis in geology: New York, John Wiley & Sons, 646 p.

Ehrlich, R., and Weinberg, B., 1970, An exact method for characterization of grain shape: Journal of Sedimentary Petrology, v. 40, p. 205–212.

Ehrlich, R., Orzeck, J. J., and Weinberg, B., 1974, Detrital quartz as a natural tracer—Fourier grain shape analysis: Journal of Sedimentary Petrology, v. 44, p. 145–150.

Full, W.E., Ehrlich, R., and Kennedy, S. K., 1984, Optimal configuration and information content of sets of frequency distributions: Journal of Sedimentary Petrology, v. 54, p. 117–126.

Galehouse, J. S., 1971, Point counting: *in* Carver, R. E., ed., Procedures in sedimentary petrology: New York, Wiley-Interscience, p. 385–407.

Ingram, R. L., 1971, Sieve analysis: *in* Carver, R. E., ed., Procedures in sedimentary petrology: New York, Wiley-Interscience, p. 49–67.

Kennedy, S. K., and Ehrlich, R., 1985, Origin of shape changes of sand and silt in a high-gradient stream system: Journal of Sedimentary Petrology, v. 55, p. 57–64.

Kuenen, P. H., 1959, Experimental abrasion. 3. Fluviatile action on sand: American Journal of Science, v. 257, p. 172–180.

Mazzullo, J. M., and Magenheimer, S., 1987, The original shapes of quartz sand grains: Journal of Sedimentary Petrology, v. 57, p. 479–487.

Moss, A. J., 1966, Origin, shaping and significance of quartz sand grains: Journal of the Geological Society of Australia, v. 13, p. 97–136.

Moss, A.J., 1972, Initial fluviatile fragmentation of granitic quartz: Journal of Sedimentary Petrology, v. 42, p. 905–916.

Moss, A. J., and Green, P., 1976, Sand and silt grains: Predetermination of their formation and properties by microfractures in quartz: Journal of the Geological Society of Australia, v. 22, p. 485–495.

Moss, A. J., Walker, P. H., and Hutka, J., 1973, Fragmentation of granitic quartz in water: Sedimentology, v. 20, p. 489–511.

Osborne, R. H., and Yeh, C.-C., 1991, Fourier grain-shape analysis of coastal and inner continental-shelf sand samples: Oceanside Littoral Cell, southern Orange and San Diego Counties, southern California, *in* Osborne, R. H., ed., From shoreline to abyss: Contributions in marine geology in honor of Francis Parker Shepard: SEPM (Society for Sedimentary Geology) Special Publication No. 46, Tulsa, Oklahoma, p. 51–66.

Peterson, M., and Mazzullo, J., 1987, The effects of fluvial abrasion on the shapes of quartz sand grains: Gulf Coast Association of Geological Societies Transactions, v. 37, p. 443–446.

Rogers, J.J.W., Krueger, W. C., and Krog, M.,1963, Sizes of naturally abraded materials: Journal of Sedimentary Petrology, v. 33, p. 628–632.

Sammis, C. G., King, G.C.P., and Biegel, R. C., 1987, The kinematics of gouge deformation: Pure and Applied Geophysics, v. 125, p. 777–812.

Schultz, D. J., 1980, The effects of hydrofluoric acid on quartz grain shape—Fourier grain shape analysis: Journal of Sedimentary Petrology, v. 50, p. 644–645.

Schumm, S. A., Mosley, M. P., and Weaver, W. E., 1987, Experimental fluvial geomorphology: New York, John Wiley & Sons, 413 p.

Schwarcz, H. P., and Shane, K. C., 1969, Measurement of particle shape by Fourier analysis: Sedimentology, v. 13, p. 213–231.

Streckeisen, A. L., 1976, To each plutonic rock its proper name: Earth Science Reviews, v. 12, p. 1–33.

Van der Plas, L., and Tobi, A. C., 1965, A chart for judging the reliability of point counting results: American Journal of Science, v. 263, p. 87–90.

Wentworth, C. K., 1992, A scale of grade and class terms for clastic sediments: Journal of Geology, v. 30, p. 377–392.

MANUSCRIPT ACCEPTED BY THE SOCIETY DECEMBER 24, 1992

Printed in U.S.A.

Geological Society of America
Special Paper 284
1993

Beach sands of Elba Island, Tuscany, Italy: Roundness study and evidence of provenance

M. Dane Picard
Department of Geology and Geophysics, University of Utah, Salt Lake City, Utah 84112
Earle F. McBride
Department of Geological Sciences, University of Texas at Austin, Austin, Texas 78713

ABSTRACT

We studied the composition and roundness of medium sand from 18 small beaches of Elba Island. Six are pocket beaches less than 100 m long; the longest is 1.3 km long. The drainage basins of streams that supply the beaches are all less than 25 km^2; most are less than 5 km^2. Beach sands range widely in composition owing to diverse source terrane.

For the various drainage basins, comparison of the outcrop areas of the different types of bedrock with the compositions of beach sand grains yields the following conclusions: (1) the relative area of granodiorite outcrop is accurately represented by the amount of quartz + feldspar + quartzofeldspathic rock fragments in all beaches, although plagioclase in beach sand is significantly reduced relative to K-feldspar; (2) the relative areas of outcrop of ophiolitic and limestone bedrock are accurately represented by beach sand in pocket beaches, but are only moderately represented (ophiolitic rocks) or poorly represented (limestone) in other beaches; (3) the relative area of outcrop of metamorphic rocks is poorly represented in beach sand (metamorphic rock fragments + polycrystalline quartz) except in one anomalous beach supplied in part by mine tailings; (4) the relative area of bedded chert outcrop is poorly represented in beach sand because bedded chert does not break down into sand-size grains; and (5) shale bedrock is not represented or is only marginally represented in beach sand.

For Elba beaches in general, the order of increasing roundness of grains, and thus increasing rate of abrasion, is: quartz < plagioclase < K-feldspar and igneous rock fragments (quartzo-feldspathic) < serpentine < metamorphic rock fragments < carbonate rock fragments (CRFs). There are no significant differences in roundness with beach length for quartz, K-feldspar, or CRFs. There are also no significant differences in roundness for the same three grain types from beaches of different size drainage basins, which indicates there is no perceptible rounding of grains by streams.

First-cycle monocrystalline quartz grains of medium sand are unrounded, although coarser grains show minor blunting of edges. The roundness of quartz, K-feldspar, and CRFs are all greater on the eastern, more protected part of the island. This reflects a significant proportion of recycled quartz and K-feldspar in the eastern beaches, but CRFs may undergo more rounding in beaches of low to moderate wave activity than in high-energy beaches.

Picard, M. D., and McBride, E. F., 1993, Beach sands of Elba Island, Tuscany, Italy: Roundness study and evidence of provenance, *in* Johnsson, M. J., and Basu, A., eds., Processes Controlling the Composition of Clastic Sediments: Boulder, Colorado, Geological Society of America Special Paper 284.

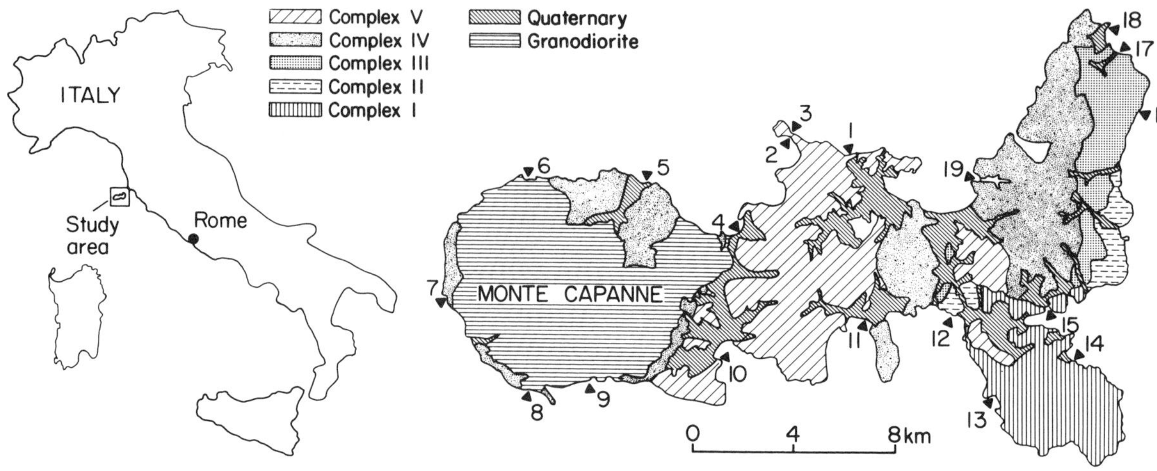

Figure 1. Generalized geologic map of Elba showing numbered sample localities. Complex I = Carboniferous and Triassic marble, dolomite, and quartzite; Complex II = Carboniferous to Jurassic metasedimentary rocks; Complex III = Carboniferous to Jurassic limestone, argillite, and sandstone; Complex IV = Jurassic and Cretaceous ophiolitic rocks, bedded chert, limestone, and argillite; Complex V = Cretaceous and Paleogene limestone, sandstone, and shale. Modified from Trevisan and Marinelli (1967). Numbers 1 through 18 indicate beach sites.

INTRODUCTION

Elba, the largest island in the Tuscan archipelago, northwestern Italy (Fig. 1), has a wide variety of bedrock exposed and a variety of sizes of beaches. We studied the composition and roundness of sand on 18 beaches (Picard and McBride, 1990). Our primary purpose was to see how well the sand composition of each beach reflects the outcrop area of different bedrock types within the drainage area of each beach. We also studied the roundness of different types of grains to assess whether grain roundness in beaches is influenced by beach width, size of the drainage basin that supplies the beach, or degree of wave activity of the beach. Although the Elba beach sands have no preservation potential in the stratigraphic record, they provide an opportunity to examine the effects of abrasion rounding and loss of grains in the geologically short time of the Holocene. Probably all sand in the present beaches accumulated since the Holocene stillstand approximately 5,000 yr ago.

The composition of sand grains carried by streams and deposited along their reaches and at their mouths is a function of bedrock composition, weathering, and modification of detritus by abrasion and selective sorting. Weathering is greatly affected by climate and relief (Krynine, 1950; Folk, 1968; Basu, 1976, 1985; Grantham, 1986; Grantham and Velbel, 1988; Ibbeken and Schleyer, 1991). In this work we compare the outcrop areas of different bedrock types with the modal composition of beach sands. We have not studied the regolith directly. This study adds to the small but growing literature on predepositional factors that control sand composition.

In area, Elba is about 223 km^2, the third largest Italian island after Sicily and Sardinia. It is 27 km long from east to west and reaches a maximum breadth of 18 km on its eastern side. The island has moderately rugged relief: the highest part of the island is 1,019 m above sea level and more than 70% of the island is more than 200 m above sea level. The land is highest on the west in the granodiorite massif of Monte Capanne (1,019 m). The centeral part of the island includes hills and small coastal plains, and the eastern part is formed of two series of hills that culminate in Cima del Monte (516 m) and Monte Calamita (413 m). The mean annual precipitation is between 75 and 100 cm, and the mean monthly temperature ranges from 8° to 26° C (Wallen, 1977).

The island is complex geologically (Trevisan and Marinelli, 1967; Perrin, 1974). It is composed of five different stratigraphic sequences or complexes, chiefly of sedimentary rock but including some ophiolitic rocks and low-grade metasedimentary rocks, that were juxtaposed by faulting during the Apennine orogeny (Oligocene-Miocene) (Fig. 1). The rocks are mostly Mesozoic, but Carboniferous and Permian rocks make up part of Complex I. Complexes IV and V have been intruded by silicic rocks (granodiorite and minor aplite and granite) as young as 7 Ma that have produced thermal metamorphism in the western third of the island. Tertiary hydrothermal activity in the eastern third of the island developed some important metal sulfide deposits and their oxidized alteration products. The sedimentary rocks are dominated by carbonate rocks and shale-argillite, but include sandstone, nodular and bedded chert, and evaporites as well. Unconsolidated Quaternary alluvium covers about 5% of the is-

TABLE 1. SIZE OF BEACHES, DRAINAGE BASINS, AND PERCENT OF BEDROCK EXPOSED IN DRAINAGE BASINS

Beach	Length of Beach (m)	Area of Basin (km^2)	Granodiorite	Ophiolitic Rocks	Metamorphic Rocks	Carbonate Rocks	Shale	Sand and Sandstone	Chert
1	100	0.2	90					10	
2	<100	Pocket	60		5	30	5		
3	<100	Pocket	60		5	30	5		
4	800	3	63			6		31	
5	800	11.8	70		16			14	
6	<100	1.6	100						
7	<100	Pocket	10	60	30				
8	<100	Pocket		50	25	25			
9	200	2.1	79	9	3			9	
10	1,300	25.4	62		8	12		18	
11	1,000	7.8	31		10	31	4	24	
12	500	6.7	9	17	22	40	1	11	
13	200	1.3			85	5		10	
14	150	0.8			90			10	
15	200	3.1		27	18			39	16
16	500	2.9		2	41	28	13	14*	2
17	200*	2.8		1		75	11	13	
18	<100*	Pocket				10	40	50	

*Beach 16 is contaminated by mine tailings; beaches 17 and 18 are thin veneers on bedrock.

land, chiefly in alluvial plains in the central part. Spoils from mining operations veneer small parts of the eastern third of the island.

The island has 147 km of coastline formed predominately of gulfs and promontories on which sandy beaches alternate with high, rocky headlands. We sampled the beaches at 18 localities, in two instances collecting two samples at a locality. The beaches we sampled are small; six are pocket beaches less than 100 m long that are supplied only by wave attack of adjacent headlands, by slope wash, or by one or two short-headed (<2 km long) ephemeral streams (beaches 2, 3, 6–8, 18). Six beaches are from 100 to 300 m long (beaches 1, 9,13–15, 17), and the six largest are from 0.5 to 1.3 km long (beaches 4, 5, 10–12, 16: Table 1). The drainage basins of ephemeral streams that supply sand to the beaches range in size from 0.2 to 25.4 km^2, although only four drainage basins exceed 5 km^2. Two beaches (beaches 17, 18) in the protected northeast corner of the island have only a thin veneer of sand above bedrock.

Beach 16 has sands of two mineralogical modes controlled hydraulically by wave sorting: one mode dominated by black opaques and the other by metamorphic rock fragments. Mine tailings are abundant in the source area of this beach and are the source of many opaque minerals. Also, about 2.5 km to the south there is a large open-pit iron mine.

We determined the sizes of drainage basins and the percentages of outcropping rock types within each by planimetry using the 1967 geologic map of Elba by Trevisan and Marinelli (1967). Sedimentary formations composed of interbedded rocks such as limestone with interbedded shale were estimated to be 75% limestone and 25% shale based on observation of outcrops. Bedrock was divided into the following categories: granodiorite (including minor quartz monzonite and aplite), metamorphic rocks, carbonate rocks, shale, sandstone (including Quaternary and deposits), chert, and ophiolitic rocks (serpentine, basalt, gabbro) (Table 1). We neither corrected map areas for the slope of the terrain nor compensated for differential erosion of bedrock (cf. Grantham and Velbel, 1988; Ibbeken and Schleyer, 1991).

PROCEDURES

Sand was collected from pristine parts of the high-tide berms at 18 localities around the island. Two samples were taken from beach 1 to test for uniformity of composition, and two samples were taken from beach 12 to show the effect of mineral segregation by waves. From each sample, a split of the 1 to 2 ϕ size fraction (medium sand) was made, impregnated in blue-dyed epoxy resin, and thin sectioned. Thin sections were stained for K-feldspar with sodium cobaltinitrite.

A total number of 300 grains per sample were counted (Table 2). Each grain was assigned to one of 16 categories: monocrystalline quartz, K-feldspar, plagioclase, coarse igneous rock fragments (quartzo-feldspathic grains), fine igneous rock fragments, metamorphic rock fragments, serpentine (including minor pyroxene-pyroxenite and pyroxene-serpentine intergrowths), volcanic rock fragments, unidentified rock fragments, chert, shale (including argillite), sandstone (including siltstone), carbonate rock fragments (including spar, micrite, and intraclasts), skeletal grains, and miscellaneous (biotite and heavy minerals). Unidentified rock fragments are grains with nondiagnostic, fine-grained textures, or with alteration products that obscure the grains. We divided the thin sections for point counting between us; we also looked at all the thin sections together. Both of us point-counted

TABLE 2. COMPOSITION, GRAIN ROUNDNESS, AND SHAPE DATA*

Sample	Mqtz	Pqtz	Kfeld	Plag	IRF-C	IRF-F	MRF	Serp	VRF	URF	Chert
1A	34.0	0.6	12.0	4.6	16.3	27.6	0.6	0.0	0.0	2.6	0.0
1B	35.0	1.3	12.6	5.0	15.0	28.6	0.3	0.0	0.0	1.3	0.0
2	16.3	5.0	9.0	2.6	12.6	8.6	0.3	0.0	0.0	0.6	6.3
3	17.3	7.0	7.0	4.3	7.6	2.0	0.6	0.0	0.0	2.0	7.3
4	42.3	5.3	22.6	5.6	7.3	2.0	3.6	0.0	0.0	1.3	0.0
5	45.6	5.0	18.6	4.6	8.6	1.0	7.0	0.0	0.0	3.0	0.0
6	47.0	1.3	21.6	13.3	7.6	0.6	1.6	0.0	0.0	1.3	0.0
7	1.3	0.3	0.3	0.6	9.3	0.0	11.6	61.6	0.0	7.3	0.0
8	10.3	1.6	7.0	5.0	0.0	0.0	11.6	21.3	1.0	10.6	1.0
9	46.6	4.0	26.6	6.0	10.3	0.0	0.0	0.0	0.0	0.0	0.0
10	32.6	5.3	34.6	6.6	5.0	0.0	3.6	0.0	0.0	2.3	0.0
11	56.3	5.0	11.0	8.0	2.3	1.6	3.0	0.6	2.3	2.6	0.0
12	52.3	21.0	4.3	4.0	3.6	3.3	4.0	0.0	1.0	2.0	0.0
13	12.6	9.0	14.0	1.3	1.6	4.3	10.0	0.3	1.0	1.3	0.0
14	29.6	20.3	11.3	5.6	4.6	1.0	10.3	0.0	1.0	5.0	0.0
15	18.3	11.0	6.0	4.0	7.6	3.6	8.3	2.0	6.3	17.3	6.0
16A	0.0	0.3	0.0	0.0	1.0	0.0	0.0	0.6	0.0	0.0	0.6
16B	8.0	4.0	3.3	0.0	0.3	0.0	62.6	4.3	0.0	2.6	2.6
17	28.3	20.0	5.3	0.3	1.6	3.0	7.3	0.6	1.6	7.3	6.0
18	5.3	1.0	1.0	0.3	0.3	0.0	1.0	0.0	0.0	5.6	2.6

several of the same thin sections to ensure similar point-count techniques. Two separate samples from beach 1 show the uniformity of modal compositions (samples 1A, 1B: Table 2).

We used the conventional point-count method instead of the Gazzi-Dickinson method (Gazzi, 1966; Dickinson, 1970). Grains were counted only if the cross-hairs fell on a part of a grain exposed at the upper surface of a thin section. This method of counting using the plane-section technique averts the volume error that can result if the full width of grains embedded in transparent epoxy resin is included in the counting method (cf. Harrell, 1983).

We estimated the roundness of monocrystalline quartz, plagioclase, K-feldspar, coarse igneous rock fragments, serpentine, and metamorphic and carbonate rock fragments from thin sections by comparing grain outlines with Powers's (1953) roundness images and Folk's (1968) roundness categories. Thirty or more grains of each type were estimated for roundness if the number of available grains were adequate. All roundness estimates were made by the senior author. Grain shapes and roundness were also examined under a binocular microscope and in scanning electron microscope images. The grain types we selected to study are those present in the majority of the beaches, and thus are useful for comparison.

The roundness and sphericity of K-feldspar grains was also determined by digitizing on an image analyzer the outline of grains visible in a petrographic microscope (Mazzullo and Kennedy, 1985) and analyzing the shapes using the Fourier series in closed form (Ehrlich and Weinberg, 1970). Harmonics 2 and 19 can be used to describe the sphericity and roundness, respectively, of grains in two dimensions (Haines and Mazzullo, 1988). Fifty grains from each of 12 beaches were analyzed (Table 1).

SAND COMPOSITION

The Elba beach sands have a broad range in modal composition, which reflects the complex bedrock composition of the island. Extremes range from the black sand of beach 16 (sample 16A), which is 97% opaque minerals, to sand composed almost entirely of granodiorite detritus ($Q_{49}F_{36}R_{15}$: beach 6), to sand in which CRF and skeletal grains, mostly reworked Mesozoic particles, make up 80% of the sand (beach 18), to sand composed more than 50% of ophiolite detritus (pyroxene, serpentine, and unidentified rock fragments: beach 7). Table 2 gives the results of the modal analyses at each beach.

Distribution of specific grains

Quartz. High in beaches (5, 6, 9) fed by streams draining granodiorite of the Monte Capanne massif and at localities (1, 4, 10–12) of the central part of Elba, where granodiorite, aplite, and turbidite sandstone crop out.

Polycrystalline quartz. Exceeds 10% in beaches 12, 14, 15, and 17. These are beaches fed by streams draining schist and hornfels.

K-feldspar. High in beaches fed primarily by granodiorite and secondarily by metamorphic rocks. Exceeds 15% at beaches 4–6, 9, and 10. The presence of K-feldspar in beach 8, composed of detritus from ophiolites, limestones and sandstones, indicates that some of it is recycled from older sandstone.

Plagioclase. Exceeds 5% in beaches 4, 6, 9, and 10, all fed chiefly from granodiorite, and in beach 14, which is fed by abundant metamorphic rocks. Plagioclase is also abundant in

TABLE 2. COMPOSITION, GRAIN ROUNDNESS, AND SHAPE DATA (continued)

Sample	Shale	SST	CRF	Skel	Misc	Total	Mat. Index	Har. 2 Spher.	Har. 19 Roundness	Kfeld Rnd	Qtz Rnd
1A	1.0	0.0	0.0	0.0	0.0	99.3	53.5	0.2250	0.0054	3.3	1.6
1B	0.6	0.0	0.0	0.0	0.0	99.7	57.2	0.2250	0.0060	3.1	1.3
2	0.0	2.0	30.0	6.0	0.3	99.6	38.3	0.2460	0.0070	3.0	
3	0.0	3.0	30.0	11.0	0.6	99.7	46.4	0.2000	0.0060		1.2
4	0.0	5.0	2.3	0.0	2.6	99.9	91.0	0.2260	0.0048	2.5	1.4
5	0.0	0.0	2.0	0.0	4.3	99.7	103.0	0.2470	0.0059	2.7	0.9
6	0.0	0.0	2.3	1.3	1.6	99.5	94.3	0.2210	0.0071	2.5	0.8
7	0.9	0.0	1.6	2.3	2.4	99.5	1.6			1.7	0.7
8	0.0	4.0	22.3	1.3	2.6	99.6	14.9	0.2210	0.0064	3.0	0.9
9	0.0	0.0	0.0	0.0	6.3	99.8	102.8	0.2110	0.0063	2.7	0.8
10	0.0	3.0	0.0	0.0	6.6	99.6	61.4	0.2190	0.0068	2.9	1.3
11	0.0	0.0	0.3	3.0	3.6	99.6	160.0	0.2230	0.0040	3.7	1.1
12	0.3	1.7	1.0	0.0	1.0	99.5	279.8	0.2130	0.0048	3.5	1.6
13	0.0	2.3	31.0	10.6	0.3	99.6	27.7	0.2150	0.0053	3.3	2.6
14	0.0	1.0	2.3	0.0	7.6	99.6	100.4	0.2290	0.0043	3.6	2.5
15	0.6	0.3	3.0	0.0	5.3	99.6	54.9				
16A	0.3	0.0	0.0	0.0	97.0	99.8	0.9				
16B	0.0	0.0	1.3	0.3	10.3	99.6	17.2				
17	0.6	0.0	14.6	2.0	1.0	99.5	120.1	0.2150	0.0050	4.1	1.3
18	1.0	0.0	76.6	4.0	1.0	99.7	9.8				

Note: Sample numbers refer to beaches shown in Figure 1; replicate samples were taken at beach 1; two samples at beach 16 include a heavy mineral placer and normal sand. Harmonic 2 characterizes grain sphericity, and harmonic 19 characterizes grain roundness of K-feldspar.

*Abbreviations: Mqtz = monocrystalline quartz; Pqtz = polycrystalline quartz; Kfeld = K-feldspar; Plag = plagioclase; IRF-C = coarse-grained igneous rock fragments; IRF-F = fine-grained igneous rock fragments; MRF = metamorphic rock fragments; Serp = serpentine and serpentine-pyroxene intergrowths; VRF = volcanic rock fragments; URF = unidentified rock fragments; Shale = shale and argillite; SSt = sandstone; CRF = carbonate rock fragments; Skel = skeletal grains; Misc = miscellaneous = heavy minerals and mica; Mat. Index = maturity index; Har. 2 = harmonic 2; Har. 19 = harmonic 19; K-spar Rnd = Powers-Folk roundness for K-feldspar; Qtz Rnd = Powers-Folk roundness for monocrystalline quartz.

beach 11, which is anomalous because only a small exposure of granodiorite contributes sand to the beach. Plagioclase exceeds K-feldspar in abundance in granodiorite bedrock, whereas K-feldspar exceeds plagioclase in abundance in beach sand. This switch in relative abundance of feldspar species reflects either the greater susceptibility of plagioclase to weathering or the selective fracturing of plagioclase into grains smaller than we studied. The latter premise is favored in view of the slight amount of chemical weathering in the Mediterranean climate of Elba and is compatible with the observations in Spain of Tortosa et al. (1989).

Igneous rock fragments (IRFs). These quartzo-feldspathic grains are essentially all granodiorite fragments, and they are abundant where granodiorite is the dominant beach source rock. However, IRFs exceed 20% only at beaches 1 and 2, in the north-central part of the island. They have no correlation with K-feldspar abundance. IRFs are not particularly abundant in other beaches derived chiefly from granodiorite (beaches 4–6), but these beaches are all higher energy beaches than those in localities 1 and 2. This suggests that quartzo-feldspathic grains break down into separate quartz and feldspar grains more readily under strong wave action.

Metamorphic rock fragments (MRFs). These grains exceed 10% at beaches 7, 8, 13, 14, and 16, all beaches fed from areas with at least 40% metamorphic rocks in the drainage basin. The exception is beach 8, whose drainage basin has only 25% metamorphic bedrock. Metamorphic bedrock is consistently underrepresented in all beaches, except where mine tailings provided a ready source of detritus. Our data are consistent with those of previous studies, which show that detritus from metamorphic bedrock breaks down rapidly during weathering and river transport (Cameron and Blatt, 1971; Cleary and Connolly, 1971), and suggest that wave abrasion is also highly destructive to these rock fragments. The latter conclusion was documented for gravel-size clasts of gneiss in Calabria, Italy, where gneiss fragments were reduced in abundance by 56% during only 7 km of beach transport (Ibbeken and Schleyer, 1991).

Serpentine. Serpentine exceeds 15% in beaches 7 and 8 and exceeds 4% in beach 16. All three beaches have source areas with abundant ophiolite bedrock. Pyroxene-serpentine aggregates are also a major component (46%) of beach 7.

Volcanic rock fragments (VRFs). These grains exceed 1% only at beaches 11 and 15. The source area of beach 15 has abundant basalts in ophiolitic rocks, but beach 11 is anomalous.

Unidentified rock fragments (URFs). These grains are

more abundant (mean = 3.8%) and widespread than VRFs, but they are most abundant in beaches where VRFs are abundant. As in a previous study that tracks both VRFs and URFs, we believe that many URFs are actually volcanic in origin (McBride and Picard, 1987). Not all URFs have such an origin, because some are also present in beaches without basaltic bedrock.

Chert. Chert exceeds 5% of beaches 2, 3, 15, and 17. Its main source in beaches 2, 3, and 17 is nodules of chert in limestone bedrock; in beach 15 the source is bedded radiolarian chert. In spite of its durability, bedded chert is underrepresented in beach sand derived from it. Beach 15 has only 6% chert sand grains derived from 16% chert outcrop area. As we found in Liguria, Italy (McBride and Picard, 1987), bedded chert readily breaks down to gravel-size clasts whose size and shape is controlled largely by the fracture network. However, because of its toughness, resistance to chemical weathering, and absence of internal bedding planes, the bedded chert does not readily break down into sand-size particles, and thus is underrepresented in stream sand derived from it (McBride and Picard, 1987). We suspect that the same explanation holds for Elba. We have no quantitative data on the abundance of chert nodules in limestone bedrock, but it is our observation that nodular chert is much better represented by chert sand than is bedded chert.

Shale/Claystone. Shale grains are present in about half the beach sands but nowhere exceed 1%. Shale is relatively abundant in two pocket beaches fed by adjacent coastal bedrock (beaches 7, 18) and in beaches with small drainage basins that tap shale bedrock (e.g., beaches 1, 12, 15). The abundance of shale sand grains in beaches is a poor indicator of the outcrop area of shale in the source area, a finding that we also made from a Ligurian beach (McBride and Picard, 1987).

Sandstone. Sandstone sand grains are present also in about half the beaches sampled, in which they average 2.5%.

Carbonate rock fragments. (CRFs). CRFs are present in all but four beaches and make up more than 15% of the sand in six samples. They comprise 77% of the sand in beach 18, a pocket beach with coastal exposures of limestone. CRFs underrepresent the area of carbonate rock outcrop in drainage basins.

Skeletal grains. Skeletal grains include modern marine skeletal remains and fossils recycled from Mesozoic limestones. They exceed 3% in beaches 2, 3, 13, and 18, all beaches with abundant CRFs. Using the criteria of Zuffa (1985), we determined that most of the skeletal grains are recycled grains. This conclusion is also supported by the observation that most of the skeletal grains are calcite, not aragonite.

Miscellaneous grains. Biotite and nonopaque heavy minerals are present in thin sections of most beaches sampled, and exceed 5% in beaches 9, 10, and 14–16 (sample 16B). The first four beaches listed have abundant ophiolites or metamorphic rocks in the source areas, whereas beach 16B contains heavy minerals concentrated by waves from sand containing abundant mine tailings. Opaque heavy minerals are abundant only in beach 16 (sample 16A), where they have also been concentrated by waves from sand derived in part from mine tailings.

SAND MATURITY VALUES

One measure of variation in beach sand composition is the mineralogical maturity of the samples. Sand maturity values reflect the relative abundance of the more durable grain types. In 1957 (p. 509), Pettijohn introduced the mineralogical maturity index as:

$$MI = \frac{\text{quartz} + \text{chert}}{\text{feldspar} + \text{rock fragments}} \times 100, \tag{1}$$

where the grains are estimated in volume percent. We (McBride and Picard, 1987, p. 1024) more recently introduced a modified maturity index (MMI) to give recognition to heavy minerals, micas, and skeletal grains, which also have implications for mineralogical maturity. The MMI is defined as:

$$MMI = \frac{\text{quartz} + \text{chert}}{\text{other grains}} \times 100. \tag{2}$$

There is a large variation in the maturity of the Elba beach sands, which range from 1 to 280 (Table 2). The most immature sands are on the black sand beach (16), at Capo Castello (18) on the northeast corner of the island—where there is little quartz and the sand is dominantly carbonate grains—and on the west (7, 8), where pocket beach sand is derived only from metamorphosed ophiolitic rocks. The most mature sand (MMI of 280 at beach 12), which is about equal in maturity to the average sandstone (~300), occurs where streams drain Carboniferous and Permo-Triassic sandstone terrains. Elba beach sands are more immature than some mainland beach sands of Italy (Gazzi et al., 1973; McBride and Picard, 1987; also unpublished data), but are comparable in maturity to some short-headed mountainous streams in Italy (McBride and Picard, 1987; Ibbeken and Schleyer, 1991). Elba beach sands are considerably more immature than beaches on broad passive continental margins (e.g., Hayes, 1965; Potter, 1986), although passive margins include carbonate beaches.

MODAL COMPOSITION VERSUS BEDROCK OUTCROP AREA

Procedure

We attempted to assess how well the composition of beach sand reflects the relative area of exposure of several distinctive types of bedrock within the drainage basin that supplies each beach. Our approach was to compare the abundance of a distinctive grain type in beach sand (i.e., chert) with the relative area of exposure of a distinctive rock type (i.e., bedded chert) or of a group of detrital grains in beach sand that could logically be considered to be derived from a particular bedrock. For example, monocrystalline quartz, feldspar, IRFs, and half the polycrystalline quartz are considered to be derived from granodiorite bedrock; serpentine, serpentine-pyroxene intergrowths, pyroxenite, VRFs and URFs, and opaques are considered to be derived from ophiolite bedrock. We qualitatively assessed correlation of sand

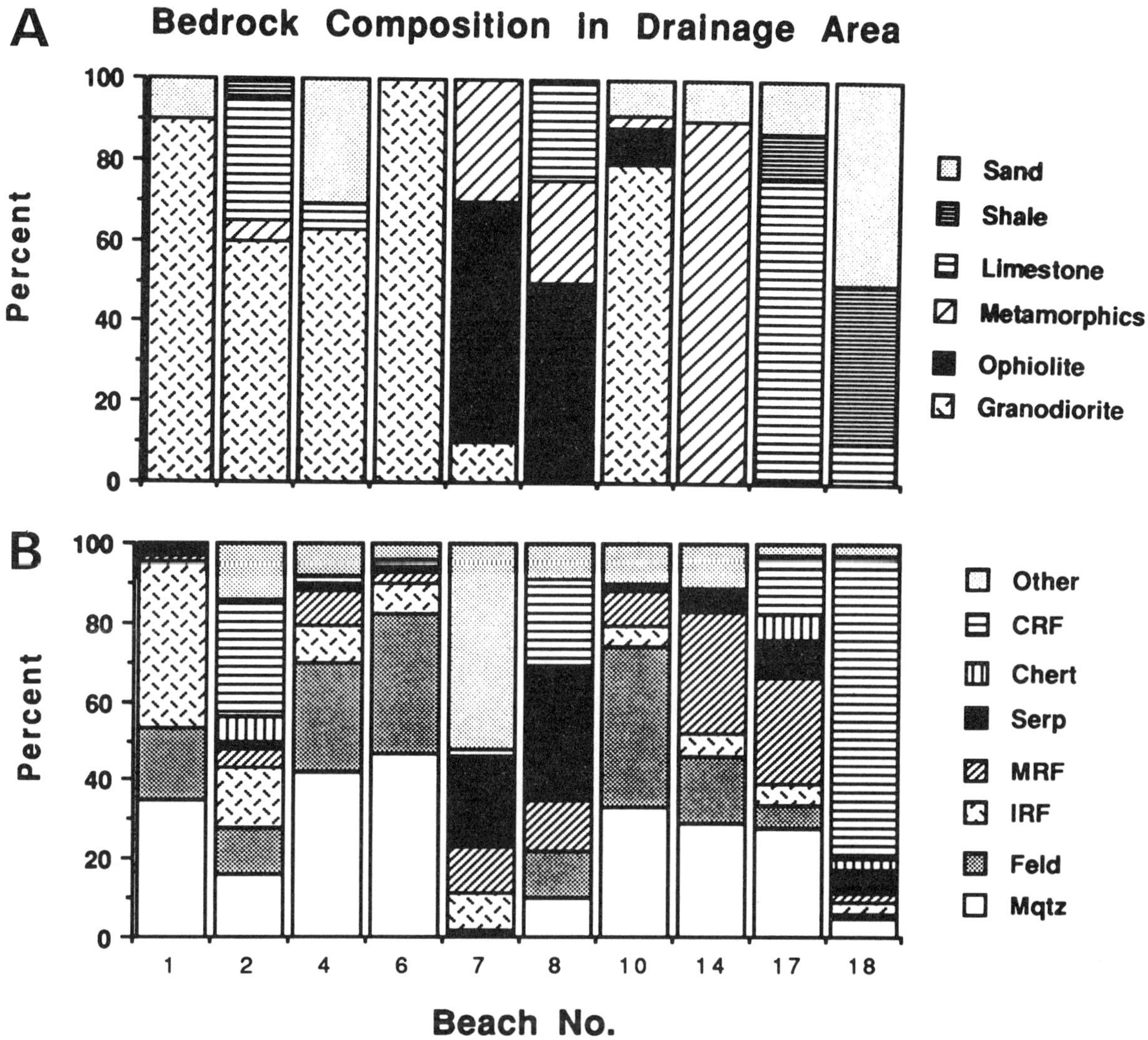

Figure 2. Comparison of bedrock outcrop area. A, With beach sand; B, composition for 10 beaches. (See Fig. 1 for locations.)

composition with outcrop area by examination of data in Tables 1 and 2, and by analyzing the graphic presentation of these data for 10 drainage basins (Fig. 2). We also assessed the degree of correlation quantitatively by calculating correlation coefficients for the area of outcrop of a particular type of bedrock versus percentage of sand grains derived from that bedrock where we had sufficient data to do so. Thus, we correlated 13 basins/sands containing metamorphic components, 12 containing granodiorite components, 11 containing limestone components, and 6 containing ophiolite components. Correlation coefficients of percentage data are of uncertain reliability because the data are percentages and are not open-ended. However, the coefficients should provide an estimate of relative ranking.

General conclusions

The abundance of granodiorite detritus in beach sands of all types of beaches faithfully represents the relative abundance of granodiorite bedrock exposure in each basin (Figs. 2, 3A). The correlation coefficient for percentage exposure of granodiorite in a drainage basin versus sand grains of inferred granodiorite provenance (monocrystalline quartz + 1/2 polycrystalline quartz + feldspar + IRFs) is 0.61; for sand grains omitting quartz grains, it is 0.85 (Fig. 3A). Metamorphic bedrock outcrop area is underrepresented in all beaches, whereas ophiolitic and limestone bedrock areas are underrepresented in beach sands except in pocket beaches with coastal exposures of them. The correlation coefficient for metamorphic bedrock versus metamorphic rock fragments (+ 1/2 the polycrystalline quartz sand) is 0.40 but is 0.76 if a spurious sample is excluded (Fig. 3B), r = 0.90 for ophiolite bedrock versus ophiolite-derived sand components, and r = –0.31 for limestone bedrock versus carbonate rock fragments (ignoring skeletal grains). Bedded chert outcrop is underrepresented in all beaches, whereas inspection suggests that nodular chert in carbonate rocks is adequately represented in beach sand. Shale bedrock is essentially unrepresented in beach sand.

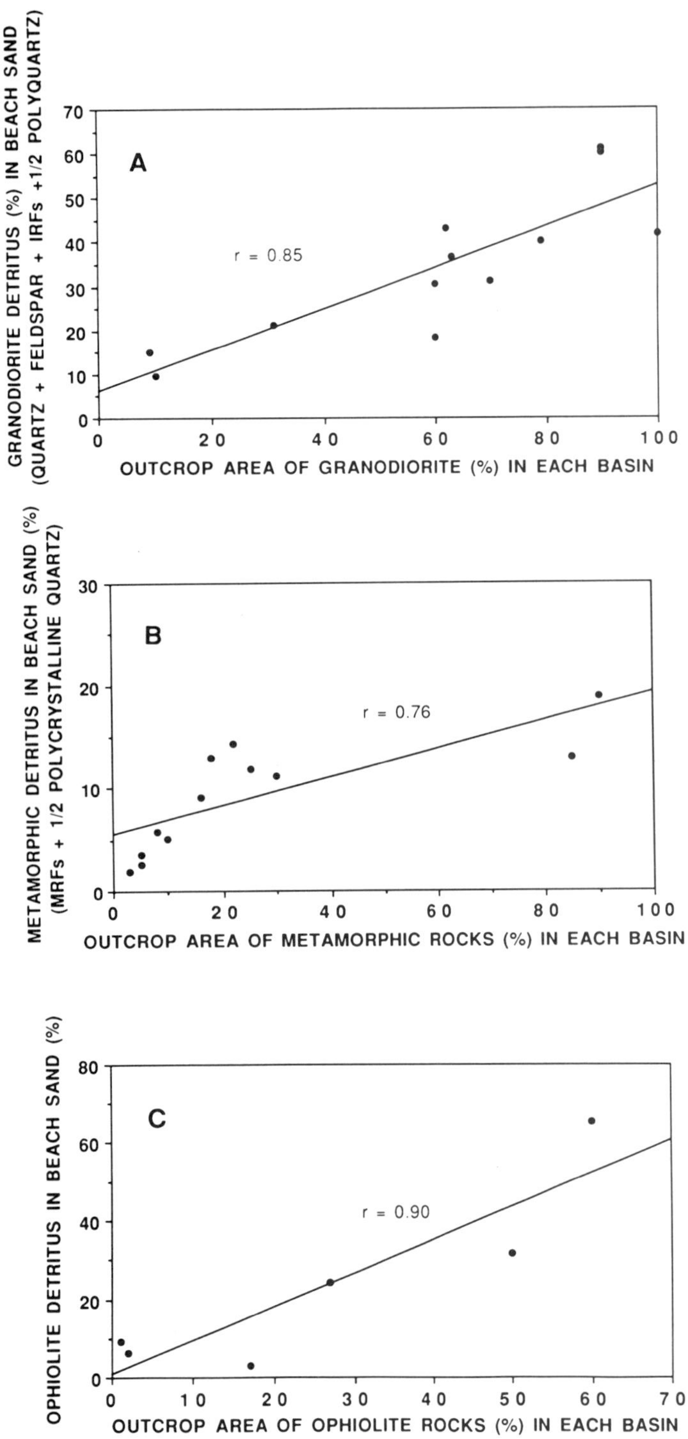

Figure 3. Cross-plots of outcrop area of various bedrock types in each basin versus the percent of sand in each beach that reflects the particular bedrock. A, Granodiorite bedrock versus detritus; B, metamorphic bedrock versus detritus (omitting the spurious data for beach 16); C, ophiolite bedrock versus detritus.

TABLE 3. COMPARISON OF ROUNDNESS BY COMPOSITION AT VARIOUS BEACHES

Grain Type	Beach					
	1	6	7	8	10	17
Monocrystalline quartz	1.3	0.8	0.7	1.5	1.3	1.3
Plagioclase	1.9	1.6	....	1.7	2.1	....
K-feldspar	3.4	2.5	1.7	3.0	2.9	4.1
Igneous rock fragments	3.3	....	....	....	3.0	....
Serpentine	....	....	3.8	3.3	....	....
Metamorphic rock fragments	....	....	....	4.0	3.8	....
Carbonate rock fragments	4.5	3.6	3.7	4.3	....	4.2

ROUNDNESS STUDY

Roundness trends

At six localities we measured roundness for all of the major constituents of the sands in order to compare variations in grain abrasion with variations in mineral and rock composition. These data (Table 3) show that variations in roundness are closely related to mineral hardness in single mineral grains or to the dominant minerals in rock fragments. In order of increasing roundness, the grains are monocrystalline quartz, plagioclase feldspar, K-feldspar and IRFs about equal, serpentine, MRFs, and CRFs. The igneous rock fragments are dominantly particles eroded from granodiorite terrains with abundant K-feldspar and are therefore similar to K-feldspar in their degree of roundness. The greater rounding of K-feldspar at locality 17 compared with serpentine is attributed to the presence of recycled K-feldspar grains from sedimentary sources.

Variations in the roundness of grains of carbonate rocks (micrite and spar), serpentine, and potassium feldspar were studied because these grains are most likely to show whether abrasion has occurred in transit to the beaches and on them. In addition, while determining the grain types of the samples, we noted a variation in roundness of some monocrystalline quartz grains. We then studied this variation at all localities. Monocrystalline quartz grains of medium sand size that are interpreted to be first cycle in origin are unrounded. Such grains display conchoidal fractures and possess sharp grain edges that are barely blunted (Fig. 4A). First-cycle monocrystalline quartz grains of coarser size possess slightly blunted edges that are characteristic of the first stages of rounding (Fig. 4B). There are a few moderately well rounded quartz grains that are probably first cycle in origin, but these are phenocrysts whose roundness is probably the result of resorbtion in magma (Fig. 4C).

Micrite grains are consistently better rounded than carbonate spar grains. This is true for each beach for which we have nine or more grains of each type for comparison (Table 4). The average roundness of micrite is 0.7 roundness units greater than the average roundness of spar grains, and the relationship is significant ($p < 0.001$). However, the reason for the difference in roundness is not clear. Spar grains are rounded equally on all edges and show no evidence of having recently been cleaved. It is possible that spar grains are initially more angular than micrite and that this difference is retained during their abrasion history.

Sphericity values (harmonic 2: Table 2) determined by digitizing grain shapes have a remarkably low range of values that correspond to less than 1 Rittenhouse sphericity value (cf. Fig. 3 of Haines and Mazzullo, 1988). The mean harmonic 2 value is 0.22, which is equivalent to a Rittenhouse sphericity value of 0.8.

Roundness of K-feldspar as measured by Fourier harmonic 19 ranges from 0.004 to 0.007, which is equivalent to Krumbein roundness values of 0.0 to 0.4, the highly angular class. Roundness expressed by harmonic 19 has a moderate correlation with our visually determined values (r = 0.65). Roundness of K-feldspar determined by Fourier analysis also has a moderate correlation (r = 0.61) with our visually determined roundness of quartz. Our visual roundness values for quartz and feldspar, however, have a slightly greater correlation (r = 0.69). The data are compatible with the interpretation that our visually determined roundness values can be used quantitatively.

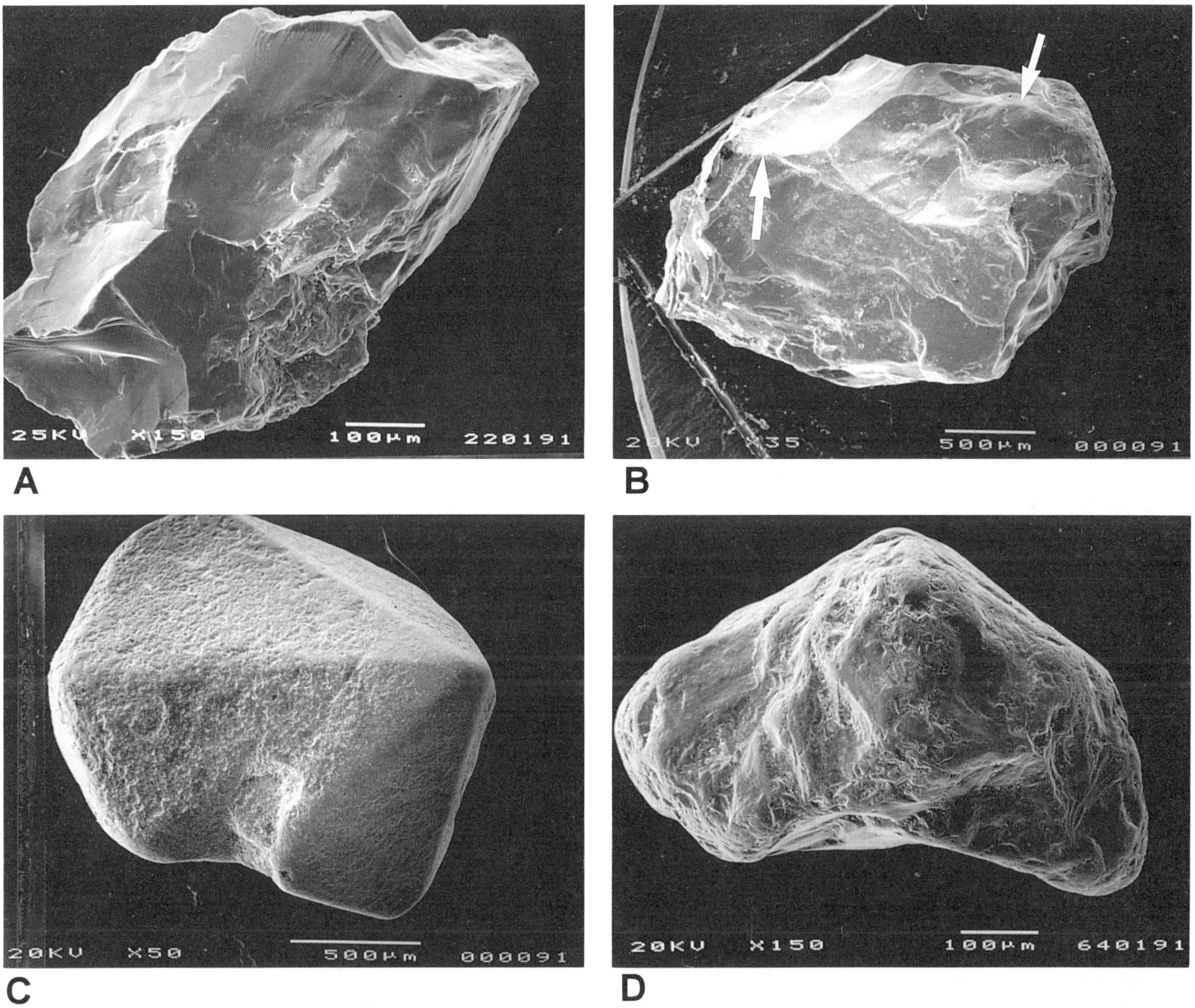

Figure 4. Scanning electron microscope images of quartz grains showing variations in shape and roundness. A, First-cycle, medium sand grain showing no rounding; beach 6. B, First-cycle, very coarse sand grain showing minor rounding of edges (arrows); beach 6. C, First-cycle, medium sand grain with moderate degree of rounding; beach 6. The roundness of this pyramidal quartz phenocryst is interpreted to result from resorption in a magma during crystallization and not from wave abrasion. D, recycled medium sand grain showing high degree of rounding; beach 15.

TABLE 4. ROUNDNESS OF SPAR AND MICRITE GRAINS

Beach	Spar Number Measured	Spar Mean	Spar SD*	Micrite Number Measured	Micrite Mean	Micrite SD*
3	14	3.8	0.61	23	4.0	0.66
7	9	3.3	0.44	9	4.2	0.71
8	25	3.2	0.45	25	3.9	0.66
13	6	3.7	0.41	9	4.6	0.60
15	10	3.9	0.73	10	4.4	0.60
17	26	4.1	0.56	7	4.6	0.35
18	10	3.6	0.33	31	4.1	0.71
	100	3.6		114	4.3	

*SD = Standard Deviation.

Roundness versus beach length

We divided the beaches into three classes based on their lengths, and we looked for trends in the roundness of monocrystalline quartz, K-feldspar, and CRFs with beach size. The smallest beaches are the pocket beaches, all less than 100 m long (beaches 2, 3, 6–8, 18); the intermediate size beaches are from 100 to 300 m long (beaches 1, 9, 13–15, 17); and the largest beaches are longer than 500 m (beaches 4, 5, 10–12, 16). There is no correlation between grain roundness and beach length as evaluated by Student's t-test at the 95% confidence level.

Roundness versus size of drainage basin

We divided the beaches into three groups on the basis of drainage basin size: <1 km^2 (beaches 1–3, 7, 8, 14, 18), from 1 to 3.1 km^2 (beaches 4, 6, 9, 13, 15–17), and >3.1 km^2 (beaches 5, 10–12). For the three grain types, there is no correlation between grain roundness and drainage basin size. This suggests that on Elba there is no measurable increase in grain roundness during transport by short-headed streams.

Roundness versus geographic location of beaches

We divided the beaches into three geographic groups: western (beaches 5–9), central (beaches 1–4, 10, 11), and eastern (beaches 12–18) (Fig. 5). The western side of Elba faces the greatest expanse of the Mediterranean and has the highest energy beaches, whereas the eastern part of the island is the most protected (Aiello et al., 1976). Thus, wave energy, in general, decreases from west to east. The roundness of all three grain types (monocrystalline quartz, K-feldspar, and CRFs) is significantly greater ($p < 0.05$ for quartz and K-feldspar, $p < 0.06$ for CRF) in the east than the west, and most increase progressively from west to central to east. For quartz and K-feldspar, we attribute this trend to the presence of greater amounts of recycled and better rounded detrital grains in the east (Fig. 4D). Sedimentary rock and low-grade metasedimentary rock increase in relative outcrop abundance from west to east on the island. However, the reason for the eastward increase in roundness of CRFs is not clear.

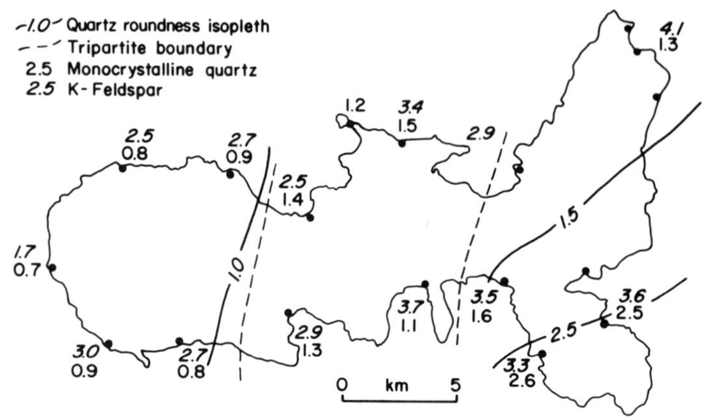

Figure 5. Average roundness values (Powers-Folk) for monocrystalline quartz and K-feldspar in Elba beaches. Isopleths are for quartz. Dashed lines define the three geographic divisions of the island for this study.

Either CRFs are recycled in greater abundance toward the east also, or they round more efficiently under gentler wave action in the eastern beaches. Few of the CRFs are single crystals of calcite that might break and cleave under more vigorous wave action, so this possibility is rejected.

Serpentine shows no geographic trend in roundness in the five beaches for which we have data (Fig. 6).

CONCLUSIONS

Comparison of the outcrop areas of different types of bedrock within drainage basins with the modal composition of the beach sands supplied by the basins yields the following conclusions:

1. The relative area of granodiorite outcrop is accurately represented by the amount of quartz + feldspar + IRFs in all beaches, although plagioclase is significantly reduced in beach sand relative to K-feldspar.

2. The relative areas of outcrop of ophiolitic and limestone bedrock are accurately represented by beach sand in pocket beaches, but are only moderately represented (ophiolitic rocks) or poorly represented (limestone) in other beaches.

3. The relative area of outcrop of metamorphic rocks is poorly represented in beach sand (MRFs + polycrystalline quartz), except in one anomalous beach supplied in part by mine tailings.

4. The relative area of bedded chert outcrop is poorly represented in beach sand because bedded chert does not readily break down to sand-size particles.

5. Shale bedrock is not represented or is only marginally represented in beach sand. For Elba beaches in general, the order of increasing roundness of grains, and thus increasing rate of abrasion, is quartz < plagioclase < K-feldspar and IRFs about equal < serpentine < MRFs < and CRFs. There are no significant differences in roundness with beach size for quartz, K-feldspar, or CRFs. There are also no significant differences in roundness for the same three grain types from beaches of different size drainage basins, which indicates there is no perceptible

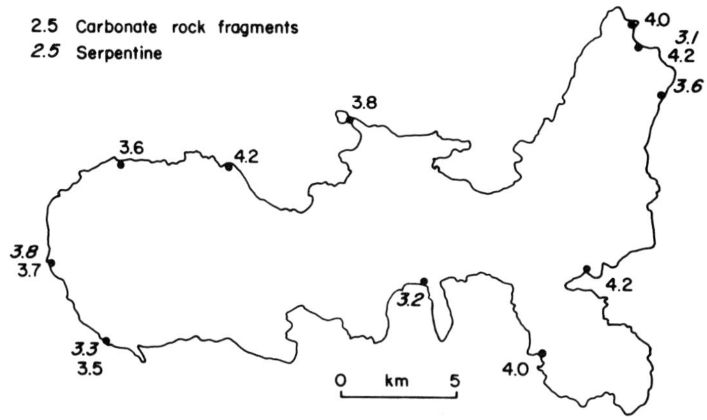

Figure 6. Average roundness values (Powers-Folk) for carbonate rock fragments and serpentine in Elba beaches.

rounding of grains by streams. First-cycle monocrystalline quartz grains of medium sand size are unrounded, although incipient rounding exists on coarser quartz grains. Quartz, K-feldspar, and CRFs are all better rounded on the eastern, more protected part of the island. We attribute this trend in quartz and K-feldspar to the greater abundance of recycled and better-rounded grains to the east. Either there are many recycled CRFs in the eastern beaches also, or CRFs round more efficiently under gentler wave action.

ACKNOWLEDGMENTS

The Geology Foundation of the University of Texas at Austin and Conoco, Inc., provided partial funding for this study. We thank the Istituto di Geologia, Università degli Studi di Modena, for its support of our studies in northern Italy. Laura Bagwell and James Mazzullo provided facilities and counsel for the Fourier shape study. We appreciate greatly the helpful reviews of the manuscript by Patricia Cowley, R. L. Folk, Robert Ehrlich, Emanuela Molinaroli, Craig Sanders, and C. A. Suczek. Rosemany Brant and Jeanette Stubbe typed several drafts of the manuscript, and Jeff Horowitz did the drafting.

REFERENCES CITED

Aiello, E., and 10 others, 1976, Il trasporto litoraneo lungo la costa toscana tra la foce del Fiume Magra ed i Monti dell' Uccellina: Bolletino della Società Geologica Italiana, v. 94, p. 1519–1571.

Basu, A., 1976, Petrology of Holocene fluvial sand derived from plutonic source rocks: Implications to paleoclimatic interpretation: Journal of Sedimentary Petrology, v. 46, p. 649–709.

Basu, A., 1985, Influence of climate and relief on composition of sands released at source areas, *in* Zuffa, G. G., ed., Provenance of arenites; NATO ASI Series: Dordrecht, Riedel Publishing Co., p. 1–15.

Cameron, K. L., and Blatt, H., 1971, Durabilities of sand size schist and "volcanic" rock fragments during fluvial transport, Elk Creek, Black Hills, South Dakota: Journal of Sedimentary Petrology, v. 41, p. 565–576.

Cleary, W. J., and Connolly, J. R., 1971, Distribution and genesis of quartz in a piedmont-coastal plain environment: Geological Society of America Bulletin, v. 82, p. 2755–2765.

Dickinson, W. R., 1970, Interpreting detrital modes of graywacke and arkose: Journal of Sedimentary Petrology, v. 40, p. 695–707.

Ehrlich, R., and Weinberg, B., 1970, An exact method for the characterization of grain shape: Journal of Sedimentary Petrology, v. 40, p. 205–212.

Folk, R. L., 1968, Petrology of sedimentary rocks: Austin, TX, Hemphill Publishing Co., 184 p.

Gazzi, P., 1966, Le arenarie del flysch supracretacico dell'Appennino modenense; correlazioni con il flysch di Monghidoro: Mineralogia et Petrologia Acta, v. 16, p. 69–97.

Gazzi, P., Zuffa, G. G., Grandolfi, G., and Paganelli, L., 1973, Provenienza e dispersione litoranea della sabbie delle spiagge adriatiche fra le foci dell' Isonzo e del Foglia: Inquadramento regionale: Memoria della Società Geologica Italiana, v. 11, p. 1–37.

Grantham, J. H., 1986, The influence of climate and relief on lithic fragment abundance in modern fluvial sands of the southern Blue Ridge Mountains, North Carolina [abs.]: Society of Economic Paleontologists and Mineralogists, Annual Midyear Meeting, Abstracts, v. 3, p. 46–47.

Grantham, J. H., and Velbel, M. A., 1988, The influence of climate and topography on rock fragment abundance in modern fluvial sands of the southern Blue Ridge Mountains, North Carolina: Journal of Sedimentary Petrology, v. 55, p. 219–227.

Haines, J., and Mazzullo, J., 1988, The original shapes of quartz silt grains: A test of the validity of the use of quartz grain shape analysis to determine the sources of terrigenous silt in marine sedimentary rocks: Marine Geology, v. 78, p. 227–240.

Harrell, J. A., 1983, Grain size and shape distributions, grain packing, and pore geometry within sand laminae: Characterization and methodologies [Ph.D. dissertation]: Cincinnati, University of Cincinnati, 567 p.

Hayes, M. O., 1965, Sedimentation in a semiarid, wave-dominated coast (south Texas), with emphasis on hurricane effects [Ph.D. dissertation]: Austin, University of Texas, 350 p.

Ibbeken, H., and Schleyer, R., 1991, Source and sediment: Berlin, Springer-Verlag, 286 p.

Krynine, P. D., 1950, Petrology and stratigraphy of the Triassic sedimentary rocks of Connecticut: Connecticut State Geological and Natural History Survey Bulletin 73, 247 p.

Mazzullo, J., and Kennedy, S., 1985, Automated measurement of the nominal sectional diameters of individual sedimentary particles: Journal of Sedimentary Petrology, v. 55, p. 593–595.

McBride, E. F., and Picard, M. D., 1987, Downstream changes in composition, roundness, and gravel size in a short-headed, high-gradient stream, northwestern Italy: Journal of Sedimentary Petrology, v. 57, p. 1018–1026.

Perrin, M., 1974, Contribution à l'étude géoglogique de l'île d'Elba (Italie): Thèse pour l'obtention du grade du Docteur ès Sciences Naturalles U.E.F. des Sciences de la Terre de l'Amenagement Regionale: Université de Caen, 763 p. [*non-vide*]

Pettijohn, F. J., 1957, Sedimentary rocks (second edition): New York, Harper and Brothers, 718 p.

Picard, M. D., and McBride, E. F., 1990, Variation in sand composition and roundness on the beaches of Elba, Tuscany, Italy [abs.]: International Sedimentological Congress, 13th, Nottingham, Abstracts, p. 346.

Potter, P. E., 1986, South America and a few grains of sand: Part 1. Beach sands: Journal of Geology, v. 94, p. 301–319.

Powers, M. C., 1953, A new roundness scale for sedimentary particles: Journal of Sedimentary Petrology, v. 23, p. 117–119.

Tortosa, A., Palomares, M., and Arribas, J., 1989, Caracterización composicional de los depositos arenosos actuales generados en el Sistema Central: Estudios Geológicos, v. 45, p. 205–215.

Trevisan, L., and Marinelli, G., eds., 1967, Carta Geologica del' isola d'Elba: Pisa, Consiglio Nazionale delle Richerche.

Wallen, C. C., 1977, Climates of central and southern Europe, *in* World survey of climatology: New York, Elsevier Scientific Publishing Co., v. 6, 278 p.

Zuffa, G. G., 1985, Optical analyses of arenites: Influence of methodology on compositional results, *in* Zuffa, G. G., ed., Provenance of arenites: Dordrecht, Riedel Publishing Co., p. 165–190.

Manuscript Accepted by the Society December 24, 1992

Printed in U.S.A.

Geological Society of America
Special Paper 284
1993

Sedimentary recycling in a temperate climate drainage basin (Senio River, north-central Italy): Composition of source rock, soil profiles, and fluvial deposits

William Cavazza
Department of Mineralogical Sciences, University of Bologna, Piazza di Porta San Donato 1, 40126 Bologna, Italy
Gian Gaspare Zuffa, Claudio Camporesi, and Carlo Ferretti
Department of Geological Sciences, University of Bologna, Via Zamboni 67, 40127 Bologna, Italy

ABSTRACT

The Senio River drainage basin (northern Apennines, Italy) provides a good opportunity to study the effects of sedimentary recycling in a temperate climate. In fact, the bedrock of its upper portion is made of a single and homogeneous sedimentary unit, the Marnoso-arenacea Formation (Miocene), which consists of graded beds composed each of sandstone, siltstone, and mudstone.

Composition of the soil sand-fraction within the upper portion of the Senio River drainage basin reflects the sandstone/siltstone composition of the underlying Marnoso-arenacea Formation. Thus, pedogenic processes are generally not strong enough to affect siliciclastic sand grains formed by the alteration of the corresponding source lithology. Carbonate bioclasts are virtually the only sand-size detritus released from the poorly indurated mudstone of the Marnoso-arenacea Formation (i.e., T_e divisions of turbidite beds and hemipelagic beds).

Siliciclastic sand-fractions of fluvial sediments and soils are very similar, indicating that only minor modifications occurred during transport. On the other hand, a large population of carbonate clasts (as much as 35% of the sand fraction) is produced by the breakage of penecontemporaneous carbonate concretions generated mostly by vegetal activity along the upper reaches of the Senio River and its tributaries. In spite of their abundance, these carbonate grains are rapidly lost during transport because of their fragile and chemically labile nature; this implies that their preservation potential in the rock record is low.

Marnoso-arenacea–derived sandstone and siltstone rock fragments are the only siliciclastic components of the detritus present along the Senio River which are modified during transport, decreasing progressively downstream probably because of breakage.

The most useful indicators of sedimentary recycling within the sediments of the Senio River are the siltstone/sandstone lithic fragments, mono- and polycrystalline calcite grains due to breakage of the Marnoso-arenacea sandstone cement, and the mostly unworn fossil tests derived from the Marnoso-arenacea mudstones.

This study confirms that composition of the sand shed off tectonically active source-areas composed of sedimentary rocks in conditions of temperate climate and high/moderate relief and short transport reflects the sediment source-rock, except for poorly indurated mudstone source rocks.

Cavazza, W., Zuffa, G. G., Camporesi, C., and Ferretti, C., 1993, Sedimentary recycling in a temperate climate drainage basin (Senio River, north-central Italy): Composition of source rock, soil profiles, and fluvial deposits, *in* Johnsson, M. J., and Basu, A., eds., Processes Controlling the Composition of Clastic Sediments: Boulder, Colorado, Geological Society of America Special Paper 284.

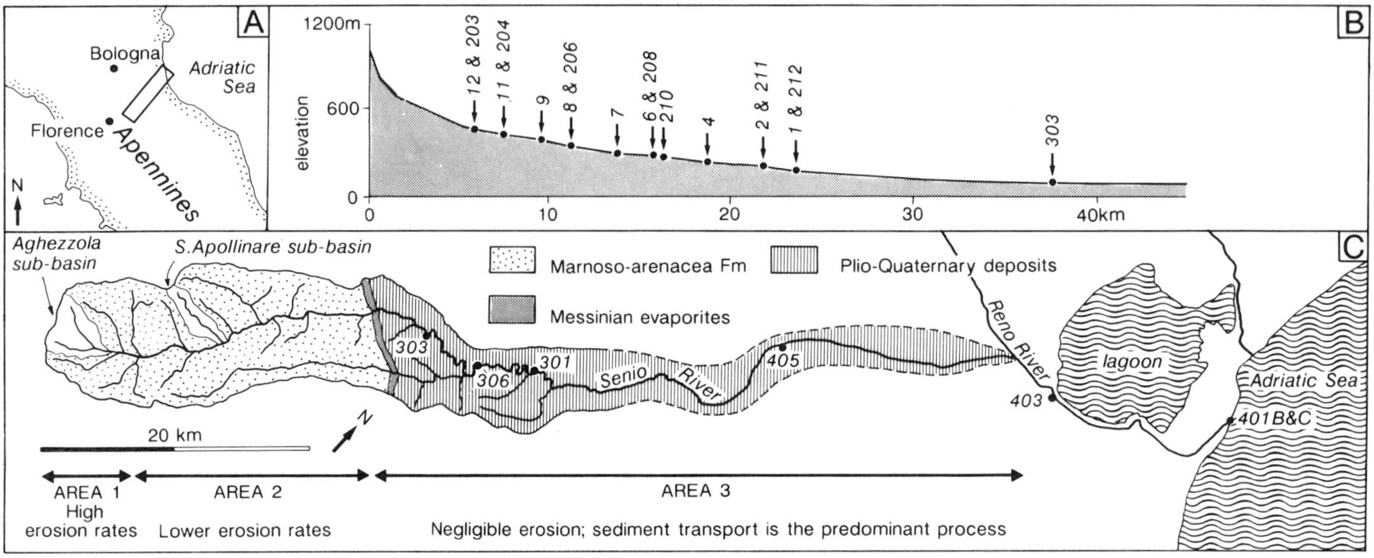

Figure 1. A, Location of the study area in north-central Italy. B, Elevation profile of the Senio River; numbers indicate location of samples. C, Senio River drainage basin (enlargement of the area outlined in A). The patterns indicate bedrock lithologies, the numbers indicate the location of samples taken from the distal Senio River and from the Reno River.

INTRODUCTION

Reliable provenance reconstructions cannot be obtained in a simplistic fashion from petrographic analysis of the framework grains of an arenite because the different physical and chemical characteristics of the mineral grains may induce dramatic selection processes during weathering, transport, deposition, and diagenesis of the sedimentary particles. In addition, multicyclic detritus derived from the erosion of preexisting sandstones can be mistaken for first-cycle detritus, complicating further the correct reconstruction of provenance. For these reasons, it is important to adopt an actualistic approach and study present-day natural systems in order to quantify the effects of these limiting factors on provenance studies. Several studies have focused on the interplay among some of the following parameters of present-day fluvial drainage basins: topography, climate, source area lithology, soil mineralogy, composition, and durability of fluvial sediments during transport (e.g., Cameron and Blatt, 1971; Basu, 1976; Breyer and Bart, 1978; Mack, 1981, Suttner et al., 1981; Grantham and Velbel, 1988; Johnsson et al., 1988; DeCelles and Hertel; 1989).

Sedimentary rocks crop out on 66% of Earth's land surface. Zuffa (1987), recalculating the data of Blatt and Jones (1975), concluded that the amount of sandstone-derived polycyclic grains in the average sandstone is about 30%. It is therefore essential to discriminate between first-cycle and second-cycle detritus for correct provenance interpretations. A few studies list the most evident petrographic indicators of sandstone recycling, including abraded overgrowths, sedimentary lithics, poor roundness sorting, high zircon-tourmaline-rutile (ZTR) index in shallowly buried sandstones (for a review, see Blatt, 1967; Folk, 1974; Zuffa, 1987; Fig. 7). However, they fail to address the problem in a quantitative way. In this work, we study the second-cycle sediment budget of the temperate-climate drainage basin of the Senio River in the northern Apennines (Italy). The peculiarity of the upper portion of this basin is due to the presence of a single sedimentary source rock, the Marnoso-arenacea Formation (Miocene). The goals of this work are to: (1) compare sandstone composition of the bedrock with the sand fraction of the soils developed on top of it, in order to study pedogenic modifications; (2) study sand composition along the Senio River and its tributaries in order to track any compositional modification during transport; (3) identify the most reliable indicators of recycling; and (4) study a population of penecontemporaneous carbonate concretions generated mostly by vegetal activity along the upper reaches of the Senio River and its tributaries.

STUDY AREA

The study area comprises the drainage basin of the Senio River in the northern Apennines of Italy (Fig. 1). This river has its source close to the divide of the northern Apennines (a collisional orogen composed of a northeast-verging fold-thrust belt) and flows northeastward to the Po Plain (the present-day expression of the adjacent foreland basin). The upper half of the river course is in high-relief terrain, where human impact on the river has been minor. The lower half is within the Po Plain where artificial banks have been built, transforming long tracts of the river into a straight channel. After a total course of about 80 km, the Senio River flows into the Reno River which, in turn, debouches into the Adriatic Sea. This research was particularly focused on the upper portion of the drainage basin, which is underlain entirely by a single sedimentary unit, the Marnoso-arenacea Formation.

TABLE 1. CLIMATIC AND HYDROLOGIC CHARACTERISTICS OF THE SENIO RIVER*

Mean annual temperature	11.0°C
Mean monthly temperature	Between 2.1°C (January) and 24.0°C (July)
Mean annual rainfall	1,053.9 mm
Mean monthly rainfall	Ranging between 42.1 mm (July) and 133.4 mm (November)
Mean annual discharge	3.13 m^3/sec
Mean monthly discharge	Ranging between 6.65 m^3/sec (February) and 0.21 m^3/sec (August)
Mean monthly values of daily maximum discharge	Ranging between 27.92 m^3/sec (February) and 2.10 m^3/sec (July)

*Data from the period 1921–1971 (Ciccacci and others, 1980).

The lower reaches of the river cut across progressively younger Messinian evaporites and fine-grained Plio-Pleistocene siliciclastic deposits.

The study area is characterized by a warm temperate climate with prolonged summers and mild winters (Ciccacci et al., 1980). Climatic and hydrologic characteristics of the upper half of the Senio River drainage basin are reported in Table 1. Following the soil taxonomy proposed by the U.S. Department of Agriculture (1975), the most common types of soils present in the upper half of the Senio River drainage basin are identified as Typic and Lithic Dystrochrepts, Typic and District Eutrochrepts, and Lithic Udorthents (Casalicchio et al., 1979).

The Senio River drainage basin can be divided into three areas characterized by different topography, erosion rates, and soil development (Fig. 1). This subdivision is based on the analysis of aerophotographs and soil and land-use maps, and on reconnaissance work. Going downstream from southwest to northeast, they are:

Area 1. This area comprises the steep headlands of the drainage basin, close to the Apenninic divide (approximately the initial 12 km of the river course). Erosion rates are high (more than 20% of this area is represented by outcrops or scree); soils are thin and poorly developed. Tributaries to the Senio River cut deeply into the Marnoso-arenacea Formation.

Area 2. This area extends from approximately 12 to 25 km downstream. It is characterized by lower relief and slower erosion rates. Soils are thicker and generally more stable because of lower gradient topography. Tributaries to the Senio River cut deeply into soils and secondarily into the Marnoso-arenacea Formation.

Area 3. This area extends from 25 km to the confluence with the Reno River. This portion of the drainage basin is characterized by subdued topography and by major human modification of the river (artificial banks, irrigation channels, etc.). Detrital input to the river along this tract is negligible because of the absence of tributaries; sediment transport, rather than sediment production, is the prevailing process.

Our effort was concentrated on the upper reaches of the drainage basin (areas 1 and 2) where erosion and clast production are maximum; area 3 was considered only for sediment modifications during transport. In order to compare the sediment supplied by areas 1 and 2 to the Senio River, we studied in detail two small tributary subbasins (Aghezzola and S. Apollinare; Fig. 1) as representative of areas 1 and 2, respectively. The subbasins were selected after examination of soil maps and aerophotographs of the whole Senio River drainage basin and reconnaissance work. Table 2 illustrates the type of vegetation cover and soil present in the Aghezzola and S. Apollinare subbasins, based on fieldwork and mapping of color aerophotographs. We will supply maps of the two subbasins (at the scale of 1:10,000) on request.

SAMPLING

A total of 61 samples were taken from the study area. The samples can be divided into three groups: 30 soil samples, taken from the drainage areas of the Aghezzola and S. Apollinare creeks; 8 stream samples, taken from the beds of the Aghezzola and S. Apollinare creeks; and 23 river samples taken along the Senio and the Reno riverbeds.

Eighty-nine percent of all samples (54 of 61) were taken from areas 1 and 2, where the bedrock is formed by the Marnoso-arenacea Formation. The remaining seven samples ("distal Senio River" and "Reno River" of Table 4) were taken in order to better track the compositional variations occurring during transport of the detritus.

Soil samples included all types of soils present in the study area. Each soil sample was obtained by mixing three separate subsamples taken 5 to 10 m apart at a depth of 15 to 45 cm below the ground surface. Subsamples were mixed in order to minimize local variations of source rock lithology (sandstone, siltstone and mudstone). Samples from the Aghezzola and S. Apollinare Creeks and from the Senio and Reno Rivers were taken from active areas of the riverbeds, escavating to a depth of 20 cm. Older, heavily vegetated fluvial terraces were avoided.

All samples were dried, weighed, wet-sieved, and decanted to separate the sand fraction. The sand fraction of soil samples varies between 6.4 and 23.6% of the total weight, averaging 12.9%. The sand fraction of the samples taken along the water courses (Aghezzola and S. Apollinare Creeks, Senio and Reno Rivers) varies between 17.9 and 98.0%, averaging 43.6%. The sand fraction was then split, impregnated with epoxy, thin sectioned, and stained for potassium feldspar.

POINT-COUNTING TECHNIQUE

Point counting was performed according to the Gazzi-Dickinson method (Gazzi, 1966; Dickinson, 1970). This method, in order to minimize the dependence of sandstone composition on grain size, separates coarse-grained lithic fragments (crystals greater than 0.0625 mm) from fine-grained lithic fragments (crystals smaller than 0.0625 mm). During transport, the former tend to break into individual crystals, whereas the latter tend to break

TABLE 2. MORPHO-VEGETATIONAL UNITS AND THEIR RELATIVE AREAL ABUNDANCE IN AGHEZZOLA AND S. APOLLINARE SUBBASINS

Unit	Definition	Aghezzola (%)	S. Apollinare (%)
Outcrop	More than 65% of the area is represented by rock outcrops. Soil is virtually absent. Local eluvial covers are present. Intense erosion	5.7	1.1
Detritus ± vegetation	Characterized by common eluvial covers and by mostly bushy and grassy vegetation. Outcrop area is less than 35%. Thin local soils. Intense erosion	14.5	5.8
Sparse wood	Sparse trees and bushes; remaining area is covered by grass. Thin widespread soils. Moderate erosion	5.9	8.9
Dense wood	Dense tree cover mostly composed of oaks, elms, and chestnut trees. Thick undergrowth. Includes afforestation areas. Moderate to deep soils. Erosion is more intense in steep Aghezzola-type terrains than in lower altitude S. Apollinare-type terrains	64.4	65.7
Orchard	Mostly peach trees and vineyards. Deep soils. Moderate erosion	–	1.0
Other farmland	Cultivated fields without trees and mostly nonirrigated. Deep soils. Moderate erosion	2.4	6.0
Grassland	Grassy noncultivated areas, including pastures. Thin to moderate soils. Little to moderate erosion, depending on relief	6.9	11.0
Urban area	Houses and their surroundings, roads, etc.	0.2	0.5
Total area		11.2 km^2	6.4 km^2

into smaller fine-grained lithics. Therefore, coarse-grained lithics are assigned to the mineral beneath the cross-hair, and their lithic nature recorded separately. Three hundred grains per thin section were counted and classified in 36 modal classes. Table 3 explains the petrographic parameters adopted for point counting and Table 4 shows the recalculated point-counting data.

In order to recognize the detritus generated by each interval of the Marnoso-arenacea turbidite beds, point counting was preceded by the microscopic analysis of the Marnoso-arenacea sandstone samples studied by Gandolfi and others (1983) and by Cavazza and Gandolfi (1992), and of new samples taken from sandstone, siltstone, and mudstone divisions of the Marnoso-arenacea turbidite beds and from hemipelagic beds.

SEDIMENT SOURCE ROCK

The Marnoso-arenacea Formation is composed of submarine-fan and basin-plain deposits sedimented in the Periadriatic foreland basin during the Miocene epoch (Ricci Lucchi, 1985). In the study area, this unit is represented by interfingering depositional lobe and basin-plain deposits with average sandstone/shale ratio between ½ and 2. Single turbidite beds comprise, from bottom to top, three rock types: sandstone, siltstone, and mudstone overlain by thin hemipelagic beds.

Marnoso-arenacea sandstones are composed of quartz and feldspar with subordinate lithics. Average sandstone composition in the study area is $Q_{56}F_{33}L_{11}$ (Fig. 2), based on the data of Gandolfi et al. (1983). Most common lithic fragments are gneisses, granitoid rocks, and phyllites, with subordinate carbonate and volcanic rocks. Sandstones and siltstones of the Marnoso-arenacea Formation are tightly cemented by ferroan calcite derived from interbedded pelites during burial diagenesis (Fontana et al., 1986).

The analyses of numerous sandstone samples taken from the study area during this research and preexisting data indicate that overall sandstone composition of this formation is similar throughout the study area.

SOIL SAND-FRACTION

Sand compositions of the Aghezzola and S. Apollinare soil samples are virtually indistinguishable (Table 4); therefore, the

TABLE 3. EXPLANATION OF PETROGRAPHIC PARAMETERS

Ancient noncarbonate (AN)	Q	Qs	= Monocrystalline quartz grains
		Op	= Polycrystalline quartz grains (chert excluded)
		Qrf	= Quartz crystals within rock fragments*
	F	Ks	= Monocrystalline potassium feldspar
		Krf	= Potassium feldspar crystals within rock fragments*
		Ps	= Monocrystalline plagioclase feldspar grains
		Prf	= Plagioclase feldspar crystals within rock fragments*
	L	V	= Volcanic and hypabyssal lithic grains
		Se	= Serpentinite and serpentine-schist lithic grains
		Chs	= Chlorite-schist lithic grains
		Sl	= Slate and phyllite lithic grains
		Arg	= Argillite lithic grains
		Ss	= Matrix within sandstone lithic grains
		Ch	= Chert lithic grains
		Sc	= Lithic grains of carbonate-cemented siltstone
		Snc	= Lithic grains of noncarbonate-cemented siltstone
		Ms	= Monocrystalline phyllosilicate grains
		Mrf	= Phyllosilicate crystals within rock fragments
		OM	= Other monocrystalline grains and other crystals within rock fragments
		Gl	= Detrital glauconite grains
		Io	= Detrital iron-oxide grains
Ancient carbonate (AC)		Spm	= Monocrystalline sparite grains
		Spss	= Sparite cement within sandstone lithic grains
		Bis	= Bioclasts
		Bsp	= Biosparite
		Bmc	= Biomicrite lithic grains
		Rfm	= Micritic-microsparitic lithic grains
		Spf	= Sparitic lithic grains
Penecontemporaneous carbonate (PC)		Bi	= Bioclasts
		Pe	= Rhizocretions and peloid grains
Others		Lc	= Limeclasts (carbonate grains of uncertain age)
		Cx	= Carbonate replacements of unidentifiable grains
		Br	= Fragments of bricks and other building materials
		Vom	= Fragments of vegetal organic matter
		Al	= Altered grains
		U	= Unidentified grains

*Including sandstone rock fragments.

two sample subpopulations have been grouped together. Average soil sand-fraction composition is shown in Figure 2. A qualitative comparison between soil sand-fraction and Marnoso-arenacea rock types (i.e., sandstone, siltstone, and mudstone) shows that sandstone and siltstone are represented in the soil. On the contrary, mudstone clasts are virtually absent, indicating that the uppermost, clayey T_e Bouma divisions of the turbidite beds and the overlying hemipelagic beds do not yield any significant sand-size *siliciclastic* detritus, as also shown by petrographic analysis of base-of-outcrop detritus. *Carbonate* fossils are the only significant sand-size products of the mudstone interval.

After subtracting the siltstone lithics from the point-counting results, composition of soil sand-fraction is virtually identical to the average sandstone composition of the bedrock (Fig. 2), as determined by Gandolfi et al. (1983). This compositional equivalence indicates that framework components of the Marnoso-arenacea sandstone beds are not significantly affected by pedogenesis.

As mentioned earlier, clasts derived from the silty intervals of the source rock are present in the soils. Unfortunately, we have no data on the exact amount of siltstone lithics released from these intervals and, therefore, it is impossible to quantify the pedogenic loss, if any, of these grains.

STREAM AND RIVER SANDS

During the study of the sand fraction of stream samples (i.e., those samples taken from the Aghezzola and S. Apollinare Creeks) and of Senio River samples, a large amount of peloidal carbonate clasts was found (PC component of Fig. 3). They are

TABLE 4. PERCENTAGES OF RECALCULATED POINT-COUNT DATA

Location	Aghezzola Soils													
Sample	110	117	123	108	111	113	120	122	109	127	128	105	112	119
Qs	7.3	9.0	13.0	25.1	17.4	14.7	23.9	18.7	25.0	34.0	23.2	23.2	26.0	14.7
Qp	0.7	1.3	3.0	8.3	6.3	4.7	6.0	4.0	7.7	7.3	3.3	3.7	6.3	2.3
Qrf	3.3	8.3	1.7	4.3	10.4	7.0	9.0	5.7	11.3	5.0	5.7	11.7	10.3	3.7
Ks	0.7	1.0	3.0	8.0	2.0	2.7	5.0	1.7	3.0	7.7	4.3	4.3	3.7	1.7
Krf	0.3	1.0	0.3	2.0	1.3	1.7	1.7	0.7	2.7	1.3	1.7	2.7	1.7	1.0
Ps	3.3	3.3	7.7	15.3	8.3	5.0	13.7	11.0	14.3	10.7	13.7	8.7	10.0	8.0
Prf	2.0	3.3	0.7	7.0	7.0	1.7	6.7	6.0	7.7	1.7	3.7	5.3	3.7	1.3
V	0.3	0.7	0.3	3.7	0.3	0.7	1.3	0.3	1.3	0.7	0.7	1.3	1.7	tr
Se	-	-	-	1.0	-	-	-	-	-	-	-	-	tr	-
Chs	-	-	-	-	-	-	-	-	-	-	-	-	-	-
Sl	-	0.3	1.7	2.3	1.3	1.3	3.3	2.0	2.3	4.7	5.0	1.7	2.3	-
Arg	0.7	2.7	2.3	2.7	-	0.7	2.0	2.0	1.0	3.3	3.0	0.7	3.3	3.3
Ss	1.3	3.7	2.3	0.3	2.7	2.3	3.0	2.7	1.7	0.7	2.0	1.7	1.7	1.3
Ch	tr	-	tr	0.3	-	tr	-	0.3	tr	-	-	tr	-	tr
Sc	19.3	22.7	15.7	1.7	16.8	20.7	4.7	14.7	6.0	4.3	6.7	8.7	5.0	9.7
Snc	2.0	1.0	1.7	1.0	-	2.3	3.7	1.3	2.0	1.7	0.7	1.3	1.3	1.3
Ms	2.7	1.0	2.3	4.0	2.3	3.0	4.7	5.3	5.0	7.0	3.0	3.3	4.0	3.3
Mrf	1.7	1.0	0.7	1.7	1.3	1.0	1.0	1.0	1.0	1.0	2.3	1.0	0.7	-
Om	tr	tr	tr	-	tr	tr	0.3	tr	tr	tr	tr	0.7	tr	tr
Gl	-	-	tr	tr	-	-	-	-	tr	tr	-	-	tr	tr
Io	2.0	2.7	3.0	3.0	1.0	3.3	2.3	1.3	0.7	1.3	0.7	1.3	1.0	1.0
Spm	1.0	2.0	0.7	-	2.3	2.0	-	2.0	2.0	2.7	2.0	2.0	1.3	0.3
Spss	2.3	2.3	3.3	-	6.7	6.3	1.0	3.3	1.0	1.3	2.7	2.7	3.3	0.7
Bis	23.8	12.7	28.6	3.7	5.0	9.3	3.3	8.0	2.0	2.0	8.0	7.3	6.3	34.0
Bsp	0.7	1.3	0.7	-	-	0.7	-	1.0	-	-	-	1.0	0.7	2.0
Bmc	0.3	0.7	-	-	0.3	-	-	-	-	-	-	-	-	-
Rfm	17.0	13.0	4.0	1.3	3.0	4.3	2.7	4.7	-	-	4.0	1.7	3.7	4.7
Spf	7.3	5.0	3.0	1.3	4.3	4.0	0.7	2.3	2.0	1.3	2.7	3.7	2.0	5.7
Bi	-	-	-	-	-	-	-	-	-	-	-	-	-	-
Pe	-	-	-	-	-	-	-	-	-	-	0.3	-	-	-
Lc	-	-	-	-	-	-	-	-	-	-	-	-	-	-
Cx	-	-	-	-	-	0.3	-	-	-	-	-	-	-	-
Br	-	-	-	1.3	-	-	-	-	tr	0.3	0.3	0.3	-	-
Vom	-	-	-	-	-	-	-	-	-	-	-	-	-	-
Al	-	-	-	0.7	-	0.3	-	-	0.3	-	0.3	-	-	-
U	-	-	0.3	-	-	-	-	-	-	-	-	-	-	-
Total	100.0	100.0	100.0	100.0	100.0	100.0	100.0	100.0	100.0	100.0	100.0	100.0	100.0	100.0

TABLE 4. PERCENTAGES OF RECALCULATED POINT-COUNT DATA (continued)

Location	S. Apollinare Soils															
Sample	57	58	69	54	65	72	55	61	71	74	51	66	73	64	67	70
Qs	25.1	28.4	16.1	17.8	21.6	22.4	28.9	29.8	14.0	23.7	28.1	27.2	20.4	26.1	15.9	30.0
Qp	8.7	8.3	4.7	6.7	6.7	9.0	6.7	12.7	3.7	6.3	7.7	3.0	6.7	9.7	5.7	5.7
Qrf	8.3	5.7	6.0	10.7	9.0	8.3	14.7	11.7	3.3	9.3	9.7	14.6	10.3	11.3	8.0	5.7
Ks	4.3	4.3	2.3	7.7	3.0	3.7	7.7	7.3	3.0	4.3	4.3	5.8	5.3	8.3	4.0	8.3
Krf	2.7	3.0	2.3	5.3	3.0	3.3	2.0	3.3	1.7	3.0	2.0	3.3	3.3	2.7	3.7	2.7
Ps	10.3	13.8	6.0	11.1	14.0	12.1	16.1	18.4	7.7	13.3	14.4	12.5	13.1	14.3	9.3	16.3
Prf	5.3	7.0	2.3	5.0	5.0	7.7	7.0	6.0	5.0	6.3	5.3	5.1	7.3	7.7	4.3	4.3
V	0.7	1.0	2.0	1.3	1.3	0.3	1.3	1.3	1.3	0.7	0.7	0.3	0.3	1.0	1.3	0.3
Se	0.3	1.0	0.3	-	0.3	tr	1.0	-	0.7	0.3	0.3	1.2	tr	tr	0.6	tr
Chs	-	tr	tr	tr	0.3	-	-	0.3	-	0.3	-	0.3	-	-	-	0.3
Sl	1.7	2.3	1.3	1.7	0.3	0.7	2.3	2.7	1.7	2.0	2.0	0.3	1.3	2.0	1.0	1.0
Arg	1.0	0.3	1.0	0.7	1.0	0.3	0.3	0.7	1.7	1.7	1.3	0.6	1.3	0.3	0.3	2.7
Ss	3.3	0.3	1.0	5.3	3.0	2.0	0.7	1.0	3.3	2.0	1.3	2.4	2.7	1.7	2.7	1.0
Ch	0.3	-	0.3	0.3	1.3	1.0	tr	0.3	1.0	2.0	1.3	tr	0.3	0.3	1.7	0.7
Sc	6.0	1.0	11.7	7.0	6.3	8.3	1.3	-	9.7	6.0	3.7	3.3	10.4	2.0	12.3	4.0
Snc	1.3	-	1.3	0.3	0.7	-	1.7	0.3	0.3	0.3	1.7	0.3	-	1.0	0.3	-
Ms	5.3	5.3	2.7	3.3	4.3	4.0	0.3	2.3	2.0	3.3	4.0	3.0	4.7	5.3	2.0	2.7
Mrf	2.7	2.0	1.0	3.3	1.3	2.3	1.0	1.0	2.0	1.7	1.3	2.4	1.3	2.3	0.7	1.3
Om	2.0	1.3	0.3	0.3	0.7	1.0	tr	0.3	1.7	0.3	1.0	1.2	1.0	0.3	1.0	1.3
Gl	tr	tr	-	tr	0.3	tr	-	0.3	tr	tr	tr	tr	tr	tr	-	tr
Io	0.3	0.3	1.0	1.0	0.7	-	1.0	tr	0.7	0.3	0.3	0.3	0.3	-	0.3	0.7
Spm	1.7	3.0	2.3	1.7	2.7	2.0	0.7	-	1.0	0.7	1.3	1.5	0.3	0.7	0.7	2.3
Spss	2.0	1.0	1.3	2.3	4.3	1.3	1.3	-	0.3	1.0	1.3	2.7	2.7	1.0	2.3	1.7
Bis	1.3	4.0	13.8	2.0	1.3	2.7	1.7	-	15.2	4.7	2.0	0.6	1.0	0.7	1.7	3.3
Bsp	0.7	1.0	4.0	0.3	0.3	1.3	-	-	6.3	1.0	0.3	0.9	0.7	-	-	0.7
Bmc	0.7	1.7	0.7	-	-	0.7	-	-	1.3	2.3	-	-	2.7	-	-	0.3
Rfm	2.7	2.3	9.3	2.3	3.7	3.7	1.0	-	4.0	2.3	1.3	3.9	1.3	0.3	15.9	1.7
Spf	0.7	0.7	4.0	1.3	2.0	1.3	-	-	5.7	-	1.7	1.2	-	-	2.0	1.0
Bi	-	-	-	-	0.3	-	-	-	-	-	-	-	-	-	-	-
Pe	-	-	-	-	-	-	-	-	0.3	-	-	-	-	-		
Lc	-	-	0.3	-	-	-	-	-	-	-	0.3	1.2	0.3	-	1.7	-
Cx	0.3	0.7	0.7	0.7	1.0	-	1.0	-	0.7	-	0.7	-	-	-	0.3	-
Br	-	-	-	-	-	-	-	-	-	0.3	-	-	-	-	-	-
Vom	-	-	-	-	-	-	-	-	-	-	-	-	-	-		
Al	-	0.3	-	0.3	0.3	0.3	-	0.3	0.7	0.3	0.7	-	0.7	0.3	-	-
U	0.3	-	-	0.3	-	0.3	0.3	-	-	0.3	-	0.9	0.3	0.7	0.3	-
Total	100.0	100.0	100.0	100.0	100.0	100.0	100.0	100.0	100.0	100.0	100.0	100.0	100.0	100.0	100.0	100.0

TABLE 4. PERCENTAGES OF RECALCULATED POINT-COUNT DATA (continued)

Location	Aghezzola (13-102) and S Apollinare Streams (56-68B)								Senio River 1990								
Sample	13	118	121	103	102	56	68A	68B	1	2	4	6	7	8	9	11	12
Qs	8.3	6.3	10.0	5.3	10.7	28.9	26.8	24.7	12.0	12.0	7.3	8.3	6.3	8.3	7.7	5.6	4.3
Qp	4.0	1.7	2.0	1.0	1.3	8.7	6.0	5.7	5.3	4.7	1.7	3.3	3.3	5.3	3.7	2.5	3.7
Qrf	10.0	8.3	15.3	10.3	12.7	8.3	6.3	6.3	8.7	7.0	1.7	3.7	4.3	10.3	10.0	4.6	6.0
Ks	1.7	1.3	2.3	1.0	1.3	5.0	5.0	4.3	2.7	3.7	1.3	0.7	1.0	1.7	1.7	0.3	0.7
Krf	3.7	0.7	1.7	3.7	2.7	2.3	1.0	1.7	2.3	1.3	1.3	1.7	2.3	1.0	2.7	1.7	1.0
Ps	2.7	4.7	3.7	4.7	4.3	15.7	10.0	9.7	5.0	7.0	4.3	6.7	3.0	5.7	3.0	1.0	1.7
Prf	5.3	4.3	5.0	5.0	4.0	5.3	4.0	4.0	6.3	4.7	0.7	2.7	3.3	7.3	7.0	4.6	4.3
V	0.7	0.7	-	tr	0.3	0.3	tr	0.3	0.7	0.3	tr	0.3	0.3	1.3	1.7	0.3	1.0
Se	0.3	-	-	-	-	0.7	0.3	tr	0.3	0.7	-	tr	0.3	tr	0.3	tr	-
Chs	-	-	-	-	-	tr	-	0.3	-	-	tr	tr	tr	0.3	tr	tr	-
Sl	1.3	2.0	2.0	0.3	1.3	2.0	1.7	1.7	2.0	0.7	0.3	0.3	0.3	1.7	0.3	0.3	0.7
Arg	0.7	-	1.7	tr	tr	0.3	-	-	0.3	0.3	tr	0.3	tr	-	-	-	1.0
Ss	5.7	5.0	3.0	5.0	3.7	1.0	1.0	0.3	5.7	2.0	0.3	2.7	2.7	5.3	7.3	2.8	7.7
Ch	tr	-	-	-	-	0.3	0.7	1.0	1.0	1.3	0.3	0.3	1.0	1.7	0.3	tr	1.0
Sc	30.5	21.2	15.0	13.8	23.7	4.7	3.3	2.3	11.0	10.7	-	7.0	13.1	16.4	26.7	8.4	17.6
Snc	1.3	3.7	2.0	1.3	tr	0.3	-	-	1.0	0.3	-	0.3	0.3	1.0	-	0.3	0.7
Ms	0.3	0.7	2.7	2.3	tr	1.7	4.7	4.0	2.3	2.0	0.7	0.3	-	0.7	-	0.7	0.7
Mrf	2.7	3.7	2.3	3.3	1.3	1.3	0.7	1.7	2.3	2.0	-	0.3	1.0	2.3	1.7	1.0	1.0
Om	0.3	0.3	tr	tr	0.3	1.0	0.3	0.7	tr	tr	tr	-	0.3	0.7	0.3	0.3	-
Gl	-	tr	0.3	tr	tr	0.3	tr	tr	tr	0.3	0.3	-	tr	0.3	tr	tr	-
Io	-	1.0	tr	0.3	0.7	-	tr	-	-	0.3	tr	-	-	0.3	-	tr	-
Spm	1.0	1.0	0.3	0.3	0.3	0.7	1.0	0.7	-	0.3	-	0.3	-	1.7	1.0	0.3	0.3
Spss	1.3	6.3	8.0	7.7	4.3	1.3	0.3	0.7	0.3	0.7	1.0	0.7	2.3	1.7	2.7	1.0	3.7
Bis	6.0	14.0	5.0	3.7	3.7	2.0	7.3	6.0	2.0	2.3	0.7	0.7	1.0	1.3	2.0	2.1	3.0
Bsp	1.3	0.3	0.7	1.0	0.7	0.3	0.7	1.3	-	1.7	-	1.0	0.7	3.0	0.3	0.7	0.3
Bmc	0.3	1.7	0.3	0.3	-	-	0.3	0.3	-	-	-	-	-	-	0.3	-	-
Rfm	2.3	7.7	1.7	4.3	3.7	2.7	1.3	1.7	3.7	5.3	2.0	6.0	6.3	3.3	4.7	1.7	2.0
Spf	0.7	2.7	5.0	3.7	4.0	1.7	2.0	1.0	2.7	2.7	1.0	2.0	1.3	-	2.0	0.3	1.0
Bi	-	-	-	-	-	-	0.3	-	-	-	0.3	-	0.3	-	-	-	-
Pe	5.3	0.7	10.0	21.1	14.7	1.0	12.3	18.3	20.1	23.8	72.9	46.1	41.7	14.1	6.0	57.8	34.3
Lc	1.7	-	-	0.3	0.3	1.3	2.0	1.0	0.7	1.3	1.0	4.0	3.3	2.3	4.0	1.7	2.0
Cx	-	-	-	-	-	0.3	0.7	-	0.3	0.3	0.3	0.3	0.3	-	1.0	-	-
Br	-	-	-	-	-	0.3	-	-	0.3	-	-	-	-	-	0.7	-	-
Vom	0.3	-	-	0.3	-	-	tr	tr	-	tr	tr	-	-	0.3	0.3	-	-
Al	-	-	-	-	-	-	-	-	0.3	-	0.3	-	-	0.7	0.3	-	-
U	0.3	-	-	-	-	0.3	-	0.3	0.7	0.3	0.3	-	-	-	0.3	-	0.3
Total	100.0	100.0	100.0	100.0	100.0	100.0	100.0	100.0	100.0	100.0	100.0	100.0	100.0	100.0	100.0	100.0	100.0

TABLE 4. PERCENTAGES OF RECALCULATED POINT-COUNT DATA (continued)

Location	Senio River 1991							Distal Senio River				Reno River		
Sample	203	204	206	208	210	211	212	303	306	301	405	403	401B	401C
Qs	12.0	8.7	8.0	11.0	10.3	9.7	14.7	13.7	13.8	14.4	14.9	18.3	26.2	29.4
Qp	4.7	5.0	1.7	4.3	5.7	4.0	4.0	7.0	6.2	6.0	6.0	7.0	5.7	5.0
Qrf	5.7	9.3	10.2	8.7	6.7	9.3	8.0	10.0	9.2	10.1	11.0	8.7	5.0	3.7
Ks	0.7	1.7	1.7	1.3	1.7	2.0	2.3	1.0	3.1	2.7	3.7	5.7	4.3	2.0
Krf	2.7	2.3	2.7	3.7	1.7	1.0	2.0	2.0	3.1	4.0	3.7	2.7	2.3	0.3
Ps	4.0	3.7	3.3	4.7	4.3	4.0	3.3	5.7	5.3	5.0	6.7	8.3	10.7	8.3
Prf	6.3	5.3	8.3	4.7	5.0	2.7	3.3	6.3	5.2	6.3	6.3	5.3	2.3	4.3
V	0.3	1.0	1.7	2.3	0.3	0.3	1.3	0.7	1.1	2.3	1.7	0.7	1.0	1.0
Se	-	-	-	-	-	0.3	-	0.3	0.4	tr	0.7	-	-	-
Chs	-	-	-	-	-	0.3	-	-	-	-	0.3	-	-	-
Sl	2.3	2.0	1.7	3.3	2.7	3.7	4.0	7.3	5.3	7.7	9.3	4.7	2.0	5.3
Arg	0.7	1.3	0.7	0.7	3.7	0.3	-	-	0.4	0.3	1.0	3.0	2.3	4.7
Ss	6.0	4.7	3.7	4.0	3.3	3.0	4.7	1.7	1.6	2.3	1.7	1.0	0.7	0.3
Ch	-	-	0.7	0.3	0.3	tr	tr	0.7	0.2	0.3	-	0.7	0.7	0.3
Sc	15.9	13.1	19.1	14.7	21.8	18.1	20.8	17.3	13.6	9.7	9.0	4.7	4.0	3.7
Snc	tr	1.3	0.7	0.3	0.7	-	-	0.7	1.1	0.3	1.0	1.7	2.3	2.0
Ms	0.7	1.0	0.7	1.3	0.3	1.3	1.7	0.3	1.6	1.0	1.3	1.3	3.0	2.7
Mrf	3.3	2.3	1.7	0.3	1.3	0.7	2.0	0.7	0.6	2.3	1.7	1.0	0.3	0.3
Om	tr	0.3	tr	0.7	0.3	tr	0.3	tr	4.4	0.3	0.7	tr	1.3	tr
Gl	-	tr	tr	tr	tr	tr	tr	-	0.2	-	-	tr	tr	tr
Io	0.3	0.3	0.7	0.3	1.3	tr	tr	0.3	0.4	0.3	tr	tr	tr	0.3
Spm	0.7	1.0	1.7	1.7	0.7	0.7	2.0	1.0	1.3	1.7	1.7	3.3	6.7	5.7
Spss	8.7	7.3	8.7	6.7	7.0	7.7	4.3	2.0	1.8	1.7	1.0	3.3	1.3	1.0
Bis	5.7	2.7	7.0	3.7	4.3	5.3	6.7	7.7	5.8	6.3	2.0	3.3	5.0	5.0
Bsp	0.7	-	-	0.7	1.0	1.3	-	1.0	0.2	0.7	tr	-	-	-
Bmc	-	-	-	-	0.3	0.3	0.3	-	0.2	1.0	-	0.7	0.3	-
Rfm	2.3	4.7	3.0	4.3	3.3	2.3	3.0	6.0	8.2	8.3	8.0	7.3	5.3	5.7
Spf	4.3	4.0	6.0	2.7	2.7	3.0	2.3	3.0	4.4	3.0	5.0	6.7	7.0	5.0
Bi	-	-	-	-	-	-	-	tr	0.2	0.3	-	0.3	-	3.0
Pe	12.0	17.0	6.3	12.7	9.3	17.8	9.0	3.3	0.7	1.7	1.3	0.3	-	-
Lc	-	-	-	0.3	-	-	-	-	-	-	-	-	-	-
Cx	-	-	-	-	-	-	-	-	-	-	-	-	-	-
Br	-	-	-	tr	-	-	-	-	-	-	-	-	-	-
Vom	-	-	-	0.3	-	0.3	-	-	-	-	-	-	-	-
Al	-	-	-	0.3	-	0.3	-	0.3	0.4	-	0.3	-	0.3	0.7
U	-	-	-	-	-	0.3	-	-	-	-	-	-	-	0.3
Total	100.0	100.0	100.0	100.0	100.0	100.0	100.0	100.0	100.0	100.0	100.0	100.0	100.0	100.0

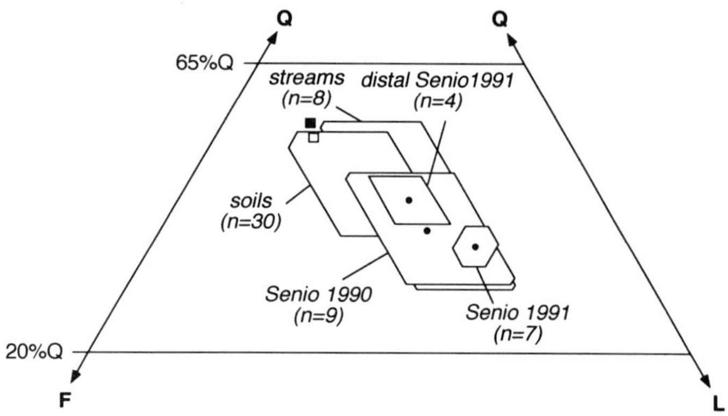

Figure 2. QFL triangular plot showing sand compositions of soil samples, stream samples (Aghezzola Creek and S. Apollinare Creek combined), Senio River samples (1990 and 1991 sampling seasons), and distal Senio River samples. Dots and polygons indicate means and standard deviations, respectively. For explanation of vertices, see Table 3. Open square indicates the mean sand composition of the 30 soil samples from the study area, recalculated excluding categories Sc, Snc, and Ss of Table 3. This recalculation allows comparison with the mean composition of the Marnoso-arenacea sandstone in the study area (solid square; data from Gandolfi and others, 1983).

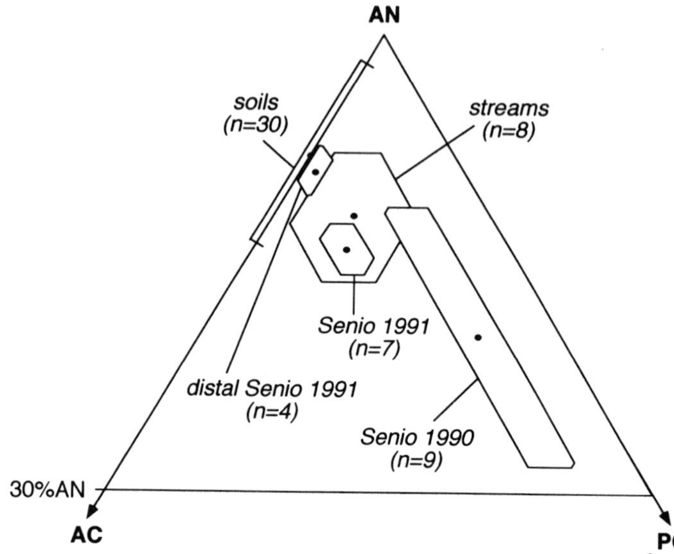

Figure 3. AN-AC-PC triangular plot showing sand compositions of soil samples, stream samples (Aghezzola Creek and S. Apollinare Creek combined), Senio River samples (1990 and 1991 sampling seasons), and distal Senio River samples. Dots and polygons indicate means and standard deviations, respectively. For explanation of vertices, see Table 3.

different from the scarce carbonate clasts present in the framework of the Marnoso-arenacea sandstones, both morphologically and texturally. These carbonate clasts are described in the following section.

Penecontemporaneous carbonate grains

Description. Peloidal carbonate grains are light gray in color and are composed of low-Mg calcitic micrite and microsparite (determined from X-ray diffraction and from energy dispersive X-ray analyses at the scanning electron microscope with a porous, spongy fabric; concentric structures may be present. Scattered siliciclastic grains and vegetal organic matter (thin roots, algal filaments, etc.) with well-preserved textures are common within these grains. Three major types of calcite peloids have been recognized: thick rhizocretions around plant roots; concretions around algal filaments; and grain coatings (Fig. 4).

1. Rhizocretions. This type represents about 70% of all peloids. In thin section, it appears as a sectioned tubule (Fig. 4A, B). The center of the tubule was originally occupied by the root, but it is now generally empty except for local traces of vegetal organic matter. The wall of the tubule can be as thick as 1 mm, and it is made of micrite-microsparite with a characteristic spongy appearance due to abundant voids; this peculiar texture is easily recognizable even in broken grains without root remains. The inner surface of the tubule may show precise casts of the root cortical structures.

2. Concretions around algal filaments. This type is characterized by bundles of tubules less than 150 μm in diameter, more characteristically between 10 and 30 μm (Fig. 4D, Fig. 5). It represents about 10% of all peloids.

3. Grain coatings. This type consists of micritic and, rarely, microsparitic calcite coating detrital grains of various types. Compared to the other types of peloids, they are finer grained and texturally more homogeneous (Fig. 4C). In a few cases, they display concentric layering. They comprise about 20% of all peloids.

Peloidal carbonate clasts are common in the samples taken from the tributaries (average 10.4% of the total sand fraction) and in the samples taken from the Senio River in 1990 (35.1%), and 1991 (12.0%). Yet they are rare in the samples taken from the distal Senio River and from the Reno River (1.0% combined average) and virtually absent in the Aghezzola and S. Apollinare soil samples (Fig. 3).

Interpretation. Based on field and laboratory evidences, these grains have been interpreted as penecontemporaneous concretions developed around plant roots, algal filaments, and sedimentary grains. They have been found only in water-saturated sediments along the Senio River and its tributaries, whereas they are altogether absent in all other areas of the drainage basin.

Type 2 concretions and most type 1 rhizoliths were generated in a dense network of small-diameter algal filaments and plant rootlets commonly oriented horizontally. This is a virtually exclusive characteristic of those plants living in lagoonal or paludal environments or, more generally, in those environments where phreatic waters are at modest depth (Cohen, 1982; Mount and Cohen, 1984).

Carbonate organo-sedimentary structures of intrabasinal origin are produced by the activity and decay of the root systems of plants (Klappa, 1980), algae, and fungi (Schneider, 1977). They are the result of the buildup and/or the replacement around or inside plant roots, algal filaments, or mycelial strands of fungi

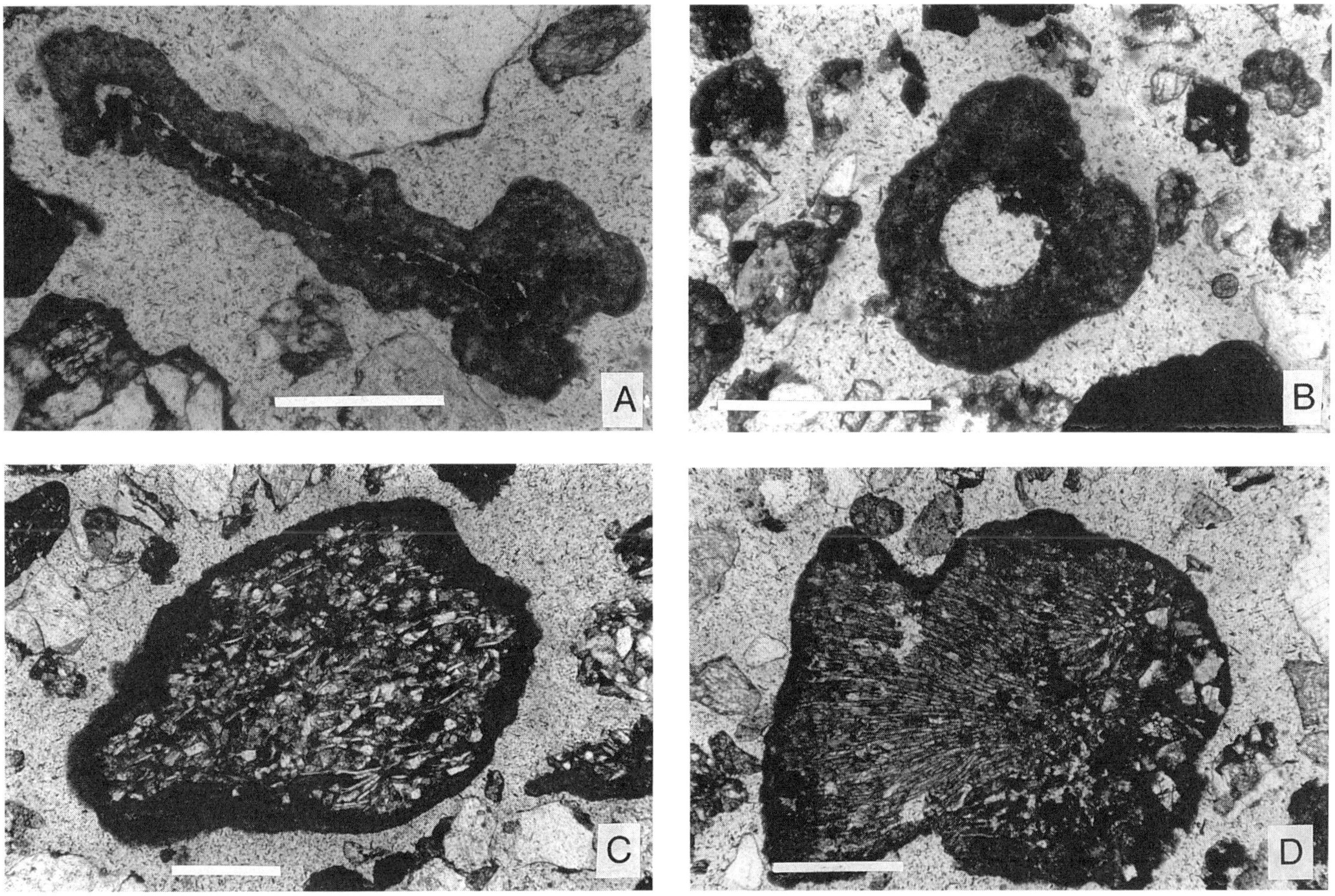

Figure 4. Photomicrographs of penecontemporaneous calcite concretions developed along the Senio riverbed. Scale bar represents 0.5 mm for all photographs. A and B show micritic-microsparitic calcite concretions around thin plant roots (rhizocretions). A, Longitudinal section showing completely preserved root. B, Transversal section. C, Calcite coatings on a siltstone lithic fragment. D, Carbonate concretion with fluidal texture due to precipitation of calcite around algal filaments.

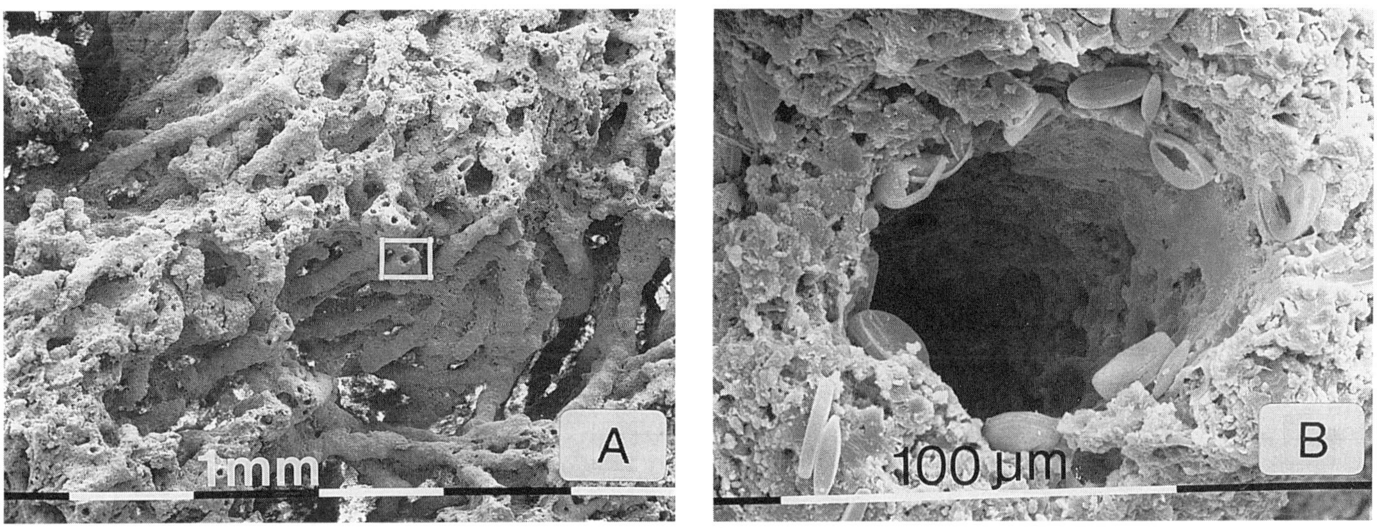

Figure 5. SEM photographs of calcite concretion around algal filaments. White outlined box in A indicates the area enlarged in B. Note diatom tests lining the filament casts.

and/or the replacement of these features by mineral substances. The development process of these carbonates in correspondence to root systems involves complex interactions among microorganisms (bacteria, fungi, blue-green algae), organic acids produced by the activity and decay of the roots, and the continuous flow of pellicular waters due to the absorption activity of roots and to evapotranspiration (Klappa, 1980; Semeniuk and Meagher, 1981; Mount and Cohen, 1984). Furthermore, some plants do not absorb calcium, facilitating its precipitation around their roots (Keller and Frederickson, 1952).

Type 3 concretions do not show clear-cut evidence of vegetal activity and are difficult to interpret as either biologically or chemically induced.

Carbonate peloids were found during the August 1990 sampling campaign. It was decided to replicate the sampling along the Senio River one year later, in August 1991, to measure any variation. The 1990 samples have an average of 35.1% of carbonate peloids, whereas the 1991 samples have only 12.0% of peloids in the total sand-fraction. The analysis of monthly mean rainfall within the upper reaches of the Senio River indicates that the 11 months preceding the 1990 sampling were characterized by unusually low rainfall (total of 444 mm of rain versus an average annual rainfall of 1,054 mm), whereas the 11 months preceding the 1991 sampling were characterized by normal rainfall (1,018 mm). Integration of these meteorological data with sedimentologic observations made periodically along the Senio River and its tributaries during the years 1990–1992 suggests that the development of the calcite concretions was induced by the unusually long period of low rainfall from October 1989 to August 1990. During this time interval, the stagnant conditions were conducive to the proliferation of vegetal life on the riverbed, as observed during the 1990 sampling season. On the contrary, an average year is characterized by several floods between November and April (Ciccacci et al., 1980), which have significant effects on the Senio riverbed, redistributing the sediment and eradicating the vegetation along the riverbed. During the floods, most calcite concretions are eroded and transported downstream.

As to the source of the calcium necessary for precipitation of the concretions, sediment-water interactions in the acidic soils of the Senio River drainage basin (Gisotti, 1988) induce the release of Ca^{2+} ions from calcite-cemented siliciclastic grains and from marl grains into the percolating waters. Rock-water interaction between interstitial waters and bedrock provides an additional source of calcium. Percolating waters, thus enriched in calcium content, flow downvalley toward the trunk of the Senio River, where plant/fungal activity induces carbonate precipitation around plant roots, algal filaments, and mycelial strands.

Stream sands

The average ancient noncarbonate/ancient carbonate ratio of soil and stream sands is almost identical (Fig. 3); the only difference between the two sample populations is represented by the amount of penecontemporaneous carbonates (PC), i.e., modern carbonate concretions and bioclasts. PC grains are virtually absent in the soil samples, whereas they range between 0.7 and 21% in the samples from the Aghezzola and S. Apollinare Creeks (Table 4).

QFL compositions of stream and soil sands overlap considerably (Fig. 2); the differences between the two sample populations have no statistical significance. The average composition of stream samples has a large standard deviation, particularly in terms of lithic content. This can be explained considering that the detritus carried downstream by the Aghezzola and the S. Apollinare Creeks is not derived exclusively from the soils, but also from outcrops of the Marnoso-arenacea Formation, locally contributing large amounts of unstable lithic fragments to the creeks. This is particularly true for the Aghezzola Creek, which cuts deeply into the bedrock.

River sands

The samples collected along the proximal segment of the Senio River in August 1990 and 1991 have very similar siliciclastic compositions, whereas their carbonate compositions are quite different. In fact, on a standard QFL diagram (Fig. 2), the two populations of samples display overlapping compositions, the 1991 samples showing a smaller standard deviation. Conversely, on the AN-AC-PC ternary plot (Fig. 3), the 1990 samples are significantly richer in penecontemporaneous carbonate clasts (35.1 versus 12.0%), although their amount varies considerably, producing a large standard deviation. As discussed earlier, the abundance of biogenic carbonate concretions in the 1990 samples was induced by local conditions related to the low-energy hydrologic regime of that year. Conversely, the 1991 samples contain relatively minor amounts of carbonate concretions and have smaller standard deviations, suggesting that 1991 floods redistributed and homogenized the river sediments irrespective of preexisting local conditions.

Compared to the 1991 proximal samples, the samples collected along the distal part of the Senio River in the same year are significantly poorer in penecontemporaneous carbonate grains (PC in Fig. 3) and fine-grained lithic fragments (L in Fig. 2). The depletion in carbonate grains in the samples from the distal Senio River (average 1.7% of the total sand fraction), as well as in those from the Reno River (0.1%), is due to their fragility and chemical instability; sediment dilution due to the 1991 floods represents an additional explanation. Since average grain size of proximal and distal samples is relatively constant, the downstream depletion in lithics can be explained as due to progressive breakage of the mechanically unstable siltstone grains during transport whose amount, therefore, decreases in the sand population. In fact, the average grain size of the siltstone grains progressively decreases downstream, as shown by microscopic analysis.

A significant modification affecting the siliciclastic component of the Senio River sand-fraction during transport involves the amount of sandstone lithics. They decrease very rapidly during the first 20 to 25 km and tends to remain relatively constant

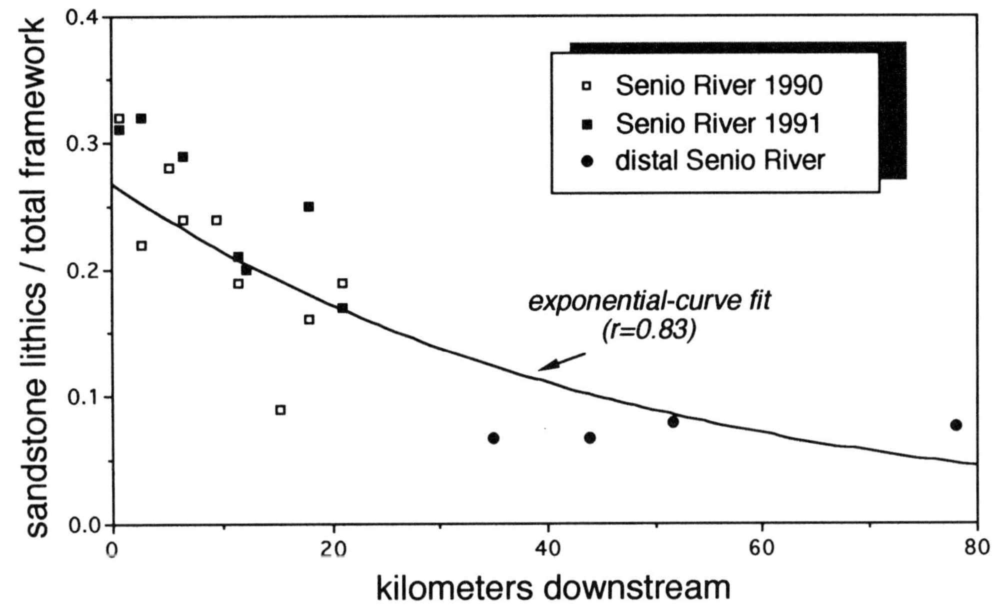

Figure 6. Scattergram showing the marked decrease in the amount of sandstone lithics on the total of the framework grains of sand samples taken along the Senio River. The relative amount of sandstone grains remains virtually constant after the first 30 km of transport.

during the remaining transport (Fig. 6). The lower reaches of the river cut across progressively younger Messinian evaporites and fine-grained Plio-Pleistocene siliciclastic deposits. Nevertheless, no sand-sized detrital contribution from these source rocks has been found in the more distal samples.

SEDIMENTARY RECYCLING

Sedimentary recycling produces different effects on provenance reconstructions based on the study of sandstone petrology, depending on whether the recycled source rock is a mudrock, a sandstone, or a carbonate rock. Mudrocks, unless extremely indurated, have little chance of producing durable sand grains capable of reaching the final site of deposition. Sandstones mostly release quartz, feldspars, and lithic fragments that can be mistaken for first-cycle contributions from igneous or metamorphic source rocks. Carbonate rocks may pass into solution or generate carbonate sands, depending mostly on climatic factors. These considerations lead to the conclusion that some source rocks are systematically overlooked because of their low preservation potential during the sedimentary cycle (e.g., Graham et al., 1986), whereas other rock types yield, if not correctly interpreted, false information on nonexisting source area lithologies. Zuffa (1987) discussed these matters and presented a list of criteria for the identification of multicycle detritus.

The Marnoso-arenacea Formation, being composed of graded beds, provides a wide range of source rock types (sandstone, siltstone, and mudstone), and therefore represents a good starting point for the study of sediment recycling. The various types of sand-sized detritus derived from erosion of specific tur-

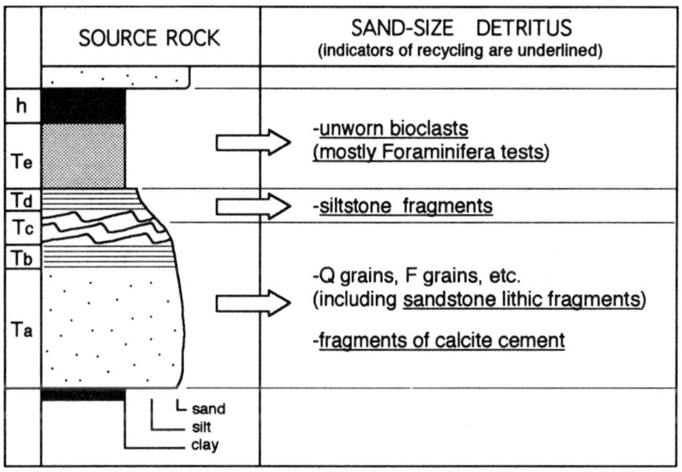

Figure 7. Relationships between sediment source rock and sand-size detritus generated from its erosion in the study area. Note that Bouma's T_e interval of the Marnoso-arenacea turbidite beds and the overlying hemipelagic beds (h) do not contribute sand-sized detritus, except for Foraminifera tests which hence, in the context of a provenance reconstruction, may represent the only indicators of the existence of a fine-grained, mud-producing source lithology.

bidite bed intervals of this unit are shown in Figure 7. Results of this study indicate that, in terms of the specific type of detritus generated, the Marnoso-arenacea beds can be divided into three parts: (1) The sandstone interval generates most of the sand-size detritus. Fortunately, not all generated grains are monocrystalline quartz and feldspar, or lithic fragments derived from crystalline rocks; some will be secondary sandstone rock fragments, thus providing direct evidence of sandstone recycling. Significant amounts of calcite sparite grains, like those derived from the cement of the Marnoso-arenacea sandstones, can also be of general use as indicators of the recycling of calcite-cemented arenites. (2) The siltstone interval releases siltstone lithic fragments, but our study indicates that an undetermined amount of these grains may be lost to pedogenic processes; this suggests that even small amounts of calcite-cemented siltstone lithics in ancient sandstones may indicate the presence of significant quantities of siltstone in the source area. (3) Unabraded fossil tests are the only quantitatively significant sand-sized products of the mudstone intervals; their occurrence in the framework composition of ancient sandstones may indicate the presence of mudrock in the source area.

CONCLUSIONS

Considering the entire Senio River drainage basin and its sediment budget, the results of this study can be summarized as follows:

1. The mudstone component of the Marnoso-arenacea Formation (representing about 30% of the stratigraphic section) does not contribute sandy detritus except for unabraded tests of planktic Foraminifera, which thus represent the only indicators of the existence of this fine-grained lithology in the source area. On the other hand, the sandstone and siltstone intervals are adequately represented in the soil sand-fraction.

2. Pedogenic effects within the upper reaches of the temperate-climate drainage basin of the Senio River are insufficient to affect the soil sand-fraction derived from the Marnoso-arenacea sandstone intervals. The likely pedogenic loss of the unstable sand fraction derived from the siltstone intervals has not been quantified.

3. Marnoso-arenacea-derived sandstone and siltstone rock fragments are the only siliciclastic components of the detritus present along the Senio River which are modified during transport, decreasing progressively downstream, probably because of breakage.

4. Relatively minor annual rainfall variations can affect significantly the composition of the sand fraction of the Senio River and induce the development of a generation of penecontemporaneous carbonate concretions. These grains could be useful paleoclimatic indicators in the study of the rock record, but this study demonstrates their extremely low preservation potential.

5. Direct evidences of sedimentary recycling of the Marnoso-arenacea Formation are the sandstone and siltstone rock fragments, the monocrystalline and polycrystalline calcite grains derived from its cement, and the sand-size fossil tests derived from T_e Bouma divisions of the turbidite beds and from the hemipelagic beds (Fig. 7). Similar criteria can be used as indicators of sandstone recycling in provenance studies of ancient clastic units.

These conclusions confirm that the detritus shed off tectonically active source areas composed of sedimentary rocks in conditions of temperate climate and high/moderate relief reflects the sediment source rock (except for poorly indurated mudstone), and is hardly modified during short transport. If these boundary conditions are met, sandstone detrital modes can be safely utilized for provenance reconstructions of the rock record.

ACKNOWLEDGMENTS

R. L. Folk, E. F. McBride, and R. T. Patterson provided comments on an earlier version of the manuscript. Critical reviews by R. M. Cox, P. G. DeCelles, and M. Johnsson are gratefully acknowledged. G. Gasparotto and G. Valdrè operated the SEM-EDAX. Technical assistance during XRD analysis was provided by F. Tateo. Unpublished climatological data were supplies by ERSA, Servizio Meteorologico Regionale, Bologna. Funding for this project was provided by Consiglio Nazionale delle Ricerche.

REFERENCES CITED

Basu, A., 1976, Petrology of Holocene fluvial sands derived from plutonic source rocks: Implications to provenance interpretations: Journal of Sedimentary Petrology, v. 46, p. 694–709.

Blatt, H., 1967, Provenance determinations and recycling of sediments: Journal of Sedimentary Petrology, v. 37, p. 1031–1044.

Blatt, H., and Jones, R. L., 1975, Proportions of exposed igneous, metamorphic, and sedimentary rocks: Geological Society of America Bulletin, v. 86, p. 1085–1088.

Breyer, J. A., and Bart, H. A., 1978, The composition of fluvial sands in a temperate semi-arid region: Journal of Sedimentary Petrology, v. 48, p. 1311–1320.

Cameron, K. L., and Blatt, H., 1971, Durabilities of sand-size schists and volcanic rock fragments during fluvial transport, Elk Creek, Black Hills, South Dakota: Journal of Sedimentary Petrology, v. 41, p. 565–576.

Casalicchio, G., Giorgi, G., Guermandi, M., Pignone, R., and Vianello, G., 1979, Soil map of Emilia-Romagna: Regione Emilia-Romagna, scale 1:200,000.

Cavazza, W., and Gandolfi, G., 1992, Diagenetic processes along a basin-wide marker bed as a function of burial depth: Journal of Sedimentary Petrology, v. 62, p. 261–272.

Ciccacci, S., Fredi, P., Lupia Palmieri, E., and Pugliese, F., 1980, Contributo dell'analisi geomorfica quantitativa alla valuzione dell'entità dell'erosione nei bacini fluviali: Bollettino Società Geologica Italiana, v. 99, p. 455–516.

Cohen, A. S., 1982, Paleoenvironments of root casts from the Koobi Fora Formation, Kenya: Journal of Sedimentary Petrology, v. 52, p. 401–414.

DeCelles, P. G., and Hertel, F., 1989, Petrology of fluvial sands from the Amazonian foreland basin, Peru and Bolivia: Geological Society of America Bulletin, v. 101, p. 1552–1562.

Dickinson, W. R., 1970, Interpreting detrital modes of graywacke and arkose: Journal of Sedimentary Petrology, v. 40, p. 695–707.

Folk, R. L., 1974, Petrology of sedimentary rocks: Austin, Tx., Hemphill's Bookstore, 182 p.

Fontana, D., McBride, E. F., and Kugler, R., 1986, Diagenesis and porosity evolution of submarine-fan and basin-plain sandstones, Marnoso-arenacea Formation, Northern Apennines, Italy: Canadian Petroleum Geologists Bulletin, v. 34, p. 313–328.

Gandolfi, G., Paganelli, L., and Zuffa, G. G., 1983, Petrology and dispersal pattern in the Marnoso-arenacea Formation (Miocene, Northern Apennines): Journal of Sedimentary Petrology, v. 53, p. 493–507.

Gazzi, P., 1966, Le arenarie del flysch sopracretaceo dell'Appennino modenese; correlazioni con il flysch di Monghidoro: Mineralogica et Petrographica Acta, v. 12, p. 69–97.

Gisotti, G., 1988, Principi di geopedologia: Bologna, Calderini, 278 p.

Graham, S. A., and 14 others, 1986, Provenance modelling as a technique for analyzing source terrane evolution and controls on foreland sedimentation, in Allen, P. A., and Homewood, P., eds., Foreland basins: International Association of Sedimentologists Special Publication 8, p. 425–436.

Grantham, J. H., and Velbel, M. A., 1988, The influence of climate and topography on rock-fragment abundance in modern fluvial sands of the southern Blue Ridge Mountains, North Carolina: Journal of Sedimentary Petrology, v. 58, p. 219–227.

Johnsson, M. J., Stallard, R. F., and Meade, R. H., 1988, First-cycle quartz arenites in the Orinoco river basin, Venezuela and Colombia: Journal of Geology, v. 96, p. 263–277.

Keller, W. D., and Frederickson, A. F., 1952, Role of plants and colloidal acids in the mechanism of weathering: American Journal of Science, v. 250, p. 594–608.

Klappa, C. F., 1980, Rhizoliths in terrestrial carbonates: classification, recognition, genesis and significance: Sedimentology, v. 27, p. 613–629.

Mack, G. H., 1981, Composition of modern stream sand in a humid climate derived from a low-grade metamorphic and sedimentary foreland fold-thrust belt of north Georgia: Journal of Sedimentary Petrology, v. 51, p. 1247–1258.

Mount, J. F., and Cohen, A. S., 1984, Petrology and geochemistry of rhizoliths from Plio-Pleistocene fluvial and marginal lacustrine deposits, East lake Turkana, Kenya: Journal of Sedimentary Petrology, v. 54, p. 263–275.

Ricci Lucchi, F., 1985, Marnoso-arenacea turbidite system, Italy, *in* Bouma, A. H., Normark, W. R., and Barnes, N. E., Submarine fans and related turbidite systems: Berlin, Springer Verlag, p. 209–216.

Schneider, J., 1977, Carbonate construction and decomposition by epilithic and endolithic micro-organisms in salt- and freshwater, *in* Frey, R. W., ed., Fossil algae: Berlin, Springer Verlag, p. 248–260.

Semeniuk, V., and Meagher, T. D., 1981, Calcrete in capillary coastal dunes in southwestern Australia: Journal of Sedimentary Petrology, v. 51, p. 47–68.

Suttner, L. J., Basu, A., and Mack, G. H., 1981, Climate and origin of quartz arenites: Journal of Sedimentary Petrology, v. 51, p. 1235–1246.

U.S. Department of Agriculture, 1975, Soil taxonomy: Washington, D.C., U.S. Government Printing Office.

Zuffa, G. G., 1987, Unravelling hinterland and offshore palaeogeography from deep-water arenites, *in* Leggett, J. K., and Zuffa, G. G., eds., Marine clastic sedimentology: London, Graham and Trotman, p. 39–61.

MANUSCRIPT ACCEPTED BY THE SOCIETY DECEMBER 24, 1992

Printed in U.S.A.

Geological Society of America
Special Paper 284
1993

Variability in sandstone composition as a function of depositional environment in coarse-grained delta systems

Suzanne Kairo* and Lee J. Suttner
Department of Geological Sciences, Indiana University, Bloomington, Indiana 47405
Prodip K. Dutta
Department of Geography and Geology, Indiana State University, Terre Haute, Indiana 47809

ABSTRACT

Differences in framework composition of first-cycle sandstones within coarse-grained delta systems of the Fountain and Minturn Formations (Pennsylvanian) near Colorado Springs and McCoy, Colorado, largely are a function of variable mechanical disaggregation and hydrodynamic sorting characteristic of different depositional environments within the deltas. Modification of composition occurred in spite of deposition in a tectonically active setting where rates of sediment supply and burial were relatively high.

Within a wave-dominated delta in the Fountain Formation near Colorado Springs foreshore sandstones are most mature ($Q_{69}F_{28}R_3$) and offshore/transition sandstones are most immature ($Q_{50}F_{48}R_2$). Differences in maturity reflect shoreline reworking processes. Feldspar is mechanically broken and abraded in the foreshore due to swash/backwash processes. Smaller grains of feldspar are winnowed from the foreshore and transported in suspension to the offshore during storms. The average composition of shoreface sandstone ($Q_{62}F_{34}R_4$) closely resembles the composition of its precursor alluvial sandstone ($Q_{61}F_{35}R_4$), suggesting most shoreface sand was derived directly from the alluvial channels and underwent little or no compositional change. However, the overall variability in the composition of shoreface sandstones is greater than that of alluvial sandstones, suggesting a more complex derivational history. Some samples of shoreface sandstone are enriched in feldspar, presumably derived directly by winnowing from the foreshore; other samples are enriched in quartz, indicating total reworking of coarser mature sand from the foreshore.

The Minturn Formation near McCoy contains facies of river-dominated deltas, fed by both meandering- and braided-river systems. Differences in depositional processes between the two systems caused distinctly different modification of the framework composition of sandstones deposited in the two systems. Sand delivered by meandering-fluvial systems presumably formed under more intense weathering conditions and contains up to 8% fewer rock fragments and as much as 12% more feldspar than braided-fluvial sandstones. Compositions of fluvial sandstones were subsequently modified by marine processes and the compositions of the sandstones in the braided and meandering fluviodeltaic systems diverged even further. The primary variation in composition is reflected by a decrease in abundance of rock fragments, with resulting enrichment of quartz and feldspar. In both systems, beach facies are very similar in

*Present address: Exxon Production Research Company, P.O. Box 2189, Houston, Texas 77252.

Kairo, S., Suttner, L. J., and Dutta, P. K., 1993, Variability in sandstone composition as a function of depositional environment in coarse-grained delta systems, *in* Johnsson, M. J., and Basu, A., eds., Processes Controlling the Composition of Clastic Sediments: Boulder, Colorado, Geological Society of America Special Paper 284.

composition to their parent alluvial sand, suggesting that sand in the Minturn deltas had low residence time in the beach and that the beaches experienced low wave activity. Overall, the marine-influenced facies of meandering-fluviodeltaic origin are more distinct in composition than those of braided-fluviodeltaic origin. Although variation in grain size between facies accounts for most of the observed difference in composition of sandstones of the meandering-fluviodeltaic system, weak correlation between grain size and QFR composition indicates that grain size by itself cannot explain all of the compositional variation in sandstones of the braided-fluviodeltaic system.

Differences in compositional modification between the wave-dominated (Fountain) and river-dominated (Minturn) coarse-grained deltas were due largely to subtle differences in the composition and grain size of alluvial sand brought to the shoreline and in both the *rigor* and *duration* of reworking in the marine environments of the two types of deltas. Largely as a consequence of the latter, the composition of beach sandstones of the Fountain was more distinctly modified relative to other facies in the Fountain. Also, more modification occurred in Fountain beach sandstones than in shoreline sandstones of the Minturn Formation where full-scale development of beaches and their long-term existence were prevented because of the dominance of fluvial processes on the delta plains and delta fronts.

INTRODUCTION

To what extent and in what manner do environmental processes modify the framework composition of sand and sandstone? In this study we attempt to answer this fundamental question through petrographic analysis of facies-related compositional differences within wave- and river-dominated coarse-grained deltas in the Fountain and Minturn Formations (Pennsylvanian) in Colorado.

Coarse-grained deltas (McPherson et al., 1987) are ideal systems in which to study the interrelationship between composition and depositional process because they display high variability and close proximity of depositional subenvironments. Pennsylvanian coarse-grained deltas in the Fountain and Minturn Formations are especially suited for study because the multiple eustatic changes in sea level characteristic of Pennsylvanian time and the tectonically active depositional settings of both formations have resulted in repeated vertical juxtaposition of a variety of alluvial, nearshore, and offshore facies. This permits compositional analysis of samples of the same facies from different stratigraphic levels. Moreover, these coarse-grained deltas consist of first-cycle, compositionally immature detritus capable of maximum response to process-induced change.

Sandstone composition records effects of multiple processes including derivation, transport, deposition, and diagenesis (Suttner, 1974) (Fig. 1). Many previous studies in sandstone petrology have focused on the influence of provenance and transportational history on composition (see references cited in Suttner, 1989, and Johnsson et al., 1991). A number of other studies have established a link between physical attributes of grains, physical processes, and sand or sandstone composition (e.g., Rubey, 1933; Lowright et al., 1972; Odom et al., 1976; Slingerland, 1977; Doyle et al., 1983; Burroughs, 1985), but with the exception of Odom et al. (1976), these studies dealt primarily with heavy minerals and micas. Relationships between depositional environment and sand composition have also been demonstrated (e.g., Shukis and Ethridge, 1975; Davis and Ethridge, 1975; Ethridge, 1977; Mack, 1978; Franzinelli and Potter, 1983; Winn et al., 1984), but only Ethridge (1977), Mack (1978), and Winn et al. (1984) were able to evaluate the effects of long-term mechanical and chemical modification because they analyzed sandstones rather than Holocene sands.

Our study differs from previous studies in that it investigates the effects of depositional environment on composition of immature detritus under conditions of relatively high rates of sediment supply, and it evaluates the relative importance of grain size, hydrodynamics, and original composition of sediment in wave- and river-dominated delta systems, as well as in braided- and meandering-fluviodeltaic settings. A better understanding of facies-related compositional variation in the framework composition of sandstones is important to both theoretical and applied petrologic study. Framework-grain compositions can play an important role in the diagenetic modification of sandstones, which in turn influences the quality of sandstone reservoirs. If framework grain composition varies significantly with facies, these variations should be included in geochemical fluid-flux models and be an integral part of studies that attempt to understand and predict facies-related reservoir quality.

GEOLOGIC SETTING

Early in Pennsylvanian time, Precambrian-cored uplifts developed in central Colorado, New Mexico, and Texas, probably in response to continental collision and suturing in the vicinity of the Ouachita-Marathon orogen, several hundred kilometers to the southeast (Kluth and Coney, 1981). Collectively, these uplifts are referred to as the Ancestral Rocky Mountains (Fig. 2). A complex association of alluvial fan, alluvial plain, and coarse-

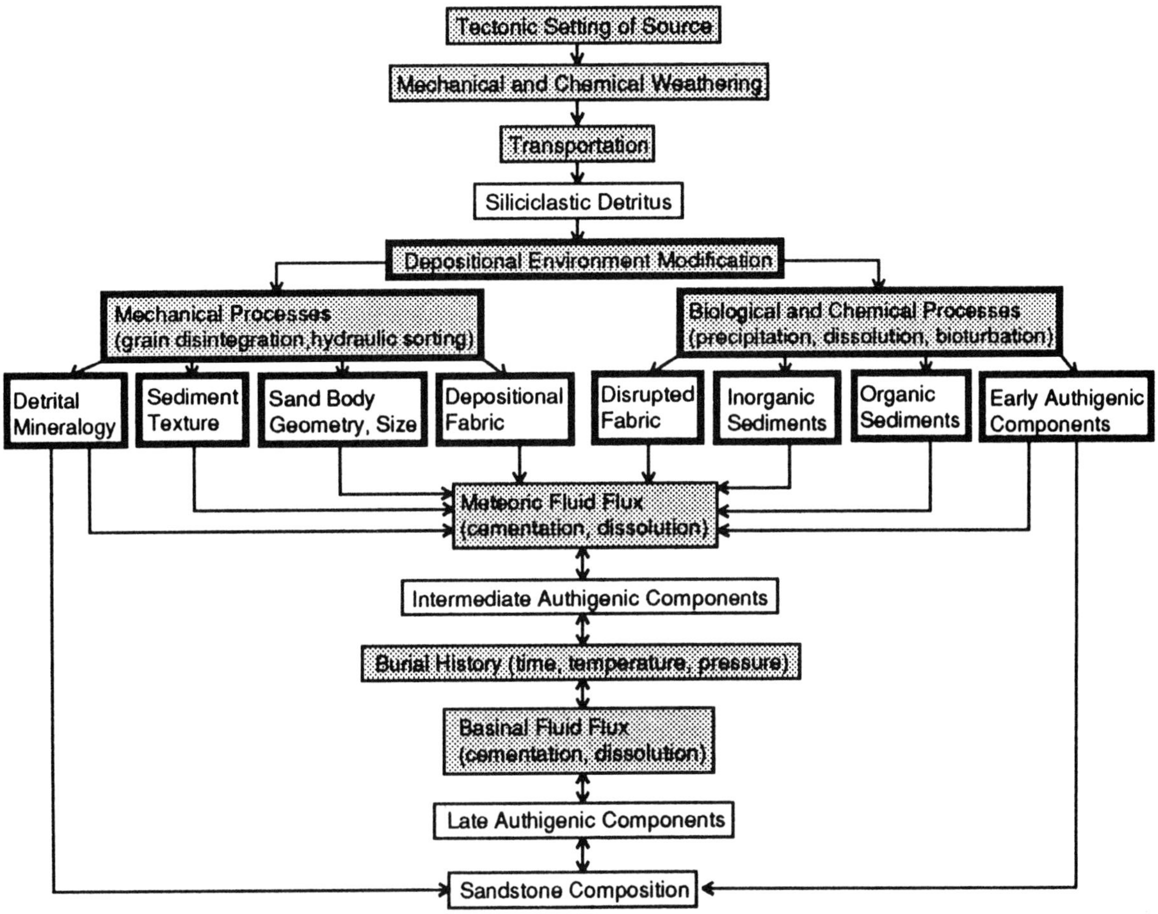

Figure 1. Processes (shaded boxes) and responses (white boxes) in the evolution of a sandstone's composition. Mechanical disintegration, hydraulic sorting, biological processes, and chemical processes are environmentally sensitive and impart a unique signature on detritus and early authigenic cements. Fluid-flux, in turn, may cause dissolution or precipitation of authigenic components in the sandstone. Subsequently, during burial diagenesis, the detrital mineralogy and texture interact with basinal fluid fluxes, and the composition of sandstone is modified by complex reaction-transport feedback processes.

grained delta systems developed along the flanks of the uplifts (Mallory, 1975; Wilson, 1975). During Atokan time, sediments of the Fountain Formation were deposited along the east side of the Ancestral Frontrangia Uplift in the Denver basin. During the Desmoinesian, the Minturn Formation accumulated on the west side of the uplift in the Eagle basin.

The Fountain and Minturn were deposited in relatively similar geological settings. Both formations were deposited under conditions of relatively rapid sediment supply and subsidence at a time when glacio-eustacy was common and widely affected sedimentation patterns. Both formations were studied in close proximity to the ancestral uplifts, and the study localities were chosen so as to maximize the number of marine and nonmarine facies available for study in a small area, while minimizing transport distance. It is unlikely that sediments in the sandstones studied were transported more than 50 km, and were most likely deposited within 25 km of their sources. The source rocks of both formations were Precambrian crystalline rocks of the Ancestral Front Range; the Fountain source rock primarily was coarse-grained granite and granite gneiss, and the provenance for the Minturn included compositionally similar but fine- to medium-grained granite and granodiorite, along with minor schist.

Although the general depositional settings of the Fountain and Minturn were essentially the same, specific environments of deposition were somewhat different. Samples of the Fountain Formation analyzed in this study were collected from outcrops near Colorado Springs. Here the Fountain was deposited in a relatively small ($\approx$200 km^2) embayment on the western margin of the Pennsylvanian interior sea. Paleogeographic reconstructions place this area in a near-equatorial position (Scotese and McKerrow, 1990), presumably in the path of prevailing easterly tradewinds (Mack and Suttner, 1977) (Fig. 2). Facies in the Fountain reflect these conditions in that well-developed wave-modified shorelines are represented in the marine facies (foreshore, shoreface, and offshore), which are intercalated with alluvial facies in the lower part of the formation. Marine and nonmarine units are systematically stacked in a series of progradational facies successions, reflecting shoreline movement asso-

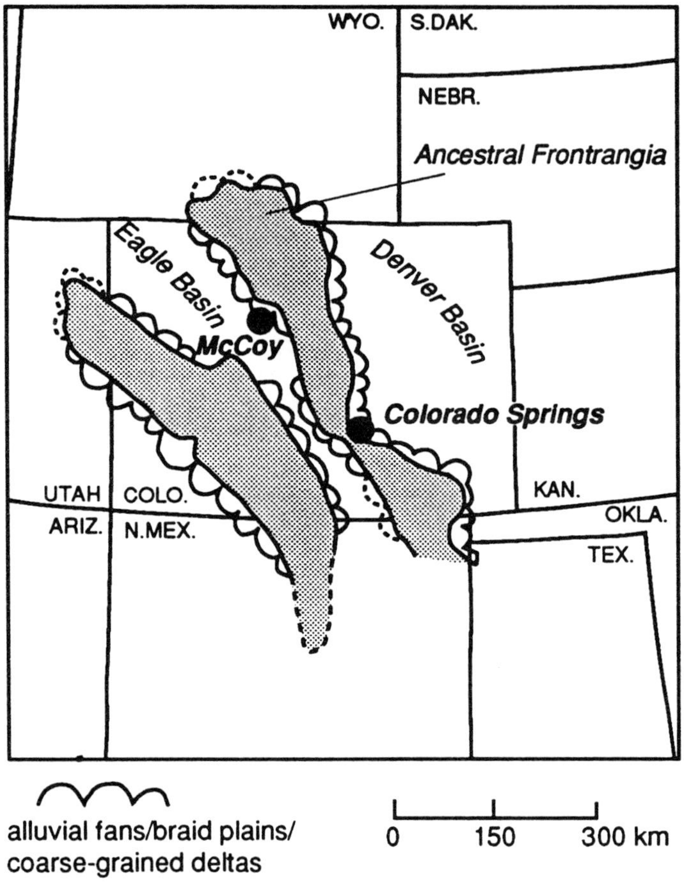

Figure 2. Distribution of the Ancestral Rocky Mountain uplifts and generalized Pennsylvanian paleogeography of Colorado and surrounding states. After Mallory, 1958. Sites of deposition of the Fountain and Minturn Formations were within 10° north of the Permo-Pennsylvanian equator.

ciated with a complex interplay of Pennsylvanian eustacy, variable sediment supply, and basin subsidence.

In contrast to the wave-dominated, coarse-grained delta deposits of the Fountain, coarse-grained delta facies in the Minturn Formation near McCoy, Colorado, reflect deposition in a river-dominated system. The Minturn was deposited on the west side of the Ancestral Frontrangia uplift in a more restricted interior basin effectively shielded from high wave-energy processes (Fig. 2). However, like the Fountain, the fluviodeltaic systems of the Minturn experienced multiple shoreline shifts that resulted in the interbedding of transgressive and regressive deposits of fluvial, marginal marine, shallow-marine, and delta-front settings. Two types of fluvial systems—nonchannelized braided and channelized meandering—fed the Gilbert-type deltas in the Minturn Formation near McCoy.

In summary, both the Fountain and Minturn Formations were deposited in comparable tectonic and geologic settings, and both formations contain immature, arkosic, first-cycle detritus, which was transported essentially the same distances from comparable sources. Consequently, differences in composition of sandstones studied from both units should reflect modification due to environmental processes. Because both formations contain a diverse assemblage of facies representative of a variety of depositional processes, the Fountain and Minturn provide an ideal framework in which to study sandstone composition as a function of physical processes intrinsic to different depositional environments.

SEDIMENTOLOGY

Three types of coarse-grained delta systems are represented in the Fountain and Minturn Formations. The Fountain Formation comprises deposits of a braided, coarse-grained, wave-dominated, fluviodeltaic system. Although the interbedding of alluvial and marine deposits distinguishes the Fountain as a fluviodeltaic system, no distinctive deltaic facies are preserved. Instead, alluvial, foreshore, shoreface, and offshore-transition facies characterize the formation, suggesting that the deltaic shoreline was reworked by wave processes with great efficiency in order to give rise to well-formed beach sequences. In contrast, the Minturn Formation includes deposits of both braided and meandering coarse-grained fluviodeltaic systems, which were dominated by fluvial processes. It differs significantly from the Fountain in that it includes shoreline and transitional deposits closely associated with the construction and reworking of Gilbert-type deltas. Fluvial, shoreline, deltaic, shallow-marine, transitional, and offshore environments characterize the Minturn Formation.

Details of the internal stratigraphy and sedimentology of the Minturn and Fountain Formations are described in the following sections, along with specific aspects of the sedimentary facies considered in this study. In this study we use the term facies to refer to a sedimentologic unit with a specific set of lithologic and stratigraphic characteristics (e.g., grain size, sorting, color, sedimentary structures, bed geometry, and associated rock types). In addition, based on a number of criteria, each facies has an inferred depositional origin.

Fountain Formation

Introduction. Samples of Fountain sandstone analyzed in this study were collected from 19 measured sections in and around Garden of the Gods park near the west edge of Colorado Springs. During Atokan time the embayment of the interior sea that existed in this area was bounded on the south and southwest by highlands elevated along the Ancestral Ute Pass fault. At least 15 progradational cycles (parasequences?) formed during Fountain deposition in response to changes in sea level, and variable rates of both subsidence and sediment delivery from the highlands into the embayment. Our samples come from the uppermost eight cycles (cycles 9-16, Fig. 3) within the measured sections that are located along an onshore-to-offshore (south-north) transect within the ancient embayment. Except for cycles 12 and 13, grain-size trends, paleocurrent data, and facies geometry indicate that the average orientation of the shoreline during Fountain

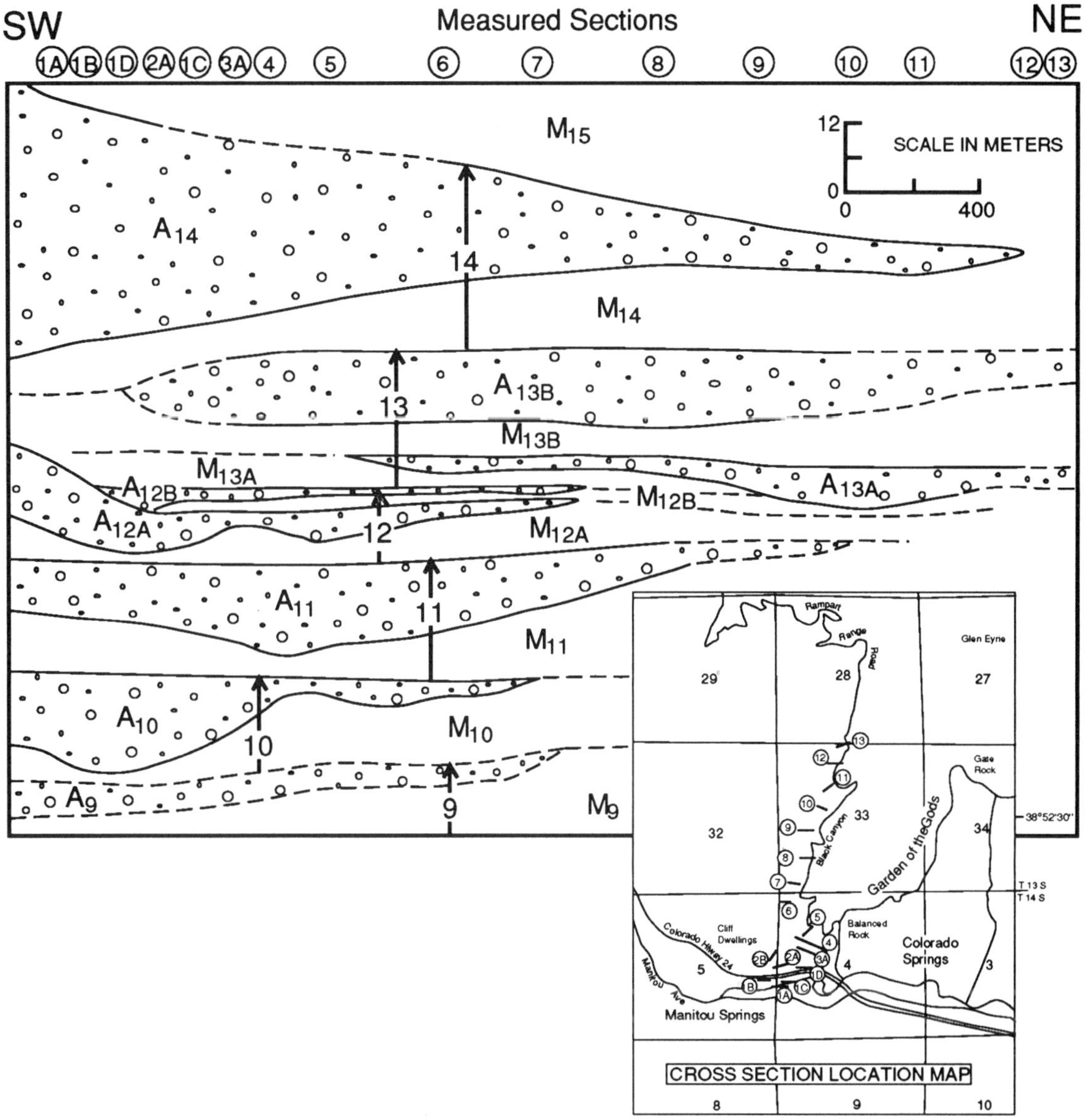

Figure 3. Cross section schematically illustrating the vertical and lateral geometric relations of the upper seven progradational cycles in the Fountain Formation in the study area near Colorado Springs. Numbers across top of cross section correspond to measured section locations, as plotted on the inset map that is taken from the Manitou Springs 7.5-min topographic quadrangle. More specific locations of measured sections are shown in Maples and Suttner (1990, Fig. 3). Letters A and M refer, respectively, to the alluvial and marine portions of each cycle. Alluvial portions of cycles 9, 10, 11, 14, and 15 thicken to the southwest at the expense of the thickness of the associated marine intervals of the cycles. Thickness gradients suggest a southwest-to-northeast onshore-to-offshore environmental trend for these cycles. These geometric relations are obscure or absent for cycles 12A through 13B. This fact, coupled with paleocurrent data and other indicators of shoreline orientation in cycles 12A through 13B, suggests a more west-to-east onshore-to-offshore gradient during cycle deposition.

deposition was about N60°W. Paleocurrents and geometry of facies in cycles 12 and 13 indicate that sediment was delivered from the west to a shoreline oriented approximately N10°W, and abundant cobble conglomerate is repeatedly interbedded in the marine and nonmarine facies of these cycles. The higher percentage of conglomeratic material in the marine facies of cycles 12 and 13, relative to within the marine facies of the other cycles, suggests a more proximal and/or higher relief source for sediment in these cycles. The abundance of conglomerates throughout this interval also suggests that sediment was supplied at a high rate. Both of these factors could have influenced the relative degree of compositional maturation of sand from these cycles.

Detailed description and interpretation of individual facies making up the progradational cycles are given in Maples and Suttner (1990). Summary descriptions and interpretations follow.

Facies description and interpretation. The systematic vertical facies succession, which is typical of an ideal progradational cycle in the Fountain Formation, consists of—from base to top—offshore/transition, shoreface, foreshore, and alluvial facies (Fig. 4). The base of each cycle is a conglomerate formed as a lag on a transgressive erosive surface. The conglomerate is overlain by red, greenish gray, or mottled red and greenish gray bioturbated mudstone interbedded with hummocky, cross-bedded, bioturbated, fine-grained sandstone. The mudstone is interpreted as the product of suspension settling from quiet water in the offshore-transition zone below fair-weather wave base; the fine-grained sandstone was deposited in the same setting, but during waning stages of storms.

Shoreface sandstones, which are gradational with the underlying offshore-transition facies, are poorly sorted, medium to very coarse grained, and planar cross-stratified. The sandstones contain rare grains of glauconite. They are interbedded with laterally continuous pebble to cobble conglomerates. Both the sandstone and conglomerate facies were deposited on the lower part of the beach between mean low-water level and fair-weather wave base. Cross-bedding orientation indicates that shoreface sands predominately were transported by longshore and offshore-directed rip currents. In contrast, the coarse clasts making up the conglomerates were delivered to the lower shoreface by storm-generated waves during the destructive phases of fair-weather–storm beach cycles.

The marine portion of each progradational sequence is capped by 1- to 3-m-thick sets of low-angle (<10°) cross-bedded sandstone representing accretion deposits on the upper beach or foreshore. Stratification in this facies is well defined by inversely graded couplets of laminae made up of coarse and very coarse (or medium and coarse) sand. Such couplets have been observed forming by a process of shear sorting during backwash on the upper parts of modern beaches (Clifton, 1969).

Complete progradational cycles in the Fountain are capped by very poorly sorted and weakly stratified medium- to coarse-grained sandstone, conglomeratic sandstone, and conglomerate. Individual lithofacies are highly lenticular and represent both channel-fill and bar deposits of a braided alluvial system.

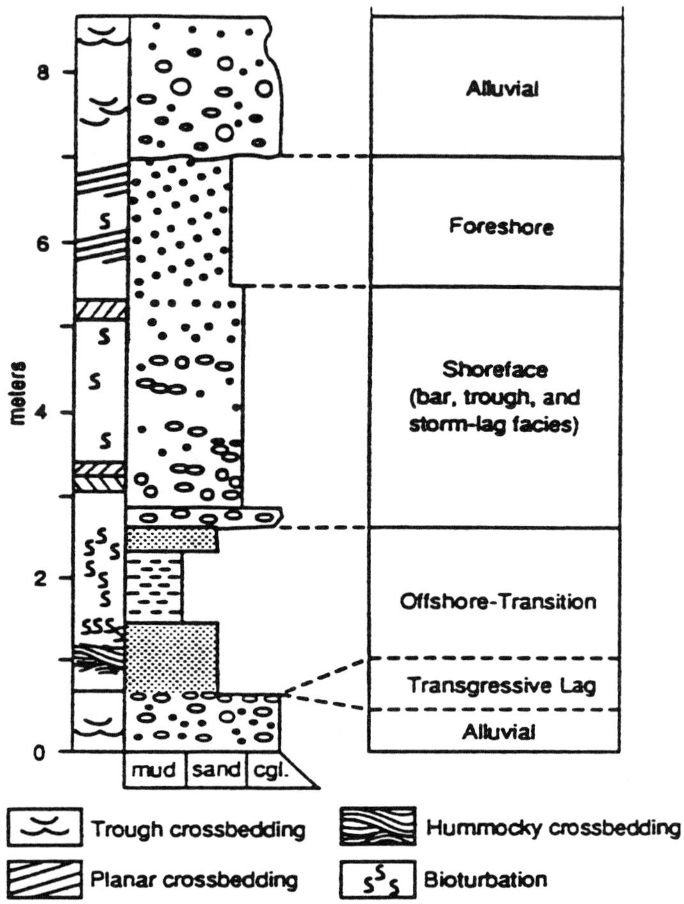

Figure 4. Typical vertical succession of facies generated by progradation during deposition of the Fountain Formation. Sixteen of these progradational cycles, or closely related derivatives of the cycles, have been mapped in the Fountain Formation west of Colorado Springs.

Minturn Formation

Introduction. Samples of Minturn sandstone analyzed in this study were collected from six composite measured sections all located within 5 km of McCoy, Colorado (Fig. 5). Samples span a 200-m-stratigraphic interval consisting of eight sequences (in the sense of Van Wagoner et al., 1990).

Stevens (1958) and Chronic and Stevens (1958) provided the first comprehensive description of stratigraphy and facies within the Minturn Formation near McCoy and provided early interpretations of paleogeography from the paleontology and general lithological characteristics of the formation in that area. Walker and Harms (unpublished field guides) provided a more detailed documentation of facies within Units 7 through 12 of Stevens (1958) and Chronic and Stevens (1958). K. J. Houck (personal and written communications, 1991, 1992) most recently modified and supplemented Chronic and Stevens' work with a detailed study of the sedimentology and stratigraphy of Units 3 through 6 of Stevens (1958) and Chronic and Stevens

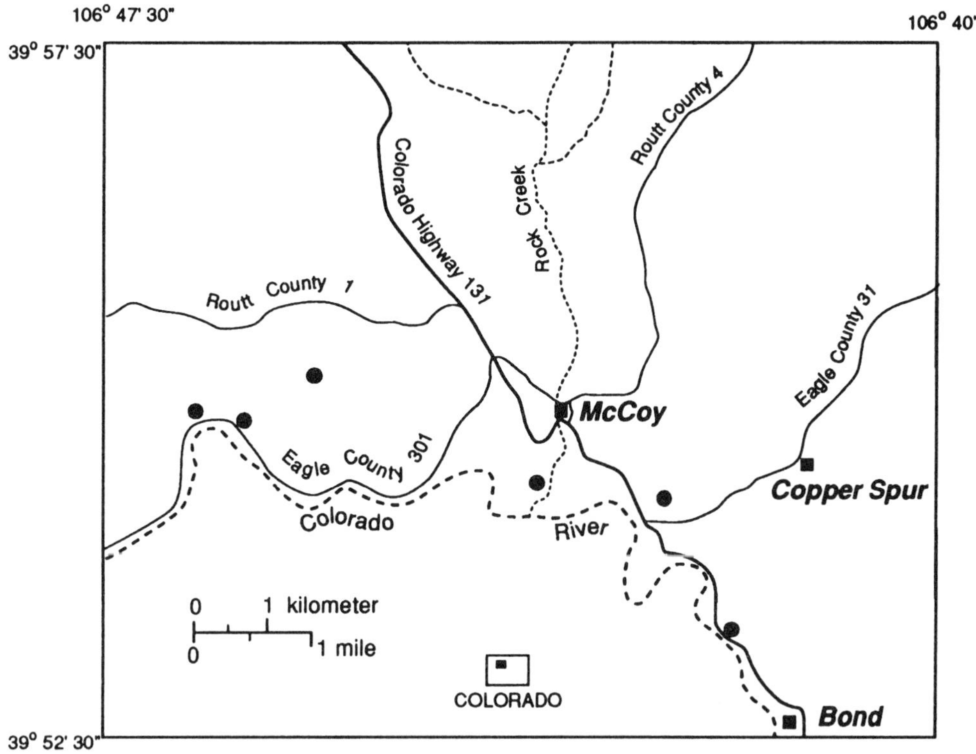

Figure 5. Location map of the Minturn study area near McCoy, Colorado. Dots designate locations of sampled measured sections. Modified from Houck, 1991.

(1958). Units 3 through 6 compose the middle third of the Minturn near McCoy and are the units sampled for this study.

Unlike facies of the Fountain Formation, facies in the Minturn Formation are not arranged in systematic vertical facies successions. Much variety in type and distribution of facies exists, including marine and nonmarine facies associated with both braided-fluviodeltaic (BFD) and meandering-fluviodeltaic (MFD) systems. Represented in the Minturn at McCoy are upper delta-plain, lower delta-plain, delta-slope, prodelta, shoreline, transitional, and shallow-marine environments. The deposits of MFD systems of the Minturn Formation differ significantly from deposits of BFD systems in that MFD systems have substantial organic matter preserved as carbonized imprints. Although deposits of BFD and MFD systems are interstratified on a scale of tens of meters, the two systems likely were not contemporary. This is reflected by the fact that BFD and MFD systems are separated by sequence boundaries (K. J. Houck, personal and written communications, 1991, 1992).

Facies of the MFD systems contain virtually no conglomerate and, overall, contain a greater proportion of fine-grained facies than the BFD systems. Within deposits of both the BFD and MFD systems, facies assemblages are recognized as characteristic of either progradational or retrogradational shorelines; K. J. Houck (personal and written communications, 1991, 1992) suggests that the two show little or no interfingering or facies transitions with each other. Progradational facies assemblages, associated with either transgressions or regressions, are dominated by fluvial or delta-plain facies, and characteristically display lateral facies transitions over distances of tens of meters. Retrogradational facies assemblages are interpreted to signal the onset of regional transgressive events, because they occur as open-marine facies capping progradational deposits over the entire study area. Retrogradational facies assemblages characteristically include limestones with open-marine fauna, and individual beds are significantly more laterally continuous than beds in progradational facies assemblages, with bedsets commonly extending for at least hundreds of meters. Specific characteristics and interpretations of all facies in the interval studied are described by Kairo (1992) and K. J. Houck (personal and written communications, 1991, 1992), and the facies in this study are described below.

Description and interpretation of deposits of braided-fluviodeltaic origin. Compositions of sandstones from both progradational and retrogradational facies of Minturn BFD systems were studied. Sandstones from deposits of *braided-fluvial channels, braided-distributary channels, beach,* and *delta-front sandsheets* were studied as representative of progradational intervals.

Braided-fluvial-channel deposits occur in isolated and amalgamated channelforms of varying size. The braided-fluvial-channel facies includes moderately to poorly sorted, medium to granular sandstones, and minor pebble to cobble conglomerates that are commonly cross-bedded and less commonly planar cross-bedded. Braided-distributary-channel sandstones are very similar to fluvial-channel sandstones in grain size and occurrence, but are distinguished from fluvial sandstones by the presence of interbedded sandy limestones bearing fragments of marine fossils. Also, they commonly occur in smaller channelforms than fluvial-channel sandstones, and are less commonly amalgamated. Beach sandstones are similar to distributary-channel sandstones in texture and occurrence, but are not trough cross-bedded. Rather, they have crudely developed low-angle planar cross-lamination or parallel horizontal lamination, suggestive of wave reworking in a swash zone. Like the distributary channels, beach sandstones are usually associated with sandy marine limestones. Although distributary-channel and beach deposits occur in progradational successions, fossils in the interbedded sandy limestones clearly indicate that the sandstones were deposited in environments exposed to occasional marine influxes. Other marine deposits associated with braided progradational intervals are rare sandsheets, interpreted as marine because they are interbedded with fossiliferous marine limestones. The marine sandsheets considered in this study are moderately sorted, fine- to medium-grained sandstones that occur as parallel-sided, planar-laminated beds or hummocky to swaley cross-stratified beds, all of which are encased in silty shale. This facies probably originated as storms reworked delta-front deposits into sheets or laterally extensive beds.

Sandstones from washover-fan and tidal-channel facies were studied as representatives of retrogradational braided intervals. Washover-fan deposits consist of medium to coarse, moderately to well-sorted sandstones in thin (1 to 2 m) lens-shaped bodies that consist mainly of steep (approximately angle of repose) planar cross-beds. Amalgamated washover-fan bodies stack up to several meters in thickness and extend laterally 100 m or more. Washover fans form when shoreline sediments are reworked from a barrier island and redeposited as a lens or wedge, which progrades landward into the standing water of lagoons.

Tidal-channel sandstones occur in 1- to 2-m-thick, several-meter-wide, channelform bodies and are found above or interbedded with washover-fan sandstones and other lagoonal deposits. In comparison to washover-fan sandstones, they may be of equal or finer grain size, are less well sorted, and have poorly developed trough cross-beds or indistinct bedding. We interpret these to be deposits of the channels through which fluvial effluents, storm surges, or tide waters were drained from back-beach areas.

Description and interpretation of deposits of meandering-fluviodeltaic origin. Sandstone facies representative of progradational successions of Minturn MFD systems include deposits of progradational fluvial and proximal distributary channels, distal distributary channels, progradational marine bars and sandsheets, and turbidites. Meandering-fluvial-channel and proximal distributary-channel facies, considered together here as meandering-fluvial-channel facies, are characterized by moderately to poorly sorted, medium- to coarse-grained sandstones that occur in trough cross-bedded channelforms. Larger channelforms (more than 100 m wide, 3 to 10 m thick) commonly contain lateral-accretion surfaces. Smaller fluvial-channel sand bodies (tens of meters wide) occur as individual channelforms.

Distal distributary-channel sandstones consist of fine- to medium-grained, moderately to well-sorted, crude trough cross-bedded sandstones that contain common to abundant detrital mica and commonly display soft-sediment deformation features. They usually occur in sand bodies less than 1 m thick and tens of meters wide, and are encased in brown, gray, or black shales.

Shallow-marine bars and sandsheets associated with progradational units include two types of deposits: nonreworked delta-front bars and bottomset beds, and storm-reworked shelf and interdistributary sandsheets. Both types of deposits consist of fine- to medium-grained sandstones with abundant detrital mica, and both have sheet-like geometries that extend laterally for many tens of meters. Delta-front sandsheets occur in 5- to 10-cm-thick beds that are usually bioturbated, are rarely parallel laminated, and stack in upward coarsening and thickening bedsets as much as 5 m thick. These sands were probably emplaced when sediments cascaded down Gilbert-type delta foresets and spread out into sheet-like deposits in distal areas in front of the delta, or were deposited as bars at the mouths of distributary channels. Shelf and interdistributary sandsheets occur in 25- to 100-cm-thick hummocky and swaley cross-beds with 1- to 3-cm-thick foresets. We interpret them to have formed when storm currents reworked sediments of the delta platform.

Turbidite facies occur at the bases of progradational successions in the MFD system deposits, but their occurrence is restricted to locations of broad, incised channels associated with underlying sequence boundaries (K. J. Houck, personal and written communications, 1991, 1992). Turbidite sandstones are fine grained with abundant silt-sized grains, and occur as 5- to 10-cm-thick bedsets that fine upward to shales. Beds stack to thicknesses as much as 15 m. Amalgamated turbidite-sandstone bodies can extend laterally for hundreds of meters, but pinch out near the edges of incised channel margins. Turbidite sandstones include planar laminated and rippled beds, with climbing ripples also common. The tops of some turbidite deposits are reworked into microhummocky cross-stratified beds as thick as a few centimeters. Because hummocky cross-bedding is considered to form under combined flow conditions during storms (Nottvedt and Kreisa, 1987), these turbidites are interpreted as prodelta turbidites laid down just above storm-weather wave base.

Beach, marine bar, and marine sandsheet deposits were studied as representatives of retrogradational facies of MFD systems. Beach sandstones consist of moderately to poorly sorted, coarse-grained, and rare medium-grained sandstones occurring in 5- to 100-cm-thick beds that stack in upward thickening successions. Lower beds of successions are structureless and bioturbated. Beds at the tops of successions have crudely developed parallel, subhorizontal to gently inclined laminations. Beach sandstones commonly contain bivalve macrofossils and are interbedded with silty fossiliferous limestones. These beaches were not of open coastline shoreface-foreshore character. Rather, we interpret them to have formed as washover sheets or coarse-grained deltaic deposits that were occasionally subject to swash-zone wave activity.

Shelf or lagoonal bars and sandsheets that occur in retrogradational successions of Minturn MFD deposits are fine-grained, moderately to well-sorted sandstones in less than 1-m-thick parallel-sided beds. Beds are either structureless or laminated and contain abundant detrital mica. We interpret them to have formed when storm- or flood-generated suspended loads were deposited.

PETROGRAPHIC METHODS

Fountain Formation

One hundred fourteen samples from the Fountain Formation were analyzed with the traditional point-counting method in order to determine the composition of framework grains. The Gazzi-Dickinson counting technique was not used because information about rock fragment survivability is lost when discrete minerals in crystalline rock fragments are counted according to this technique (Suttner and Basu, 1985). A counting grid was employed to traverse the thin section, but only the composition of the medium-sand grains that fell on the grid points was counted. The composition of at least 300 medium-sand grains was obtained in most counts, except for samples in which medium-sand grains were rare. For these generally fine- and very fine-grained sandstones, the composition of fine-sand grains was tallied. Original compositions of altered grains were interpreted from relict features and added to the appropriate categories.

Minturn Formation

The compositions of 59 sandstone samples from the Minturn Formation were petrographically determined by modal analysis using the traditional counting technique. For each sample 300 to 400 points were counted on an evenly spaced 1-mm grid. At least 250 grid points fell on framework grains and the balance on cement and pore space. Original compositions of altered grains and pseudomatrix were interpreted whenever possible and added to the appropriate categories.

To better understand the effects of grain-size variations on the composition of samples, for every point intersecting a grain, a Wentworth grain size was recorded. Because larger grains were intersected by a larger number of grid points than smaller grains, the number of points for each grain size does not represent the number of grains in that size fraction. Rather, grain-size information is representative of the proportion of the area of the thin section made up of grains of a particular size. In this way an area-percent grain-size distribution can be calculated and assumed to be similar to a volume-percent grain-size distribution. No grain-size conversion factors were employed.

Discussion of point-count methods

Fountain and Minturn point-count methods were designed for two separate studies with different approaches to the same problem, specifically, how to determine type and amount of variation in composition of sandstone from different facies. Because the pont-count methods were different, Fountain and Minturn data sets are not entirely cross-comparable.

The Fountain point-count procedure *deemphasizes* effects of grain size by studying the variation in composition from only one size fraction (medium sand). The main disadvantage of this method is that the compositional information obtained might not be representative of the overall composition of a sandstone, if the average or most abundant grain size of the sandstone facies of interest is not medium sand. Nonetheless, it provides controlled information on how a particular size fraction of the sand is modified from facies to facies.

Minturn data were collected in a way that would *quantify* the effects of grain size on compositional modification. The particular method used was chosen to obtain information about framework-grain and cement compositions, porosity abundance, and grain-size distributions all in the same point count. We emphasize that the method employed was a point count, not a grain count. The count provides compositional information based on the percentage of the surface of the thin section occupied by grains of a particular size range.

SANDSTONE FRAMEWORK COMPOSITION

Fountain Formation

Description. Fountain sandstones generally are arkosic to subarkosic, and rarely quartz arenitic (Table 1). They fall into two textural categories: poorly to moderately sorted medium- to coarse-grained sandstones and well-sorted fine-grained sandstones. Common framework grain types include monocrystalline quartz, monocrystalline feldspar, polycrystalline quartz, polycrystalline feldspar, and polymineralic plutonic rock fragments. Feldspar types include microcline, orthoclase, and plagioclase. Many feldspar crystals have been partially replaced by calcite or by phyllosilicates, but relict grain boundaries and unaltered portions of grains usually allow identification of the original grain type. Rock fragments generally are composed of quartz and feldspar and may contain crystals of muscovite and biotite mica. Well-rounded glauconite grains and detrital mica flakes are minor components of some of the sandstones.

Fountain sandstones contain, on average, less than 4% rock fragments. Consequently, variation in composition between facies conveniently can be expressed by quartz/feldspar (Q/F) ratios. Maximum compositional immaturity exists in the offshore-transition sandstone, where the average Q/F is approximately 1. The most compositionally mature sandstones are from the foreshore facies and have an average Q/F of 2.5, but some samples are nearly pure quartz arenites. Alluvial and shoreface sandstones

TABLE 1. SUMMARY OF COMPOSITION OF FOUNTAIN AND MINTURN SANDSTONE FACIES

Facies	Number of Samples	Quartz	Feldspar	Rock Fragments*	Average Q:F:R
		% Range			
		I. Fountain Formation			
A. Cycles 9-16					
Alluvial	37	52–75	22–44	0–12	61:35:4
Foreshore	38	51–98	1–47	0–11	69:28:3
Shoreface	36	46–95	2–52	0–18	62:34:4
Offshore/transition	3	43–55	43–56	1–5	50:48:2
B. Cycles 12, 13					
Alluvial	13	54–66	32–44	1–12	59:37:4
Foreshore	11	51–66	31–47	1—7	58:38:4
Shoreface	15	46–73	21–52	0–18	59:37:4
Offshore-transition	2	43-55	44–56	1	49:50:1
C. Cycles 9-11, 14-16					
Alluvial	24	52–75	22–43	0–9	62:34:4
Foreshore	27	51–98	1–47	0–11	74:23:3
Shoreface	20	51–95	2–47	0–11	66:31:3
Offshore-transition	1	52	43	5	52:43:5
		II. Minturn Formation			
A. Braided system					
Fluvial	6	17–30	11–24	49–67	22:18:60
Distributary	8	19–36	15–28	37–63	29:22:49
Beach	3	17–27	22–30	52–54	22:25:53
Marine sandstone (prograd.)	3	28–37	19–30	38–53	33:22:45
Tidal channel	4	24–48	14–32	27–61	36:21:43
Washover fan	6	18–46	13–25	37–66	32:19:49
B. Meandering System					
Fluvial	7	14–25	25–36	43–59	20:30:50
Distributary	5	29–42	32–49	16–37	34:42:24
Marine sandstone (prograd.)	4	22–49	37–13	11–14	35:40:25
Beach	4	16–28	19–41	31–55	24:30:46
Marine sandstone (retro.)	2	33–36	52–55	9–15	35:53:12
Turbidite	4	61–73	20–34	1–22	13:27:10

*Rock Fragments category includes both fine- and coarse-grained crystalline rock fragments and polycrystalline quartz. Percentages are normalized to 100 percent following subtraction of accessory minerals and detrital micas.

have nearly identical mean compositions; the average Q/F for the alluvial sandstones is 1.7; for the shoreface sandstone, the average is 1.8. However, the range in Q/F for the shoreface sandstone substantially exceeds the range for alluvial sandstone.

When the field of variation (Ingersoll and Suczek, 1979) for the mean modal composition for each facies is plotted on a ternary diagram, substantial overlap in the alluvial, shoreface, and foreshore fields exists (Fig. 6A). Offshore-transition samples, however, are compositionally distinct from the other facies. The overlap diminishes if samples from cycles 12 and 13, noted earlier as having an origin different than the other cycles, are not included in the plots (Fig. 6C). When the mean values for samples from each facies from these cycles are plotted by themselves, it is clear that maturation trends evident in samples from the cycles 9 through 11 and 14 through 16 are not present in the samples from cycles 12 and 13 (Fig. 6B). For example, foreshore samples from cycles 9 through 11 and 14 through 16 are significantly enriched in quartz relative to the same facies from cycles 12 and 13 (Fig. 6C). Unfortunately, only two samples of offshore sandstone from cycles 12 and 13 were analyzed for this study, so cross-comparison for the offshore facies from the different cycles cannot be made. However, overall, marine sandstones from cycles 12 and 13 are virtually identical in composition to the alluvial sandstones of the same age (Fig. 6B).

Interpretation. Fundamental to the validity of interpretation of changes in composition of Fountain sandstone as a function of environmental processes is that the alluvial sandstones in all cycles are compositionally uniform. This is, in fact, true (Fig. 6). Accordingly, departures in the compositions of marine sandstones from the composition of alluvial sandstones are not due to

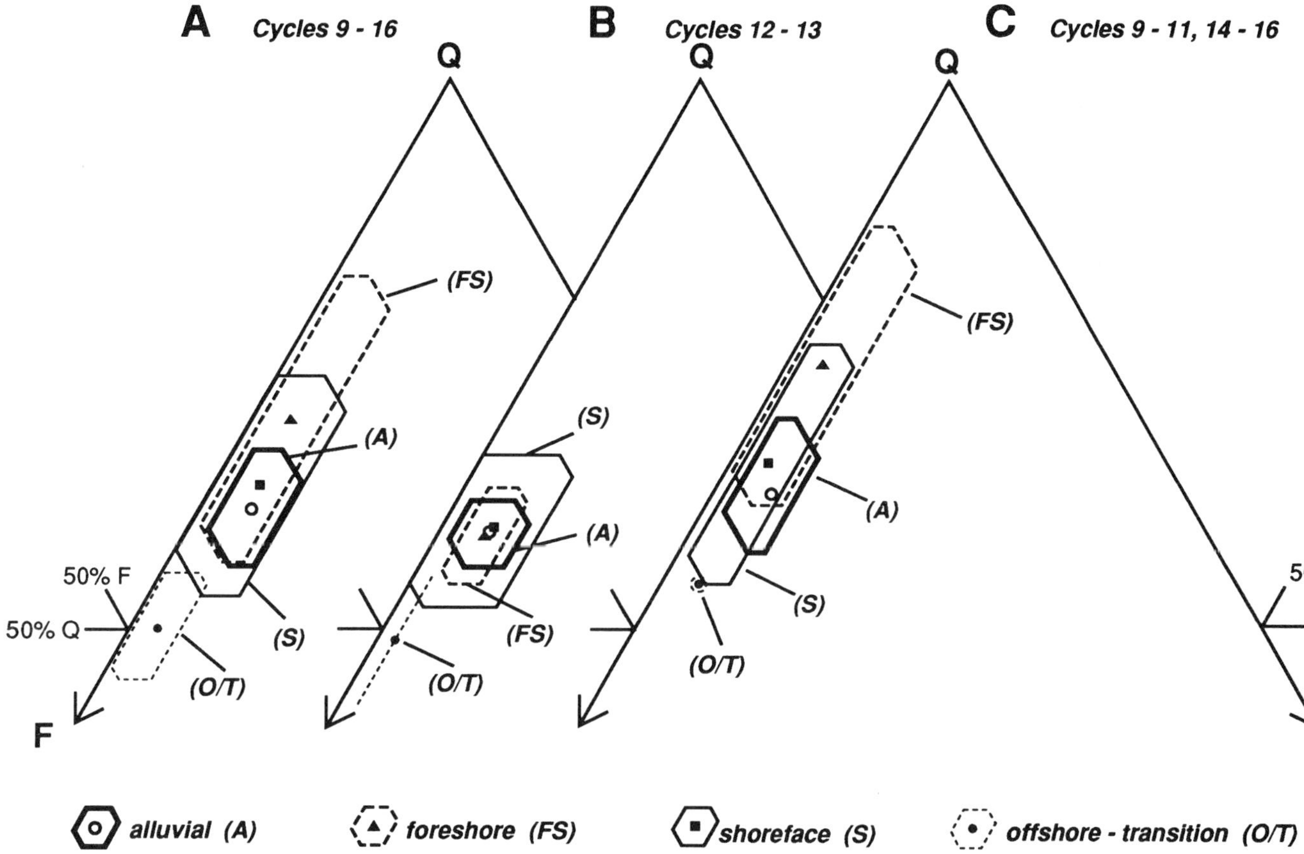

Figure 6. Ternary diagrams showing the average content of quartz (Q), feldspar (F), and rock fragment (R) in Fountain sandstone facies. A, Values for samples from all cycles. B, Values for samples from cycles 12 and 13. C, values for cycles 9 through 11 and 14 through 16. On these and subsequent diagrams, the hexagon symbols represent fields of variation calculated using the technique of Ingersoll and Suczek (1979).

variation through time and space in composition of sand delivered by rivers to the shoreline. Instead, these departures must be caused by variable degrees of winnowing and enrichment of quartz and destruction of feldspar in the different shoreline and offshore hydrodynamic regimes. This relationship demonstrates that environmental processes are effective in modifying the composition of sand even in coarse-grained delta settings characterized by relatively high rates of sediment supply.

Overall, no significant net loss in the number of feldspar grains within the marine system is apparent. In other words, the depletion of feldspar in one environment (e.g., foreshore) is balanced by the enrichment somewhere else (e.g., offshore-transition environment). This suggests that chemical dissolution of feldspar is an insignificant process. It also implies that change in the quartz/feldspar ratio caused by a particular process in one hydrodynamic zone likely will bring about change in some other zone as sand is reworked from one environment to another.

The influence of modification of sand composition in one environment on the composition of sand elsewhere in the system is manifested most clearly in the foreshore and offshore-transition facies. On most beaches, maximum cumulative destructional energy should be focused in the foreshore or upper beach where continuous swash/backwash effectively breaks feldspar down into finer sizes, thereby permitting it to be winnowed and carried away. This would account for the relative depletion of feldspar in the foreshore facies. Some of this feldspar must be moved into the shoreface, in order to account for minor enrichment of feldspar in the shoreface relative to the precursor alluvial sand, but most must bypass the shoreface, probably in suspension, ultimately to be deposited below fair-weather wave base in the offshore-transition zone. Storm-related mechanisms for the transport of sand eroded from beaches into deeper water environments have been proposed by Hayes (1967).

Causes of modification of the composition of sand in the shoreface are more varied than in the foreshore and offshore. One might conclude that the overall similarity in composition of the alluvial and shoreface sand indicates that the shoreface sand was derived directly from the alluvial system and experienced little subsequent modification. However, closer inspection of the distribution of individual sample compositions in Figure 7A reveals that some shoreface samples are enriched, but others depleted in feldspar with respect to the alluvial precursor sand. This, of

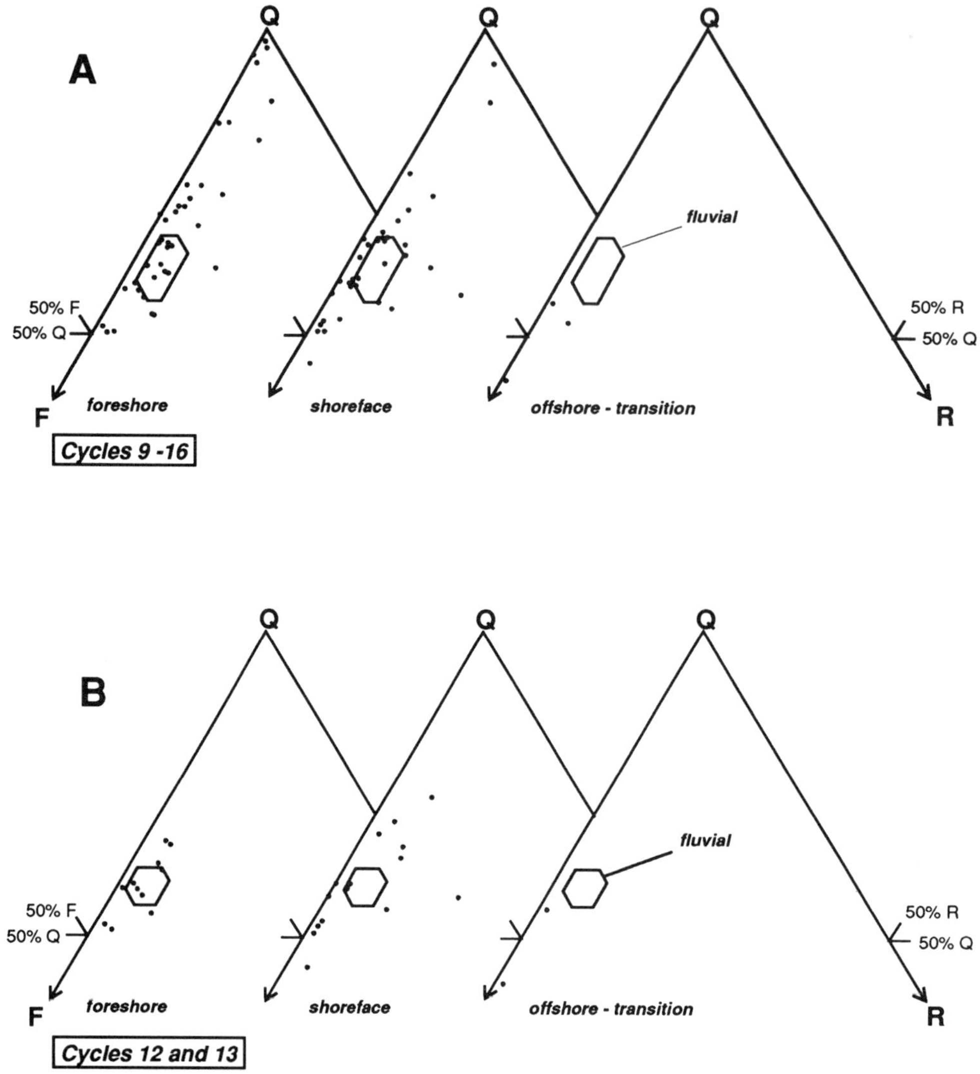

Figure 7. Ternary diagrams showing the composition of individual samples of Fountain marine-influenced sandstone facies in cycles 9-16 (A), and cycles 12 and 13 (B). Field of variation shown is based on composition of associated alluvial samples from the same cycles. Showing the alluvial field of variation permits visual evaluation of how marine sandstones differ in composition from their alluvial precursors.

course, accounts for the enalrged field of variation for the shoreface samples relative to the alluvial field (Fig. 6). It also indicates that some shoreface sand must be reworked in bulk directly from the more mature foreshore sand, but under different conditions some shoreface sand must receive feldspar that was winnowed from the foreshore.

Many of the shoreface samples have compositions nearly identical to the parent alluvial sand. Although this does not necessarily require that all or most of the shoreface sand passed directly from the alluvial feeder channels into the shoreface zone with no compositional modification, this is the most simple explanation of the observation. Theoretically, all of the shoreface sand could have been reworked from the foreshore, but a delicate balance in the shoreface between the processes of reworking of quartz-rich foreshore sand and feldspar enrichment could have restored the shoreface sand to the original composition it had when it left the feeder channel. However, maintaining this balance through several cycles of shoreface development would be highly fortuitous. Therefore, we conclude that substantial shoreface sand must have undergone little compositional change between the time it left the mouths of its feeder channels and its ultimate burial.

Intuitively, one would expect that the magnitude of compositional modification sand experiences in the shallow-marine regime should be directly related to both the amount of wave/tide energy acting on the sand per unit time and the total length of time the sand resides at or near the sediment-water interface. The latter should be a function of rate of sediment supply and accumulation. The similarity in composition of alluvial sandstones in

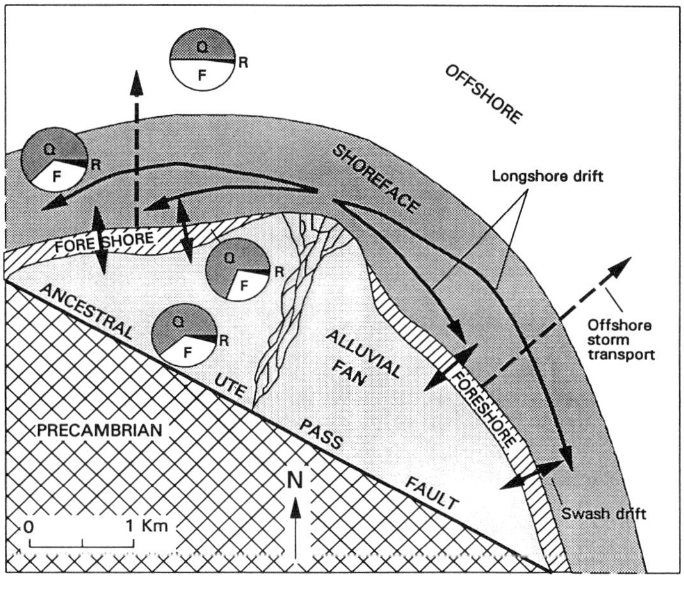

Figure 8. Schematic representation of processes affecting sandstone composition in the Fountain Formation. Pie diagrams illustrate variation in composition as a function of environment. Sediment was derived from a Precambrian source terrane located southwest of the ancestral Ute Pass Fault and transported northeastward across a broad fan or alluvial plain. Sand initially bypassed the foreshore enroute to a shoreface zone. Later it was reworked by swash processes in the foreshore. Feldspar, winnowed from foreshore sand, was transported into the offshore-transition environment during storms. These processes enriched the foreshore sand in quartz and the offshore-transition sand in feldspar.

cycles 12 and 13 (Fig. 7B) implies that sand in cycles 12 and 13 underwent insignificant compositional change between the time it was derived from the feeder rivers and the time it ultimately was buried. As noted earlier, the geometry of these cycles, facies and paleocurrent trends, and grain size, all indicate that sediment in cycles 12 and 13 was derived from a more proximal if not higher relief source. Consequently, both rates of sediment supply and burial likely were greater during deposition of these cycles and the sand did not experience a long-enough period of mechanical reworking to cause alteration of its original composition. We are puzzled by the fact that a high percentage of shoreface samples from cycles 12 and 13 plot outside of their associated alluvial field (Fig. 7B), but perhaps this also is reflecting the relatively complicated history of derivation of shoreface sand as discussed above.

Figure 8 schematically illustrates and summarizes the above discussion of effects of hydrodynamic processes on grain survivability in the wave-dominated coarse-grained delta of the Fountain Formation. Longshore currents transport sand from the river mouths to beaches located downcurrent. Modification of sand composition during this longshore drift was not significant, as is evident from the near-similarity in composition of the shoreface sand and its precursor alluvial sand. However, at some time sand moved by longshore currents must have been transported landward onto the upper beach face by wave-induced currents associated with the shoaling breaker and surf zones. Once this sand resided in the foreshore, significant maturation occurred and medium- to coarse-sand sized feldspar grains were mechanically broken down during swash/backwash movement. Subsequent reworking of subordinate amounts of both quartz-rich and feldspar-rich sand into the shoreface by high-amplitude waves during storms would explain the rare quartz- and feldspar-enriched sandstones of shoreface origin.

The relative enrichment of feldspar in the offshore-transition facies partly is an artifact of grain size, because this facies is dominantly silt to fine sand and it has long been known that feldspar is concentrated in these fine grain sizes (Odom et al., 1976). However, the enrichment was enhanced through winnowing of smaller feldspar grains from the shoreface and foreshore and transport of these grains in suspension into the offshore-transition setting during or shortly after major storms (Fig. 8).

In spite of the fact that sand in the Fountain system initially lacked sizeable amounts of rock fragments, thereby limiting compositional change to the quartz-feldspar grain population, a process-induced compositional signature, albeit simple in character, was imparted to the sediment.

MINTURN FORMATION

Description. Average Minturn sandstone is feldspathic and medium to coarse grained. The most common grain types include monocrystalline and polycrystalline quartz, monocrystalline and polycrystalline feldspar, and plutonic rock fragments made up of quartz, feldspar, and minor mica. Both sodium-rich plagioclase and potassium feldspars are present. Less common grain types include quartz-mica-schist fragments, chlorite-schist fragments, and fragments of carbonate fossils. Present in varying proportions are grains altered to phyllosilicates and iron oxide, as well as grains of detrital mica, including muscovite, biotite, and chlorite.

Average compositions of all facies studied in the Minturn are summarized in Table 1 and are shown in the ternary diagram in Figure 9. Similarities and differences between the compositions of various facies are reflected in the quartz (Q), feldspar (F), and rock fragment (R) abundances of the facies as normalized to (Q + F + R) and are illustrated on a ternary diagram (Fig. 9). Comparison of the average relative abundances of Q, F, and R in Minturn sandstones illustrates that sandstones from environments associated with the BFD systems form a separate population from sandstones associated with the MFD systems (Fig. 9). The difference is defined by the ratio of feldspar to rock fragments (F/R) and the ratio of quartz to feldspar (Q/F) in sandstones of the two types of depositional systems.

Among sandstones of both the BFD and MFD depositional systems, fluvial sandstones contain the largest percentage of rock fragments of all facies in the systems. All other facies, on average, have either more quartz, more feldspar, or more of both of these components than fluvial sandstones of their respective systems. In the deposits of BFD systems, the average composition of beach sandstones most closely approximates the average composition of the fluvial-channel facies, with an equal amount of quartz, 7%

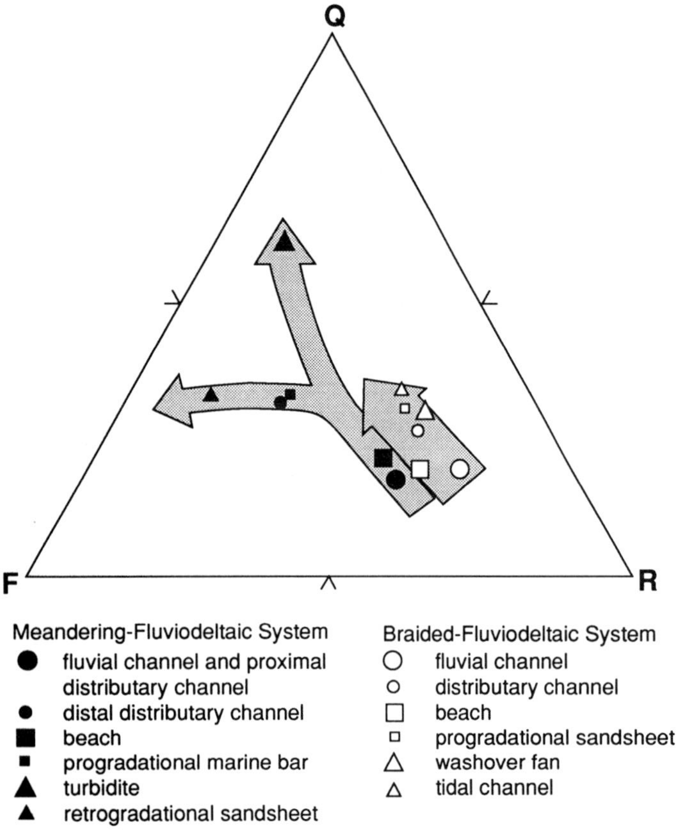

Figure 9. Ternary plot of the average quartz (Q) feldspar (F), and rock fragment (R) percentages for all facies studied in both the braided- (open symbols) and meandering- (solid symbols) fluviodeltaic systems of the Minturn Formation. Broad shaded arrows schematically represent paths of compositional maturation for sandstone from the two systems.

more feldspar, and 7% fewer rock fragments than fluvial-channel sandstones. All other sandstone facies in the deposits of BFD systems contain an average of 7 to 14% more quartz than braided-fluvial sandstones. These other facies have only 4% more feldspar than the fluvial facies but 11 to 17% fewer rock fragments. Almost all marine and shoreline facies of MFD systems contain more quartz and feldspar but fewer rock fragments than sandstones of the meandering-fluvial channel/proximal distributary-channel facies of that system. The only exceptions are the turbidite facies, which average 3% less feldspar than the meandering-fluvial channel facies, and the beach facies, which has an average feldspar composition identical to that of the fluvial-channel facies. As in the case of beach deposits of the BFD systems, the average composition of sandstones from the beach facies of the MFD systems is closest to the average composition of the fluvial-channel facies. Distal distributary-channel sandstones, on the average, have 14% more quartz, 12% more feldspar, and 26% fewer rock fragments than the meandering-fluvial channel/proximal distributary-channel facies.

Progradational marine bars and sandsheets deposited by MFD systems have essentially the same average composition as distal distributary sandstones, whereas transgressive-marine sandsheets have the same amount of quartz, but 11% more feldspar and 12% fewer rock fragments than distal-distributary channel facies. The average composition of transgressive-marine sandsheets is considerably different than the average fluvial composition, with 15% more quartz, 23% more feldspar, and 38% fewer rock fragments than the fluvial-channel facies. Turbidite sandstones also have a considerably different average composition than the average fluvial-channel sandstone, with 43% more quartz, 3% less feldspar, and 40% fewer rock fragments than the fluvial-channel facies.

When the field of variation (Ingersoll and Suczek, 1979) for the mean modal composition for each facies is plotted on a ternary diagram, the differences in composition among facies of the BFD systems are not as distinct as the differences in composition among facies of MFD systems (Fig. 10). The fields of variation for facies from MFD systems show some overlap (Fig. 10A), but compositions of several facies are different enough to plot as nonoverlapping fields. Facies of BFD systems show substantial overlap (Fig. 10B), implying that a particular composition could be representative of several different facies.

Interpretation of the differences in the composition of sandstones from braided- and meandering-fluviodeltaic systems. Perhaps the most striking aspect of Minturn sandstone composition is that the average compositions of sandstone facies of the BFD system plot as a separate population from the average compositions of sandstone facies of the MFD systems as defined by the ranges in the F/R and Q/F values of the two populations of samples. This is best exemplified by the nonoverlapping fields of variation for the composition of sandstones deposited in braided- and meandering-fluvial channels (Fig. 11).

Differences in composition between facies in the BFD and MFD systems likely do not reflect changes through time in the composition of source rocks that provided sands for the two systems. Even though deposits associated with the BFD and MFD depositional systems occur at different stratigraphic levels, deposits of the two types of systems are repeatedly interbedded within the formation. Regardless of their stratigraphic position, BFD and MFD sandstones plot as two separate populations on QFR diagrams. Moreover, there is no apparent difference in composition of sand-sized rock fragments and pebbles in sandstones from the BFD and MFD systems, suggesting that the sources of sand for both systems were similar. Also, rock fragments in sandstones from both systems are primarily fine- to medium-grained granitoids (crystal diameters, <5.00 mm). Plutonic rocks in the source area may have been made up of numerous intrusions of variable composition, but a source area consisting primarily of granite-like rocks probably would not give rise to two distinctly different populations of framework grains repeatedly through time by simple erosional unroofing.

Another explanation for the difference in composition of sandstones associated with braided- and meandering-fluvial channels may be related to the grain size of sediment carried by the two systems. It is well established that sand and sandstone composition varies with grain size; feldspar is enriched in sands

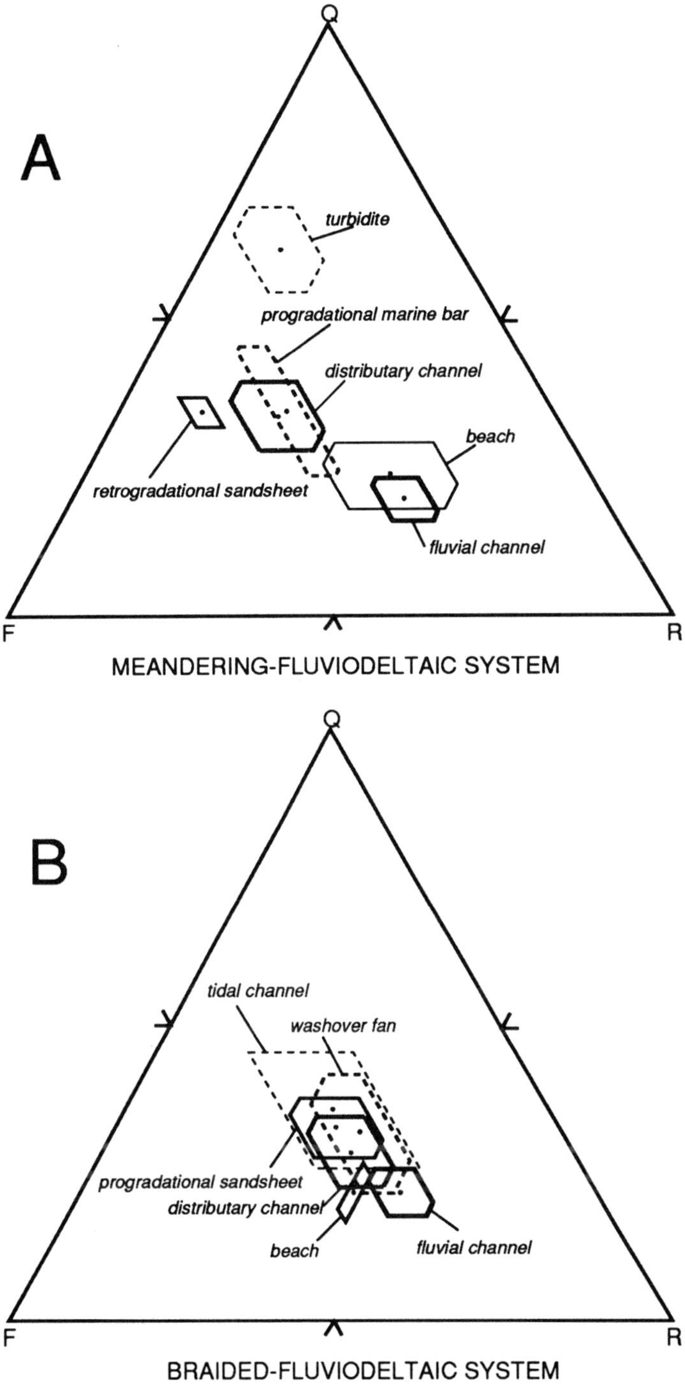

Figure 10. Ternary plots of quartz, feldspar, and rock fragment percentages for sandstone facies from the meandering- (A) and braided- (B) fluviodeltaic systems of the Minturn Formation. Dots mark the average quartz, feldspar, and rock fragment percentages for each facies.

and sandstones of dominantly fine grain size and rock fragments are enriched in coarse-sand-sized detritus (e.g., Allen, 1962, Hunter, 1967; Okada, 1966, Basu, 1976, Odom and others, 1976). Deposits of BFD and associated marine systems generally have a greater number of coarser grained facies than MFD and associated marine system deposits, which have several fine- to

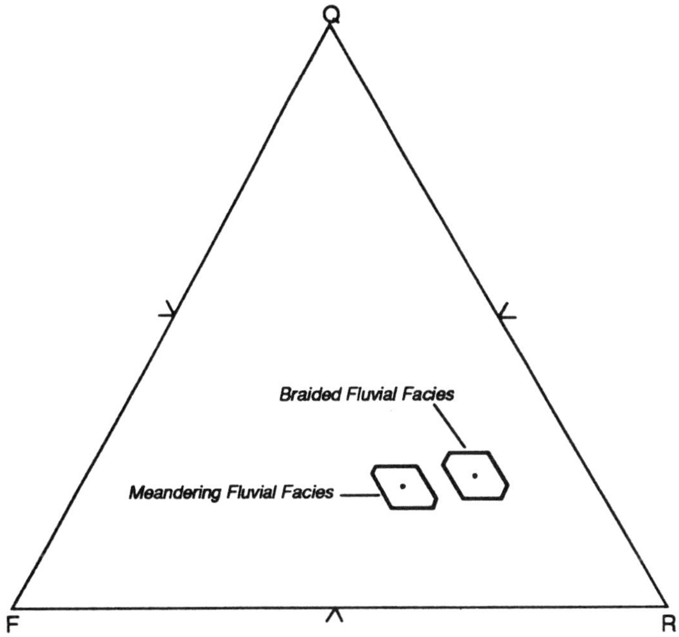

Figure 11. Ternary diagram of the average QFR composition of sandstones deposited in meandering- and braided-fluvial channels.

medium-grained sandstone facies. Grain size might explain the distinctly lower F/R ratio of braided-fluvial sandstones and the distinctly higher Q/F ratio of meandering-fluvial sandstones. To test this possibility, F/R and Q/F ratios of sandstones from the fluvial facies of both types of systems were compared to the grain size of the sample as represented by the percentage of the sample composed of grains greater than or equal to coarse sand (Fig. 12). As expected, in sandstones of both braided- and meandering-fluvial systems, the F/R ratio generally decreases with grain size. However, F/R ratios of samples from braided-fluvial-channel deposits are always greater than the F/R ratios of samples of similar grain size (i.e., containing similar percentages of grains larger than or equal to coarse sand size) from meandering fluvial channel deposits. The cross-plot of Q/F ratios with grain size in Figure 13 shows that the Q/F ratios for braided-fluvial sandstones also are consistently greater than those for meandering-fluvial sandstones. In addition, Q/F is not covariant with grain size in either braided-fluvial-channel or meandering-fluvial-channel sandstones. Both of these observations suggest that differences in composition between the two populations of sandstones are not a result of the grain-size difference between the two systems.

Because climate influences composition of sediment derived from parent rocks, climatic fluctuations may have been the cause of the difference between compositions of sandstones in the two types of depositional systems. Based on palynology, climate in parts of the Ancestral Rockies during the Desmoinesian fluctuated between cool-dry and warm-wet conditions (Rueger, 1984). This climatic fluctuation may have been the driving force behind alternate development of braided-fluvial systems with associated

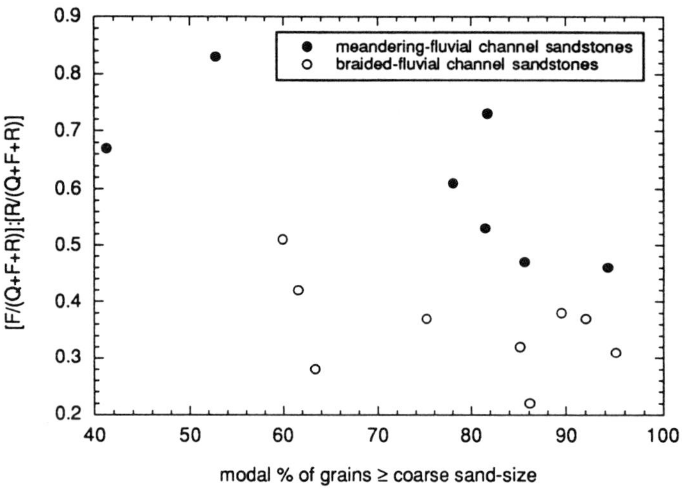

Figure 12. Plot of the ratio of feldspar to rock fragments (both normalized to Q + F + R) versus the modal percentage of grains greater than or equal to coarse sand-size in braided- and meandering-fluvial channel sandstones.

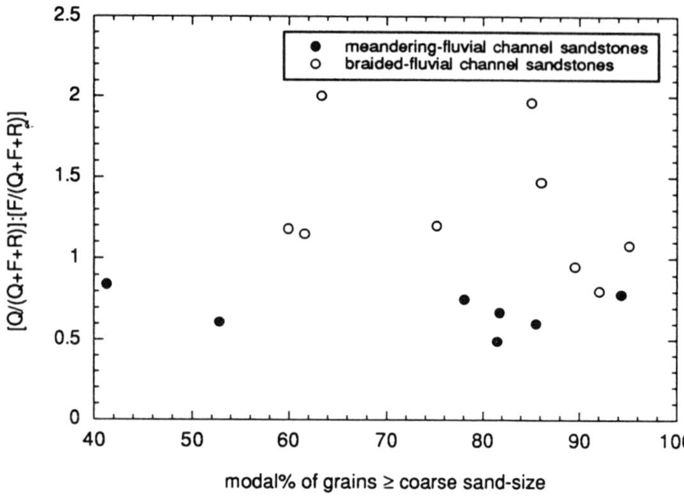

Figure 13. Plot of the ratio of quartz to feldspar (both normalized to Q + F + R) versus the modal percentage of grains greater than or equal to coarse sand-size in braided- and meandering-fluvial channel sandstones.

limestones (semi-arid to seasonally wet climate) and meandering-fluvial systems with associated black, coaly shales (warm-humid climate). Fluctuations in climate could have caused repeated variations in the kinds of sediments produced from source rocks. For example, Basu (1976) demonstrated that composition of sand derived from weathering of granitic plutonic rocks in a humid climate will differ from that of sand produced from granites in a semi-arid climate. He found that in humid-temperate climates, rock fragments disaggregated into monocrystalline grains of their component minerals so that rock fragments were more abundant in sediments produced in semi-arid climates. This would explain enrichment of rock fragments in Minturn sandstones associated with the braided-fluviodeltaic systems. However, contrary to the predictions from Basu's model, quartz is not enriched in facies associated with the MFD systems, which were most likely deposited in a relatively wet climate. The absence of quartz enrichment can be explained if chemical weathering of source rocks was transport limited (Johnsson, 1992; Johnsson et al., 1988). In a transport-limited system, the degree of chemical weathering of detritus is limited by rates of erosion (e.g., faster rates of erosion, less chemical weathering). High relative rates of runoff and high relief also increase the abundance of rock fragments introduced to detritus (Grantham and Velbel, 1988). If, in the active tectonic setting of the Ancestral Rocky Mountains, chemical weathering of crystalline rocks was transport limited, weathering may have been adequate to disaggregate rock fragments along crystal boundaries, thereby producing more monomineralic grains. But, weathering may not have been intense enough to remove labile minerals by dissolution. Given the granitic to granodioritic composition of source rocks, simple disaggregation of rock fragments along crystal boundaries would produce monomineralic feldspar in amounts greater than or equal to the quartz produced, providing a simple means for enrichment of feldspar in MFD facies relative to BFD facies. Even though climate may have been an important factor in modifying the composition of Minturn sandstones, the effect of depositional environment is superimposed on the compositional differences present among sandstones of the MFD and BFD systems. This is evident from differences between the compositions of various sedimentary facies within the MFD and BFD systems. The ternary plots in Figures 9 and 10 illustrate that, on the average, the composition of sandstones of marine and shoreline facies in both MFD and BFD systems are, to greater or lesser degree, modified from the composition of their associated fluvial facies. The compositions evolve away from the rock fragments pole and away from the compositions of fluvial sandstones along trends indicated by the shaded arrows in Figure 9. All modification is represented by a decrease in the amount of rock fragments, but whether this decrease is compensated for by an increase in quartz, feldspar, or both, varies from system to system and from facies to facies. Presumably, sandstones deposited in shoreline and marine subenvironments of the Minturn fluviodeltaic systems were ultimately delivered to the shoreline via fluvial channels, so the trends in compositional variation must reflect the effects of processes within the depositional environment. The differences in the degree of compositional modification that occurred in MFD and BFD depositional environments may reflect differences in processes of transport and deposition.

Interpretation of variations in compositions of meandering-fluviodeltaic facies. Apparently grain-size sorting by processes of transportation and deposition exerts an influence on the composition of sandstones deposited in MFD systems. As expected, the finer grained sedimentary deposits (turbidites, retrogradational sandsheets, shallow-marine bars and sandsheets, and distal distributary-channels) are depleted in rock fragments relative to the coarser grained sedimentary facies (meandering-fluvial channel and beach sandstones). This variation in framework grain composition probably resulted from processes that were active in the Minturn depositional environments. As rock fragments passed through the meandering-fluvial channel and shoreline settings, they would have become disaggregated into monocrystalline grains composed of the various components of the rock fragments. These finer grained monomineralic grains would have in turn become concentrated in finer grained facies by processes of grain-size sorting. The results of grain disaggregation and grain-size sorting are exemplified by the compositions of retrogradational shallow-marine sandsheets and turbidites, which are the most fine-grained facies of MFD systems and also the most modified from the compositions of fluvial sandstones.

Sandstones in the retrogradational shallow-marine sandsheets of the MFD systems contain abundant silt and fine sand and, overall, are much finer grained than most other facies from the same system. As expected because of their finer grain size, these sandstones are depleted in rock fragments with respect to other facies. The enrichment of feldspar in the fine sand- and silt-sized fractions of sandstones was demonstrated by Odom et al. (1976), who proposed that it reflected the tendency of feldspars to cleave into finer grain sizes during mechanical destruction. Such destruction of grains in the fluvial and shoreline environments could result in enrichment of very fine feldspar grains. This process of feldspar destruction might be enhanced in transgressive deposits because previously deposited grains would undergo a second stage of mechanical destruction as sediments of the shoreline and fluvial system were reworked by transgressing seas. Simple size sorting of very fine sand by marine processes could concentrate the very fine feldspars in MFD shallow-marine facies in the same fashion, as suggested for enrichment of feldspar in offshore-transition facies of the Fountain Formation.

Turbidite sandstones of Minturn MFD systems also are fine grained and, as expected, contain few to no rock fragments, but, surprisingly, the retrogradational shallow-marine sandsheet and turbidite facies have different amounts of feldspar despite their similar grain size. Turbidite sandstones in the Minturn have abundant quartz, which is generally not known to be enriched in the silt to fine sand-sized fraction of sandstones unless directly supplied from a source in this grain size. This may have been the case for the Minturn turbidites. Turbidite sandstones always lie directly above erosional surfaces interpreted as sequence boundaries (i.e., unconformities), and the deposits underlying the sequence boundaries might have had a more mature composition due to chemical alteration during the periods of erosion and nondeposition represented by the sequence boundaries. Feldspar could have been chemically destroyed and fine-grained quartz residually enriched by weathering during soil development. Sequence boundaries forming the base of MFD successions have not been analyzed for evidence of chemical alteration, so this explanation is speculative.

Possibly grain-size similarities are the cause of the compositional similarities of beach and fluvial sandstones, and the compositional similarities between distal distributary, progradational bar, and sandsheet sandstones. However, a more rigorous analysis of the grain size and composition of sedimentary facies than that presented here is necessary to support this hypothesis.

Interpretation of variations in compositions of braided-fluviodeltaic facies. Although much of the compositional variation in facies of the MFD systems seems to be strongly linked to grain size, grain size appears to correlate only slightly with compositional variations in BFD system facies. This weak relationship between grain size and composition is explicitly shown in Figure 14. Beside each point in Figure 14, a number gives the percentage of that sample made up of grains greater than or equal in size to coarse sand. The samples reflect a broad range in grain size, with samples being composed of 4 to 97% coarse sand and granular grains. In general, samples with the smallest percentages of coarse grains have the smallest percentages of rock fragments, but grain size decreases nonsystematically away from the rock fragment pole. Relatively coarse samples plot adjacent to relatively fine samples. Perhaps the effect of grain size is weak because the facies of the braided system tend to be more poorly sorted due to the more irregular discharge and higher energy nature of braided-fluvial deposition. Poor sorting probably would inhibit compositional variation dependent on concentration of a particular grain size, because a mix of grain sizes would in effect dilute the influence on composition of any one size fraction.

If grain size plays any role in the nature of modification of sands in the braided-fluviodeltaic system, its influence, at best, is subtle or nonsystematic. The control of grain size is demonstrated only when grain size and composition of samples from the same facies (i.e., distributary channel, beach, progradational sandsheet, washover-fan, and tidal channel) are compared to the compositional field of the fluvial facies. A tendency exists for the coarser samples to fall within the fluvial composition field, whereas samples having a smaller percentage of coarse grains plot outside the fluvial field. Even though this comparison suggests that associated facies have compositions reflecting grain-size modification of fluvial sediment, in the samples analyzed for this study modifications do not result in a systematic change that is uniform from one facies to another.

Comparison of Figures 9 and 10 reveals that the differences among the compositions of BFD sandstone facies are not as distinct as the compositional differences among facies of MFD systems (Fig. 10). However, when the compositions of shoreline and marine sandstones from BFD systems are plotted on a QFR ternary diagram and compared to the composition of their associated fluvial sandstones, it is apparent that marine and shoreline sandstones of BFD systems are enriched in quartz and feldspar and depleted in rock fragments relative to the fluvial sandstones. Physical and/or chemical sedimentary processes in the depositional environment presumably modified the composition of sediments delivered to the system by braided fluvial channels. Mechanical disaggregation of rock fragments into monomineralic quartz and feldspar reduced the amount of rock fragments while increasing the amounts of quartz and feldspar in sediments of the BFD systems.

Trends in the compositional modification of Minturn BFD

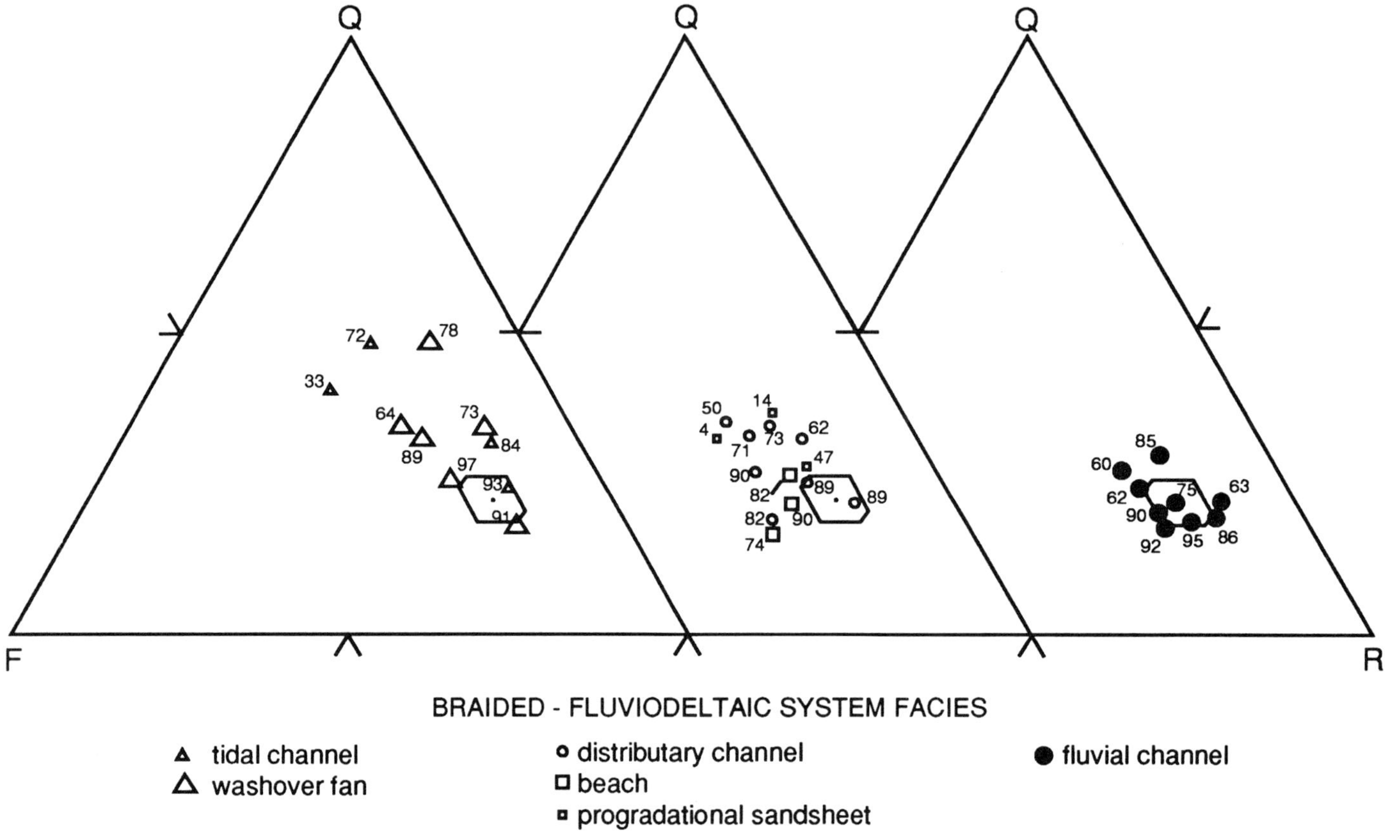

Figure 14. Ternary plots of the quartz, feldspar, and rock fragment compositions of individual samples from fluvial, deltaic, shoreline, and marine deposits of braided-fluviodeltaic systems. Facies include fluvial channel, distributary channel, progradational sandsheet, beach, tidal channel, and washover-fan facies. Adjacent to each point are values representing the modal percentage of each sample made up of grains equal to or coarser than coarse sand. (For example, a value of 90 indicates the composition was determined from a sample composed of 90% coarse sand.) The hexagon symbol represents field of variation for associated braided-fluvial channel sandstones. Showing the fluvial field of variation permits visual evaluation of how marine sandstones differ in composition from their fluvial precursor.

sandstones are reflected primarily by variations in quartz abundance and subordinately by variations in feldspar abundance (Fig. 14). The enrichment of quartz relative to feldspar in BFD facies is probably a function of both grain size of facies and durability of quartz relative to feldspar. Following the course of mechanical destruction, feldspar would have been broken into finer grain sizes and winnowed from facies deposited under high-energy depositional conditions. Fine-grained feldspar would have been enriched and deposited in fine- to medium-grained facies under lower energy depositional conditions. Qualitative analysis of additional fine-grained marine and fluvial facies from BFD deposits suggests these facies do have a relatively high feldspar content, supporting the viability of this explanation.

Washover-fan and tidal-channel sandstones are the most enriched in quartz of all BFD system facies. This enrichment may be related to the fact that these facies occur in retrogradational intervals and were formed from reworking of delta-platform sediments. As a result of repeated reworking and redeposition on a transgressed surface, sediments deposited in retrogradational facies should record more physical destruction of rock fragments and feldspar. Conversely, sandstones interpreted as beach, distributary-channel, and progradational sandsheet deposits occur in progradational intervals of the Minturn BFD systems that were characterized by higher rates of sediment supply. With a higher rate of sediment supply, less opportunity existed for shoreline or marine processes to rework sands delivered by braided fluvial channels.

DISCUSSION

Fluviodeltaic sandstones in both the Fountain and Minturn Formations display evidence of environment-related compositional change despite tectonically induced high rates of sediment accumulation and burial, but the specific nature of change is not identical for the two formations. This is because the path of compositional change that sand follows depends on three conditions: the composition and texture of the "starting" detritus, the degree of size sorting during transport and deposition, and the rigor and duration of reworking by marine processes. Even though both the Fountain and Minturn accumulated in comparable geologic settings, the factors that influence compositional change were not identical for both formations and account for differences in the maturation patterns we have described.

The most significant difference in maturation patterns of sandstones from the Fountain and Minturn results from differences in the "starting" detritus. Sandstones of the Minturn contain significantly more rock fragments than Fountain sandstone, reflecting the higher percentage of fine-grained plutonic rocks in the Minturn source area. In contrast, the coarsely crystalline parent rocks for the Fountain produced mostly monocrystalline detritus. Consequently, Fountain sand began with a more restricted composition, and modification pathways as expressed on a ternary diagram are simple and essentially confined to the quartz-feldspar tieline. Trends of modification represented in facies of the Minturn, on the other hand, are primarily characterized by a depletion of rock fragments, accompanied by an enrichment of quartz and/or feldspar.

The second most significant influence on modification of detritus results from the degree of grain-size sorting during transport and deposition. The most pronounced segregation of sandstone facies on the basis of framework composition occurs among the facies of the MFD systems of the Minturn Formation. Less differentiation exists among facies of the Fountain Formation and the various facies of the BFD system of the Minturn. Because the wave-dominated Fountain delta likely also was fed by high-discharge braided streams, we are not surprised that the degree of differentiation is comparable for facies of the Fountain and the Minturn BFD systems. Based on our comparison of the MFD and BFD systems, the amount of compositional difference among facies of the same association apparently is partially dependent on the magnitude of grain-size difference within a facies and between facies. In the meandering rivers, relatively uniform flow conditions probably promoted more efficient grain-size sorting. Furthermore, the discharge of sediment to the mouths of the deltas fed by the meandering rivers likely was less than for the braided rivers. Together, these two factors should have enhanced the effectiveness of concentration of different size fractions in the various marine and transitional environments at the lower end of the delta plain and on the delta front, assuming other hydrodynamic factors were equal. This would explain the relatively sharp difference in composition as a function of grain size in Minturn sandstones from facies within the MFD system and the absence of such differences for the BFD systems in both the Fountain and Minturn.

The third influence on the degree of compositional maturation relates to the rigor and duration of reworking processes active at the shorelines of the Fountain and Minturn depositional systems. The progradational foreshore facies of the Fountain Formation is most enriched in quartz as a result of reworking by waves in the swash zone. Beach deposits in the Minturn, however, show only minor evidence of reworking in a swash zone. They do not exhibit well-developed shoreface-foreshore facies, nor do they show a maturation trend. Shoreline facies in the Minturn that *do* show a maturation trend are the retrogradational-shoreline deposits associated within the BFD intervals. These facies, including deposits of tidal channels and washover fans, are the most consistently quartzose of all the BFD facies of the Minturn. Their maturation is attributed to a lengthy period of reworking associated with transgression of delta and shoreline facies, rather than fair-weather wave reworking of shoreline sediment, as was the case in the Fountain. We conclude from this comparison that maturation of detritus in the Fountain was primarily a function of rigor of reworking, whereas maturation in Minturn shoreline facies was primarily a function of duration of reworking.

CONCLUSIONS

1. First-cycle sandstones from ancient coarse-grained deltas in the Fountain and Minturn Formations have been compositionally modified by hydrodynamic processes characteristic of different environments in the deltas. Both the Fountain and Minturn were deposited in an active tectonic setting characterized by high rates of sedimentation and burial. Compositional modification likely would have been significantly greater in deltas located in more passive settings.

2. In the wave-dominated delta of the Fountain Formation, foreshore sand was enriched in quartz through mechanical breakdown and winnowing of feldspar. During storms, this feldspar was concentrated in the offshore/transition zone. Most shoreface sandstone resembles its precursor alluvial sandstone in composition, suggesting that little modification occurred during longshore transport. However, some samples of shoreface sandstone are either sharply enriched or depleted in feldspar and had a more complicated origin involving reworking of foreshore sand.

3. Variations in composition of the Minturn sandstone are more complex than those in the Fountain because of the broader range of environments associated with the river-dominated deltas in the Minturn. Some of the deltas were fed by braided and some by meandering rivers, and each river type carried a distinctive mix of quartz, feldspar, and rock fragments. Consequently, change in composition of the sand caused by hydrodynamic processes in the transitional and marine environments was different for each delta type.

4. Meandering rivers in the Minturn carried a finer grained, better sorted load than the braided rivers, which primarily accounts for the lower percentage of rock fragments in the meandering-fluvial-channel sandstones. Compositional modification of sediment in the meandering-fluviodeltaic system increases from the nonmarine to the offshore marine environments, resulting from deposition of rock fragment–rich, coarse-grained detritus upstream in the fluvial channels, and deposition of more quartz- and feldspar-rich detritus in the marine and transitional environments where lower transport energies prevailed. The most distal of the meandering-fluviodeltaic facies, the prodelta turbidite deposits, are the most modified and the most compositionally

mature, owing to a combination of chemical and mechanical differentiation.

5. Compositional differentiation in the Minturn braided-fluviodeltaic system is not as distinct as in the meandering system, primarily owing to a lower quality of grain-size sorting and less consistency in mechanical reworking in this system. Although compositional fields of the facies of the braided-fluviodeltaic system overlap significantly, substantial facies-related modification occurred. As a result, distributary-channel and marine sandstones are more quartzose than fluvial sandstones.

6. Among both the braided-deltaic and meandering-deltaic system deposits, retrogradational-shoreline and marine facies are compositionally more modified than the progradational facies. Samples from retrogradational braided-deltaic shoreline facies (washover fans and tidal channels), may contain up to 20% more quartz and 11% more feldspar than the braided fluvial sandstones. Retrogradational shallow-marine sandstones associated with the meandering-fluvial system contain up to 14% more quartz and 21% more feldspar than meandering-fluvial sandstones.

7. The specific nature and magnitude of compositional change that sand undergoes as it passes through various environments in coarse-grained deltas are influenced by three factors: the sand's starting composition, the degree of size sorting of the sand that occurs in the different environments, and the rigor and duration of hydrodynamic processes that cause the physical breakdown of grains. Each of these factors differed to some extent both within and between the Fountain and Minturn facies associations, and this explains the diagnostic differences observed in the associations.

ACKNOWLEDGMENTS

Acknowledgment is made to Donors of the Petroleum Research Fund, administered by the American Chemical Society, for support of the research on the Fountain Formation. The Department of Geological Sciences at Indiana University is acknowledged for its support of field study of the Minturn Formation and for providing thin sections of Minturn samples. Exxon Production Research Company also produced a number of the Minturn thin sections used in this study. Wes Boberg is thanked for his assistance in processing point-count data, and we express our gratitude to Patty Byrum for typing numerous drafts of the manuscript. We are especially indebted to Karen Houck for permitting us to work in her dissertation study area near McCoy and for freely sharing her observations and interpretations of the Minturn with us prior to their publication. She, as well as Ray Ingersoll and Mark Johnsson, provided fair and constructive reviews of the manuscript, for which we are also grateful. It is a pleasure to be a part of a scientific community in which one's peers are so professional and conscientious in meeting their responsibilities as reviewers.

REFERENCES CITED

Allen, J.R.L., 1962, Petrology, origin and deposition of the highest Lower Old Red Sandstone of Shropshire, England: Journal of Sedimentary Petrology, v. 32, p. 657–697.

Basu, A., 1976, Petrology of Holocene fluvial sand derived from plutonic source rocks: Implications to paleoclimatic interpretation: Journal of Sedimentary Petrology, v. 40, p. 694–709.

Burroughs, W. A., 1985, The hydraulic equivalence of mica—Discussion: Journal of Sedimentary Petrology, v. 55, p. 291–294.

Chronic, J., and Stevens, C., 1958, Pennsylvanian paleogeography in the McCoy Area, Eagle County, *in* Symposium on Pennsylvanian rocks of Colorado and adjacent areas: Rocky Mountain Association of Geologists Field Conference Guidebook, p. 86–90.

Clifton, H. E., 1969, Beach cementation: Nature and origin: Marine Geology, v. 7, p. 553–559.

Davis, D. K., and Ethridge, F. C., 1975, Sandstone composition and depositional environments: American Association of Petroleum Geologists Bulletin, v. 59, p. 234–264.

Doyle, L. J., Carder, K. L., and Steward, R. G., 1983, The hydraulic equivalence of mica: Journal of Sedimentary Petrology, v. 53, p. 643–648.

Ethridge, F. G., 1977, Petrology, transport, and environment in isochronous upper Devonian sandstone and siltstone units, New York: Journal of Sedimentary Petrology, v. 47, p. 53–65.

Franzinelli, E., and Potter, P. E., 1983, Petrology, chemistry and texture of modern river sands, Amazon River system: Journal of Geology, v. 91, p. 23–39.

Grantham, J. H., and Velbel, M. A., 1988, The influence of climate and topography on rock fragment abundance in modern fluvial sands of the southern Blue Ridge Mountains, North Carolina: Journal of Sedimentary Petrology, v. 58, p. 219–227.

Hayes, M. O., 1967, Hurricanes as geological agents: case studies of Hurricanes Carla, 1961, and Cindy, 1963: Austin, TX; Report of Investigations, Bureau of Economic Geology, v. 61, 54 p.

Houck, K. J., 1991, Structural control on the distribution of sedimentary facies in the Pennsylvanian Minturn Formation of north-central Colorado: U.S. Geological Survey Bulletin 1787, p. Y1–Y33.

Hunter, R. E., 1967, The petrography of some Illinois Pleistocene and Recent sands: Sedimentary Geology, v. 1, p. 57–75.

Ingersoll, R. V., and Suczek, C. A., 1979, Petrology and provenance of Neogene sand from Nicobar and Bengal Fans: DSOP sites 211 and 218: Journal of Sedimentary Petrology, v. 49, p. 1217–1228.

Johnsson, M. J., Stallard, R. F., and Meade, R. H., 1988, First-cycle quartz arenites in the Orinoco River Basin, Venezuela and Columbia: Journal of Geology, v. 96, p. 263–277.

Johnsson, M. J., Stallard, R. F., and Lundberg, N., 1991, Controls on the composition of fluvial sands from a tropical weathering environment: Sands of the Orinoco River drainage basin, Venezuela and Columbia: Geological Society of America Bulletin, v. 103, p. 1622–1647.

Kairo, S., 1992, Facies-related mineralogic and geochemical variations in sandstones from the Minturn Formation (Pennsylvanian), north-central Colorado [Ph.D. thesis]: Bloomington, Indiana University, 376 p.

Kluth, C. F., and Coney, P. V., 1981, Plate tectonics of the Ancestral Rocky Mountains: Geology, v. 9, p. 10–15.

Lowright, R., Williams, E. G., and Dachille, F., 1972, An analysis of the factors controlling deviations in hydraulic equivalence of some modern sands: Journal of Sedimentary Petrology, v. 42, p. 643–645.

Mack, G. H., 1978, The survivability of labile light mineral grains in fluvial, eolian, and littoral environments; The Permian Cutler–Cedar Mesa Formations, Moab, Utah: Sedimentology, v. 25, p. 587–604.

Mack, G. H., and Suttner, L. J., 1977, Paleoclimate interpretation from a petrographic comparison of Holocene sands and the Fountain Formation (Pennsylvanian) in the Colorado Front Range: Journal of Sedimentary Petrology, v. 47, p. 89–100.

Mallory, W. W., 1958, Pennsylvanian coarse arkosic redbeds and associated mountains in Colorado, *in* Symposium on Pennsylvanian rocks of Colorado and adjacent areas: Rocky Mountain Association of Geologists Field Conference Guidebook, p. 17–20.

Mallory, W. W., 1975, Middle southern Rocky Mountains, northern Colorado Plateau and eastern Great Basin, *in* McKee, E. D., and Crosby, E. J., Paleotectonic investigations of the Pennsylvanian System in the United States: U.S. Geological Survey Professional Paper 853, p. 265–278.

Maples, C. G., and Suttner, L. J., 1990, Trace fossils and marine-nonmarine cyclicity in the Fountain Formation (Pennsylvanian: Morrowan-Atokan) near Manitou Springs, Colorado: Journal of Paleontology, v. 64, p. 859–880.

McPherson, J. G., Shanmugam, G., and Moiola, R. J., 1987, Fan-deltas and braid deltas: Varieties of coarse-grained deltas: Geological Society of America Bulletin, v. 99, p. 331–340.

Nottvedt, A., and Kreisa, R. D., 1987, Model for combined flow origin of hummocky cross-stratification: Geology, v. 15, p. 357–361.

Odom, I. E., Doe, T. W., and Dott, R. H., 1976, Nature of feldspar-grain size relations in some quartz-rich sandstones: Journal of Sedimentary Petrology, v. 46, p. 862–870.

Okada, H., 1966, Non-graywacke "turbidite" sandstones in the Welsh geosyncline: Sedimentology, v. 7, p. 211–232.

Rubey, W. W., 1933, The size-distribution of heavy minerals within a water-laid sandstone: Journal of Sedimentary Petrology, v. 3, p. 3–29.

Rueger, B. F., 1984, Geology and palynology of the Paradox Formation (Desmoinesian), southeastern Utah [M.S. thesis]: Boulder, University of Colorado, 192 p.

Scotese, C. R., and McKerrow, W. S., 1990, Revised world maps and introduction, *in* McKerrow, W. S., and Scotese, C. R., eds., Paleozoic paleogeography and biogeography: Geological Society of London Memoir No. 12, p. 1–21.

Shukis, P. S., and Ethridge, F. G., 1975, A petrographic reconnaissance of sand sized sediment, upper Saint Francis River, southeast Missouri: Journal of Sedimentary Petrology, v. 45, p. 115–127.

Slingerland, R. L., 1977, The effects of entrainment on the hydraulic equivalence relationships of light and heavy minerals in sands: Journal of Sedimentary Petrology, v. 47, p. 753–770.

Stevens, C. H., 1958, Stratigraphy and paleontology of the McCoy, Colorado, area [M.S. thesis]: Boulder, University of Colorado, 189 p.

Suttner, L. J., 1974, Sedimentary petrographic provinces: An evaluation, *in* Ross, C. A., ed., Paleogeographic provinces and provinciality: Society of Economic Paleontologists and Mineralogists Special Publication 21, p. 75–84.

Suttner, L. J., 1989, Recent advances in the study of the detrital mineralogy of sand and sandstone: Implications for teaching: Journal of Geological Education, v. 37, p. 235–240.

Suttner, L. J., and Basu, A., 1985, The effect of grain size on detrital modes: A test of the Gazzi-Dickinson point-counting method—A discussion: Journal of Sedimentary Petrology, v. 55, p. 616–617.

Van Wagoner, J. C., Mitchum, R. M., Campion, K. M., and Rahmanian, V. D., 1990, Siliciclastic sequence stratigraphy in well logs, cores and outcrops: Concepts for high-resolution correlation of time and facies: American Association of Petroleum Geologists Bulletin, Methods in Exploration Series, No. 7, 55 p.

Wilson, R. F., 1975, Eastern Colorado, *in* McKee, E. D., and Crosby, E. J., Paleotectonic investigations of the Pennsylvanian system in the United States: U.S. Geological Survey Professional Paper 853, Part I, p. 265–278.

Winn, R. D., Jr., Stonecipher, S. A., and Bishop, M. G., 1984, Sorting and wave abrasion: Controls on composition and diagenesis in Lower Frontier sandstones, southwestern Wyoming: American Association of Petroleum Geologists Bulletin, v. 68, p. 268–284.

MANUSCRIPT ACCEPTED BY THE SOCIETY DECEMBER 24, 1992

Printed in U.S.A.

Geological Society of America
Special Paper 284
1993

The effects of basin asymmetry on sand composition: Examples from Lake Tanganyika, Africa

Michael J. Soreghan and Andrew S. Cohen
Department of Geosciences, University of Arizona, Tucson, Arizona 85721

ABSTRACT

Distinctive compositional trends occur in sand suites collected from structurally diverse segments of Lake Tanganyika (East African Rift System). Climate (subhumid, tropical) and source lithology (mixed granitic and metasedimentary rocks) are generally invariant between the source areas studied. Therefore, differences in sand composition relate directly to differences in compositional modification during transport and deposition of the sands. Sand composition along contrasting segments of the lake basin can be related to the strong structural asymmetry of the basin.

We point-counted medium sands from river mouth, deltaic, beach, and shallow-water environments of four structurally distinct, lake-margin settings: (1) escarpment margin (Pemba, Zaire site); (2) hinged margin (Nyanza Lac, Burundi, and Rumonge, Burundi sites); (3) accommodation-zone margin (Magara, Burundi site); and (4) axial margin (Ruzizi River site, Burundi). These four margin-types are typical of the structural morphology in strongly asymmetric half-graben rift basins. Using the Gazzi-Dickinson point counting methodology, margins two through four cluster tightly in QFL space, suggesting similar source lithology. Samples from the Pemba site (escarpment margin) are significantly depleted in feldspar. Because of the uncertainties in interpreting this difference, the escarpment margin sands were not included in subsequent analyses.

Variations in sand composition occur between the remaining sites when using the traditional point-counting methodology. The suite of fluvial sands entering the lake basin along the accommodation-zone margin (Magara site) averages 38% rock fragments compared to an average of 15 to 20% and 18% rock fragments for the suite of fluvial sands entering the basin along the hinged or axial margins, respectively. This is a fundamental contrast that correlates directly to differences in drainage basin size.

Reworking at the depositional site produces additional compositional differences across the structural margins. In comparing the sample suites for each site, it is the variability in the data that is more important than average composition in highlighting these differences. Minimal compositional modification occurs at the depositional site along both the accommodation-zone margin and the axial margin. In contrast, significant environmental overprinting occurs along the hinged margin sites. Limited compositional modification along the accommodation-zone margin results from the steep depositional gradient and the very narrow high-energy zone which limits reworking of the sediment. The limited reworking and the consequent minimal compositional modification along the axial margin are primarily a function of the high sedimentation rate. The compositional modification (enrichment in quartz) along the hinged margin is a function of low sediment accumulation rates coupled with low subsidence rates, leaving large regions of shallow-water substrate exposed to long periods of reworking. Chemical

Soreghan, M. J., and Cohen, A. S., 1993, The effects of basin asymmetry on sand compositions: Examples from Lake Tanganyika, Africa, *in* Johnsson, M. J., and Basu, A., eds., Processes Controlling the Composition of Clastic Sediments: Boulder, Colorado, Geological Society of America Special Paper 284.

alteration of labile phases also occurs along the hinged margin through carbonate replacement of feldspar.

Our results suggest that, with careful statistical analysis, it is possible to use sandstone composition to distinguish between tectonic margins typical of half-graben basins. Using this approach in ancient studies, it is necessary to collect from several coeval facies among different study sites and to constrain source lithology and climate. In addition, these results are most appropriate for humid-climate basins, where weathering processes are most extreme.

INTRODUCTION

The structure of a basin imparts a strong control on sedimentation patterns (Miall, 1981). Sedimentary controls such as the duration of weathering at the sediment source, basin and hinterland morphology, sediment residence time during transit to the depositional site, and even the nature of the depositional and transport systems are all influenced by basin tectonics. These same sedimentary processes are commonly included within the "environmental" variable of Suttner (1974). Other variables that make up the complex equation that governs sand composition include source lithology, climate, and diagenesis. An important goal in sedimentary petrology is to isolate the effects of these variables in order to understand their relative importance in controlling sandstone composition. Ultimately, this will allow geologists to interpret many facets of the paleogeology of the basin/hinterland (Basu, 1985).

Isolating the effects of variables that influence sand composition is made difficult, however, because the variables are not independent; in fact, recent studies have moved toward exploring the interrelationship between variables. For example, it is now well established that climate can alter the imprint on sand composition imparted by source lithology (Basu, 1976; Suttner et al., 1981; Potter, 1986; Suttner and Dutta, 1986; Johnsson et al., 1988, 1991). In addition, the magnitude of the climatic effect is, to a certain extent, dependent on the environment of deposition and residence time of sediment in the transport and depositional systems (Basu, 1985; Grantham and Velbel, 1988; Johnsson et al., 1988, 1991). One link that has not received much attention, however, is the relationship between the structure of a basin and its influence on the sedimentary or "environmental" variables that control sand composition.

The goal of this study is to document the variability in sand composition within a modern basin that reflects changes in the sedimentary variables induced by basin structure. Lake Tanganyika, Africa, is an excellent region to undertake such a study for a number of reasons. First, by studying sands within a modern basin we can document directly the variability in both source lithology and climate. Second, we can eliminate the effects of burial diagenesis. Finally, Lake Tanganyika, which is part of the East African Rift, exhibits a highly asymmetric structural morphology (half-graben) attributable to nonuniform extension. This allows us to contrast the sedimentary variables and consequent changes in sand composition between structurally distinct margins of the lake-basin.

BACKGROUND

Previous work

Both transport and depositional processes may play important roles in altering the composition of weathered parent material (Bradley, 1970; Pittman, 1970; Cameron and Blatt, 1971; Davies and Ethridge, 1975; Self, 1975; Ethridge, 1977; Breyer and Bart, 1978; Mack, 1978; Potter, 1978; Grantham and Velbel, 1988). Most previous studies have attempted to isolate an aspect of the total "environmental" variable in order to identify the signature of that particular aspect of the transport or depositional system as recorded by the composition of the detritus. For example, the influence of transport processes on sand composition has been defined by the length or gradient of the drainage system. It appears that high gradients alter the composition of sands within the fluvial environment by decreasing the abundance of rock fragments (Cameron and Blatt, 1971; Mack, 1978), although the nature of the parent material is also an important variable (Mann and Cavaroc, 1973; Mack, 1981). In low-gradient fluvial systems, distance of transport plays an important role only in humid to subhumid climates; in fact, residence time or sediment "parking" outside the transportation "system" appears to be a more important factor in modifying sand composition (Potter, 1978; Suttner et al., 1981; Basu, 1985; Johnsson et al., 1988, 1991; Grantham and Velbel, 1988; Savage and Potter, 1991). Within high-energy zones of certain depositional systems, hydraulic sorting and physical breakdown of grains can be an important modifier of sand composition (Davies and Ethridge, 1975; Mack, 1978). In these studies the authors try to link variation in sand composition directly to the internal processes of the depositional system.

This study differs from those listed above because we attempt to relate the variation observed in the composition of sands directly to basin tectonics. In attempting to document the link between sand composition and basin structure, we integrate, rather than differentiate, the individual components of the "environmental" variable in order to relate variations in sand composition to the total set of depositional and transport processes that are controlled by structure. In this approach we utilize basin structure as a "master variable" that underlies the first-order variables (e.g., slope, relief, residence time) that control sand composition. If we can better quantify this link we can ultimately utilize sand composition not only for inferring provenance of the hinterland, but also for interpreting the paleostructure of the basin.

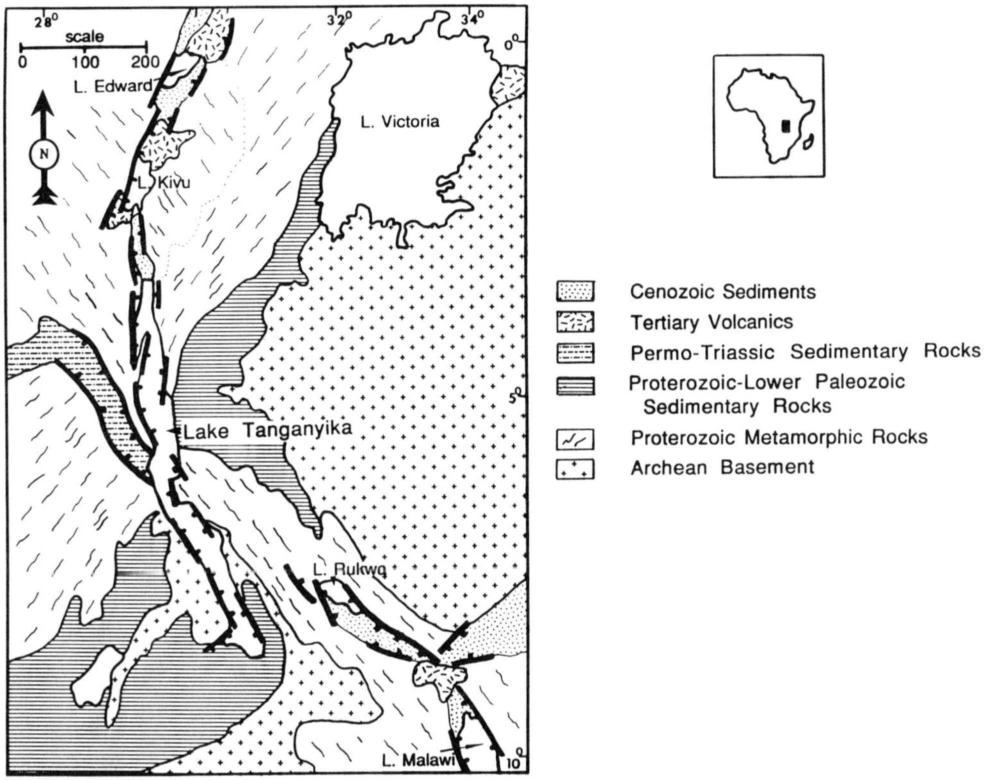

Figure 1. Generalized geological map of western arm of East African Rift System showing location of Lake Tanganyika. Thick, hachured lines represent Cenozoic normal faults. Thin, wavy dashes represent structural trends in Proterozoic mobile belts. Note that the rift structure generally avoids the Archean craton. Modified from McConnell (1972) and Ebinger (1989).

Setting of Lake Tanganyika

Tectonics and geology. Lake Tanganyika lies within the western arm of the East African Rift System. Figure 1 outlines the geology and generalized tectonic grain of both the Cenozoic rift structure and Precambrian mobile belts. The tectonic and geologic evolution of this portion of the rift has been well documented (McConnell, 1972; Ebinger, 1989; Sander and Rosendahl, 1989; Tiercelin, 1991; Tiercelin and Mondeguer, 1991).

Ten half-graben structural subbasins with alternating polarity lie along the rift axis and together constitute the Lake Tanganyika basin (Fig. 2) (Rosendahl et al., 1986; Morley, 1988; Ebinger, 1989; Sander and Rosendahl, 1989). These basins, which are 80 to 160 km long and 30 to 60 km wide, are decoupled by complex structural horsts, termed accommodation zones (Fig. 2), which typically trend at a high angle to the rift axis (Rosendahl et al., 1986; Morley et al., 1990). This structural geometry produces four lake-margin types (Fig. 3): (1) hinged margins, which occur along the hanging wall of individual half-grabens; (2) escarpment margins, which occur along the footwall of individual half-grabens; (3) accommodation-zone margins, which lie across the structural horsts between half-grabens; and (4) axial margins, which occur at the end of the lake-basin and are oriented perpendicular to the rift axis.

Climate and hydrology. Lake Tanganyika lies between 3° to 9°S latitude, within the present subtropical humid zone. Average air temperature ranges from 17° to 29°C; precipitation averages approximately 90 cm/yr and falls primarily in a single rainy season from October through March (Coulter and Spigel, 1991).

Lake Tanganyika is permanently anoxic below water depths of 90 to 250 m, depending on the season and location within the lake (Kufferath, 1952; Hecky and Degens, 1973). Because of the long fetch of the lake, normal wave base can be as deep as 3 to 4 m. Storm wave base is less well known, but current structures are observed in 10-m water depth. Wind-driven longshore currents form an important component of the circulation system; permanent currents formed by upwelling have been documented (Coulter, 1963).

Influence of basin structure on sedimentary processes

In both modern and ancient half-grabens, structural asymmetry results in asymmetry of sedimentation patterns (Turner-Peterson and Smoot, 1985; Gore, 1988; Hamblin and Rust, 1989; Cohen, 1990; Anadón et al., 1991). Leeder and Gawthrope (1987) concluded that differences in sedimentary controls across half-grabens relate directly to the spatial variability of two fundamental structural controls: subsidence and slope. Sedimenta-

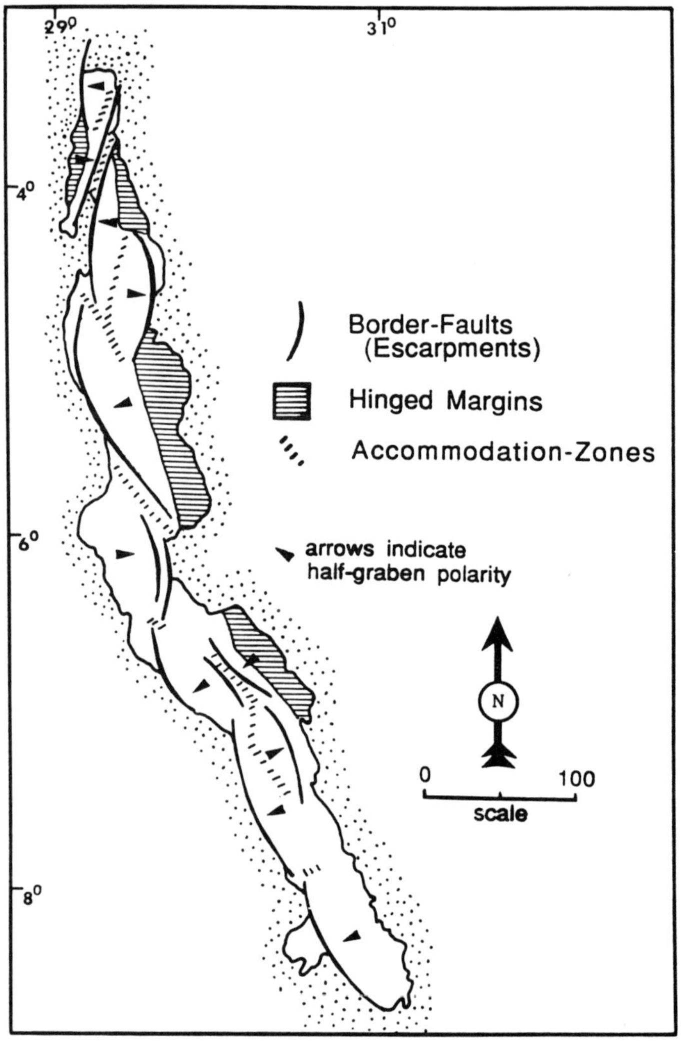

Figure 2. Generalized tectonic map of Lake Tanganyika. Discrete, segmented border faults produce a series of alternating half-graben basins (arrows point toward border faults). See text for discussion. Modified from Rosendahl et al. (1986), and Morley (1988).

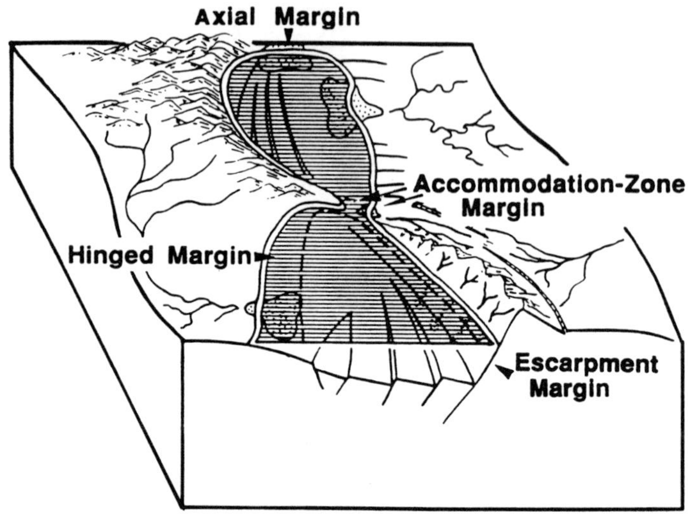

Figure 3. Idealized block diagram showing structural margin types created by alternating, half-graben morphology. Accommodation-zone, hinged, and escarpment margins are all structurally controlled; axial margin is controlled by depositional processes. Sedimentary processes along these margins vary because of significant differences in the structural characteristics among margin types.

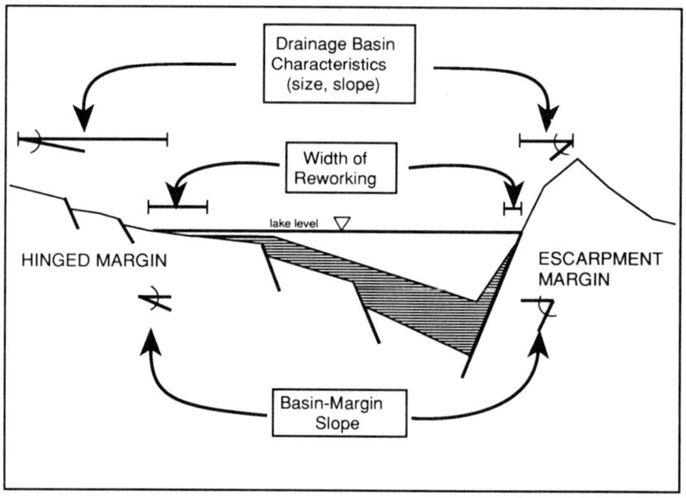

Figure 4. Schematic representation of some of the important sedimentary processes that vary across a half-graben basin. Tectonic slope and non-uniform tectonic subsidence create the asymmetry in sedimentary processes.

tion patterns in both the hinterland and depositional site vary as a consequence of these two tectonic controls (Fig. 4).

Drainages originating in the uplifted flanks of border faults typically drain away from the rift basin (Baker, 1986; Frostick and Reid, 1989; Cohen, 1990). Drainages that do debouch into the rift trough are dominantly axial streams or streams that begin by draining away from the rift and are channeled back toward the rift by synthetic or antithetic faults (Baker, 1986; Frostick and Reid, 1989). Within the Lake Tanganyika basin, drainage basin characteristics relate directly to the asymmetric structure (Fig. 5). For example, small, steep drainages occur adjacent to escarpment margins, whereas larger streams occur along axial or hinged margins. Accommodation-zone margin streams are highly variable in both size and gradient, but are generally smaller and steeper than axial or hinged margin streams (Fig. 5). Differences in the drainage characteristics also influence the littoral environments (Table 1); for example, most of the large deltas within Lake Tanganyika occur on shoaling or axial margins.

At the depositional site, differences in nearshore energy, widths of coarse-grained facies, size and spacing of deltas, and depositional gradients along different segments of rift lakes are all controlled by basin asymmetry (Table 1) (Cohen, 1990; Soreghan and Cohen, 1991).

We predict that these same variables controlled by basin structure also influence sand composition. Variations in sand composition can be affected by two processes: weathering and transport mechanisms that affect composition before the sands

TABLE 1. FACIES CHARACTERISTICS ALONG STRUCTURAL MARGINS OF LAKE TANGANYIKA

Margin Type	Facies Characteristics	Sedimentary Controls
Hinged Margin	Shore-parallel facies. Poorly to well-developed deltas. Well-sorted sands dominated as much as two kilometers offshore. Carbonate sand dominant lithology	Large, low-gradient drainages. Low-gradient, bench-like bathymetric profile. Wide high-energy zone. Low subsidence rates; low net sedimentation rates
Axial Margin	Asymmetric facies pattern. Typically large, well-developed, progradational delta. Sands common along shoreface, delta mouth and abandoned delta lobe. Clastic sand/silt dominant lithology	Large, low-gradient drainages. Low-gradient, asymmetric bathymetric profile. Wide high-energy zone. Moderate subsidence rates; high net sedimentation rates
Accommodation-Zone Margin	Discontinuous, shore-parallel facies. Widely spaced, nonprograding fan deltas. Poorly sorted sands confined to beach, fan-delta mouths. Clastic silt-dominant lithology; biogenic sediment also common	Small to intermediate, high-gradient drainages. Steep, complex bathymetric profile. Narrow high-energy zone. Low subsidence rates; moderate to very low net sedimentation rates
Escarpment Margin	Discontinuous, shore-parallel facies. Widely spaced, nonprograding fan-deltas. Poorly sorted sands confined to beach, fan-delta mouths. Biogenic sediment-dominant lithology; clastic sand/silt also common	Small to intermediate, high-gradient drainages. Steep, cliff-like bathymetric profile. Very narrow high-energy zone. High subsidence rates; high net sedimentation rates

enter the lake, and subsequent reworking that occurs within the depositional site.

METHODS

There were 71 samples collected either by hand (using SCUBA) or with a modified Eckman grab sampler from northern Lake Tanganyika during field seasons in 1985, 1986, and 1989. We concentrated our sampling in four settings that represent each of the four tectonic lake margins described above (Fig. 6: Ruzizi Delta, Burundi; Magara, Burundi; Nyanza Lac, Burundi; and Pemba, Zaire). Samples collected south of the Rumonge Fault (Fig. 6) lie along a diffuse boundary between the accommodation zone and the hinged margin to the south. Because the morphological characteristics of this region differed significantly from the region north of the Rumonge Fault, we separated the two Magara populations. The samples from south of the fault are included with the hinged margin samples, representing a second population for this margin type. These samples are referred to as the Rumonge site samples.

At each locality we collected samples and categorized them into one of four environments: (1) river-mouth samples, collected within the stream channel directly at the mouth of the river, or in some cases a few kilometers upstream; (2) beach samples, collected along the shoreface and spaced at least 1 km apart across the study sites; (3) delta samples, collected offshore of streams that exhibited well-developed delta deposits; and (4) shallow-water samples, collected offshore and away from direct influence of the sediment plume emanating from nearby river mouths. We attempted to analyze an equal number of samples from each site and each environment, although in a few cases this was not possible because of the minimal extent of certain environments along certain sites.

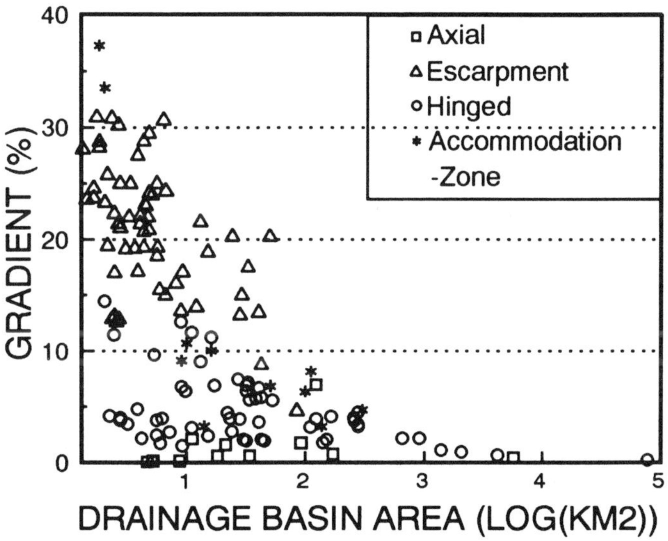

Figure 5. Plot of stream gradient (in percent) versus the log of drainage basin area (square kilometers) for all streams greater than 1 km^2 entering Lake Tanganyika along the three northernmost half-grabens (Fig. 2). Streams are coded according to the structural margin type on which they occur. Note that there is an inverse relationship between stream gradient and drainage basin area.

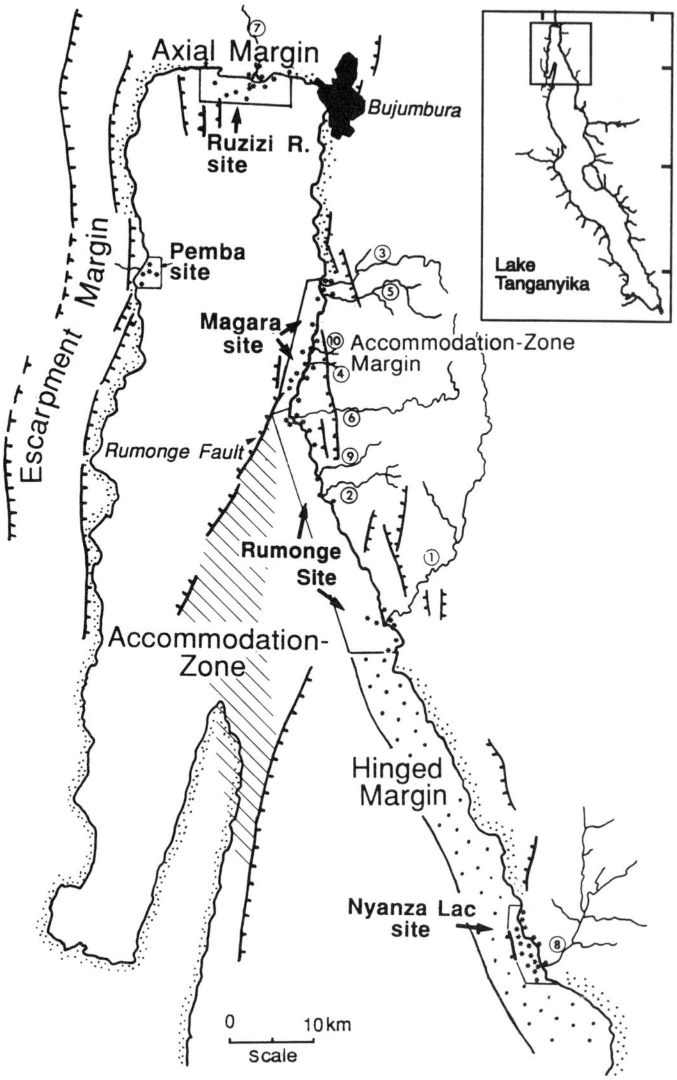

Figure 6. Location map outlining the study sites and rivers discussed in text. Boxed regions outline areas where sampling was conducted along each margin type (sites). Small asterisks within boxed regions represent sample sites. Asterisks representing beach and river-mouth sands located on shoreward side of lake-margin line. Numbered streams keyed to Table 3. Hachured lines represent normal faults and are taken from Ebinger (1989). Boundary between Magara site (accommodation-zone margin) and Rumonge site (hinged margin) is the Rumonge Fault.

TABLE 2. GRAIN CATEGORIES AND PARAMETERS

Abbreviation	Definition of Grain Type
Qm	Monocrystalline quartz
$Qp_{2\text{-}3}$	Polycrystalline quartz with two or three grains
$Qp_{>4}$	Polycrystalline quartz with more than four grains
Qr	Quartz grain within a rock fragment*
P	Plagioclase feldspar grain
Pr	Plagioclase feldspar grain within a rock fragment*
K	Potassium feldspar grain
Kr	Potassium feldspar grain within a rock fragment*
M	Mica grain
Mr	Mica grain within a rock fragment*
D	Accessory mineral grain
Dr	Accessory mineral grain within a rock fragment*
L	Lithic grain defined by Gazzi Dickinson methodology†
Int	Intraclasts; both clastic and carbonate mud

Note: Recalculated parameters for analyses:

Ratios using traditional methodology:

Q_t/Q_t+F_t+R

F_t/Q_t+F_t+R

R/O_t+F_t+R

Ratios using Gazzi-Dickinson methodology:

$Q_{GD}/Q_{GD}+F_{GD}+L$

$F_{GD}/Q_{GD}+F_{GD}+L$

$L/Q_{GD}+F_{GD}+L$

where

$Q_t = Qm+Qp_{2\text{-}3}+Qp_{>4}$

$Q_{GD} = Qm+Qp_{2\text{-}3}+Qp_{>4}+Qr$

$F_t = P+K$

$F_{GD} = P+Pr+K+Kr$

$R = Qr+Pr+Kr+Mr+Dr+L$

$L = L$

*Rock fragments as defined by Suttner and others (1981); see text for details.

†L category includes metasedimentary, metavolcanic, volcanic, and other lithics.

We prepared thin sections from the 1.25 to 1.5 ϕ size fraction (medium sand) of the sieved samples. A representative thin section for each sample was prepared by carefully pouring warmed epoxy over a homogenized split of the sample contained in small trays. The warmed epoxy was of sufficiently low density to eliminate any separation and biasing due to density contrasts. Once the epoxy hardened, the "artificial rock" was ground and mounted using standard methods. All thin sections were stained with sodium cobaltinitrate to assist in identification of alkali-feldspars.

We counted 400 points for each sample, with a spacing sufficient to cover the whole slide. Both the Gazzi-Dickinson (GD) and traditional (T) methods were employed simultaneously in counting (Dickinson, 1970; Suttner et al., 1981; Ingersoll et al., 1984). In the traditional methodology a rock fragment was defined in the sense of Suttner and others (1981; see also Ingersoll et al., 1984): a multicrystal grain with either two sand-sized phases, or a grain in which no phase comprises more than 90% of the clast area. The number of points varies on recasting the data for either QFL (GD method) or QFR (T-method) diagrams, since multigrain fragments that are counted as mica or accessory minerals in the GD method are counted as rock fragments in the T-method. Categories utilized in classifying grains are presented in Table 2.

In order to test for differences in composition of the different populations (sand suites), we performed a series of one-way analysis of variance (ANOVA) on each of the variables: quartz, feldspar, and rock fragments/lithics. For these tests, our null hypothesis was that the means of the populations (sand suites) were not different above a confidence level chosen at 0.01 (i.e., $\alpha = 0.01$). In testing the null hypothesis, if the calculated ratio of

TABLE 3. CHARACTERISTICS AND SOURCE LITHOLOGY OF DRAINAGE BASINS SAMPLED IN STUDY

River	Location (see Fig. 6)	Size (km^2)	Gradient (%)	Source Lithology Occurring within Drainage Basins: Holo-Pleist Sedimentary Rocks* (%)	Miocene-Pliocene Volcanics (%)	PC Meta-sedimentary Rocks (%)	PC Granite-Gneiss (%)
Dama	1	295	2.96	4	0	40	56
Gonya	2	35	7.92	3	0	60	37
Karonga	3	52	6.81	2	0	38	60
Kavagambwe	4	1.6	37.2	0	0	0	100
Kirasa	5	111	8.15	0.5	0	12.5	87
Ruzibazi	6	165	4.13	1	0	48	51
Ruzizi	7	5,709	423	23	22	21	34
Rwaba	8	285	2.79	33	0	0	67
Shange	9	15	8.56	2	0	0	98
Unnamed	10	6	33.5	0	0	10	90

Note: Relative percent of each source lithology calculated using a polar planimeter by tracing the drainage area of each basin over geologic maps of Theunissen (1986), Claessens and Theunissen (1988), and Ebinger (1989). Source lithology categories represent the dominant (but not exclusive) lithology within each measured map unit.

*Primarily unconsolidated floodplain deposits, but also represents fault blocks of Pleistocene lacustrine strata.

sum of squares (F-ratio) of the populations was larger than the critical value at the confidence level of $p = 0.01$, we rejected the null hypothesis (indicated by $p<\alpha$) and accepted the hypothesis that the sample suites were significantly different. Otherwise, if the calculated F-ratio was less than the critical value we could not reject the null hypothesis (indicated by $p>\alpha$) and we concluded that the composition of the sand suites are not different. All statistical measures of compositional data (means, variances, analysis of variance) were calculated using arcsine transformed data (Sokal and Rohlf, 1981). This angular transformation makes variance independent of the mean, although it does not necessarily eliminate the forced inverse relationships created by constant sum of the percentage data. The calculated statistical measures are recast into percentages to facilitate comparisons.

CONTROL OF CLIMATE AND SOURCE LITHOLOGY VARIATION

In order to investigate the relationship between environmental factors that influence sand composition and the underlying control of basin structure, it is necessary to hold constant other variables that influence sand composition. Two of these variables are climate and source lithology. Within a natural laboratory the scale of Lake Tanganyika, it is impossible to hold these factors truly invariant. Nevertheless, we outline below our means of minimizing the effects of these two variables.

Climate

Situations in which microclimate can be assumed to be constant around a depositional basin are rare. As discussed above, the Lake Tanganyika basin lies within a subhumid region. Detailed climate records for the region are generally poor but suggest that rainfall varies between 838 mm/yr, at Bujumbura, Burundi, the northern edge of the lake, and 1,053 mm/yr, at Nyanza Lac, Burundi, the southernmost of the study sites (a distance of 120 km). Temperatures are warm (20° to 30°C) with little seasonal variation at any of the sites. Proxies for climate, such as vegetation, do not vary significantly over the range of the study sites. Native vegetation was dominated by *Brachystegia* woodland flora with montane forests along range crests and wet savannas on the larger floodplains; much of this has been replaced by tropical cultivation and pasture rangeland in the past 100 yr. Variations in vegetation are controlled primarily by elevation, which, as noted above, is controlled by basin structure.

Source lithology

Table 3 outlines the relative percentage of different rock types within the drainages from which river-mouth sand samples were collected. This provides one means of comparing the variability in source lithology among the sampled source regions. From Table 3, it is apparent that lithology is dominated by a mixture of two Proterozoic-age (PC) rock suites: gneissic/granitic rocks; and metasediments, primarily phyllites, schists, and quartzites. Although the relative proportion of granitic to metasedimentary rocks varies among the individual drainage basins, the only significant variation in source rock types occurs within the Ruzizi drainage basin, which contains significant Pliocene-Miocene–age volcanic rocks (Ebinger, 1989).

TABLE 4. MEAN MODAL COMPOSITIONS OF SAND SUITES CAST IN GAZZI-DICKINSON POINT-COUNTING METHODOLOGY

Sample Suites (Site)	Number of Samples	Quartz Mean (%)	Quartz SD*	Feldspar Mean (%)	Feldspar SD*	Lithics Mean (%)	Lithics SD*
Nyanza Lac	17	74.6	+5.9, -6.3	20.4	+5.6, -5.1	4.6	+4.1, -2.9
Magara	16	73.3	+5.3, -5.6	21.3	+3.7, -3.5	5.1	+3.0, -2.3
Rumonge	17	78.2	+ 6.8, -7.6	18.8	+9.1, -7.7	2.3	+1.8, -1.3
Ruzizi River	14	71.1	+4.0, -4.2	21.8	+3.7, -3.5	6.7	+3.2, -2.6
Pemba, Zaire	5	78.4	+4.5, -4.9	4.4	+3.7, -2.6	16.8	+3.5, -3.2

Note: Means and standard deviations calculated on arcsine transformed data (see text for discussion) and transformed back into percentage data for display. Consequently, rounding errorrs produce sums of less than 100%. In addition, standard deviations are asymmetric around the means when presented in percentage form.
*Standard deviation.

RESULTS

As a first step, we test the assumption of minimal source rock variation discussed above through an empirical analysis of the data. We have plotted all point-count data on a single QFL diagram, with the data cast to conform to the (GD) methodology (Fig. 7A; Table 4). Note that four of the sand suites, samples from the axial, hinged, and accommodation-zone margins, are virtually indistinguishable; these four sand suites are all from the east (Burundi) side of the lake. Both the mean and standard deviation of these four sand suites generally overlap on the ternary diagram (Fig. 7B). An ANOVA, performed on these four sand suites (populations), indicates that the compositional differences are not statistically significant ($p>\alpha$) for both quartz (Q_{GD}) and feldspar (F_{GD}), but the differences are statistically significant ($p<\alpha$) for the lithic category (L). The axial site sand suite is slightly enriched in volcanic lithic fragments derived from the volcanic source rocks present within the Ruzizi River drainage (Table 3); this may explain much of the observed variation in the lithic category. As the Gazzi-Dickinson methodology maximizes source lithology information and minimizes the effects of transport and depositional processes (Dickinson, 1970; Ingersoll et al., 1984), we infer that, at least among the four Burundi populations, source lithology variation is minimal, but not invariant.

The significant difference between the Burundi and Zaire samples apparent in Figure 7A is more enigmatic. The Zaire samples are enriched in aphanitic lithic fragments, primarily quartz-mica aggregates, and relatively depleted in feldspar, relative to the Burundi samples. Although geologic maps (Leppersonne, 1977) indicate that the rock suites are generally the same across the northern part of the basin (Fig. 1), the contrast in sand composition may be related to small-scale source rock differences within the small drainage basin adjacent to the Zaire site. However, at least in the case of the relative dearth of feldspar, the cause may be related to syndepositional alteration (i.e., environmental, not lithologic controls). The Zaire site encompasses a region recently documented as a zone of hydrothermal activity (Tiercelin et al., 1989; Tiercelin and Mondeguer, 1991). The influent hydrothermal waters are more acidic (pH = 7.8) than the lake waters (pH = 8.9) and reach 65°C; in addition, a kaolinite horizon, as much as 8 m thick, blankets the hill slope immediately adjacent to the site. This horizon is interpreted as an alteration zone formed by reaction of the bedrock and the hydrothermal waters (J. J. Tiercelin, personal communication). Syndepositional hydrothermal alteration at this site may have affected sand composition to such an extent that source lithology information is completely masked. Nevertheless, because of the uncertainties in the cause of the compositional variation between sands of the Zaire and remaining Burundi sites, we do not include the Zaire site in the subsequent analyses of environmental effects.

In conclusion, there are not significant differences in the amounts of quartz and feldspar; there are statistically significant, but minor, differences in aphanitic lithic fragments between the accommodation zone, axial and hinged margins when the data are cast in the Gazzi-Dickinson methodology. We conclude that the compositional similarity observed in the data implies that source lithology between sites is generally similar. However, when the data are cast using the traditional methodology (Table 5), there are significant compositional differences between the sites. An ANOVA, performed on the same populations (four sites), but employing the data cast in the traditional methodology, indicates that between-group differences are statistically significant ($p<<\alpha$) for both quartz (Q_t) and rock fragments (R), but are not statistically significant ($p>\alpha$) in the case of feldspar (F_t). Most of the variability resides in the ratio of monomineralic grains to rock fragments. In this case we conclude that these compositional differences, which form the basis for the discussion below, are related to environmental differences between the structurally diverse segments of the lake margin.

Effects of transport processes

The sands entering Lake Tanganyika from rivers along the study sites exhibit large differences in the relative percentage of

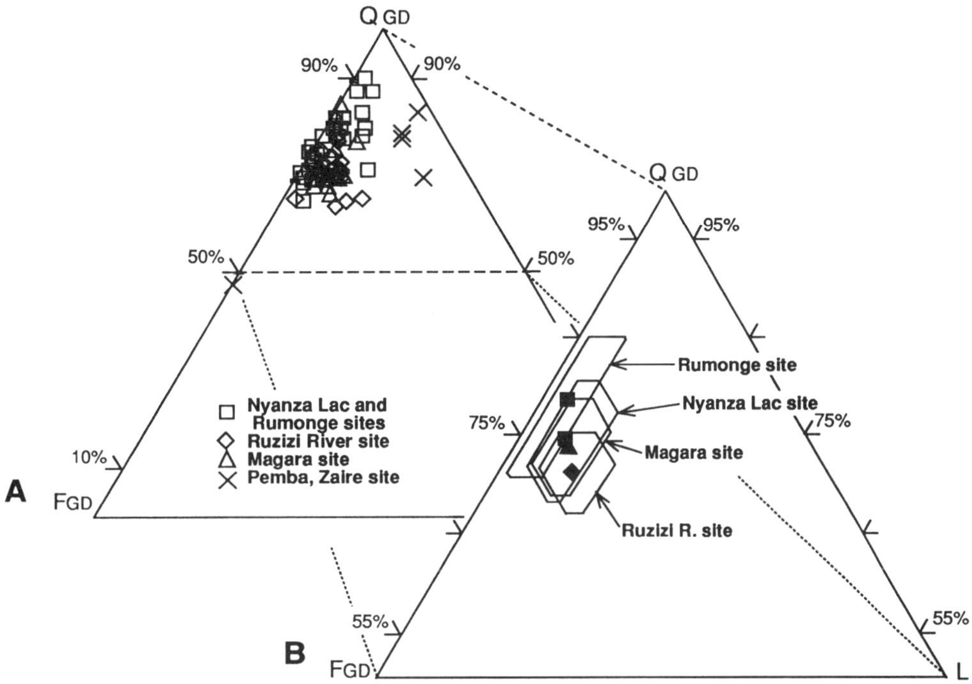

Figure 7. QFL ternary diagram (GD method) of modal sand composition for the Lake Tanganyika sand samples (grain definitions found in Table 2). A, Each point represents one sample, and samples are coded according to margin type. Note the tight clustering of four of the five sand suites. This suggests that variation in source lithology among the sites is minimal. The Pemba, Zaire, samples are outliers and are not considered in later analyses. See text for discussion. B, Enlargement of A plotting means (symbols) and standard deviations (polygons) for four of the sand suites (Magara, Nyanza Lac, Rumonge, and Ruzizi River). Note change in scale from A.

rock fragments (% R in individual samples ranges from 53 to 13%). The rock fragments are primarily quartz-feldspar(-mica) aggregates and metasedimentary clasts, mainly quartz-mica (-feldspar) aggregates. The river-mouth samples are plotted on a QFR diagram in Figure 8; note that the samples roughly lie on a line extending away from the rock-fragment pole. The samples also show distinct clustering by location. Accommodation-zone river-mouth samples consistently contain the highest percentage of rock fragments (average % QFR = 49:13:38) and exhibit large variance (Table 5), whereas shoaling and axial margin river-mouth samples contain significantly fewer rock fragments (average % QFR = 63:14:23—Nyanza Lac site; 65:18:17—Rumonge site; and 66:19:15—Ruzizi R. site) and exhibit much less variance (Table 5). An ANOVA was performed on two populations: (1) the suite of river-mouth samples from the accommodation zone; and (2) the suite of river-mouth samples from the other margins combined. The results suggest that the between-group differences are significant ($p<\alpha$) for both quartz (Q_t) and rock fragments (R), but not for feldspar (F_t).

In summary, the suite of river-mouth sands along the accommodation-zone margin are compositionally less mature relative to the suite of river-mouth sands from either the axial or hinged margins. Sands along these latter two margins, however, do not differ in composition relative to one another; both are proportionately enriched in quartz and depleted in rock fragments.

If source lithology and climate are considered approximately constant between the sites, then compositional differences among sites outlined above likely relate to weathering duration and transport processes. In Figure 9, four different compositional variables for the river-mouth samples are plotted against the corresponding area of the individual drainage basins. The ratio of total number of single grains to grains plus aggregates, the ratio of single quartz grains (Q_t in Table 2) to total quartz (Q_{GD} in Table 2) and the ratio of feldspar grains (Ft in Table 2) to total feldspar (F_{GD} in Table 2) show highly significant positive correlations to drainage area (Fig. 9, A-C). The correlation between monocrystalline quartz grains (Qm in Table 2) to total quartz is not significant (Fig. 9D). These relationships suggest that drainage-basin size is one important control on sand composition as it enters Lake Tanganyika. In Figure 9A, the four points that plot above the regression line, i.e., the "grain-rich" samples, are all from hinged margin streams. The "grain-rich" nature of these river-mouth samples may be related to the lower gradients of hinged margin streams in comparison to accommodation-zone margin streams of equivalent size. For example, the Ruzibazi River (from hinged margin) is similar in size to the Kirasa and Karonga Rivers (from accommodation-zone margin), yet the latter two are nearly

TABLE 5. MEAN MODAL COMPOSITIONS OF SAND SUITES CAST IN TRADITIONAL POINT-COUNTING METHODOLOGY

Sample Suites (Site)	Number of Samples	Quartz Mean (%)	Quartz SD*	Feldspar Mean (%)	Feldspar SD*	Rock Fragments Mean (%)	Rock Fragments SD*
Nyanza Lac	17	63.7	+7.2, -7.5	16.0	+5.2, -4.6	19.6	+7.1, -6.3
Beach	5	60.0	+4.2, -4.3	12.8	+3.8, -3.4	26.8	+7.4, -6.8
Delta	4	60.6	+9.3, -9.7	20.8	+6.8, -6.1	18.3	+5.4, -4.9
Shallow water	7	69.3	+5.1, -5.4	15.2	+2.1, -2.0	15.4	+3.8, -3.4
Rivermouth	1	63.4	- -	14.1	- -	22.5	- -
Magara	16	55.0	+10.3,-10.6	13.8	+3.2, -2.9	30.7	+11.4,-10.3
Beach	5	62.8	+5.8, -6.0	13.7	+4.3, -3.8	23.1	+6.8, -6.2
Delta	1	70.5	- -	14.7	- -	14.7	- -
Shallow water	6	51.8	+10.8,-10.9	14.3	+2.1, -2.0	33.6	+10.4, -9.7
Rivermouth	4	49.1	+11.6,-11.5	12.4	+3.5, -3.1	37.9	+14.1,-13.2
Rumonge site	17	68.8	+7.8, -8.3	16.1	+7.6, -6.3	14.7	+3.7, -3.4
Beach	6	68.0	+9.3,-10.0	16.1	+8.7, -7.1	15.5	+2.8, -2.6
Delta	4	69.8	+7.5, -8.1	16.5	+10.0, -8.0	13.1	+2.2, -2.1
Shallow water	2	78.7	+1.0, -1.0	11.3	+4.8, -4.0	9.6	+3.7, -3.2
Rivermouth	5	64.6	+4.5, -4.6	17.9	+6.2, -5.5	17.2	+3.0, -2.8
Ruzizi River	14	63.7	+4.1, -4.2	18.4	+3.5, -3.3	17.6	+4.6, -4.2
Beach	3	64.0	+2.0, -2.1	21.7	+5.2, -4.8	15.9	+5.8, -5.1
Delta	6	64.9	+4.5, -4.7	16.7	+1.6, -1.5	18.3	+4.9, -4.4
Shallow water	3	59.6	+4.6, -4.7	19.4	+2.0, -2.0	21.0	+2.7, -2.6
Rivermouth	2	65.9	+1.3, -1.4	18.9	+5.2, -4.7	15.0	+4.8, -4.2

Note: Means and standard deviations calculated on arcsine transformed data (see text for discussion) and transformed back into percentage data for display. Consequently, rounding errors produced sums of less than 100%. In addition, standard deviations are asymmetric around the means when presented in percentage form.
*Standard deviation.

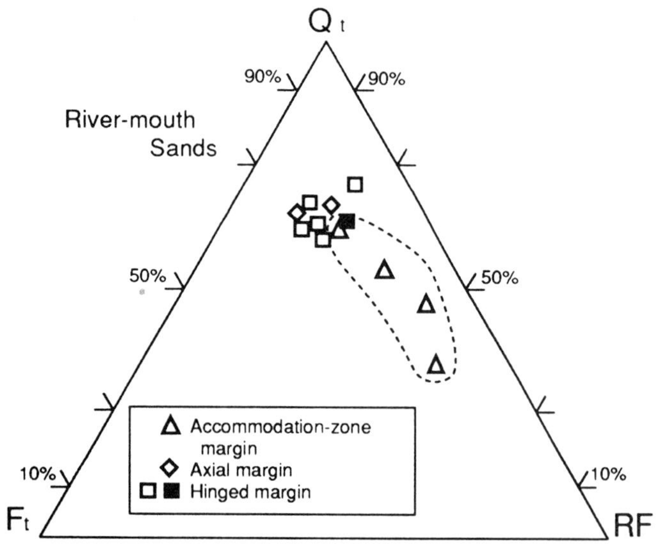

Figure 8. QFR ternary diagram (T-method) of modal sand composition for the river-mouth sand samples. Samples coded according to the tectonic margin from which they were collected. The solid square represents the single river-mouth sand from the Nyanza Lac site. Dashed line qualitatively outlines the range in composition of the accommodation-zone margin samples (Magara site).

twice as steep (Table 3). This implies that stream gradient is also an important variable in controlling sand composition. In fact, the F_t/F_{GD} ratio correlates significantly better with stream gradient ($r = 0.91$; $p<0.001$) than with drainage-basin size; samples with a higher ratio of feldspar grains to multi-grain fragments occur in rivers of lower gradients.

The importance of both size and gradient of the transportation system in modifying sand composition is not a new conclusion (Cameron and Blatt, 1971; Self, 1975; Breyer and Bart, 1978; Potter, 1978; Grantham and Velbel, 1988, Johnsson et al., 1991; Savage and Potter, 1991) although a number of workers have downplayed its importance (Breyer and Bart, 1978; Mack, 1978). Our data support the conclusion that, within a subhumid to humid climate, both the length and gradient of the river are key variables affecting sand composition. More importantly, since size and gradient of the drainage basin are directly related to basin structure (Fig. 5), the asymmetry of the basin underlies the extent of compositional modification incurred by the sediment during transport to the depositional site.

We have utilized drainage basin size as a proxy for the total mineralogic maturation potential experienced by the sediment from its formation in the hinterland to deposition at the lake margin. The maturation of sediment through this process can be

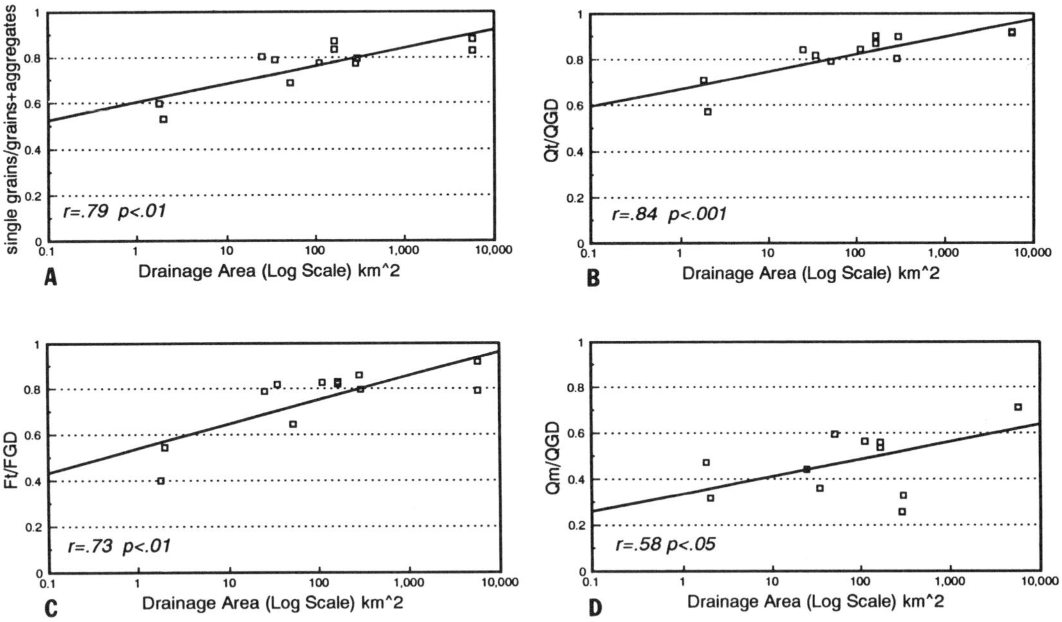

Figure 9. Plots of different compositional ratios versus log of drainage area (square kilometers). A, On Y-axis, numerator (grains) represents the summation of Qm,Qp_{2-3}, $Qp_{>4}$, P,K,M,D, whereas denominator (grains + aggregates) represents the summation of the numerator plus the summation of Qr,Pr,Kr,Mr,Dr,L (see Table 2 for explanation of abbreviations). B-D, See Table 2 for definition of numerator and denominator in ratios. Sloping lines represent logarithmic regression; residuals (r) and confidence level (p) reported on each plot.

divided into four parts: (1) nature and intensity of weathering at the source site; (2) removal rate from the weathering site; (3) transport rate to the depositional site, including periods of sediment storage; and, (4) magnitude of physical and chemical energy imparted on the sediment during transport (Johnsson et al., 1991). Rates of weathering and erosion are generally poorly constrained variables; in the case of the Lake Tanganyika basin, modern erosion rates are unquestionably accelerating due to anthropogenic effects (Bizimana and Duchafour, 1991). Nevertheless, all drainages are sourced in steep, generally uniform mountainous terrane, and most of the mountain slopes in the northern part of the lake basin are equally disturbed (A. S. Cohen, unpublished data). Therefore we assume that the rates of weathering and erosion are not significantly different among the sites. This leaves the transport system itself—factors 3 and 4 above—as the most important variable in affecting sand composition. As documented in Figure 5, the size of the basin is inversely proportional to the gradient of the drainage, so sediment within the larger drainage basins experiences a longer period of transport, not only because of the length of transport, but also because of periods of presumed sediment storage.

Effects of depositional processes

Figure 10 depicts sand composition data from all three structural margins on QFR diagrams; the mean and standard deviations for each sand suite are presented in Table 5. Included on each plot is a dashed line outlining the range of composition of river-mouth sands from that site; this range can be thought of as the initial composition of the sand before any subsequent compositional modification at the depositional site. There is considerable overlap in composition among the sites, but, as stated above, an ANOVA performed on each compositional variable (Q_t, F_t, R) indicates that the between-group differences are significant for both quartz and rock fragments, but not for feldspar. In addition, on close inspection of Figure 10, A-D and Table 5, one can discern that compositional variability (scatter on ternary diagrams) exhibit subtle differences of each site. We propose that it is the structurally controlled differences in depositional site processes that control these subtle differences in the nature of the compositional variability for each site.

Accommodation-zone margin (Magara site). The accommodation-zone margin sands are relatively quartz-poor (average % QFR = 55:14:31). Of all four sites, this sand suite exhibits the largest variance in quartz and rock-fragment abundances (Table 5). In addition, from Figure 10A it is apparent that the compositional scatter in the population as a whole is represented by the compositional scatter in the river-mouth samples alone (dashed line).

As discussed above, the large variance in the composition of river-mouth sands indicates that the effect of transport processes on sand composition is highly variable along this type of margin. The clustering of sand composition from the other environments of the sand suite within the range of composition of river-mouth sands (Fig. 10A) suggests that the magnitude of reworking at the depositional site is minimal along this margin. This is especially

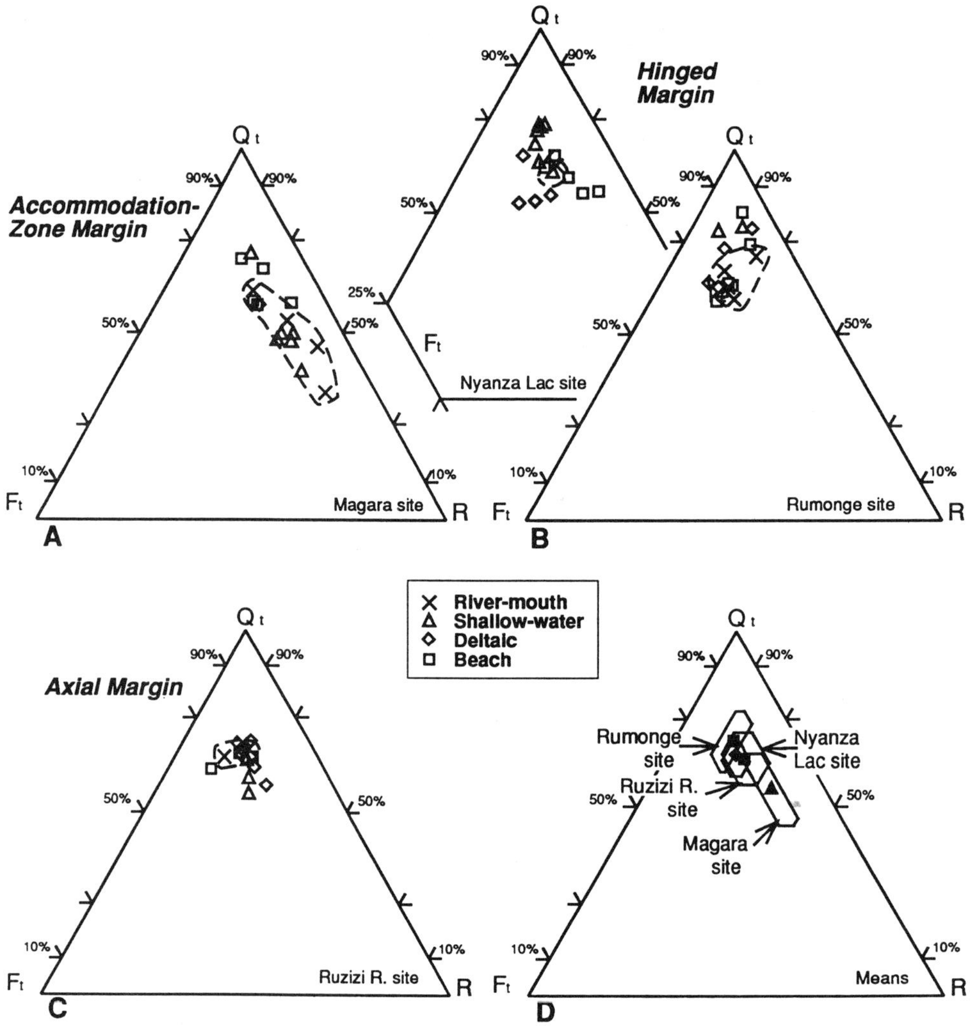

Figure 10. A-C, QFR ternary diagram (T-method) of modal sand composition for three of the tectonic margin types (escarpment margin samples not included). A, Accommodation-zone margin. B, Hinged margin. C, Axial margin. D, Means (symbols) and standard deviations (polygons) for each of the margins in A-C. Dashed lines in A-C represent the range in composition of river-mouth sands from each study site.

apparent by considering that, with one exception the shallow-water samples, all collected within a kilometer of the shore, cluster within the scatter of the river-mouth samples. Because of the structural morphology of this type of margin, gradients are very steep; consequently, the zone above wave base (zone of reworking) is very narrow, limiting the extent of sediment reworking. Grain-size data support this interpretation; the shallow-water samples are consistently poorly sorted.

The composition of beach samples along the accommodation-zone margin are slightly enriched in quartz and depleted in rock fragments compared to the composition of the shallow-water and river-mouth samples (Table 5). Furthermore, beach sands cluster more tightly than river-mouth sands, suggesting that sands are compositionally homogenized by nearshore/swash-zone reworking. A relative increase in quartz can occur through consequent breakdown of unstable grains during constant agitation and abrasion within the shoreface environment (Davies and Ethridge, 1975; Mack, 1978).

Hinged margin (Nyanza Lac and Rumonge sites). The hinged margin sand suite from both sites is plotted in Figure 10B. The composition of sands from these two sites is both quartz-rich and depleted in rock fragments relative to the accommodation-zone margin (average % QFR = 64:16:20 and 69:16:15 for the Nyanza Lac and Rumonge sand suites, respectively). Inspection of the standard deviations of the two sand suites (Table 5) suggests that variance in the composition of the hinged margin sand suite is less (except for feldspar) than the variance found in the accommodation-zone margin sand suite. In addition, for both the hinged margin sand suites, quartz displays the largest variance of the three variables. The high degree of scatter in quartz content among environments appears in Figure 10B as a rough vertical trend in the data toward the quartz pole. This trend implies an enrichment in quartz at the expense of both rock fragments and feldspar and is most distinctive in the shallow-water samples. The average quartz content of the shallow-water sands ranges from 69 (Nyanza Lac site) to 79% (Rumonge site), whereas the average quartz content of the corresponding river-mouth sands is 63 and 65%, respectively. The quartz enrichment may be caused by two possible environmental factors: physical reworking and associated mechanical destruction of unstable grains, and chemical destruction of feldspar and rock fragments.

The shallow-water samples were collected along large low-gradient benches typical of hinged margins (Cohen, 1990). These benches experience low net sedimentation rates that result from the wide spacing of the larger drainage systems, the pervasive reworking of sediments by fairweather and storm currents, and low subsidence rates (Cohen, 1990). These factors conspire to produce long residence times for the sediment at the water/sediment interface. For example, Cohen (1989) obtained radiocarbon ages as old as 3 Ka for surficial bioclasts on these benches. Previously, Mack (1978) demonstrated that the relative abundance of feldspar grains are reduced by as much as 50% in the shallow-water environment of an ancient marine basin when compared to the sands from the derivative fluvial system; he attributed this to mechanical breakdown of the feldspar during long periods of abrasion in the zone above wave base. We similarly conclude that, at least for the rock-fragment population, mechanical breakdown during extended periods of physical reworking on these morphologic benches decreases the relative amounts of rock fragments. This reworking may have been accentuated when water depths were shallower during the lake-level changes that occurred throughout the Pleistocene and Recent.

In addition, petrographic evidence suggests that feldspar grains in this environment are also undergoing chemical destruction. Much of the sediment along these margins develops a carbonate (algal) coating; extensive regions of these morphologic benches are covered by ooid shoals and a diverse suite of carbonate sediments (Cohen and Thouin, 1987). Carbonate sedimentation is much more prevalent along hinged margins in Lake Tanganyika because of the wide, shallow benches that most clastic sediment bypasses (Cohen, 1990). In the samples rich in coated grains, feldspar, especially if coated, exhibits a high degree of alteration. In some cases the grains were virtually destroyed. Therefore, we conclude that the feldspar grains along these low-gradient benches are destroyed primarily by replacement through carbonate formation, rather than by mechanical destruction.

The average composition of the beach and delta samples from these two sites is very similar to the average composition of the corresponding river-mouth samples (Table 5), suggesting that significant compositional modification is limited to the shallow-water benches described above.

Axial margin (Ruzizi River site). The suite of sands from the axial margin (average % QFR = 64:18:18) is very similar in composition to both hinged margin sand suites. However, the variance, as exhibited through the standard deviations (Table 5), is lower than in any of the other sand suites. This is manifested by the tight clustering of the data in Figure 10C; this clustering is significantly different than the scatter observed in Figure 10, A or B.

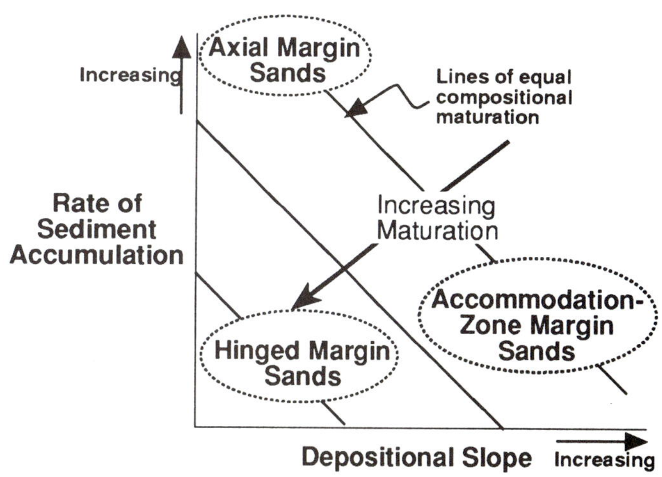

Figure 11. Schematic representation of the relationship between depositional slope and sediment accumulation rate on controlling the degree of compositional maturation of sands at the depositional site. No scale is implied. Note that the accommodation-zone and axial margin sands both plot in the region of low compositional maturation, but are controlled by very different processes. Modification that occurs in transit is not depicted in the diagram, but is significantly different between the axial and the accommodation-zone margin sands.

We interpret the relatively tight clustering of the axial margin samples to represent minimal reworking and compositional modification at the depositional site. A mechanism for limiting the amount of reworking along the axial margin may be the much higher sedimentation rate, which limits the residence time of the sediment at the sediment/water interface. Short-term sedimentation rates of 5 mm/yr have been obtained from ^{14}C dating of short cores collected just south of the Ruzizi R. site (Tiercelin and Mondeguer, 1991).

The lack of additional modification at the depositional site of the axial margin is similar to the lack of compositional modification interpreted for the accommodation-zone margin and unlike the significant modification interpreted for the hinged margin. This can be attributed to three variables: extent of modification during transport, depositional gradient, and sedimentation rate. The interaction of the last two variables is depicted in Figure 11. Note that although the axial and accommodation-zone margins experience similarly minor reworking, the extent of modification during transport creates compositionally distinct sands.

Summary of environmental processes on sand composition

Figure 12 summarizes the primary and secondary trends observed in the composition of the Lake Tanganyika sands that can be attributed to transport and depositional processes. Transport processes are controlled by two variables: drainage size (i.e., length of drainage), and gradient of the drainage system.

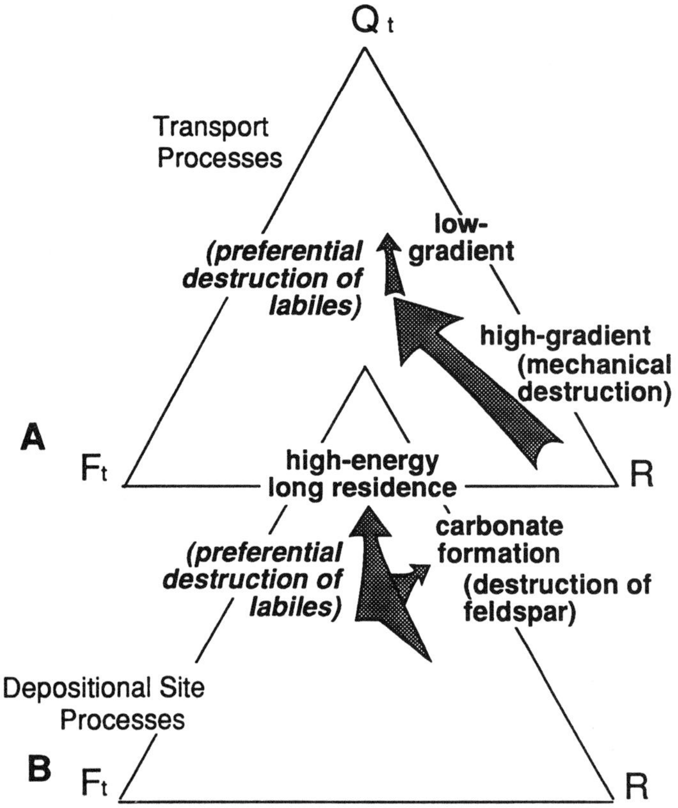

Figure 12. Summary of the trends observed in Lake Tanganyika sands that can be attributed to transport and depositional processes. A, Transport processes. B, Depositional site processes. Note that not all processes depicted in the A and B occur at each study site. The size of the arrows indicate relative magnitude of process.

Mechanical destruction primarily of rock fragments is the dominant process in the sampled drainages; the degree of disaggregation of rock fragments is proportional to the length of transport, and the resultant compositional trend is directed away from the rock-fragment pole in a QFR diagram (Fig. 12A). In the larger drainages (>200 km^2), or in drainages with relatively low gradients (<10%), preferential destruction of both feldspar and rock fragments and consequent enrichment in quartz may be a second process, although it is not well constrained by the present data.

At the depositional site (Fig. 12B), mechanical destruction of labile phases appears to occur within the high-energy beach environment only when the starting material is generally labile-rich, that is, has not undergone significant modification during transport. Preferential destruction of labile phases is dominant on shallow-water benches, where sediment residence time is long and carbonate precipitation is common. Preferential destruction of feldspar appears to be a locally important process where carbonate formation preferentially replaces feldspar.

INTERPRETING BASIN STRUCTURE FROM SAND COMPOSITION

The trends in the Lake Tanganyika data suggest that differentiation of structural morphology of an ancient asymmetric basin is feasible based on careful statistical analysis of sand composition. It is important to stress that the compositional changes are subtle, and primarily limited to the ratio of monomineralic grains and rock fragments; significant changes in mineralogic abundances in the sands were not observed. In attempting to decipher basin structure for ancient basins from sandstone composition, it is necessary to define a specific sampling strategy. Sands from different study sites (cores/outcrops) must be collected from coeval units; the largest stratigraphic interval that can be compared among sites will be determined by the variability in source lithology as it is unroofed. Sandstones from laterally and vertically distinct units within the targeted interval should be sampled to obtain sands from a range of environments (e.g., beach, deltaic, fluvial, shallow water). This will provide the requisite sand suite from which to analyze compositional trends within each study site. Finally, the data must be cast in the traditional method (Suttner et al., 1981), but we suggest the data be recorded so it can also be recast using the Gazzi-Dickinson methodology.

Figure 13 illustrates predictions of sandstone composition along different regions of an asymmetric basin similar to Lake Tanganyika where bedrock lithology and climate are uniform across the basin. Sands from the axial margin will be uniformly quartzose, with very limited scatter in the data. Hinged margin sands will also be quartz-rich; there may be no significant difference between the average composition of the axial and the hinged margins sands. However, the hinged margin samples show a more significant dispersion in the composition (primarily toward the quartz pole), since the sands along this margin experience the most drastic environmental overprinting at the depositional site. Sands from the accommodation-zone margin will exhibit an even higher degree of variation in sand composition; two distinct populations may be discernible. This scatter results from the dramatic structural transition and the consequent variability in sedimentary processes typical of this type of margin. Because of the variability in drainage morphology typical of these margins, initial composition of sands will be heterogeneous; additional depositional site reworking will vary from minimal (along the steeper parts, e.g., Magara site) to extensive (along the transition to the adjacent hinged margin, e.g., Rumonge site). Finally, although our present data are minimal for this margin type, we predict that escarpment margin sands should be consistently rich in rock fragments, with low to moderate scatter in the composition of sands from different facies. Because drainages along the escarpments are consistently small, we infer there should be less scatter in the data compared to the accommodation-zone margin samples.

In this type of analysis many other factors can complicate the relationships outlined above. Climate and source lithology must be well constrained and invariant to illuminate these patterns. In addition, the subhumid climate of Lake Tanganyika accentuates the environmental effects induced by the basin structure. Whether the above results apply to basins in more arid climates must await further data collection. Finally, mixing due to longshore currents can produce overlap between the distinct fields. An example of the effect of this last process can be seen in

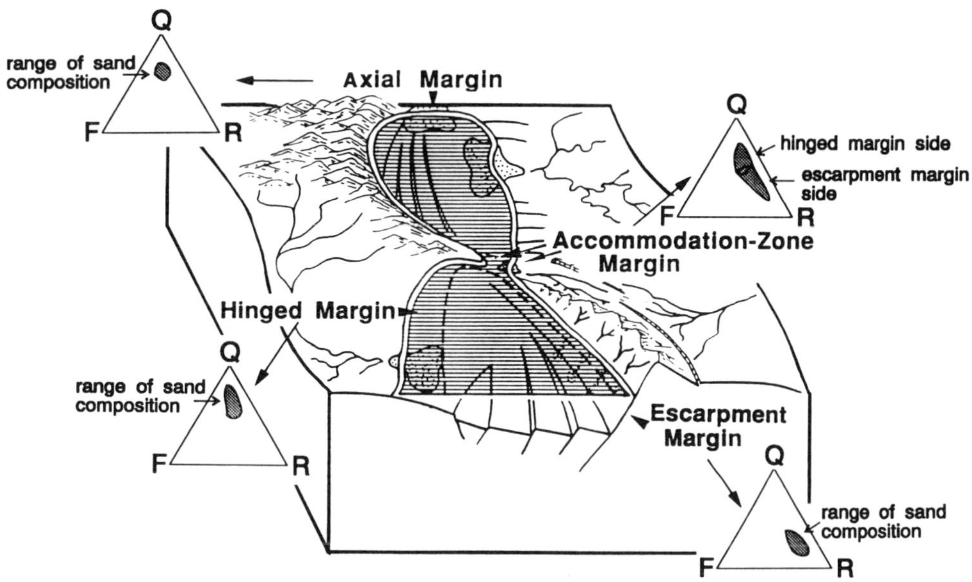

Figure 13. Prediction of range of composition for sandstone suites collected from different facies within a half-graben lake basin. Both average composition of the sands and the trends observed in the composition are important in distinguishing structural margins solely from sand composition.

Figure 10A. Note that within the sand suite for the accommodation-zone margin there is a single, anomalously quartz-rich shallow-water sample. This sample was collected immediately north of the Rumonge Fault (on the downthrown block) in 63-m water depth. We infer that this sand is derived from the adjacent hinged margin to the south and has been carried by longshore currents over the Rumonge Fault and into the deeper water. The composition of this single sample is approximately equivalent in composition to the average composition of the sand suite from the Rumonge site. This type of mixing may create complications in analyzing differences between populations, but may also provide important information about sediment transport processes.

CONCLUSIONS

1. Within the asymmetric Lake Tanganyika basin, transport and depositional processes differ as a consequence of basin asymmetry. The asymmetry in transport and reworking variables produces distinct compositional suites of sands along the four structural margins of the basin when the data are cast using the traditional point-counting method.

2. Sands entering Lake Tanganyika along the steeper, faulted margins of the lake (accommodation zone) contain, on average, twice as many rock fragments as sands entering along the hinged (unfaulted) or axial margins of the lake. The small/steep drainages along faulted margins produce only limited compositional modification dominated by mechanical destruction of rock fragments. The large/low-gradient streams of the axial and hinged margin produce more significant compositional modification.

3. Once sands reach the depositional site along the accommodation-zone margins, little additional modification occurs. Sands vary in composition primarily because they are derived from a number of streams of varying size. Only beach samples exhibit compositional modification; they are generally more quartzose than deeper water samples.

4. Hinged margin sands exhibit the greatest degree of compositional modification. Long residence time on high-energy, shallow-water benches produces an increase in quartz content relative to river-mouth sands. Destruction of unstable grains is aided by carbonate precipitation in the form of ooids, algal coatings, and beach-rock formation.

5. Axial margin sands derived from a single river (Ruzizi River) are uniformly quartzose, and exhibit very little compositional scatter. High deltaic sedimentation rates may limit the extent of depositional site overprinting.

6. The structurally induced changes in sand composition may be used to infer paleostructure in ancient, asymmetric basins. Distinction between margins along steep, faulted sides and the relatively low-gradient, unfaulted sides can be made by comparison of composition alone. Distinguishing between hinged and axial margins will be more difficult, requiring analysis of a suite of samples collected from a number of different facies. Hinged margin sands will exhibit a higher degree of scatter, primarily toward the quartz pole, betraying the long residence time of the sediment at the depositional site. Axial margin sands, however, will be more compositionally homogeneous, even when comparing different facies.

7. Changes in source lithology among different parts of the basin, highly fluctuating climatic conditions, and significant longshore currents within the lake can produce complexities in trying

to interpret paleostructure in ancient basins. In addition, no attempt has been made to predict the effects of burial diagenesis in preferentially masking the trends outlined.

ACKNOWLEDGMENTS

This work was partially funded by grants from Amoco Production Company, Conoco, Inc., Marathon Oil Company, and the H. M. Keck Foundation. We deeply appreciate the assistance of Gaspard Ntakimazi, Laurent Ntahuga, Pontien Ndabaneze, and others at the Faculté des Sciences, Université du Burundi, for logistical support while in Africa. We also thank Debra Gevirtzman, Mark Johnston, Robin Jones, Elinor Michel, Kelly West, Bouduain Hussein, and Ebode Mkoni for their valuable assistance over our various field seasons. Lynn Soreghan and William R. Dickinson provided valuable comments on earlier versions of the manuscript. H. A. Cohen, M. J. Johnsson, and C. A. Suczek provided highly rigorous reviews.

REFERENCES CITED

Anadón, P., Cabrera, L., Julia, R., and Marzo, M., 1991, Sequential arrangement and asymmetrical fill in the Miocene Rubielos de Mora Basin (northeast Spain), *in* Anadón, P., Cabrera, L., and Kelts, K., eds., Lacustrine facies analysis: International Association of Sedimentologists Special Publication 13, p. 257–278.

Baker, B. H., 1986, Tectonics and volcanism of the southern Kenya rift valley and its influence on rift sedimentation, *in* Frostick, L. E., Renaut, R. W., Reid, I., and Tiercelin, J. J., eds., Sedimentation in the East African rifts: Geological Society Special Publication 25, p. 45–58.

Basu, A., 1976, Petrology of Holocene fluvial sand derived from plutonic source rocks: Implications to paleoclimatic interpretations: Journal of Sedimentary Petrology, v. 46, p. 694–709.

Basu, A., 1985, Influence of climate and relief on compositions of sands released at source areas, *in* Zuffa, G. G., ed., Provenance of arenites: Dordrecht, D. Reidel, p. 1–18.

Bizimana, M., and Duchafour, H., 1991, Etude pour l'amenagement de bassins versants: Cas du bassin de la riviere Ntahangwa [abs.], *in* Cohen, A. S., ed., Report on the First International Conference on the Conservation and Biodiversity of Lake Tanganyika: Biodiversity Support Project, p. 43.

Bradley, W. C., 1970, Effect of weathering on abrasion of granitic gravel, Colorado River (Texas): Geological Society of America Bulletin, v. 81, p. 61–80.

Breyer, J. A., and Bart, H. A., 1978, The composition of fluvial sands in a temperate semiarid region: Journal of Sedimentary Petrology, v. 48, p. 1311–1320.

Cameron, K. L., and Blatt, H., 1971, Durabilities of sand size schist and "volcanic" rock fragments during fluvial transport, Elk Creek, Black Hills, South Dakota: Journal of Sedimentary Petrology, v. 41, p. 565–576.

Claessens, W., and Theunissen, K., 1988, Carte géologique de Burundi, Feuille Bururi: Burundi, Burundi Ministère Travaux Publique, Energie, Mines, scale 1:100,000.

Cohen, A. S., 1989, The taphonomy of gastropod shell accumulations in large lakes: An example from Lake Tanganyika, Africa: Paleobiology, v. 15, p. 26–45.

Cohen, A. S., 1990, Tectono-stratigraphic model for sedimentation in Lake Tanganyika, Africa, *in* Katz, B. J., eds., Lacustrine basin exploration—Case studies and modern analogs: American Association of Petroleum Geologists Memoir 50, p. 137–150.

Cohen, A. S., and Thouin, C., 1987, Nearshore carbonate deposits in Lake Tanganyika: Geology, v. 15, p. 414–418.

Coulter, G. W., 1963, Hydrological changes in relation to biological productivity in southern Lake Tanganyika: Limnology and Oceanography, v. 8, p. 463–477.

Coulter, G. W., and Spigel, R. H., 1991, Hydrodynamics (Chapter 3), *in* Coulter, G. W., ed., Lake Tanganyika and its life: New York, Oxford University Press, p. 49–75.

Davies, D. K., and Ethridge, F. G., 1975, Sandstone composition and depositional environment: American Association of Petroleum Geologists Bulletin, v. 59, p. 239–264.

Dickinson, W. R., 1970, Interpreting detrital modes of graywacke and arkose: Journal of Sedimentary Petrology, v. 40, p. 695–707.

Dickinson, W. R., 1985, Interpreting provenance relations from detrital modes of sandstones, *in* Zuffa, G. G., ed., Provenance of arenites: Dordrecht, D. Reidel, p. 333–361.

Ebinger, C. J., 1989, Tectonic development of the western branch of the East African Rift System: Geological Society of America Bulletin, v. 101, p. 885–903.

Ethridge, F. G., 1977, Petrology, transport, and environment in isochronous Upper Devonian sandstone and siltstone units, New York: Journal of Sedimentary Petrology, v. 47, p. 53–65.

Frostick, L., and Reid, I., 1989, Is structure the main control of river drainage and sedimentation in rifts?: Journal of African Earth Sciences, v. 8, p. 165–182.

Gore, P.J.W., 1988, Lacustrine sequences in an early Mesozoic rift basin: Culpepper Basin, Virginia, USA, *in* Fleet, A. J., Kelts, K., and Talbot, M. R., eds., Lacustrine petroleum source rocks: Geological Society Special Publication 40, p. 247–278.

Grantham, J. H., and Velbel, M. A., 1988, The influence of climate and topography on rock-fragment abundance in modern fluvial sands of the southern Blue Ridge Mountains, North Carolina: Journal of Sedimentary Petrology, v. 58, p. 219–227.

Hamblin, A. P., and Rust, B. R., 1989, Tectono-sedimentary analysis of alternate-polarity half-graben basin-fill successions: Late Devonian–Early Carboniferous Horton Group, Cape Breton Island, Nova Scotia: Basin Research, v. 2, p. 239–255.

Hecky, R. E., and Degens, E. T., 1973, Late Pleistocene–Holocene chemical stratigraphy and paleolimnology of the rift valley lakes of central Africa: Woods Hole Oceanographic Institution Technical Report 73–28, 93 p.

Ingersoll, R. V., Bullard, T. F., Ford, R. L., Grimm, J. P., Pickle, J. D., and Sares, S. W., 1984, The effect of grain size on detrital modes: A test of the Gazzi-Dickinson point-counting method: Journal of Sedimentary Petrology, v. 54, p. 103–116.

Johnsson, M. J., Stallard, R. F., and Meade, R. H., 1988, First-cycle quartz arenites in the Orinoco River Basin, Venezuela and Columbia: Journal of Geology, v. 96, p. 263a–278.

Johnsson, M. J., Stallard, R. F., and Lundberg, N., 1991, Controls on the composition of fluvial sands from a tropical weathering environment: Sands of the Orinoco River drainage basin, Venezuela and Columbia: Geological Society of America Bulletin, v. 103, p. 1622–1647.

Kufferath, J., 1952, Le milieu biochemique, exploration hydrobiologique du Lac Tanganyika—Résultats scientifiques: Institut Royale des Sciences Naturelles de Belgique, v. 1, p. 29–47.

Leeder, M., and Gawthorpe, R., 1987, Sedimentary models for extensional tilt-block/half-graben basins, *in* Coward, M. P., Dewey, J. F., and Hancock, P. L., eds., Continental extensional tectonics: Geological Society of London Special Publication 28, p. 139–152.

Leppersonne, J., 1977, Carte géologique du Zaire: Tervueren, Belgium, Musée Royal de l'Afrique Centrale, scale 1:2,000,000.

Mack, G. H., 1978, The survivability of labile light-mineral grains in fluvial, aeolian, and littoral marine environments: The Permian Cutler and Cedar Mesa Formations, Moab, Utah: Sedimentology, v. 25, p. 587–604.

Mack, G. H., 1981, Composition of modern stream sand in a humid climate derived from a low-grade metamorphic and sedimentary foreland fold-thrust belt of North Georgia: Journal of Sedimentary Petrology, v. 51, p. 1247–1258.

Mann, W. R., and Cavaroc, V. V., 1973, Composition of sand released from three source areas under humid, low relief weathering in the North Carolina

Piedmont: Journal of Sedimentary Petrology, v. 43, p. 870–881.
McConnell, R. B., 1972, Geological development of the rift-system of eastern Africa: Geological Society of America Bulletin, v. 83, p. 2549–2572.
Miall, A. D., 1981, Alluvial sedimentary basins: Tectonic setting and basin architecture, *in* Miall, A. D., eds., Sedimentation and tectonics in alluvial basins: Geological Association of Canada Special Paper 23, p. 1–33.
Morley, C. K., 1988, Variable extension in Lake Tanganyika: Tectonics, v. 7, p. 785–801.
Morley, C. K., Nelson, R. A., Patton, T. L., and Munn, S. G., 1990, Transfer zones in the East African Rift System and their relevance to hydrocarbon exploration in rifts: American Association of Petroleum Geologists Bulletin, v. 74, p. 1234–1253.
Pittman, E. D., 1970, Plagioclase feldspars as an indicator of provenance in sedimentary rocks: Journal of Sedimentary Petrology, v. 40, p. 591–598.
Potter, P. E., 1978, Petrology and chemistry of modern big river sands: Journal of Geology, v. 86, p. 423–449.
Potter, P. E., 1986, South America and a few grains of sand: Part 1—Beach sands: Journal of Geology, v. 94, p. 301–319.
Rosendahl, B. R., and 7 others, 1986, Structural expressions of rifting: Lessons from Lake Tanganyika, *in* Frostick, L. E., Renault, R. W., Reid, I., and Tiercelin, J. J., eds., Sedimentation in the East African rifts: Geological Society Special Publication 25, p. 29–44.
Sander, S., and Rosendahl, B. R., 1989, The geometry of rifting in Lake Tanganyika, East Africa: Journal of African Earth Sciences, v. 8, p. 323–354.
Savage, K. M., and Potter, P. E., 1991, Petrology of modern sands of the Rios Guaviare and Inirida, southern Columbia: Tropical climate and sand composition: Journal of Geology, v. 99, p. 287–298.
Self, R. P., 1975, Petrologic changes in fluvial sediments in the Rio Nautla drainage basin, Veracruz, Mexico: Journal of Sedimentary Petrology, v. 45, p. 140–149.
Sokal, R. R., and Rohlf, F. J., 1981, Biometry (second edition): New York, W.H. Freeman, 859 p.
Soreghan, M. J., and Cohen, A. S., 1991, Facies distributions within contrasting structural components of a rift lake: Lake Tanganyika, Africa [abs.]: American Association of Petroleum Geologists Bulletin, v. 75, p. 674.
Suttner, L. J., 1974, Sedimentary petrographic provinces: An evaluation, *in* Ross, C. A., ed., Paleogeographic provinces and provinciality: Society of Economic Paleontologists and Mineralogists Special Publication 21, p. 75–84.
Suttner, L. J., and Dutta, P. K., 1986, Alluviual sandstone composition and paleoclimate, I. Framework mineralogy: Journal of Sedimentary Petrology, v. 56, p. 329–345.
Suttner, L. J., Basu, A., and Mack, G. H., 1981, Climate and the origin of quartz arenites: Journal of Sedimentary Petrology, v. 51, p. 1235–1246.
Theunissen, K., 1986, Carte géologique de Burundi, Feuille Rumonge: Burundi, Burundi Ministère Travaux Publique, Energie, Mines, scale 1:100,000.
Tiercelin, J. J., 1991, Natural resources in the lacustrine facies of the Cenozoic rift basins of East Africa, *in* Anadón, P., Cabrera, L., and Kelts, K., eds., Lacustrine facies analysis: International Association of Sedimentologists Special Publication 13, p. 3–38.
Tiercelin, J. J., and Mondeguer, A., 1991, The geology of the Tanganyika Trough (Chapter 2), *in* Coulter, G. W., ed., Lake Tanganyika and its life: New York, Oxford University Press, p. 7–48.
Tiercelin, J. J., Thouin, C., Tchibangu, K., and Mondeguer, A., 1989, Discovery of sublacustrine hydrothermal activity and associated massive sulfides and hydrocarbons in the north Tanganyika Trough, East African Rift: Geology, v. 17, p. 969–1064.
Turner-Peterson, C. E., and Smoot, J., 1985, New thoughts on facies relationships in the Triassic Stockton and Lockatong Formations, Pennsylvania and New Jersey, *in* Robinson, G. R., and Froelich, A. J., eds., Proceedings, 2nd, U.S. Geological Survey Workshop on the Early Mesozoic Basins of the Eastern United States: U.S. Geological Survey 946, p. 10–16.

Manuscript Accepted by the Society December 24, 1992

Printed in U.S.A.

Geological Society of America
Special Paper 284
1993

Contribution of Saharan dust to the Central Mediterranean Basin

Emanuela Molinaroli
Dipartimento di Scienze Ambientali, Università di Venezia, Dorso Duro 2137, 30123 Venezia, Italy
Stefano Guerzoni
Istituto di Geologia Marina, C.N.R., via Zamboni 65, Bologna, Italy
Giancarlo Rampazzo
Dipartimento di Scienze Ambientali, Università di Venezia, Dorso Duro 2137, 30123 Venezia, Italy

ABSTRACT

The Mediterranean Basin receives sedimentary contributions from several sources with different geological characteristics. The most important natural source of dust is the Saharan desert. Data from aerosol dust samples collected in a remote site in the Central Mediterranean Basin were used to investigate incursions of Saharan dust during a seven-month period (October through April). This work is an attempt to trace source areas and transport of dust plumes by means of grain-size, mineralogical, and chemical analyses. Two sets of aerosol samples typical of "desert-dominated" and "anthropogenic-dominated" air masses were studied. We propose a model for assessing the effects of latitude on the chemistry and mineralogy of the Central Mediterranean aerosol. Calculations and measurements of insoluble fluxes, in particular for Al (80 $\mu g/cm^2 \cdot yr^{-1}$) and Ca ($28 \mu g/cm^2 \cdot yr^{-1}$), were performed, and results were compared with fluxes from the literature calculated in other stations in the Mediterranean Basin.

INTRODUCTION

Eolian dust deposits are prominent in many parts of the world, both on continents and in oceans (Pye, 1987), and play a major role in the formation of soils and deep-sea sediments. The continents are the primary sources of mineral aerosols, particularly in arid and semi-arid regions. Estimates of the amount of mineral aerosol generated in the particle-size range capable of long-range transport are in the order of hundreds to thousands of million tons annually (Peterson and Junge, 1971; Robinson and Robbins, 1968; Schütz, 1980). Fluxes of atmospheric solid particles across the air-sea interface can be of geological, pedological, and ecological importance. Eolian dust can also affect the composition of ice in glaciers (De Angelis and Gaudichet, 1991; Waghenbach and Geis, 1989).

The Mediterranean Basin receives sedimentary material from many sources with different geological characteristics. Saharan dust originating from North Africa is the major source of natural atmospheric particles over the North Atlantic and Mediterranean. The eolian transport of desert aerosol is also responsible for the episodes of "red snow" or "red rain" described in various parts of Europe (Prodi and Fea, 1979; Bücher et al., 1983; Bergametti, 1987; Pye, 1987). In the following discussion, the generic term "Saharan" refers mainly to northwestern Africa (southern regions of Morocco, Algeria, Tunisia, and Libya), which is considered to be the dominant dust source for the Central Mediterranean Basin. In this work, we use the term "desert dominated" to include all samples made up of dust abounding in Saharan particles. The term "anthropogenic dominated" is used to include all samples made of dust abounding in anthropogenic particles, with some contributions of maritime and local particles.

Although much time and study have been devoted to the transport of Saharan dust to the west (over the North Atlantic), very little is known about its transport to the north (over the Mediterranean). The transport of Saharan dust over the Mediterranean region is poorly documented, although a few authors have studied one or two dust events from the meteorological, mineralogical, or chemical points of view (Tomadin et al., 1984; Bücher et al., 1983; Ganor and Mamane, 1982; Chester et al., 1984; Lefèvre et al., 1986; Loye-Pilot et al., 1986; Dulac et al., 1987).

Molinaroli, E., Guerzoni, S., and Rampazzo, G., 1993, Contribution of Saharan dust to the Central Mediterranean Basin, *in* Johnsson, M. J., and Basu, A., eds., Processes Controlling the Composition of Clastic Sediments: Boulder, Colorado, Geological Society of America Special Paper 284.

Because these studies are limited in time, major questions remain about the frequency of Saharan dust events in this region, and also about their source regions, transport processes, and deposition patterns. The elemental chemistry of particulate material in the Mediterranean is essentially controlled by dilution of European background material (anthropogenic-dominated) with desert components (Chester et al., 1981). This diversity of sources is reflected in the composition of the aerosols; mineralogical markers can also be used to recognize typical *mineral assemblages* of the source areas, and to differentiate Saharan dusts from other crustal components (Tomadin and Lenaz, 1989). It would appear that latitudinal control exists over the distribution of mineral aerosols in the Mediterranean basin, with a general increase to the south (Guerzoni and others, 1989).

Furthermore, relatively few compositional data are available that may be used to characterize the desert-derived components and so identify actual incidents of the incursion of dust over the Central Mediterranean. In an attempt to investigate such incursions of Saharan dust, a series of aerosol collections were made in Sardinia.

In the present work, we would note the differences between Saharan dust and other dusts from different sources, collected as a series of aerosols over a seven-month period (October 1990 through April 1991). To reach this goal we used new and more complete way to analyze aerosol samples. Grain-size, mineralogical, and chemical parameters were employed for a better characterization of the aerosol. The two sets of aerosols for which data are reported here are representative of desert-dominated and anthropogenic-dominated air masses, and should offer ideal sample sets for evaluation of the effects of latitudinal controls on the chemistry of the Central Mediterranean aerosol. Moreover, our purpose was to investigate the amount and flux of Saharan dust over the Mediterranean Sea; thus we present some calculations and measurements, along with discussion, of fluxes and dry (aerosol dust deposited without rain) and wet (aerosol dust deposited during rain) deposition of trace metals to the Central Mediterranean.

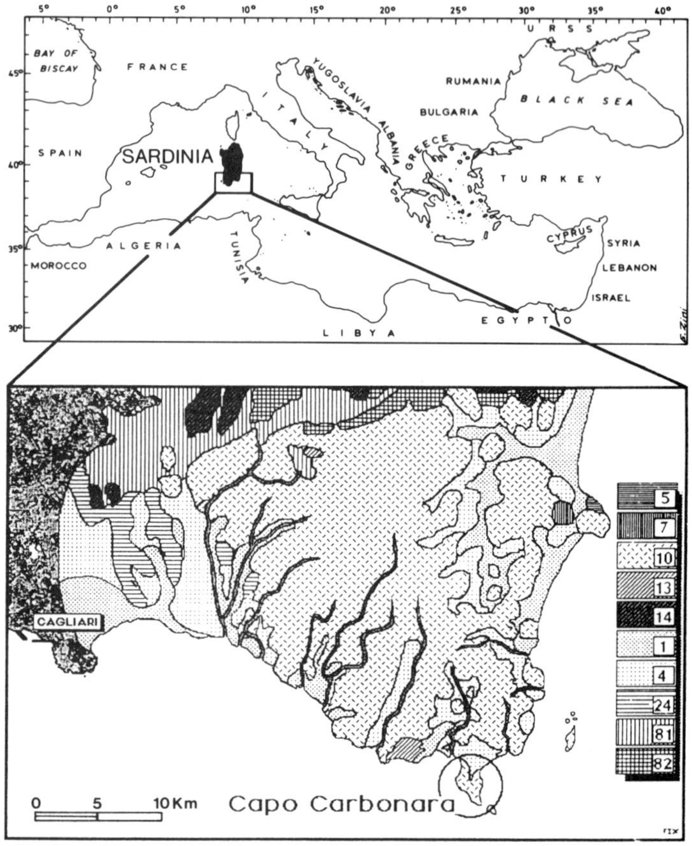

Figure 1. Geographical map of Mediterranean Sea, and location of sampling station with geological sketch. Key to numbered column: 1 = colluvial, alluvial, and fluviolacustrine deposits, modern beach sediments: Holocene; 4 = colluvial, alluvial, fluviolacustrine, and fluvioglacial deposits: Pleistocene; 5 = alkaline and subalkaline basalts, trachybasalts, basanites, tephrites and sodic trachyandesites, andesitic basalts: Neogene; 7 = rhyodacites, rhyolites, and liparites: Miocene; 10 = granites and granodiorites: Paleozoic; 13 = gabbros: Paleozoic; 14 = rhyodacites and rhyolites (ignimbrites): Permo-Carboniferous; 24 = marls, commonly with chert, of pelagic facies: Middle-Lower Miocene; 81 = marine deposits, mainly arenaceous units: Silurian-Ordovician; 82 = marine deposits, mainly shaly units, sometimes carbonaceous: Silurian-Ordovician.

METHODS

Sampling

Aerosol samples were collected at Capo Carbonara, a remote coastal station in southeastern Sardinia (Fig. 1), between October 1990 and April 1991. The sampling sites are 200 km north of the Saharan desert and more than 400 km from continental Italy (northeast) and France (northwest). The most abundant rocks in the station area are Paleozoic granitic plutons (Fig. 1). The principal winds are from the west and east-southeast, and the station is located far from local sources of pollutants. Sampling was carried out using a high-volume Sierra Andersen dry sampler, with filters made of 10-μm monofilament fibers located at upper level. Sampling lasted 24 to 72 hr.

In this preliminary study we have considered six samples: two with prevailing winds from the west, two with prevailing winds from the east-southeast, one with mixed wind directions, and one with prevailing winds from the northeast-east.

Analytical techniques

Aerosols were removed from the filters by washing in water treated with "Milli-Q water system" in the laboratory. Water-soluble particles were removed from the samples.

Particle-size analyses of the insoluble particulate matter were carried out using a Model TA II Coulter Counter, linked to an automatic data acquisition system (Boldrin et al., 1986) for particle counting. Orifice tubings of 50 and 102 μm were used to investigate the size domain after a preliminary scansion during which no longer particles were detected.

TABLE 1. COLLECTED DATA FOR PARTICULATE MATTER SAMPLED AT CAPO CARBONARA*

Sample Number	Air-mass Origin	Wind (m/sec)	TSP (μg/m³)	MSP (μg/m³)	Insol. (%)
9†	West	12	31.8	3.3	10
22§	West	9	23.3	0.9	4
3**	E-SE	6	78.7	40.6	52
42**	S-SE	5	97.8	47.4	49
21§	Mixed	8	31.8	0.8	3
28§	NE-E	7	11.3	0.8	7

*TSP = total suspended particles; MSP = mineral suspended particles; Insol. = insoluable fraction.
†Anthropogenic-dominated particles with contribution of local particles.
§Anthropogenic-dominated particles with contribution of maritime particles.
**Desert-dominated particles.

Mineralogical studies were performed by x-ray diffraction on powders and oriented-particle mounts. For convenient identification and systematic morphological analysis of the different minerals, a scanning electron microscope (SEM) fitted with an x-ray energy dispersive system (EDS) was used. Major elements were analyzed on a flame atomic absorption spectrophotometer; trace elements were run using a graphite furnace. Analytical precision and procedures are presented in detail elsewhere (Guerzoni et al., 1987, 1991).

RESULTS AND DISCUSSION

Central Mediterranean southern trades (CMST) aerosol population

Cyclonic activity in the Mediterranean Basin is most intense during late winter and spring, when blocking anticyclones are frequently observed over the Atlantic. During these periods air masses commonly come from West Europe. Sometimes (~15% of yearly frequency), a deep Atlantic trough reaches North Africa and causes a strong flow from the south over the Mediterranean. Dust storms moving from the Sahara to the Mediterranean generally occur between October and May (Bücher, 1989). Thus, meteorological conditions make concentrations of crustal material in the aerosols fluctuate. There is also evidence of the presence of dust "pulses" during the collection period.

The amounts of total suspended particles (TSP) and mineral suspended particles (MSP) are reported in Table 1, together with air-mass provenance and wind speed. Concentrations of TSP values range from 11.3 $\mu g/m^3$, for samples 9, 22, 21, and 28, to more than 97.8 $\mu g/m^3$ for samples 3 and 42. The average TSP content in samples 9, 22, 21, and 28 is 24.6 $\mu g/m^3$, whereas during the southern wind episodes, samples 3 and 42, the average mass loading increases up to 88 $\mu g/m^3$. Accordingly, the MSP fraction is 0.8 to 3.3 $\mu g/m^3$ (for samples 22, 21, 28, and 9), constituting 3 to 10% of the total mass of dust, for the former group and 40 to 47.6 $\mu g/m^3$ (for samples 3 and 42), i.e., 49 to 52% of the total mass of dust transported during southerly episodes. Therefore, four of the samples represent anthropogenic-dominated aerosols, and two, represent desert-dominated aerosols.

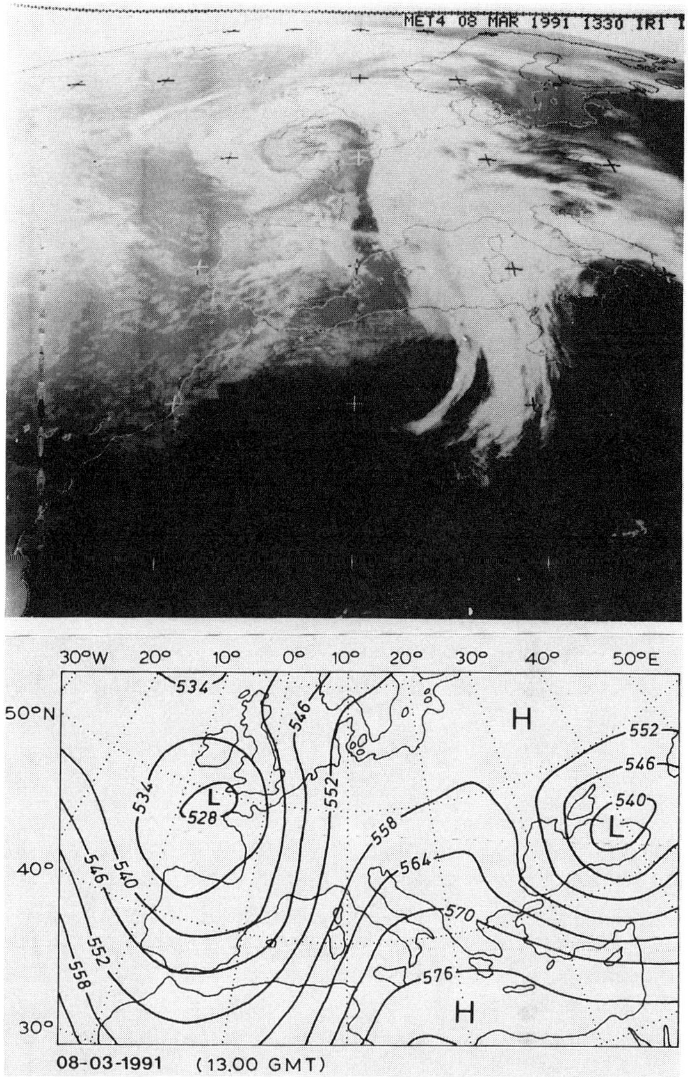

Figure 2. IR *Meteosat* photo (top) and synoptic meteorological situation at 500 mb (bottom) of southern transport episode of March 1991.

Both desert-dominated episodes were related to the position of the sub-tropical anticyclone over the Northeastern Atlantic and to the presence of polar air-masses over both the Atlantic Ocean and the continent. The first was recorded in October 1990 and the second in March 1991; the latter is presented in Figure 2, which shows a *Meteosat* photograph and the synoptic meteorological situation at 500 mb.

Here we compare our values of TSP during desert-dominated events with three TSP concentrations with similar meteorological conditions: 92 $\mu g/m^3$ (Lefèvre et al., 1986) along the southern coast of Sicily influenced by Saharan sources during August 1984, back-air trajectories were predominantly westerly;

72 μg/m^3 (Correggiari et al., 1989) south of 40°N latitude on board ship during February 1983, back-air trajectories were predominantly southerly clearly indicates the unique contribution from North Africa; and 100 μg/m^3 (Chester et al., 1984) Central Tyrrhenian during October 1979, back-air trajectories were all predominantly southerly and indicate that the winds had crossed part of the North African mainland.

We also compare our values of TSP during anthropogenic-dominated events with three TSP concentrations with similar meteorological conditions: 33 μg/m^3 (Lefèvre et al., 1986) along the southern coast of Sicily during July 1984, back-air trajectories were predominantly northerly; 5 μg/m^3 (Guerzoni et al., 1989) southern Tyrrhenian Sea on board ship during November 1982, back-air trajectories were predominantly northwesterly; and 11 μg/m^3 (Chester and others, 1984) Central Tyrrhenian during October 1979, back-air trajectories predominantly easterly.

Size distribution and mineralogical composition of aerosol

We have not found any previous studies reporting size analyses of aerosol samples collected at land stations. This is the first study to report the size distribution of Saharan dust from a land station.

The grain-size distribution of the investigated dust samples is between 0.63 and 102 μm, mainly falling between 0.63 and 4 μm.

Comparisons between representative cumulative curves of the samples are shown in Figure 3. Particle-size analysis mainly reveals the presence of silty clays or clayey silts. The cumulative curves indicate the median (Md), accurately shown at the 50 percentile.

The desert-dominated samples, 3 and 42, have Md values of 4.5 and 2.1 μm, at wind speeds of 6 and 5 m/s, respectively, from east-southeast; the anthropogenic-dominated samples usually have a lower Md value. Sample 22 has an Md value of 0.98 μm, at wind speeds of 9 m/s from the west; sample 21 has an Md value of 1.0 μm, at a wind speed at 8 m/s from several directions; sample 28 has an Md value of 1.4 μm, at a wind speed of 7 m/s from northeast-east; and sample 9 has an Md value of 7.2 μm, at a wind speed of 12 m/s from west. Sample 9 seems to be the coarsest dust, probably related to different transport velocity, but is also may be considered a true local dust. Sample 28 seems to be a coarse dust, and thus may contain some local materials. As indicated in Figure 3, the size distribution of the two desert-dominated aerosol is shifted toward a different population. Bimodal distribution, already reported by Prodi and Fea (1979) as a characteristic of mineral aerosols from the Sahara (D'Almeida and Scütz, 1983) was also observed in our samples. In the bimodal distribution the smaller particle mode can be assumed as the well-mixed long-range transport (>500 km) fraction of the mineral aerosol over the desert, whereas the large-particle mode represents proximal contributions (100 to 300 km) activated during the "Sirocco" wind (a typical wind transporting coarse, desert-derived dust in the low atmosphere). It also seems reasonable to attribute individual differences in the size distribution of the Saharan dust collected at Capo Carbonara to different transport times, which are in turn related to different transport velocities and the different lengths of airborne trajectories. Instead, the distribution size of the two anthropogenic-dominated aerosols, with the contribution of maritime particles, are shifted toward smaller particles.

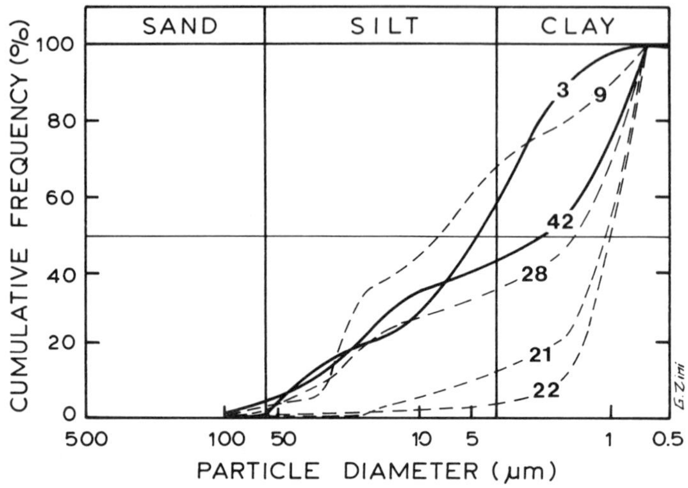

Figure 3. Cumulative grain-size curves of six dust samples collected at Capo Carbonara. Solid line indicates desert-dominated samples; broken line, anthropogenic-dominated samples.

Mineralogical analysis shows the following mineral assemblage: quartz, (the most abundant nonclay mineral), calcite, feldspars, dolomite, talc, illite, kaolinite, chlorite, and palygorskite. Our main purpose was to identify local and Saharan indicators, taking into account both morphology and particle species. We looked for mineral associations of some species generally considered as indicators of Saharan origin. In the two observed samples, the mineralogical assemblage is similar to that of other well-known Saharan dusts (Coudé-Gaussen, 1981; Pye, 1987). As in those dusts, besides a substantial proportion of quartz grains, the Capo Carbonara dusts show feldspars, kaolinite, palygorskite, and abundant dolomite. The two events of African contributions are of concern and are studied by SEM and EDS analysis. Shape and surface features of the particles are related to the processes resulting from eolation because liberation from the source and SEM/EDS analysis of particle mineralogy reveals their possible allogeny.

Some authors have found that most particles forming a desert-dust plume originate during the wind erosion process as products of mechanical breakage of particle crusts or "splashing" (Gillette, 1981). Both coarse and very fine silt were observed by SEM. The high frequency of small, chemically rounded particles and clay mineral clusters also points to in-situ erosion of already very small particles. The most probable source areas for small particles are silt and clay sediments from floodplains, and fine-grained eolian sediments probably rejuvenated by fluvial erosion (Coudé-Gaussen, 1982).

The most characteristic mineral in Saharan dusts is palygorskite, which is a very common clay in the limestone plateaus of the northern Sahara. The occurrence in our desert-dominated samples of rounded palygorskite is a very good indicator of eolian transport (Fig. 4, a–b). The size of these palygorskite particles ranged between 4 and 40 μm. Such grains are wind shaped from varied clayey substrata of the Sahara and transported beyond the desert margins (Coudé-Gaussen and Blanc, 1985).

Another characteristic mineral found was dolomite, sometimes very well preserved (Fig. 4c). Most of the quartz particles have a subrounded or rounded form, and vary in size considerably (Fig. 4, e–f). Among the altered grain types, calcite and feldspar grains commonly occur (Fig. 4h).

Mineralogical and morphological properties also characterize particles of non-Saharan origin. It appears from mineralogical observation that quartz, feldspars, and talc are the characteristic local inputs, because we did find some unusually coarse (~40 μm) angular quartz (Fig. 4d) and some very well preserved feldspars (Fig. 4g) that obviously came from the local plutons.

In Figure 5 we indicate the effects of latitude on the mineral suspended particles content of the Mediterranean aerosol. MSP aerosol values collected during southern events on board ship by our institute in the period 1981 through 1983 (Correggiari et al., 1989) were plotted versus latitude, with other samples collected in the Tyrrhenian Sea by Chester and coworkers (Chester et al., 1984). In the same figure are also plotted dust loadings of samples (averages with ranges) collected from land-based stations with southern winds, at two different latitudes: Messina, Sicily (38.1°N latitude, 6 yr of weekly sampling by National Meteorological Service; Cruciani and others, 1991) and Corsica (42.3°N latitude, 2 yr of daily sampling; Bergametti, 1987), together with our Sardinian station (39.1°N latitude). The dust loading exhibits a sharp decrease in concentration from the North Africa toward Europe. The very clear and constant latitudinal trend of dust loading in the Mediterranean atmosphere is striking, in spite of different periods and sometimes different collection techniques. This decrease was modeled using an exponential decay pattern (correlation coefficient = –0.76, p<0.001%). According to the semi-empirical exponential models developed in this study, the atmospheric concentrations of mineral aerosol halve approximately every 1.5° of latitude northward.

Elemental chemistry of insoluble aerosol

Geochemical analysis of the insoluble aerosol samples was performed. The concentrations of selected indicator elements are given in Table 2. The contents of Si, Al, Ca, Mg, and K in the Saharan samples was 10 to 30 times higher than in the others; the largest increase was observed for Ca content, associated with enrichments in calcite and dolomite. Concentrations correlated very well those of dust loadings, in tandem with major contributors of crustal elements to TSP. The Al values of the desert-dominated dust samples were high (~3,000 ng/m^3), whereas the lowest Al values (~25 ng/m^3) were observed in anthropogenic-dominated samples. Particulate Al in the atmosphere was associated almost exclusively with aluminosilicates; for this reason it is commonly employed as an indicator of the amount of crustal material in an aerosol population (Chester and others, 1981, 1984, 1986). Clearly, the high Al concentrations over the Capo Carbonara station are the result of the transport of crustal material to that region from Saharan sources. Our Al concentrations are within the same orders of magnitude of those found for the northeast Trade Winds off West Africa (e.g., Chester et al., 1979).

Table 2 also shows variations in the Al/Si ratios for the six samples. This ratio changes with particle size. Thus, anthropogenic-dominated samples 21 and 22 (the finest sampled) have an Al/Si ratio significantly lower than that of desert-dominated samples 3 and 42 (Fig. 3). This indicates the occurrence of a component of fine Si particles that are not, for the most part, associated with Al. These results indicate an important relation between geochemical and grain-size analyses as shown also in Figure 11 of Coudé-Gaussen et al., 1987.

Concentrations of other elements (Zn, Pb, Cd, and Cr) in the aerosol also show considerable variations between the two groups of samples (desert dominated and anthropogenic dominated). In order to understand the compositional differences in terms of desert and anthropogenic elemental sources, enrichment factors (EF) were used. Table 3 shows the EF of the analyzed elements, calculated employing Al as a crustal reference element and according to the equation:

$$EF_{crust} = (E/Al_{air})/(E/Al_{crust}),$$

in which E/Al_{air} and E/Al_{crust} are the concentrations of an element E and of Al in the atmosphere and in the crustal material respectively (e.g., Taylor, 1964). Enrichment factors close to one are taken to indicate that an element has a mainly crustal origin, and those greater than about 10 are considered to indicate that a substantial portion of the abundance of an element has a noncrustal source.

In the following sections, the possible origin of some of the elements is discussed, and an attempt is made to distinguish the natural from the anthropogenic signature. In terms of EF, the Mediterranean Basin aerosols can be divided into three general groups: (1) nonenriched elements, with EF generally less than 10, i.e., Si, Ca, K, Mg, and Cr; (2) intermediate elements, with EF between 10 and 100, i.e., Cr, Ni, V, Cu; and (3) enriched elements, with EF which may exceed 100, i.e., Zn, Cd, Pb (Chester et al., 1984). Table 3 shows that enriched and nonenriched elements behave differently in the two sets of samples. Nonenriched elements generally have low ranges of EF in all samples. In contrast, there are considerable variations in the EF of enriched elements (Zn, Cd, and especially Pb). However, the EF for these elements is considerably higher in the anthropogenic-dominated samples. These high values cannot be considered of crustal origin; we never found any mineralogical evidence to support this hypothesis. Thus, a clear distinction can be made between the two

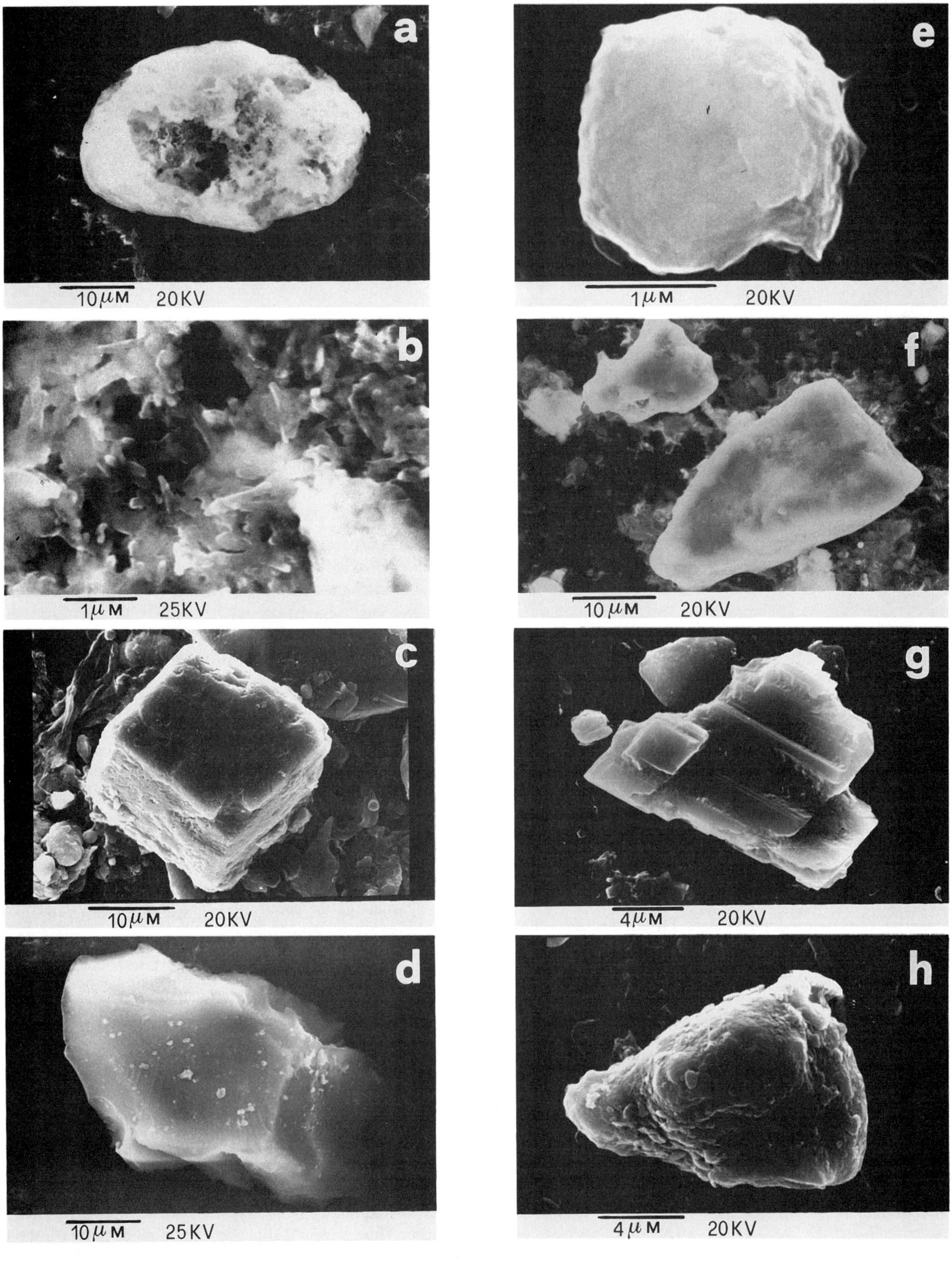
a
10μM 20KV
b
1μM 25KV
c
10μM 20KV
d
10μM 25KV
e
1μM 20KV
f
10μM 20KV
g
4μM 20KV
h
4μM 20KV

TABLE 2. CHEMICAL COMPOSITION AND Al/Si RATIO OF SAMPLES COLLECTED AT CAPO CARBONARA

Sample Number	Element (ng/m³)									Al/Si Ratio
	Si	Al	Ca	Mg	K	Zn	Pb	Cd	Cr	
9	249	101	11.1	20.6	19.4	3.9	38.7	0.028	0.37	0.34
22	139	24	1.8	3.8	7.3	0.3	11.2	0.003	0.04	0.17
3	6,910	2,827	1,210.0	681.0	653.0	22.7	7.2	0.055	5.16	0.40
42	8,490	3,414	1,259.0	662.0	604.8	29.2	40.3	0.064	3.98	0.40
21	156	26	2.7	3.9	6.0	0.3	11.4	<0.001	0.07	0.17
28	76	24	2.5	3.6	7.7	0.8	11.3	0.003	0.17	0.32

sets of samples. This result is in agreement with most previous work on the Mediterranean Sea (Dulac et al., 1987; Bergametti et al., 1988; Chester et al., 1984; Correggiari et al., 1989) and shows that aerosol composition is influenced by both pollutants and crustal sources.

We can compare our desert-dominated aerosol with other samples collected in similar meteorological situations (direct transport from desert areas) in the Atlantic and Arabian Seas. In Table 4 CMST data from this study are compared with an aerosol collected off the west coast of Africa (ANET), and with the Arabian Sea aerosol collected during the northeast monsoon. These samples are among the most desert-dominated population found in the world ocean (Chester et al., 1986). Most of the data are very similar; the only exception is Pb content, probably due to the anthropogenic Pb (very high in the Mediterranean basin) that is attached to Saharan particles.

Mean concentration and EF values of trace metals in our anthropogenic-dominated aerosols are also very similar to samples collected in Corsica and in the North Sea under winds coming from Europe (Table 5).

Insoluble atmospheric particle and trace metal fluxes

A considerable amount of data have been reported on mineral dust and trace metal fluxes over Atlantic (Buat-Menard and Chesslet, 1979) and on the qualitative eolian contribution in the Mediterranean (e.g., Chester et al., 1981; Tomadin et al., 1984). However, few such flux data are available for Central Mediterranean and from sampling sites close to the desert source.

Aerosol deposition from the atmosphere is controlled by a combination of wet and dry processes. Dry-deposition velocities of trace metals from the atmosphere vary with particle size (Arimoto and Duce, 1986). According to data reported for the deposition of trace metals to the surface of the North Sea and Western Mediterranean (Dulac et al., 1987) dry-deposition velocities for crustal elements (Al, Fe, Si) are about 1.5 cm/ and, because

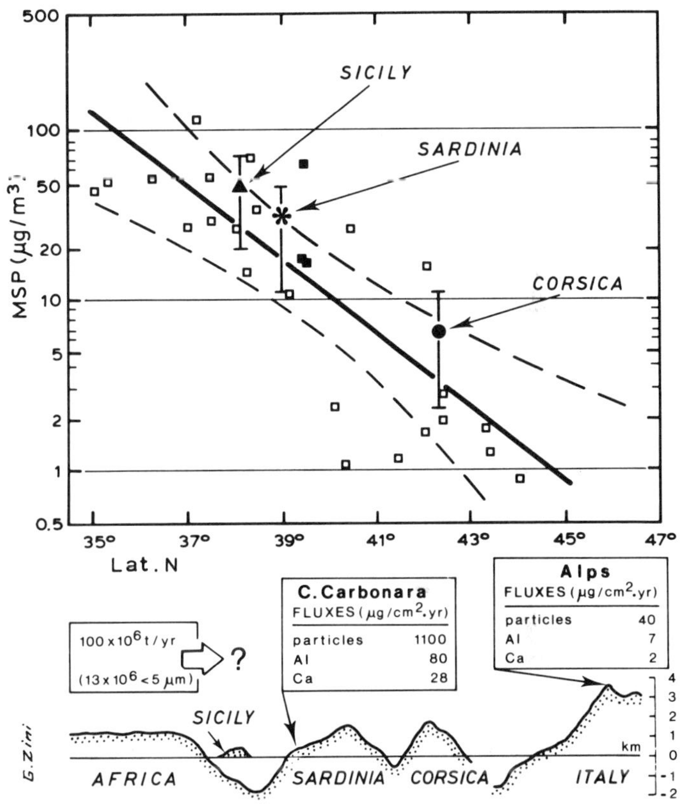

Figure 5. Mineral suspended particle values of aerosol dusts plotted against latitude. Open squares indicate samples collected on board ship by the Institute in the period 1981 through 1983 (Correggiari et al., 1989); filled squares indicate samples collected on board ship in Central Tyrrhenian Sea by Chester and coworkers (Chester et al., 1984). Filled symbols indicate average values (with ranges) from land-based station on islands: triangle represents Messina, Sicily, 5 yr of weekly sampling (Cruciani et al., 1991); asterisk represents Capo Carbonara, Sardinia, the present work; circle represents Capo Cavallo, Corsica, 18 mo of daily sampling (Bergametti et al., 1988). Lower part of plot shows sketch profile from Africa toward Europe along 9°E longitude. Left, Africa—showing amount of dust that leaves Africa northward (D'Almeida, 1986); center—Capo Carbonara, insoluble fluxes calculated in this study; right—Alps, average fluxes (40 yr) recorded in ice cores (Waghenbach and Geis, 1989; De Angelis and Gaudichet, 1991).

Figure 4. Scanning electron micrographs of tracer minerals. (a) Wind-shaped palygorskite grain; (b) close-up of (a) showing palygorskite fibers, sample 42; (c) very well preserved dolomite crystal, sample 42; (d) huge angular quartz (~40 μm), sample 22; (e) very small smooth quartz of desert type; sample 3; (f) large wind-shaped quartz grain, sample 42; (g) very well preserved angular feldspar grain, sample 28; (h) rounded and altered feldspar grain, sample 3.

TABLE 3. ENRICHMENT FACTORS OF THE STUDY SAMPLES

Sample Number	Si	Al	Ca	Mg	K	Zn	Pb	Cd	Cr
9	0.8	1	0.2	0.6	0.8	45	2,531	112	3
22	1.2	1	0.1	0.4	1.2	14	3,022	56	1
3	0.7	1	0.8	0.7	0.9	9	17	8	1
42	0.7	1	0.7	0.5	0.7	9	68	7	1
21	1.7	1	0.2	0.4	1	13	2,817	3	2
28	0.9	1	0.2	0.4	1.3	38	3,158	55	6

TABLE 4. MEAN CONCENTRATION AND EF VALUES OF TRACE METALS IN DESERT-DOMINATED AEROSOLS

Element	CM* (ng/m³)	EF	ANET† (ng/m³)	EF	AS§ (ng/m³)	EF
Ca	1,234.00	0.80	....	....	....	....
Al	3,221.00	1.00	5,925.00	1.00	1,977.00	1.00
Si	7,700.00	0.70	....	....	....	....
Cr	4.60	1.00	10.00	1.40	5.20	2.20
Zn	26.00	9.00	16.00	3.20	98.00	7.30
Cd	0.06	8.00	0.12	8.30	0.06	14.00
Pb	24.00	42.00	6.90	7.70	6.00	20.00

*Central Mediterranean (this study).
†Atlantic northeastern trade (Murphy, 1985).
§Arabian Sea (Chester et al., 1991).

TABLE 5. MEAN CONCENTRATION AND EF VALUES OF TRACE METALS IN ANTHROPOGENIC-DOMINATED AEROSOLS

Element	Sardinia* (ng/m³)	EF	Corsica† (ng/m³)	EF	North Sea§ (ng/m³)	EF
Al	44.0	1.0	200	1.0	159.0	1
Si	155.00	1.2	625	0.9	...	...
Cr	0.16	4.0	...	...	4.5	23
Zn	1.30	28.0	26	152.0	153.0	1,432
Cd	0.01	57.0	...	...	...	...
Pb	18.20	2,882.0	22	724.0	147.0	6,028

*Sardinia (this study).
†Corsica (Bergametti et al., 1988).
§North Sea (Cambray et al., 1975).

of nearness to the land, an average value of 1.5 cm/s was assumed for these elements in the Saharan aerosols. In contrast, depositional velocities for anthropogenic elements (e.g., Pb, Zn, and Cd), which are found on small particles, were reported to be less than about 0.5 cm/s. Very few data on the wet/dry deposition ratio of trace metals over the Mediterranean are available (Guerzoni et al., 1991; Guieu et al., 1991).

We used the average trace metal concentrations given in Tables 4 and 5 to calculate the dry deposition fluxes of particles insoluble Al and insoluble Ca to the sea surface over the Central Mediterranean Sea (CMS). The total atmospheric deposition was weighted to take account the relative frequencies of desert- and anthropogenic-dominated episodes (15% and 85%, respectively).

The equation to get fluxes (F) is the following:

$$F = C * V_d * time\ (sec),$$

where C is the concentration in air, and V_d the deposition velocity (1.5 and 0.5 cm/s for desert- and anthropogenic-dominated aerosol, respectively).

The desert-dominated aerosols have an average concentration of MSP of 30 $\mu g/m^3$, of Al 3,000 ng/m^3. Therefore, for MSP we multiply the mean concentration of desert-dominated events, which is 30 $\mu g/m^3$, by 1.5 cm/s and 15% of the year (equal to 4.7×10^6 sec), and we obtain a yearly mineral flux of approximately 210 $\mu g/cm^2$. The same calculations for Al and Ca give 21 and 7 $\mu g/cm^2$/year, respectively (Table 6).

The concentration values of the anthropogenic-dominated aerosols are 2 $\mu g/m^3$ for MSP and 50 and 5 ng/m^3 for Al and Ca, respectively. The related flux calculations, for the 85% of yearly frequency of this type of events (26.8×10^6 s) are 30 $\mu g/cm^2$ for particles and 1 and 0.1 $\mu g/cm^2$ for Al and Ca, respectively.

Concerning the wet and dry ratio, in the model proposed by Duce et al., 1976 it is assumed that wet deposition is approximately twice that of dry deposition, so that the total trace metal flux is three times the dry deposition flux. In our recent work in Sardinia (Guerzoni and others, 1993), we have found that dry particle deposition is 20 to 25% of the total deposition. We have estimated the total (wet + dry) fluxes being three to four times the dry one. The total (wet + dry) yearly fluxes are 750 to 1,000 $\mu g/cm^2$ for mineral particles, 65 to 90 $\mu g/cm^2$ for Al, and 20 to 30 $\mu g/cm^2$ for Ca. Contributions to sediments in the Central Mediterranean are therefore less than 1 mg/cm^2 yr, i.e., 1 cm of sediment every 1,000 yr.

The trace metal flux data of Al to the sea surface for the CMS are listed in Table 7, together with those from other Mediterranean areas and the North Sea. Mediterranean data are very consistent, and one order of magnitude higher than North Sea ones.

In Figure 5, where MSP values with latitude are indicated, fluxes calculated in Sardinia are compared to those derived from records in Alpine ice cores (45.6°N latitude: Waghenbach and Geis, 1989; De Angelis and Gaudichet, 1991), together with the amount of dust that travels every year northward from the Sahara

TABLE 6. DRY INSOLUBLE FLUXES MEASURED AT CAPO CARBONARA OF DRY AND WET FLUXES (see text for explanation)

	Aerosols		Fluxes (µg/cm²)	
	Desert-dominated	Anthropogenic-maritime/and Local Dominated	Total Dry	Total Wet + Dry
Particles	210	30.0	240.0	750–1,000
Al	21	1.0	22.0	65–90
Ca	7	0.1	7.1	20–30

(100 × 10^6 tons). The same exponential model that explains mineral decrease in the atmosphere also seems to explain the variations between the two sites in terms of fluxes. According to the semi-empirical exponential models developed in this study, the predicted atmospheric fluxes of particles, Al and Ca, would be similar (within a factor of 2) to that found in the Alpine site, 6°N of the Sardinian site.

SUMMARY AND CONCLUSION

1. Combined analysis of the grain-size, mineralogical, and chemical properties of aerosol dust is a good way to discriminate between desert-dominated and anthropogenic-dominated air masses.

2. The size distribution of desert-dominated aerosols is shifted toward larger particles and a bimodal distribution was observed. The size distribution of anthropogenic-dominated aerosol with contribution of local particles seems to be the coarsest dust. Instead, the size distribution of anthropogenic-dominated air masses with contribution of maritime particles is shifted toward smaller particles.

3. Palygorskite, dolomite, calcite, and rounded quartz are strongly suggestive of desert-dominated air masses.

4. Si, Al, Ca, Mg, and K concentrations in desert-dominated samples are about 10 to 30 times higher than in anthropogenic-dominated samples.

5. The Al/Si ratio is a good discriminator between desert- and anthropogenic-dominated samples, with high values in the former group and low ones in the latter.

TABLE 7. TRACE METAL FLUX DATA OF Al TO THE SEA SURFACE

Atmospheric Flux	Al (µg/cm²/yr)
CMST*	65–90
Corsica†	97
Gulf de Lion§	175
North Sea**	4–15

*This study.
†Bergametti, 1987.
§Guieu and others, 1991.
**GESAMP (Joint Group of Experts on the Scientific Aspect of Marine Pollution), 1989.

6. Enrichment factors for enriched elements (Zn, Cd, and especially Pb) do not exceed 100 in desert-dominated samples; they do exceed 100 many times in anthropogenic-dominated samples.

7. An exponential curve for aerosol mineral loading was calculated between 35° and 45°N latitude, and the northward decrease of MSP values fits most of the published data.

8. The same curve may also explain variations between particle fluxes calculated at Capo Carbonara (less than 1 mg/cm^2 yr) and those recorded in ice cores in the Alps.

ACKNOWLEDGMENTS

This work was partially supported by the Commission of the European Communities/Sciences and Technology Environmental Programme Contract STEP-CT N-090-0080, and by World Meteorological Organization Contract 20.028/A/CNS. This is Instituto di Geologia Marina–Consiglio Nazionale delle Ricerche scientific contribution 914. We thank the Italian Meteorological Service, at the Capo Carbonara station, for the use of its size and meteorological data during sampling. We also thank S. Rabitti for discussions and grain-size analyses, W. Landuzzi for mineralogical data, A. Ibba for SEM analyses, G. Quarantotto and Flavia Visin for chemistry work, G. Marozzi for cooperation with photographic work, and G. Zini for drawings.

REFERENCES CITED

Arimoto, R., and Duce, R. A., 1986, Dry deposition models and air/sea exchange of trace elements: Journal of Geophysical Research, v. 91, p. 2787–2792.

Bergametti, G., 1987, Apports de matiere par voie atmospherique a la Mediterranee Occidentale: Aspects geochimiques et meteorologiques [Ph.D. thesis]: Paris, University of Paris VII, 296 p.

Bergametti, G., Buat-Menard, P., and Martin, D., 1988, Trace metals in the mediterranean atmosphere, *in* Field measurement and their interpretation: Air Pollution Research Report Series 14, C.E.C., p. 88–95.

Boldrin, A., Cesca, A., and Rabitti, S., 1986, An automatic data acquisition system for Coulter Counter Model TA II: Bollettino di Oceanologia Teorica e Applicata, v. IV (3), p. 159–164.

Buat-Menard, P., and Chesslet, R., 1979, Variable influence of the atmospheric flux on the trace metals chemistry of the oceanic suspended matter: Earth and Planetary Science Letters, v. 42, p. 399–411.

Bücher, A., 1989, Fallout of Saharan dust in the northwestern Mediterranean region, *in* Leine, M., and Sarnthein, M., eds., Paleoclimatology and paleometeorology: Modern and past patterns of global atmospheric transport: Dordrecht, Kluwer Academic Publisher, p. 565–584.

Bücher, A., Dubief, J., and Lucas, C., 1983, Retombees estivales de poussieres sahariennes sur l'Europe: Revue de Geologie Dynamique et de Geographie Physique, v. 24, p. 153–165.

Cambray, R. S., Jofferies, D. S., and Toping, G., 1975, An estimate of the input of atmospheric trace elements in to the North Sea and the Clyde Sea (1972–1973): UK Atomic Energy Authority, Harwell Report AERE-R 7733.

Chester, R., Griffiths, A. G., and Hirst, J. M., 1979, The influence of soil-sized atmospheric particulates on the elemental chemistry of deep-sea sediments of northeastern Atlantic: Marine Geology, v. 32, p. 141–154.

Chester, R., Saydam, A. C., and Sharples, E. J., 1981, An approach to the assessment of local trace metal pollution in the Mediterranean marine atmosphere: Marine Pollution Bulletin, v. 12, p. 426–431.

Chester, R., Sharples, E. J., and Sanders, G. S., 1984, Saharan dust incursion over the Tyrrhenian Sea: Atmospheric Environment, v. 18, p. 929–935.

Chester, R., Murphy, K.J.T., Towner, J., and Thomas, A., 1986, The partitioning of elements in crust-dominated marine aerosols: Chemical Geology, v. 54, p. 1–15.

Chester, R., Berry, A. S., and Murphy, K.J.T., 1991, The distribution of particulate atmospheric trace metals and mineral aerosol over the Indian Ocean: Marine Chemistry, v. 34, p. 261–290.

Correggiari, A., Guerzoni, S., Lenaz, R., Quarantotto, G., and Rampazzo, G., 1989, Dust deposition in the central Mediterranean (Tyrrhenian and Adriatic Seas): Relationships with marine sediments and riverine input: Terra Nova, v. 1, p. 549–558.

Coudé-Gaussen, G., 1981, Etude détailleée d'un échantillon de poussières éoliennes prélevéau Tanezrouft le 10 décembre 1980: Recherche Gègraphic Strasbourg, v. 16/17, p. 121–130.

Coudé-Gaussen, G., 1982, Les poussières éoliennes sahariennes—mise au point: Revue de Géomorphologie Dynamique, v. 31, p. 49–69.

Coudé-Gaussen, G., and Blanc, P., 1985, Présence de grains éolisés de palygorskite dans les poussières actuelles et les sédiments récents d'origine désertique: Bulletine Societé Géologique France, v. 8, p. 571–579.

Coudé-Gaussen, G., and 6 others, 1987, Saharan dust on Fuerteventura Island (Canaries): Chemical and mineralogical characteristics, air mass trajectories, and probable sources: Journal of Geophysical Research, v. 92, p. 9753–9771.

Cruciani, L., Falasconi, L., Guerzoni, S., Rampazzo, G., and Quarantotto, G., 1991, Six years (1985–1990) of suspended particulate matter (SPM) and precipitation data at Messina station, Italy: Relationships between dust transport and rain chemistry: 2nd Workshop on Airborne Pollution of the Mediterranean Sea, Monaco, 8–12 April 1991.

D'Almeida, G. A., 1986, A model for Saharan dust transport: Journal of Climate and Applied Meteorology, v. 25, p. 903–916.

D'Almeida, G. A., and Schütz, L., 1983, Number, mass and volume distribution of mineral aerosol and soils of the Sahara: Journal of Climate Applied Meteorology, v. 22, p. 233–243.

De Angelis, M., and Gaudichet, A., 1991, Saharan dust deposition over Mont Blanc (French Alps) during the last 30 years: Tellus, v. 43B, p. 61–75.

Duce, R. A., Ray, B. J., Hoffman, G. L., and Walsh, P. R., 1976, Trace metal concentration as a function of particle size in marine aerosols from Bermuda: Geophysical Research Letter, v. 3, p. 339–342.

Dulac, F., Buat-Menard, P., Arnold, M., and Ezat, U., 1987, Atmospheric input of trace metals to the Western Mediterranean Sea: 1. Factors controlling the variability of atmospheric concentration: Journal of Geophysical Research, v. 96, p. 8437–8453.

Ganor, E., and Mamane, V., 1982, Transport of Saharan dust across the eastern Mediterranean: Atmospheric Environment, v. 10, p. 1079–1084.

GESAMP, 1989, The atmospheric input of trace species to the world ocean, in Reports and Studies, no. 38, p. 111.

Gillette, D., 1981, Production of dust that may be carried great distances, *in* Pewe, T., ed., Desert dust: Origin, characteristics, and effects on man: Geological Society of America Special Paper 186, p. 11–26.

Guerzoni, S., Rovatti, G., Molinaroli E., and Rampazzo G., 1987, Total and "selective" extraction methods for trace metals in marine sediment reference samples (Mess-1, NBS 1646): Chemistry in Ecology, v. 3, p. 39–48.

Guerzoni, S., Lenaz, R., Quarantotto, G., Rampazzo, G., and Correggiari, A., 1989, Trace metal composition of airborne particles over the Mediterranean Sea: Giornale di Geologia, v. 51, p. 117–130.

Guerzoni, S., Quarantotto, G., and Rampazzo, G., 1991, Rapporto tra input atmosferico e fluviale di metalli pesanti nei mari Adriatico e Tirreno: Atti Convegno AIOL, S. Margherita Ligure, 22–25 November 1991.

Guerzoni, S., and 6 others, 1993, Mineral atmospheric particulate from south to NW Mediterranean: Seasonal variations and characteristics: Water Pollution Research Report 28, p. 483–493.

Guieu, C., Martin, J. M., Thomas, A. J., and Elbaz-Poulichet, F., 1991, Atmospheric versus river inputs of metals to the Gulf of Lions: Marine Pollution Bulletin, v. 22, p. 176–183.

Lefèvre, R., Gaudichet, A., and De Félice, P., 1986, Caractérisation chimico-mineralogiques des flux microparticulaires dans la basse atmosphère dela Méditerranée moyenne. Permanence et fluctuation des apports atmosphériques à la sédimentation: Centre de Research Academia de Science Paris, v. 303, p. 1215–1220.

Loye-Pilot, M. D., Martin, J. M., Morelli, J., Gros, J. M., and Strauss, B., 1986, Impact des poussieres sahariennes sur l'acidite des pluies dans l'atmosphere mediterraneenne, *in* Proceedings, Fourth European Symposium, Stresa, Italy, 23–25 September 1986, p. 489–499.

Murphy, K.J.T., 1985, Trace metals chemistry of the Atlantic aerosol [Ph.D. thesis]: Liverpool, University of Liverpool, 380 p.

Peterson, J. T., and Junge, C. E., 1971, Source of particulate matter in the atmosphere, *in* Matthews, W., Kellogg, W., and Robinson, G. D., eds., Man's impact on climate: Cambridge, MIT Press, p. 310–320.

Prodi, F., and Fea, G., 1979, A case of transport and deposition of Saharan dust over the Italian Peninsula and southern Europe: Journal of Geophysical Research, 84C, p. 6951–6960.

Pye, K., 1987, Aeolian dust and dust deposition: Academic Press, 334 p.

Robinson, E., and Robbins, R. C., 1968, Sources, abundance and fate of gaseous atmospheric pollutants: Final Report Project PR-6755, Menlo Park, California, Stanford Research Institute.

Schütz, L., 1980, Long range transport of desert dust with special emphasis on the Sahara: Annals of New York Acadcmy of Sciences, v. 338, p. 515–532.

Taylor, S. R., 1964, Abundance of chemical elements in the continental crust: a new table: Geochimica et Cosmochimica Acta, v. 28, p. 1273–1285.

Tomadin, L., Lenaz, R., Landuzzi, V., Mazzucotelli, A., and Vannucci, R., 1984, Wind-blown dusts over the Central Mediterranean: Oceanologica Acta, v. 7, p. 13–23.

Tomadin, L., and Lenaz, R., 1989, Eolian dust over the Mediterranean and their contribution to the present sedimentation, *in* Leine, M., and Sarnthein, M., eds., Paleoclimatology and paleometeorology: Modern and past patterns of global atmospheric transport: Dordrecht, Kluwer Academic Publisher, p. 267–282.

Waghenbach, D., and Geis, K., 1989, The mineral dust record in a high altitude alpine glacier (Colle Gnifetti, Swiss Alps), *in* Leine, M., and Sarnthein, M., eds., Paleoclimatology and paleometerology: Modern and past patterns of global atmospheric transport: Dordrecht, Kluwer Academic Publisher, p. 543–564.

MANUSCRIPT ACCEPTED BY THE SOCIETY DECEMBER 24, 1992

Printed in U.S.A.

Geological Society of America
Special Paper 284
1993

Modern stream sands from compound crystalline sources: Composition and sand generation index

Marta Palomares and José Arribas
Departamento de Petrología y Geoquímica, Universidad Complutense de Madrid, 28040 Madrid, Spain

ABSTRACT

Compositions of modern first-cycle sands derived from granitic and metamorphic terrains in the Spanish Central System have been analyzed in order to evaluate the contributions of different bedrock types. The results of this work indicate that sand composition normally does not permit quantitative statements concerning sourceland composition. This is due to the fact that each rock type has a different potential to generate sand, dependent on such properties as its mineralogy, average crystal size, and microfabric. We introduce the concept of sand generation index (SGI), which is a relative measure of the capacity of one bedrock type to generate sand with respect to another in a compound source area. SGI of granitoid is 14 to 20 times greater than SGI of slate-schist when these rock types appear in a dual crystalline source. In the case of a gneiss + slate-schist source rock association, the SGI of gneiss is about five times greater than the SGI of slate-schist, whereas the SGI of gneiss is similar to that of granitoid in case of a dual granitoid + gneiss source. Finally, our results show that quantitative estimates of source land composition based on QFR diagrams are hazardous if the concept of the SGI is disregarded.

INTRODUCTION

As shown by Mann and Cavaroc (1973), Basu et al. (1975), Dickinson and Valloni (1980) and other workers, the study of modern sands provides for a better understanding of the influence of source area composition, tectonic setting, weathering, transport, and climate on final sand composition. An integration of the results obtained from such studies should yield a more accurate interpretation of provenance in ancient deposits. In order to infer provenance, modal compositions of sands are usually plotted on ternary diagrams, such as QFR, QFL, or Q_mFLt, subdivided into diagnostic compositional fields (e.g., Dickinson and Suczek, 1979; Pettijohn et al., 1987). However, current interpretation schemes often appear insufficient, inasmuch as they either are based on the assumption of source areas consisting of a single rock type (Folk, 1974; Pettijohn et al., 1987) or, at the other extreme, they concern the discrimination between major geotectonic settings (Dickinson, 1985; Valloni, 1985). Petrofacies of sands derived from wide drainage terrains with different source materials can be used as a valuable tool for the distinction of major provenance categories. On the contrary, on a smaller scale lithological variability in the source area critically affects sand composition (Ingersoll, 1991). At present many studies address more local problems in provenance analysis (e.g., Arribas and Arribas, 1991), where interpretations based on continental-scale data (Dickinson et al., 1983) are of more limited value. Therefore, the necessity arises to search for diagnostic criteria that will improve provenance interpretations at a smaller scale.

Little work has been done on quantifying the relative contribution of various source-rock types in a given source area to sand populations derived from that source area. Mack (1981) analyzed the compositional characteristics of modern sands derived from a mixed metamorphic and sedimentary source terrain. The results obtained suggest that this kind of approach may constitute a promising field for future research. What is the composition of a sand released from a source area that consists of two or more lithologic types in known relative proportions? What is the capacity of a certain source rock type to produce sand with

Palomares, M., and Arribas, J., 1993, Modern stream sands from compound crystalline sources: Composition and sand generation index, *in* Johnsson, M. J., and Basu, A., eds., Processes Controlling the Composition of Clastic Sediments: Boulder, Colorado, Geological Society of America Special Paper 284.

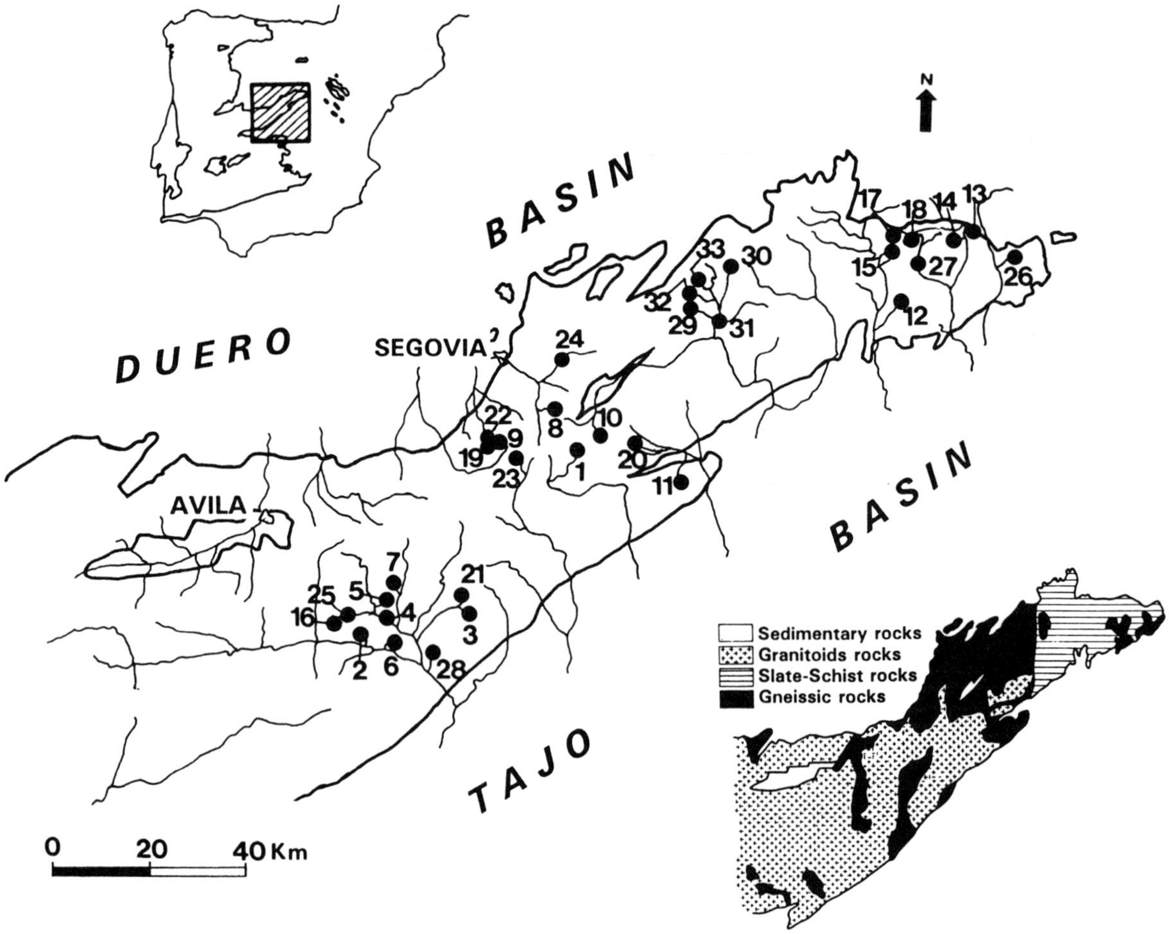

Figure 1. Location map of sampled drainage systems in the Spanish Central System. See Table 1 for identification and characterization of the individual drainage systems.

respect to another? To answer these questions, studies need to be attempted, particularly on modern sands, where source areas are well exposed and other provenance parameters (such as weathering, transport, climate) are known or can be quantitatively estimated.

The purpose of this study is to provide data concerning the composition of first-cycle sands derived from three different crystalline source rock associations, with particular emphasis on the influence of differing proportions of specific bedrock types in the source area on the final sand composition. The Spanish Central System has been selected because it provides exposures of various igneous and metamorphic rock types, as well as well-defined drainage basins with modern first-cycle stream sands. By using classical petrographic methods, the relative potential for sand generation of the different source rock types is quantitatively assessed.

CHARACTERISTICS OF THE SOURCE AREAS

The Central System is a northeast-southwest–trending mountain chain, about 400 km long, that is situated in the central part of the Iberian Peninsula (Fig. 1). The elevation increases toward the southwest, reaching a maximum of 2,592 m (Pico Almanzor). Streams drain northward (Duero Basin) and southward (Tajo Basin). The Central System is a part of the Hesperian Massif (i.e., the southernmost portion of the European Hercynian Chain), uplifted during the Alpine Cycle, and consists largely of granitoids of late Hercynian age that intrude low- to high-rank (Precambrian and Paleozoic) metamorphic basement (Fig. 1). Granitoid lithologic varieties include biotite leucogranites, adamellites, pegmatitic to aplitic leucogranites, porphyritic adamellites, and granodiorites (Aparicio et al., 1975; Casillas and Peinado, 1988). The emplacement of the granitoids is thought to be related to late orogenic regional extension (Casquet et al., 1988). Metamorphic rocks include low-rank slates and schists (Lopez Ruiz et al., 1975), as well as gneisses (Aparicio and Galán, 1980; Navidad and Peinado, 1981; Villaseca, 1984). Composition and textures of sands derived from "simple" (i.e., single lithology) sources in the same area have been studied by Tortosa et al. (1989), Palomares et al. (1990) and Tortosa et al. (1991). The present climate in the Central System is semiarid, with mean annual precipitation ranging from 500 to 1,500 mm and mean annual temperatures ranging between 6° and 14°C.

TABLE 1. GENERAL DATA CONCERNING DRAINAGE SYSTEMS CONSIDERED IN THIS STUDY

Drainage Area	Lithology	Number of Samples	Surface Area (km^2)	Maximum Length (km)	Relief Ratio*	Altitude (max.–min.) (m)
1 La Pedriza	Granitoid (Gr)	4	8.75	4.47	0.22	2,015–1,030
2 Jomillo	Granitoid (Gr)	1	0.15	0.70	0.10	690–620
3 Sobaquillo	Granitoid (Gr)	1	0.20	0.75	0.30	1,090–860
4 Fraile	Granitoid (Gr)	2	4.40	3.85	0.11	1,178–750
5 Enebrosilla	Granitoid (Gr)	4	3.85	3.15	0.10	1,071–720
6 Rosados	Granitoid (Gr)	2	0.75	0.90	0.16	831–685
7 Quintanar	Granitoid (Gr)	3	1.65	1.90	0.14	1,030–760
8 Valdeclemente	Granitoid (Gr)	1	2.06	3.50	0.27	2,182–1,220
9 Mujer Muerta	Gneiss (Gn)	4	5.60	3.60	0.24	2,196–1,320
10 Purgatorio	Gneiss (Gn)	2	1.15	1.50	0.21	1,610–1,290
11 San Pedro	Gneiss (GN)	2	1.40	2.65	0.19	1,422–910
12 Prado Redondo	Gneiss (Gn)	2	1.00	1.45	0.07	1,100–990
13 Valdecanal	Slate-schist (Sl-Sc)	3	0.70	1.27	0.13	1,280–1,110
14 Retuerta	Slate-schist (Sl-Sc)	3	2.40	2.25	0.14	1,449–1,130
15 Morequero	Slate-schist (Sl-Sc)	2	0.90	1.50	0.22	1,589–1,260
16 Pajares	Slate-schist (Sl-Sc)	1	2.81	2.00	0.11	1,320–1,100
17 Pelagallinas	Slate-schist (Sl-Sc)	2	29.00	14.50	0.05	1,852–1,125
18 Cristóbal 1	Slate-schist (Sl-Sc)	2	6.12	5.90	0.09	1,821–1,300
19 Peña Hombre	Gr + Gn	5	3.95	3.10	0.23	2,003–1,290
20 Carpintera	Gr + Gn	4	4.92	7.00	0.08	1,460–895
21 Puebla	Gr + Gn	1	8.53	4.35	0.08	1,178–810
22 Milanillos	Gr + Gn	6	11.62	6.25	0.17	2,196–1,150
23 Infierno	Gr + Gn	6	2.09	2.25	0.32	2,009–1,290
24 Cambrones	Gr + Gn	9	48.00	12.50	0.07	2,078–1,120
25 Pizarra	Gr + Sl-Sc	6	20.96	7.75	0.07	1,320–730
26 Chorrón	Gn + Sl-Sc	3	1.71	2.00	0.16	1,408–1,090
27 Cristóbal 2	Gn + Sl-Sc	5	47.27	13.90	0.05	1,821–1,050
28 Lorenzo	Gn + Sl-Sc	1	1.81	2.40	0.11	811–540
29 Cigüeñuela	Gn + Sl-Sc	1	21.21	8.10	0.10	1,854–1,030
30 Xalle	Gn + Sl-Sc	2	12.02	5.15	0.15	1,834–1,050
31 Manadero	Gn + Sl-Sc	8	79.36	13.85	0.08	2,045–960
32 Solana	Gn + Sl-Sc	1	10.71	4.55	0.14	1,833–1,190
33 Dehesa	Gn + Sl-Sc	2	6.70	3.90	0.16	1,833–1,230

*Relief (maximum-minimum elevations) divided by maximum length.

Accordingly, both mechanical and chemical weathering are moderate (Wilson, 1969).

METHODS

For a quantitative assessment of the effects of a "compound" (i.e. mixed lithology) source on sand composition, the detrital modes of sands derived from simple sources were first analyzed. The source rock types present in the Central System have been grouped into three categories: granitoids, gneisses, and slate and schists. Sampling was performed on 33 modern drainage systems (Fig. 1). Two types of systems are distinguished, according to bedrock lithology at the source (Table 1): (1) Drainage systems with simple crystalline sources (constituted by granitoids, gneisses, or slate-schists). The area of the drainage basins ranges from 0.1 to 30 km^2. They are between 700 to 2,200 m above sea level. First- or second-order streams, with lengths between 0.7 and 15 km and relief ratios ranging from 0.07 to 0.30, are typical. (2) Drainage systems with dual crystalline sources (constituted by an association of granitoid + gneiss, granitoid + slate-schist, or gneiss + slate-schist). Drainage areas are between 2 and 80 km^2 and elevations between 800 and 2,200 m above sea level. Relief is moderate, with ratios ranging from 0.05 to 0.32. Streams are of first- to sixth-order, with lengths ranging from 2 to 14 km. In the case of the samples derived from this type of drainage areas, relative proportions of the individual lithologies in a given source area have been quantified according to Mack (1981; our Fig. 2).

The stream sands studied are from 81 different sites. At several sites more than one sample has been taken in order to test the reproducibility of analytical results. To avoid contamination, subaqueous bars in the inner reaches of the stream channels were sampled. Forty-one sand samples were collected in 18 drainage systems from homogeneous source terrains (one lithology) and 60 sands were sampled in main channels and major tributaries of 15 systems from compound source terrains (Table 1). In the source

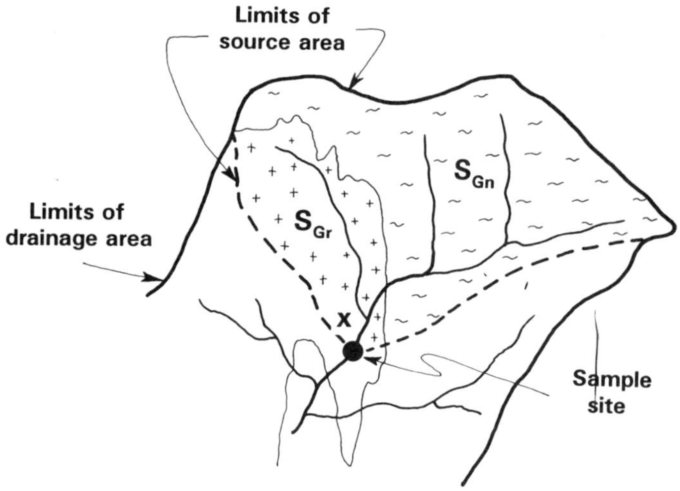

$$\text{Source rock (granitoid) ratio characteristic of sample X (Mack, 1981)} = \frac{S_{Gr}}{S_{Gr} + S_{Gn}} \times 100$$

Figure 2. Sketch of a drainage system explaining the terminology used and showing the criterion adopted to establish the relative areal proportions of the bedrock types constituting a source terrain. S_{Gr} and S_{Gn} refer to the outcrop area of granitoid and gneiss bedrock, respectively.

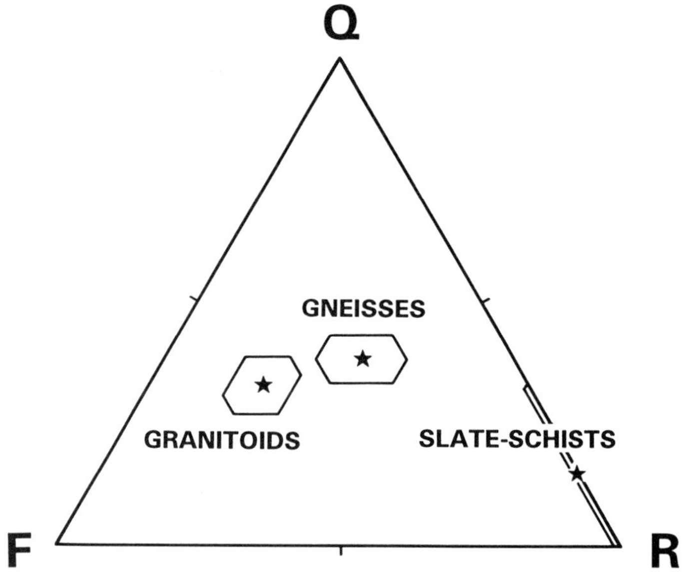

Figure 3. Mean compositions with standard deviations (1 SD) of modern medium-grained sands derived from simple sources (granitoids, gneisses, or slate-schists) in the Central System.

terrains of the latter, the relative abundance of the constituent bedrock types has a considerable range. To avoid size-composition effects, only the medium-sand fraction (0.25 to 0.5 mm) has been considered. This fraction was separated by dry sieving and cemented with epoxy resin for thin-section preparation. Four hundred points were counted per thin section (Chayes, 1956) and the data subsequently evaluated according to both the "traditional" and Gazzi-Dickinson (1966) criteria (see also Zuffa, 1980, 1985). The modal compositions listed in Table 2 have then been plotted in QFR diagrams.

SANDS DERIVED FROM SIMPLE SOURCES

Grain types

Monocrystalline quartz is the dominant quartz grain type in both sands derived from simple granitoid and gneissic sources (25% and 26%, respectively, cf. Table 2). By contrast, the Q_m/Q_p (monocrystalline quartz/polycrystalline quartz) ratio in sands proceeding from slate-schist sources is typically around 1. In sands from granitoid sources, the percentages of undulatory and nonundulatory monocrystalline quartz is similar, as outlined by Tortosa et al. (1991). Medium sand–sized polycrystalline quartz grains from granitoid and gneissic sources (abundance: 5 to 8%, respectively) consist of a few crystals, typically less than three, whereas polycrystalline quartz grains derived from slate-schists (4.6%) tend to contain more than three crystal units per grain. This is related to the original crystal size in the parent rocks.

K-Feldspar and plagioclase occur as idiomorphic and subidiomorphic grains, both twinned and untwinned. Sands derived from granitoid sources contain the highest proportions of feldspar (46%), with a clear dominance of plagioclase (K/P = 0.8). In sands released from gneissic sources feldspar content is less (27%), and K-feldspar (mainly microcline) prevails over plagioclase (K/P = 1.4). The feldspar content in sands from slate-schist terrains is very low (0.6%); both plagioclase and K-feldspar are occasionally present.

Rock fragments in sands derived from granitoid as well as those from gneissic terrains (abundance, 19.6 and 34.7%, respectively) are composed of quartz-feldspar and quartz-feldspar-mica aggregates, typically with 3 to 5 crystals per grain in the former and 4 to 10 in the latter. Labile rock fragments (L-grains; Dickinson, 1970) are most frequent in sands released from slate-schist sources (58%). Muscovite, biotite, and chlorite also occur as single grains, but are quantitatively subordinate (generally <2%) in the medium sand fraction, as micas as individual grains tend to concentrate in the size fractions finer than 0.25 mm.

Modal composition

The compositions of sands derived from slate-schist terrains plots within a narrow field parallel to the QR edge next to the R apex in the QFR diagram, with a mean of $Q_{14}F_1R_{85}$ (Fig. 3). As shown by Tortosa et al. (1989), the relative proportions of quartz and rock fragments vary as a function of the original crystal size in the parent rock. Modal compositions of sands derived from granitoid and gneissic sources are similar, with means of $Q_{33}F_{47}R_{20}$ and $Q_{38}F_{27}R_{35}$, respectively. The textural heterogeneity of gneisses (such as foliation, variable crystal size) results in

TABLE 2. MODAL COMPOSITIONS WITH STANDARD DEVIATION (1 SD) OF MEDIUM SANDS DERIVED FROM THE THREE TYPES OF COMPOUND SOURCES DISTINGUISHED*

Mixing	Lithology Percentage	Qnu†	Qu§	Qp2-3**	Qp>3‡	K-Fsp	Plag	Mica	Rc§§	L***	Q	F	R
	100 Gr	11.0 ± 8.4	14.7 ± 8.5	5.0 ± 3.5	1.8 ± 1.9	21.2 ± 5.8	25.0 ± 7.2	1.4 ± 1.2	19.6 ± 5.6	0	32.7 ± 5.7	46.2 ± 8.4	19.6 ± 5.6
Gr + Gn	30-50 Gn	21.3 ± 4.7	4.3 ± 1.0	5.4 ± 1.8	1.6 ± 1.8	19.0 ± 5.0	17.6 ± 7.3	3.3 ± 2.0	26.4 ± 7.3	0.2 ± 0.5	34.1 ± 4.0	38.2 ± 8.4	27.4 ± 7.4
	>50 Gn	18.6 ± 7.2	4.9 ± 1.9	7.3 ± 2.6	3.3 ± 2.2	21.0 ± 4.2	9.3 ± 3.0	3.2±1.6	30.0 ± 5.3	1.6 ± 1.0	36.3 ± 5.7	31.0 ± 5.2	32.7 ± 6.2
	100 Gn	20.7 ± 6.5	5.6 ± 4.5	8.3 ± 2.7	4.1 ± 2.8	16.1 ± 5.6	11.1 ± 6.1	1.9±1.0	34.7 ± 9.0	0	38.6 ± 6.7	27.2 ± 8.8	34.7 ± 9.0
	100 Gr	11.0 ± 8.4	14.7 ± 8.5	5.0 ± 3.5	1.8 ± 1.9	21.2 ± 5.8	25.0 ± 7.2	1.4±1.2	19.6 ± 5.6	0	32.7 ± 5.7	46.2 ± 8.4	19.6 ± 5.6
Gr + Sl-Sc	10-30 Gr	15.9 ± 1.2	3.9 ± 2.1	4.5 ± 1.0	1.4 ± 0.9	19.4 ± 4.3	16.2 ± 3.3	3.6±0.9	28.5 ± 4.4	4.5 ± 2.4	28.0 ± 3.0	36.4 ± 4.7	35.4 ± 4.2
	5 Gr	11.1	4.6	5.3	6.2	14.5	6.8	3.4	39.1	8.1	28.3	22.3	49.2
	100 Sl-Sc	4.9 ± 7.5	1.3 ± 1.7	1.9 ± 3.5	4.6 ± 5.2	0.5 ± 0.6	0.1 ± 0.2	0.4±0.4	26.8 ± 10	58.1 ± 20.5	13.6 ± 18	0.6 ± 0.7	85.6 ± 18.7
	100 Sl-Sc	4.9 ± 7.5	1.3 ± 1.7	1.9 ± 3.5	4.6 ± 5.2	0.5 ± 0.6	0.1 ± 0.2	0.4±0.4	26.8 ± 10	58.1 ± 20.5	13.6 ± 18	0.6 ± 0.7	85.6 ± 18.7
Gn + Sl-Sc	<40 Gn	5.6 ± 3.0	1.4 ± 0.9	4.2 ± 2.1	10.5 ± 4.7	1.8 ± 0.6	10.4 ± 3.5	2.9±2.4	52.6 ± 4.5	9.7 ± 5.1	22.5 ± 7.1	12.5 ± 3.5	64.8 ± 4.5
	>40 Gn	21.7 ± 5.8	4.3 ± 2.1	8.5 ± 2.2	2.2 ± 1.0	9.1 ± 5.1	11.2 ± 3.5	3.3±1.5	37.2 ± 5.2	2.3 ± 1.2	38.6 ± 7.0	20.2 ± 4.7	40.4 ± 5.8
	100 Gn	20.7 ± 6.5	5.4 ± 4.5	8.3 ± 2.7	4.1 ± 2.8	16.1 ± 5.6	11.1 ± 6.1	1.9±1.0	34.7 ± 9.0	0	38.6 ± 6.7	27.2 ± 8.8	34.7 ± 9.0

*Data from sands derived from simple sources (100% granitoid, 100% gneiss, or 100% slate-schist) are included to facilitate compositional trend analysis.

†Nonundulatory monocrystalline quartz.

§Undulatory monocrystalline quartz.

**Polycrystalline quartz with 2 or 3 crystal units/grain.

‡Polycrystalline quartz with more than 3 units/grain.

§§Nonlabile rock fragments.

***Labile rock fragments.

greater quantities of rock fragments in their weathered products as compared to granitoid-derived sands (Tortosa et al., 1989). The detrital mode of the sands derived from granitoid sources compares well with that of sands derived from similar source rocks under semiarid climatic conditions (Basu, 1976).

SANDS DERIVED FROM DUAL SOURCES

Granitoid + gneiss (Gr + Gn)-derived sands

The principal tendencies in compositional variation between sands derived from mixed sources are illustrated in Figures 4-6.

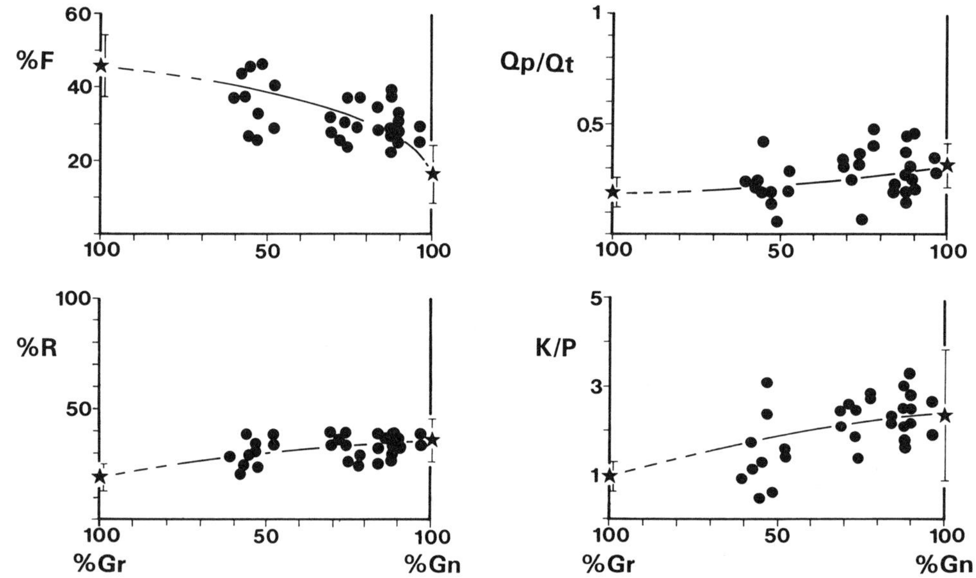

Figure 4. Compositional characteristics of mixed (granitoid + gneiss) derived sands versus percentage outcrop area of gneissic bedrock in the source terrain. Stars represent the mean compositions (with standard deviations) of sands derived from a 100% granitoid (Gr) source terrain and from a 100% gneissic (Gn) source. Qt = total quartz; Qp = polycrystalline quartz; F = total feldspar; K = K-feldspar; P = plagioclase; R = total rock fragments.

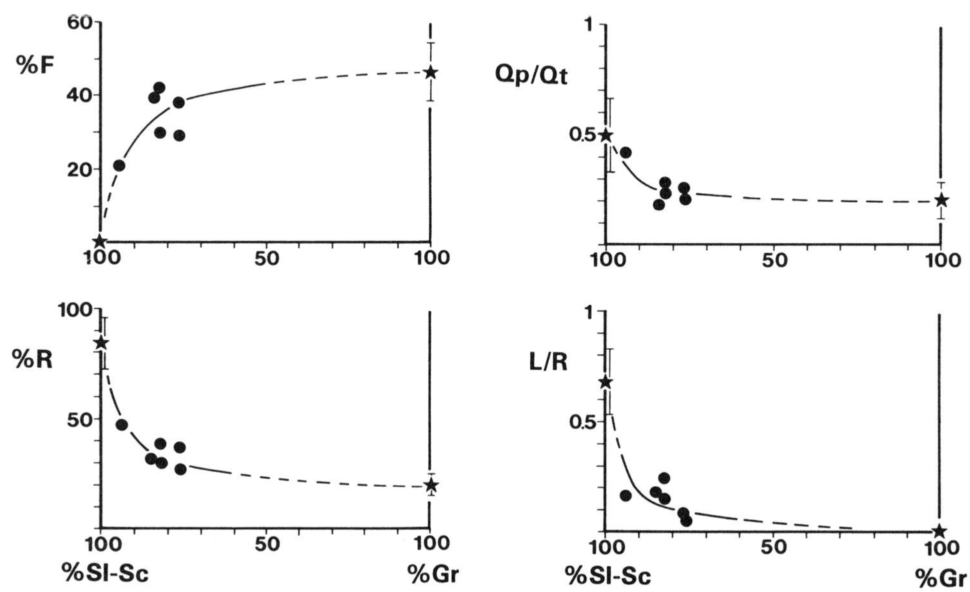

Figure 5. Compositional characteristics of mixed (granitoid + slate-schist) derived sands versus percentage outcrop area of granitoid bedrock in the source terrain. Stars represent the mean compositions (with standard deviations) of sands derived from a 100% granitoid (Gr) source terrain and from a 100% slate-schist (Sl-Sc) source. R = total rock fragments; L = labile rock fragments. See Figure 4 for definition of other symbols.

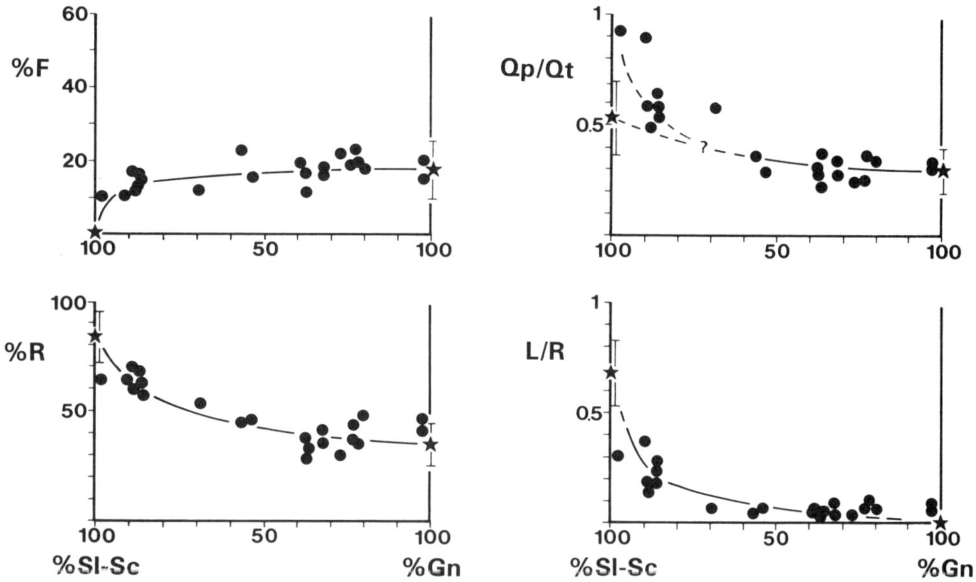

Figure 6. Compositional characteristics of mixed (gneiss + slate-schist) derived sands versus percentage outcrop area of gneiss bedrock in the source terrain. Stars represent the mean compositions (with standard deviations) of sands derived from a 100% gneiss (Gn) source terrain and from a 100% slate-schist (Sl-Sc) source. Symbols as in Figures 4 and 5.

Among sands derived from the (Gr + Gn) source terrains, the percentage of rock fragments and polycrystalline quartz increases with increasing percentage of gneiss in the source area, whereas the feldspar content decreases (Fig. 4). However, the higher plagioclase content of granitoid with respect to gneissic rocks causes the K/P ratio (Dickinson, 1985) to rise with an increase of the proportion of gneissic bedrock at the source. In order to analyze modal composition of the sands in the QFR diagram, the samples have been grouped, according to the relative abundances of the bedrock types, in (Gr_{70-50})-derived and ($Gr_{<50}$)-derived sands (Fig. 7, above). Mean modal composition of these two groups are $Q_{34}F_{39}R_{27}$ and $Q_{36}F_{30}R_{34}$, respectively. Compositions of sands derived from this type of compound source plot roughly on a straight line linking the compositions that characterize the pure (granitoid or gneiss) end member sands. Compositional trends in the diagrams in Figures 4 and 7 (above) are near-linear. Mean compositional values on the trend line in the QFR diagram are roughly proportional to the percentage of granitoid or gneiss in the source area. This is due to the similar composition and crystal size of the two parent rock types and points to their similar potential to generate sand.

Granitoid + slate-schist (Gr + Sl-Sc)–derived sands

Sources characterized by an association of (Gr + Sl-Sc) are rare in the Central System. Six samples derived from sources with a limited range in abundance in granitoid and slate-schist (Gr_5 + $Sl\text{-}Sc_{95}$ to Gr_{30} + $Sl\text{-}Sc_{70}$) have been analyzed. Nevertheless, the compositional data of these sands are particularly significant, when compared to those derived from simple (granitoid or slate-schist) sources (Fig. 5). In this case compositional trends strikingly diverge from linearity. For instance, a small increase in the abundance of granitoid bedrock produces an important increase in feldspars (K-feldspar and plagioclase) and a disproportionate diminution of labile rock fragments, thus resulting a clear dominance of granitoid-derived grain types. The mean modal composition of sands derived from (Gr_{30-10} + $Sl\text{-}Sc_{70-90}$) terrains is $Q_{28}F_{36}R_{36}$. The modal composition of one sand sample derived from a (Gr_5 + $Sl\text{-}Sc_{95}$) source is $Q_{28}F_{22}R_{50}$. Plotting these values and the means of the simple granitoid-derived and slate-schist-derived sands in the QFR diagram permits the development of the compositional trend (Fig. 7, center). It is evident from the diagram in Fig. 7 that no more than a 5% of granitoid in the source area is needed to produce sands with a modal composition near the midpoint of the trend line and that an important shift toward the mean composition of sands derived from pure granitoid terrains results when the granitoid proportion tends toward higher values.

Gneiss + slate-schist (Gn + Sl-Sc)–derived sands

Variations in abundance of the different grain types in (Gn + Sl-Sc)-derived sands show trends similar to those found in sands derived from (Gr + Sl-Sc) sources (Fig. 6, Table 2). Feldspar content markedly rises (from negligible to more than 10%) when gneiss appears as bedrock in the source area. Beyond a value of about 15% of gneissic bedrock, the feldspar content remains relatively stable. A similar, but inverted, trend is observed in the case of polycrystalline quartz grains and rock fragments. Moreover, when gneiss exceeds 20% of the source area the content of labile rock fragments in the sands becomes insignificant. On the other hand, the Q_p/Q_t ratio characterizing pure slate-schist derived

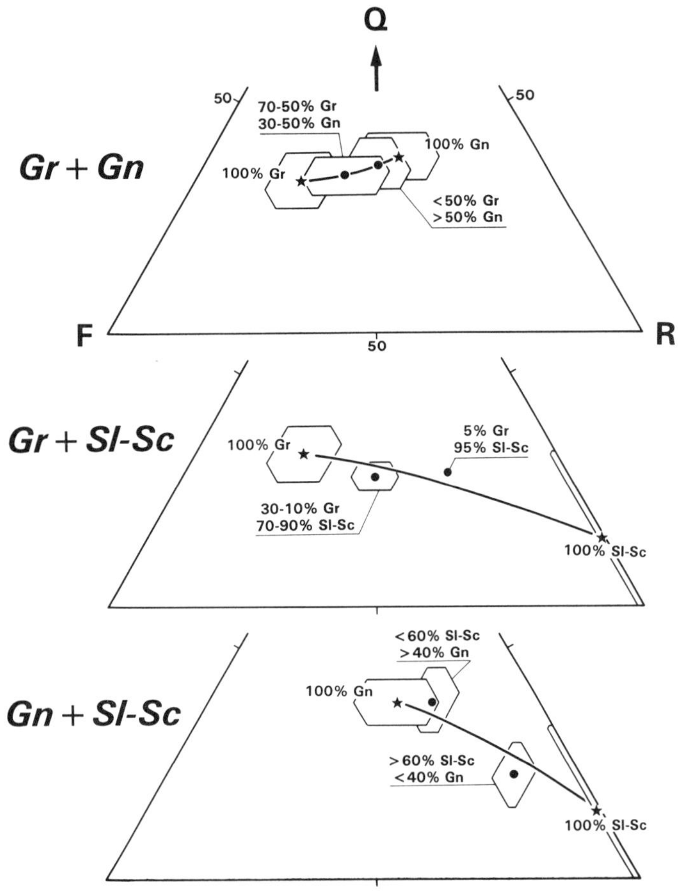

Figure 7. QFR compositions (mean and standard deviation) of sands derived from the three different source rock associations considered. Above, granitoid + gneiss (Gr + Gn); center, granitoid + slate-schist (Gr + Sl-Sc); below, gneiss + slate-schist (Gn + Sl-Sc). For clarity, compositional data concerning individual sand samples have been grouped into classes according to the relative abundance of the two bedrock types at the sources. Stars represent end member compositions of sands derived from simple (single lithology) sources (from Fig. 3).

sands is lower than expected, if compared with Q_p/Q_t values of sands from terrains with gneiss $<20\%$. However, the mean composition for slate-schist–derived sand has been obtained from data relative to 13 sands derived from slates and schists with different percentages of interbedded quartzitic rocks. This probably explains the low Q_p/Q_t index observed. Mean compositions of ($Gn_{>40}$ + $Sl\text{-}Sc_{<60}$)- and ($Gn_{<40}$ + $Sl\text{-}Sc_{>60}$)–derived sands are $Q_{38}F_{20}R_{42}$ and $Q_{27}F_{10}R_{63}$, respectively. The trend line that links the means of exclusively gneiss derived and slate-schist derived sands in the QFR diagram (Fig. 7, below) is similar in shape but shorter than that established for sands generated from (Gr + Sl-Sc) sources. This reflects the greater potential of gneiss to yield rock fragments compared to granitoid lithologies. As in (Gr + Sl-Sc)-derived sands, modal compositions are not directly proportional to the relative percentages of gneiss and slate-schist in the source area. Sands derived from terrains with no more than a 40% of gneiss have a composition similar to that of sands generated from simple gneissic sources.

INTERPRETATION

The compositional characteristics of first-cycle sands derived from the three types of dual sources considered show that no simple relation exists between sand composition and the percentage of surface area of the different bedrock types at the source. Each bedrock type (i.e., granitoid, gneiss, and slate-schist) has a different potential to produce sand-size grains. Sands derived from compound sources therefore will be dominated by grains proceeding from the rock type with the highest potential. This potential is mainly a function of the specific properties of a given rock type (Cooke and Doornkamp, 1990, p. 325). Crystal size in the parent rock controls the size of the mechanically weathered products and rocks with a mean crystal size in the range of 2.0 to 0.062 mm have a greater capacity to produce sand than others characterized by a smaller average crystal size (e.g., granitoids versus slates and schists). Another factor is rock isotropy. The virtual isotropy of plutonic rocks causes them to disintegrate preferentially along crystal boundaries, thus providing a great amount of sand-size monomineralic grains. In contrast, fabric heterogeneities (such as foliation and oriented microfabrics) typical of metamorphic rocks produce additional unstable surfaces. In fact, slates, schists, and gneisses tend to break preferentially along such surfaces, providing either coarse-clastic debris or products smaller than sand-size, with a consequent impoverishment of the sand fraction (e.g. Palomares, 1988; Palomares et al., 1990). Chemical stability and hardness of rock-forming minerals constitute a third important factor that affects the volume of sand generated. Low-stability minerals release high quantities of silt- and clay-size clasts, apart from soluble products, thus decreasing the sand generation potential of the parent rock.

In order to make allowance for the contribution of a particular bedrock constituent of a compound source area to the derivative sand, we propose the concept of sand generation index (SGI). It is convenient to express the SGI of a given bedrock type A of a dual source A + B in terms of the outcrop area of A required to produce a sand whose modal composition represents the average between the modes of the sands derived from simple A and B sources. Average modal compositions can be read from the QFR diagram (Fig. 3), and corresponding area percentages subsequently assessed from binary diagrams, such as shown in figures 4, 5, and 6. Thus,

$$SGI_{A(A+B)} = \frac{(S_A + S_B)}{S_A},$$

where $SGI_{A(A+B)}$ denotes the SGI of A referred to a dual source constituted by the association A + B; ($S_A + S_B$), the total surface area of the source terrain and S_A, the surface area occupied by A.

For example, in a source area constituted by (Gr + Sl-Sc), only about 5% outcrop area of granitoid bedrock is needed to obtain a sand with a modal composition of $Q_{28}F_{22}R_{50}$ (the average between that of an exclusively granitoid-derived and that of a pure slate-schist–derived sand). Consequently $SGI_{Gr(Gr\,+}$

TABLE 3. SAND GENERATION INDEX (SGI) AND RELATED PARAMETERS FOR GRANITOIDS, GNEISSES, AND SLATE-SCHISTS IN THE THREE BEDROCK ASSOCIATIONS CONSIDERED

Accociated Rock Types in the Source Area	Average Modal Compositions of Sands Derived from Simple Sources*	Surface Area Corresponding to Average Composition† (%)		SGI
Granitoid + gneiss	$Q_{35.5} F_{37} R_{27.5}$	Gr	40–50	2.0–2.5
		Gn	50–60	1.66–2.0
Granitoid + slate-schist	$Q_{23.5} F_{24} R_{52.5}$	Gr	5–7	14.28–20.0
		Sl-Sc	93–95	1.04–1.07
Gneiss + slate-schist	$Q_{24} F_{14} R_{60}$	Gn	15–20	5.0–6.66
		Sl-Sc	80–85	1.25–1.77

*From Figure 3.
†From Figures 4, 5, and 6.

$_{Sl\text{-}Sc)}$ is between 14 and 20 ($S_{Gr} + S_{Sl\text{-}Sc}/S_{Gr}$), whereas $SGI_{Sl\text{-}Sc(Gr + Sl\text{-}Sc)}$ is about 1 ($S_{Gr} + S_{Sl\text{-}Sc}/S_{Sl\text{-}Sc}$). In other words, granitoid rocks are 14 to 20 times more efficient in producing sand than are slates and schists. Table 3 shows the SGI values of the three bedrock types in the three associations considered in this study. It is obvious from the above definition that relative values result, rather than absolute (rock-specific) values. Granitoid has the highest SGI when associated with slate-schist. Gneiss generates four times less sand than granitoids do when associated with slate-schist. As expected, in both cases slate-schist bedrocks turn out to be a minor sand source when compared with granitoid and gneissic bedrocks. SGI values of granitoid and gneiss are similar when these two lithologies are present in the source area.

It is essential to remember that our analytical data refer to the medium-sand fraction. The SGI values listed in Table 3, therefore, only reflect the potential of a rock to generate grains in that size range. Complementary studies concerning the coarse- and fine-sand fractions will be attempted in the future. Furthermore, our results are based on sands from first- to sixth-order streams between <1 and 15 km in length, and thus the effects of transport on sand composition are negligible. Longer transport distances evidently cause the content of rock fragments and feldspar to decrease (Cameron and Blatt, 1971; Breyer and Bart, 1978; Mack, 1978), with an attendant increase in quartz grains. Accordingly, with increasing transport distances, one has to expect a progressive shift toward the Q apex of the compositional fields and trend lines illustrated in Figure 7. Humid climatic conditions are likely to produce a parallel shift (Suttner et al., 1981; Basu, 1985; Grantham and Velbel, 1988).

In spite of the above arguments, our data suggest that the model composition of sands and sandstones (in particular arkoses) may be determined by a rock type that represents only a subordinate portion of the source area and, thus, attempts to estimate sourceland composition directly from QFR plots are hazardous. Our results call for a critical revision of conventional paleogeographical and paleogeological interpretations based on detrital modes. A case in point is the lower Triassic Buntsandstein in the Iberian Range, a formation including arkoses, subarkoses, and quartzarenites, generally with less than 12% of unstable low-rank metamorphic rock fragments (slate and micaceous schists) (Arribas et al., 1985). Traditionally, provenance of these deposits has been interpreted in terms of a "principally" granitic-gneissic source, with "minor" contributions of low-rank metamorphic rocks. In view of the results of this study, such straightforward assertions are no longer warranted and, unless carefully documented, caution is recommended when using semiquantitative terms (such as dominant, principal, minor) in provenance interpretations.

CONCLUSIONS

1. The comparison of modal compositions of first-cycle sands derived from source areas consisting of one and two lithologies (simple and dual crystalline sources, respectively) provide information concerning the relative capacity of different bedrocks types (granitoid, gneiss, and slate-schist) in generating medium sand–size grains.

2. The concept of sand generation index (SGI) is introduced. SGI permits specification of the capacity of a given bedrock type to produce sand with respect to others in a compound source area.

3. SGI of a specific bedrock type, such as defined here, varies with respect to the source rock association. Thus, in case of a granitoid + slate-schist source SGI of granitoid is 14 to 20, whereas the SGI of slate-schist is about 1. By contrast, the SGI of granitoid is 2.0 to 2.5 when this rock type occurs in association with gneiss in the source area. In this case the SGI of gneiss is in the range of 1.66 to 2.00. Finally, in a gneiss + slate-schist association, the SGI of gneiss is 5.0 to 6.66, and the SGI of slate-schist is 1.25 to 1.77.

4. It is shown that the overall composition of a sand or sandstone in cases can be critically influenced by a bedrock type that constitutes only a minor portion of the source area. Quantitative estimates concerning the sourceland geology that are based exclusively on QFR plots may therefore be incorrect.

ACKNOWLEDGMENTS

We are grateful to Glen Hieshima, who kindly reported on our ideas at the 1991 annual meeting of the Geological Society of America. Abhijit Basu and Amparo Tortosa helped and encouraged us during research and in the preparation of the manuscript. The manuscript was read critically by O. Kälin and reviewed by W. C. James and S. W. Young; we greatly appreciate the helpful comments, corrections, and criticisms.

REFERENCES CITED

Aparicio, A., and Galán, E., 1980, Las características del metamorfismo hercínico de bajo y muy bajo grado en el sector oriental del Sistema Central (provincia de Guadalajara): Estudios Geológicos, v. 36, p. 75–84.

Aparicio, A., Barrera, J. L., Caraballo, J. M., Peinado, M., and Tinao, J. M., 1975, Los materiales graníticos hercínicos del Sistema Central español: Memoria del Instituto Geológico y Minero de España, v. 88, 145 p.

Arribas, J., and Arribas, E., 1991, Petrographic evidence of different provenance in two alluvial fan systems (Palaeogene of the northern Tajo Basin, Spain), *in* Morton, A. C., Todd, S. P., and Haughton, P.D.W., eds., Developments in sedimentary provenance studies: Geological Society of London Special Publication 57, p. 263–271.

Arribas, J., Marfil, R., and De La Peña, J. A., 1985, Provenance of Triassic feldspathic sandstones in the Iberian Range, Spain: Significance of quartz types: Journal of Sedimentary Petrology, v. 55, p. 864–868.

Basu, A., 1976, Petrology of Holocene fluvial sands derived from plutonic source rocks: Implications to paleoclimatic interpretation: Journal of Sedimentary Petrology, v. 46, p. 694–709.

Basu, A., 1985, Influence of climate and relief on compositions of sands released at source areas, *in* Zuffa, G. G., ed., Provenance of arenites: Dordrecht, D. Reidel, p. 1–18.

Basu, A., Young, S. W., Suttner, L. J., James, C. W., and Mack, G. H., 1975, Re-evaluation of the use of undulatory extinction and polycrystallinity in detrital quartz for provenance interpretation: Journal of Sedimentary Petrology, v. 45, p. 873–882.

Breyer, J. A. and Bart, H. A., 1978, The composition of fluvial sands in a temperate semi-arid region: Journal of Sedimentary Petrology, v. 48, p. 1311–1320.

Cameron, K. L., and Blatt, H., 1971, Durabilities of sand-sized schist and volcanic rock fragments during fluvial transport, Elk Creek, Black Hills, South Dakota: Journal of Sedimentary Petrology, v. 41, p. 565–576.

Casillas, R. and Peinado, M., 1988, Secuencias graníticas en el área de San Martín de Valdeiglesias (Sistema Central español), *in* Libro Homenaje a L.C. García de Figuerola: Geología de los granitoides y rocas asociadas del macizo hespérico: Madrid, Editorial Rueda, p. 281–292.

Casquet, C., Fuster, J. M., Gonzalez-Casado, J. M., Peinado, M., and Villaseca, C., 1988, Extensional tectonics and granite emplacement in the Spanish Central System: A discussion: Proceedings of the 5th Workshop on the European Geotraverse Project, p. 65–76.

Chayes, F., 1956, Petrographic modal analysis: New York, John Wiley & Sons, 113 p.

Cooke, R. U., and Doornkamp, J. C., 1990, Geomorphology in environmental management: Oxford, Clarendon Press, 410 p.

Dickinson, W. R., 1970, Interpreting detrital modes of graywacke and arkose: Journal of Sedimentary Petrology, v. 40, p. 695–707.

Dickinson, W. R., 1985, Interpreting provenance relations from detrital modes of sandstones, *in* Zuffa, G. G., ed., Provenance of arenites: Dordrecht, D. Reidel, p. 333–361.

Dickinson, W. R. and Suczek, C. A., 1979, Plate tectonics and sandstone compositions: American Association of Petroleum Geologists Bulletin, v. 63, p. 2164–2182.

Dickinson, W. R. and Valloni, R., 1980, Plate settings and provenance of sands in modern ocean basins: Geology, v. 8, p. 82–86.

Dickinson, W.R., and 8 others, 1983, Provenance of North American Phanerozoic sandstones in relation to tectonic setting: Geological Society of America Bulletin, v. 94, p. 222–235.

Folk, R. L., 1974, Petrology of sedimentary rocks: Austin, TX, Hemphill's, 182 p.

Gazzi, P., 1966, Le arenarie del flysch sopracretaceo dell'Appennino modenese; correlazioni con il Flysch di Monghidoro: Mineralogia et Petrographia Acta, v. 12, p. 69–97.

Grantham, J. H., and Velbel, M. A., 1988, The influence of climate and topography on rock-fragment abundance in modern fluvial sands of the Southern Blue Ridge Mountains, North Carolina: Journal of Sedimentary Petrology, v. 58, p. 219–227.

Ingersoll, R. V., 1991, Actualistic sandstone petrofacies: Effects of variable sampling scale: Geological Society of America Abstracts with Programs, p. A70.

Lopez Ruiz, J., Aparicio, A., and García Cacho, L., 1975, El metamorfismo de la sierra de Guadarrama, Sistema Central español: Memorias del Instituto Geológico y Minero de España, v. 86, 127 p.

Mack, G. H., 1978, The survivability of labile light mineral grains in fluvial, eolian and littoral marine environments: The Permian Cutler and Cedar Mesa Formation, Moab, Utah: Sedimentology, v. 25, p. 587–606.

Mack, G. H., 1981, Composition of modern stream sand in humid climate derived from a low-grade metamorphic and sedimentary foreland fold-thrust belt of north Georgia: Journal of Sedimentary Petrology, v. 51, p. 1247–1258.

Mann, W. R., and Cavaroc, V. V., 1973, Composition of sands released from three source areas under humid, low relief weathering in the North Carolina piedmont. Journal of Sedimentary Petrology, v. 43, p. 870–881.

Navidad, M., and Peinado, M., 1981, Ortogneis y metasedimentos de la formacion infrabasal al Ollo de Sapo (macizo Hiendelaencina, Guadarrama oriental): Cuadernos de Geología Ibérica, v. 7, p. 183–199.

Palomares, M., 1988, Análisis de las arenas actuales derivadas de rocas metamórficas del Sistema Central: aplicación a los estudios de procedencia. [Licenciatura thesis]: Madrid, Universidad Complutense, 191 p.

Palomares, M., Tortosa, A., and Arribas, J., 1990, Caracterización clastométrica de los depósitos detríticos de cabecera de arroyos en el Sistema Central Influencia de la litología del área fuente: Boletín de la Real Sociedad Española de Historia Natural (sección Geología), v. 85, p. 5–21.

Pettijohn, F. J., Potter, P. E., and Siever, R., 1987, Sand and sandstones, second edition: New York, Springer-Verlag, 553 p.

Suttner, L. J., Basu, A., and Mack, G. M., 1981, Climate and the origin of quartzarenites: Journal of Sedimentary Petrology, v. 51, p. 1235–1246.

Tortosa, A., Palomares, M., and Arribas, J., 1989, Caracterización composicional de los depósitos arenosos actuales generados en el Sistema Central: Estudios Geológicos, v. 45, p. 205–213.

Tortosa, A., Palomares, M., and Arribas, J., 1991, Quartz grain types in Holocene deposits from the Spanish Central System: Some problems in provenance analysis, *in* Morton, A. C., Todd, S. P., and Haughton, P.D.W., eds., Developments in sedimentary provenance studies: Geological Society of London Special Publication 57, p. 47–54.

Valloni, R., 1985, Reading provenance from modern marine sands, *in* Zuffa, G. G., ed., Provenance of arenites, Dordrecht, D. Reidel, p. 309–332.

Villaseca, C., 1984, Evolución metamórfica del sector centro septentrional de la Sierra de Guadarrama [Doctoral thesis]: Madrid, Universidad Complutense, 331 p.

Wilson, L., 1969, Les rélations entre les processus géomorphologiques et le climat moderne comme méthode de paléoclimatologie: Revue de Géographie Phisique et de Géologie Dinamique, v. 11, p. 303–314.

Zuffa, G. G., 1980, Hybrid arenites: Their composition and classification: Journal of Sedimentary Petrology, v. 50, p. 21–29.

Zuffa, G. G., 1985, Optical analyses of arenites: Influence of methodology on compositional results, *in* Zuffa, G. G., ed., Provenance of arenites, Dordrecht, D. Reidel, p. 165–189.

MANUSCRIPT ACCEPTED BY THE SOCIETY DECEMBER 24, 1992

Printed in U.S.A.

Geological Society of America
Special Paper 284
1993

Toward quantitative provenance analysis: A brief review and case study

Emanuela Molinaroli
Dipartimento di Scienze Ambientali, Università di Venezia, Dorso Duro 2137, 30123 Venezia, Italy
Abhijit Basu
Department of Geological Sciences, Indiana University, Bloomington, Indiana 47405

ABSTRACT

Standard petrographic (e.g., QFL) and chemical (e.g., bulk chemistry) methods for provenance determination successfully classify approximately 55 to 85% of the empirical data on which the methods are based. New approaches in provenance analysis include: (1) age dating of single zircon grains or rock fragments; (2) calculating the P-T-t paths of metamorphic clasts; (3) interpreting rare earth element distribution and Sm-Nd isotope systematics to infer mantle precursors; and (4) analyzing single groups of minerals; Fe-Ti oxide minerals appear to be an especially promising group of minerals because, despite extensive chemical alteration, intergrowth patterns assist in identifying ilmenite and magnetite precursors.

The goals of quantitative provenance analysis are to estimate both the proportions of source rocks represented in a body of detrital sediments and the rates at which the detritus was derived. We have quantitatively compared the distribution and the chemistry of selected heavy mineral species in Holocene sands and the sandstones of the Oligocene Renova Formation in southwestern Montana; field evidence suggests that the source rocks of these sands and sandstones were the same. Our previous statistical analysis of the chemical and textural properties of detrital opaque oxide minerals in the sands and sandstones suggested that the Renova sandstones consist of 77% igneous (mostly granodiorites and dacites) and 23% metamorphic (mostly amphibolite-facies schists and gneisses) detritus.

We have developed a first approximation of an algebraic mass balance model, using characteristic index minerals for specific parent rocks, which would calculate the proportions of parent rocks that were eroded to produce a body of detrital sediments. The largest uncertainty in the model is in the estimation of the proportions of parent rocks that must be eroded to release equal amounts of the respective index minerals, and in the relative preservation potentials of the index minerals during weathering and diagenesis. We have applied this model using garnets and zircons as two index minerals to compare the proportions of parent rocks of Holocene sands and Renova sandstones in southwestern Montana. Microprobe analyses of 574 single grains of garnet and zircon show that the compositions of detrital garnets and zircons in the Holocene sands and Renova sandstones are similar, suggesting that diagenetic alteration of these two minerals in the Renova has not been extensive. Our model calculations, based on modal analyses for zircons and garnets in 143,607 heavy mineral grains, suggest that ratios of igneous and metamorphic source rocks that were eroded to produce the Holocene sands and Renova sandstones in the study area are 81:19 and 82:18, respectively. The close correspondence of these two ratios possibly suggests that relative contributions of igneous and metamorphic sources in this locality have not changed much since the Oligocene time.

Molinaroli, E., and Basu, A., 1993, Toward quantitative provenance analysis: A brief review and case study, *in* Johnsson, M. J., and Basu, A., eds., Processes Controlling the Composition of Clastic Sediments: Boulder, Colorado, Geological Society of America Special Paper 284.

INTRODUCTION

This work briefly tracks recent advances in provenance analysis that go beyond source rock identification and attempts to estimate the relative proportions of different source rocks present in a body of sediments (e.g., Ibbeken and Schleyer, 1991; Basu and Molinaroli, 1991; Stöffler et al., 1981). With a case study, we show how statistical analysis and mass balance calculations based on petrological data may provide a preliminary step toward quantitative provenance analysis.

QUALITATIVE PROVENANCE ANALYSIS

Provenance analyses of siliciclastic sediments have advanced considerably since Pettijohn declared that "the question of provenance is one of the most difficult the sedimentary petrographer is called on to solve" (Pettijohn et al., 1972, p. 298). These advances have been based mostly on empirical evidence. For example, Dickinson et al. (1983) have shown that detrital modes of sandstones, determined by the Gazzi-Dickinson method, are useful in identifying the plate tectonic setting of the provenance of a suite of sandstones. Bhatia (1983), van de Kamp and Leake (1985), and others have shown that the chemical compositions of sandstones, especially of those associated with plate margin provenance, are useful for discriminating tectonic provenance. Earlier, Basu et al. (1975) and Young (1976) showed that quartz types in sands and sandstones can also be used to discriminate igneous and metamorphic sources. Suttner et al. (1981) and Basu (1976, 1985) showed how climate may affect sand composition, and that it may be possible to infer the relative influence of climate from the compositions of consanguineous sandstones. More recently, other techniques have been used for specific identification of source rocks. For example, Drewery et al. (1987), Cliff et al. (1991), and Girty and Wardlaw (1985) have used age dates of detrital zircons, and Graham et al. (1991) have used age dating of rock fragments in conglomerates to identify sources of sandstones. Cuthbert (1991) has calculated the P-T-t paths of metamorphic clasts in the Hornelen basin of western Norway to identify the sequence of thrust loading in source areas. McLennan et al. (1990, 1991) and Nelson and DePaolo (1988) have been attempting to characterize the crustal and mantle progenitors of the crystalline rocks that were eroded to supply the detritus of the sandstones that they study.

However, sandstones usually have multiple sources, and hybrid arenites are the rule rather than the exception (Zuffa, 1980; Mack, 1984). Quantitative estimation of the relative contributions of different source rocks or of different tectonic regimes to a suite of siliciclastic rocks was beyond the scope of the general studies mentioned above. However, it is important to identify multiple sources of detrital rocks (e.g., Zuffa et al., 1987; Garzanti, 1991). Further, new advances may be made if one can estimate the proportions of detrital material that were derived from specific source rocks, or estimate the proportions of source rocks that were eroded to give rise to a body of detrital sediments and the rates of sediment supply.

The primary variables that control the compositions of siliciclastic material before final deposition are source rocks (S), weathering (W), and transportation (T). Thus, if composition is expressed as relative abundances of quartz, feldspar, and rock fragments (QFR), then QFR = f(S,W,T), which is commonly accepted (Suttner, 1974). In a previous study we have seen that quartz populations in granitic detritus produced under humid and semi-arid climates may vary by 40%; thus, $f_{W(ST)}(Q) = \Delta Q$ {–40%} (Basu, 1976). Johnsson has noticed even higher enrichment of quartz as a result of climatic influence on weathering (M. J. Johnsson, personal communication, 1992). We have also found that, under tropical weathering, rock fragments in basaltic detritus may be reduced to clay in a very short period of time; thus, $f_{W(ST)}(R) = \Delta R$ {–100%} (Basu et al., 1986). Transportation causes mechanical degradation and feldspars concentrate in the silt fraction; thus, $f_{T(SW)}(F) = \Delta F$ {–25%} (Suttner et al., 1981). Blatt (1978) has found that a combination of short transport and chemical weathering may completely remove basaltic detritus from a stream sand. In short, QFR is as much a function of maturity as it is of source rocks (see also the discussion by Johnsson, 1990a,b). This implies that QFR, QFL, and Q_mFL_t types of analysis of unlithified sands of unspecified maturity can provide perhaps only a qualitative assessment of source rock assemblages. Although not suitable for quantitative provenance analysis, the QFL type of investigation does identify provenance of sands and sandstones (Dickinson, 1988).

We have tested the statistical reliability of the general models of Dickinson et al. (1983), van de Kamp and Leake (1985), and of Basu et al. (1975) using the very data sets with which all three models were constructed (Molinaroli et al., 1991). Our results show that the overall efficiency with which the models can classify their own data sets is no more than 85%; in particular, transitional arc, dissected arc, and undissected arc provenances are correctly identified only in 55, 63, and 70% of cases with Q_mFL_t data (Molinaroli et al., 1991). In our opinion, such low success rates with the empirical data imply that the methods should not be used to construct general discriminatory models. We mention this to set the quantitative context of the case studies reported below.

It has been shown in the last decade that the chemical compositions of bulk concentrates of detrital ilmenite are sensitive to source rocks (Darby, 1984; Darby and Tsang, 1985), although some of the "ilmenite" would have intergrown phases such as magnetite and hematite affecting the chemical composition of the bulk sample. Nearly 50% of detrital opaque Fe-Ti oxide mineral grains in the heavy residues of sands and sandstones may be polymineralic, showing intergrowth textures (Molinaroli and Basu, 1987; Riezebos, 1979). The data on the crystallographic orientation of the intergrowths, their widths, and

Figure 1. Photomicrograph of a weathered detrital Fe-Ti oxide mineral showing one direction of intergrowth between ilmenite and hematite in sand sample 8582059 collected from a stream draining principally metamorphic rocks in southwestern Montana. Length of frame, 270 μm.

Figure 2. Photomicrograph of a weathered detrital Fe-Ti oxide mineral showing three directions of intergrowth between magnetite and hematite in sand sample 8581664 collected from a stream draining principally granitic rocks in southwestern Montana. Length of frame, 270 μm.

the chemical compositions of the ilmenite phase may be combined to form a very robust provenance discriminant (Basu and Molinaroli, 1989). More recently, Grigsby (1990, 1992) has shown that chemical compositions of individual optically homogeneous magnetite and ilmenite are also sensitive to source rocks. He has further shown that the distribution of homogeneous grains and those with different types of intergrowths may also be sensitive to provenance.

In a case study with Holocene sands, we have found that the combined chemical-petrographical discriminant was successful in classifying more than 95% of the samples (Basu and Molinaroli, 1989). Most importantly, we discovered that, although weathering and diagenesis may alter much of the mineral chemistries of the intergrown phases, relict textures continue to persist in lithified sandstones. Thus, in altered Fe-Ti oxide grains, pinacoidally oriented intergrowths (one direction of intergrowth in a random section) indicate an ilmenite precursor, and octahedrally oriented intergrowths (three directions of intergrowths in a random section) indicate a magnetite precursor (Figs. 1, 2). Ilmenite and magnetite dominate the Fe-Ti oxide mineral distribution in S-type and I-type granites, respectively (Chappel and White, 1974; Ishihara, 1977). Therefore, an abundance of ilmenite or one direction of intergrowth, even if relict, in the population of detrital opaque oxide minerals likely indicates an S-type source, and an abundance of magnetite or three directions of intergrowths likely indicate an I-type source. We have found that ilmenite and grains with one set of parallel intergrowths are indeed more abundant in the detritus from deep-seated S-type granites, and that magnetite and grains with one octahedral set of intergrowths are more abundant in the detritus from shallow I-type granites (Basu and Molinaroli, 1989). In general, I-type granites are commonly associated with magmatic arcs and shallow-level plutons, and S-type granites are more common in continental blocks (Atherton and Tarney, 1979). Therefore, we suggest that intergrowth textures of detrital opaque oxide minerals may be guides to identifying magmatic arc and continental block provenances because the textures seem to survive alteration through weathering and diagenesis better than other properties of detrital Fe-Ti oxide minerals.

QUANTITATIVE PROVENANCE ANALYSIS

Quantitative provenance analysis consists of two calculations: estimating the proportions of parent rocks represented in a sediment, and determining the rates of sediment supply to and the rates of sediment trapping in a basin. In this study we attempt to perform only the first calculation, on the assumption that qualitative identification of source rocks would be possible using one or more of the standard models mentioned above. Thus, given the mass of a sediment body, we need to calculate the amounts of different parent rocks eroded in source areas. If such estimates are available, it would then be possible to reconstruct the distribution of rock types present in the source areas.

Morton (1985) has argued convincingly about the benefits of using a single durable heavy mineral for provenance studies so as to avoid or minimize "the problems caused by diagenetic and hydraulic processes" (Morton, 1991, p. 31). We describe below two approaches to quantitative provenance analysis of the intraarc sandstones of the Oligocene Renova Formation in southwestern Montana using crystallographic, chemical, and textural properties of detrital opaque oxides, garnets, and zircons.

The sandstones of the Oligocene Renova Formation that we

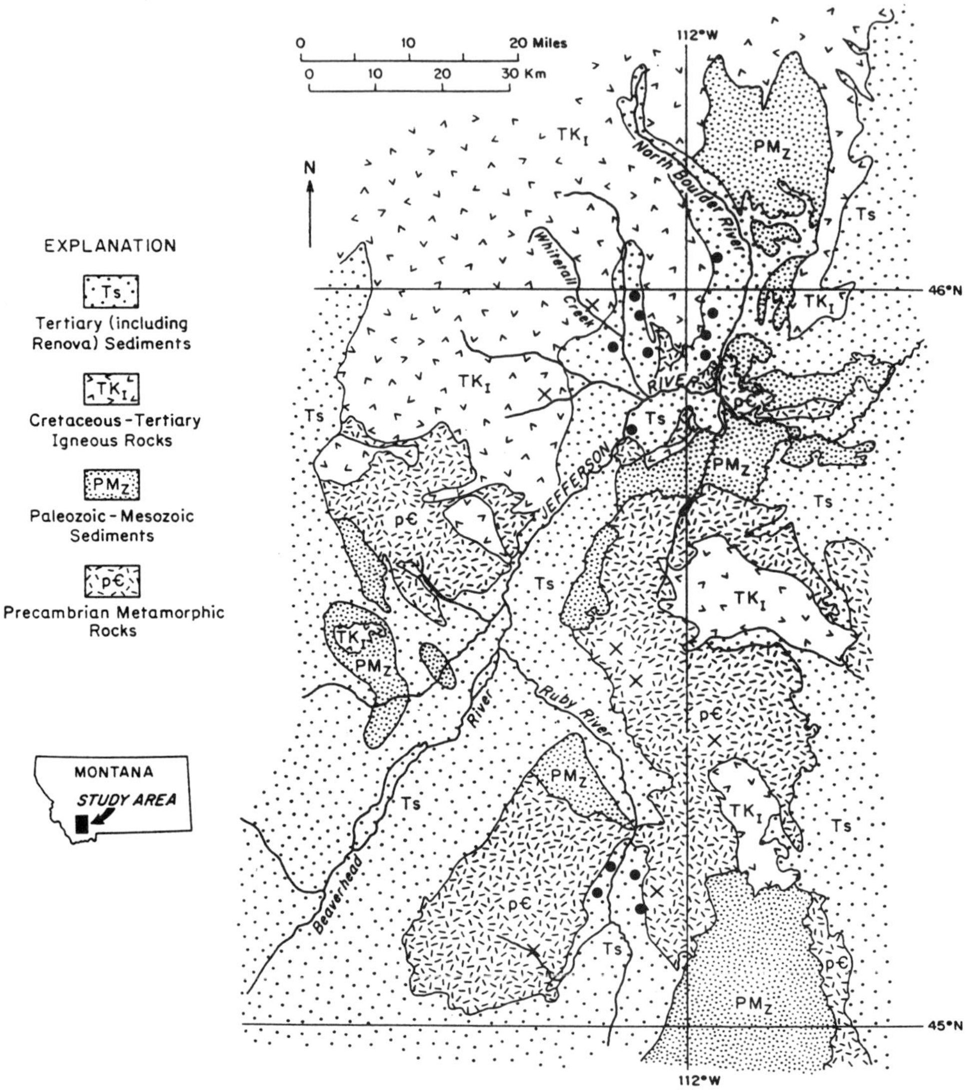

Figure 3. Sketched geological map of the study area in southwestern Montana. The distinction between the Renova and the Holocene is not shown to avoid clutter. Major sampling sites of paired Renova and Holocene samples are indicated by black dots. Locations marked with an X are sampling sites of Holocene sand only, which were used for the Fe-Ti oxide mineral study. Modified after Basu and Molinaroli, 1991.

have selected for our study generally lie unconformably as alluvial fan aprons on their source rocks (Fig. 3). The source rocks are broadly divisible into two groups. The Precambrian metamorphic source rocks consist of garnet-bearing gneisses and amphibolite-facies schists, including an anthophyllite-garnet schist. The Cretaceous-Tertiary igneous source rocks consist of the Boulder Batholith and the Elkhorn Mountain Volcanics; the rock types are mostly granodiorites, dacites, and andesites. We have also sampled the Holocene sediments from streams draining these very source rocks in the same area. Thus our samples represent the Renova-Holocene pair, and we can use the Holocene samples as the standard for quantitative provenance analysis of the Renova in this locality. A more complete account of the geology of the area and sampling plan is given in Basu and Molinaroli (1991). Suffice it to state here that the sampling plan was designed to represent all the principal Renova outcrops and all the major drainage basins in the area for collecting a representative suite of Holocene sands. To the best of our knowledge our sampling stations are distributed over the Renova outcrops and in the present-day drainage basins as uniformly as possible. For the garnet/zircon study described below, the Holocene samples were collected from streams near the Renova outcrops but avoiding possible contamination from the Renova. These Holocene sands for this garnet/zircon study were derived principally but not exclusively from igneous or metamorphic lithologies. Thus, the complete suite of Holocene sands represents a mixture of source rock lithologies, just as the sandstones of the Oligocene Renova Formation in this region are also products of mixed source rocks.

We have sampled the Holocene sands from nine localities and the Renova sandstones from outcrops along three streams (Fig. 3). We do not believe that sand from any one stream is representative of all Holocene sands and, respectively, of the region nor that sandstone from any one Renova fan is representative of all Renova sandstones. Therefore, data from all the samples in the region were averaged for comparing the Holocene and the Renova.

Laboratory methods and results

The clay fraction was removed from samples of stream sands and disaggregated sandstones by washing the samples with water. Samples were not sieved. Heavy mineral concentrates were obtained from the unsieved, washed samples by using bromoform as the heavy liquid. We have conducted petrographic and electron microprobe analyses on polished thin sections of heavy mineral concentrates from Holocene sands and Renova sandstones. The grain size (largest intercept normal to the long axis of a grain in thin section) in the concentrates ranged from about 20 μm to about 1 mm; more than 90% of garnets and zircons are in the 0.1 mm to 0.3 mm range. We have previously determined the mineralogy, and if present, the orientation and widths of the intergrowths of 24,791 Fe-Ti oxide mineral grains, and analyzed 828 grains of ilmenite, magnetite, and titaniferous hematite with an electron microprobe. Results of the Fe-Ti oxide mineral studies were given in Basu and Molinaroli (1989, 1991).

We have determined the abundances of garnet and zircon in a total population of 143,607 heavy mineral grains and have analyzed 347 garnets and 227 zircons with an electron microprobe. Results of the garnet and zircon modal analyses are given in Table 1; chemical compositions are graphically presented in Figures 4 and 5.

TABLE 1. DISTRIBUTION OF GARNETS AND ZIRCONS IN HOLOCENE FLUVIAL SANDS AND THE OLIGOCENE RENOVA SANDSTONES*

	Garnet (of total heavy mineral grains) (%)	Zircon (%)	Total HM
	HOLOCENE SANDS		
Northern areas	2.31	0.68	53,861
Southern areas	22.22	0.07	43,806
Holocene total	**9.54**	**0.40**	**97,667**
	OLIGOCENE RENOVA SANDSTONES		
Ruby River Section	21.51	0.82	33,063
N. Boulder Section	0.19	0.56	7,553
Whitetail Section	0.00	0.32	5,324
Renova total	**15.60**	**0.72**	**45,940**

*Both are derived from the same Precambrian metamorphic rocks and Cretaceous-Tertiary igneous rocks. The totals are averages of samples that represent parent rocks better than the individual areas listed.

A statistical approach

Grigsby (1990, 1992) has discovered that petrographic and chemical properties of both magnetite and ilmenite discriminate source rocks quite efficiently. We have reached a similar conclusion from our study of the detrital Fe-Ti oxide minerals, treating them as a single family of heavy minerals, in the sands of the Appalachians and the Rocky Mountains (Basu and Molinaroli, 1989). Discriminant function analysis of types of intergrowths and the chemical compositions of the detrital Fe-Ti oxide mineral population in Holocene sands in southwestern Montana correctly classifies (100% statistical success) their igneous or metamorphic sources on the basis of these properties (Basu and Molinaroli, 1991). We have not tested whether the coefficients generated from this analysis are universally applicable. But, because these Holocene sands were derived from the same source rocks as the sands of the Renova sandstones that we are investigating, it is appropriate to apply the coefficients derived from the discriminant function analysis of Holocene sands for classifying the Renova sandstones. A listing of the variables and the coefficients used is given in Appendix 1. Results show that 77% of the detrital Fe-Ti oxide minerals in the Renova were derived from igneous sources (mostly granodiorites and dacites), and the rest, 23%, were derived from mostly amphibolite-facies schists and gneisses, i.e., metamorphic sources (Basu and Molinaroli, 1991).

We may assume that the preservation potential of any one Fe-Ti oxide mineral is not dependent on its parent rocks; i.e., an ilmenite (ideally of a certain invariable composition and structure) from a granite, gneiss, basalt, or schist would have the same survivability during weathering, transportation, deposition, and diagenesis. If so, then the proportions, ϵ, of detrital Fe-Ti oxide minerals from igneous and metamorphic source rocks also represent the proportions, η, of source rocks that were eroded to produce the detritus, i.e., $\epsilon = \eta$. However, if two different index minerals or two varieties of a single mineral with two different preservation potentials are considered instead of a single mineral, the proportion of detrital abundances of these two minerals, υ, will not represent the proportion, π, of the source rocks. If the preservation potentials of these two index minerals were to differ by a constant factor, ψ, then the proportion of two different source rocks (υ) eroded to produce the detritus would be ψ/π, i.e., $\pi = \psi/\upsilon$. This concept is further discussed below.

A mass balance approach

We have developed a *metamorphic mineral/igneous mineral* ratio model for our calculations. The model requires a set of samples from known sources, used as the standard. Corresponding data on samples of unknown pedigree are used to calculate the factors by which the unknown sources differ from the known standard. This is a crucial concept in our argument. We do not calculate the absolute ratios in which two sources or two parent rocks were eroded to produce a body of detrital sediments. We calculate a nominal ratio in which two known parent rocks were

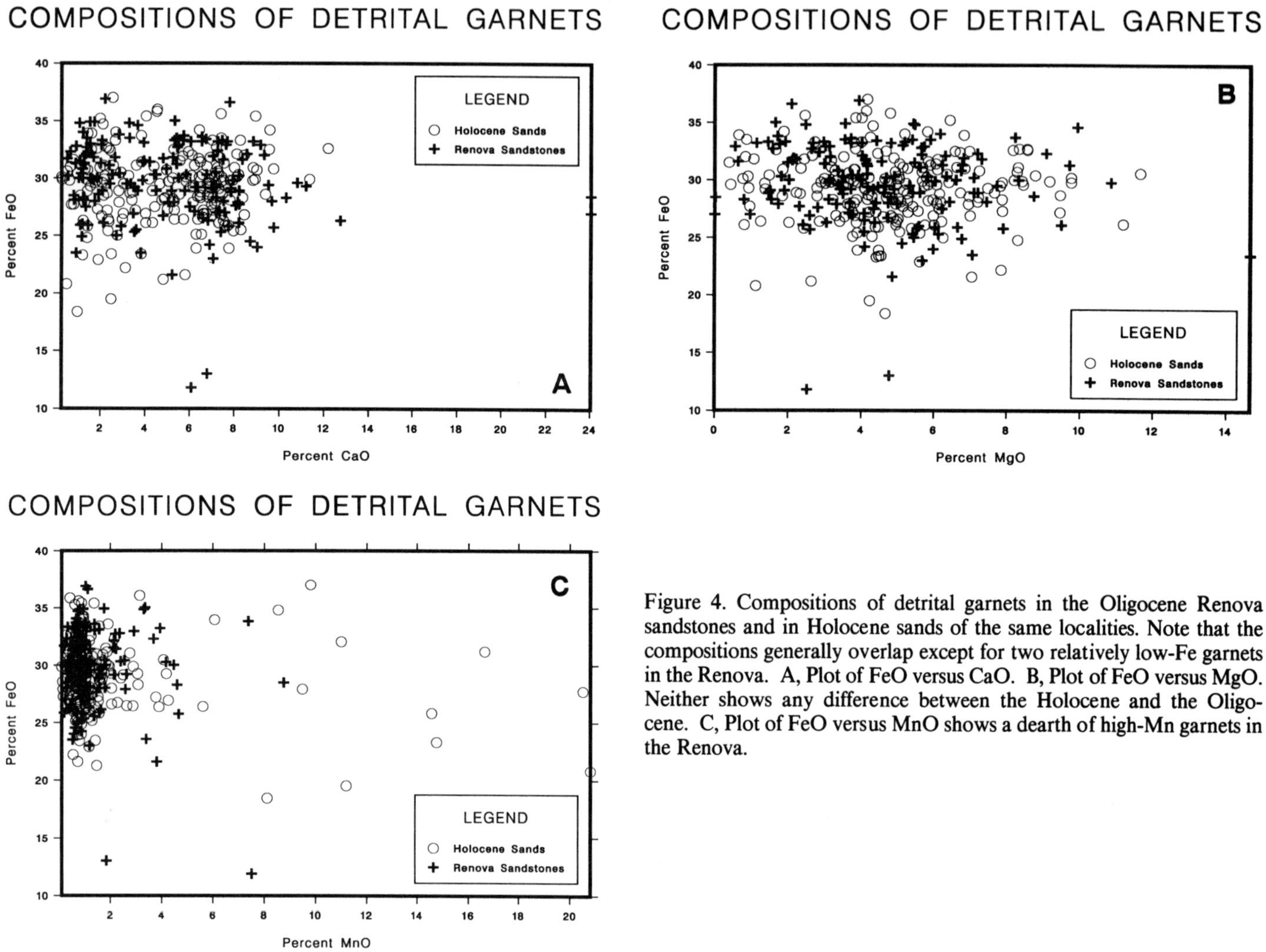

Figure 4. Compositions of detrital garnets in the Oligocene Renova sandstones and in Holocene sands of the same localities. Note that the compositions generally overlap except for two relatively low-Fe garnets in the Renova. A, Plot of FeO versus CaO. B, Plot of FeO versus MgO. Neither shows any difference between the Holocene and the Oligocene. C, Plot of FeO versus MnO shows a dearth of high-Mn garnets in the Renova.

eroded to produce the stream sands of the present time. We calculate another nominal ratio in which two similar parent rocks might have been eroded in the past to produce a body of ancient sediments. We then find the factor by which the ancient differs from the represent and evaluate the geological significance, if any, of that factor.

The selection of the index minerals poses a problem. To be useful, the chosen minerals must be abundant enough in source rocks and in sediments, and must also be exclusive enough to represent a parent rock. Because such a combination is rare in nature, a compromise has to be accepted. In order to determine the relative proportions of igneous (mostly granodiorite and dacite) and metamorphic (mostly amphibolite-facies schists and gneisses) contribution to the Renova sandstones, we have selected zircon and garnet as two index minerals representing these two sources. Garnet and zircon do not have similar deposition and preservation potentials. Zircon is much heavier than garnet, and the two also commonly differ in shape; thus, they are sorted differently. Further, zircon is by far the most durable heavy mineral, whereas garnet is more prone to chemical weathering

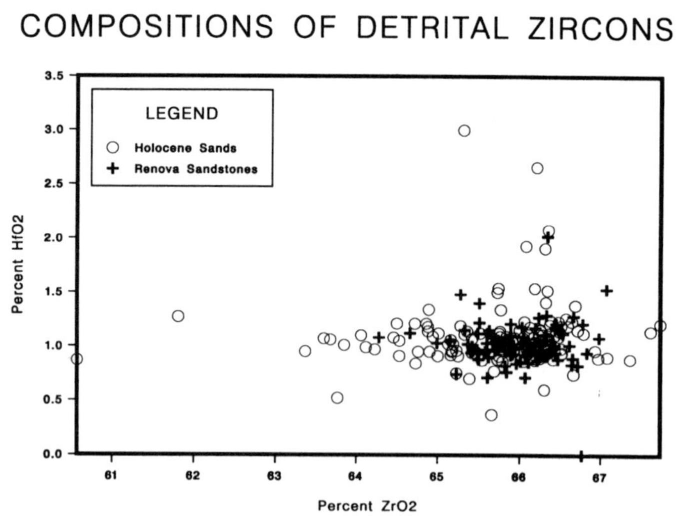

Figure 5. Compositions of detrital zircons in the Oligocene Renova sandstones and in Holocene sands of the same localities. Note that the compositions generally overlap except for two relatively low-Zr and two high-Hf zircons in the Holocene.

(cf. Orr and Folk, 1983) and diagenesis (Morton, 1985). In general, garnets may be of much larger sizes than zircons in parent rocks and may concentrate in larger size fractions, especially in immature sediments (cf. Force ane Stone, 1990). It is also true that the composition of any fraction of a sand sample is a function of the mean grain size of the sample. Because we have not sieved our samples and are using all available sizes, and because the size range of 90% of our detrital garnets and zircons is rather narrow (0.1 to 0.3 mm), we are significantly reducing the size control on the distribution of detrital garnets and zircon. Further, chemical weathering by itself renders the preservation potential of garnets lower than that of zircon. The relative abundances of these two minerals in any sediment are also dependent on their relative abundances in the parent rocks, which we do not consider explicitly. All of these factors must be taken into account if these two minerals are used to estimate the proportions of parent rocks that were eroded to give rise to a body of detrital sediments. Yet, these minerals are generally durable in the sedimentary milieu and, given that this part of the Renova may have been buried only to shallow depths (Fields et al., 1985), it is likely that variations in their abundances due to differential diagenetic alteration may be small, although not necessarily insignificant.

To check the possible influence of differential diagenetic alteration, we have plotted compositions of detrital garnets and zircons from the Holocene and the Renova in Figures 4 and 5, respectively. The plots show that, in general, compositions of garnet and zircon are not dependent on age. To be sure, there is a greater variability in the compositions of both zircon and garnets in the Holocene sands than in those from the Renova sandstones. For example, a few garnets with high Mn show up in the Holocene, whereas no high-Mn garnets are present in the Renova samples. We ascribe the greater variability in Holocene sands to minor differences in immediate source rocks and conclude that differential diagenetic alteration of garnets and zircons must have been undetectably small in the Renova. Morton (1991) has noted that Ca-rich garnets tend to be less durable than garnets of other compositions during deep burial diagenesis of North Sea sandstones. We apparently do not have that problem; perhaps the continental Renova sediments may not have been buried deeply enough to cause extensive dissolution of garnets. Similarity in the compositions of zircons and garnets in the Renova sandstones and Holocene sands is compatible with the interpretation, drawn from field relationships, that the parent rocks of the two deposits were the same.

An inexact garnet/zircon ratio model

Assumptions. In our specific case we assume that: (1) garnets are present only in the metamorphic source rocks; (2) zircons are present only in the igneous source rocks; (3) the ratio of igneous to metamorphic rock volumes yielding (after weathering, erosion, and deposition) equal volumes of zircons and garnets in the heavy mineral fraction is α, and $\alpha>1$; and (4) detrital zircons suffer little reduction because of weathering or diagenesis but the volume of detrital garnets is modified by a factor of β ($\beta<1$) relative to zircons because of weathering and diagenesis. The first two assumptions are not entirely correct because garnets may be present in igneous rocks and relict zircons do persist in metamorphic rocks. However, the Boulder Batholith, despite its heterogeneity, and the Elkhorn Mountain Volcanics are not reported to contain any garnet (Tilling, 1964; Knopf, 1957). And, except for a part of one small amphibolite body, which contains 0.2% zircon, none of the other major gneisses and schists that make up by far the bulk of the metamorphic rocks in the area contain any measurable amount of zircon (Cordua, 1973). Further, given the uncertainties in the values of α and β, these two assumptions are reasonable for our purpose.

It follows that (1) $\alpha = I_Z/M_G$, where M_G is the volume of metamorphic source rocks releasing a given volume of garnets and I_Z is the volume of igneous source rocks releasing the same volume of zircons; (2) the volume of detrital garnets in a sediment is $N_{SG} = \beta N_{RG}$, where N_{RG} is the volume of garnets released from a unit volume of metamorphic source rocks; (3) the volume of detrital zircons in a sediment is $N_{SZ} = N_{RZ}$, where N_{RZ} is the volume of zircons released from a unit volume of igneous source rocks; (4) $N_{RG} = k_g M_G$, where k_g is a constant; and (5) $N_{RZ} = k_z I_Z$, where k_z is a constant. Note that k_g and k_z reflect the efficiency with which garnets and zircons are released from respective parent rocks. The relative efficiency k_g/k_z is obtianed by setting $N_{RG} = N_{RZ}$, and is equal to α.

Thus, equal amounts of garnet-bearing (~metamorphic) and zircon-bearing (~igneous) source rocks would produce detrital garnets and zircons in the ratio of $\alpha\beta$. Stated another way, igneous and metamorphic source rocks in the ratio of $\alpha\beta$ would produce equal amounts of detrital zircons and garnets. This relationship, of course, would hold only if the abundance of garnet in igneous source rocks and that of zircon in metamorphic source rocks were negligibly small.

In this first approximation model we have allowed two flaws that need to be adequately addressed in future. First, we have not explicitly included a variable that reflects the abundance of each index mineral in its parent rock; instead, we have implicitly included such a variable in our factor α. Second, a more complete model would require that we expand our factor β such that changes in the abundance of any index mineral during source rock weathering, abrasion during transport, weathering during transport, selective sorting, selective diagenesis, etc. are treated separately and explicitly. In short, the factor $\alpha\beta$ includes many uncertainties that need to be addressed as we progress. Ralph Hunter (personal communication, 1992) has kindly expanded a part of our model to include the volume proportion of garnets and zircons in source rocks and to separate the effects of weathering and transport, which we reproduce with minor editorial modification in Appendix 2. Because our present goal is to compare the Holocene with the Renova, comparable systematic errors in both the Holocene and the Renova should cancel each other out.

In any case, our model specifies that if a unit of metamorphic source rocks provides 1% garnet in a sediment, $\alpha\beta$ units of igneous source rocks would be necessary to provide 1% zircon in the sediment. On the basis of our general experience with the source rocks and weathering products in southwestern Montana, we estimate, somewhat arbitrarily, the factor $\alpha\beta$ to be somewhere between 10 and 10^3; we assume that $\alpha\beta = 10^2$ for a first approximation type of calculation, but a better estimate obviously is necessary. Thus, if 1 cm^3 of detrital garnet is obtained from 1 km^3 of metamorphic rocks, it takes 100 km^3 of igneous rocks to produce 1 cm^3 of detrital zircon. In our calculations, let M_G = the volume of Precambrian metamorphic source rocks; I_Z = the volume of Cretaceous-Tertiary igneous source rocks; Gt% = the percentage of garnet in the heavy mineral fraction; and Zc% = the percentage of zircon in the heavy mineral fraction.

For a sediment of mixed provenance, the proportion of Precambrian metamorphic source rocks = $M_G/(M_G + I_Z) \approx$ Gt % / (Gt % + $\alpha\beta$Zc %), where $\alpha\beta$ = 100 (estimated).

Results and implications. Results of the model calculations with our data on the distribution of garnets and zircons in the heavy mineral fraction of Holocene sands and Renova sandstones are given in Table 2. Because we believe that the distribution of sample sites represent Holocene and Renova deposits better than the averages of the different stream sections, the totals in Table 2 are averages of the samples in their respective groups.

TABLE 2. CALCULATED PROPORTIONS OF PRECAMBRIAN METAMORPHIC ROCKS AND CRETACEOUS-TERTIARY IGNEOUS ROCKS THAT WERE ERODED TO PRODUCE THE HOLOCENE FLUVIAL SANDS AND THE OLIGOCENE RENOVA SANDSTONES IN THE STUDY AREA*

	$M_G/(M_G + I_Z)$
HOLOCENE SANDS	
Northern areas	0.03
Southern areas	0.76
Holocene total	**0.19**
OLIGOCENE RENOVA SANDSTONES	
Ruby River Section	0.21
N. Boulder Section	0.003
Whitetail Section	0.00
Renova total	**0.18**

*The totals are averages of samples that represent parent rocks better than the individual areas listed.

Model calculations thus show that 19% of metamorphic rocks and 81% of igneous rocks were eroded to produce the Holocene sands in the study area. In the Oligocene, 18% of metamorphic rocks and 82% of igneous rocks were eroded to deposit the Renova sandstones in the same area. These proportions are virtually identical, suggesting that the overall proportion of igneous and metamorphic contribution to Renova sands and Holocene sands in southwestern Montana is not much different. In detail, however, different proportions of igneous and metamorphic detritus were delivered to different fans in different locations within the area. For example, 100% of the Renova sandstones in the Whitetail Creek section were probably derived from igneous source rocks. We emphasize that the geological importance of the results of our model calculations is not the absolute proportions of source rock contribution to the Holocene and the Renova, but the importance is in the fact that the overall partitioning of parent-rock contributions to the Holocene and the Renova are so similar.

DISCUSSION AND CONCLUSION

It is important to note that it is necessary to select a pair of proximal sands that have identical source rocks for the present model to be applicable to comparative studies. The probability that garnets, zircons, and Fe-Ti oxide minerals in the Renova have not suffered extensive diagenesis also aids in assuming that the factor $\alpha\beta$ is more or less the same for the Holocene sands and the Renova sandstones in the study area. It is by coincidence, and because of the assumption that $\alpha\beta$ = 100, that the results of the mass balance approach using a garnet/zircon ratio model inadvertently turn out to be similar to those of the statistical approach using detrital Fe-Ti oxide minerals (82:18 and 77:23, respectively, for igneous:metamorphic contributions). Many studies have shown that major mineral distributions in sands and sandstones are suitable for gross identification of provenance (Dickinson, 1988). Our studies of Fe-Ti oxide minerals, garnets, and zircon suggest that minor mineral distributions apparently are more useful in quantitative provenance analysis.

We emphasize that this is a preliminary attempt to construct a simple algebraic model, which by no means provides a universal solution. Our immediate goal is to put forth a proposal that may generate discussion and new ideas regarding quantitative provenance analysis. We hope that some day a model will emerge that will allow us to calculate the amounts, mass or volume, of different parent rock that were eroded to produce the detritus in a body of a clastic sedimentary rock.

ACKNOWLEDGMENTS

This research was supported in part by the Indiana University Foundation and by a fellowship from NATO-CNR (Italy) (to E.M.). We thank Ian Steele for assisting with electron probe microanalyses at the University of Chicago, and Jean Olson for collecting some of the samples used in this study. Thanks also to Lorie Canada, Barbara Hill, and James Tolen, who assisted in preparation of the manuscript. Comments from R. L. Folk, J. Grigsby, M. J. Johnsson, R. Hunter, and J. B. Maynard on a substantially different version of this study are gratefully acknowledged; the present manuscript was reviewed and edited by M. J. Johnsson and E. Merino, whose assistance is much appreciated.

APPENDIX 1. PARTIAL OUTPUT FROM SPSS

Discriminant analysis on groups defined by IGN and MET
296 (unweighted) cases were processed.
252 of these were excluded from the analysis.
252 had out-of-range group codes.
44 (unweighted) cases will be used in the analysis.

Variables: Al, Cr, Mg, Mn, Pe, and Ti are the respective oxide weight percents in single mineral phases.
W1, W2, and W3 are minimum widths of intergrowth lamellae—<2 μm, 2–10 μm, and >10 μm respectively—in polyphase Fe-Ti oxide grains
D1, D2, and D3 are the numbers of directions of intergrowth present— 1, 2, and 3 respectively—in polyphase Fe-Ti oxide grains

Number of cases by group

Group	Number of Cases Unweighted	Weighted Label
IGN	14	14.0
MET	30	30.0
Total	44	44.0

The following two variables failed the tolerance test:

Variable	Groups Variance	Tolerance	Minimum Tolerance
W3	92.6571	0.0001608	0.0000884
D2	0.000000E+00	0.0000000	0.0000000

Classification Function Coefficients
(Fisher's Linear Discriminant Functions)

Group =	IGN	MET
W1	22.54062	25.66077
W2	27.16624	32.53332
D1	-3.823051	-4.747657
D3	-7.856502	-9.341378
Mg	78.59168	111.8906
Al	20.65249	22.09746
Ti	11.99410	11.53383
Cr	-27.07180	-36.76480
Mn	17.56615	22.74184
Fe	9.045995	8.866997
(constant)	-1241.041	-1502.833

Canonical Discriminant Functions

Function	Eigenvalue	Percent of Variance	Cumulative Percent	Canonical Correlation
1*	15.23439	100.00	100.00	0.9687117

After Function	Wilks' Lambda	Chi-Squared	D.F.
0	0.0615976	103.12	10

*The one canonical discriminant functions remaining in the analysis.

Standardized Canonical Discriminant Function Coefficients

	Func. 1
W1	4.63022
W2	8.27802
D1	-3.96351
D3	-5.36539
Mg	1.53598
Al	0.05709
Ti	-0.25310
Cr	-0.02891
Mn	1.94527
Fe	-0.13608

Structural Matrix:
Pooled within-group correlations between discriminating variables and canonical discriminant functions (variables ordered by size of correlation within function):

	Func 1
WI	-0.23772
D3	-0.20616
W2	0.16677
W3	0.07831
D1	0.06937
Al	-0.06748
Cr	-0.05217
Mg	-0.04289
Mn	-0.02225
Fe	0.00663
Ti	-0.00580

Unstandardized canonical discriminant function coefficients:

	Func 1
W1	0.3810977
W2	0.6555405
D1	-0.1129321
D3	-0.1813641
Mg	4.067165
Al	0.1764894
Ti	-0.5621773E-01
Cr	-1.183912
Mn	0.6321628
Fe	-0.2186295E-01
(constant)	-33.46405

Canonical discriminant functions evaluated at group means (Group centroids):

Group	Func 1
IGN	-5.58223
MET	2.60504

Test of equality of group covariance matrices using box's m. The ranks and natural logarithms of determinants printed are those of the group covariance matrices.

Group Label	Rank	Log Determinant
IGN	9	(singular)
MET	8	(singular)
Pooled within-groups Covariance matrix	10	11.956718

No test can be performed without at least two nonsingular group covariance matrices.

Actual Group	Number of Cases	Predicted Group Membership IGN	MET
Group IGN	14	14	0
		100.0%	0.0%
Group MET	30	0	30
		0.0%	100.0%
Ungrouped cases	252	193	59
		76.6%	23.4%

Percent of "grouped" cases correctly classified: 100.00%

Classification process summary:
296 cases were processed.
0 cases were excluded for missing or out-of-range group codes.
0 cases had at least one missing discriminating variable.
296 cases were used for printed output.

APPENDIX 2. ALTERNATE MODEL

Let:

G be an abbreviation for garnet and GSR for garnet-bearing source rock,

Z be an abbreviation for zircon and ZSR for zircon-bearing source rock,

V_{GSR} be defined as the total volume of garnet-bearing source rock eroded to form a sedimentary deposit,

V_{ZSR} be defined as the total volume of zircon-bearing source rock eroded to form a sedimentary deposit,

N_G be defined as the volumetric proportion of garnet in GSR, i.e., as the volume of garnet in a unit volume of GSR,

N_Z be defined similarly to N_G.

Note that the volume of garnet in the total eroded volume of GSR is $N_G V_{GSR}$ and that the volume of zircon in the total eroded volume of ZSR is $N_Z V_{ZSR}$.

Let:

N_{RG} be defined as the volume of garnet released by weathering of unit volume of GSR,

G_G be defined by the equation $N_{RG} = C_G N_G$; then, C_G may be called the "release factor" for garnet,

N_{RZ} and C_Z be defined similarly to N_{RG} and C_G.

Note that the volume of garnet left after weathering of the total eroded volume of GSR is $N_{RG} V_{GSR} = C_G N_G V_{GSR}$, and that the volume of zircon left after weathering of the total eroded volume of ZSR is $N_{RZ} V_{ZSR} = C_Z N_Z V_{ZSR}$.

Let:

N_{SG} be defined as the volume of garnet deposited after weathering and erosion of unit volume of GSR,

B_G be defined by the equation $N_{SG} = B_G N_{RG}$; then B_G may be called the "attrition factor" for garnet,

N_{SZ} and B_Z be defined similarly to N_{SG} and B_G,

V_{SG} be defined as the total volume of garnet in the deposit resulting from the weathering and erosion of the source rock,

V_{SZ} be defined similarly to V_{SG}.

Therefore,

$$V_{SG} = N_{SG} V_{GSR} = B_G N_{RG} V_{GSR} = B_G C_G N_G V_{GSR}$$

and,

$$V_{SZ} = N_{SZ} V_{ZSR} = B_Z R_{RZ} V_{ZSR} = B_Z C_Z N_Z V_{ZSR}.$$

Let:

R be defined as the ratio of garnet to zircon in the deposit, then $R = V_{SG}/V_{SZ}$,

R_{SR} be defined as the ratio of garnet-bearing source rocks to zircon-bearing source rocks,

Then $R_{SR} = V_{GSR}/V_{ZSR}$.

Substituting,

$$R_{SR} = \{B_Z C_Z N_Z / B_G C_G N_G\} R.$$

Converting from ratios to percentages,

$R = P_{SG}/P_{SZ}$, where P_{SG} is the percent of garnet in the deposit and P_{SZ} is the percent of zircon in the deposit.

Given that $R_{SR} = P_{GSR}/P_{ZSR}$ and $P_{GSR} + P_{ZSR} = 1$, where P_{GSR} is the percent of eroded garnet-bearing source rock and P_{ZSR} is the percent of eroded zircon-bearing source rock, we get by substituting,

$$P_{GSR} = P_{SR}/(1 + R_{SR}) = P_{SR}) = P_{SG}/\{P_{SG} + (f_G/f_Z) P_{SZ}\},$$

where $f_G = B_G C_G N_G$ and $f_Z = B_Z C_Z N_Z$.

Therefore, this alternate model is equivalent to the original model if the ratio

$$B_G C_G N_G / B_Z C_Z N_Z = \alpha\beta.$$

Note that, in the original model, no zircon is supposed to be lost in the sedimentary processes, i.e., $B_Z = 1$, but that garnet suffers some loss such that $B_G = \beta$.

Therefore, the models are equivalent if $C_G N_G / C_Z N_Z = \alpha$.

Presumably $C_Z = 1$, just like B_Z, and the condition of equivalence is given by:

$$C_G N_G / N_Z = \alpha.$$

In essence, the factors C_G/C_Z and N_G/N_Z are compressed in the original model into a single factor, α.

REFERENCES CITED

Atherton, M. P., and Tarney, J., 1979, Origin of granitic batholiths: Orpington, UK, Shiva Publishing, 147 p.

Basu, A., 1976, Petrology of Holocene fluvial sands derived from plutonic source rocks: Implications to paleoclimatic interpretation: Journal of Sedimentary Petrology, v. 46, p. 694–709.

Basu, A., 1985, Influence of climate and relief on compositions of sands released at source areas, *in* Zuffa, G. G., ed., Provenance of arenites: Dordrecht, Reidel, p. 1–18.

Basu, A., and Molinaroli, E., 1989, Provenance characteristics of detrital opaque Fe-Ti oxide minerals: Journal of Sedimentary Petrology, v. 59, p. 922–934.

Basu, A., and Molinaroli, E., 1991, Reliability and application of detrital opaque Fe-Ti oxide minerals in provenance determination, *in* Morton, A. C., Todd, S. P., and Haughton, P.D.W., eds., Developments in sedimentary provenance studies: Geological Society of London Special Publication 57, p. 55–65.

Basu, A., Young, S. W., Suttner, L. J., James, W. C., and Mack, G., 1975, Re-evaluation of the use of undulatory extinction and polycrystallinity in detrital quartz for provenance interpretation: Journal of Sedimentary Petrology, v. 45, p. 873–882.

Basu, A., Brown, M., Elzea, J., and Muffler, S., 1986, Siliciclastic sediments derived from the tropical weathering of a basalt in eastern India: Geological Society of America Abstracts with Programs, v. 18, p. 279.

Bhatia, M. R., 1983, Plate tectonics and the geochemical composition of sandstones: Journal of Geology, v. 91, p. 611–627.

Blatt, H., 1978, Sediment dispersal from Vogelsberg basalt, Hessen, West Germany: Geologische Rundschau, v. 67, p. 1009–1015.

Chappell, B. W., and White, A. J., 1974, Two contrasting granite types: Pacific Geology, v. 8, p. 173–174.

Cliff, R. A., Drewery, S. E., and Leeder, M. R., 1991, Sourcelands for the Carboniferous Pennine river system: Constraints from sedimentary evidence and U-Pb geochronology using zircon and monazite, *in* Morton, A. C., Todd, S. P., and Haughton, P.D.W., eds., Developments in sedimentary provenance studies: Geological Society of London Special Publication 57, p. 137–159.

Cordua, W. S., 1973, Precambrian Geology of the Southern Tobacco Root Mountains, Madison County, Montana [Ph.D. dissertation]: Bloomington, Indiana University, 257 p.

Cuthbert, S. J., 1991, Evolution of the Devonian Hornelen Basin, west Norway: New constraints from petrological studies of metamorphic clasts, *in* Morton, A. C., Todd, S. P., and Haughton, P.D.W., eds., Developments in sedimentary provenance studies: Geological Society of London Special Publication 57, p. 343–360.

Darby, D. A., 1984, Trace elements in ilmenite: A way to discriminate provenance or age in coastal sands: Geological Society of America Bulletin, v. 95, p. 1208–1218.

Darby, D. A., and Tsang, Y. W., 1987, Variation in ilmenite element composition within and among drainage basins: Implications for provenance: Journal of Sedimentary Petrology, v. 57, p. 831–838.

Dickinson, W. R., 1988, Provenance and sediment dispersal in relation to paleotectonics and paleogeography of sedimentary basins (chapter 1), *in* Kleinspehn, K. L., and Paola, C., eds., New perspectives in basin analysis: New York, Springer-Verlag, p. 3–25.

Dickinson, W. R., and 8 others, 1983, Provenance of North American Phanerozoic sandstones in relation to tectonic setting: Geological Society of America Bulletin, v. 94, p. 222–235.

Drewery, S., Cliff, R. A., and Leeder, M. R., 1987, Provenance of Carboniferous sandstones from U-Pb dating of detrital zircons: Nature, v. 325, p. 51–53.

Fields, R. W., Rasmussen, D. L., Tabrum, A. R., and Nichols, R., 1985, Cenozoic rocks of the intermontane basins of western Montana and eastern Idaho: A summary, in Flores, R. S., and Kaplan, S. S., eds., Cenozoic paleogeography of the west-central United States: Denver, Rocky Mountain Section, Society of Economic Paleontologists and Mineralogists, p. 9–36.

Force, E. R., and Stone, B. D., 1990, Heavy-mineral dispersal and deposition in sandy deltas of glacial Lake Quinebaug, Connecticut: U.S. Geological Survey Bulletin 1874, 21 p.

Garzanti, E., 1991, Non-carbonate intrabasinal grains in arenites: Their recognition, significance, and relationship to eustatic cycles and tectonic setting: Journal of Sedimentary Petrology, v. 61, p. 959–975.

Girty, G. H., and Wardlaw, M. S., 1985, Petrology and provenance of pre–Late Devonian sandstones, Shoo Fly Complex, northern Sierra Nevada, California: Geological Society of America Bulletin, v. 96, p. 516–521.

Graham, J. R., Wrafter, J. P., Daly, J. S., and Menuge, J. F., 1991, A local source for the Ordovician Derryveeny Formation, western Ireland: Implications for the Connemara Dalradian, *in* Morton, A. C., Todd, S. P., and Haughton, P.D.W., eds., Developments in sedimentary provenance studies: Geological Society of London Special Publication 57, p. 199–213.

Grigsby, J. D., 1990, Detrital magnetite as a provenance indicator: Journal of Sedimentary Petrology, v. 60, p. 940–951.

Grigsby, J. D., 1992, Chemical fingerprinting in detrital ilmenite: A viable alternative in provenance research: Journal of Sedimentary Petrology, v. 62, p. 331–337.

Ibbeken, H., and Schleyer, R., 1991, Source and sediment: Berlin, Springer-Verlag, p. 286.

Ishihara, S., 1977, The magnetite-series and ilmenite-series granitic rocks: Mining Geology (Tokyo), v. 27, p. 293–305.

Johnsson, M. J., 1990a, Tectonic versus chemical-weathering controls on the compositionn of fluvial sands in tropical environments: Sedimentology, v. 37, p. 713–826.

Johnsson, M. J., 1990a, Tectonic versus chemical-weathering controls on the composition of fluvial sands in tropical environments: Sedimentology, v. 37, p. 713–726.

Johnsson, M. J., 1990b, Chemical weathering of fluvial sediments during alluvial storage: The Macuapanim Island point bar, Solimões River, Brazil: Journal of Sedimentary Petrology, v. 60, p. 827–842.

Knopf, A., 1957, The Boulder Batholith of Montana: American Journal of Science, v. 255, p. 81–103.

Mack, G. H., 1984, Exceptions to the relationship between plate tectonics and sandstone composition: Journal of Sedimentary Petrology, v. 54, p. 212–220.

McLennan, S. M., and Taylor, S. R., 1991, Sedimentary rocks and crustal evolution: Tectonic settings and secular trends: Journal of Geology, v. 99, p. 1–22.

McLennan, S. M., Taylor, S. R., McCulloch, M. T., and Maynard, J. B., 1990, Geochemical and Nd-Sr isotopic composition of deep sea turbidites: Crustal evolution and plate tectonic associations: Geochimica et Cosmochimica Acta, v. 54, p. 2015–2050.

Molinaroli, E., and Basu, A., 1987, Studio di minerali opoachi in sabbie fluviali oloceniche e nelle corispondenti rocce madri di zone sottoposte a climi diversi (Montagne Rocciose e Monti Appalachi in U.S.A.): Rendiconti della Societa Italiana di Mineralogia e Petrologia, v. 42, p. 271–283.

Molinaroli, E., Blom, M., and Basu, A., 1991, Methods of provenance determination tested with discriminant function analysis: Journal of Sedimentary Petrology, v. 61, p. 900–908.

Morton, A., 1985, Heavy minerals in provenance interpretation, *in* Zuffa, G. G., ed., Provenance of arenites: Dordrecht, Reidel, p. 249–277.

Morton, A. C., 1991, Geochemical studies of detrital heavy minerals and their application to provenance research, *in* Morton, A. C., Todd, S. P., and Haughton, P.D.W., eds., Developments in sedimentary provenance studies: Geological Society of London Special Publication 57, p. 31–45.

Nelson, B. K., and DePaolo, D. J., 1988, Comparison of isotopic and petrographic provenance indicators in sediments from Tertiary continental basins in New Mexico: Journal of Sedimentary Petrology, v. 58, p. 348–357.

Orr, E. D., and Folk, R. L., 1983, New scents on the chattermark trail: Weathering enhances obscure microfractures: Journal of Sedimentary Petrology, v. 53, p. 121–129.

Pettijohn, F. P., Potter, P. E., and Siever, R., 1972, Sand and sandstones: New York, Springer-Verlag, p. 553.

Riezebos, P. A., 1979, Compositional downstream variation of opaque and translucent heavy residues in some modern Rio Magdalena sands (Colombia): Sedimentary Geology, v. 24, p. 197–225.

Stöffler, D., Ostertag, R., Reimold, W. U., Borchardt, R., Malley, J., and Rehfeldt, A., 1981, Distribution and provenance of lunar highland rock types at North Ray Crater, Apollo 16: Proceedings Lunar and Planetary Science Conference, v. 12b, p. 185–207.

Suttner, L. J., 1974, Sedimentary petrographic provinces: An evaluation: Society of Economic Paleontologists and Mineralogists Special Publication 21, p. 75–84.

Suttner, L. J., Basu, A., and Mack, G. H., 1981, Climate and the origin of quartz arenites: Journal of Sedimentary Petrology, v. 51, p. 1235–1246.

Tilling, R. I., 1964, Variation in modes and norms of an "homogeneous" pluton of the Boulder Batholith, Montana: U.S. Geological Survey Professional Paper 501-D, p. D8–D13.

van de Kamp, P. C., and Leake, B. E., 1985, Petrography and geochemistry of feldspathic and mafic sediments of the Northeastern Pacific Margin: Transactions of the Royal Society of Edinburgh, Earth Sciences, v. 76, p. 411–449.

Young, S. W., 1976, Petrographic textures of detrital polycrystalline quartz as an aid to interpreting crystalline source rocks: Journal of Sedimentary Petrology, v. 46, p. 595–603.

Zuffa, G. G., 1980, Hybrid arenites: Their composition and classification: Journal of Sedimentary Petrology, v. 50, p. 21–29.

Zuffa, G. G., 1987, Unravelling hinterland and offshore palaeogeography from deep-water arenites, *in* Legget, J. K., and Zuffa, G. G., eds., Marine clastic sedimentology: London, Graham and Trotman, p. 39–61.

MANUSCRIPT ACCEPTED BY THE SOCIETY DECEMBER 24, 1992

Printed in U.S.A.

Index

[Italic page numbers indicate major references]

N

O

P

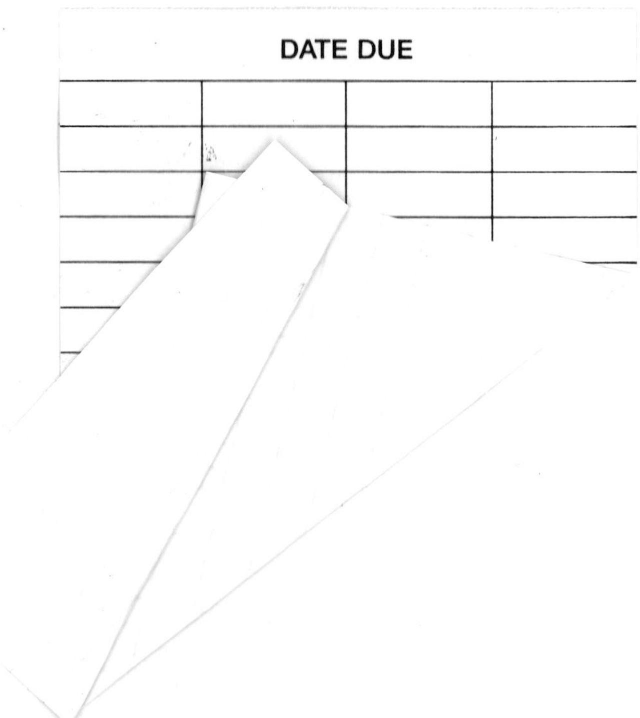

Typeset by WESType Publishing Services, Inc., Boulder, Colorado
Printed in U.S.A. by Malloy Lithographing, Inc., Ann Arbor, Michigan